Industrial
Noise and
Hearing
Conservation

Industrial Noise and Hearing Conservation

Edited by
Julian B. Olishifski, P.E.
Earl R. Harford, Ph.D.

NATIONAL SAFETY COUNCIL
Chicago, Ill. 60611

International Standard Book Number: 0–87912–085–1
Library of Congress Catalog Card Number: 75–11313
Printed in the United States of America
20M107508   Stock No. 151.17

Contents

CONTENTS

Appendices

NSC Occupational Safety Series

The National Safety Council Occupational Safety Series is dedicated to the publication of current and vital information about occupational safety. Other books in this series are:

> *Accident Prevention Manual for Industrial Operations,*
> *Communications for the Safety Professional,*
> *Fundamentals of Industrial Hygiene,*
> *Supervisors Safety Manual,*
> *Supervisors Guide to Human Relations,*
> *Motor Fleet Safety Manual.*

All are designed to help the reader identify accident problems, establish priorities, gather and analyze data to help identify problems, and develop methods and procedures that will eliminate or decrease the seriousness of accident problems—to mitigate injury and minimize economic loss resulting from accidents.

Audiovisuals and instructors guides are available for use with many of these books. For details, contact the National Safety Council.

FRANK E. McELROY, P.E.
Director, Technical Publications
National Safety Council

Foreword

Noise is perhaps the most common pollutant in industry. Thousands of workers in many different industries are exposed to relatively high sound pressure levels. Occupational safety and health regulations in many states, as well as the Federal Government, require that employees be protected from exposure to high noise levels in order to preserve their hearing. The larger companies, employing industrial hygienists, have been alert for a number of years to the hazards of exposure to high noise levels. However, many smaller industries do not employ industrial hygienists and may not generally be aware of the adverse effects of noise. Since industrial hygienists are in serious short supply, it becomes imperative that supervisors, safety directors, or others assigned duties for protection of employee health and safety become familiar with the basic principles of noise control and conservation of hearing.

During recent years, regulations to control employee noise exposures have come into existence, and workers' compensation benefits to employees suffering noise induced hearing loss have been increasing rapidly. It is unlikely that there will be any relaxation in these regulations and trends, and effective noise control and hearing-conservation programs are now vitally important to employers.

This book will serve as a valuable reference and fill a serious need. It covers the fundamental principles necessary for the establishment of an effective, on-going hearing-conservation program. Subjects are covered in some detail by professionals who are recognized specialists in their respective fields of hearing conservation and noise control.

This material is presented to assist supervisors, safety professionals, and others to recognize, evaluate, and control industrial noise problems. Alternate methods of noise control are presented in terms readily understandable to the nonprofessional. It is recommended that industrial hygienists or consultants be employed for the control of complex noise problems.

BURL E. McGARRAHAN
Member, Board of Directors National Safety Council
Member, Inter-Industry Noise Study Steering Committee

Dayton, Ohio

Preface

The primary purpose of this book is to provide a reference for persons who have either an interest in or a direct responsibility for the preservation of hearing of those exposed to high levels of industrial noise. Thus, it is intended to be of use to engineers, executives, audiometric technicians, industrial hygienists, nurses, audiologists, physicians, and safety personnel. It should also be appropriate as a self-teaching text as well as a reference for training courses in industrial noise and hearing conservation. With this in mind, bibliographies have been included in many chapters for those who wish to pursue specific topics in greater depth.

This book is the result of material developed for several short courses and workshops given for occupational safety and health personnel. Originally, a relatively short manual—one containing information specific to the organization and operation of an industrial noise control and hearing-conservation program—was envisaged. As time progressed, however, the need for a more comprehensive text became evident, and as a result, some theoretical concepts and practical material have been included.

An effective industrial hearing-conservation program is one that successfully reduces hazardous noise at its source. Consequently, considerable information has been included to assist in dealing with noise reduction. Although some of the subject matter may be meaningful only to those having an engineering background, we hope that most of the information in this book will be of value and useful to the general reader.

Industrial noise problems can be extremely complex; there is no standard solution applicable to all situations. As a reasonable minimum, an industrial hearing-conservation program should include noise surveys, engineering and administrative controls, audiometry, and hearing protection for those employees working in areas where noise cannot be reduced at its source. Each of these basic ingredients of an industrial hearing-conservation program is discussed in this book.

The book is organized into six major parts, each of which is meant to stand alone as a reference source. For that reason, we have permitted a certain amount of redundancy to occur.

Part one includes chapters on the fundamental aspects of the noise problem, and the physical and subjective aspects of sound. A background is provided which lays the groundwork for understanding how sound energy is produced, transmitted, and measured.

Part two of the manual is concerned with the measurement of sound. Basic information is given on the various types of instruments available to measure sound levels and describes how to use the instruments properly to obtain valid measurements as well as where and when to take sound measurements.

Part three describes the effects of noise on man. The anatomy, physiology, and common pathologies of the ear are discussed along with the medical aspects of hearing conservation, for a full understanding of the impact of noise on man. To make this section more meaningful, illustrations are included of the handicapping effects of hearing impairment, as well as examples of the extra-auditory effects of noise. Because noise and hearing conservation have become a matter of government concern, federal regulations and guidelines for noise exposure are discussed in this part, in addition to workers' compensation and medicolegal factors. Also included in part three is one of the more important aspects of a hearing-conservation program—the methods used to evaluate the extent of noise exposure.

Part four deals with the control of noise and noise exposure. Although noise problems vary, the basic principles of noise control, problem-solving techniques, and the examples of engineering noise control measures given here are general enough to have wide application. To augment the basics, specific information is covered in vibration measurements and control. An important part of noise control is prevention, and with that in mind, a chapter has been included dealing with sound level specifications for the purchase of equipment and machinery. Logically, if the noise cannot be reduced, the worker must be protected. The last chapter in this part covers personal hearing-protective devices.

Part five is directed specifically to persons responsible for conducting hearing tests for industrial personnel. Discussed in detail are the fundamental concepts of air-conduction, puretone audiometry, audiometer calibration, and techniques for the determination of a hearing threshold. Manual and automatic audiometry, interpretation of audiograms, and an explanation of the testing problems a technician might encounter during an audiometric examination are described.

Part six brings the book to a logical conclusion with a description of an effective industrial hearing-conservation program. Included are chapters on the roles of the nurse, the audiologist, and the physician in the implementation of a successful program.

One of the most difficult parts of getting any project started is finding sources of help and information. For this reason, we have included in the final chapter of this book a comprehensive, annotated bibliography, and listings of professional and service organizations, standards, governmental agencies, training courses, and audiovisuals.

Finally, extensive appendices are included to assist the reader. These appendices use excerpts from the Federal Occupational Noise Regulations, Federal Environmental Noise Regulations, Intersociety Committee Guidelines for Noise Control, a glossary, and a review of mathematics. An extensive index is included to assist the reader in locating information in this text.

Many charts and tables have been included to assist the reader in better understanding the subject matter.

The authors were carefully selected on the basis of their expertise and years of experience in their respective areas of specialization. The editors are grateful to each of them for their valuable contribution to this text. We hope that this book will at least fulfill a portion of our expectations for its application.

The information and recommendations contained in this book have been compiled from sources believed to be reliable and to represent the best current opinion on the subject. No warranty, guarantee, or representation is made by the National Safety Council as to the absolute correctness or sufficiency of any representation contained in this and other publications, and the National Safety Council assumes no responsibility in connection therewith; nor can it be assumed that all acceptable safety measures are contained in this (and other) publications, or that other or additional measures may not be required under particular or exceptional conditions or circumstances.

This manual will be revised periodically. Contributions and comments from readers are invited.

JULIAN B. OLISHIFSKI, P.E.
EARL R. HARFORD, PH.D.

Contributors

Many of the chapter authors listed below are also instructors in National Safety Council Training courses on "Industrial Noise Control" or "Industrial Audiometric Technician Training."

Production of this book was handled by the following: editing by Robert Pedroza and Frank McElroy; proofreading by Harry Sharp and Jim Weging; manuscript typing by Muriel Perry and Florence Kanter. Drawings were by Edwin Huff; cover art and book design was by Vito DePinto of First Impressions.

Chapter authors:

Edwin L. Alpaugh, P.E.
Industrial Hygienist
International Harvester Co.
Chicago, Illinois
Chapter 6

Charles L. Cheever, B.S.
Industrial Hygiene Engineer
Argonne National Laboratory
Argonne, Illinois
Chapter 15

Jack D. Clemis, M.D., F.A.C.S., F.R.C.S.(c)
Associate Professor of Otolaryngology and Maxillofacial Surgery
Northwestern University

Otologic Professional Associates
Chicago, Illinois
Chapter 7

Alexander Cohen, Ph.D.
Chief, Behavioral & Motivational Factors Branch
National Institute for Occupational Safety and Health
Cincinnati, Ohio
Chapter 10

Meyer S. Fox, M.D.
Clinical Professor
University of Wisconsin Medical School
Consultant in Industrial Otology
Milwaukee, Wisconsin
Chapters 8 and 13

Earl R. Harford, Ph.D.*
Professor of Audiology
Director, Hearing Clinic
Northwestern University
Evanston, Illinois
Chapters 21, 24, 26, 27, and 28

Bill Ihde, P.E.
Consultant in Acoustics
Industrial Noise Services, Inc.
Hinsdale, Illinois
Chapters 5 and 17

Herbert H. Jones, M.S.
Professor, Department of Industrial
Safety and Hygiene
Central Missouri State University
Warrensburg, Missouri
Chapter 11

George W. Kamperman, P.E.
Kamperman Associates, Inc.
Downers Grove, Illinois
Chapter 16

Richard LaJeunesse, B.S., P.E.
Design Engineer
Time Data Division of General
Radio Co.
Palo Alto, California
Chapter 4

Noel D. Matkin, Ph.D.
Professor of Audiology
Northwestern University
Evanston, Illinois
Chapters 22 and 26

Hayes A. Newby, Ph.D.
Chairman, Department of Hearing
and Speech Sciences
University of Maryland
Edgewater, Maryland
Chapter 31

Julian B. Olishifski, M.S., P.E.
Director of Industrial Hygiene
National Safety Council
Adjunct Assistant Professor of Occupational and Environmental
Medicine
University of Illinois
Chicago, Illinois
*Chapters 1, 2, 3, 14,
18, 19, 29, and 33*

Bill Reich
President
Reich Associates, Inc.
Acton, Massachusetts
Chapter 23

**Donna M. Reichmuth, R.N.,
C.O.H.N.**
Division Health Consultant
Liberty Mutual Insurance Co.
Chicago, Illinois
Chapter 30

William F. Rintelmann, Ph.D.
Professor and Chairman of Audiology
University of Pennsylvania Medical School
Philadelphia, Pennsylvania
Chapter 25

---

* Dr. Earl R. Harford's position and address, after January 1, 1976, will be Director, Bill Wilkerson Hearing and Speech Center, Nashville, Tenn., and Professor and Chairman of Hearing and Speech Sciences, Vanderbilt University, Nashville, Tenn.

CONTRIBUTORS

Joseph Sataloff, M.D., D.Sc.
Professor of Otology
Thomas Jefferson University
Scientific Member
U.S. Dept. of Labor Standards Advisory Committee on Noise
Philadelphia, Pennsylvania
Chapters 9 and 32

F. A. Van Atta, Ph.D.
Senior Scientist
Occupational Safety & Health Administration
Special Programs
U.S. Department of Labor
Washington, D.C.
Chapter 12

Richard L. Swift, B.S.
Product Line Manager
Mine Safety Appliance Co.
Pittsburgh, Pennsylvania
Chapter 20

Part 1 Introduction

Sources (Measured at operator/ listener distance from source)	Aural Effect	Sound Level In Decibels
Shotgun blast Jet plane at take-off Firecrackers, exploding	Human ear Pain Threshold	140
Rock Music (amplified) Hockey game crowd Thunder, severe Pneumatic jackhammer	Uncomfortably Loud	120
Powered lawn mower Tractor, farm type Subway train (interior) Motorcycle Snowmobile Cocktail party (100 guests)	Extremely Loud	100
Window air-conditioner Crowded restaurant Diesel-powered truck/tractor	Moderately Loud	80
Singing birds Normal conversation	Quiet	60
Rustle of leaves Faucet, dripping Light rainfall	Very Quiet	40
Whisper	Just Audible	10

FIG. 1–1.—Levels of some familiar sounds.

Chapter One

Industrial Noise and Hearing Conservation

By Julian B. Olishifski, P.E.

The recognition, evaluation, and control of noise problems are introduced in this chapter. Basically, they include: (*a*) making an appraisal of the extent of the noise problem, (*b*) setting objectives for a noise control program, and (*c*) measuring, analyzing, and controlling industrial noise exposures.

Man and his machines can be noisy. When there is a lot of space, there is a chance for noise to dissipate. But when man and his machines are grouped together, the chance for the noise levels to dissipate goes down. Excessive noise levels have, therefore, penetrated virtually every aspect of modern life and, generally speaking, the levels are increasing. Soon 75 percent of the U.S. population will be clustered in major metropolitan areas—and overall noise levels will tend to increase as noise sources rise both in number and in geographical density. As population increases, so does the proliferation of more noise producing equipment—there are more cars, busses, and trucks; more tractors and other farm equipment; more and larger machinery; jet aircraft; even loud electrical musical instruments. More highways, airports, and airplanes make sure that noise will be even more widely distributed. In some cases, noise produces annoyance; in the more severe situations, it results in loss of hearing in those people who are exposed to it.

What is noise?

Noise is any undesired sound. At times the same sound may be pleasing; at other times, it may be annoying or harmful. A stereo set

3

playing loud music on the first floor of a house may have a pleasant psychological effect on some persons, but, to someone trying to sleep in an upstairs bedroom, it may be disturbing.

Musical sounds are usually the result of regular disturbances of the ambient air pressure; irregular and erratic disturbances are usually unpleasant to hear and are termed noise. Thus noise, in a general sense, is more or less any random disturbance. At any location there is always some noise; it may come from many sources near or far. It may be reflected from walls, ceilings, and floors. This composite all-encompassing noise associated with a given environment is called *ambient noise*.

Useless sound can also be called noise. It carries no information; instead, it tends to interfere with our ability to receive and interpret useful sound. In many cases it is difficult to decide whether a sound is information or noise. Often it is both. For example, the sound of a machine conveys information to the operator. It lets him know whether it is running normally. But, to an adjacent operator tending another machine, the sound coming from his coworker's machine could be considered noise. Noise is sound. But it is not *just* sound; it has additional characteristics. First, noise is sound that is subjective, that is, it must be heard before a value judgement can be passed and a determination made that the sound heard is indeed noise. Noise, unlike sound which can exist alone, must be heard by someone if it is to exist. The second characteristic of sound as noise is that it must be judged as an unwanted, irregular, and erratic (random) disturbance, or useless sound (meaning that it carries no information and tends to interfere with the reception and interpretation of useful sound).

A certain degree of environmental quietness is desirable in itself. People in general do not like to live in the immediate vicinity of an airfield, or near roads with heavy traffic, or near other noisy places. For that reason the noise levels in the immediate vicinity have a considerable impact on the price of nearby property.

Public awareness

In addition to the increased number of noise sources, public awareness of noise levels has been increasing dramatically. Newspapers, magazines, and television commentators describe undesirable noise levels. Community groups get involved and demand that government officials do something to curb excessive noise levels.

Noise in homes is rising, as more tools and appliances are being used and, as their power is increased, so has the noise level. The combination of electronic amplification equipment and rock music has probably affected the hearing of a whole generation of young listeners. Noise levels in apartments and private dwellings, particularly in kitchens, are beginning to approach those in factories (see Figure 1-1).

In choosing home appliances, such as refrigerators, vacuum cleaners, washing machines, air conditioning systems, and oil burners, the buyer should definitely consider the noise levels of the items. The control of product noise is therefore important not only from an annoyance and health point of view but also from an economic point of view. Within a few years, product noise standards will almost certainly exist.

Municipalities are moving to curtail construction noise in the streets and to insert acoustic standards in building codes. There may soon be stringent standards—monitoring of sound levels will be required in periodic inspections. (See Appendix B.)

Industrial noise problems

Industrial noise problems are extremely complex. There is no standard program applicable to all situations. In view of changes in enforcement and compensation aspects, however, industry should consider and evaluate its noise problems and take steps toward the establishment of effective hearing-conservation programs.

As a minimum, a hearing-conservation program in industry should include:

1. Noise exposure evaluation
2. Engineering control of noise
3. Audiometric examinations of employees
4. Hearing protection for those employees working in areas where noise cannot be controlled through engineering methods

These topics will all be discussed in this book. But before doing so, there are other aspects that must be introduced—economics, workers' compensation, noise regulations, the effects of noise, and the characteristics and the measurement of noise.

Economic aspects

Noise control expenditures on products and materials will become even greater with the current and expected increase in antinoise legislation. A tremendous increase has occurred in the sales of sound measuring and analysis instrumentation over the past few years. The market for noise limiting products encompasses virtually all of industry. Primary application for such products are surface vehicles and aircraft, machinery, industrial plants, consumer products (including home appliances and recreational vehicles), and building and construction equipment. Methods for achieving noise reduction include vibration isolation, partitions and barriers, lined ducts, mufflers, enclosures, and acoustic absorption. Part Four of this book covers this topic in great detail.

Noise control techniques are being incorporated into products during the design stage. Machine tool buyers represent one group who currently specify maximum noise levels in their purchase orders. Equipment can be designed with lower noise levels, but there may be performance tradeoffs of weight, size, power consumption, and perhaps increased maintenance costs. These new, quieter products will probably weigh more, be bigger and bulkier, cost more, and be more difficult and expensive to service and maintain.

To attain quieter products, the engineer must be prepared to trade off, to some degree, many of the design goals which have been achieved in response to market demands. Lightweight, low cost, portability, ease of operation, and simplicity of maintenance should not be cast aside lightly. Price increases may be inevitable and the cost/benefit relationship should be examined in each case. In addition to paying for higher original equipment costs, the user (both as a consumer and as a taxpayer) will also bear the burden of increased direct and indirect costs.

Laboratory solutions to noise problems sometimes cannot be applied to products presently in the marketplace because the noise control costs may exceed the value of the products. Because noise control efforts must meet this test of the marketplace, it is imperative to evaluate the costs to assure a successful noise control program. Unfortunately, few studies of the cost of noise control have appeared in published literature.

Compensation aspects

The growing interest of employers in hearing loss claims due to industrial noise exposure has been stimulated by the trend toward coverage of partial loss of hearing under state compensation laws. Along with the OSHA noise control regulations, new and revised compensation laws can be expected. Compensation laws that cover loss of hearing due to noise exposure have been enacted in many states. Compensation is being awarded in other states even though hearing loss is not specifically defined in the compensation laws.

That workers in noisy occupations develop a greater than average degree of hearing loss has been known for more than 100 years. Prior to 1950, however, hearing loss was not regarded as a significant factor in compensation awards. Some claims occurred, but these were primarily due to traumatic injuries from such causes as explosions, concussions, blows to the head, and foreign objects or infections in the ears. Since 1950, a number of states have passed legislation or revised existing laws to include noise induced partial hearing loss as a basis for workers' compensation benefits. For more details see Chapter 13, Worker's Compensation and Medicolegal Factors.

At the present time, approximately 36 states recognize occupational hearing loss either specifically or as an interpretation of the existing laws. In some states, however, a total loss of hearing must be incurred before compensation will be paid.

It has been estimated that 1.7 million workers in the U.S. between 50 and 59 years of age have enough hearing loss to be awarded compensation. Assuming that only 10 percent of these persons file for compensation and that the average claim amounts to $3000, the potential cost to industry could run to over $500 million.

Estimates show that 14 percent of the working population is employed where the noise level is in excess of 90 dBA* (see Table 1-A). In the present state of knowledge, no one can predict which individuals will incur a hearing loss; but if enough individuals are placed in an environment where the predominant noise level exceeds 90 dBA, some of them will incur a hearing impairment *in excess* of that due to presbycusis (the loss of hearing due to aging). The number of workers subjected to noise levels that may impair their hearing probably exceeds those exposed to any other significant occupational hazard.

Until the development of the audiometer in recent years, there was no means of measuring with appreciable accuracy the degree of hearing loss. Now partial hearing losses are easily measurable by use of commercially available instruments. This is discussed in detail in Part Five —Industrial Audiometry.

A partial hearing loss due to excessive noise exposure may occur in a few months or longer depending upon the duration of exposure and the levels. However, a person with a partial hearing loss may still be able to perform the primary job function although with greater difficulty. However, problems can arise when speech communication, such as signals or directions, is misunderstood and results in an unsafe action.

---

* The term dBA is explained in the next chapter.

TABLE 1-A

NOISE EXPOSURES ABOVE 90 dBA IN MANUFACTURING

Code	Number of Plants In Sample	Total Number of Employees in Sample	Number Located in Areas 90 dBA and Above	Percent of Work Force Exposed	Total Work Force	Number Projected to Be Located in Areas 90 dBA and Over
Textile Mill Products	23	12,764	5,634	44.1	963,300	424,815
Petroleum and Coal Products	16	20,493	5,875	28.6	192,800	55,140
Lumber and Wood Products	14	5,654	1,460	25.8	601,000	155,058
Food and Kindred Products	17	23,690	5,959	25.1	1,898,600	476,549
Furniture and Fixtures	11	10,374	1,849	17.8	465,400	82,841
Fabricated Metal Products	56	41,371	7,079	17.1	1,335,000	228,285
Stone, Clay and Glass Products	5	2,502	416	16.6	643,800	106,870
Primary Metal Industries	51	71,208	11,001	15.4	1,190,000	183,260
Rubber and Plastic Products	4	7,671	1,105	14.4	589,500	84,888
Transportation Equipment	46	199,212	23,445	11.7	1,705,500	199,543
Electrical Equipment and Supplies	7	8,790	973	11.0	1,778,100	195,591
Chemicals and Allied Products	8	3,081	324	10.5	1,014,400	106,512
Apparel and Other Textile Products	1	50	5	10.0	1,353,100	*
Paper and Allied Products	21	14,997	1,385	9.2	687,400	63,240
Ordnance and Accessories	12	39,403	3,480	8.8	193,900	17,063
Instruments and Related Products	6	3,254	193	5.9	433,800	25,594
Machinery Except Electrical	38	25,016	1,144	4.5	1,768,000	79,560
Printing and Publishing	5	5,597	237	4.2	1,085,900	45,607
Total	341**	504,427	71,564	14.1	16,999,500	2,533,416

Source: NIOSH Criteria Document.

* Insufficient data for projection
** 2709 questionnaires were sent to the manufacturing industries listed, of which 1559 were returned. 341 of these respondents answered this question.

Noise control regulations

Noise problems have been investigated by scientists and engineers for more than 80 years. As long as environmental noise problems were being dealt with in an abstract professional sense, they could be considered a scientific curiosity or an arena in which a consensus of scientific views prevailed. But, in dealing with noise regulations enforceable by penalties and other sanctions, a multidiscipline approach is essential. The regulation of environmental noise levels entails consideration of the extremely complex interrelationship of scientific, engineering, technical, social, economic, and political spheres. In addition, the skills and knowledge of scientists, physicians, lawyers, economists, and political scientists must be utilized. More important, the views of those most affected by these noise standards— those for whom protection is being provided and those who are to be regulated—should be brought into the standards-making process.

Federal occupational noise exposure regulations

Historical development. The federal regulation of occupational noise exposure started with the rules issued under the authority of the Walsh-Healey Public Contracts Act. Those regulations required that occupational noise exposure must be reasonably controlled to minimize fatigue and the probability of accidents. The federal occupational noise exposure regulations were originally written to apply to contractors under the Walsh-Healey Public Contracts Act and the McNamara-O'Hara Service Contracts Act. Under the Williams-Steiger Occupational Safety and Health Act of 1970, the Bureau of Labor Standards was replaced by the Occupational Safety and Health Administration (OSHA).

The National Institute for Occupational Safety and Health (NIOSH) was established within the Department of Health, Education and Welfare by the Occupational Safety and Health Act of 1970 to conduct research and to recommend new occupational safety and health standards. These recommendations are transmitted to the Department of Labor, which has the responsibility for the final setting, promulgation, and enforcement of the standards.

On August 14, 1972, NIOSH provided the Department of Labor with a criteria package (HSM* 73–11001), *Occupational Exposure to Noise.*

---

* Health Services and Mental Health Administration. See Appendix A-3.

9

Thereupon, the Assistant Secretary of Labor determined that a standard advisory committee on noise should be appointed. The purpose of this OSHA Advisory Committee was to obtain and evaluate additional recommendations from labor, management, government, and independent experts. The Committee considered written and oral comments directed to it by interested parties. It transmitted its recommendations for a revised standard to the Occupational Safety and Health Administration (OSHA) on December 20, 1973.

The 1974 OSHA standard limits an employee's exposure level to 90 dBA, calculated as an eight-hour, time-weighted average. NIOSH indicated a need for reducing the eight-hour exposure level to 85 dBA, but was unable to recommend a specific time period after which the 85 dBA noise level should become effective for all industry due to the unavailability of sufficient data relating to the technological feasibility of this level. (See Appendix A.)

The Environmental Protection Agency (EPA), reviewed the risk of hearing impairment and recommended that OSHA reduce the limit at least to 85 dBA. The EPA, which reviewed the draft standard proposal under the authority of the Noise Control Act of 1972, further recommended that additional studies be undertaken to explore reducing the permissible level still further at some future date. The proposed revision to the rules for occupational noise exposure were published in the *Federal Register* on October 24, 1974. (For further information, see Chapter 12, Federal Regulations and Guidelines, and Appendix A.)

Environmental noise regulations

The Noise Control Act of 1972 established the Office of Noise Abatement and Control (ONAC) within the U.S. Environmental Protection Agency. The following discussion presents the background of this legislation and a general consideration of the provisions of the Act.

The Noise Control Act authorizes the Environmental Protection Agency (EPA) to head a national noise abatement program and to perform the following:

• Promulgate and enforce noise emission regulations for machines that make excessive noise.

• Promulgate noise emission regulations for railroads and interstate motor carriers (to be enforced by the Department of Transportation).

• Study and make recommendations on the control of aircraft and airport noise.

10

- Coordinate other federal agencies' activities in noise control.

In setting noise standards or limits for machines, the EPA considers the following:

- How is the machine used: singly or in combination with other noise sources?

- What is the degree of noise reduction achievable through the application of the best available technology?

- What are the costs of compliance with the standard?

In addition to the considerations mentioned, the EPA has an obligation to inform the public of the hazards of noise and of the progress being made in noise control.

The Noise Control Act singled out construction equipment, transportation equipment (including recreational vehicles), motors and engines, and electrical and electronic equipment for noise regulation. EPA is also authorized to list other products which are major sources of noise and to update the list periodically. EPA deals in performance, rather than in specifications and standards. Thus, the machine manufacturer may design his machine any way he sees fit as long as in operation it complies with the EPA noise regulations. EPA is also empowered to promulgate labeling regulations covering excessively noisy machinery and noise reduction products. The law gives EPA review authority, but no veto power, over all proposed federal noise standards. (See Appendix B-1.)

Criteria and standards

Although the words *criteria* and *standards* are synonymous in professional usage (both being a measure by which something is judged), in the environmental sense, *criteria* are a recitation of factual information as to cause-and effect relationships without making judgments as to what is desired or desirable. Data exist which demonstrate that noise evokes a significant subjective physiological and psychological response pattern. However, the mere recitation of data, no matter how scientifically valid, is not acceptable to the courts, affected industries, or the public.

A need exists to identify the basis of environmental standards and to provide adequate statements for reasons behind decisions. Therefore, the general standard development process must include a credible chain of reasoning leading to the conclusion that the proposed standards are the best available, given the accumulation of data upon which these standards are based (see Chapter 11).

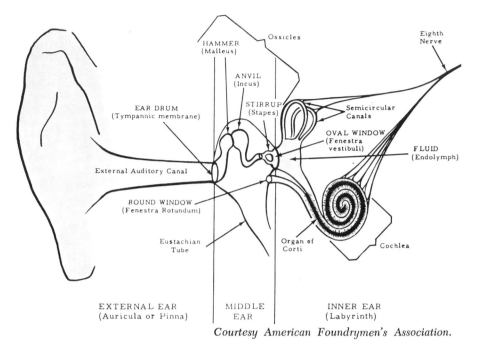

Courtesy American Foundrymen's Association.

Fig. 1–2.—Shown here is a schematic diagram showing the transmission of sound waves to the inner ear.

It must be clearly stated how and why the particular standard was selected. It is also necessary to demonstrate what alternatives were developed, with all the relevant information as to cost, impact, projected results, and the like. How these alternatives were examined individually and then tested against the others to produce the decision in the form of proposed standards should be clearly shown.

A regulatory program usually implies some form of legislation. Legislation regulating noise is nothing new in this country—many states and major cities have general and qualitative laws against excessive or unwarranted noise. But such laws are very difficult to enforce, because determining whether a violation has occurred is a matter of subjective judgment and enforcement officials are reticent to enforce qualitative laws.

Noise regulations can and should be objective, quantitative laws. Criteria are well established for those noise levels that are objectionable to a majority of people. Equipment and measurement procedures for

12

determining when noise standards are exceeded are readily available, and their reliability is also well established.

Not only should laws be quantitative, but they must be specific as to the source of noise as well. A realistic law must recognize that a garden lawn mower is inherently less noisy than a jackhammer and it should assign limitations appropriate to each piece of equipment. To establish one absolute noise level not to be exceeded, regardless of the noise source, simply is not feasible, nor would it really solve the noise problem, because the annoying characteristics of noise depend on the background noise as well as the intruding noise.

In addition to the source, the *area* where the noise source is used needs regulation. A jackhammer on a country road bothers no one except the operator (and he can wear muffs). But if the jackhammer is used within occupied buildings it may be exceedingly annoying to the occupants.

Another requirement for effective noise legislation is that it be consistent with current technology, while still recognizing that further reductions of the noise level in the future may be desirable. The noise regulations should begin by a gradual phasing out of old and noisy devices that create the peak disturbances. Newer, quieter devices can then be phased in without imposing too great a burden on manufacturers or operators of these devices.

Control of noise will require a public awareness both of the problem and of the potentials for solving it. Research and development have an important role to play in seeking new and improved ways of controlling noise. But what is primarily needed is to apply, on a broad scale, the knowledge that is already available.

Effects of noise

Clear evidence supports the following statements about the effects on people of exposure to noise of sufficient intensity and duration:

• Exposure to noise can result in temporary hearing losses, and repeated exposure can lead to permanent hearing loss.

• Noise can interfere with speech communication and the perception of other auditory signals.

• Noise can disturb sleep.

• Noise can be a source of annoyance.

• Noise can interfere with the performance of complicated tasks and can especially disturb performance when speech communication or response to auditory signals is demanded.

• Noise can adversely influence mood and disturb relaxation.

Understanding the relationships between the human auditory system and the principles of sound generation, propagation, and measurement is confusing unless the terminology and technical references are clearly defined. To clarify the subject, the next section is devoted to a short explanation of how the human auditory system works. (See Chapter 7 for more details.)

How we hear

The organ of hearing consists of three distinct compartments or regions, each of which has its own special part to play in the mechanism of hearing. These regions are termed the outer (external), middle, and inner (internal) ear. When sound enters the ear, the waves pass along the ear canal to the eardrum; each wave causes a sympathetic vibration of the eardrum. The eardrum conducts these vibrations to three tiny bones called ossicles—the three smallest bones in the body.

The ossicles transmit the vibrations to a fluid contained in a tiny, snail-shaped structure called the cochlea. Within the cochlea are microscopic nerve endings that respond to these fluid vibrations. The neural impulses created by the nerve endings are transmitted to the brain where they are interpreted as sound (see Figure 1–2).

Hearing loss

Loss of hearing may be defined as any reduction in the ability to hear from that of a normal person. Such loss may be classified into two general categories:

• *Temporary hearing loss* from exposure to loud noises; normal hearing usually returns after a rest period. The recovery period may be minutes, hours, days, or even longer, depending upon the individual and the severity and length of exposure.

• *Permanent hearing loss* may occur as a result of the aging process, diseases, injury, or exposure to loud noises for extended periods of time. The hearing loss associated with exposure to industrial noise is commonly referred to as noise induced hearing loss. This type of hearing loss is the result of nerve or hair cell destruction in the hearing organ and it is *not* reversible. However, such hearing losses usually are only partial.

Total hearing loss is most frequently associated with disease or traumatic injury.

The most common levels of industrial noise exposures are well below the threshold of pain. There is a wide range of noise levels and exposure times which may cause an impairment of hearing in exposed individuals.

Individual susceptibility may also be a factor. Noise induced permanent hearing loss is first evident in a reduction in the ability to hear the higher frequency sounds. As the exposure continues, the reduction progresses to the frequency of sounds in the speech range. Research studies indicate that any permanent effect on the hearing organ is unlikely unless it is preceded by a temporary threshold shift of the hearing level. This shift can be detected by an appropriate program of monitoring audiometry, which should be used to alert management that there is a risk of permanent damage to the hearing organs of some employees.

Effects similar to, or perhaps identical with, those produced by industrial noise are also produced by other agents, or result from other causes. Certain drugs used for the treatment of disease may reduce the hearing sensitivity of the ear.

Another complicating factor in the industrial noise picture is the loss in hearing sensitivity that takes place as people grow older. Technically, the decrease in hearing sensitivity accompanying the aging process is known as *presbycusis*. Hearing loss due to both noise exposure and presbycusis is the result of nerve or hair cell destruction or deterioration, and both are permanent.

In general, hearing losses are of two major types: conductive and sensori-neural. The accompanying typical audiograms (Figures 1–3 and 1–4) show the nature of these types. The conductive type is not caused by sustained noise exposure. The sensori-neural type may be caused by noise exposure as well as other causes, including presbycusis.

There are numerous infectious diseases that may produce loss of hearing, which may be confused with a loss resulting from exposure to industrial noise. It is important that these other factors be given appropriate consideration when evaluating the effects of industrial noise on man.

Speech interference

Speech interferences is an easily demonstrable effect of noise exposure. For situations such as listening to the radio or conversing, speech interference can be a distinct source of annoyance. For situations in which

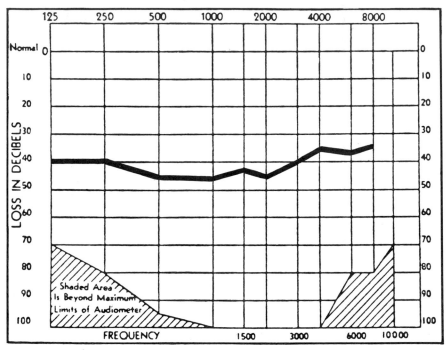

Fig. 1–3.—This audiogram shows the typical configuration associated with conductive or middle ear hearing impairment.

verbal communication is critical—commands to a crane operator who is hoisting or lowering loads, for example—speech interference can be a very real hazard.

Annoyance

Psychologic (emotional) reactions of a person to a sound involve a multiplicity of factors which vary with the characteristics of sound—its appropriateness, unexpectedness, interference with speech communication, and intermittancy, as well as its intensity and frequency. The quality of the noise rather than the quantity is usually the deciding factor in influencing the emotional reaction to noise.

No doubt the most widespread reaction to noise is that of annoyance. Certain characteristics of sound appear more annoying than others. These characteristics are:

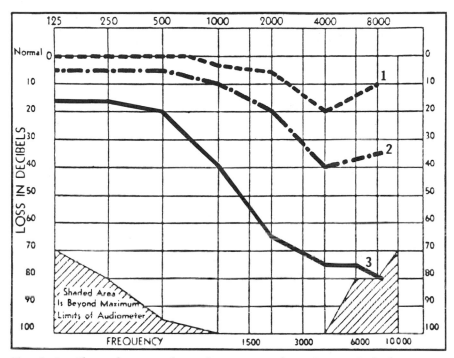

Fig. 1–4.—Shown here are the audiometric configurations typical of sensory-neural hearing losses. Curve 1 early, curve 2 intermediate, curve 3 advanced.

• *Loudness.* The more intense, louder noises are considered more annoying.

• *Pitch.* A high-pitch noise is generally more annoying than a low-pitch noise of equal loudness.

• *Intermittancy and irregularity.* Sound that occurs randomly or varies in intensity or frequency appears to be more annoying than continuous or unchanging sounds.

• *Localization.* A sound which appears to change its relative location to the listener is more annoying than a stationary source.

• *Noise level.* Annoyance can increase as the level of the noise increases above the background noise level.

• *Exposure time.* Length of exposure is obviously an important factor in the degree of annoyance.

• *Control.* Actually, the lack of control. It is very frustrating and annoying to be subjected to a loud noise and not be able to do something about it. Conversely, the degree of annoyance will be less if the person being subjected to the noise has it in his power to reduce the noise level (turn the volume down on the radio), move away from the source, or turn the offending source off.

For more details on annoyance, see Chapters 3 and 10.

Measurement and analysis

To be able to control noise effectively, it is necessary that the noise be measured objectively, according to recognized standardized procedures, and that the measured results be evaluated against accepted criteria. This section will describe briefly the properties of sound and some of the measurement methods.

Sound

Sound is a form of energy that travels through air (or anything else) in waves, by the alternate compression and rarefaction of the elastic medium. Sound waves are, therefore, a series of pressure variations that radiate from a source as spherical shells. Sound becomes *noise* when it is unwanted. As each wave—first a compression, then a rarefaction— encounters an object, it exerts a force—a push, then a pull—on the object. This is why sound waves can break a glass or cause a window screen to vibrate. Sound waves have three significant characteristics to the human listener—pitch, loudness, and quality (or timbre) (see Figure 1–5).

Pitch is related to frequency which expresses the number of times the air pressure changes in a given time. Musicians ordinarily use the term *pitch*, engineers the term *frequency*, expressed as cycles per second and abbreviated as: cycles/sec, cps, c/s, or simply "cycles." Recently, the term "hertz," abbreviated Hz and preferred in current usage, has been adopted to represent one cycle per second. Thus, a sound source that causes the pressure to change through a complete cycle 2000 times a second is said to have a frequency of 2000 cps or 2000 Hz (or 2 kHz, kilohertz).

Pitch is related to the noise source vibration frequency—the greater the number of vibrations, the higher the pitch. A low or a high note,

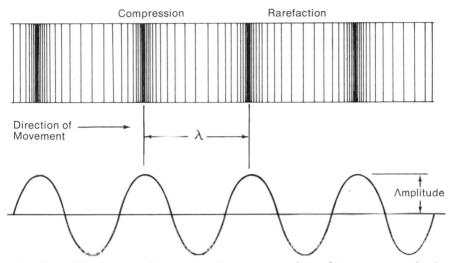

Fig. 1–5.—Illustration of how vibrations are transformed into waves. A vibrating source produces compression of the air molecules. Sound waves are a series of pressure variations that radiate outward from a source.

the bass or treble of the piano, or a bass or soprano voice, all mean that the sounds are of low or high pitch, respectively.

Loudness is a measure of the intensity of the sound wave as perceived by the human ear. The intensity of a sound wave may be compared with how hard a sound wave hits an object, a characteristic which can be measured precisely with instruments. But the loudness heard by a human ear is slightly different from the purely objective physical measurements. The ear hears sound at intermediate frequencies better than sound at very low or very high frequencies.

The relative loudness of a sound depends upon the frequency and the intensity. The ear is unsurpassed in the vast range of intensities to which it responds, as well as in its extreme sensitivity to faint signals. It is, however, relatively insensitive to changes in intensity. Intensity must be increased roughly 26 percent before the ear can sense a change in loudness.

Sound pressure. Sound intensity can be expressed as pressure. The sound pressure response range of the ear (from 4 to 4 ten-millionths

19

b/sq ft) means that sound level measurements over that range involve ratios of up to 10 million to 1. Because such ratios are awkward to handle arithmetically, the term decibel, which represents relative intensity, was borrowed from electrical communication engineering. The decibel (abbreviated "dB") is a dimensionless unit that expresses the ratio of two numerical values on a logarithmic scale with reference to base level. (For more information on the decibel, see Chapter 2.)

Quality is the peculiarity or distinctiveness of a sound apart from its pitch and is also called timbre. It makes middle C played on a piano distinguishable from the same note played on a trumpet, clarinet, or violin. A tone, unless it is pure, will have one or more harmonics superimposed on it. The harmonics alter the tone, and impart richness and color to it. Sounds of the same predominant vibration frequency—that is, of the same pitch—may differ greatly in quality. The quality of the sound wave depends upon the number and character of the simple harmonic waves of which it is composed.

Sound measuring instruments

The most commonly used instrument for measuring noise is the sound level meter. The typical sound level meter electronically weights the amplitudes of the various frequencies approximately in accordance with a person's hearing sensitivity and sums the resulting weighted spectrum to obtain a single number. Typically the sound level meter contains three different response weighting networks: A, B, and C networks (see Figure 1–6).

The most commonly used scale on the sound level meter is the A-scale, since it has been found to account fairly well, although not perfectly, for man's perception of sound. When using the sound level meter on A-weighting, the quantity obtained is the A-weighted sound level. Its unit is the decibel (dB), often popularly referred to as dBA.

Sound measuring instruments are designed to provide a readout in decibels. A sound level meter is composed basically of a microphone, an electronic amplifier, an indicating meter calibrated to read in decibels, and attenuating networks to alter the range. This meter can measure total sound pressure variations if the sound is steady; that is, if its fluctuations are not rapid, such as from an impact.

Octave-band analyzers are used to measure the sound level in a specified frequency band. Usually, the output of the sound level meter is fed into the analyzer, then by manipulation of the filters, the level of

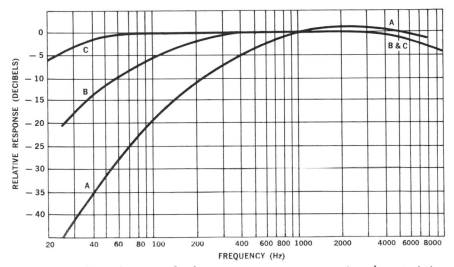

FIG. 1–6.—Shown here are the frequency response attenuation characteristics for the A, B, and C networks.

sound in each octave band is read.

Other instruments used to measure properties of sound waves include sophisticated analyzers with filter networks for measuring one-third octave bands, impact sound level meters, dosimeters, and vibration level meters. Instruments are discussed in detail in Part Two—Measurement of Sound.

What measurements to make. Noise measurements range from a single sound level to a detailed analysis showing hundreds of components of a complex vibration (see Figure 1–7, next page).

The number of measurements taken and the type of instruments needed depends on the information that is required. If the problem is one of checking compliance with a certain noise specification, the particular measurement required is reasonably clear and only some guidance as to choice of instruments and their use is needed. But, if the goal is to reduce the noise produced by industrial operations in general, the situation is more complex and careful attention to the acoustic environment is essential.

In principle, it is impossible to assemble a noise measurement system

Fig. 1–7.—Noise measurements may range from a single measurement to a detailed analysis of the various octave-band levels that may be present.

that does not in some way distort or alter the sound field being studied, partly because the sound measuring instruments interfere with the sound field. It is essential that the observer have an understanding of the sound field under observation and a basic understanding of what is going on within the instrumentation system.

Measurement of the noise field may require using different types of sound level measuring instruments. These measurements must be repeated as changes in noise producing equipment or operating procedures occur. Calibration checks of the instruments should be made before, during, and after the sound level survey.

For many noise measurement situations, the A-weighted sound level, expressed in units of dBA, provides very useful information. The A-weighted sound level is obtained by electronically filtering or weighting a sound signal to give greater emphasis to the speech frequencies and lesser emphasis to the lower frequencies. (As mentioned earlier, this parallels the way that the human ear hears the sound spectrum.)

The use of the dBA scale for preliminary noise measurement greatly simplifies the collection of sound level survey data. Detailed sound level survey and octave-band analysis data are necessary to provide sufficient information to determine the proper remedial measures in noise control procedures.

Analysis of noise exposure

The critical factors in the analysis of noise exposures are (a) A-weighted sound level, (b) frequency composition or spectrum of the noise, and (c) duration and distribution of noise exposure during a typical workday.

It is currently believed that exposure of unprotected ears to sound levels in excess of 115 dBA is hazardous and should be avoided. Exposure to sound levels below 70 dBA can be assumed to be safe and not produce any permanent hearing loss. The majority of industrial noise exposures fall within this 45 dBA range and additional information is required to reach a decision as to the degree of hazard.

It would be very helpful to know the predominant frequencies present or contributions from each of the frequency bands that make up the overall level. It is currently believed that noise energy having predominant frequencies above 500 Hz has a greater potential for causing hearing loss than noise energy concentrated in the low-frequency regions. It is also believed that noises, having a sharp peak in a narrow-frequency band (such as a pure tone), present a greater hazard to hearing than those noises having a continuous distribution of energy across a broad frequency range.

The incidence of noise induced hearing losses is directly related to the total exposure time. In addition, it is believed that intermittent exposures are far less damaging to the ear than continuous exposures, even if the sound pressure levels for the intermittent exposures are considerably higher, because the rest periods between noise exposures allow the ear to recuperate.

At the present time, the deleterious effects of noise exposure and the energy content of the noise cannot be directly equated. Doubling the energy content does not produce double the hearing loss. In general, the greater the total energy content of the noise, the shorter the time of

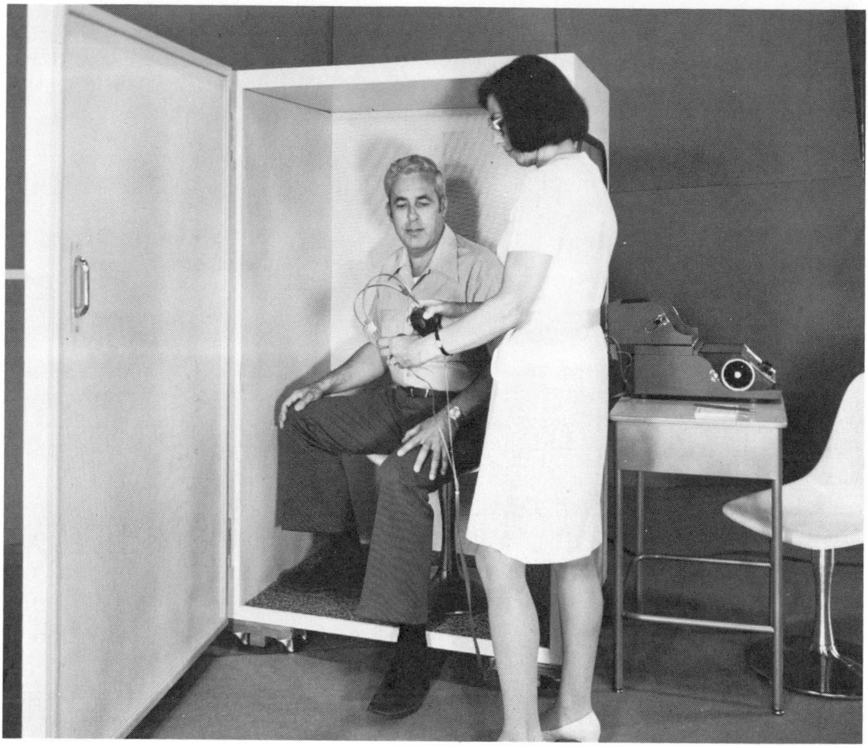

Fig. 1–8.—The audiometric techncian should describe to the subject what he should do during the audiometric test.

exposure required to produce the same amount of hearing loss, but the exact relation between time and energy is not known.

Another factor that should be considered in the analysis of noise exposures is the type of noise, for instance impact noise, such as that generated by drop hammers and punch presses, or steady-state noise, such as from turbines or fans. Impact noise is a sharp burst of sound and sophisticated instrumentation is necessary to determine the peak levels for this type of noise. Additional research needs to be done to fully define the effects of impact noise upon the ears.

The total exposure during a normal work life must be known to arrive at a valid judgment of how noise will affect hearing loss. Instruments such as noise dosimeters can be used to determine the exposure pattern of a particular individual. Instruments such as sound level meters can be used to determine the noise exposure at a given instant in time (that is, during the time the test is being taken). An exposure pattern can be established using a series of such tests and the work history of the individual.

Industrial audiometry

Industrial audiometry is an important part of the hearing conservation program and is the subject of Part Five—Industrial Audiometry. Briefly, the objectives are to accomplish the following:

1. Obtain a baseline audiogram of an individual's hearing ability at the time of the preemployment examination
2. Detect significant hearing threshold shifts in exposed employees during the course of their employment
3. Provide a record of an employee's hearing acuity
4. Check on the effectiveness of noise control measures by measuring the hearing thresholds of exposed employees
5. Comply with government regulations

An audiometer is required for assessing an individual's hearing ability. It consists of an oscillator which produces pure tone sounds at predetermined frequencies, an attenuator which controls the intensity of the sound or tone produced, a presenter switch, and earphones through which the person whose hearing is being tested hears the tone (see Figure 1–8).

Threshold audiometry. Threshold audiometry, as its name implies, is used to determine the auditory threshold for a given stimulus. Measurements of hearing are made to determine the presence of abnormal function of the ear. Before hearing can be described as abnormal, a reference point, or normal value, must be designated.

The standard values of the average normal hearing threshold for pure tones are expressed in decibels of sound pressure level greater than the reference sound pressure level, 0.0002 dyne/square centimeter (20 micropascals). The quantity that is of interest, however, is not the sound pressure level of the normal hearing threshold but rather the magnitude of departure from this standard reference threshold. Audiometers are calibrated to read "zero" when the sound pressure level of the output signal is equal to the sound pressure level of the reference standard. Departures from reference can then be read directly from the setting of the audiometer attenuator dial.

Hearing threshold levels are those intensities at which sound or a tone can just be heard. The term "air conduction" describes the air path by which the test sounds generated at the earphones are conducted through air to stimulate the eardrum.

The record of measured hearing thresholds is called a *threshold audiogram*. Audiometric tests can also be recorded in the form of audiograms on which are plotted both sound intensity (in decibels) and frequency (in hertz). A sample audiogram is shown in Figure 1–3, p. 16.

There are at least four sources of variability which may affect the results of pure tone threshold tests—namely, variability in the test environment, in the subject's performance, in the tester's performance, and in the test equipment.

Variability in the test environment is caused mainly by ambient noise in the hearing testing area. Unless test areas are specially treated to reduce ambient noise below specified values, the measured hearing threshold of persons with normal hearing may be masked by the background noise. Because the more common ambient noises in test environments are predominantly low-frequency sounds, most of the masking occurs at 1000 Hz and below. For more details, see Chapter 27.

Ambient noise in the testing environment is perhaps the most frequent environmental cause of variations in audiograms, but it is by no means the only cause. The subject must be comfortable so that he will not become restless and squirm or fidget. Humidity, temperature, and the size and decor of the test room are all factors that can influence comfort, and it is important to pay attention to these details in order to provide optimum conditions for obtaining good audiometric measurements. See Chapter 28.

Variability in test results can also be caused by malfunctioning equipment. Audiometers must be calibrated periodically if they are to produce accurate test results. The calibration should include both the electronic circuits and the earphones, and should be repeated at intervals not longer than one year and more frequently when erratic operation or erroneous test results are suspected. A biological check of the audiometer, made by testing the hearing of persons with established audiometric patterns, should be performed periodically before use. (See Chapter 22 for more details.)

Data form

The medical form used in audiometric testing programs should include all basic data relating to the hearing evaluation. Hearing threshold values, noise exposure history, and pertinent medical history items should be accurately recorded each time an employee's hearing is tested. During preemployment examination, it is essential to obtain a thorough history, both medical and occupational. This includes any previous ear disease, exposure to noise, and family history of deafness. Some typical forms used in industry are shown in Figure 1–9. The employee is identified by name, social security number, sex, and age. Additional information includes the date and time of test (day of week, time of day), conditions of test, and the name of the examiner.

Who should be examined

Preemployment hearing threshold tests should be taken by all job applicants, not just those who are to work in noisy areas. This will establish a baseline hearing threshold for each employee for future comparison. Preemployment hearing tests are essential if a company is to protect itself from being found liable for preexisting hearing loss incurred elsewhere. If an employee is hired with hearing damage and he is subsequently exposed to high noise levels, the company may be liable for all of his hearing loss, unless it can be proven that the employee had a preexisting hearing loss.

If a preemployment audiogram shows any unusual irregularity, particularly an abrupt loss beginning at 2000 Hz, the applicant should be referred to a physician for otological examination and placement evaluation. The audiometric technician should be sure that at least 14 hours have elapsed since the subject's last noise exposure to levels in excess of 80 dBA, whenever an audiogram is performed. This requirement may be met by having the employee wear hearing protectors that reduce the noise exposure level to less than 80 dBA.

Periodic followup hearing tests should be given to persons stationed in areas where noise exposures exceed the suggested hearing conservation criteria. Not all employees have the same susceptibility to hearing loss. What may be a permissible level for some may cause hearing damage to others. This is primarily why periodic audiograms should be made for each employee working in sound levels that exceed acceptable values.

The schedule for periodic retesting of hearing depends largely upon the employee's noise exposure. Assuming that a record of the worker's hearing status was established at the time of employment or placement,

(*Text continues on page 30.*)

AUDIOLOGICAL EXAMINATIONS

NAME		DATE OF BIRTH	OCCUPATION		SHIFT	I.D.NO.

Occupational History (Beginning with last previous, working back to first job.)

DEPT.

	EMPLOYER	CITY	DUTIES	LENGTH OF SERVICE	NOISE EXPOSURE	EAR PROTECTORS
1.					☐ YES ☐ NO	☐ YES ☐ NO
2.					☐ YES ☐ NO	☐ YES ☐ NO
3.					☐ YES ☐ NO	☐ YES ☐ NO
4.					☐ YES ☐ NO	☐ YES ☐ NO

MILITARY SERVICE	TIME SERVED	BRANCH			(OTHER)	EXPOSURE TO GUNFIRE & NOISE
		☐ ARMY ☐ NAVY ☐ MARINES ☐ AIR FORCE				☐ YES ☐ NO

CHECK IF YOU HAVE OR HAD ANY OF THE FOLLOWING:

MEDICAL HISTORY

☐ ALLERGY ☐ DIZZINESS ☐ EARACHES ☐ FREQUENT COLDS ☐ INFLUENZA ☐ MEASLES

☐ MUMPS ☐ RINGING EARS ☐ RUNNING EARS ☐ SCARLET FEVER ☐ WHOOPING COUGH ☐ OTHERS

SERIOUS MEDICAL AILMENTS	SURGERY (TYPE OF OPERATIONS)	INDUSTRIAL INJURIES OR DISEASE

HEAD INJURIES	HEARING IMPAIRMENT IN FAMILY	WORKER'S EVALUATION OF HEARING STATUS
		☐ GOOD ☐ FAIR ☐ POOR

EDUCATION (HIGHEST GRADE COMPLETED)	PREVIOUS HEARING TEST (DATE & COMPANY)	DATE LAST WORKED

TECHNICIAN	STATION	COMMENTS	DATE	A.M. P.M.	DAY OF WEEK

FIG. 1-9.—The audiometric data form should include all basic data relating to the hearing evaluation.

1703 Recording Audiometer
Grason-Stadler a GR company

NAME______________________________ NO.________

AGE____ SEX____ DATE____ TIME____ BY____ AUDIOMETER NO.____

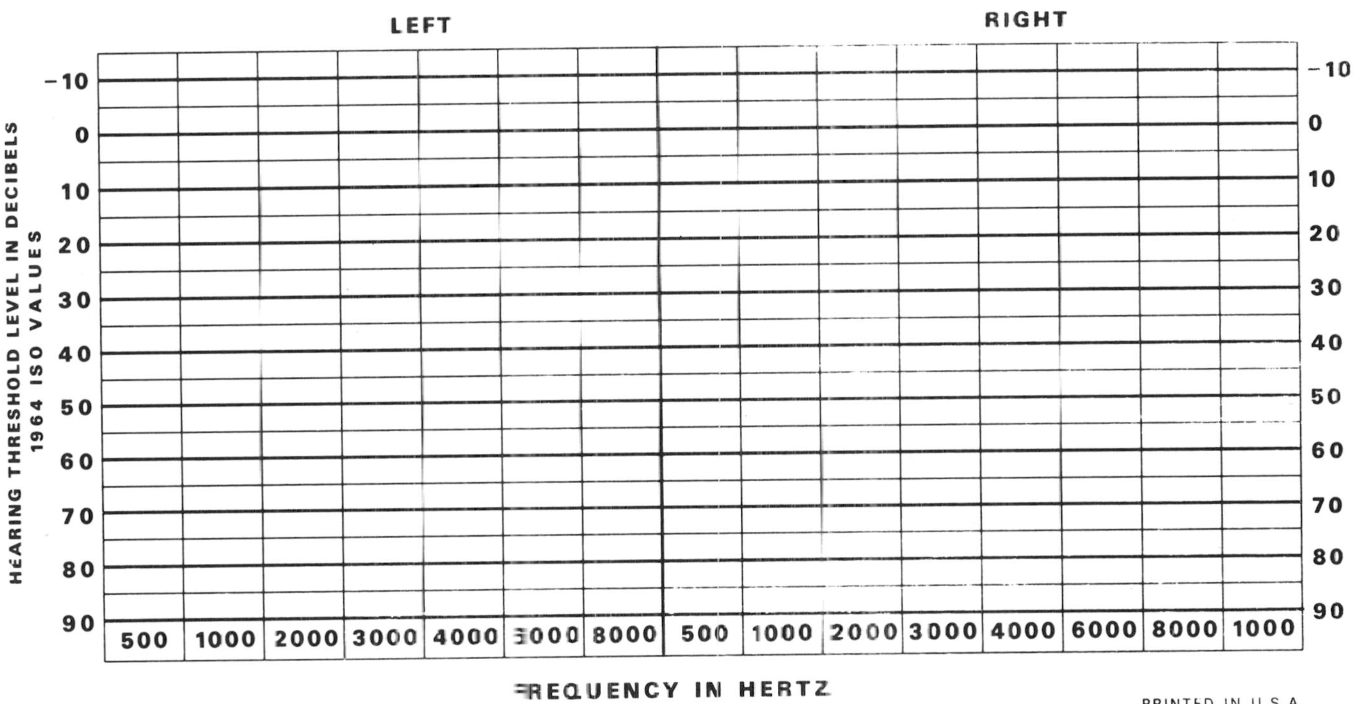
LEFT
RIGHT
HEARING THRESHOLD LEVEL IN DECIBELS
1964 ISO VALUES
-10
0
10
20
30
40
50
60
70
80
90
500
1000
2000
3000
4000
8000
500
1000
2000
3000
4000
6000
8000
1000
FREQUENCY IN HERTZ

CHART No.1703-9102

PRINTED IN U.S.A.

the first reexamination should be made from nine to twelve months after placement. If no significant threshold shifts (greater than 10 dB) relative to the preplacement audiogram are noted, subsequent followup tests can be made at yearly intervals. Where the noise exposure is relatively low, the interval between followup tests can be increased, based upon conditions of exposure and medical judgment considering results of previous audiograms and clinical examination.

Noise exposure is by no means the only reason for a change in an individual's audiogram. When a change in hearing status is confirmed, a determination must be made as to the cause. Improper placement of earphones and excessive ambient noise in the test room would certainly affect audiogram results. Physiological changes such as age would also affect audiogram results. Motivation of the individual and his attitudes toward the test can affect performance.

The industrial audiometric program can identify persons having hearing threshold changes not related to noise exposure. These workers should be referred to their family physician for diagnosis and treatment. However, when workers having threshold shifts related to noise exposure are identified, the subsequent procedure could be followed:

1. Check the fit of hearing protection, if worn by the worker

2. Repeat or start educational sessions to encourage wearing hearing protection if it has not been worn

3. Investigate the noise levels in the work area particularly if a previous sound level survey failed to reveal noise hazards

Noise exposure information, correlated with audiometric test results, is needed to arrive at intelligent decisions about a firm's hearing conservation program. Should all hearing tests and medical opinion point to a progressive deterioration of an individual's hearing, the safety professional can insist that hearing protection equipment be worn, and that the individual's time of exposure to excessive noise be controlled.

Noise control programs

Accurate information defining the extent, duration, and distribution of noise levels would be needed to provide a solution to a noise problem. The degree of noise reduction required is obtained by comparing the measured levels with acceptable noise criteria. The next step is to consider various noise control measures such as engineering design, limiting the time of exposure, or the use of personal protective devices to achieve

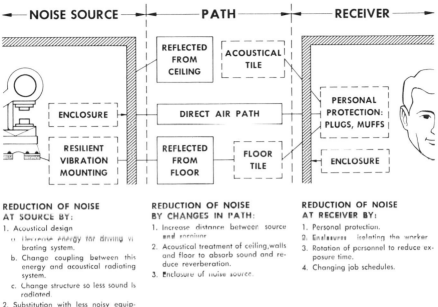

REDUCTION OF NOISE AT SOURCE BY:

1. Acoustical design
   a. Decrease energy for driving vibrating system.
   b. Change coupling between this energy and acoustical radiating system.
   c. Change structure so less sound is radiated.
2. Substitution with less noisy equipment.
3. Change in method of processing.

REDUCTION OF NOISE BY CHANGES IN PATH:

1. Increase distance between source and receiver.
2. Acoustical treatment of ceiling, walls and floor to absorb sound and reduce reverberation.
3. Enclosure of noise source.

REDUCTION OF NOISE AT RECEIVER BY:

1. Personal protection.
2. Enclosures isolating the worker
3. Rotation of personnel to reduce exposure time.
4. Changing job schedules.

FIG. 1–10.—Every noise problem breaks down into three component parts, (*a*) a source radiating sound energy, (*b*) a path along which the sound energy travels, (*c*) a receiver such as the human ear.

the desired level of reduction.

Every noise problem breaks down into three component parts (*a*) a source radiating sound energy (*b*) a path along which the sound energy travels and (*c*) a receiver such as the human ear (Figure 1–10). The "system" approach to noise problem analysis and control will assist in understanding both the problem and the changes that will be necessary for noise reduction. If each part of the system—source, path, and receiver —is put in its proper perspective, the overall problem will be greatly simplified. To help translate these principles into practical terms, specific examples of controlling industrial noise exposure are outlined in the next section.

Source

Noise control often can be designed into the equipment so that little or no compromise in the design goals is required. Noise control undertaken on existing equipment usually is more difficult. Engineering control of industrial noise problems requires the skill of individuals who are highly proficient in this field of knowledge.

Noise control strategies need careful objective analysis on both a technical and economic basis.

Complete redesign would have product designers immediately consider noise level as a primary product specification for design of all new products. Full replacement of the product population would eventually take place depending on the service life of each product. Many designers feel this approach will minimize cost increases associated with noise control measures. Existing product modifications would require manufacturers to modify or replace existing products to lower the noise levels of noisy equipment.

The existing equipment within any plant was probably selected on the basis that it was the most economical and efficient method of producing the product. Careful acoustical design can result in quieter equipment that would also be more economical to operate than noisier equipment. Examples of control at the source are the substitution of quieter machines, the use of vibration-isolation mountings, and the reduction of the external surface areas of the vibrating parts as much as possible. Machines mounted directly on floors and walls may cause them to vibrate, resulting in sound radiation. The proper use of machine mountings insulates the machines and reduces the transmission of vibrations to the floors and walls.

Although substitution of less noisy machines may have limited application, there are certain areas in which substitution has potential application. Examples include using "squeeze" type equipment in the place of drop hammers, welding in place of riveting, and chemical cleaning of metal instead of high-speed polishing and grinding.

Noise path

Because the desired amount of noise reduction cannot always be achieved by good acoustic design at the source, modification of the noise path and receiver must be considered.

Noise reduction along the path can be accomplished in many ways— by shielding or enclosing the source, by increasing the distance between source and receiver, or by placing a shield between the source and the receiver. Examples of noise reduction along the path are baffles and the enclosure of noise producing equipment to minimize transmission of noise to areas occupied by employees. Use of acoustical material on walls, ceilings, and floors to absorb sound waves and to reduce reverberations can result in significant noise reduction.

Noise produced by a source travels out in all directions. When the sound waves encounter solid objects, such as machinery or walls, they are reflected. Thus, the total noise level within the room depends upon both the direct and reflected noise. The application of sound absorption material to walls can reduce the reflected noise level in the room.

Enclosures

Noise entering or leaving an area can best be prevented by sealing all outlets. In extreme cases, double structures can be used. Special treatment, including the use of steel and lead panels, is available to prevent noise leakage in certain cases. Gaskets around doors also can reduce noise transmission from one space to another. Inside a building, carpets, drapery, and acoustical tiles can provide some reduction of the noise levels in the building, although such internal treatments do not affect the noise intrusion from outside. In situations where the number of operators is small and the process is such that the work task can be confined to a limited area, isolation of workers in a separate noise insulated room provides effective control (see Figure 1–11, next page).

Checklist. A checklist of some of the most effective approaches to noise control which can be used by any versatile engineer in evaluating and attempting to solve some of the simpler noise problems is presented here.

- Reduce impact or acceleration.

- Reduce unbalanced forces.

- Reduce large radiating areas (sounding boards).

- Reduce resonance effects in mechanical and acoustical systems and couplings.

- Use acoustic enclosures where possible to contain the noise completely.

- Use acoustic barriers to shield or deflect or help absorb the noise, or partial enclosures to contain the noise if complete enclosures are impractical.

- Reduce or eliminate noise leakage paths.

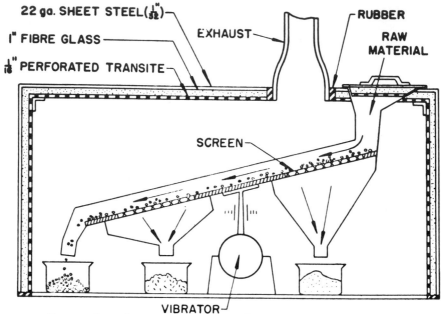

22 ga. SHEET STEEL($\frac{1}{32}$")
I" FIBRE GLASS
$\frac{1}{16}$" PERFORATED TRANSITE
EXHAUST
RUBBER
RAW MATERIAL
SCREEN
VIBRATOR

From Industrial Noise Manual, 2nd ed., *American Industrial Hygiene Assn.*

FIG. 1–11.—A total enclosure for a screen sifter provides sufficient noise attenuation.

- Use acoustic absorption material to absorb sound energy inside confined spaces and in sound control ducts or passageways.

- Use mufflers or attenuators to reduce noise in gas-flow paths.

- Provide greater distances between the noise sources and the noise receivers.

- Use vibration isolation mounts to isolate a vibration source from a noise radiating surface (the sounding board).

- Use flexible connections (electrical, piping, ducting, structural, etc.) between the isolated source and its base supporting structure.

- Use vibration-damping materials to reduce noise radiation from thin surfaces.

- Seek less noisy methods of performing the same function.

The final success of a noise reduction project usually depends on the ingenuity with which these basic approaches can be applied without decreasing maximum use and accessibility of the machine or other noise source that is being quieted.

Noise control at the ear of the receiver would include administrative controls or regulation of exposure time and the use of personal hearing-protective devices.

Administrative controls. Administrative controls involve a reduction of the noise exposure time by changing job schedules or rotating personnel to reduce the amount of time spent in noisy areas. Remote control (TV monitors) and control rooms can also be used to reduce the time of noise exposure by allowing the employee to monitor a particular process or to perform an operation away from the noise producing area.

Personal protection. There are some operations in industry that cannot be quieted by engineering methods. Earplugs or muffs worn by the worker can provide effective protection in such cases. Numerous reports have shown that such devices can provide up to 25 to 30 dB of protection if properly worn (see Figure 1–12).

Of the two general types of hearing protection—plugs and muffs—plugs are designed to occlude the ear canal and muffs are designed to cover the external ear. See details in Chapter 20, Personal Hearing-Protective Devices.

The choice of plugs or muffs or both depends in part on the work situation. Will the employee's head be confined to a work space so small there is no room for muffs? Must he wear a safety hat in addition to ear protection? And so forth. There are advantages and disadvantages to the use of either plugs or muffs, and before a choice is made between the two, all the circumstances of a particular job should be considered.

Fitting and indocrination. An employee's ears should be examined and his hearing tested at the time he is fitted with hearing protectors. Plugs should be fitted individually for each ear. If the ear canals are not the same size or shape, they may require plugs of different size. To promote the acceptance of earplugs, an employee should be allowed to choose from two or three different makes at the time he is fitted (see Figure 1–13).

As with other kinds of personal protection (safety hats, safety glasses, safety shoes, or respirators), it may be difficult to convince employees

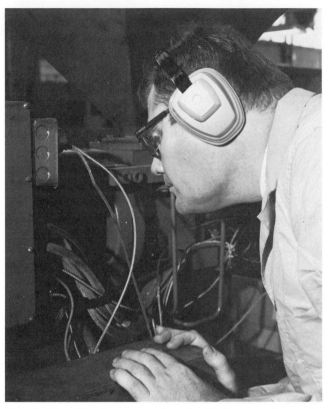

Fig. 1–12.—Hearing protection should be worn by workers at those operations where the noise source cannot be controlled by engineering methods.

that they should wear hearing protectors. Successful personal protection programs are based on thorough indoctrination of personnel. An employee *must* be impressed with the importance of hearing protection and the benefits to be gained from its consistent use. He should know the following facts:

• Good protection depends on a good seal between the surfaces of the skin and the surface of the hearing protector. A very small leak can destroy the effectiveness of the protection. Because protectors have a tendency to work loose as a result of talking or chewing, they must be reseated from time to time during the workday.

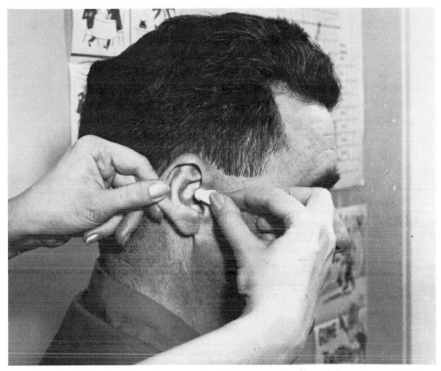

FIG. 1–13.—The employee's ears should be examined and hearing protection should be fitted individually for each ear.

• A good seal cannot be obtained without some initial discomfort.

• There will be no untoward reactions as a result of the use of hearing protectors if they are kept reasonably clean.

• A properly designed and fitted, clean ear protector will cause no more difficulty than a pair of safety goggles.

• The use of hearing protection will not make it more difficult to understand speech or to hear warning signals when worn in a noisy environment.

Most of the available hearing protectors, when correctly fitted, provide about the same amount of protection. The best hearing protector, therefore, is the one that is accepted by the employee and works properly. Properly fitted protectors can be worn continuously by most persons and will provide adequate protection against most industrial noise exposures. The average worker will wear his hearing protection properly if he understands the problem through education by qualified personnel. Therefore, education of the employee is a major facet in an effective hearing conservation program.

The effectiveness of a hearing-conservation program depends upon the cooperation of employers, employees, and others concerned. Management's responsibility in this type of program includes noise control measures, provision of hearing protection equipment where it is required, and informing employees of the benefits to be derived from a hearing conservation program.

It is the employee's responsibility to make proper use of the protective equipment provided by management. It is also the employee's responsibility to observe any rules or regulations in the use of equipment to minimize the noise level exposure.

Sources of help. The safety specialist who believes that he has an industrial noise problem should consult a competent acoustical engineer or industrial hygienist. Such help may be obtained from the insurance carrier, the National Safety Council, or private consultants.

Bibliography

American Industrial Hygiene Association, *Industrial Noise Manual,* 2nd ed. Akron, Ohio: AIHA, 1966.

Background for Loss of Hearing Claims. American Mutual Insurance Alliance, Chicago, Ill., 1964.

Criteria for Background Noise in the Audiometer Room S3.1–1960 (R-1971). American National Standards Institute, New York, N.Y.

"Guidelines for Noise Exposure Control," *American Hygiene Association Journal,* Oct. 1967.

Harris, C. M. ed. *Handbook of Noise Control.* New York, N.Y.: McGraw-Hill Book Co., 1957.

Hosey, A. D. and Powell, C. H., eds. *Industrial Noise—A Guide to its Evaluation and Control.* (Public Health Service Publication No. 1572, U.S. Dept. HEW), Washington, D.C.: U.S. Government Printing Office, 1967.

Industrial Noise and Hearing Protection, Revised. Employers Insurance of Wausau, Wausau, Wis., 1967.

Petersen, A. P. G. and Gross, Jr., E. E. *Handbook of Noise Measurement,* 7th ed. Concord, Mass.: General Radio Co., 1972.

Chapter Two Physical Characteristics of Sound

By Julian B. Olishifski, P.E.

The elementary principles and physical laws that characterize the measurement, production, transmission, and absorption of sound are discussed in this chapter. Because the chief sources of sounds in air are vibrations of solid objects, the measurement of vibration will also be discussed.

Definition of sound

Sound may be defined in two ways—objectively and subjectively.

• *Objectively,* sound is a form of wave motion due to pressure alteration or particle displacement in an elastic medium.

• *Subjectively,* sound is a sensory experience in the brain.

Sound requires a medium for propagation—it can be gas, liquid, or solid. Unlike electromagnetic waves, sound cannot travel through a vacuum. Thus sound at a particular point is a rapid variation in the pressure of the medium at that point around a steady-state value. In air, the steady-state pressure is atmospheric pressure. (Of course, the average atmospheric pressure does change, but this change is slow enough to be considered constant compared to the rapid pressure variation of sound.)[1]

Sound is a relatively simple form of energy, causing variations in pressure and alterations in direction of molecular movement within media. Sound, like all objective things, exists whether or not any living thing hears it or is affected by it. Noise, on the other hand, is "undesired sound." What is "unwanted" is a subjective determination that must be made by an individual. Chapter 3 discusses this in more detail.

The term *sound* is usually applied to the form of energy that produces a sensation of hearing in humans, while *vibration* usually refers to the nonaudible acoustic phenomena which are recognized by the tactile experience of touch or feeling. However, there is no essential physical difference between the sonic and vibratory forms of sound energy.

The generation and propagation of sound is easily visualized by means of a simple model. Consider a plate suspended in mid air by a string (see Figure 2–1). When struck, the plate vibrates rapidly back and

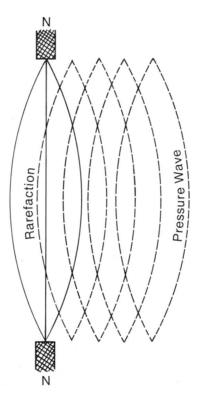

FIG. 2–1.—As the plate moves, it compresses the air in the direction of its motion and when it reverses direction, it produces a partial vacuum or rarefaction.

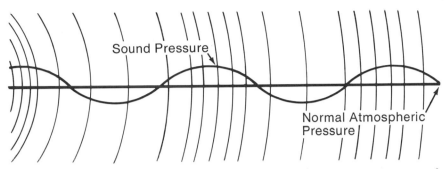

FIG. 2–2.—Air is an elastic medium and behaves as if it were a succession of adjoining particles. The resulting motion of the medium is known as wave motion and the instantaneous form of the disturbance is called a sound wave.

forth. As the plate moves, it compresses the air in the direction of its motion, and when it reverses direction, it leaves a partial vacuum or rarefaction of the air. These alternate compressions and rarefactions cause small but repeated fluctuations in atmospheric pressure which extend outward from the plate. When these pressure variations strike an eardrum, they cause it to vibrate in response to the changes in pressure. The disturbance of the eardrum is translated into a neural sensation and is carried to the brain where it is interpreted as sound.

Sound is invariably produced by vibratory motion of some sort. The sounding body may vibrate transversely, longitudinally, or perform torsional vibrations around an axis. In any case, it must act upon some medium to produce vibrations characteristic of sound. Any type of vibration may be a source of sound, but by definition only longitudinal vibration of the conducting medium is a sound wave.[1]

Sound waves

The elastic medium through which sound is transmitted behaves as if it were a succession of adjoining particles between which are balanced forces of attraction and repulsion. If one of these particles is set in motion, its neighbor follows with the same motion, though lagging behind the first in time (see Figure 2–2). The resulting motion of the medium is known as wave motion, and the instantaneous form of the disturbance as a whole is called a *wave*. This disturbance (initiated at some point) is transmitted (propagated) to other points in a predictable manner determined by the physical properties of the medium existing between

41

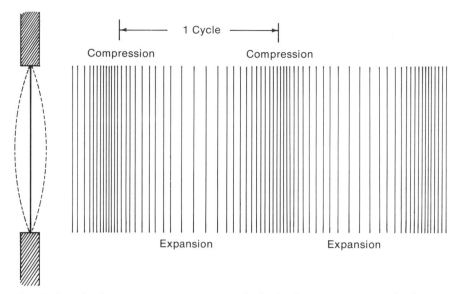

FIG. 2–3.—A plane wave is a wave in which the fronts are perpendicular to the direction of travel.

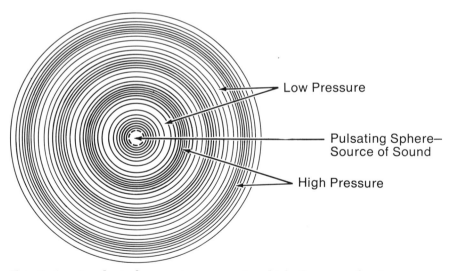

FIG. 2–4.—A spherical wave is a wave in which the wave fronts are concentric spheres.

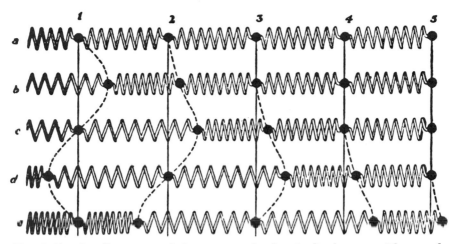

Fig. 2-5.—An illustration of the motion of a longitudinal wave. The metal balls represent the vibrating air particle and the coiled spring represents the elastic medium. The particle in column one successively passes through a compression (condensation) and expansion (rarefaction) phase as it moves about the midpoint.

the points of observation.

Sound waves are a particular form of a general class of waves known as *elastic waves*, which can occur in media having the properties of mass (inertia) and elasticity. Air is a good example of this type media. The elasticity tends to pull a displaced particle back to its original position as a spring would, and because of its inertia, the displaced particle can transfer momentum to an adjoining particle. Because air possesses both inertia and elasticity, a sound wave can be propagated in air.

Sound wave propagation is of two major types—*plane waves* and *spherical waves*. These names are derived from the geometric description of the wave front—the area occupied by the forward part of the pressure disturbance.

• *A plane wave* results, in general, if the source is large, if the measuring point is at a long distance from the source, and if there are no reflecting surfaces of any kind between the source and the measuring point (see Figure 2-3).

• *The spherical wave* generally comes from a pulsating sphere where the point of measurement is reasonably close to the sphere and there

43

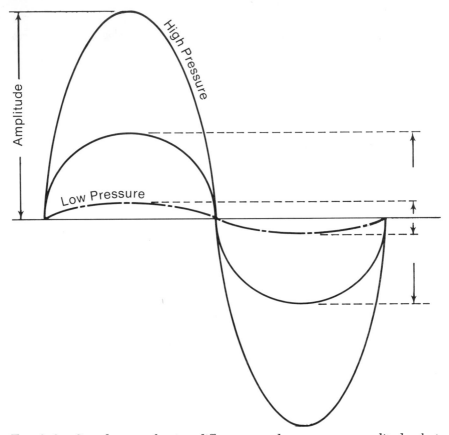

FIG. 2–6.—Sound waves having different sound pressure or amplitudes but the same frequencies are shown here.

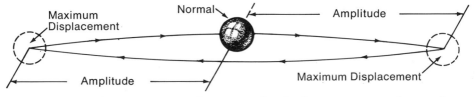

FIG. 2–7.—Relative positions of an air molecule during one complete cycle of motion.

are no reflecting surfaces in the vicinity of the source or the point of measurement (see Figure 2–4). The maximum pressure would lie on the surface of a sphere with the center of the sphere located at the pulsating source. A reasonably small section of the forward wave front at some distance from the source would occupy an area approaching a flat plane.

As described earlier, a sound wave is characterized as a succession of condensations and rarefactions of the particles of the conducting medium. As another illustration (see Figure 2–5), consider a series of metal balls supported by long threads and connected in line by elastic springs. If one ball is moved, its neighbor moves the same way a little later, and this disturbance is transmitted down the line of metal balls. A study of the model would show that the advancing wave is marked by a compression or condensation of the elastic spring between the metal balls followed by a rarefaction, which is indicated by the extended spring.

The distance between the two nearest metal balls that are moving in the same direction at the same phase at the same time is called the *wavelength*. The number of times this complete cycle of compression and extension occurs per second is called *frequency* and is measured in cycles per second, called hertz (Hz).

One sound wave may have three times the frequency and one third the amplitude of another sound wave. However, if both waves cross their respective zero positions in the same direction at the same time, they are said to be *in phase* (see Figure 2–6).

Frequency. A vibrating body disturbs air particles in the same manner as described in the illustration of the metal balls and produces a variation in normal atmospheric pressure. When this disturbance or pressure variation reaches the eardrums it is translated into a sensation that is interpreted as sound. The number of pressure variations (or vibrations) that occur per second determines the frequency. One complete vibration of the object corresponds to one complete cycle of pressure change.

Frequency is the number of times per second that a point on the sound source is displaced from its position of equilibrium, rebounds through the equilibrium position to a maximum displacement opposite in direction to the initial displacement, and then returns to its equilibrium position. In other words, frequency is the number of times per second a vibrating body traces out one complete cycle of motion (see

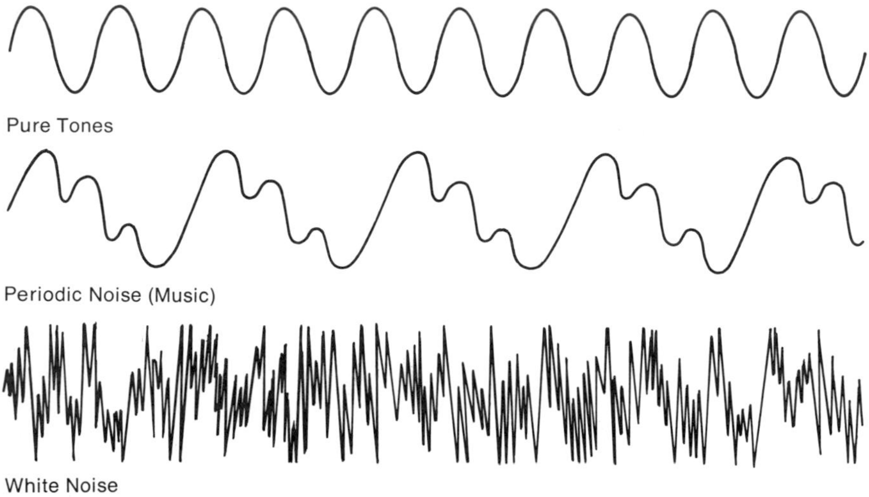

Pure Tones

Periodic Noise (Music)

White Noise

FIG. 2–8.—Representation of waveforms of pure tones, music, and noise.

Figure 2–7). The time required for each cycle is known as the period of the wave, and is simply the reciprocal of the frequency. For a frequency of 100 Hz, the period of vibration is 1/100 or 0.01 second.

Pure tone. The simplest type of sound, called a pure tone, can be described as having only a single frequency. Sounds, as encountered in nature, rarely consist of a single frequency. Music, speech, and noise are each composed of many frequencies. The frequencies comprising speech are found principally between 300 and 3000 Hz (see Figure 2–8).

The range of frequencies that concern people are classified as the *sonic* range (20 to 20,000 Hz), the *ultrasonic* range (above 20,000 Hz), and the *infrasonic* range (below 20 Hz).[2]

The problems of hearing and noise exposure are normally associated with the sonic range. Ultrasonic cleaners are common sources which generate ultrasonic energy. Machines with unbalanced parts that rotate or vibrate very slowly, as well as earthquakes or tremors, are common sources of infrasonic waves.

The medium and the speed of sound. Since a sound wave represents a pressure disturbance in a medium, and since this pressure disturbance moves through the medium, there must clearly be a certain speed of propagation of this pressure disturbance. The speed of sound within any medium depends upon the compressibility and the density of the medium.

In a homogeneous medium, sounds of all frequencies travel at the same speed. The speed of sound at 70 F (21 C) is approximately 1130 ft/sec (344 m/sec) in air, 4700 ft/sec (1433 m/sec) in water, 13,000 ft/sec (3962 m/sec) in wood, and 16,500 ft/sec (5179 m/sec) in steel.[1]

The distance that a sound wave travels in one period or cycle is called the wavelength of the sound. The Greek letter λ (lambda) is always used to express wavelength. The velocity of a sound wave is always equal to the product of the wavelength and the frequency:

$$c = \lambda f$$

where   c — speed, in feet or meters per second
          λ = wavelength in feet or meters
          f — frequency in hertz

It is important to note that the speed of sound is dependent only upon the properties of the medium and is constant at a given temperature. Because wavelength times frequency equals a constant, the wavelength decreases as the frequency increases, and vice versa. If the frequency were 1000 Hz, the wavelength in air would be 1.13 feet (0.34 meter). Likewise, a 100 Hz wave moves a distance of 11.3 feet (3.4 meters) in one cycle when the speed of sound is 1130 ft/sec (344.4 m/sec).

Wavelength. Wavelength is an important property of sound. For example, sound waves having a wavelength much larger than the size of an obstacle are little affected by the presence of that obstacle; the sound waves will bend around it. If a 100 Hz (11.3 ft) sound wave passes through a picket fence, it is disturbed only a little; the sound wave travels on as if the fence were not present at all. This bending of the sound around obstacles is called *diffraction*.

If the wavelength of the sound is small in comparison with the size of the obstacle, (these would be high-frequency sounds), the sound will be reflected or scattered in many directions and the obstacle will cast a shadow. Actually, some sound is diffracted into the "shadow" and there is significant reflection of the sound. As a consequence of diffraction, a wall is little value as a shield against sound of low frequency (long wavelength), but it can be an effective barrier against high-frequency (short wavelength) sound (see Figure 2–9).

Root-mean-square amplitude. One way of describing the magnitude

47

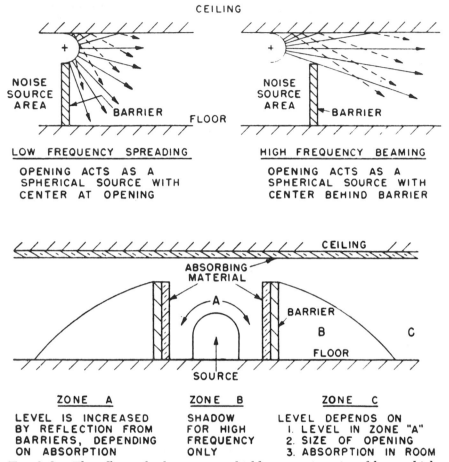

FIG. 2–9.—The effects of a barrier as a shield to contain noise of low or high frequency is shown in this illustration.

of a wave motion is the peak or maximum value called the amplitude. For some purposes, other ways of describing the magnitude of a wave are more convenient or significant (see Figure 2–10). A very useful kind of mean value of a wave is its root-mean-square amplitude that is, the square root of the mean squared displacements during one period. For a sine wave, the rms amplitude is 0.707 of the maximum value. The rms value is also known as the *effective value*.[3]

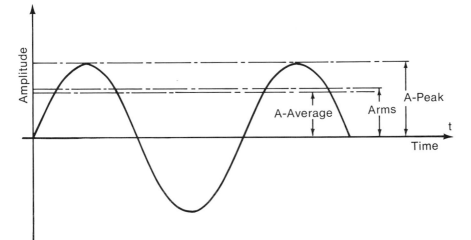

Fɪɢ. 2–10.—The peak and RMS values of a sound wave are shown in this illustration.

Most sounds met with in daily life are not purely sinusoidal. They vary with time, frequency, and magnitude. Simple mathematical expressions just do not exist for such complex sounds. So to characterize the magnitude of the signal, amplitude density is used instead of amplitude, because different amplitude values occur with a certain "density" when the sound is studied statistically over a certain period of time (see Figure 2–11).

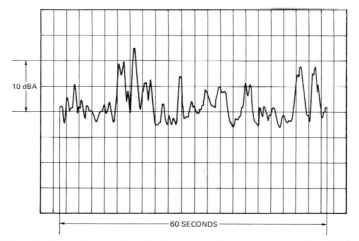

Fɪɢ. 2–11.—The amplitude density is the frequency with which certain values of the amplitude occur in a specific time period.

Sound pressure

Air consists of a large number of molecules or very small spheres distributed uniformly within space, each molecule moving in random motion. If a portion of space were enclosed in a box, first with no air and then with the box filled with air at normal room temperatures and pressures, it would be found that the air exerts a force on the surfaces of the box amounting to about 14.7 lb/in², or in metric units, about 100,000 newtons/m² (1.013×10^5 N/m² or 1.013×10^5 pascals). This force per unit area is normally called *atmospheric pressure* and is directly related to the density, or number of molecules, of air in the box.

The instantaneous pressure recorded in a sound wave is usually a small variation from the normal atmospheric pressure. This small change from atmospheric pressure is called the excess sound pressure, or simply sound pressure.

Reference pressure. In order to measure sound pressures it is necessary to have a reference pressure. Because sound pressures are very small variations in atmospheric pressure, it is convenient to measure sound pressure in the same unit. Atmospheric pressure is measured in various units—in millimeters of mercury (760 mm Hg = 1 atmosphere), pounds per square inch (14.7), millibars (1000), dynes per square centimeter (1,000,000), or newtons per square meter (100,000). Newtons per square meter (known in SI units as pascals) is extensively used in acoustics to denote sound pressure, and this unit will be used throughout this book for that purpose.

The measured sound pressures have little significance in themselves; they are almost always compared with some base or reference, and they are usually quoted as "levels re (that reference)." The units and reference levels are derived from common, readily observed phenomena.

The weakest sound pressure that can be perceived as sound is a very small quantity; however, the range of sound pressure perceived as sound is extremely large.

The zero level of sound pressure, for example, is not a true physical zero (that is, the absence of any pressure in excess of what would have existed with no sound energy present in the medium); rather, it approximates the average threshold of audibility at about 1000 Hz for humans.

Audible sound pressures range between the threshold of hearing and the threshold of feeling, between 0.00002 N/m² and 20 N/m². This

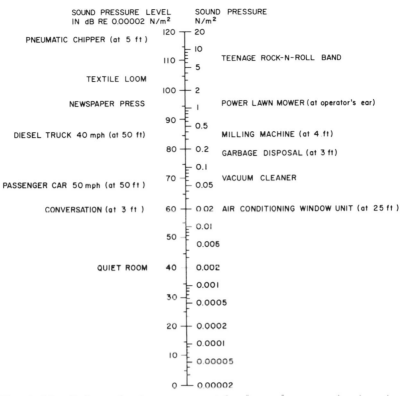

SOUND PRESSURE LEVEL IN dB RE 0.00002 N/m² · SOUND PRESSURE N/m²

PNEUMATIC CHIPPER (at 5 ft) — 120 — 20

— 10

110 — TEENAGE ROCK-N-ROLL BAND

— 5

TEXTILE LOOM

100 — 2

NEWSPAPER PRESS — 1 — POWER LAWN MOWER (at operator's ear)

90 — 0.5

DIESEL TRUCK 40 mph (at 50 ft) — MILLING MACHINE (at 4 ft)

80 — 0.2 — GARBAGE DISPOSAL (at 3 ft)

— 0.1

70 — VACUUM CLEANER

— 0.05

PASSENGER CAR 50 mph (at 50 ft)

CONVERSATION (at 3 ft) — 60 — 0.02 — AIR CONDITIONING WINDOW UNIT (at 25 ft)

— 0.01

50 — 0.005

QUIET ROOM — 40 — 0.002

— 0.001

30 — 0.0005

20 — 0.0002

— 0.0001

10 — 0.00005

0 — 0.00002

Fig. 2–12.—Relationship between a weighted sound pressure levels and decibels (dB) and sound pressure in N/m.²

corresponds to a ratio of 1 to 10,000,000 (see Figure 2–12). In acoustics, there is just as much interest in observing the effects of small changes near the threshold of hearing as there is in observing the effects of small changes near the upper end of the scale, it would be impossible to construct a linear scale with the required range. An analogous situation (which might be more meaningful) would be a scale ranging from one inch to 16,000 miles—the same ruler to be used to measure changes of a few inches or changes of a few miles.

Decibels

Selecting a practical scale for sound measurements involves two problems. The first is caused by the tremendous range of sound pressures that are encountered, and the second problem arises from the non-linear manner in which the ear responds. The ear tends to respond logarithmically to the intensity of an acoustical stimulus.

Both of these problems can be solved by the use of a logarithmic or decibel scale. One characteristic of the decibel scale is that it is possible to show, on an ordinary sheet of graph paper, a large range of sound pressures in such a manner that the small variations are as accurately portrayed as are the large variations.

By definition, the decibel (abbreviated as dB) is a dimensionless unit used to express the logarithm of the ratio of a measured quantity to a reference quantity.[2]

$$dB = 10 \log_{10} \frac{P_1}{P_o}$$

Decibels are not linear units like miles or pounds. Rather, they are representative points on a sharply rising curve. Thus, while 10 decibels is 10 times greater than one decibel, 20 decibels is 100 times greater (10×10), 30 decibels is 1000 times greater ($10 \times 10 \times 10$) and so on. Sound equal to one hundred decibels, therefore, is 10 billion times as intense (that is, it represents 10 billion times as much acoustic energy) as one decibel. The reason for such a complicated scale is simply that the human ear detects a wide range of acoustic energy.

Because decibels are logarithmic units, they cannot be added or subtracted arithmetically. In fact, if the intensity of a sound is doubled, there is a corresponding increase of only three decibels, not double the number. For example, if one machine produced a sound of 90 dB and a second identical machine was placed adjacent to the first, the combined sound would be 93 dB, not 180 dB. On the decibel scale, zero is the threshold of hearing, and 130 decibels is the threshold of pain.

The reference base for the decibel was chosen as the lower limit of hearing or the threshold of audibility for a person with good hearing. This threshold (or reference base) is 4 ten-millionths lb/sq ft; in the metric system, this is 0.00002 newtons per square meter (or 20 micro-newtons per square meter, $20\mu N/m^2$).

The maximum sound the ear can sense is about 4 lb/sq ft, or 200 N/m^2—actually ¼ of normal atmospheric pressure, or 140 on the decibel scale. From 0 to 140 decibels, sounds may be ranked from the threshold of acute hearing to the threshold of maximum auditory response (see Figure 2–13).

The gentle rustle of leaves, for example, is rated at 20 decibels, while

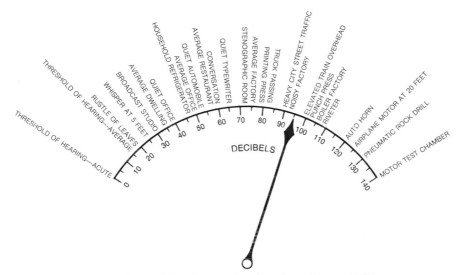

FIG. 2-13.—Typical sound levels associated with various activities.

a typical office has about 50 decibels of background noise. Moderate traffic noise ranges around 70 decibels; a police whistle hits 80. Subways and elevated trains rank just below thunder at 100 decibels. At just above 120 decibels the ear begins to feel pain. Sound levels are usually measured at their source, thus their decibel rating decreases as the distance from that source increases. These ratings should, therefore, be regarded as averages and should be used primarily for comparative purposes.

Decibel notation. The decibel notation system has two important features:

1. Scale compression and expansion
2. Computational simplicity

In the field of acoustics, the range of possible values is very large. Sound pressure levels as low as zero dB and as high as 180 dB have been measured. This is a range of one billion to one. It is most inconvenient to try to work with a variable having such a wide range on a linear scale. On the other hand, a logarithmic scale conveniently compresses such a range.

Associated with this advantage of scale compression provided by decibel notation are, in general, a reduction of resolution near the top of the range of value of interest, and an improvement near the bottom of the range. In general, the decibel scale provides a precision which is proportional to the value of a quantity—as the value grows, the precision also grows. The linear scale provides a precision which increases with the value of a quantity—as the value grows, the precision also increases.

Another major advantage of the use of decibel notation is its computational simplicity. Because decibels are logarithms, data multiplications are made by additions, and data divisions are made by substractions.

Now to go into more detail. In acoustics, the general relationship for a sound level, in the specialized logarithmic sense of the word, is

$$\text{Level} = \log_r \frac{q}{q_0}$$

where q is the quantity whose level is being obtained, q_0 is the reference quantity of like kind, and r is the base of the logarithm. Thus, the level of a quantity is the logarithm of the ratio of that quantity to a reference quantity. The base of the logarithm and the reference quantity must be specified. Since there are many kinds of levels, the type of level involved must be specified.[2]

The decibel is a unit of level when $r = 10^{1/10}$ or $\sqrt[10]{10}$ and the quantities q and q_0 are in some idealized situation proportional to power. That is, the decibel is used as a unit of level of quantities such as sound pressure squared. Strictly, the decibel is thus a unit of "pressure-squared" level, but the usual convention is to shorten this to pressure level. With a little juggling in accordance with the rules of logarithms, it follows that

$$L = 10 \log_{10} \frac{q}{q_0} \text{ dB}$$

Another unit of level, which is seldom used, is the bel, for which $r = 10$. Thus $10 \text{ dB} = 1 \text{ bel}$.

The usual unit of sound pressure level is the decibel (abbreviated dB). It is of such nature that doubling any sound pressure corresponds to an increase of sound pressure level of 6 dB. The sound pressure level corresponding to a sound pressure of 1 microbar is 74 dB; that corresponding to 2 microbars is 80 dB. A change of sound pressure by a factor of 10 corresponds to a change in sound pressure level of 20 dB.

The sound pressure level L_p in decibels, corresponding to a sound pressure p, is defined by

$$L_p = 10 \log \frac{p^2}{p_o^2} = 20 \log \frac{p}{p_o} \, dB$$

where p_o is a reference pressure. Here as elsewhere in this manual the common logarithm to the base 10 is to be understood unless there is a specific notation to the contrary.

Sound level measurement

A sound level meter is designed to measure sound pressure level. To assist in obtaining reasonable uniformity between different instruments, ANSI has established a standard S1.4–1971 to which sound level meters should conform. The current ANSI standard for sound level meters requires that three response characteristics be provided in the instrument (see Figure 1–6, p. 21). These three responses are obtained by weighting networks designated as A, B, and C.

Since the sound level meter reading obtained depends on the weighting characteristics used, that characteristic must be specified. Readings using each of the three weighting networks provide some indication of the frequency distribution of the noise. If the sound level is essentially the same on all three networks, the sound energy consists of frequencies above 600 Hz. If the sound level is greater on the C-network than on the A-network by several decibels, much of the noise is probably below 600 Hz, because the A network discriminates quite severely against the low frequencies.

Octave bands. An octave-band analysis indicates how the sound energy is distributed over the audible range of frequencies (see Figure 2–14). When the sound to be measured consists of a number of tones which are spread over many octaves, it may be necessary to determine the sound energy distribution according to frequency. The most practical and widely used instrument for this purpose is the octave-band analyzer.

In an octave-band analysis, the acoustic energy is electronically separated into various frequency bands, in this case, octave bands, each of which covers a 2-to-1 range of frequencies. The analysis yields a series of levels, one for each band, called "band levels," or, for octave bands, "octave-band levels." It is apparent that the band in which a level reading is obtained must be specified, if the information is to be of value.

Bandwidth. Electrical filters are used to reject signals below the

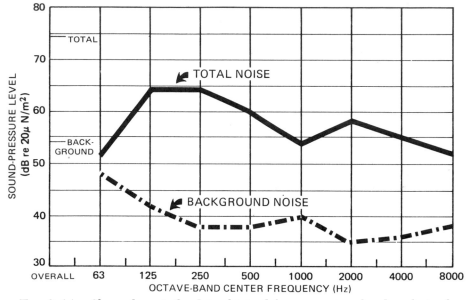

FIG. 2–14.—Shown here is the data obtained from an octave-band analysis of a noise source showing the total noise and the background noise levels when the noise source is not in operation.

lower cutoff frequency and above the upper cutoff frequency. Signals between the upper- and lower-limit frequencies are passed by the filter which is consequently called a band-pass filter. The difference between the cutoff frequencies is the bandwidth (see Figure 2–15).

The preferred series ANSI S1.6–1966 of octave bands of acoustic measurements cover the audible range in ten bands. The center frequencies of these bands are 31.5, 63, 125, 250, 500, 1000, 2000, 4000, 8000, and 16,000 Hz. The frequency range of any one of these bands represents a ratio of 2 to 1; for example, the effective range for the 1000 Hz octave band extends from 707 to 1414 Hz.

When a graph is made of octave-band sound level measurements, the frequency scale is commonly divided into equal intervals along the horizontal x-axis. The sound level in each band is plotted as a point on each of the frequency positions along the vertical or y-axis. Adjacent points are then connected by straight lines. An example of a plot of this type is given in Figure 2–14.[1]

56

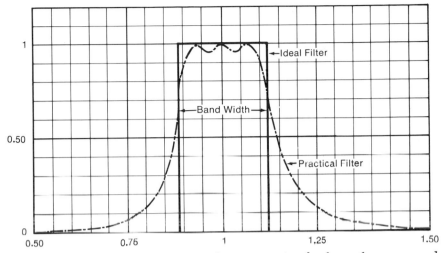

FIG. 2–15.—Electrical filters are used to reject signals above the upper and below the lower cutoff frequency. The difference between the cutoff frequencies is the bandwidth.

Fluctuating levels

One of the dominant characteristics of environmental noise at any location is that it fluctuates considerably from relative quiet at one instant to loud the next. Thus, noise exposure must be described by an approach that takes time into account. This can be achieved by providing information depicting the cumulative distribution of sound levels; that is, by showing the percent of time during the observation period that each sound level is exceeded. Noise levels are often specified in terms of levels exceeded 10 percent of the time, 50 percent of the time, and 90 percent of the time.

The sound level exceeded 10 percent of the time, expressed as L_{10}, gives an approximate measure of the higher level and short duration noise. A measure of the median sound level is given by L_{50} and represents the level exceeded 50 percent of the time. The residual sound level is approximated by L_{90}, which is the sound level exceeded 90 percent of the time.

The energy mean noise level (L_{eq}). A measure accounting for both the duration and the magnitude of all the sounds occurring during a given period is the *average sound level,* sometimes called the equivalent continuous noise level. It is the continuous level that is equivalent in terms of noise energy content to the actual fluctuating noise existing at a location during the observation period. It is also called the *energy mean noise level* (L_{eq}). By definition, L_{eq} is the level of the steady-state continuous noise having the same energy as the actual time-varying noise. In terms of assessing the effect of noise on humans, L_{eq} is an important measure of environmental noise.[4]

Types of noise. Noise is further differentiated into steady-state, fluctuating, intermittent, and impulsive noise.

• *Steady-state noise* is noise whose quality and intensity is practically constant (varying less than ±5 dB) over an appreciable period of time.

• *Fluctuating noise* is noise whose intensity rises or falls more than 5 dB.

• *Intermittent noise* is discontinuous and differs from impulsive noise in being of longer duration and less specifically defined.

• *Impulsive noise* is transient, like a gunshot. The impulse must be less than 500 milliseconds (0.5 second) duration and have a magnitude (change in sound pressure level) of at least 40 dB within that time. A single impulse may be heard as a discrete event occurring in otherwise quiet conditions, or it may be superimposed upon a background of steady-state or fluctuating noise. It may be characterized by the following basic parameters:

1. Peak sound pressure level
2. Duration (in milliseconds or microseconds)
3. Rise and decay time
4. Types of waveform (time-course)
5. Number of impulses

Two types of impulses, "A" and "B," are shown in Figure 2–16. In the

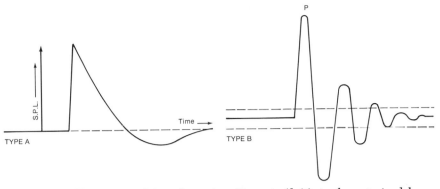

Fig. 2–16.—Two types of impulse noise. Type A (*left*) is characterized by a rapid rise to a peak SPL followed by a decay to negligible magnitude. The type B impulse displays a negative pressure wave after the peak.[4]

Type A impulse, there is a rapid rise to a peak SPL followed by a decay to a negligible magnitude. In the Type B impulse or event, a subsequent negative pressure wave of much smaller magnitude occurs. The duration is taken as being the time for the envelope to decay to a value 20 dB below the peak. It is important to note that impulse noises can be distinguished as to type and properly measured only by oscillographic techniques.[4] See Chapter 4 for more details.

Sound fields

Sound can be thought of as a field, just as electromagnetic waves are fields. There are many common types of sound fields, such as the near field, far field, free field, plane sound field, spherical sound field, reverberant sound field, and the diffuse sound field (see Figure 2–17). A brief introduction was given under Sound Waves, earlier in this chapter.

Any vibrating object will radiate sound in a gaseous medium such as air; the propagation of sound waves takes place in the form of density variations in the medium in the direction of propagation, as explained earlier in this chapter. The most common method of measuring these density variations is to measure the associated variations in pressure, that is, to measure the sound pressure.

Plane field. When the microphone is relatively far away from the sound source, the sound field may appear to be a plane field, in which the sound pressure is constant in any plane perpendicular to the direc-

59

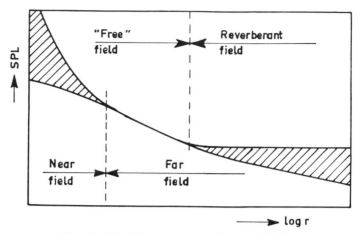

FIG. 2–17.—Various types of sound fields.

tion of propagation (see Figure 2–3).[2]

Spherical field. A spherical field is characterized by a wave propagating radially away from a small pulsating sphere (a point source) (see Figure 2–4). A very important relationship between the emitted sound power from a point source and the sound pressure existing at a distance, R, from the source can be derived.[3]

Assuming that the emitted sound power, E, spreads out from the source in the form of spheres with a continuously increasing radius, R, the sound intensity passing through the surface of a sphere or radius, R, must equal the emitted power, E, divided by the area of the surface (area $= 4\pi R^2$):

$$I = \frac{E}{4\pi R^2}$$

This relationship is the so-called inverse-distance law (sometimes also called the "inverse-square law" for obvious reasons), and governs sound radiation in the acoustic "far" field of a sound source (see Figure 2–18).

Free field. Sound in a homogeneous space propagates outward from a source in all directions and consequently the sound pressure decreases with the square of the distance from the source. If such a point (or spherical) source is in the air far from any other objects, including the ground, the sound pressure produced by the source is the same in every

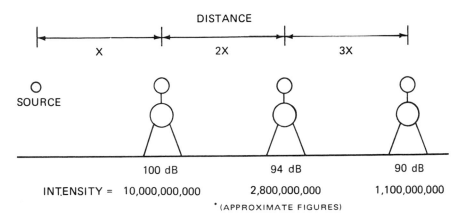

NOISE CONTROL BY DISTANCE

Fig. 2–18.—In the far field as the distance from the noise source increases, the sound pressure decreases.

direction at equal distances from the point source. Furthermore, the sound pressure is halved for each doubling of distance from the point. This change is usually expressed as a decrease in sound pressure level of 6 dB. The sound field produced under these idealized conditions is called a *free sound field* because it is uniform, free from all bounding surfaces, and undisturbed by other sources of sound.

For certain tests, free-field conditions are necessary and often outdoor measurements are impractical. For such tests, special rooms have been built in which the sound-absorptive construction of the walls, floor, and ceiling is such that practically no sound is reflected from them. These are called *free-field rooms* or *anechoic* (*echo-free*) rooms. Points of measurement are commonly referred to as *far field* and *near field*.

A measuring point is considered to be in the far field if, by increasing the distance of the measuring point from a small source, the pressure measurement change follows the inverse square law, that is, as the distance from the source is doubled, the measured sound pressure level decreases 6 dB. An engineering rule of thumb commonly used to specify a measuring position as being in the far field is that the measuring point should be at least several wavelengths from the source.

The near field of a source is defined as that region where changes in pressure amplitudes do not obey the inverse square law. Normally, such positions are within a few wavelengths of the source. A transitional region clearly exists between the near field and the far field as one increases the distance from the source to the measuring point. Unfortunately, the mathematical description of the wave in this transitional region is far from simple. In practice one tries to avoid making measurements in the transitional region, but in real life it is sometimes impossible to avoid making measurements within this ill-defined space.

Two regions of radiation from a sound source have been briefly mentioned, the acoustic near field and the acoustic far field, both with no sound-reflecting obstacles. When sound-reflecting objects are introduced into the sound field, the wave picture changes completely because of the reflections. There is now not only the original sound wave but also a reflected sound wave traveling in the opposite direction to the original one. The sound pressure in the field is then, at any instant, the combination of the pressure due to the original wave and the pressure(s) due to the reflected wave(s).

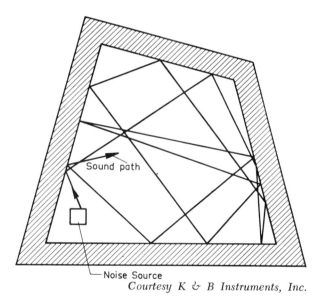

Courtesy K & B Instruments, Inc.

Fig. 2–19.—The sound from a noise source in a reverberant field is reflected by the walls, floor, and ceiling.

Reverberant field. The sound that a noise source radiates in a room is reflected by the walls, floor, and ceiling. The reflected sound will again be reflected when it strikes another boundary, with some absorption of energy at each reflection. The net result is that the intensity of the sound is different from what it would be if the reflecting surfaces were not there (see Figure 2–19).

Close to the source of sound there is little effect from these reflections, since the direct sound dominates. But far from the source, unless the boundaries are very absorbing, the reflected sound dominates, and this region is called the *reverberant field.*

The sound pressure level in this region depends on the acoustic power radiated, the size of the room, and the acoustic absorption characteristics of the materials in the room. These factors and the directivity characteristics of the source also determine the region over which the transition between reverberant and direct sound occurs.

Standing waves. A second effect of reflected sound is that the measured level does not decrease steadily as the measuring position is moved away from the sound source. Variations of up to 10 dB are common and, in particular situations, much more may be found. These variations usually show the following characteristic—as the measuring microphone is moved away from the source, the measured sound level decreases to a minimum, rises again to a maximum, decreases to a minimum again, etc. These patterns are called *standing waves.*

They are noticeable mainly when the sound source has strong frequency components in the vicinity of one of the very many possible resonances of the room. See also Chapter 5.

Acoustic power

The power passing through a unit area is a measure of sound intensity. Because it is difficult to measure sound intensity directly, most acoustical measurements are made of the effective sound pressure or square root of the mean square sound pressure. The sound power of the source can be computed from the relation:

$$W = I_{avg} 4\pi R^2 \text{ watts}$$

Where I_{avg} is the average sound intensity at a distance R from a sound source whose acoustical power is W as shown in Figure 2–20. The quantity $4\pi R^2$ is the area of the sphere over the surface of which the intensity is averaged.

Note that the intensity will decrease with the square of the distance. As we move away from the source, the power per unit area passing any given position decreases because the total power must diverge with

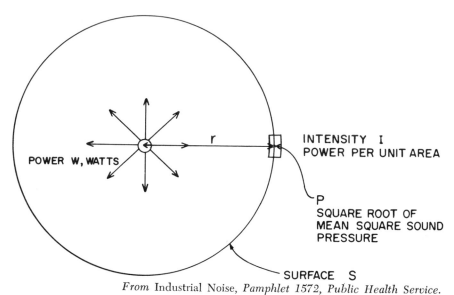

From Industrial Noise, *Pamphlet 1572, Public Health Service.*

Fɪɢ. 2–20.—A single nondirectional sound source radiating W watts producing a sound intensity, I, in watts per unit area.

distance. We call the power passing through a unit area the *intensity.*

For a spherical surface S surrounding a simple sound source, the intensity is the same at every point on the surface. Thus, the total power W (watts) is equal to the *intensity*, I (watts per square meter), multiplied by the total area of the surface S. That is:

$$W = IS \quad \text{watts}$$

Sound power levels should not be confused with sound pressure levels, which are also expressed in decibels. We can see this distinction readily by remembering that the sound power level describes the total acoustic power radiated by a source. On the other hand, the sound pressure level at a given point depends upon the distance from the source, losses in the intervening air, room effects (if indoors), and other factors.

Measurement of acoustic power. A noise rating is intended to make a prediction of the noise level that an apparatus will produce when installed. In order for the rating to be adequate for this purpose, the total acoustic power radiated by a source and the acoustic directivity pattern of the source should be included as part of the rating.

For example, an air compressor may be rated by the manufacturer as producing a noise level of 85 dB at a distance of five feet. This level may have been calculated by an averaging of a few sound level readings five feet from the compressor. When it is installed and the level is measured, the new level may be 90 dB at five feet. Naturally, the purchaser feels that he should complain because in his opinion the machine was incorrectly rated. The difference of 5 dB *may* have been caused by incorrect installation, but usually such a difference is a result of the acoustical characteristics of the factory space. By the use of an adequate rating system and a knowledge of acoustical room characteristics, it would have been possible to predict this effect (see Figure 2–21).

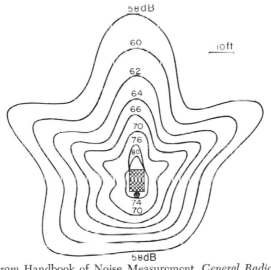

From Handbook of Noise Measurement, *General Radio Co.*

Fig. 2–21.—Contours of equal sound pressure level around a noise source.

Another part of this problem is the prediction of levels at places in the factory other than at the measurement distance. For example, the nearest worker may be 20 feet away, and the levels at that distance are now more important. A knowledge of the acoustic power radiated and of the acoustical characteristics of the factory space is needed to predict the probable sound level at this new location.

Directivity factor. It is common for sound sources to radiate more sound in one direction than in others. Low-frequency (long-wavelength) sound emanating from a source is more uniformly radiated in all directions. In general, a sound source that is small in comparison with the wavelength of sound it produces tends to be an omnidirectional source; a sound source large in comparison with the wavelength of the sound produced is likely to be directional.

The directivity factor is a term often employed to describe how directional a source is. The *directivity factor*, Q, is defined as the ratio of the mean square sound pressure at some fixed distance and specified direction, to the mean square sound pressure at the same distance but averaged over all directions from the source. The distance must be sufficiently great so that the sound appears to come from a single point called the *acoustic center* of the source. The directivity factor for the specified direction is:

$$Q = \frac{p_d^2}{p_{avg}^2}$$

Noise control measures

The general principles used to reduce structure-borne sound are:

• Isolate the source of vibration from a sound radiating surface

• Place a sound barrier between the noise source and the equipment or machine exterior

• Damp the amplitude of vibration of the sound radiating surface

Vibration-isolation materials make use of a resilient mounting to separate the energy source from the vibrating surface. Products that reduce transmission of sound by conduction include: high-density glass wool materials, steel springs, elastomers, cork, and felt.

Sound barriers use a material capable of retarding airborne sound transmission. They can be air-impervious materials and include: gypsum board, hardboard or plywood, plastic sheeting, lead-loaded vinyl sheeting, and metal sheet or heavy foil.

Panel damping consists of the application of a material with high internal damping properties. The effect of such treatment is to reduce the ring or tinny sound of a metal surface.

Transmission loss. Normally, part of the problem of noise control is how to prohibit sound from traveling from one place to another. A wall separating a noise source from an area desired to be quiet can be characterized by its transmission loss (TL) rating. Transmission loss is a property of the wall itself, and is defined as the ratio of the sound energy transmitted through the wall to the sound energy striking the wall.

$$TL = \frac{\text{incident energy}}{\text{transmitted energy}}$$

When sound waves impinge upon a wall, the following occur:

1. The fluctuating pressure exerted on the wall causes it to vibrate like a diaphragm. Sound is then radiated by the vibrating wall into the space on both sides of the wall.

2. Depending principally on the nature of the surface, a considerable proportion of the sound energy will be reflected, some will be absorbed, and the remaining will be transmitted.

When subjected to some disturbance, such as a sound pressure wave,

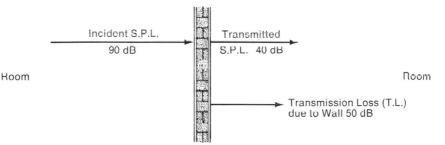

Fig. 2–22.—Transmission loss is a property of the material and is defined as the ratio of the sound energy transmitted through the barrier to the sound energy striking the barrier.

any building structure—such as a wall, ceiling, floor, or door—will vibrate. It will vibrate more readily at certain frequencies, the lowest of which is known as the *natural frequency* of the structure. Because the natural frequencies of floors and walls tend to be low, it is much more difficult to insulate against the transmission of noise that consists of predominantly low frequencies (see Figure 2–22).

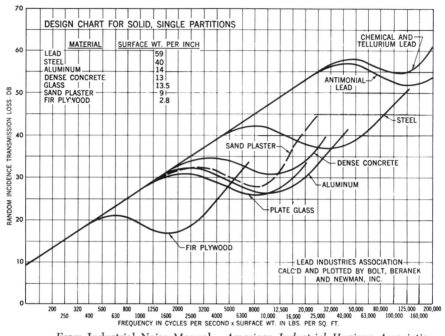

From Industrial Noise Manual. *American Industrial Hygiene Association.*

FIG. 2–23.—Theoretical transmission loss as a function of mass and frequency.

Slope. The transmission loss in structural elements such as walls and floors always varies directly with the frequency of the incident sound wave, the amount of necessary insulation rising as the frequency rises. This is called the "slope" of the transmission loss and is expressed as "x dB per octave." The transmission loss in a brick wall increases by approximately 5 dB with every octave frequency increase. Other forms of construction will have a higher or lower slope. The transmission loss of a single partition or structure depends primarily on the weight of the partition or structure. This is known as the *mass law* and is illustrated in Figure 2–23.

The new single classification of rating walls, called *sound transmission class* (STC), relates to the frequency characteristics of a wall. Frequency contours of TL have been agreed upon and in general these TL values, in decibels, rise from 125 to about 350 Hz at a rate of approximately 10 dB/octave and then change to a rate of about 3 dB/octave for

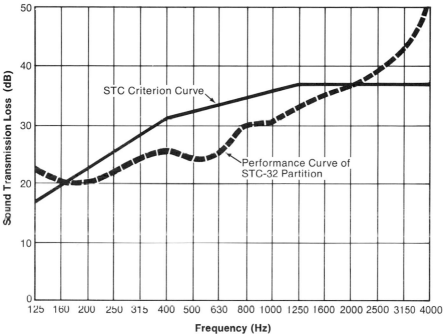

Fig. 2–24.—The single classification of rating sound barriers is called the sound transmission class (STC). Shown here is the (STC) criteria specification and the performance curves for a partition.

frequencies of 350 Hz to 1500 Hz and remain uniform with frequencies above 1500 Hz. The STC rating of a wall is given as the uniform (flat) value of this frequency contour for the higher frequencies. Thus a wall with an STC value of 45 dB has transmission loss values (actually attenuation values) which equal or exceed the frequency characteristics of the standard contour which carries the "45 dB" rating (see Figure 2–24).

Noise transmission paths should receive very careful consideration before any noise control procedures are adopted. The primary sound transmission path cannot be identified by listening. If the sound energy is being transmitted by a partition and also by a ceiling, the sound which arrives at the listener first determines the apparent direction of the sound. A listener in the receiving room will consider that the sound is coming through the partition (because that is the shortest route) although the ceiling may, in fact, be the main acoustic transmission path.

Sound absorption. Absorption of sound by a material means that the sound is not being reflected from the surface of the material. Generally speaking, any material that readily absorbs sound pressure waves just as readily transmits them, allows them to pass through. Any material or surface which reflects sound pressure waves is also a good insulating material. In other words, the insulating material prevents the sound waves from entering or passing through to a designated area.

No single material can be both an efficient absorbent and a good insulator. Covering the surface of a wall with acoustic tiles or foam rubber, for example, will make no difference to the transmission loss in that wall. It should be borne in mind that the reduction of sound by absorption (that is, the reduction of reverberant sound) can never be greater than 10 dB, and is often much less.

The *absorption coefficient* of any material represents the ratio of absorbed energy to the energy incident upon the material. If the material is a perfect reflector, the absorption coefficient is zero; and if the material is a perfect absorber, the absorption coefficient is one. Absorption of sound is accomplished by generation of heat which results from the movement of air particles against the surface of the absorbing material.

The unit of sound absorption is the *sabin*. One sabin per 0.093 m^2 (1 ft^2) is total absorption. If one wishes to have a high absorption coefficient for low-frequency sounds, the depth of the absorptive material must be great or the material must be spaced away from the wall. This is one of the major problems in absorbing low-frequency sound.

Resonance. When a small object—such as a watch or a piano wire—is isolated, it produces relatively little sound. But, when it is in contact with a large acoustic conductor—such as the sounding board of a piano—there is a transfer of sound energy that results in considerable amplification. Machinery and equipment act the same way if insulators or isolators are not provided to prevent the noise from being transmitted and considerably amplified.

Vibration

This section deals with vibration as a characteristic of a sound source that can be measured in the attempt to control noise. Vibration is a term used to describe the movement of a body. The motion may be simple like that of a pendulum, or it may be complex like a ride in the "whip" at an amusement park. The motion of the vibrating body may produce sound when the rate of vibration is in the audible range.

Many mechanical vibrations lie in the frequency range of 1 to 2000 Hz (corresponding to rotational speeds of 60 to 120,000 rpm). In some specialized fields, however, both lower and higher frequencies are important. For example, in seismological work, vibration studies of frequencies may extend down to a small fraction of a hertz, while in loudspeaker cone design, frequencies up to 20,000 Hz are being studied.

Types of motion

Vibration problems occur in many devices and operations and usually involve more than one type of vibratory motion. Vibration can be measured in terms of *displacement, velocity, acceleration,* and *jerk.*

Displacement. The easiest measurement to understand is displacement, that is, the magnitude of motion of a body. When the rate of motion (frequency of vibration) is low, the displacement can be measured directly with a dial gage micrometer; if high, an ordinary distance scale. Displacement measurements are significant in the study of deformation and bending of structures.

Velocity is considered to be the best single criterion for use in scheduling preventive maintenance operations of rotating machinery. Peak-to-peak displacement has also been used for this purpose, but the amplitude of displacement selected as a desirable upper limit varies markedly with rotational speed.

Velocity is the time rate of change of displacement, so that for sinusoidal vibration the velocity is proportional to displacement and to frequency of vibration. In most machines, however, relatively small parts are vibrating at relatively low frequencies. This situation may be compared to a small loudspeaker without a baffle. At low frequencies, the air may be pumped back and forth from one side of the cone to the other at a very high velocity, but without building up much pressure or radiating much sound energy because of the very low air load. Under these conditions, vibration acceleration measurement provides a better measure of the amount of noise radiated than does a velocity measurement.

Acceleration. In many cases of mechanical vibration, and especially where mechanical failure is a consideration, the actual forces set up in the vibrating parts are important factors. The acceleration of a given mass is proportional to the applied force; this produces a reacting force equal

71

but opposite in direction. The component members of a vibrating structure, therefore, exert forces on the total structure that are a function of the masses and the accelerations of the vibrating parts. For this reason, acceleration measurements are important when vibrations are severe enough to cause actual mechanical failure.

Acceleration is the time rate of change of velocity, so that for a sinusoidal vibration, it is proportional to the displacement and to the square of the frequency or to the velocity and the frequency.

Jerk is the time rate of change of acceleration. At low frequencies this change is related to riding comfort of autos and elevators. It is also important for determining load tiedown in planes, trains, and trucks.

Vibration is discussed in Chapter 18 of this Manual.

Summary

The basic principles of sound generation were discussed in this chapter. To minimize, limit, or prevent excessive exposure to noise requires some knowledge of the elementary principals and physical laws that characterize the measurement, production, transmission, and absorption of sound.

Objectively, sound is a form of wave motion due to pressure alternation or particle displacement in an elastic medium. Sound is invariably produced by vibratory motion of some sort in a gaseous, liquid, or solid body. Any type of vibration may be a source of sound, but only longitudinal vibration of the conducting medium constitutes sound waves.

A sound wave may be characterized as a succession of condensations and rarefactions of the particles of the conducting medium. When this disturbance or pressure variation reaches the eardrum, it is translated into the sensation we interpret as sound. The number of pressure vibrations or complete cycles, which occur per second, determines the frequency expressed in cycles per second or hertz (Hz).

One complete vibration of the object corresponds to one complete cycle of pressure change. The distance that a sound wave travels in one period or cycle is called the wavelength of the sound.

Because of the tremendous range of sound pressures involved in noise measurements, it has become customary to deal with sound pressure levels instead of sound pressure.

The decibel notation system is very useful for dealing with sound measurements. The decibel is not a unit of measurement, like an inch or a watt, it is the logarithm of a nondimensional ratio of two powers or two power-like quantities. In general, the decibel scale provides a precision that is proportional to the value of the quantity; a linear scale provides a precision that increases with the value of the quantity.

Most commonly used sound measuring instruments are designed to respond to sound pressure changes. When the sound to be measured is complex, consisting of a number of tones spread over many octaves, it may be necessary to determine the sound pressure distribution according to frequency. An octave band analysis indicates how the sound energy is distributed over the audible range of frequencies.

References

1. American Industrial Hygiene Association. *Industrial Noise Manual*, 2nd ed. Detroit, Mich.: A.I.H.A., 1966.
2. Petersen, A. P. G. and Gross, Jr., E. E. *Handbook of Noise Measurement*. West Concord, Mass.: General Radio Co., 1972.
3. Hosey, A. D. and Powell, C. H. eds. *Industrial Noise—A Guide To Its Evaluation and Control*, U.S. Public Health Service Publication No. 1572. Washington, D.C.: U.S. Government Printing Office, 1967.
4. U.S. Environmental Protection Agency. *Public Health and Welfare Criteria for Noise*. Washington, D.C.: U.S. Government Printing Office, 1973.

Bibliography

Beranek, L. L. ed. *Noise and Vibration Control*. New York, N.Y.: McGraw-Hill Book Co., 1971.

Harris, C. M. ed. *Handbook of Noise Control*. New York, N.Y.: McGraw-Hill Book Co., 1957.

Keast, D. N. *Measurements In Mechanical Dynamics*. New York, N.Y.: McGraw-Hill Book Co., 1967.

Chapter Three Subjective Aspects of Sound

By Julian B. Olishifski, P.E.

Subjectively, sound may be defined as "a stimulus that produces a sensory response in the brain." The perception of sound resulting in the sensation called "hearing" is the principal sensory response; however, under certain conditions, additional subjective sensations, ranging from pressure in the chest cavity to actual pain in the ears, can be produced. There are certain effects of sounds that appear to be universal for all people. The effects include (a) the masking of wanted sounds, particularly speech, (b) auditory fatigue and damage to hearing, (c) excessive loudness, and (d) annoyance.

Noise usually is sound that bears no information and whose intensity usually varies randomly in time. The word *noise* is often used to mean "sound that is unwanted by the listener." Because it is unpleasant, it interferes with the perception of wanted sound, or it is physiologically harmful.[1]

Noise does not necessarily have any particular physical characteristic to distinguish it from wanted sound.

All the definitions and measurement techniques that are described in the first two chapters about sound apply equally to noise. No instrument can distinguish between a sound and a noise—only human reaction can. A variety of methods have been devised which try to relate physical measurements of sound to the human perception. The purpose of this chapter is to outline the factors involved and to summarize the important subjective aspects of sound.

Perception of sound

The perception of sound is a very complicated process. Even though the basic works of a number of investigators have greatly helped to clarify its functioning, many details are still not completely understood.

The hearing process

From a vibrating sound source, the external ear directs the vibrations to the eardrum, multiplies them by means of small bones arranged as levers in the middle ear, and transmits the vibrations through a fluid to nerve endings within the inner ear. These nerves transmit an impulse to the brain, which, in a fraction of a second, analyzes and translates the impulse into a mental or physical response.

The vibration of the eardrum is mechanically transferred via the middle ear to the inner ear. Systematic stimulation of the nerve endings takes place along the basilar membrane of the cochlea in the inner ear. (See Chapter 7 for a more detailed explanation.) Sounds of various frequencies are analyzed by the basilar membrane at different distances from the oval window.

The nerve pulses produced in the organ of Corti along the basilar membrane are dependent on the fluid vibrations and the excitation of the nerve endings. To produce a single pulse, a certain excitation level has to be exceeded, and in this way the "limits of hearing" may be explained, at least, to a certain extent.

Threshold of hearing. A great deal of effort has gone into the development of systems which relate physical measurements of sound to subjective human response. In order to get reliable measures, experimenters had to simplify the conditions under which people react to sounds. Some of the conditions that have to be controlled and specified are: the acoustical environment of the observer, particularly the background or ambient sound level; the method of presenting the auditory signals, including the order of presentation, duration, frequency, intensity, the selection of the observers, the instructions to the observers, the experience of the observers in the specific test procedure, the normal hearing characteristics of the observers, the method of getting the responses, and the method of handling the data. The details of audiometric testing are given in Part Five—Industrial Audiometry.

Human awareness of sound is usually a function of three measurable physical qualities: (See Figure 3–1.)

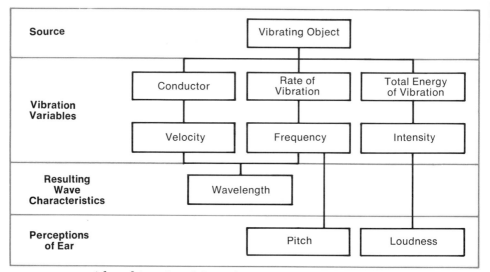

Adapted from Sound Control Construction, *United States Gypsum Co., 1972.*

FIG. 3–1.—Various terms used to describe sound are shown. A vibrating object will produce sound waves which can be characterized by their frequency, velocity, intensity, and wave length.

- Sound level (in decibels), which relates to loudness

- Frequency (in hertz) relates to pitch

- Duration (from microseconds to hours) indicates how long a sound persists.

Sound level. The variations in air pressure give sound the quality we call "loudness." The human ear is sensitive to sound pressure variations as low as 20 micronewtons per square meter ($20~\mu\mathrm{N/m^2}$). At the other end of the scale, the ear responds to sound pressures up to 20 newtons per square meter—10 million times the pressure of the weakest sound heard. Because the extremes are so great, it has become customary to express noise magnitude in decibels which are logarithmic ratios comparing pressures of interest to a reference pressure. The reference pressure commonly used in noise measurement is 20 micronewtons per square meter.[3]

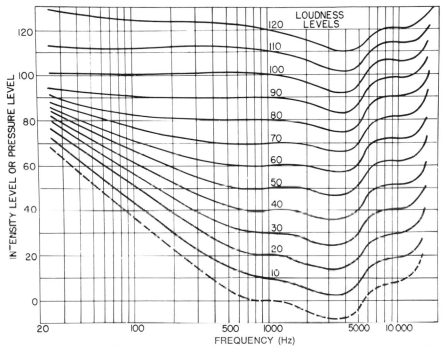

Fig. 3–2.—Free-field equal-loudness contours of pure tones. Because the human ear is more sensitive to the higher frequencies of sound, changing the frequency of a sound changes its relative loudness. These are also called Fletcher-Munson contours.

Frequency. While loudness depends primarily on sound pressure, it is also affected by frequency. Pitch is most closely related to frequency. The reason for this is that the human ear is more sensitive to high frequencies than to low. Frequency is defined as the number of times the pressure variations occur in a second, usually expressed as hertz (Hz). Human beings, generally, have the ability to hear sound from 20 to 20,000 Hz. (See Figure 3–2.)

The upper limit of frequency at which airborne sounds can be heard depends primarily on the condition of a person's hearing and on the intensity of the sound. This upper limit is usually quoted as being somewhere between 16,000 and 20,000 Hz. For most practical purposes the actual figure is not important. It is important, however, to realize that

77

it is in this upper frequency region where most people lose sensitivity as they grow older. This aging effect is called presbycusis.

The complete hearing process seems to consist of a number of separate processes that, in themselves, are fairly complicated and no simple relationship exists between the physical measurement of a sound pressure level and the human perception of the sound. The loudness of a certain pure tone may be judged to sound different compared to that of another pure tone, even though the sound pressure level is the same in both cases. Sound pressure levels, therefore, are only a part of the story and can be deceiving. For example, enjoyable sounds can be louder to the ear than annoying ones. The fundamental problem is that the quantities to be measured must include man's reaction to the sound—a reaction that may be determined by such varied factors as the time of day, his relative physical comfort and discomfort, characteristics of the sound, and attitude toward the person or device generating the sound.[3]

How loud is "loud"?

In the course of time, various loudness level rating methods have been suggested and a number of different criteria for tolerable noise levels have been proposed. A complete physical description of sound must include its frequency spectrum, its overall sound pressure level, and the variation of both of these quantities with time. Loudness is the human subjective response to sound pressure and intensity. At any given frequency, the loudness varies directly as the sound pressure and intensity, but not in a simple, straight-line manner.[3]

The physical characteristics of a sound as measured by an instrument, and the "noisiness" of a sound as a subjective characteristic may bear no relationship one to the other. A sound level meter cannot distinguish between a pleasant sound and an unpleasant one—it merely measures pressures and frequencies. A human reaction is required to differentiate between a pleasant sound and noise. Loudness is not merely a question of sound pressure level. A sound which has a constant sound pressure can be made to appear quieter, or louder, by changing its frequency.

Equal-loudness contours. Experiments to determine the response of the human ear to sound were reported to Fletcher and Munson in 1933 (see References). A reference tone and a test tone were presented alternately to the test subjects (young men), who were asked to adjust the level of the test tone until it sounded as loud to them as the reference (1000 Hz) tone. The results of these experiments with many test subjects yielded the now familiar Fletcher-Munson or equal-loudness contours (see Figure 3–2).[3]

The contours represent the sound pressure level necessary at each frequency to produce the same loudness response in the average listener. The nonlinearity of the ear is exhibited by the changing contour shapes as the sound pressure level is increased, a phenomenon that is particularly noticeable at low frequencies. The lower, dashed curve indicates the so-called "threshold of hearing" that represents the sound pressure level necessary to produce the sensation of hearing in the average listener. The actual threshold varies as much as ± 10 dB among normal individuals.

Loudness level. The loudness level of any sound is defined as the sound pressure level of a standard sound which appears to a significant number of observers to be as loud as the unknown.

PHONS. Loudness level is measured in *phons*—the loudness level of any sound in phons is equal to the sound pressure level in dB of an equally loud standard sound. If a large number of observers compared a 100 Hz tone with a 1000 Hz tone, they would probably judge the two tones to be equally loud, only when the 100 Hz tone had a higher sound pressure level than the 1000 Hz tone.

To determine whether one sound is louder, equally loud, or less loud than another, a statistically significant number of people would have to compare the sounds. Similarly, to determine the loudness of a sound, a significant number of people should compare an unknown sound with a standard one. The accepted standard sound is a 1000 Hz pure tone. The loudness level in phons of 1000 Hz tone by definition is the same as the sound pressure level in decibels.

A sound which is judged to be as loud as a 40 dB 1000 Hz tone has a loudness level of 40 phons. A factor of two in loudness does not correspond to double the number of phons. Over most of the audible range (that is, for loudness levels of 40 phons and greater), the corresponding increment is 10 phons.

SONES. Experimenters have asked observers to make judgments of the loudness ratio of sounds; that is, to state when one sound is twice, four times, one half (or whatever) as loud as another. The resultant judgments depend on the sound pressure level and to a considerable extent on how the problem was presented to the observer. Based upon the results of these tests, several scales of loudness have been devised. These scales rate sounds from "soft" to "loud" in units of sones. As a reference, the loudness of a 1000 Hz tone with a sound pressure level of 40 dB re 20 μN/m^2 (a loudness level of 40 phons) is taken to be 1 sone. A tone that sounds twice as loud has a loudness of 2 sones. For loudness levels of 40 phons or greater, the relationship between the numerical values of loudness level L_s (in phons) and loudness S (in sones) is given by:[4]

$$S = 2^{(L_s-40)/10}$$

Directivity

Although the two ears receive the sound signals simultaneously, each ear receives the signal at a slightly different level and phase, that is, earlier or later in time. The brain apparently analyzes the signals from each ear, integrates them, and extracts information from them. Location of the source of a sound is determined for low-pitched tones by detecting the difference in phase between signals arriving at each ear. Individuals can determine the direction of a low-frequency tone to an accuracy of about 10 degrees. At low frequencies little sound is blocked by the head; the wavelength is large and sound is diffracted around the head.

For sounds of higher pitch (where the wavelength is equal to or smaller than the size of the head), the head acts as a barrier, so that the ear in the acoustic shadow receives much less sound energy. Above 1000 Hz, sound localization is governed primarily by intensity rather than phase differences. Thus, the source can be located with a fair degree of accuracy and can be characterized, even in the presence of a constant sound background.

Effects of noise exposure

The effects of human exposure to noise include:

- Hearing impairment

- Speech interference

- Annoyance

- Physiological reactions

Effects of noise on humans can be classified in various ways. For example, the effects can be treated in the context of health or medical problems owing to their underlying biological basis. For example noise induced hearing loss involves damage to the structure of the hearing organ. In contrast speech interference and annoyance can be considered to be nonbiological problems because they involve no pathology.

Hearing impairment

The ear is especially adapted and most responsive to the pressure changes underlying airborne sounds or noise. The outer and middle ear structures are rarely damaged by exposure to intense sound energy, although explosive sounds or blasts can rupture the eardrum and possibly dislodge the ossicular chain. More commonly, excessive exposure produces hearing loss that involves injury to the hair cells of the inner ear.

Temporary threshold shift (TTS) can be produced by a brief exposure to high-level sound. TTS is greatest immediately after exposure to excessive noise and progressively diminishes with increasing rest time, as the ear recovers from the apparent noise overstimulation. A noise capable of causing significant TTS with brief exposures is probably capable of causing a significant permanent threshold shift (PTS) upon prolonged or recurrent exposures.

Permanent threshold shift resembles TTS except that the recovery of hearing is less than total.

Important variables in the development of temporary and permanent hearing threshold changes include the following:[5]

1. Sound level. Sound levels must exceed 60 to 80 dBA before the typical person will experience TTS.

2. Frequency distribution of sound. Sounds having most of their energy in the speech frequencies are more potent in causing a threshold shift than are sounds having most of their energy below the speech frequencies.

3. Duration of sound. The longer the sound lasts, the greater the amount of threshold shift.

4. Temporal distribution of sound exposure. The number and length of quiet periods between periods of sound influences the potentiality of threshold shift.

5. Individual differences in tolerance of sound may vary greatly among individuals.

6. Type of sound—steady-state, intermittent, impulse, or impact. The

tolerance to peak sound pressure is greatly reduced by increasing the rise time and/or burst duration of the sound.

Masking

The ability of sound to interfere with speech communication is well known. Speech communication is essential in employee training, in giving and understanding orders, in giving warning signals, and in other phases of work. It should be pointed out that many noises that are not intense enough to cause hearing loss can interfere seriously with speech communication.[6]

Here is why this happens. Ordinary speech consists of a complicated sequence of sounds whose pressure level and spectral or frequency distribution constantly vary. Because of this, some speech sounds will be masked by a specific steady noise while other speech sounds will not. Because the energy of speech sounds in different frequency regions fluctuates from moment to moment, a sound that is masked at one instant could be clearly perceptible the next.

It is not usually necessary for the listener to hear all the speech sounds in a sentence because ordinary speech is very redundant—that is, it contains repetitive information. The listener decodes speech sounds by a synthesizing process, only partly understood at present, that depends not only on the acoustic cues but also on his knowledge of the language and of the context in which the speech sound occurs.[5]

If the ear is simultaneously exposed to two different sounds, the louder sound drowns out the softer one, shifts its apparent location, or changes its quality. This phenomenon is called *masking*—the louder sound is said to mask the other sound. The degree of masking also depends upon the difference in frequency between the two sounds. The shift in hearing threshold is greatest near the frequency of the masking tone and is different for pure tones and for bands of noise of the same overall level.[6]

Acoustic privacy. It should be pointed out that not all masking is bad. A noise that can be ignored may be able to blot out an annoying one. Indeed, offices can be made so quiet that everyone can hear the talking, phone conversations, and other sounds produced by everyone else—in which case speech sounds become "unwanted noise." Only when a particular sound becomes sufficiently loud can it be detected over the background. The background sounds whatever their nature tend to mask wanted sounds until the level exceeds the background.

For acoustic privacy, therefore, a moderate amount of background noise may be desirable. If an office area has been made too quiet, a low level of noise may have to be reintroduced, so that its sound level permits ordinary conversation at 10 feet or less but requires raising the voice in order to be heard at greater distances. The optimum noise level is seldom, if ever, complete silence.

Privacy is somewhat like noise—a subjective thing, meaning something different to almost everyone. Typical high-privacy spaces include offices for attorneys, doctors, psychologists and other counselors, editors, craftsmen, top management, and the like. In acoustical design, privacy usually refers to sufficient attenuation (reduction) of sound so other sounds do not intrude. When it is impossible or impractical to provide enough isolation between spaces by means of a barrier, it is occasionally helpful to raise the background level within a space so that intruding sound from adjacent areas is no longer intelligible—that is, it is masked. This may be done by adding masking sound in the form of steady, broadband, neutral sound, such as air-diffuser noise, splashing fountains, and other *white noise*. Music may not be effective, since it contains many pauses and has relatively pure, discrete tones.[6]

Masking is a useful, but a complex and limited procedure that normally should be attempted only by experienced professionals. It cannot be used in spaces where low background levels are required.

Speech interference

It was found that many variables influence the interference of communication from speaker to listener. In addition to the masking noise that is present at the listener's ear, all of the following can be important:[7]

- The voice characteristics of the speaker

- The transmission path from speaker to listener

- The relative spatial locations of the speaker, noise source, and listener

- The noise level at the speaker's and at the listener's ear

- The presence or absence of reverberation

- The hearing ability of the listener

Special measures for rating or predicting the interfering effects of noise have been developed. These take into account the acoustic energy found within those frequency bands of noise which encompass the critical speech frequency range. These measures are used in setting the acoustic requirements for offices or other living spaces where speech and other forms of sound reception serve important functions.[8]

Fortunately, man has an unusual ability to discriminate from among the elements of a complex sound. From these, he can discern discrete, particular signals of interest to him. Lip reading or using facial or body gestures can also aid spoken communication. Almost everyone has some small amount of skill at lip reading and body language interpretation.

In describing speech interference, the noise concerned can be defined either in terms of its specific spectrum and level or in terms of any number of summarizing schemes. In addition to the average A-weighted sound level, the two most generally used alternative methods of characterizing noises in respect to their speech-masking abilities are:[7]

The Articulation Index (AI)

The Speech Interference Level (SIL)

The articulation index (AI) recognizes that certain frequencies in the masking noise are more effective in masking than other frequencies. The AI is determined as follows:[7]

1. The frequency range in which significant speech energy exists (250 to 7000 Hz) is divided into 20 bands, each of which contributes 1/20 of the total intelligibility of speech.

2. The difference between the average speech level and the average noise level (that is, the signal-to-noise ratio) is determined for each of these bands.

3. The numbers are combined to give a single index.

By predicting how much masking of specific speech sounds will occur, the AI estimates the intelligibility of speech at a given level in a specific noise. Simplified procedures for estimating the AI from measurements of octave-band levels have also been developed. Although the AI is to date the most accurate measurement used to predict the effects of noise on speech intelligibility, it is difficult to use and interpret.

The speech interference level (SIL), introduced as a simplified substitute for the AI, is an indication of only the average general masking capability of the noise. Contributions to intelligibility by the lowest and highest frequencies are ignored. As originally formulated, it was defined

TABLE 3-A
SPEECH INTERFERENCE LEVELS
(in dB re 20μN/m^2)

Distance (Feet)	Voice Level			
	Normal	Raised	Very Loud	Shouting
1	70	76	82	88
3	60	66	72	78
6	54	60	66	72
12	48	54	60	66

From Handbook of Noise Measurement, 7th ed. *General Radio Co.,* 1972.

as the arithmetical average of the sound pressure levels (SPL) in the 600 to 1200, 1200 to 2400, and 2400 to 4800 Hz octave bands. The modern version of the speech interference level is the average of the SPL's in the three octave bands centered at 500, 1000, and 2000 Hz.[8]

Many variations of SIL have been developed so that a shorthand descriptive notation is now used. SIL (.5, 1, 2) is the average of the SPL's of the three octave bands centered at 500, 1000, and 2000 Hz; SIL (.25, .5, 1, 2) includes the 250 Hz band in the average. The original SIL would be SIL (.85, 1.7, 3.4) in this notation.

Noise level, vocal effort, and distance. Because much speech is spoken at a reasonably constant level in ordinary surroundings, it is possible to tabulate many of the empirical facts about average speech communication. Table 3-A indicates that speech spoken at a normal level can be heard at a distance of three feet when the background noise has an SIL (.85, 1.7, 3.4) of 55 dB. When voice level is increased from this "normal" to "raised," "very loud," and (sustained) "shouting," each higher step represents a four-fold increase in vocal output or a 6 dB increase in acoustic output. If the voice rises 6 dB for each step, then, as a first approximation, the noise can also increase by the same amount without changing the intelligibility of the speech.

The speech level will drop 6 dB if the distance is doubled from the source. If the listener is 6 ft from a speaker, the speech level at his ears will have dropped to 6 dB less than what it was at 3 ft, and the noise that will permit normal conversation will also be 6 dB lower.

Adequate communication in higher noise levels than those indicated in Table 3-A can occur if the possible messages are known to be restricted. This factor accounts for the success of communication in many industrial situations with high levels of noise. Failure may occur, however, when an important but unpredictable message must be communicated.[7]

The A-weighted sound level is a useful index of the masking ability of a noise. The A-weighting filter network emphasizes the median of the speech frequencies, as do the various SIL's. However, in contrast to most SIL schemes, A-weighting does not completely ignore the lower frequencies.

Table 3-A applies only to reasonably steady sound. Intermittent sound impulses will, of course, mask certain signals only while they are present, and sound fluctuating in level will provide variable degrees of masking. Again, speech is redundant enough so that an isolated one-second burst of sound is unlikely to produce much disruption of the communication process. However, the probability of such a disruption occurring grows with both the frequency and the duration of the sound bursts.

While the A-weighted noise level is an adequate measure of many noises, some noise situations demand a more detailed analysis. This is particularly true of noises that consist almost exclusively of either low frequencies or high frequencies—for example the rumble of low rpm engines or the hiss of compressed air being released. The SIL and dBA values of any two noises will ordinarily not be the same (since each value will depend on the spectral energy distribution). Attempts have been made to determine an average conversion (SIL to dBA) number for a more or less vaguely defined "average" noise. The A-weighted sound levels that will permit conversation are shown in Figure 3–3.

Signal-to-noise ratio is obtained by dividing the desired signal level by the noise level. When the signal-to-noise ratio becomes sufficiently large, the signal is detected and information becomes available. Either the signal (the wanted sound) must be presented at a suitable level above the noise level, or the noise level must be reduced below the signal level—or both. For intelligibility, a signal-to-noise ratio of 10 dB (that is, the signal sound pressure level 10 dB above the noise level) is sufficient; however, this may not mean that a comfortable or pleasant situation exists. Shouting above the noise may make the message understandable, but the cost of hearing fatigue, discomfort, or annoyance should be considered.[3]

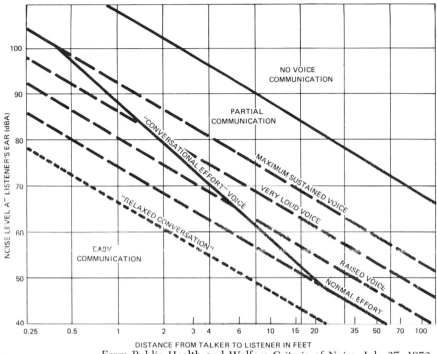

NOISE LEVEL AT LISTENER'S EAR (dBA)

NO VOICE COMMUNICATION

PARTIAL COMMUNICATION

EASY COMMUNICATION

"CONVERSATIONAL EFFORT" VOICE

"RELAXED CONVERSATION"

MAXIMUM SUSTAINED VOICE

VERY LOUD VOICE

RAISED VOICE

NORMAL EFFORT

DISTANCE FROM TALKER TO LISTENER IN FEET

From Public Health and Welfare Criteria of Noise, July 27, 1973.
U.S. Environmental Protection Agency.

FIG. 3-3.—Distance at which ordinary speech can be understood (as a function of A-weighted sound levels of masking noise in the outdoor environment).

Annoyance

The noisiness or annoyance of a sound depends on the subjective reaction to the sound. Psychoacoustic experiments have shown that recognition and identification of a specific sound characteristic can have a substantial emotional effect on the listener. Annoyance to noise is a psychosocial response. Annoyance has its roots in the unpleasantness of noise and the disruption by noise of ongoing activities.

The degree of annoyance and whether that annoyance leads to complaints or produces rejection of a noise source are dependent upon many factors. Individual responses of people to noise are often studied in the laboratory. Usually, these studies involve judgments of individual noise events in controlled environments. Such studies have been helpful in isolating some of the factors contributing to annoyance by noise.[7]

Some characteristics of sound seem to be more annoying than others. These parallel the discussion of the characteristics of sound that were discussed in Chapter 2.

1. *Loudness.* The more intense (and consequently louder) noises are considered more annoying.

2. *Pitch.* A high-pitch noise (that is, one containing predominantly frequencies above 1500 Hz) is more annoying than a low-pitch noise of equal loudness.

3. *Intermittency and irregularity.* A sound that occurs randomly or varies in intensity or frequency is believed more annoying than one which is continuous and unchanging.

4. *Localization.* A sound that repeatedly seems to change in its location relative to the listener is less preferred than one that remains stationary.

A measure of noise that describes its annoyance value has recently been developed. This measure called *perceived noise level, PndB,* is derived from calculations based upon the octave band intensity levels of the noise together with data showing equal annoyance ratings for different octave bands of noise. While of some value, perceived noise levels and other annoyance measures based upon single judgments of the noise stimulus are expected to have only limited usefulness in gaging the "complaint potential" of a noise. This is due to the many non-acoustical considerations that enter into such judgments.

Some of these factors are cited below with examples to illustrate each of them:[1]

• *The sound has unpleasant associations.* The annoyance caused by the intrusion of aircraft noise into communities around airports is based, in part, upon the residents' fear of the planes crashing into their homes.

• *The sound is inappropriate to the activity at hand.* Music tolerated during waking hours may be annoying during sleeping hours.

• *The sound is unnecessary.* People may complain of the noise made by the neighbor's pets but not by the delivery trucks in the same neighborhood.

• *The sound has an advantage associated with it.* The comfort derived from air conditioning outweighs the noise of such units. Similarly, the economic value of nearby plants to a community may balance out the noise produced by the plants. Annoyance due to military aircraft noise may be offset by the assurance against surprise attack by an enemy.

• *Individual tolerance to noise.* Some individuals complain about all kinds of noise as well as other types of nuisances, others do not.

Even though many details of the hearing process are still not well understood, the concept of loudness seems to have been fairly well agreed upon and seems to be predictable from the measurement of sound pressure levels. The concepts of annoyance and noisiness, on the other hand, are not so firmly established and seem, at least at present, not to be predictable from measured physical data, at least not in any general sense. Annoyance has a great many psychological aspects which make it difficult to define and measure in any simple manner.

Certain general trends in the annoyance effects of noise might be found even though they cannot be physically scaled. These were described at the beginning of this section on annoyance factors—loudness, pitch, intermittency, and localization.

A further effect concerns localization of sound. When a large office has acoustically hard walls, floor, and ceiling, the room is said to be "live" or reverberant. The noise from any office machinery is reflected back and forth, and the workers are immersed in the noise with the feeling that it comes from everywhere. If the office is heavily treated with absorbing material, the reflected sound is reduced, and the workers then feel that the noise is coming directly from the machine. This localized noise seems to be less annoying.

Physiological Effects

Noise can elicit many different physiological responses. However, no clear evidence indicates that the continued activation of these responses leads to irreversible changes and permanent health effects. It is important, therefore, to consider not only the more overt effects of noise, such as hearing loss and the masking of speech, but the more subtle effects which noise can produce. These nonauditory effects can be merely transitory or, in some cases, longlasting. They can take place without conscious knowledge of their occurrence. For more details, see Chapter 10.

Pain. There are two general types of aural pain or discomfort. The first type is caused by the stretching of the tympanic membrane tissues in response to large amplitude sound waves. Although there is a fairly wide range of individual variability, especially for high-frequency stimuli, the threshold of pain for normal ears is approximately 135 to 140 dB SPL. This pain threshold is essentially independent of frequency, and it can affect people with a hearing loss as well as those with normal hearing since it is not a function of the ear's sensori-neural system.

A second type of aural discomfort occurs as a result of abnormal function in the cochlea or inner ear. Frequently, noise induced hearing losses are accompanied by a condition called auditory recruitment, which is defined as an abnormal increase in loudness perception. In some cases, the condition is severe, and it can lead to considerably lower thresholds of aural discomfort or pain. Consequently, sound levels of only moderate intensity can occasionally be quite uncomfortable to individuals experiencing auditory recruitment.

Startle reflex. Noise induced orienting reflexes serve to locate the source of a sudden sound and, in combination with the startle reflex, prepare the individual to take appropriate action in the event danger is present. Apart from possibly increasing the chance of an accident in some situations, there are no clear indications that the effects are harmful since these effects are of short duration and do not cause long-time body changes.[7]

Effect on sleep. Noise can interfere with sleep; the problem of relating noise exposure level to quality of sleep is, however, difficult. Even noise of a very moderate level can change the patterns of sleep, but the determination of the significance of these changes is still an open question. Noise exposure may cause fatigue, irritability, or insomnia in some individuals, but the quantitative evidence in this regard is unclear. No firm relationship between noise and these factors can be established at this time.

Stress. Noise exposure can be presumed to cause general stress by itself or in conjunction with other stressors. Neither the relationship between noise exposure and stress nor the threshold noise level or duration at which stress may appear has been resolved.

Short and infrequent periods of stress are usually harmless if there is an opportunity for the body to recuperate within a brief period after exposure. Long-term stress, however, poses a potential danger to the health of an individual; this attitude is based on extensive experimental work on animals. Major factors that have not yet been resolved are (a) the point at which a noise becomes a stressing agent in man, and (b) what amount of exposure is necessary to cause long-lasting or permanent physiological changes.[7]

The concept that stress is universally bad and unhealthy is misleading. At certain times, some stressing agents and stressful situations might be construed as necessary (such as alerting, orienting, or motivating). Thus, although it is plausible that noise can be detrimental as a stressing agent, there is insufficient data to state unequivocally that noise as a stressor is sufficiently severe to cause serious troublesome reactions.[7]

Vibration and Infrasound

Infrasound occurs in nature at relatively low intensities. Sources of natural infrasonic frequencies are: earthquakes, volcanic eruptions, winds, air turbulence, thunder, large waterfalls, and impact of waves on beaches. Man-made infrasound waves usually occur at higher intensity levels than those found in nature.

Generally, people exposed to high levels of infrasound complain of disorientation, nausea, and general feelings of discomfort. Responses generally are mostly of a nonspecific nature, resembling reactions to mild stress or alarm. Physical effects of excessive vibration include disorientation, nausea, damage to tissues, and other phenomena. The average vibration levels found in industry are well below those capable of causing such dramatic effects.[7]

Threshold and tolerance levels vary with frequency (as is true of audible sound, but in quite a different way). There is still little agreement on whether people respond to acceleration, velocity, or displacement, or to a combination of all three. The circumstances under which people are subjected to vibration appear to influence strongly their response to the vibration. In transportation equipment, for example, much more vibration is tolerated than is generally accepted in stationary environments.[8] (See Figure 3–4.)

A variety of bizarre sensations have been reported during exposure to airborne infrasonic waves, such as fluttering or pulsating sensations. There is some evidence that intense infrasound (120 dB sound pressure level or above) can stimulate the vestibular system (the inner ear balance mechanism), and lead to disequilibrium if the stimulation is intense enough. The data available suggests that infrasonics do not pose a serious problem to the hearing mechanism when intensities are below 130 dB SPL (which is generally the case); however, where high intensities (above 130 dB SPL) are present, a serious hazard may exist.

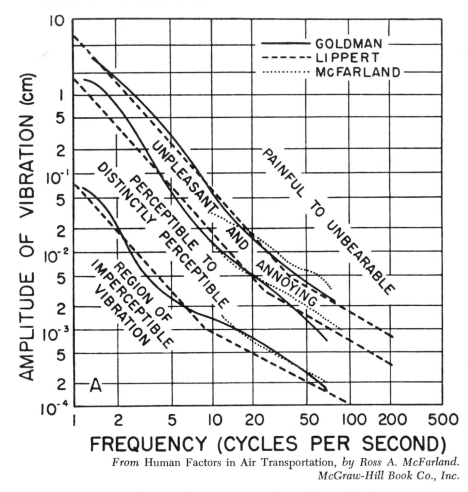

From Human Factors in Air Transportation, *by Ross A. McFarland.*
McGraw-Hill Book Co., Inc.

Fig. 3–4.—Threshold values for perception, discomfort, and tolerance to sinusoidal vibration at a single frequency are plotted on logarithmic coordinates of amplitude (in centimeters) versus frequency (cycles per second). Amplitudes are vector values (deviations each side of an average value).

The most common objection to vibration in buildings and other normal human environments is the audible effects; the surfaces and components of the environments vibrate strongly enough to turn them into secondary sound sources, often amplifying the original sound source appreciably.

Ultrasonics

Ultrasonic frequencies (above 20,000 Hz) are produced by a variety of equipment in industry, such as high-speed drills and cleaning devices. Ultrasonic waves became recognized as a potential health problem with the advent of jet engines. A number of persons working in the vicinity of jet engines reported symptoms of excessive fatigue, nausea, and headache. These responses resemble those found during stress. The problem, however, is difficult to study because of two factors:

1. Ultrasonic waves are highly absorbed by air, and, therefore, are of significance only near a source.

2. Airborne ultrasonics from ordinary sources are often accompanied by broadband noise and by subharmonics, both of which fall into the audible range.

In one study the hearing threshold levels of employees were measured in the frequency range 2000 to 12,000 Hz before and after exposure to the noise over a working day. No significant temporary threshold shifts were detected. On the assumption that noise exposure not severe enough to cause a temporary threshold shift cannot produce premanent damage, it was concluded that hearing damage due to exposure to the noise from industrial ultrasonic devices is unlikely. Exposure to high levels of ultrasonic devices is unlikely. Exposure to high levels of ultrasound (above 105 dB SPL) may have some effects on man. It is important to recognize, however, that a hazard also arises from exposure to the high levels of components in the audible range that often accompany ultrasonic waves.

Summary

Because the process of hearing is complicated, no simple relationship exists between physical measurement of sound pressure and human perception of sound. The loudness of a pure tone may be judged to sound different when compared with another pure tone even though the sound pressure level is the same in both cases. The awareness of sound is usually a function of three measurable physical variables—sound level, frequency, and duration. While loudness depends primarily on sound pressure, it is also affected by frequency because the ear is more sensitive to high frequencies than to low. A sound that has a constant sound pressure can be made to appear quieter or louder by changing its frequency.

Measures for rating or predicting the interfering effects of sound have been developed for offices and other living spaces where speech and other forms of sound reception serve important functions. These measures take into account the acoustic energy found within the critical speech frequency range.

Noisiness is subjective because it depends on a person's reaction to a sound. Recognition and identification of a specific sound characteristic can substantially affect a listener because annoyance is a psychosocial response—it depends on how unpleasant a noise is and how much it disrupts ongoing activities.

Noise exposure can be presumed to cause general stress by itself or in conjunction with other stressors. Short and infrequent periods of stress are usually harmless if there is an opportunity to recuperate within a brief period after the exposure.

Yet to be resolved are (a) the point at which noise becomes a stressing agent in man, and (b) what amount of exposure is necessary to cause long-lasting or permanent physiological damage.

References

1. Kryter, K. D. *The Effects of Noise on Man.* New York, N.Y.: Academic Press, Inc., 1970.
2. Newby, H. A. *Audiology,* 3d ed. New York, N.Y.: Appleton-Century-Crofts, 1972.
3. Petersen, A. P. G., and Gross, Jr., E. E. *Handbook of Noise Measurement,* 7th ed. West Concord, Mass.: General Radio Co., 1972.
4. *Acoustics Handbook.* Palo Alto., Calif.: Hewlett-Packard Co., 1968.
5. *Criteria for a Recommended Standard . . . Occupational Exposure to Noise.* U.S. Department of Health, Education, and Welfare, Cincinnati, Ohio, 1972.
6. Broch, J. T. *Acoustic Noise Measurements.* Cleveland, Ohio: Bruel & Kjaer, 1971.
7. U.S. Environmental Protection Agency. *Public Health and Welfare Criteria for Noise.* Washington, D.C.: U.S. Government Printing Office, 1973.
8. Beranek, L. L. ed. *Noise and Vibration Control.* New York, N.Y.: McGraw-Hill Book Co., 1971.

Bibliography

American Academy of Ophthalmology and Otolaryngology. *Guide for Conservation of Hearing.* Rochester, Minn.: AAOO, 1970.
Sataloff, J., and Michael, P. *Hearing Conservation.* Springfield, Ill.: Charles C Thomas, 1973.
Ward, W. D., and Fricke, F. E., eds. *Noise as a Public Hazard.* Washington, D.C.: The American Speech and Hearing Association, 1969.

Part 2 Measurement of Sound

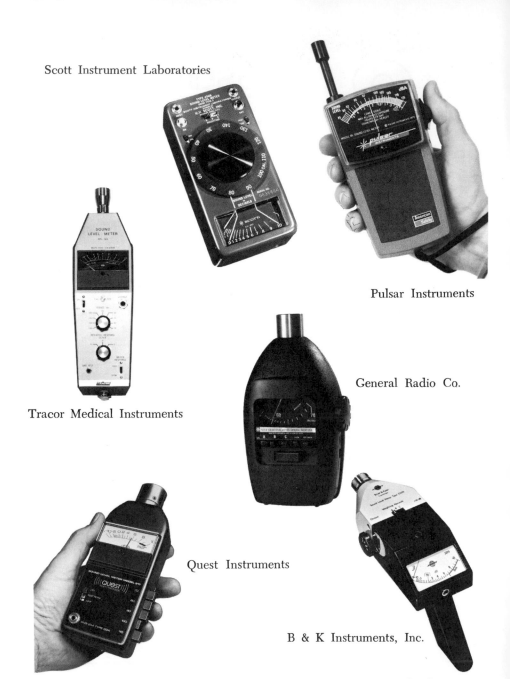

Scott Instrument Laboratories

Pulsar Instruments

General Radio Co.

Tracor Medical Instruments

Quest Instruments

B & K Instruments, Inc.

FIG. 4–1.—The basic sound measuring instrument is the sound level meter. Some of the many available are shown here.

96

Chapter Four   Instrumentation for Sound Measurement

By Richard La Jeunesse

The basic sound measuring instrument—the sound level meter—consists of a calibrated microphone, amplifier, attenuator, weighting networks, and an indicating meter (see Figure 4–1). When additional or different sound measurement information is required, other instruments such as a frequency analyzer, graphic level recorder, noise exposure monitor, impact noise analyzer, oscilloscope, noise dosimeter, and tape recorder are used (see Table 4-A and later illustrations in this chapter).

Industrial noise is usually not a simple pure tone, but a complex sound made up of many tones of various pitches and fluctuating sound levels. Two different sounds may indicate the same numerical reading on the dial of a sound level meter, but one sound may be apparently louder to the average ear. The frequency analyzer can be used to evaluate this loudness characteristic of sounds or to identify offensive components of the noise source.

Often mechanical vibrations are an integral part of the noise problem under study; in which case, a vibration measurement should be made. By replacing the microphone with a vibration pickup, some sound level meters can be used to obtain this measurement. The meter response is converted to units of displacement, velocity, or acceleration in an auxiliary control box.

All sound measuring instruments depend on a transducer to transform air pressure variations or structure-borne vibrations into an electrical signal. Once this signal is obtained, auxiliary electronic instruments are

97

TABLE 4-A. SOUND- AND VIBRATION-MEASURING INSTRUMENTS

	Calibration	Transducers & Accessories	Preamplifier	Sound & Vibration Meters, Tape Recorder, Impact Meter	Analyzers/ Filters	Graphic Recorders & Display Units
Acoustics	Sound level calibrator	Ceramic microphones		Precision sound level meter and analyzer (octave)		
		Electret microphones	Preamplifier (optional)	Sound level meter		Oscilloscope
		Condenser microphones	Preamplifier with bias supply	Noise exposure meter	Sound & vibration analyzer	Graphic level recorder
		Couplers	Multichannel amplifier	Impact noise analyzer	Wave analyzer	
		Cables		Cassette data recorder	Real-time analyzer	d-c recorder
		Windscreens				Storage display unit
		Tripod				
Vibration	Vibration calibrator	Accelerometers pickups	Preamplifier (optional)	Vibration meter	Universal filter	
		Cables	Multichannel amplifier	Cassette data recorder		
		Control boxes to use with SLM or analyzers				

Adapted from Handbook of Noise Measurement, 7th ed. General Radio Co. 1972.

used to measure certain characteristics of the electric signal or to store it for later measurement and evaluation.

• The sound level meter is specifically designed for acoustic measurement and is the most basic instrument for measuring noise. It includes a calibrated microphone, an amplifier, and a metering circuit to indicate the output of the amplifier. The sound level meter is most frequently used when the level of a noise is to be measured directly at the source.

• The frequency analyzer (also called an octave-band analyzer) is used to measure the sound pressure level in different frequency bands.

• The tape recorder is used to record sounds and to store the information for future measurement and analysis.

• The graphic level recorder provides a permanent chart record of the sound level as a function of time; or, if used with an octave-band analyzer, it provides a record of the frequency spectrum.

• The noise exposure monitor or dosimeter accumulates or integrates information regarding sound levels during a specific period of time.

• The oscilloscope displays the instantaneous waveform of the electrical signal generated by the transducer, and in combination with a camera attachment, a permanent record of the signal display on the scope can be made.

• The impact noise analyzer provides information on the peak characteristics of impact-type sound pressure variations, such as those produced by punch presses and drop hammers.

The particular noise level or characteristic may be measured by a number of different instruments; however, the one or combination chosen depends on the information desired.

First, the transducer for picking up sound waves will be discussed, then each electronic instrument will be examined in detail.

Microphones

For airborne sound waves, the transducer is a microphone; while, for structure-borne vibration, the transducer is a vibration pickup. Although the medium for transfer of energy for each type of transducer is different, the type of output (an electrical signal) of each transducer is the same. To convert a sound pressure variation to an electrical signal, the ideal microphone should meet the following design requirements:

TABLE 4-B. MICROPHONE CHARACTERISTICS

Characteristics	Piezoelectric		Condenser		Dynamic
	Crystal	Ceramic	Air-Dielectric	Electret	
Advantages	Self-generating, simple associated conversion instruments. Relatively rugged. Inexpensive Higher sound levels	High sensitivity, wide frequency range. Many sizes available.	Inexpensive. No need for independent polarizing supply. Rugged. Self-generating	Self-generating. Simple associated conversion instruments. High sensitivity. Insensitive to high humidity. Inexpensive. Fits special applications	
Limitations	Sensitive to vibration and Humidity Low Sensitivity		Expensive. Subject to Breakdown in high humidity	Sensitive to high temperature	Sensitive to magnetic fields. Relatively large size. Limited frequency range. Heavy and awkward.
Dynamic range (dB)	80–200 (Ammonium Dihydrogen Phosphate-ADP)	24–160	24–200	24–200	24–140
Frequency response (Hz)	20–10,000	Flat over usable range 20–20,000	20–20,000	20–20,000	50–8000
Temperature range	Below 115 F (46.1 C)	–22 to 185 F (–30 to +85 C)	–22 to 194 F (–30 to +90 C) including Preamp	–22 to 176 F (–30 to 80 C)	–22 to 392 F (–30 to 200 C)
Cable corrections	Correction necessary when using long cables (10 ft or more) because of temperature effect.	Recommendation of manufacturer should be followed.	Preamp required	Same as ceramic	Low impedance makes unit ideal for use with long cable. No correction required.

NOTE: The mere adherence of a microphone to one of the construction types shown here does not guarantee the performance shown in this table.

100

- Sensitivity within the desired range of interest

- A flat frequency response over the measurement range

- A suitable dynamic range

- A dynamic pressure response independent of the static pressure level

- Minimum diffraction of the sound field (that is, the dimensions of the microphone should be small in relation to the shortest wavelength of the sound wave that is being measured)

- The directivity pattern or variation in response with the angle of incidence of the sound wave should be known and accounted for

- A low noise floor or high signal-to-noise ratio

- Low output impedance

- A high acoustic impedance compared to air (and thus absorbs no acoustic energy)

- Essentially independent of the environment (that is, stable with respect to time, temperature, humidity, etc.)

These requirements form a basis for the evaluation of microphone performance. The relative importance of each of the requirements depends on the design parameters and measurement goals of the specific sound measurement system. (See Table 4-B for microphone characteristics.)

Sensitivity

The sensitivity of a microphone is a measure of its electrical output for a given acoustic input. The electric output is commonly given in volts at the output terminals with no electric load on the microphone. The acoustic input is usually taken as a "1 microbar" sound pressure level at the microphone (that is, a level that would read 74 dB re 20 μnewton/m^2). Unfortunately, microphone sensitivity is also expressed in decibels (re 1 v/μbar), and these dB bear no mathematical relation with the dB on a sound level meter. Moreover, they are *negative* decibels, so the larger the dB in numerical value, the lower the sensitivity. As a rule of thumb, a −60 dB microphone will produce 1 millivolt output from a 74 dB sound level. A −80 dB microphone would produce 1/10 millivolt from the same 74 dB level.

Sound level meters are calibrated to be used with a microphone of a particular sensitivity, and the instrument reading must be corrected if a microphone of different sensitivity is used.

Frequency response

The frequency response is related to the physical size of the microphone. The smaller the physical dimensions of the microphone, the higher is its upper frequency limit and the lesser are the effects of directivity. On the other hand, the sensitivity of the microphone usually is reduced as its dimensions decrease so that very low noise levels can be measured only with the larger size microphones. At times, compromises concerning these various factors will have to be made.

The frequency response also depends upon the angle of incidence between the sound wave and the microphone diaphragm. Most microphones are built to be omnidirectional (no one direction is favored), and they should respond equally well for all directions of sound incidence at frequencies below 1000 Hz.

Dynamic range

Consideration should be given to the dynamic range or the maximum or minimum sound pressure levels that will be measured. When very low noise levels—for example, audiometric test locations or nighttime ambient noise levels—are to be measured, then the minimum sound pressure level that a microphone can measure may be the determining factor.

The choice of microphones for measuring extremely intense noise levels, such as those produced by jet or rocket engines, will depend upon the maximum sound pressure level that a particular microphone can measure without excessive distortion or failure. No single microphone can satisfactorily measure extremely high and extremely low sound levels. A microphone designed to measure very low noise levels has a limited maximum sound pressure level response. Conversely, high-level microphones are not suitable for measuring low noise levels.

As the amplitude of the sound pressure variation increases (as the sound gets louder), the output of a microphone increases correspondingly. At a sufficiently high sound level, however, the microphone output will no longer increase proportionately to the sound pressure; and the response is then said to be nonlinear. This departure from linearity is usually gradual, and only a certain amount can be tolerated.

Directivity pattern

The variation in response of the microphone with the angle of inci-

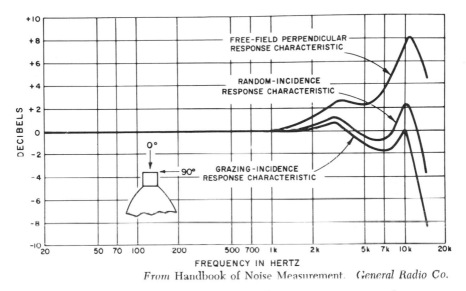

From Handbook of Noise Measurement. *General Radio Co.*

Fɪɢ. 4–2.—Typical response characteristics of a 1 in. ceramic microphone.

dence of a sound wave is called the directivity pattern and is ordinarily shown graphically (see Figure 4–2). Recommendations of the supplier should be followed when orienting the microphone for a noise measurement.

Noise floor

A microphone can generate an electrical noise signal by thermal agitation in the microphone and the input circuit of the amplifier. This inherent instrument noise usually sets the lower limit of sound pressure level that can be measured by a particular sound measurement system.

Impedance mismatch

Microphones of different types vary greatly in electrical output impedance. A microphone of high electrical impedance should not ordinarily be connected directly to a sound measuring instrument (or a cable) having a low impedance input because a loss in sensitivity will result.

Environmental effects

The response of a microphone varies with temperature. A knowledge of the temperature coefficient of response of a microphone is therefore desirable, if the microphone is to be used at temperatures appreciably different from ordinary room temperatures.

The maximum temperatures at which microphones will continue to operate satisfactorily range from 100 to about 500 F (38 to 260 C). At normal room temperatures, all microphones will work satisfactorily; however, at temperatures above 300 F (150 C), only a specially designed microphone meets the requirements. At the higher temperatures the weakest link in the sound measuring system is not likely to be the microphone but the associated preamplifier, cabling, and electronic equipment.

Humidity can also affect the operation of many microphones. Air dielectric condenser microphones may be particularly susceptible because of internal electrical leakage caused by moisture on the insulators. This can be minimized by keeping the microphone at a temperature higher than the ambient temperature using the heat generated by the microphone preamplifier.

Types of microphones

There are a variety of microphones available to perform specialized functions. For intelligible voice communication by telephone, a microphone of only limited frequency range is required; it must be rugged and sensitive and be able to operate without a preamplifier. The carbon granule microphone fulfills these requirements. The radio and television industries require a microphone that has a broad frequency range for good fidelity and is highly directional to reduce interference from unwanted sounds. The ribbon microphone handles this job well.

Although microphones such as these play an important role in their specific area and do indeed convert sound pressure to voltage, their relatively loose operating characteristics make them totally unsuitable for accurate sound pressure level measurements. Microphones used for accurate sound pressure level measurements include the piezoelectric, condenser, and dynamic microphones.

Piezoelectric. The term "piezoelectric" means the generation of electric current in crystals by applied mechanical stress. These piezoelectric crystals may be natural such as quartz, lithium sulfate, and tourmaline, or man-made by adding impurities to the natural minerals. Man-made crystals are barium titanates and lead zirconate titanates.

• *Crystal microphones* have a good frequency response, and no external voltage or current is required. However, their use is limited in sound measuring instruments because of their susceptibility to high temperatures and humidity and poor aging characteristics.

• *Ceramic microphones* have a smooth frequency response and are extremely rugged and relatively unaffected by normal temperature and humidity changes. The microphone can be mounted directly on the measuring instrument or separately with connection by extension cable when it is necessary to avoid the effects of the observer and the instru-

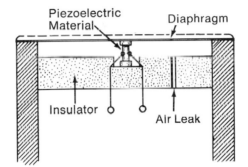

FIG. 4–3.—Typical piezoelectric microphone construction.

ment case (see Figure 4–3). It is also the least expensive type of microphone for sound measurement.

Unlike the air-dielectric condenser microphone, the ceramic microphone requires no polarization voltage, an advantage when designing instruments for portable operation. However, the upper frequency limit of the ceramic microphone is not as high as that of the condenser type, since fewer sizes are available (for high-frequency work).

Condenser microphones. The condenser, electrostatic, or capacitor microphone is widely used for accurate sound level measurement. The variation of the electrical capacitance, in response to changes in sound level in this type of microphone, is used to control an electrical signal. These microphones generally have excellent response characteristics and are available in several sizes.

Currently two types of condenser microphones are available for general use—the air-dielectric and the electret. The two types have nearly iden-

105

tical response characteristics and accuracy, but differ greatly in complexity and cost.

• *Air-dielectric.* The physical construction of a condenser microphone is shown in Figure 4–4. The sensing membrane (diaphragm) is one plate

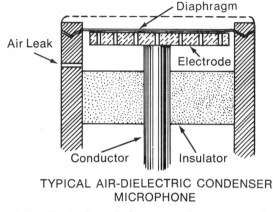

TYPICAL AIR-DIELECTRIC CONDENSER
MICROPHONE

FIG. 4–4.—Typical air-dielectric condenser microphone.

of the condenser or capacitor; the polarization electrode is the other. The membrane is attached to the housing, which is at ground potential; the polarization voltage is applied to the polarization electrode through a contact. An insulator supports the polarization electrode as well as insulating it from the housing.

The displacement of the membrane from its rest position is proportional to the applied sound pressure. The voltage across the capacitor having a given charge is proportional to the distance between the plates. The charge is supplied by the polarization voltage; and if the polarization voltage source has an impedance high enough to prevent significant current flow (even at the lowest frequency), then the charge on the condenser can be considered constant.

The air-dielectric condenser microphone has a high sensitivity and a flat frequency response (± 1 dB from 20 Hz to 20,000 Hz for the ½-in. configuration).

• *The electret condenser microphone* is identical in principle to the air-dielectric type, deriving its electrical output from changes in distance between two polarized plates (see Figure 4–5). However, the electret

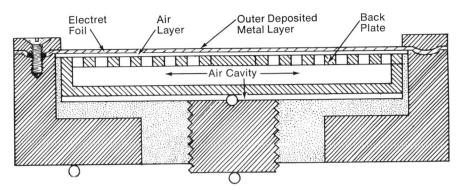

Electret Foil   Air Layer   Outer Deposited Metal Layer   Back Plate

Air Cavity

FIG. 4–5.—Typical construction of an electret microphone.

microphone employs a permanently charged plastic material as the dielectric between the two plates, eliminating the need for a separate polarizing voltage supply and providing a more rugged configuration.

Since the manufacturing process is simpler and easier to control than that required for the air-dielectric types, the electret-condenser microphone is relatively inexpensive—especially in the smaller sizes. Its sensitivity and distortion characteristics lie between the air-dielectric and ceramic types (see Table 4-B).

Dynamic microphone. This microphone consists of a thin diaphragm of aluminum or plastic to which is fastened a coil projecting into an annular gap in a magnetic field (see Figure 4–6). When the diaphragm moves, a voltage is induced in the coil. An electrical transformer is customarily used to step up the voltage from the dynamic microphone before voltage is sent to the amplifier, giving the microphone an overall sensitivity equal to the better ceramic microphones. However, its overall frequency response, especially at lower frequencies, is much poorer than any of the other types described here.

The dynamic microphone is particularly sensitive to extraneous alternating magnetic fields, such as are present near power transformers, motors, and other electrical devices. The transformer used with the microphone may be similarly affected unless it is well shielded. Dynamic microphones are seldom used now that the other types are commonly available at comparatively low prices.

Sound level meters

The basic instrument used to measure sound pressure variations in air

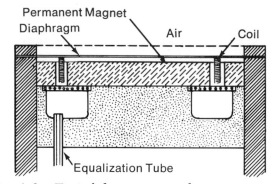

Fɪɢ. 4–6.—Typical dynamic microphone construction.

is the sound level meter. This instrument contains a microphone, an amplifier with a calibrated attenuator, a set of frequency response networks (weighting networks), and an indicating meter (see Figures 4–1 and 4–7).

The sound level meter is a sensitive electronic voltmeter that measures the electrical signal from a microphone which is ordinarily supplied with and attached to the instrument. The alternating current electrical signal from the microphone is amplified sufficiently so that, after conversion to direct current by means of a rectifier, the signal can deflect a needle on an indicating meter. An attenuator controls the overall amplification of the instrument. The response vs. frequency characteristics of the amplified signal are controlled by electrical circuits called weighting networks.

Before an instrument can be called a sound level meter (according to American National Standard S1.4–1971, *Specification for Sound Level Meters*), it must be able to measure the root-mean-square (rms) level of the sound pressure; have a response which meets a tight specification not only in smoothness of response, but also in linearity and attenuator accuracy; and contain weighting networks, labeled A, B, and C (see Table 4-C).

Sound level meters have a measurement range of about 40 dB to 140 dB (re 20.0 μN/m^2) without special accessory equipment. Special microphones permit measurement of lower or considerably higher sound levels. An amplified electrical output signal of the microphone is usually provided on the sound level meter for hook-up to other instruments for recording and analysis. The sound level meter is basically designed to be

TABLE 4-C

COMPARISONS OF SPECIFICATIONS FOR SOUND
LEVEL METERS BASED ON ANSI S1.4-1971

Type	Type			
	1	*2*	*3*	*S*
Type	*Precision*	*General Purpose*	*Survey*	*Special Purpose*
Weighting networks required	A, B & C	A, B & C	A, B & C	One network (more optional)
Total tolerance limit				Consistent with sound level meter to which it is qualified (i.e. type 1, 2 or 3)
@ 1000 Hz dBA dBB dBC	±1.0 dB ±1.0 dB ±1.0 dB	+2.0 dB ±2.0 dB ±1.5 dB	+3.0 dB ±3.0 dB +2.5 dB	
@ 2000 Hz dBA dBB dBC	±1.0 dB ±1.0 dB ±1.0 dB	±3.0 dB +3.0 dB ±2.5 dB	±4.0 dB ±4.0 dB ±3.5 dB	
@ 4000 Hz dBA dBB dBC	±1.0 dB ±1.0 dB ±1.0 dB	+5.5, −4.5 dB +5.5, −4.5 dB +5.0, −4.0 dB	±5.5 dB ±5.5 dB +5.0 dB	
Dynamic range	40 dB Below maximum scale reading	40 dB Below maximum scale reading	Not stated	Not stated
Sensitivity calibration	External acoustic calibrator	External acoustic calibrator	External acoustic calibrator or insert voltage method	Not stated
Gain calibration	Internal electrical check	Internal electrical check	Specified by manufacturer	Not stated

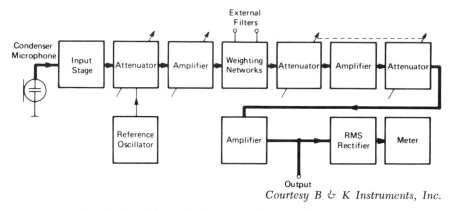

Courtesy B & K Instruments, Inc.

Fig. 4–7.—Schematic diagram of a sound level meter.

a device for field use and as such should be reliable, rugged, reasonably stable under battery operation, and light in weight.

Basic elements of a sound level meter

The **microphone** responds to sound pressure variations and produces an electrical signal which is processed by the sound level meter. (The microphone was discussed earlier in this chapter.)

The **amplifier** in a sound lever meter must have a high available gain in order to measure the low voltage signal from a microphone in a quiet location. It should have a wide frequency range, usually on the order of 20 to 20,000 Hz. The range of greatest interest in noise measurements is 50 to 6000 Hz. The inherent electronic noise floor and hum level of the amplifier must be low.

Attenuators. Sound level meters are used for measuring sounds that differ greatly in level. A small portion of this range is covered by the relative deflection of the needle on the indicating meter; the rest of the range is covered by an adjustable attenuator, which is an electrical resistance network inserted in the amplifier to produce known ranges of signal level. To simplify use, it is customary to have the attenuator adjustable in steps of 10 dB.

In some instances, the attenuator may be split into sections between various amplifying stages in order to improve the signal-to-noise ratio of the instrument and to limit the dynamic range.

Weighting networks. The sound level meter response at various frequencies can be controlled by electrical weighting networks (Figure 4–8). The response curves for these particular networks have been established in ANSI S1.4–1971. One of the networks, called "C-weighting," is intended to apply to a uniform response over the frequency range from 25 to 8000 Hz. Changes in the electronic circuit are sometimes made to compensate for the response of particular microphones, so that the net response is uniform (flat) within the tolerance allowed by the standards. The C-weighting is generally used when the sound level meter supplies a signal to an auxiliary instrument for a more detailed analysis.

The other networks, called A- and B-weighting, have a response that decreases with decreasing frequency, as shown in Figure 4–8. These responses approximate the equal loudness contours for pure tones; the A-weighting corresponds to the 40 dB contour and the B-weighting corresponds to the 70 dB contour. (See Chapter 3 for discussion of equal loudness contours.) Instrument readings taken using these networks are called *weighted sound levels in decibels,* not sound *pressure* levels.

Metering system. After the electrical signal from the microphone is

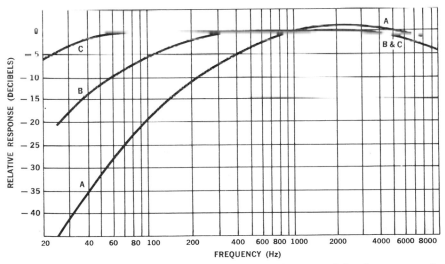

Fig. 4–8.—Frequency-response characteristics of a sound level meter with A-, B-, and C-weighting.

111

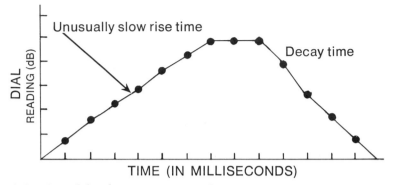

Fig. 4–9.—Sound level meter response showing unusually slow rise time.

amplified and sent through the attenuators and weighting networks, it is used to drive a metering circuit. This meter face displays a value that is proportional to the electrical signal applied to it. The American National Standard S1.4–1971 on sound level meters specifies that the rms value of the signal should be indicated. This requirement corresponds to adding up the different components of the sound wave on an energy basis. For sound measurements, the rms value is a useful indication of the general energy content.

Since the indicating needle on the dial of a sound level meter cannot follow rapidly changing variations in sound pressure, a running average of the rectified output of the metering circuit is shown. The average time (or response speed) is determined by the meter ballistics and by response circuits (chosen by a switch).

• *The fast response* is specified by ANSI S1.4–1971, which states that if a 1000 Hz signal of 0.2 second duration is applied, the maximum reading for a Type 1 (precision) instrument shall be 0 to 2 decibels less than the reading for a steady signal of the same frequency and amplitude. For Type 2 (general purpose) and Type 3 (survey) instruments, the maximum reading shall be 0 to 4 decibels less. Also, if a signal between 125 and 8000 Hz is applied and held constant, the maximum reading shall exceed the final steady reading by 0 to 1.1 decibels. ANSI S1.4–1971 further specifies that the decay time be essentially the same as the rise time. (See Figure 4–9.)

• *Slow response characteristics* are specified by ANSI S1.4–1971, which states that if a 1000 Hz signal of 0.5 second duration is applied, the maximum reading for a Type 1 instrument shall be 3 to 5 decibels less than the reading for a steady signal of the same frequency and amplitude. For Type 2 and Type 3 instruments, the maximum reading shall be 2 to 6 decibels less. Also, if a signal between 63 and 8000 Hz is applied and held constant, the maximum reading shall exceed the final steady reading by 0 to 1.6 decibels. The steady *slow* reading for any signal between 31.5 and 8000 Hz shall not differ from the corresponding steady *fast* reading by more than 0.1 dB for Types 1 and 2, and 0.3 dB for Type 3 sound level meters. This slower response is helpful in obtaining an average value, when measuring a noise level that fluctuates over a range of 4 dB or more.

The scale of the indicating meter is usually graduated in steps of one decibel to cover the full range of 20 dB between attenuated steps. It is common practice to label the full-scale value as "+10.0 dB" and the lower 10 dB value below zero as "−10 dB"; so that, when operating from −10 to +10.0 dB, the indicated dial reading is added directly to the setting of the attenuator to obtain the sound level measurement. Some of the later models display a full 50 dB measuring range on a single scale.

Output connection. Many sound level meters have an output connection that supplies an amplified electrical signal from the microphone. This amplified output is used to provide a signal to other instruments, for example, graphic level recorders, oscillographs, or spectrum analyzers. In order for this output signal to be of greatest value, it should be an accurate reproduction of the microphone signal with a minimum of distortion and inherent electronic noise.

The output signal level is usually in the order of one volt when the indicating meter is deflected to a full-scale value. This output is adequate to drive most frequency analyzers and graphic level recorders. The maximum output before serious nonlinear distortion begins is ordinarily about 10 dB above full scale. This overload capacity is included to take care of signals that have a large peak value, but only a moderate rms value. The overload capacity is seriously affected by the condition of the batteries and vacuum tubes (if they are used).

Noise floor

The internal electrical noise produced by the instrument is important when the amplified signal is to be analyzed. Over most of the range of a sound level meter, the *internal electrical instrument noise floor* is ordinarily 40 dB or so below the full scale output level.

The internal electrical noise floor of most instruments is such that broadband measurements of sound levels below about 24 dB are not possible, particularly at high frequencies. When the sound level meter is used to measure levels at the low end of its dynamic range in conjunction with a frequency analyzer, it is often possible to obtain erroneous readings—readings of the internal electrical noise floor of the sound level meter rather than the sound field being studied. At the high end of its dynamic range, in the presence of sound pressure levels above 140 to 150 dB, the sound level meter is generally limited by *microphonic excitation* (energy that bypasses the microphone and excites the instrument itself).

Specifications

The American National Standards establish design objectives, so that all sound level meters meeting a given standard will provide reasonably similar readings. The current standard on sound level meters, S1.4–1971, defines the response-frequency characteristics, the indicating instrument dynamics, and the calibration requirements for the instruments. Because standards are continually being revised to keep current with improved technology, the user should familiarize himself with the latest standard that specifies performance requirements of sound measuring equipment. For most general purposes, however, equipment that meets the current American National Standard should suffice. If more detailed calibration data on a particular instrument system is desired, this is usually available from the manufacturer at extra cost.

Frequency analyzers

In some cases the information provided by a sound level meter may be inadequate for a complete understanding of the problem. The sound level or number of decibels indicated by a sound level meter tells very little about the frequency distribution of that sound. By judicious use of the weighting networks of a sound level meter, one can learn something about the predominant frequencies present, but this is only a rough approximation.

For many industrial noise problems, it is necessary to use some type of analyzer to determine where the energy lies in the frequency spectrum. This is especially true if engineering control of noise problems is planned, because industrial noise is always complex and is made up of various frequency components at various sound intensities.

In order to analyze sound, two types of frequency analyzers are used:

- Constant percentage bandwidth (octave-band and narrow-band analyzers)

- Constant bandwidth (heterodyne analyzers)

The constant percentage analyzer has a bandwidth which is a fixed percentage of the band center frequency. Thus, at low frequencies the bandwidth will be only a few hertz wide, while at high frequencies the bandwidth will include a wider range of frequencies.

The constant bandwidth analyzer has a fixed bandwidth commonly between 3 and 200 Hz wide. The constant bandwidth analyzer is seldom used as a portable instrument to obtain information on noise control problems in the field because of the transient nature of industrial noise and the relatively long time required to gather the data.

Portable constant percentage analyzers are generally used in the field to obtain useful information in noise control problems. Different models of these frequency analyzers vary in cost, complexity, and ease of operation. Which one to use is determined by the problem and the amount and type of information needed to solve the problem. In general, as more information is required, a more selective analyzer must be used and proportionately more time is required to gather the information. Many laws and measurement standards require octave-band analysis of the noise source.

Octave-band analyzers

The octave-band analyzer is generally used for simplified analysis of noises having complex spectra. Octave-band analyzers are widely used for analysis of community and industrial noise problems.

In an octave-band analysis, the acoustic energy spectrum is electronically separated by filters into frequency bands (octave bands). The higher frequency limits of each octave band at rolloff is twice the lower frequency band limit (see Figure 4–10). This frequency analysis yields a series of measurements, one for each band, called "band levels" or, for octave bands, "octave-band levels." The band width must be specified if the information is to be of value.

The preferred series of octave bands for acoustic measurements (ANSI S1.11–1966) cover the audible range in ten bands. The center frequencies of these bands are 31.5, 63, 125, 250, 500, 1000, 2000, 4000, 8000, and 16,000 Hz. The lower frequency limit of any one of these

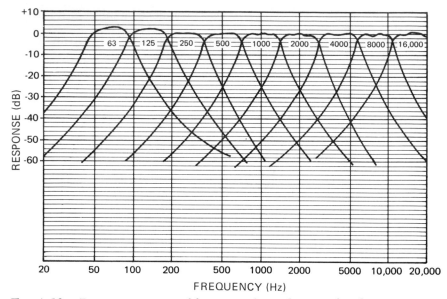

Fig. 4–10.—Response in terms of frequency for each octave band.

bands is 0.707 times the center frequency; for example, the effective bandwidth for the 1000 Hz octave band extends from 707 to 1414 Hz. (This was discussed in Chapter 2.)

Another series of octave bands has been widely used in the past. The older bands were a 75 Hz low-pass filter (all frequencies below 75 Hz) and the octave bands of 75 to 150, 150 to 300, 300 to 600, 600 to 1200, 1200 to 2400, 2400 to 4800, and 4800 to 9600 Hz. The term *low pass* is used to designate a filter that passes signals whose frequency is lower than a designated frequency, and similarly, a *high-pass* filter passes all signals whose frequency is higher than a certain value. This older series of octave bands is still specified in a number of test codes, and the published data obtained with this older series is extensive.

When the results of octave-band sound level measurements are plotted on a graph, the frequency scale (x-axis or abscissa) is commonly divided into equal intervals between the position designated for each band and the position for the band adjacent to it in frequency. The sound level in each band is plotted as a point on each of these positions along the y-axis or ordinate. Adjacent points are then connected by straight lines. An example of a plot of this type is given in Figure 4–11.

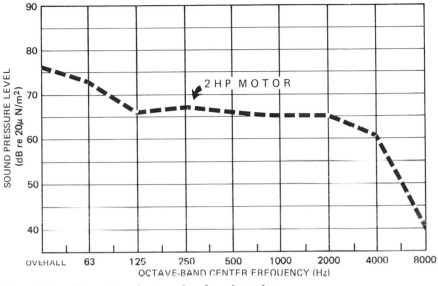

Fig. 4–11.—Examples of octave-band analysis data.

A typical octave-band analyzer might have the internal arrangement indicated in Figure 4–12. The input signal is fed directly to the overall position or to one of the filters and then to a 10 dB-step attenuator. The output of the attenuator goes to an amplifier stage whose gain may be set during calibration. Finally, a meter and an output jack similar to those on the sound level meter are fed through output amplifiers. It is usually possible to vary the damping in the meter with a switch marked FAST and SLOW, as on the sound level meter.

Narrow-band analyzers

One-third octave-band analysis is now widely used, particularly for checking compliance with military noise and vibration specifications. It is most often used with a graphic level recorder to give a graph of the energy distribution of the noise and vibration as a function of frequency.

The actual effective band for a one-third octave filter at 1000 Hz extends from about 891 to 1122 Hz; that is, the bandwidth is about 23 percent of the center frequency. Frequency analyzers with one-tenth octave bands (about 7 percent in width) and others with 1 percent bands are also available.

For a more detailed analysis of the distribution of sound energy as a

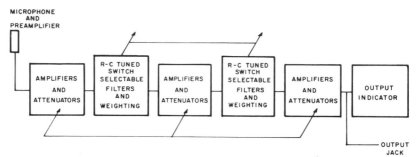

Fig. 4–12.—Block diagram of a typical octave-band analyzer.

function of frequency, the one-tenth octave-band analyzer can be used. This choice is based on the fact that ten narrow-band filters can be arranged effectively to cover a 10-to-1 frequency range. The preferred center frequencies (ANSI S1.11–1966) would be 100, 125, 160, 200, 250, 315, 400, 500, 630, and 800 Hz. The next higher 10-to-1 set would start with 1000 Hz as the center frequency and continue by multiplying each number in the preceding series by 10 (1000, 1250, 1600, 2000 . . .). Similarly, lower preferred frequencies are obtained by a division of 10, 100, and so on. For practical reasons, the usual frequency range for one-tenth octave-band noise analysis is 25 to 20,000 Hz. In some specialized problems, these narrow-band systems are essential for use in noise and especially in vibration measurements, where the frequency components are very closely spaced.

Some octave-band filters are built into sound level meters and use the electronics of the sound level meter. Others simply plug into the output of the sound level meter and indicate the band levels on their indicating dials.

Measuring vibration

Vibration analysis may be helpful, when studying noise problems associated with the generation of sound. The vibration meter is similar in operation to the sound level meter, but instead of a microphone, the vibration meter has a transducer or pickup that produces a signal when held against a vibrating source (see Figure 4–13). The signal is amplified and displayed on the indicating meter to show the vibrating source's displacement, velocity, acceleration, or jerk. (These terms are discussed in Chapter 2 and will be further amplified in Chapter 18.)

When supplied with an analyzer (octave, one-third octave, or what-

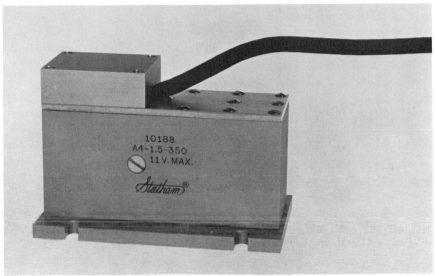

FIG. 4–13.—A typical vibration transducer or accelerometer.

ever), the output at discrete frequency bands can be determined.

Vibration meters are calibrated directly in terms of peak, peak-to-peak, or average displacement (in mils), velocity (inches per second), acceleration (inches per second per second), and jerk (inches per second per second per second).

Other models of vibration meters may indicate the same quantities in metric units—millimeters, meters per second, meters per second per second, and meters per second per second per second. If the vibration pickup used is the acceleration type, integrating and differentiating circuits are built in as part of the amplifier.

Some vibration meters are direct reading for acceleration, velocity, and displacement from 2 to 2000 Hz, and direct reading for jerk from 1 to 20 Hz. For velocity and displacement measurements, the upper end of the frequency range is limited to about 2000 Hz.

Vibration measurements can be made with sound level meters when a vibration pickup is substituted for the microphone. An auxiliary control box, which is connected between the meter and the pickup, converts the signal so that the meter indicates velocity and displacement as well as acceleration (see Figure 4–14). The combination of pickup and

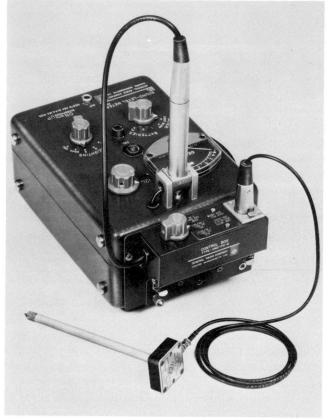

Courtesy General Radio Co.

FIG. 4–14.—A sound level meter converted to a vibration meter by adding a control box and a pickup.

control box, called a *vibration pickup system,* provides a convenient and inexpensive way for owners of sound level meters to make vibration measurements within the audio-frequency range. However, the sound level meter circuits respond down to only 20 Hz, and consequently the combination is not suitable for measuring lower frequency vibrations. The vibration meter should be used where low frequencies are important.

The sound level meter is calibrated in decibels, which must be converted to vibration amplitude, velocity, or acceleration. The calibration chart supplied with each vibration pickup system gives the proper conversion factors for that system when it is used with a particular sound level meter.

Auxiliary instruments

Noise exposure monitors

The noise exposure monitor measures and accumulates data on sound levels over a specific time period. Continuous noise monitoring instruments provide a measurement that can be related to the *noise dose* to which a person is exposed.

Different types of noise integrators are available which are designed to indicate the total noise exposure during a specified time interval (see Figure 4–15). Each of these instruments uses a different approach— one type accumulates the total amount of sound energy to which a worker is exposed for a workday, another type measures the length of time that a specified decibel level is exceeded, and a third measures the rate at which sound energy impinges on the exposed person for selected short-term periods of time.

When the noise level fluctuates significantly during the workday, the manual computation technique to determine the total integrated noise exposure becomes difficult and time consuming. Use of an instrument that automatically measures and computes the total noise exposure is a more economical method when compared to the manhours involved in collecting data and making the computations.

Most of these automatic monitors measure the noise exposure and compute how much of the daily maximum legal exposure has been accumulated. On one unit, the digital readout displays the percentage of the exposure limit that has been accumulated and the duration of the measurement and also indicates whether the instantaneous (115 dBA) and impact (140 dB peak) levels have been exceeded.

On other models a stripchart recorder can be connected to produce a permanent plot of the sound level *vs.* time. The noise exposure monitor can be placed in a central location while the data are gathered by remote microphones.

Noise dosimeters

Sound level meters can be used to accurately measure the noise level at a given point at the time of the observation. They are not designed to electronically sum up the noise exposures of mobile individuals exposed to changing noise levels during their work shift.

FIG. 4–15.—A sound level integrator is a fully automatic instrument that computes the cumulative daily noise average.

Courtesy Columbia Research Laboratories, Inc.

When a person moves away from a given sound source, the sound level at his ears usually decreases. That is why the microphone of the sound level meter must be held near the person if it is to measure accurately what sound actually reaches his ear. A sufficient number of observations would be required to develop a sound level map of the area. Then to estimate an employee's noise exposure, a time and motion study would be required to determine just how long the employee spent in each sound level area during each day.

This problem is minimized by use of the *noise dosimeter.* (The sound level measured by the dosimeter differs from that measured on a sound level meter, depending on the relative positions of the microphone, the employee, and the noise source, because of body shielding and reflection.)

One type of noise dosimeter, including a built-in microphone, is worn in the employee's shirt pocket (see Figure 4–16). All sound levels between 90 and 115 dBA are accumulated continuously during the 8 hour work period in a semiconductor memory circuit.

A readout device is used to obtain a reading of the noise exposure level experienced by the worker wearing the dosimeter and to calibrate the individual monitor units. When the monitor is inserted into the socket of the readout device, the worker's daily noise exposure is dis-

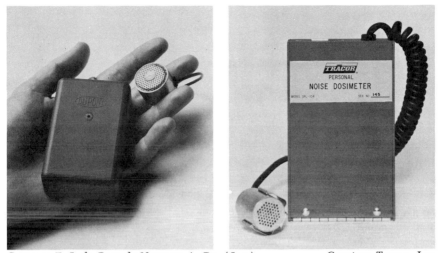

Fig. 4–16.—A noise dosimeter worn by the worker stores information on the daily noise levels a worker is exposed to.

played as a percentage of the allowable limit (see Figure 4–17). The monitor can be reset after the exposure information is read from the readout device, or it can be returned to the worker for further accumulation. The final readout is the sum of all the exposure ratios that were measured and calculated by the monitor during the work period.

Any dosimeter used should meet the accuracy requirements specified by the American National Standards Institute for general purpose sound level meters (S1.4-1971 Type 2), should be capable of handling the accumulation specifications of OSHA, and be capable of acoustical calibration in the field.

Short-duration noise measurement

It is important to recognize the distinction between "impulse" and "impact" measurements of short-duration noise (such as produced by punch presses, drop hammers, and the like).

● The term *impulse* is used to describe a special meter response (*i.e.*, time constant) on a sound level meter. This meter response should have a fast rise time and slow decay time to allow a more accurate measure of the energy in a short-duration noise pulse than would the normal "fast" speed in standard sound level meters.

123

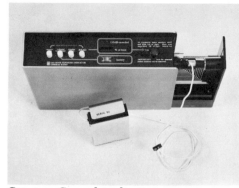

Courtesy General Radio Co.

Fɪɢ. 4–17.—Some dosimeter storage units plug into a readout unit.

Courtesy Willson Products Div., ESB Inc.

• "True peak" sound levels can only be measured with an oscilloscope or an *impact analyzer*. The term *impact* describes a meter response with a rise time of not more than 35 milliseconds to peak intensity for noise whose duration is not more than 500 milliseconds. The meter should have a holding circuit that maintains the peak reading indefinitely (until reset by the operator).

A sound level meter need not incorporate impulse or impact measurement capability to meet ANSI S1.4–1971 requirements; therefore, most sound level meters do not contain either. The few meters that do contain these features are the "top of the line" units.

Impact-noise analyzer

The instrument generally used for studying impact-type noise is the impact-noise analyzer which can be operated directly from the output of a sound level meter to measure the peak level (see Figure 4–18).

Through the use of electronic circuits, three characteristics are measured by the impact-noise analyzer—a peak instantaneous level, an average level, and continuous indication of peak level. The duration of the impact can be estimated from the difference between peak instantaneous level and average level. Any one of the three characteristics can be switch-selected for presentation on the display dial. A reset position of the selector switch restores the impact-noise analyzer to its initial preexposure condition.

Impact noise analyzers use an amplifier of very low output impedance to drive a rectifier which charges a capacitor, in a fraction of a millisecond, to the peak value of the signal from the microphone. This value is stored electronically in the capacitor for subsequent display on a voltmeter. At the same time that the capacitor is charged (peak signal value), a type of time averaged value is obtained by using a longer charging time for a second rectifier-capacitor system driven by the amplifier. The difference in level between the peak value and the time-averaged value is a measure of the time duration of the impact.

Oscilloscope

The display on an oscilloscope tube face can be used to analyze the electrical signal from a sound level or a vibration meter (see Figure 4–19). The oscilloscope has a cathode-ray tube similar to a television tube. The electron beam signal is deflected so that the trace on the display

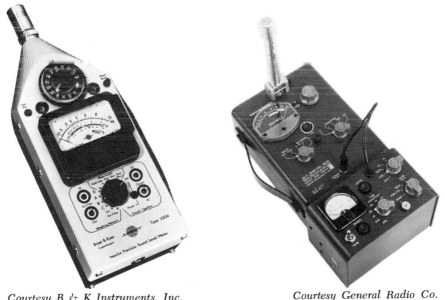

Courtesy B & K Instruments, Inc.   *Courtesy General Radio Co.*

FIG. 4–18.—Two meters for measuring impact sound.

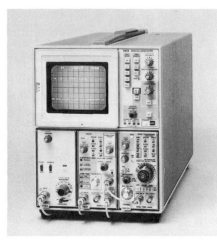

FIG. 4–19.—An oscilloscope is especially useful—not only can the amplitude and frequency be displayed visually, but a camera can be attached and a permanent record made.

Courtesy of Tektronix, Inc.

screen moves at a uniform rate towards the right in a horizontal direction (the x-axis). When the trace reaches the far right edge of the display screen, the sweeping signal quickly returns it to the left, and the pattern is repeated. The electrical signal from the sound level or vibration meter to be displayed is applied so as to produce a deflection of the trace in the vertical direction (called the y-axis). The combined motion results in a display of the signal as a trace of the instantaneous amplitude of the sound wave as a function of time.

The oscilloscope can be calibrated to measure the peak amplitude of a sound wave. In addition, the oscilloscope makes possible the study of the instantaneous values of a vibratory motion. In contrast with the sound and vibration analyzer and other wave analyzers that present information in terms of frequency, the oscilloscope presents information as a function of *time* which is often of great assistance in the solution of vibration problems. Because the oscilloscope presents information instantly and continuously and because its frequency response is not a limiting factor, it is useful in the study of vibration problems.

For sound and vibration measurements, an oscilloscope with slow sweep rates, a long-persistence screen, and a d-c amplifier is recommended. Many oscilloscopes have provision for a camera to make possible a permanent record of the wave shape being studied.

Multipurpose instruments

Development of miniaturization of the components in electronic equipment has led to the availability of multipurpose instruments for noise measurement. It is now possible to get the most often used characteristics of a sound level meter, octave-band analyzer, and impact analyzer in a single package—at a lower price than the combination of individual instruments.

In addition to meeting all requirements for both sound level meters (ANSI S1.4–1971, Type 1) and octave-band filters (ANSI S1.11–1966), the multipurpose instrument contains electronic circuitry for the measurement of both impact and impulsive noise (see Figure 4–20).

The instrument's meter response is linear in dB (vs. logarithmic, as found on most other types), and outputs are available both for graphic level (a-c) recording and for simple, less expensive d-c recording.

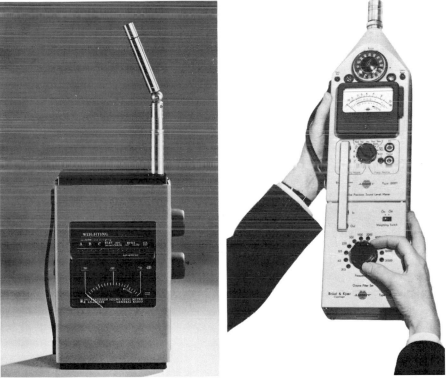

Courtesy General Radio Co. *Courtesy B & K Instruments, Inc.*

Fɪɢ. 4–20.—The multipurpose instruments shown here can be used as sound level meters, octave-band analyzers and impact/impulse noise meters.

Courtesy General Radio Co.

Courtesy Pulsar Instruments, Inc.

FIG. 4–21.—Graphic level recorders are particularly valuable when a permanent record is desirable.

Graphic level recorder

The graphic level recorder produces a permanent chart record of the level of an applied electrical signal (see Figure 4–21). For noise and vibration measurements, this signal is usually obtained from the output of a sound level meter or band analyzer.

Many noise sources produce signals whose instantaneous levels vary unpredictably with time. These signals contain many frequency components with amplitudes that vary randomly. A fan, for example, may produce a seemingly constant overall sound level, but a detailed study of the noise signal would confirm its randomly varying nature. Such random signals present measurement problems, and if the sound levels are not averaged over a sufficiently long period, individual measurements will differ greatly from each other and from a long-time average measurement.

The true level of a randomly varying noise can be estimated by averaging several independent measurements taken sufficiently far apart so that the random signal that affected one measurement does not infiuence the next. Thus, it takes time to ensure accuracy; and, conversely, accuracy will be sacrificed if the time devoted to the measurement is reduced.

The graphic level recorder can be used to record the noise levels in industry, the sound levels near highways, airports, residences, or the vibration levels of building floors or walls. To facilitate observations over a long period, the recorder can be left unattended for most of the time. The resulting information is much more extensive than that obtainable from a few isolated readings with a sound level meter.

The graphic level recorder, when connected to the output of a sound level meter and properly set for sensitivity, records sound pressure level as a function of time. It can also be used to record other variables.

Used with an analyzer, the graphic level recorder can plot the frequency spectrum of a sound source (the curve of amplitude *vs.* frequency) or of a vibrating object (its displacement, velocity, or acceleration *vs.* frequency). Mechanical linkages and special chart papers reproduce the frequency scale of the analyzer at the recorder.

The combination of the graphic level recorder and the sound and vibration analyzer produces permanent records that are essential for checking compliance with specifications, and at least one manufacturer offers this combination in a single package.

Magnetic tape recorder

The magnetic tape recorder, which stores an electrical signal as variations in the magnetic state of finely divided iron oxide particles coated on one side of a plastic tape, is very useful in measuring noise levels. When it is difficult to analyze a sound at its place of origin, a recording can be made and stored until such time as the analysis can be made. The recorded tape is played back on the recorder and measurements on the output signal can be performed in the laboratory.

The uniformity of response *vs.* frequency for a tape recorder is determined by tape speed, magnetic gap length, gap alignment, nature of head and tape contact, type of tape, equalization, and amplifier response. If the tape does not move by the magnetic gaps at a uniform speed, the reproduced signal is modified by fluctuations in the relative time scale. If a pure tone is recorded, these fluctuations produce a variation in output frequency that is called "flutter" or "wow." Unless such fluctuations are very small, they are audible when certain types of sounds are reproduced. For general application to noise measurements, therefore, only a better-quality tape recorder should be used.

The inherent instrument noise in a magnetic tape recorder depends on many factors, such as amplifier design and adjustment, bias oscillator distortion, stray magnetic fields, and type of tape used.

The maximum level at which essentially linear recording of the signal can be made is set by such factors as type of tape used, bias oscillator signal, amplifier operation, equalization, and the frequency distribution of the energy of the signal to be recorded. This distortion level sets an upper limit and the inherent noise level sets a lower limit to the range of signal levels that can be handled satisfactorily. This range is sometimes called dynamic range, and in a high-quality tape recorder it is 50 dB or more, which is adequate for most sound recording purposes.

Calibration

Prior to use, a sound measuring system should always be checked for accuracy. This is especially important when one is combining more than one instrument into a sound measuring system. The two common methods of calibration are internal and external with a secondary standard.

Electrical calibration

Many sound level meters and octave-band analyzers have a "calibrate" position, which permits comparing the dial reading of the instrument to the electronic calibration that was set at the time of manufacture. This assures that all of the electrical circuits are operating properly. *An electrical calibration does not, however, check the transducer* (microphone or vibration pickup). These must be checked by external means.

Acoustical calibration

An acoustic calibrator in common use today comprises a small, stabilized, and rugged loudspeaker system mounted in an acoustic enclosure, which fits over the microphone of the sound level meter or analyzer (Figure 4–22). The chamber is so designed that the acoustic coupling between the calibrator loudspeaker and microphone is fixed and measurements can readily be repeated. The sound level produced by the loudspeaker in the calibrator is high enough, so that instrument readings are unaffected by normal background noises.

The calibrator includes its own battery-operated oscillator to drive a loudspeaker that supplies sound at a known level and frequency. This device permits a quick check of the performance of a sound level meter or octave-band analyzer.

Calibration of tape recorders

Before making any noise or vibration recordings, a suitable acoustic calibrator should be energized and placed over the microphone. With the attenuator set to give the recording level recommended by the manu-

FIG. 4–22.—A sound level calibrator supplies sound at a known level and frequency to permit a performance check on a sound level meter or octave-band analyzer.

Courtesy of General Radio Co.

facturer, a brief recording of this calibration sound should be made. The calibrator is then removed and the noises of interest are recorded. It is important to note all changes in attenuator or volume control setting while the recording is made.

The level of a recorded sound on playback is determined by comparing its level with that of the calibration tone. Corrections should be made for any changes of the stepped volume control settings. Tape recorders with automatic or continuously variable volume control are not suitable for noise measurements.

If high noise peaks are to be recorded, there is danger of overloading the recorder. To avoid this difficulty, it is necessary to reduce the recorder gain considerably below that level used when recording continuous noises. Recordings of the impact noises should be made at several attenuator settings with about 10 dB between successive steps. This procedure assures that at least some of the recordings will be made without overloading on the peaks.

Vibration calibrator

The vibration calibrator is a small, single-frequency calibrator used for checking the overall operation of a resiliently supported cylindrical mass; it is driven by a small, transistorized, electromechanical oscillator mounted within the cylinder. To calibrate an accelerometer, the level control is adjusted for a meter reading corresponding to the mass added to the moving system of the calibrator. The accelerometer is then being driven at an acceleration of one g ($1\ g = 32$ ft/sec^2) at 100 Hz. The excursion of the calibrator can be adjusted for one g acceleration with any pickup weighing up to 300 grams.

To obtain valid and reliable results, an accurately calibrated acoustical or vibration measurement system is essential. With accurate calibration, the consistency of comparison measurements can be improved and a closer approach to an allowed performance specification is possible. Careful attention to measurement technique will result in more accurate measurements.

Summary

The basic sound measuring instruments have been described in this chapter. Much of the value of these instruments stems from their ability to obtain reliable data under field conditions with a minimum of time and effort.

Sound measuring instruments are electronic devices and, like other electronic devices, they can be damaged over a period of time. The day-to-day changes may go unnoticed unless the user is aware of certain problems that can exist. No matter which type of instrument—regardless of the make or model—the user must know his instrument intimately before he starts taking noise measurements.

The most important component in a sound measuring system is the transducer (a microphone for airborne sound and a vibration pickup for structure-borne vibration). Because the microphone must translate the noise into precise electrical pulses for measurement, the selection of the right microphone for the right job is imperative.

The accuracy, validity, and reliability of sound measurements depends in large part on how the instruments are used. One must be thoroughly familiar with an instrument to make full use of its capabilities. He can then decide what results are seriously limited by the instruments and how they are used. The material in this chapter should help the reader become familiar with the capabilities of the various instruments. A thorough study of the literature supplied by the manufacturer of a particular instrument is essential.

Special types of sound measuring instruments are available depending upon the data required and the sound being measured. Money spent for a high-quality sound measuring system, in order to achieve accurate results, will have been spent in vain if the instruments, individually and together, are not calibrated periodically. Remember, in order to get valid and reliable data, frequent, accurate calibration is essential. In the same vein, modern solid state and printed-circuit card instruments have a high reliability, but much depends on how they are used. Take care of your instruments. Become acquainted with every facet of their operation. Consult and follow the directions in the manuals provided. Don't guess; find out the right way and follow it.

Bibliography

Beranek, L. L. ed. *Noise and Vibration Control*. New York N.Y.: McGraw-Hill Book Company, 1971.

———— ed. *Noise Reduction*. New York, N.Y.: McGraw-Hill Book Company, 1960.

Harris, C. M. ed. *Handbook of Noise Control*. New York, N.Y.: McGraw-Hill Book Company, 1957.

Hosey, A. D. and Powell, C. H., eds. *Industrial Noise—A Guide to its Evaluation and Control*. (Public Health Service Publication No. 1572, U.S. Dept. HEW), Washington, D.C.: U.S. Government Printing Office, 1967.

American Industrial Hygiene Association, *Industrial Noise Manual*, 2nd ed. Akron, Ohio: AIHA, 1966.

Petersen, A. P. G. and Gross, Jr., E. E. *Handbook of Noise Measurement*, 7th ed. Concord, Mass.: General Radio Co., 1972.

Chapter Five

Sound Measurement Techniques

By William Ihde, P.E.

Sound measurement requirements range from a single observation of the overall sound pressure level to a detailed narrow-band analysis of the individual components of a complex sound or vibration. The choice of which measurements to make and which instruments to use depends upon the information needed to solve a particular noise problem.

If the problem is simply checking sound levels for compliance with a standard or specification, the needed sound measurement is reasonably clear—only some guidance as to instrument selection and use is needed. But, if the problem involves analysis of a complex noise such as that produced by many typical industrial operations, the situation is more complicated and so is the selection on the proper type of instrument.

The choice of instrument depends on the particular noise measurement problem—what type of noise is involved, what time is available for measurement, what is the aim of the investigation, and what measurement accuracy is required?

For example, when the noise is more or less steady and the aim of the investigation is to obtain an estimate of the A-weighted sound level, the readout provided by the sound level meter may be quite satisfactory. If knowledge of the A-weighted sound level does not provide sufficient information, the noise can next be analyzed with an octave-band analyzer to determine the spectral distribution and the frequency of its important components.

If the noise is transient or intermittent, or is rapidly fluctuating, the

134

use of graphic level recording is normally preferred to a visual meter reading. Graphic level recording may also be desirable because the analysis can be performed and recorded automatically. The noise may be recorded on magnetic tape and later analyzed with an octave-band analyzer. The magnetic tape recording also provides a permanent record of the noise which subsequently may be subjected to a more detailed type of analysis.

Impulse or impact noises have high-level peaks of very short duration and the standard sound level meter cannot be used to obtain a satisfactory measurement. Oscilloscopes in conjunction with sound level meters and special peak-reading impact-noise analyzers are used for this purpose.

A large variety of noise measurement equipment is available for a wide range of application. The portability, accuracy, required frequency response, dynamic range, and the degree of resolution required determine the final choice of sound measuring instrument systems for any particular application.

Altering the sound field

In principle, it is impossible to assemble a noise measurement system that does not in some way distort or alter the sound field being studied, partly because any measurement requires some interference with the observed process. It is essential that the observer understand the phenomena under observation and what is going on within the sound or vibration instrumentation system.

Factors not normally associated with a sound field can affect the accuracy of measurement. For the most part, this interference is caused by objects introduced into a sound field in order to measure it. Such objects include the microphone, its supporting structure, associated measuring instruments, and even the observer himself. Other equipment (such as machines, walls, fixtures, and the like), although they do affect the sound field, are not considered interference factors because they are permanent structures in the sound field.

The microphone is the only part of the measuring system that must actually be located in the sound field, and since its dimensions are often significant in relation to the wavelength of the sound signal being measured, it does affect the nature of the sound field.

The basic procedure for measuring the sound level at a given point is to locate the microphone at that point and read the sound level meter. Preliminary exploration of the sound field is usually necessary to determine that the point selected is the correct one.

Practical descriptions of techniques for sound measurement are given in this chapter. Detailed information on the actual manipulation of the individual instrument controls can be found in the instruction books that are furnished with the individual instruments.

This chapter will supply information about the instrument or combination of instruments and accessories that would be suitable for obtaining sound level measurements. The choice and use of microphones and auxiliary apparatus, the effects of extraneous influences, ambient conditions, background noise, instrument precautions, and the calibration of the instruments will be discussed.

Choice of microphones

The microphones supplied with typical sound level meters are suitable for most sound measurements. However, for very high sound levels, for high-temperature areas, and for accurate sound measurements above 12,000 Hz, special microphones may be needed. Types of microphones were discussed in the previous chapter. The performance of typical microphones, as well as their limitations and some of the problems encountered with their use, is discussed in the following sections.

Measurement of low sound levels

Microphones used to measure low sound levels must have low "self-noise" characteristics, and produce an output voltage sufficient to override the electrical circuit noise generated by the amplifier in the sound level meter. Sound levels down to about 24 dB can be measured with microphones that meet ANSI S1.4–1971 Type 1 specifications (a sensitivity of ±0.5 dB from 22.4 to 11,200 Hz). An octave-band analyzer may have a measurable band level lower than 24 dB, because the internal noise in a selected band may be less than the total electrical circuit noise.

Frequency range

Ceramic- and condenser-type microphones are well suited for measuring low-frequency noise. With either of these microphones, measurements may be made at very low frequencies if special amplifiers are provided. The primary requirements for accurate measurement of high-frequency sounds are small microphones and a uniform frequency response of the microphone at high frequencies.

For measuring typical factory noise levels, the high-frequency characteristic is not too important because most machinery noises do not include many high-frequency components. Even for those sounds that do include significant energy at high frequencies, the decrease in response required at very high frequencies for the standard weightings means that the important noise energy is generally well within the range of the regular microphone furnished with typical sound level meters.

Ambient conditions

Humidity. Long exposure of any microphone to very high or very low humidities should be avoided. Ceramic microphones are very resistant to extremes of humidity. The Rochelle-salt crystal microphone has been replaced by the ceramic microphone which is much more rugged for field use. Also, the Rochelle-salt type does not meet the ANSI S1.4–1971 standard for sound level meter microphones.

Condenser microphones, although not damaged by exposure to high humidity, may be seriously affected unless proper precautions are taken. The exposed insulating surface in the condenser microphone usually has been treated to maintain low electrical leakage even under conditions of high humidity. In spite of this precaution, the electrical leakage may become excessive under some conditions, and it is advisable to keep the microphone at a temperature higher than the ambient temperature.

Temperature. Although most industrial noise measurements are made indoors at average room temperatures, some measurement conditions expose the microphone to much higher or lower temperatures. When these conditions are encountered, it is essential to know the temperature limitation of the equipment.

Most ceramic microphones will withstand temperatures of -20 F to 212 F (-30 C to 100 C) without damage. The maximum safe operating temperature for most condenser microphones is about 400 F (204 C) (see Table 4-B in Chapter 4).

Although the microphone itself must be exposed to the sound field, it is usually possible through the use of long cables to place the sound level meter at a location with more reasonable temperatures. The operation of the sound level meter at extreme temperatures is limited by the temperature limitations of the batteries, so that operation at temperatures above 130 F (55 C) or below -10 F (-23 C) is not ordinarily possible without special batteries.

Since microphones are usually calibrated at normal room temperatures, a microphone operated at extremely high or low temperatures will have a different sensitivity and a correction factor should be

Fig. 5–1.—A microphone windscreen can reduce the effects of ambient wind noise and protect the microphone diaphragm in oily, misty, and dusty environments.

Courtesy General Radio Co.

applied. The correction for sensitivity for most ceramic microphones is about −0.005 dB per degree Fahrenheit (−0.009 dB per degree C), so that for most purposes the correction can be neglected. Condenser microphones have a temperature coefficient of sensitivity of about −0.02 dB per degree Fahrenheit (−0.036 dB per degree C).

Magnetic fields. Dynamic microphones are sometimes used for measurement purposes because they are readily used with long cables. However, when dynamic microphones are used, care must be taken to avoid hum pickup, which is the induction of undesired electrical signals from the external magnetic field of equipment such as transformers, motors, and generators. Ceramic and condenser microphones are relatively free from this undesirable effect.

Microphone location

Using cables

For the most accurate sound measurements, only the microphone should be put into the sound field, and the measuring instrument and the observer should not be near the point where the sound level is to be measured. For this reason and also for situations when it is impossible or impractical for the observer to be near the microphone, an extension cable is ordinarily used to connect the microphone to the instrument. If the microphone is attached directly to a preamplifier, long cables can be used with little loss in sensitivity. The instruction manuals for the particular microphone or instrument should be consulted.

Wind screens

The microphone should be kept out of the direct path of strong air currents.   Turbulent air movement past the microphone diaphragm produces a low-frequency noise signal. This wind noise may seriously upset the measurement, particularly when the microphone has a good low-frequency response. If it is not possible to avoid rapid air currents near the microphone, a wind screen should be used (see Figure 5–1). This wind screen can be obtained from the instrument manufacturer or it can be made of a single layer of silk cloth stretched across a wire frame that encloses the microphone. The frame should be much larger than the microphone.

Directional effects

Most microphones used for sound measurements are designed to be essentially omnidirectional at low frequencies; that is, the response of the microphone is independent of the direction of arrival of the sound wave. However, at high frequencies, the size of the microphone is comparable to the wavelength of sound, and a microphone will show directional effects. The extent of these effects is indicated by the frequency response characteristics shown in Figure 5–2. The direction of travel of the sound wave should be considered in positioning the microphone so that the response to the incident sound is as uniform as possible.

When the sound level is measured in a reverberant room at a point that is not close to a noise source, the sound arrives at the microphone from many different directions. The orientation of the microphone is not

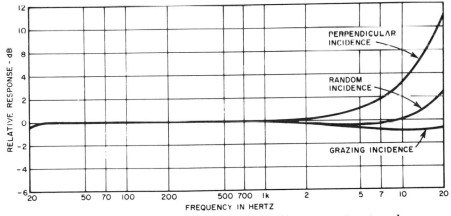

Fig. 5–2.—Typical frequency response for a ½ in. ceramic microphone.

139

critical under these conditions, and the response is assumed to be random incidence. It is usually desirable to avoid having the microphone pointing at a nearby hard surface from which high-frequency sounds can be reflected to arrive perpendicular (0 degree incidence) to the plane of the diaphragm in the microphone. If this condition cannot be avoided, the possibility for error from this effect can be reduced by placing acoustic absorbing material on the reflecting surface.

When measurements are made in a reverberant room at varying distances from a noise source, the microphone should generally be oriented so that a line joining the cylindrical axis of the microphone and the source is at an angle of about 70 degrees (tipped toward the noise source). When the microphone is located near the noise source, most of the sound comes directly from the source, and a 70 degree incidence response applies. On the other hand, near the boundaries of the room, the incidence is more nearly random, and the orientation for random-incidence response applies. These two response curves are nearly the same so that there is little change in the effective response characteristic as the microphone is moved about the room. This desirable result would not be obtained if the cylindrical axis of the microphone were pointed at the noise source.

Type of measurement

In general, the microphone location is determined by the type of measurement to be made. For example, the noise of a particular machine is usually measured at the normal operator's work position with the microphone placed near the machine according to the specific rules of a test code or standard. Or if the characteristics of the machine as a noise source are desired, a comparatively large number of measurements will have to be made. It is important to explore the noise field before deciding on definite locations for the position of the microphone. Many measurement locations may be necessary for specifying the noise field, particularly if the apparatus produces a noise that is highly directional. The possible effects of obstacles in interfering with the distribution of sound waves, particularly at high frequencies, should be kept in mind during this exploration.

If the noise level is measured for calculation of the speech interference level or for determination of hearing damage risk levels, it is important to explore the noise field to make sure that the noise measurements are representative.

All hearing damage risk measurements should be made without anyone being in the immediate vicinity of the microphone. The microphone

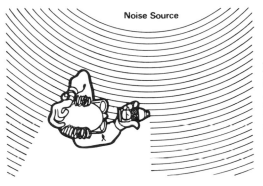

Noise Source

FIG. 5–3.—A microphone should be pointed in a direction perpendicular to the noise path, keeping the body out of the path.

should be positioned about where the operator's ears would normally be.

When the acoustical environment for hearing damage risk is being measured, no change should be made in the usual location of equipment that is normally present. The sound field should be carefully explored to make sure that the selected location for the microphone is not in an acoustic shadow cast by a nearby object or is not in a minimum of the directivity pattern of the noise source.

Because the observer can affect the instrument readings if he is close to the microphone when measurements are made near a noise source, it is advisable for the observer to stand well to the side of the direct path between the source and the microphone.

For very precise measurements in a very dead room, the instruments and the observer should be in another room with only the noise source, the microphone, the extension cable, and a minimum of supporting structure in the dead room.

When making handheld measurements, care must be taken to position the microphone of the sound level meter properly. The unit should be held in front of and away from the body so that the diaphragm of the microphone does not face the noise source directly (see Figure 5–3). The significance of the microphone position, as well as that of the observer, is shown in Figure 5–4. In position A (the recommended position), the noise energy impinges upon the microphone at a 90-degree angle of incidence; while in position B, the microphone faces the noise source (angle of incidence is 0 degree). It is apparent that if the sound level meter is held properly, little error in reading of the overall level will occur for most sounds.

The meter case itself may also disturb the sound field at the microphone. For most typical noises, there will be little error in measuring the overall level if the microphone is allowed to remain on the instrument. When an octave-band analyzer is used with the sound level meter, however, it is advisable to minimize reflections by separating the microphone from the instruments and using an extension cable.

Measurements at noise source

To determine the sound coming directly from the noise source, measurements should be made in a room containing a considerable amount of absorption, or outdoors. The reason for this is that when a noisy device is tested for its acoustic output, the space in which it is tested can have a significant effect on the results.

The measurement room used for evaluating a noise source should be sufficiently well treated so that no appreciable standing wave patterns exist. When a sound wave is reflected back toward the noise source from a large surface such as the wall of a room, alternate addition and cancellation of the sound wave may occur, resulting in maximum and minimum sound level readings as the microphone is moved along a line between the source and the reflecting surface. The resulting patterns are called *standing waves.* A standing wave pattern, however, should not be confused with the normal decrease in sound level with increasing distance or with the directivity pattern of the source. To check for standing waves, move the microphone back and forth along a direct line pointing toward the sound source.

Objects reflect sound waves; consequently, all unnecessary objects should be removed from the measurement area. If it is impractical to follow this principle, the objects should usually be covered with sound absorbent material. In general, no objects, including the observer, should be close to the microphone.

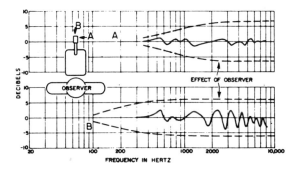

Fig. 5–4.—Frequency response of a sound level meter with and without an observer present.

Another troublesome effect may be sympathetic vibrations from nearby objects. A large, thin, metal panel if undamped can readily be set into vibration either by airborne sound or by vibration transmitted through the structure. One way of checking that the sympathetic effect is not present to any important degree is to measure the decrease in sound level as a function of the distance from the noise source. The sound meter reading should decrease 6 dB for each doubling of the distance from the source when measurements are made in the far field. This procedure also checks for the effects of reflections.

Background noise

The presence of extraneous background noise may affect the desired sound level readings. Frequently, when measuring the level of a noise source, the noise produced by other machinery masks the noise which is to be measured. In such cases, either the noise source being investigated has to be moved to a quieter place or sound level readings should be taken when the machinery producing the background noise is shut down.

To be accurate, the measurement of a noise source should determine only the direct airborne sound from the source without any appreciable contribution from noise produced by other sources. As a test to determine that this requirement has been met, the *Method for the Physical Measurement of Sound*, ANSI S1.2–1962 (par. 2.5), specifies the following:

> "If the increase in the sound pressure level in any given band, with the sound source operating, compared to the ambient sound pressure level alone is 10 dB or more, the sound pressure level due to both the sound source and ambient sound is essentially the sound pressure level due to the sound source. This is the preferred criterion."

If the background noise level and the apparatus noise level are steady, a correction may be applied to the measured data according to Figure 5–5.

Here's how to make the correction. After the test position for the microphone has been selected (as just discussed), the ambient background noise level alone is measured at the test position with the noise source (to be measured) not in operation. Then the sound level of the total sound is measured including the ambient background noise and the noise source. The measured difference in sound level determines the correction to be used. If the difference is less than 3 dB, the level obtained by use of the correction factor should be regarded as only an approximation and not as an accurate measurement. If the difference is greater than 10 dB, the background noise has virtually no effect; and the reading is the desired sound level of that machine.

An example of a situation intermediate between these two is as follows. The background noise level with the machine not operating is 80.0 dB, and the total noise with the machine under test operating is 85.0 dB. The correction from Figure 5–5, for a 5.0 dB difference, is 1.7 dB, so that the corrected level is $85.0 - 1.7 = 83.3$ dB. Another way of stating this is that the machine under test generates 83.3 dB and the background noise level contributes an additional 1.7 dB to make up the measured 85.0 dB reading.

If the difference in decibel readings between background noise level and total noise level is slight, first try to reduce the background noise level at its point of origin. Next, work on the transmission path between the source and the point of measurement. This step may mean simply closing doors and windows if the source is outside the room or it may mean erecting barriers, applying acoustical treatment to the room, and opening doors and windows if the source is in the room.

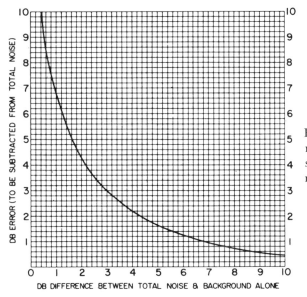

Fig. 5–5.—Background noise correction for sound level measurements.

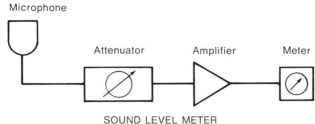

Microphone

Attenuator   Amplifier   Meter

SOUND LEVEL METER

Fig. 5–6.—A simple schematic showing the basic components of a sound level meter.

The third step is to magnify the difference in sound levels by the method of measurement. It may be possible to select a point closer to the apparatus, or to explore the background noise field to find a measuring position that minimizes the magnitude of background noise.

Auxiliary equipment

The sound level meter indicates only the overall sound level within the range of frequencies to which the instrument responds (see Figure 5–6). For more detailed noise studies concerned with noise sources and possible engineering methods of control, more information is required—for example, the relative noise level in various frequency bands or the level of the highest components. Octave-band analysis, the simplest and most common type of analysis, can be performed in a minimum of time. This analysis can be made on an octave-band analyzer, although many analyzers capable of more elaborate analysis are also adapted for octave-band measurements.

With a true "white" noise, that is, where the energy is distributed equally across the octave bands, the reading on adjacent octave bands will increase by 3 dB per band as the bands are switched toward the higher frequencies. This will often give some clue as to the type of noise being measured in a series of adjacent octave bands; that is, whether it is more or less randomly distributed or composed of discrete frequency components.

Octave-band analyzer

An octave-band analyzer is equipped with an indicating dial similar to that found on the sound level meter (see Figure 5–7). Before use, it is necessary to adjust the analyzer sensitivity with the band filters out of the circuit. Thereafter, individual band readings can be obtained for

145

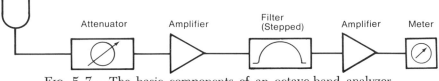

FIG. 5–7.—The basic components of an octave-band analyzer.

each octave band by turning the range selector. Operation of all octave-band analyzers is essentially similar. Follow the manufacturer's instructions regarding attenuator settings and battery checks to make sure measurements are accurate.

When the level in any octave band is unusually low, an auxiliary gain control or attenuator on the octave-band analyzer can be used to provide additional sensitivity. When an octave filter set is connected to a sound level meter, do not turn up the sensitivity of the sound level meter itself to obtain low-level readings on the analyzer (see Figure 5–8). This will overload the input stage of the sound level meter and cause a serious error. On combined sound level meters and analyzers, the manufacturer's instructions should be followed closely.

The sound level in any band is the sum of the readings of the attenuator setting on the analyzer and the indicating dial on the analyzer. Obtaining a reading with the analyzer is accomplished in exactly the same manner as with the sound level meter. The passband control of the octave-band analyzer is adjusted sequentially to each of the octave bands and the dial reading noted. The attenuator on the analyzer, which increases the overall sensitivity, is adjusted to obtain a satisfactory on-scale reading in each octave band.

The additional sensitivity available on the octave-band analyzer does not increase the overall range except on individual bands. The internal noise in the electronic circuits will limit the use of the extra sensitivity for wide band measurements and will also limit the accuracy to which a low-level band can be measured.

Recorders

Two types of recorders are of interest to those making sound measurements. One is the magnetic tape sound recorder (see Figure 5–9); the other is the graphic level recorder (see Figure 5–10) that automatically prints out a record of the readings of the sound instruments it is monitoring. Recorders are often used with sound analyzers to provide permanent records of sound levels *vs.* time. When the graphic level recorder is properly coupled to a continuously tunable analyzer, automatic analyses may be performed.

Before using a recorder with sound measuring equipment, the sensitivity of the recorder should be adjusted so that the readings or stored information will be on-scale of the sound level meter or analyzer. This is most easily accomplished with a calibration tone applied to the microphone or input circuit of the sound level meter. The sensitivity of the recorder is then adjusted so that the maximum reading is obtained when the indicating instrument on the sound level meter or analyzer reads full scale.

A magnetic tape recorder can be used in noise recording measurements. For measurements where a portable unit is required, a cassette tape recorder, such as that shown in Figure 5–9, can be used. The signal to be stored must be supplied to the recorder as an electrical signal obtained from a high-quality microphone. When measurements are to be made on the stored signal, the recorded tape is played back on the recorder and measurements are made on the electrical output signal.

Fig. 5–8.—Some sound level meters can be converted into octave-band analyzers by attaching an octave-band filter set.

Courtesy B & K Instruments, Inc.

To perform these functions, the magnetic tape recorder should include an accurate step attenuator, an amplifier with high gain and high input impedance, and a transient overload indicator with good response down to 15 Hz.

It is important that a noise level reference be established when a magnetic tape recording is made. This can be done in several ways—one of them is to note the overall reading of the sound level meter in decibels along with a notation of the source of noise, microphone location, microphone orientation (and other variables) on the noise survey form.

If the overall sound pressure level was 88 dB at the time of recording, the playback gain control of the tape recorder, when analyzing the noise at a later time, should be adjusted so that the analyzer will read 88 dB in the overall position.

A second method is to record the overall sound pressure level and the other pertinent data by speaking into the microphone and recording the data on the tape. This eliminates the danger that the measurement records and the tape can become separated or mixed because of some unforeseen circumstances.

FIG. 5–9.—A magnetic tape cassette recorder being used with a sound level meter to record the sound for later evaluation.

Courtesy General Radio Co.

The magnetic tape recorder can be used to perform the following functions in the field of noise measurement:

• *Keep reproducible records* of progressive changes in a sound level. These changes may be a result of the application of successive noise control procedures.

• *Record a noise for analysis by a number of methods* when the particular approach to be used is not at first obvious and it is not convenient or possible to repeatedly measure the original source.

• *Record a noise in the field for subsequent detailed laboratory study,* where complex instrumentation systems can be used. Frequency analyses may be obtained by playing back the tape recordings and passing the signal through any desired analyzer.

• *Record a sound that varies with time.* Samples can then be selected from the tape recording for analysis to get the change in spectrum as a function of time.

• *Monitor sound levels on tape over long periods of time* to catch intermittent sounds, which can then be separated out for analysis.

• *Record sounds that are erratic or intermittent* to aid in tracking down sources.

• *Permit subjective comparison* among sounds recorded on tape at different times. The subjective judgment can then be made by groups listening to the different sounds.

• *Permit observation of the subjective effects* of altering a signal, for example, by filtering, clipping, or adding noise.

When a noise signal, recorded on a magnetic tape recorder, is to be studied, it is customary to take short samples for analysis. These samples are cut from the full recording and formed into loops that can be run continuously in the recorder. Using short samples directly limits the fineness of detail possible in the subsequent analysis and also limits the accuracy with which one can determine the actual level in a particular band.

This limitation of accuracy results from the fact that the maximum time during which independent information can be obtained is the sample duration. If the noise is sufficiently uniform with time, a longer sample can be used to obtain increased accuracy, or measurements on a number of samples can be averaged.

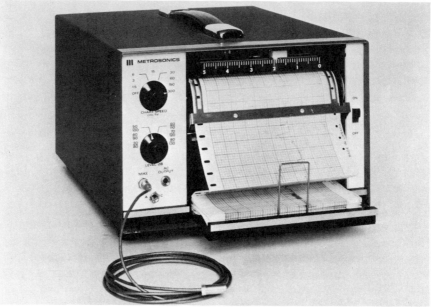

Courtesy Metrosonics.

FIG. 5–10.—A graphic level recorder can provide a permanent graphic record of the sound levels with regard to time.

Most of these applications apply to vibration signals as well.

The graphic level recorder is an instrument that produces a permanent chart record of the level of an applied signal (Figure 5–10). For noise and vibration measurements, this signal is usually obtained from the output of a sound level meter, vibration meter, or an analyzer. It is used either to record level as a function of time—as when used with the sound level meter—or as a function of frequency when used with an analyzer.

Impact noise

Many times the noise being measured is of the pulsed or transient variety, such as noise from a drop forge, riveting gun, or a foundry shakeout. When this noise is measured, the rms (root mean square), not the peak value, of the noise is indicated by the sound level meter. This rms value for the same noise peak can vary according to the time width of the noise pulse. For this reason it is useful to measure the peak noise value in addition to the rms or average value. For intermittent noises where the level varies very rapidly with time, or for impulse (impact) noises where the peak-to-rms ratio is quite high, one must resort to the use of a cathode-ray oscilloscope or an impact-noise analyzer.

The reason that sound level meters and octave-band analyzers should not be used for the measurement of intermittent and impulse noises is that their meter response is not fast enough. Even on the fast scale of a sound level meter or octave-band analyzer, the meter requires approximately 0.2 second to attain a final dial reading of the incoming signal. In other words, a transient noise that reaches its peak and decays in a time shorter than 0.2 second would not be correctly indicated on even the fast scale. The slow-response scale requires that the noise be present for even a longer period of time. The primary purpose of the slow scale is to perform an averaging of fluctuating noise levels.

Cathode-ray oscilloscope. When measuring impulse-type noises, it is necessary to use an instrument that can measure the noise peaks. One way to measure the peak noise is by feeding the output signal of the sound level meter into an oscilloscope that has been calibrated to the peak voltage output.

Impact-noise analyzer. Although the oscilloscope provides a useful means for observing the complete waveform of an impulse or transient signal, it is a bulky instrument and is suitable primarily for laboratory use. Of more practical use in the field is an impact-noise analyzer, several of which are commercially available. These analyzers simplify the procedure of measuring peak sound pressure level and are especially designed for this purpose. The response time is of the order of 0.0001 second. These analyzers are simple, compact, and very satisfactory for use in the field (see Figure 4–18, page 125).

Storage of the signal in the electronic circuit of the impact-noise analyzer is sufficiently long so that peak readings can be taken directly for most impulses by merely observing the maximum needle deflection indicating the maximum sound pressure level reached by the noise. Another characteristic which can be measured is the time-average level, which is a measure of the sound level maintained over a stated time period. The difference between the peak level and the time-average level is a measure of the time duration of the sound wave. Assuming impact noise to be an exponentially decaying random noise, the impact decay time is given in terms of the ratio of the decay time and the time used in obtaining the average level.

The impact-noise analyzer has an amplifier whose output charges a capacitor and shows this charge on a meter. The analyzer stores this peak noise measurement information so that it can be easily read following the event. By this technique the dynamics of the meter do not influence the reading. The instructions provided by the instrument manufacturer should be referred to for detailed measurement procedures.

Explanation of terms. Note that in any discussion of impact or intermittent bursts of noise, it is well to understand the terms "peak," "rms," and "crest factor." Understanding these terms helps us comprehend the electronic engineers' problem of finding an instrument that provides meaningful data.

PEAK. Peak values of the output signal of sound level meters and octave-band analyzers can be displayed on oscilloscopes. These peak (or maximum) values can be stored and read out on special circuits.

Rms or root mean square is more difficult to describe, and, to the uninitiated, even to visualize. It is defined as "the square root of the mean value of the squares of the instantaneous values of a varying quantity," whether it be sound pressure level, voltage level, or vibration level. In electrical circuits, the rms value can be described as equivalent to the power that would be dissipated in a straight resistive load by a d-c voltage.

In an electrical circuit, the power dissipated in a resistance is equal to the product of the voltage and the current through it. If the voltage that is applied to the resistor is suddenly terminated, the current will stop flowing at the same time. Because the resistor then will have power applied to it intermittently, this will result in less power absorbed by the resistor. If the duty cycle is 50 percent, the resistor would dissipate 50 percent of the power.

In order to compare this pulsed voltage to an equivalent d-c voltage, engineers have mathematically derived an expression which is designated the rms voltage. This varying voltage will always be less than the peak value of the voltage, except in the limiting case when the pulsed voltage is equal to the d-c voltage.

CREST FACTOR. In a burst or pulsed situation, the narrower the pulse the higher the peak voltage must be to maintain equal power compared to d-c; this poses great difficulty to the electronics engineer who is trying to indicate rms voltage on his meter. He must allow in his electronic circuitry enough reserve to take the peak voltage without overloading, and still have his instrument read on-scale. This ability to handle large peak-to-rms voltage ratios is called "crest factor capability," and in most high-quality instruments the ratio is anywhere from 3-to-1 to 5-to-1. Many environmental sounds have crest factors that are considerably in excess of this range. For example, a hand clap can have a crest factor of 30-to-1 or more.

A way to measure this type of sound is to use an oscilloscope with storage capability that responds to peak signals or to use an instrument called the impact-noise analyzer.

If an impact-noise analyzer is not available, it is possible to get approximate data for many impact noise sources by using a standard sound level meter. The C-weighted, fast network is used to see if the impact noise burst will cause the indicating needle to exceed 125 dB. If it does not, it usually means that the impact will not exceed 140 dB if the same noise burst is measured using an impact-noise analyzer.

Real-time analyzers

Another approach to analyzing sound measurements either on a level *vs.* time basis, or for analysis of impact- or burst-type noises, is an instrument which may be best described as a real-time analyzer (see Figure 5–11).

"Real time" is an expression that is used more and more in this space age to describe what is happening at this very instant in time. A tape recorder uses stored time as a frame of reference since the recorded sound occurred at some time previous to the playback. Real time describes a period of time which is occurring at the same time you are observing and analyzing the results. A typical example of a real-time situation is the telecast of the astronauts on the moon performing their experiments. Another example would be an oscilloscope display of the noise pattern of a diesel engine while it is operating.

The so-called real-time analyzers used in sound and vibration studies are not strictly real time since they do store information. But most of them sample and store the data and then read it out so fast that they are considered to be real time.

It might be asked how these units could be used in an industrial situation since they are very expensive. The approach here must, of course, be an economic one. An engineering specialist is well paid; if he must collect a great amount of information to satisfy governmental regulations, this might prove very expensive, if not impractical, to do manually.

FIG. 5–11.—The real-time analyzer is a sophisticated instrument which can monitor an entire plant, gather, analyze, and printout data applicable to a specific problem.

Courtesy General Radio Co.

A real-time analyzer, together with some ancillary equipment, can monitor a whole plant, analyze and record the noise measurements, and have it read out in useful data applying specifically to the problem. While a system may run as high as $50,000 to $70,000, it could not only save many hours of engineering time, but also provide data that would be impractical to get otherwise.

Another application of a real-time analyzer is to get an analysis of a short burst of noise while it is happening or very shortly afterward. This would be particularly useful if the engineer were interested in controlling the noise. Some of the analyzers have sampling times as short as ⅛ of a second.

Some of the larger companies who have complex noise problems may be very interested in this approach to noise analysis. If so, they should contact the manufacturers of these analyzers and get an engineering evaluation of their problem. They might find that this elaborate system will provide the most economical means of acquiring the data they need.

Instrument precautions

Circuit noise

When low noise levels are to be measured, the internal electrical circuit noise generated by resistors and transistors of the instrument may contribute to the measured level. This effect is usually noticeable in the range below 40 dB when the microphone is used on the end of a very long cable. If the microphone is mounted directly on the sound level meter, the level at which the internal instrument noise effect may be important is below 30 dB if C-weighting is used, or even lower if the A- or B-weighting is used. A correction can be made for this circuit noise, if necessary, using the same procedure as outlined for background noise. If the circuit noise is comparable to the noise being measured, some improvement in the measurement can usually be obtained by use of an octave-band analyzer. The circuit noise in each band should be checked also to see if correction is necessary. Whenever low noise levels must be measured and extension cables are used, a preamplifier should be used at the microphone.

Hum pickup

When noise levels are to be measured near electrical equipment, the electromagnetic field may affect the measurements of the sound measuring system. Because of the directional character of the electromagnetic field, the orientation of the instruments with respect to the equipment should be varied to see if there is any significant change in the indicated sound level. If an analyzer is used, it should be tuned to the electrical power supply frequency to determine its effect on the dial reading. If no analyzer is included, the C-weighting should be used in this test to make the effect of hum most noticeable, and a good quality pair of earphones with tight-fitting ear cushions should be used to listen to the output of the sound level meter.

If hum pickup is present, the instruments can usually be moved away from the source of the electromagnetic field, or, alternatively, a proper orientation is usually sufficient to reduce the pickup to a negligible value.

When sound measuring instruments requiring connection to electrical supply lines are used as part of the measuring setup, a check should be made at 120 Hz as well as 60 Hz for hum. This hum may be in the instruments, or it may appear as a result of the interconnection of different instruments. Sometimes reversal of the power plug connection to the electrical supply line helps to reduce the hum.

Microphonics

All vacuum tubes and some transistors are affected by mechanical vibration that can produce an undesired signal. This effect, called microphonics, is usually not experienced until the airborne sound levels are well above 100 dB. Microphonics can occur if the instruments are placed on supports that transmit vibrations directly to the instrument case.

The usual test for microphonics is to disconnect the transducer or microphone and observe whether or not the residual signal is generated internally within the instrument. The instruments can also be hand held to see whether or not the vibrations are transmitted to the instrument case through the supports or if it is the airborne sound that causes the problem.

Possible solutions for microphonic problems are as follows:

• Place the instrument on soft rubber pads.

• Remove the instruments from the strong noise field, place them in another room and interconnect them with long cables.

• Place sound barriers between the instruments and the sound source.

The microphone is also affected by mechanical vibration in that the output of the microphone is dependent on the airborne and solid-borne vibrations that are impressed upon it. The effects of the solid-borne vibrations can usually be discounted in the standard, sensitive microphones because of the type of construction used; but these vibrations have great effect on the low-sensitivity microphones used in the measurement of high sound levels. A mechanically soft mounting should generally be used for such a microphone in order to avoid trouble from these vibrations. Often merely suspending the microphone by means of its connecting cable is adequate.

Readout

The measured sound level is the sum of the attenuator settings and the reading of the indicating needle on the dial. With some sound level meters, there is more than one attenuator and the reading of all must be added together. The instructions of the manufacturer should be followed closely to obtain the correct sound level reading.

Ballistics

Shaded-pole pieces are generally used in the galvanometer movements of sound level meters in order to obtain a decibel scale which is substantially linear. Therefore, the ballistic characteristics of the indicating instrument are not constant over its entire range of movement. This sometimes results in a difference between readings on adjacent attenuator settings where one reading is taken at the upper end of the instrument scale and another reading is taken at the lower end of the scale. Where such a difference is noticed, it is desirable to use the lower attenuator setting, which gives deflection at the high end of the scale, because the ballistic characteristics are more closely controlled over that portion of the scale.

Meter pointer fluctuations

Two ballistic characteristics are provided for the indicating needle on the sound level meter—a FAST and a SLOW position. Most industrial sounds vary widely in intensity and do not give a constant meter reading. The reading may fluctuate over a range of a few decibels and sometimes over a range of many decibels, particularly when measurements are made at low frequencies. The maximum and minimum readings should usually be noted. These levels can be entered on the data sheets as "85 to 91 dB" or "88 + 3 dB."

When an average sound pressure level is desired and the fluctuations are less than 6 dB, a simple average of the maximum and minimum level is usually taken. If the range of fluctuation is greater than 6 dB, the average sound pressure level is usually taken to be three decibels below the maximum level. In selecting this maximum level, it is also customary to ignore any unusually high levels that occur infrequently.

The slow meter speed should be used to obtain an average reading when the fluctuations on the fast position are more than 3 or 4 dB. On steady sounds, the reading of the meter will be the same for either the slow or fast position; while on fluctuating sounds, the slow position provides a longer time-average reading.

When one observes the needle fluctuations for a time and estimates an average, one can remember and use only a small portion of the total observed behavior. The observations are not independent because of the finite time required for the pointer to assume a new value. In the fast position of the meter, allow about a half second between observa-

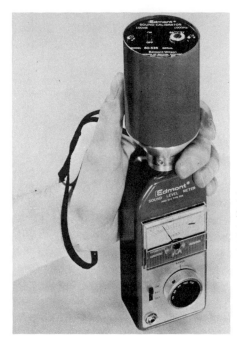

Fig. 5–12.—The acoustic calibrator fits over the sound level meter microphone and produces a fixed sound pressure to which the sound level meter is adjusted.

Courtesy Edmont-Wilson

tions; in the slow position, an interval of one to two seconds is desirable.

If an octave-band analyzer is used, the extent of the fluctuations also depends on the bandwidth. The narrower the band, the greater are the fluctuations, and the longer is the meter averaging time required for a satisfactory estimate of the level. A relatively simple principle is involved here. The narrow band is used to get fineness of detail. But the finer the detail that is desired, the more time is needed to obtain a result with a certain degree of confidence.

One can find many examples of an overall sound level that fluctuates over many decibels. One example is the background noise present in private offices; for C-weighting in the slow meter position, one can commonly find fluctuations of three or more decibels. The fluctuation may be a pure tone that corresponds to a band that is only a few hertz wide rather than the complete range of 200 to 10,000 Hz. This is because the energy in the sound is concentrated in the low frequencies over a relatively narrow band. The fluctuations reflect only the relation between the equivalent frequency band of the signal applied to the metering circuit and the averaging time of the circuit. Whether the energy is concentrated in a narrow band by means of an electrical circuit in the analyzer or by the source and path to the microphone is immaterial.

Calibration

Satisfactory noise measurements depend on the use of measuring equipment that is kept in proper operating condition. Although most instruments are reliable and stable, in time the performance of the instruments may change. In order to be sure that any important changes will be discovered and corrected, certain simple checks should usually be made before and after any set of noise measurements.

Some sound level meters and analyzers have built-in electrical calibration circuits for checking amplifier gain. In each case the gain of the amplifier is compared with the attenuation of a stable, resistive attenuator. This assures the operator that all of the electrical circuits are operating properly. This method does not, however, check the transducer (microphone or vibration pickup). These must be checked by external means. Since this check does not include the microphone, it is possible for the overall acoustical calibration to be off even though the electrical system checks satisfactorily. Although most microphones supplied with sound level meters are highly stable, if they are not abused, an acoustic calibration of the total noise measurement system is desirable.

Acoustical calibration

A simple check of the sensitivity of the microphone on a sound level meter is to use a spare microphone of similar characteristics. The spare microphone can be used to take readings of a known stable noise source to indicate whether there has been a change in sensitivity of the measurement system.

To check on the sensitivity of the entire sound measurement system, an acoustic calibrator may be employed. This device produces a fixed sound pressure in an enclosure that is designed to fit over the microphone of a sound level meter (see Figure 5–12). Usually the sound pressure is generated by a small loudspeaker. In addition to its advantage of portability, the acoustic calibrator assures that calibration checks will always be made with the same type of signal.

The procedure for the acoustic calibration of a sound level meter or analyzer is to place the calibrator over the microphone, and then adjust the dial reading on the sound level meter or octave-band analyzer to coincide with the known acoustic output of the calibrator.

Another type of external calibration is the microphone reciprocity calibrator. This instrument uses the reciprocity technique, which is preferred for the laboratory calibration of standard microphones.

Comparison of test results

The duration of the noise source may impose several restraints on the kind and amount of information that can be obtained. Some noise sources may not be under the control of the operator, or the source may not emit sound waves uniformly over an extended period of time.

Because of inherent variability of random noise, analyses of distinct samples of the same noise will not yield identical results. Unless this inherent variability is appreciated, one can be led into rejecting useful data, rejecting a useful analysis system, or placing too much reliance on a particular noise measurement.

If measurements are made of the same noise with two different sound level meters, the readings may differ significantly. The discussion in this chapter should have indicated most of the possible sources of this discrepancy. Differences in the microphone characteristics are usually the chief cause of this discrepancy. For example, if one sound level meter uses a dynamic microphone and another uses a ceramic microphone and if the noise contains strong low-frequency components, great differences in instrument readings can occur because of the generally poorer low-frequency response of the dynamic microphone.

Another factor that can contribute to this discrepancy concerns the average level. For purposes of meeting certain tolerances, the average level of an instrument made by one manufacturer may be set slightly different from that made by another.

If the instruments are not operating properly or if standing waves are not averaged out, serious discrepancies can, of course, be expected.

Sound surveys that include only weighted readings are very easy to make. The Occupational Safety and Health Act regulations provide the limits of what is considered a reasonable risk. Taking an engineering survey, however, which includes the octave-band analysis, does require a knowledge of the principles just discussed. An appreciation of the decibel and how to work with it is very important. Intelligent use and operation of sound measuring instruments will avoid the collection of meaningless data.

Bibliography

American National Standard, S1.4–1971. New York, N.Y.: American National Standards Institute, 1971.

Beranek, L. L. *Acoustic Measurements.* New York, N.Y.: John Wiley & Sons, Inc. 1949.

———. *Acoustics.* New York, N.Y.: McGraw-Hill Book Co., 1971.

Beranek, L. L., ed. *Noise and Vibration Control.* New York, N.Y.: McGraw-Hill Book Co., Inc. 1971.

Harris, C. M., ed. *Handbook of Noise Control.* New York, N.Y.: McGraw-Hill Book Co., Inc. 1957.

Petersen, A. P. G. and Cross, Jr., E. E. *Handbook of Noise Measurement,* 7th ed. West Concord, Mass.: General Radio Co. 1972.

Chapter Six

Sound Survey Techniques

By Edwin L. Alpaugh, P.E.

The procedures, methods, and techniques involved in sound surveys to obtain valid sound measurements are described in this chapter. Proper evaluation of a noise problem depends upon accurate, reliable, and reproducible sound level measurements. The location, number, and type of sound measurements to be made depend upon the criteria being used to evaluate tnat particular noise problem.

The type of sound involved, the amount of information desired, and the amount of time available to obtain the information will determine the type of sound survey that will have to be made. The amount and kind of information that is needed about a noise source will depend upon the use that will be made of that information.

If the sound level measurements are to be used to estimate potential damage to hearing, A-weighted sound level readings may be adequate. If, however, the data is to be used to redesign a machine to eliminate or minimize excessive sound (noise) levels, octave- or narrow-band analysis of the noise will be required.

The amount of time required for taking sound measurements may be related to the operating schedule of noise sources as well as to economic factors. For instance, a four hour sound survey might be adequate to evaluate one specific job, but possibly a forty hour or longer sound survey would be required to obtain sufficient information on all operations. The amount and kind of information required will determine the amount of time needed, either in taking sound measurements in the field or analyzing the data. The number of different measurements may vary—it depends upon how often certain measurements are to be repeated, the number of sound sources to be measured, and how much detail is required.

Sound survey

Classification

Sound surveys are classified according to their purpose or need:

- Simple screening survey
- Detailed general sound survey
- Special purpose sound survey

The simple screening survey should be made with a sound level meter that meets ANSI S1.4-1971 requirements for Type 1 or 2. The survey usually involves taking a number of measurements of A-weighted or overall sound pressure levels in an area for rapid evaluation of the problem.

The detailed general sound survey generally involves the use of an octave-band analyzer in order to determine the spectral distribution of the sound energy with regard to frequency. For industrial hygiene noise control activities, an octave-band analyzer normally would be used to measure the sound level in the various frequency bands.

The special purpose sound survey usually requires more specialized equipment. Many industrial noise problems can be surveyed directly with the sound level meter and octave-band analyzer. However, an impulse or impact sound such as that made by a forging hammer cannot be measured accurately with a sound level meter alone. An impact-noise analyzer, in conjunction with a sound level meter, can be used to obtain measurements of peak noise levels. A cathode-ray oscilloscope and magnetic tape recorder also may be used in a special purpose sound survey. (These instruments were described in Chapter 4.)

Industrial hygienists often refer to their work as the recognition, evaluation, and control of occupational health hazards. Excessive exposure to high sound levels over a period of time can cause partial loss of hearing. Thus noise becomes an industrial hygiene problem and must be evaluated using accepted industrial hygiene techniques.

Unlike some other industrial hygiene problems, however, recognition of a noise problem is relatively easy, and ordinarily there is no question about the need for sound level measurements. Complications arise when an attempt is made to evaluate the exposure to noise in terms of hearing damage risk, or compliance with criteria even though the threshold limit values for noise offer a guideline in this respect. Control of excessive exposure to noise may range from very simple to extremely complex engineering measures, or in some cases, there may be no feasible solution other than use of personal hearing protection devices.

Types of sound surveys

As a general rule, sound level surveys are conducted for one or more of the following reasons:

- Establish hearing conservation or damage risk criteria

- Determine annoyance or speech interference levels

- Evaluate compliance with established noise criteria and specifications

- Obtain data for engineering studies or controls

Hearing conservation. Surveys, made to evaluate possible detrimental effects on hearing acuity, are those most commonly made by safety professionals. A survey of this type is reasonably easy to make if the limitations and operating characteristics of the sound level measuring instruments are understood by the observer conducting the survey. If the noise "on" and "off" times and levels are not too complex, an accurate time study relating noise on-time to noise level for individual workers can be very helpful. This permits a reliable estimate of the total exposure time. A meaningful hearing conservation survey can be difficult, however, if the noise level varies greatly from hour to hour and day to day.

In order to estimate the degree of exposure to noise, two variables must be measured:

- Noise levels in the hearing zone of employees

- The duration of noise exposure experienced by each employee at each level

Sound levels should be measured using only those instruments which meet American National Standard S1.4-1971 for Type 1 or 2 sound level meters. Measurements should be made with the instrument set to

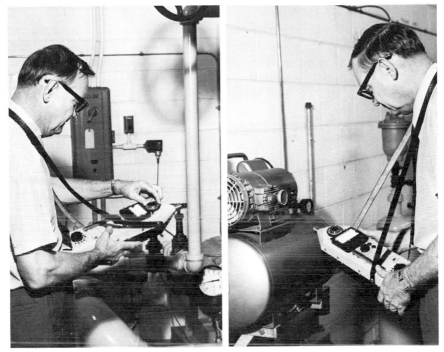

A pump.   An air compressor.

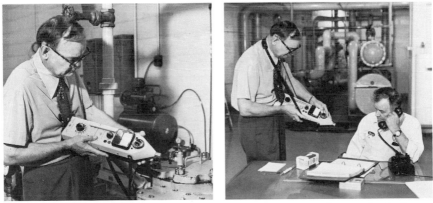

A gear housing.   Maintenance foreman's desk.

Fig. 6–1.—Taking a sound survey using an octave-band analyzer.

operate at slow response on the A-weighted network. The instrument should be acoustically calibrated before and after noise measurements are made.

The weighting circuit used to obtain A-scale measurements (referred to as dBA) discriminates against low-frequency sounds somewhat as the human ear does, and for most industrial noise situations A-scale measurements are as effective as octave-band measurements with respect to damage risk criteria. On the other hand, A-scale data alone is not adequate for the evaluation of most noise problems to determine the proper types of engineering controls, as will be discussed a little later in this chapter. When the noise being measured is not broadband or wide spectrum and the sound energy is concentrated in rather narrow peaks, an octave-band analysis should also be made.

Prior to the concept of using A-scale measurements for evaluating hearing damage risk, a hearing conservation survey required measurement of the overall sound pressure level and the sound level of each of several octave bands. This required the use of an octave-band analyzer in order to determine the sound level in each octave band.

The various problem areas in the plant should be investigated carefully to determine the exact location and number of sound measurements necessary to obtain a representative pattern of employee exposure (see Figure 6–1). Work areas in which it is necessary to speak loudly to be understood are logical places to take sound measurements.

A hearing conservation survey requires more than just a compilation of noise measurements. Other details that must be recorded are the on-off time noise exposure patterns, the kinds of noise control measures being used, and the time-weighted average level of noise exposure of employees.

Regulations, promulgated under the Occupational Safety and Health Act, state the allowable exposure time at various sound levels. In many industrial situations, noise levels vary in amplitude as variations in manufacturing processes occur and it is necessary to determine the time pattern of noise exposure during the manufacturing process.

There are several ways to estimate or evaluate time patterns of noise exposure. One is to relate the employee's time pattern of exposure to the noise level recorded on a graphic level recorder. By knowing the noise level and movement of personnel into and out of the area it is possible to estimate the individual noise exposure of an employee by referring to the time vs. level plot of the recorder and correlating it with the time pattern of movement of personnel.

Fig. 6–2.—An employee wearing an audio dosimeter to measure the noise level-time exposure.

Courtesy Unico Environmental Instruments, Inc.

The more common approach is to monitor the noise level over a period of time and calculate the equivalent noise exposure as compared to allowable levels for eight hours. In such cases, it is necessary to estimate the work periods and relate them to their respective levels in order to evaluate equivalent exposures. Noise exposure meters are now available which consist of a small wearable dosimeter (described in Chapter 4) that indicates the actual noise exposure as compared to the percent allowable exposure for an equivalent eight hour day (see Figure 6–2). Such personal noise monitors can give a reasonably good measure of the actual noise exposures over a period of time. A personal monitor can also be used as an area monitor, if desired.

Annoyance or speech interference. The survey to measure speech interference and annoyance levels is relatively easy to perform but somewhat difficult to evaluate. This is especially true for annoyance levels because of the extreme variation of tolerance levels in individuals and their reaction to noise. Typical areas to be covered in this type of survey are conference rooms and private offices, and those areas where noise levels may interfere with telephone use. Guidelines are available for evaluating annoyance levels for community noises, noise levels in occupied spaces, and speech interference. (See Chapter 3 for more details.)

167

The employees can provide helpful information to assist in locating those work areas where the noise levels may interfere with speech. If there is difficulty in carrying on a normal conversation at a close distance, this indicates there may be need for further study to determine if the noise levels are high enough to have adverse effects on hearing. Complaints from employees about head noises, ringing in the ears, and speech or music having a muffled sound at the end of the work day are very definite signs of potential problems and call for a noise survey and evaluation of exposures in the work area.

Compliance with established noise criteria. New equipment purchase orders listing specifications for acceptable operating noise levels are becoming relatively common. These specifications are usually based on A-weighted sound levels, maximum overall sound levels, or maximum octave-band levels at certain specified measuring points.

Engineering groups and trade associations have prepared standardized test codes describing techniques for measuring the noise levels generated by machines or specific pieces of equipment. These codes are often referenced as a part of a purchase specification and provide details on the measurement procedure to be used in checking the equipment for compliance with maximum noise requirement. (For further details see Chapter 19, Sound Level Specifications.)

State environmental agencies and some cities regulate the maximum allowable noise levels at the lot lines or boundaries of a manufacturing plant. These regulations may be based on A-weighted sound levels or octave-band levels. Since the noise producing operations in a typical factory can vary considerably through the day-and-night cycles of a manufacturing operation, monitoring the noise levels with a graphic recorder may be required.

The extensive use of window air conditioning units or central units with outdoor heat exchangers has led to regulations restricting noise levels in residential neighborhoods. The usual reference measurement is the A-weighted sound level at a specified location.

Some cities and towns have passed laws setting maximum limits on the noise a motor vehicle can make. Many of these laws are stated in terms of the A-weighted sound level at a specified measuring point. Further details of the specific requirements in a particular locality should be obtained from the governing body that has jurisdiction.

Many products and devices can be examined for noise output at the time they are assembled and tested. However, while noise measurements on the production line are often possible, they can hardly ever be done in an ideal manner. Precision acoustical testing usually requires a large, isolated, echo-free space, which ordinarily would not be available as part of the production line. Useful noise measurements, however, can still be made with relatively simple procedures, although the accuracy of the measurement may be significantly reduced.

To minimize equipment noise levels, the various noise producing mechanisms contained in a specific device or product must first be identified—a complex and difficult task. Many machines, both old and new, have complicated combinations of forcing motions, complex structural vibrations, and resonating surfaces as primary noise sources. A useful technique to pinpoint the noise sources for quality control purposes depends on a vibration measurement that has been correlated with airborne sound level measurements. For example, airborne sound level measurements may show that a noisy product or device produces noise peaks in one or two octave bands and a vibration measurement may then be taken to determine which levels correspond to these noise peaks. Some exploration of the vibration levels of the various surfaces of the device under test will be necessary to find the critical areas. The vibration tests may then be used to predict sound levels of the finished product.

Background noise may seriously interfere with noise studies of individual machines. However, background noise may be much less during lunch periods or shift changes, and noise measurements of a particular machine may then be practical. Tape recordings of the noise sources can be made during the relatively quiet background-noise periods and the recorded noise levels of the noise source can then be analyzed in the laboratory at a later, more convenient time. (See Chapter 5.)

Engineering surveys. Because the purpose of an engineering noise survey is to determine feasible means to control noise at its source or along its path, such surveys are nearly always more complex than those made for hearing conservation. As a general rule, an engineering noise survey requires a wide selection of noise measuring equipment as well as a good knowledge of acoustics by the individuals using the equipment. It is not unusual for noise problems to require extensive engineering noise surveys to determine the primary noise sources. Octave-band sound level measurements are generally made to determine the type and amount of control necessary to reduce the noise to an acceptable level (see Figure 6–3). This may require many measure-

FIG. 6–3.—Performing an engineering noise survey prior to determining control methods.

ments to locate and identify the noise generating sources.

Surveys also can be made to determine the effectiveness of noise control measures. Periodic surveys made after the control measures are instituted will indicate if they are operating effectively and efficiently.

Noise and vibration levels are obvious factors to consider in the planning and layout of manufacturing operations. Conference rooms, private offices, and audiometric testing rooms may be much more expensive to design and build if they must be placed adjacent to noisy manufacturing operations and, at the same time, have low background noise within the enclosed area.

A careful engineering study of sound and vibration levels is needed for best use of existing facilities. Useful preliminary information can be obtained with a sound level meter, but an octave-band analysis of the sounds and vibration levels can provide more complete and useful data. The cost of acoustically isolating low-frequency noise is much greater than isolating high-frequency noise; therefore, it is necessary to know the levels of low-frequency sounds and vibrations when making cost studies.

Engineering surveys usually require detailed information about the noise source, such as its sound pressure output, its directional characteristics, and the variation in sound level as measured radially away from the source of noise. Specialized noise measuring equipment—such as a narrow-band analyzer, impact-noise analyzer, high quality tape recorder, graphic level recorder, oscilloscope, or vibration measuring equipment and stroboscopes—is often required to obtain enough information for evaluating a complex engineering noise problem.

Types of sounds

The type of sound to be measured in a survey is an important factor in determining the equipment necessary for accurate measurements. Sound (and thus noise) may be classified as follows:

• *Broadband sound* (wideband sound) may vary in overall sound level, and the energy may be spread over a wide range of frequencies. Examples of broadband sounds are the sounds generated in textile mill weave rooms, piston-type aircraft engines, and air moving through a duct.

• *Narrow-band sounds* are those sounds having their energy concentrated in a narrow frequency range. For instance, sounds produced by circular saws, generators, sirens, air compressors, planers, transformers, and jet engine compressors are classified as narrow band.

• *Impulse or impact sound* is characterized as a single event of short duration and is created, for example, by explosions, drop hammers, punch presses, pistol shots, doors slamming, and similar operations.

• *Repeated impulse or impact sound* refers to a series of repeated events such as that generated by chipping and jack hammers, riveting operations, and foundry shakeouts.

• *Transient sound,* best exemplified by a passing train or low aircraft.

Within any of these classifications, additional characteristics may exist which set a sound apart from others within the same category; a sound may be continuous, intermittent, cyclic, or transient. Other characteristics may include pure tone or beats, varying speed sound sources, and random noise.

Continuous and intermittent sound

If the variations of sound level involve higher readings at intervals of one second or less, the higher sound level reading can be considered to be the continuous level. This means that when the indicating needle on the dial of the sound level meter, on the A-scale at slow response, generally reads 88 dBA with intermittent levels of 92 dBA at intervals of one second or less, the 92 dBA reading should be considered as the continous noise level at that measurement position.

Intermittent sounds whose duration is *greater* than one second should, insofar as practical, be measured as to level and duration. The total duration of the intermittent sounds at peak intensity during a day should be used to determine whether the noise exposure to employees is within permissible limits. These intermittent sound levels, which can be measured with a sound level meter, should not be confused with impulse sounds of very short duration resulting from impacts or explosions.

Typically during sound surveys of industrial operations, impact noise may be mixed with broadband noise, or there may be work areas where impact noise predominates. It should be emphasized that efforts to measure impact noise with an ordinary sound level meter may give completely inaccurate readings. The true peak levels of impact noise will be much higher than the maximum sound level meter readings. Consequently, impact-noise analyzers should be used to measure sudden sharp impact or impulse-type noise. However, if the impact is repeated at a rate of about 200 per minute, a sound level meter can be used to give an *estimate* of the overall noise level.

Pure tone and beats

Examples of sounds that display a steady indicated sound level are transformer hum and the sound from some rotating electrical machinery. When the combined sound produced by several similar machines is measured, the indicated level is also constant, unless the speed of the machines is such that some of the major sound components are only a few cycles apart in frequency. In this situation, an audible beat—a periodic rise and fall in sound level—occurs, and the indicated sound level meter reading also rises and falls.

It is also possible to obtain sound level readings differing by several decibels due to standing waves simply by moving the sound level meter microphone nearer, or farther away from the sound source.

Varying speed sound sources

Machinery that operates at varying speeds usually produces a noise that fluctuates in level. If machinery speed varies periodically, the sound level may also vary periodically. If the machine speed varies erratically, the sound level will also vary erratically, and the behavior may be similar to that of random noise.

Random noise

The indicated dial reading of the sound level of a random noise, such as that produced by some jets, blowers, or combustion processes, will be erratic. Most industrial noise sources have enough random energy components so that the indicated sound level will fluctuate noticeably. The extent of this fluctuation depends upon the nature of the sound source. The fluctuations in sound level are ordinarily not a result of erratic behavior of sound measuring equipment, but rather reflect irregularities in the noise production process. This process often can be considered as a combination of many sources that produce sound at random time intervals.

The noise level and problems in older mill-type factory buildings with high ceilings, wooden floors, and open doors and windows will differ from that found in modern air conditioned plants where windows have been eliminated and production is carried on through the operation of a great number of closely spaced machines, conveyors, and other auxiliary equipment. The noise problems and their solutions would be considerably different in the two cases.

Typical noise levels

Typical examples of noise levels common to a number of industrial operations are shown in Figure 6–4 and Table 6-A. Although based on a survey made several years ago, the results are still valid and give some idea of the range of noise levels that might be found in similar environments.[1] In addition, the octave-band sound pressure levels of a few widely used machines are also shown. The shaded area shows the range of observations.

The loudest operations shown are chipping and riveting on large steel tanks and aircraft assemblies. Similar operations on smaller or more massive pieces, like small castings or concrete, are grouped with other pneumatic power tools such as drills and wrenches. Note that the noise is more a function of the work than of the tool, although many air-actuated tools are in themselves quite noisy. A chipping hammer, for example, used on large steel plates causes vibrations with large amplitudes. Massive castings or concrete structure radiate less.

The range of observations for saws includes circular wood saws and stone saws. A circular wood saw measured close to 90 dB in every octave band; stone saws were about 85 dB.

The graph for planers includes operations on wood and stone. A
(*Text continues on page 176.*)

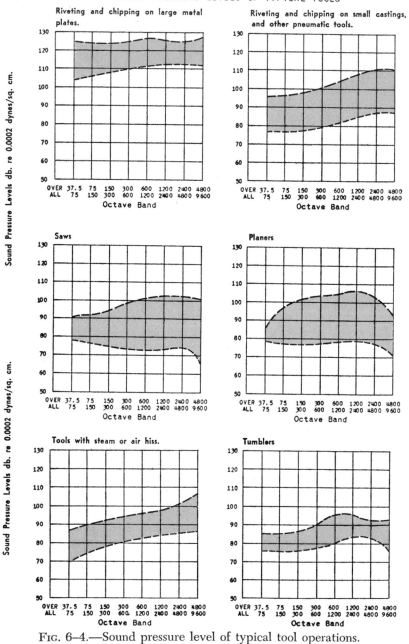

FIG. 6–4.—Sound pressure level of typical tool operations.

SOUND PRESSURE LEVELS OF TYPICAL TOOLS

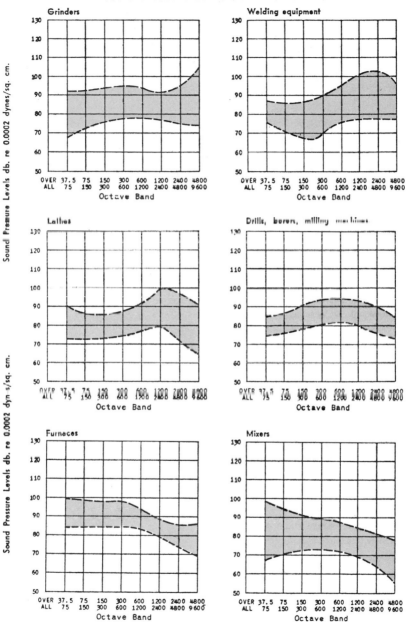

Sound Pressure Levels db. re 0.0002 dynes/sq. cm.

Grinders

Welding equipment

Lathes

Drills, borers, milling machines

Furnaces

Mixers

Octave Band

Adapted from A Noise Survey of Manufacturing Industries, *H. B. Karplus and G. L. Bonvallet.*[1]

TABLE 6-A

FOUNDARY HEARING ZONE NOISE LEVELS
Measured 2 to 5 Ft From a Machine

Operation	Overall Noise Level
Shakeout, 50 sq ft (½–1 cu ft castings)	114
Shakeout, 40 sq ft (½–1 cu ft castings)	105
Shakeout, 10 ft x 10 ft (2–4 cu ft castings)	98
Shakeout, 8 ft x 5 ft (1 cu ft castings)	108
Pneumatic chipper, 100 psi (1–3 cu ft castings)	110
Pneumatic chipper, 100 psi (4–6 ft castings)	112
Push-up machine, 14 in. x 15 in. sand molding	103
Jolt squeeze machine, 13 in. sand molding	100
Pneumatic ram, 100 psi, sand mold ramming	90
Core blower, making sand cores	116
Tumbler, 36 in. x 72 in., small castings	100
Stand grinder, 36 in. small castings	96
Pre-mix burner, 2 in. outlet, gas, crucible heat	94
Electric furnace 30 ton, early in run	105
Air hoist, 2000 lb, pneumatic	112
Sand slinger, air rush and sand noise	102
Scarfing, acetylene weld, equipment	91

NOTE. The noise levels shown here are indicative only and will not necessarily be representative of a particular situation since the noise levels from two identical machines will vary due to maintenance policies, differences in equipment, casting size, noise control, and many other factors. The exposure may be steady or intermittent, or both, depending on the operation, and may include a number of different types of noises such as those due to air, impact, combustion, arcing, or friction. Foundry noise survey measurements indicate that the use of compressed air, mechanical vibrating equipment, and tumblers accounts for a large percentage of the noise sources. These sources should be studied first in order to determine where control will be most effective.

wide range of sound levels is produced. Planing a flat surface on fairly soft stone is the least noisy, whereas planing a concave stone surface was fairly close to the upper limit of the range. Finish planing on pine was close to the lower limit, whereas planing operations (making a deep cut on hardwood) were found close to the upper limit.

Steam and air hiss, encountered in many operations, is shown in Figure 6–4. Tumblers, in which small castings are knocked together in a deburring operation, grinding, welding, and machine shop operations (like lathe work, drilling, milling, and boring) are, in general, less noisy than the other operations shown in the graphs.

Furnaces and mixers are examples of machines having downward-sloping sound spectra. Furnaces have fairly high overall levels, but are not usually judged to be very loud because of the low intensity in the high-frequency region. Mixers are the least noisy class of machines listed.

Noise sources in industrial operations are many and varied; some of the more typical are as follows:

Noise Source	*Type of Noise*
Punch presses Riveting Forging	Metal-to-metal contact
Portable tools such as Chipping hammers Grinders	Tool or wheel contacting metal and compressed air exhaust
Blow-off hoses Air hoists	Discharge of air exhaust
Moving machine parts that fit poorly, or not well secured to supporting members	Vibration
High volume of in-rushing air Burners Compressors	Roar due to air turbulence
Vibrating walls, roofs, and other structures—when, for example, driven by an out-of-balance fan	Sympathetic vibration

Instruments

The type of noise to be measured influences the selection of instruments to be used (see Chapter 4). For example, a tape recorder for data storage usually is necessary to obtain data on noise generating equipment that operates intermittently. On the other hand, the noise levels of equipment or machinery operating continuously producing a steady broadband noise can be easily measured with a sound level meter or an octave-band analyzer. Yet neither of these noise measuring systems should be used for the measurement of impact or impulse noise.

In many cases it is advisable to run preliminary noise surveys before making detailed measurements with more sophisticated instruments. These preliminary measurements can be made by safety professionals with survey-type equipment. The data then can be sent to specialists such as acoustical engineers or industrial hygienists for their evaluation and recommendations as to what further measurements should be taken.

The actual taking of sound measurements is not too difficult, provided that the sound levels are not too erratic. The type of sound measuring instrument necessary to obtain a reasonably detailed picture of the noise exposure involved depends on the information that is required. For preliminary surveys, an ANSI Type 2 sound level meter is essential.

The Type 2 survey meter does not have all the features of a Type 1 sound level meter but is still accurate enough for a preliminary survey. Although their physical dimensions may be rather small, these sound level measuring instruments are complex electronic devices, and frequent calibration checks are necessary to make sure they are functioning properly.

A good set of headphones used with a sound level meter can be of great value in making a noise survey. Listening to the noise levels through earphones frequently can provide a good evaluation of the noise source as well as determining the condition of the instrument.

A sound level measuring instrument should meet the following minimum requirements:

• The dials should be clear and easy to read and the calibration marks legible—a crude measurement is of limited value.

• The instrument's accuracy, response, and other characteristics should be fully predictable and completely supported by readily available specifications and data.

• The instrument should be rugged, reliable, and simple to operate.

• The instrument should be fully portable and convenient to use—preferably held in one hand (so that the other hand can record data).

• The instrument should have a sufficient dynamic range and frequency response to cover the acoustical region of interest.

• The instrument should be reasonably unaffected by temperature, humidity, vibration, and the electrical or magnetic fields normally encountered in noise surveys. (See Chapter 5 for a discussion of these effects.)

The American National Standard S1.4-1971 is frequently cited by both manufacturers and purchasers of sound level meters as a minimum standard.

The typical sound level meter consists of a calibrated microphone, an amplifier, and a metering circuit to indicate the output of the amplifier. There is generally a provision for making the amplifier output available for other auxiliary instruments or for listening with earphones. (See details in Chapter 4.)

The calibrated microphone is one of the most important parts of the instrument. The two types of microphones that are the most popular at the present time are the ceramic microphone and the condenser microphone. The advantage of the ceramic microphone is its simplified associated electronics and its reliability. The advantage of the condenser microphone is in the smooth response which makes it much more uniform at higher frequencies. (See details in the previous chapter.)

The octave-band analyzer may have the filter set as a separate instrument or may be combined in the same case with the sound level meter. The analyzer has an attenuator and cutoff filter or filters permitting the selection of the octave bands to be measured. If they are separate instruments, the analyzer and the sound level meter can be connected by a cable.

The sound level obtained in any octave band is the sum, observing the appropriate signs, of the attenuator setting of the octave-band analyzer and the needle reading on the dial of the indicating meter.

If more detail concerning the noise source is needed, a one-third octave-band analyzer can be used. The narrower the bandwidth of the frequency analyzer the more accurately one can determine the peak components of the noise spectrum. Narrow-band analyzers are necessary to locate pure tone components or sharp narrow band peaks.

If it is important to know accurately the sound levels at different frequencies, the type of frequency analyzer used should be chosen carefully. However, one should keep in mind that the narrower the bandwidth of the frequency analyzer the greater the amount of time required to collect noise measurements at the source.

For a particular noise source, the simplest and quickest reading is the A-weighted sound level as read on a sound level meter. A single set of octave-band level readings for this same source would take about eight times as long, and a one-third octave-band analysis would take at least 24 times as long to take as a single A-weighted reading.

A question that should be asked—is the additional detail obtained by use of a one-third octave-band analyzer really necessary? The answer, of course, depends upon the information needed. For many noise problems, an octave-band analysis will be all that is needed. For example, where the noise is fairly continuous and does not exhibit any sharp peaks and dips in sound level or contain any pure tone components—octave-band measurements may be completely satisfactory. In cases where the noise is being measured to determine its potential for producing hearing damage, A-weighted sound levels and knowledge that impact noise is not present would be sufficient. Usually an experienced listener armed with some knowledge of the noise source can ascertain this much supplementary detail by ear.

However, many noise problems require more detail than that provided by an octave-band analysis. For example, in attempting to redesign a complicated piece of machinery to make it quieter, one can make very good use of detailed information about the distribution of noise energy provided by a narrow-band analysis, particularly in isolating the various components of the noise source that contribute to its total noise output.

Calibration of equipment

Whether the sound level meter or the octave-band analyzer is used, the presurvey operational steps are the same. They seem simple but are quite important. Before leaving the office or laboratory, make sure the instruments are operating properly. Check the batteries and electrical circuit calibration as described in the instrument manual furnished by the manufacturer. Make sure the microphone is functioning properly by using an acoustical calibrating device. Make sure the calibrator is specifically designed for the microphone being used. A calibrator is a device that impresses a noise of known intensity and frequency upon the microphone and is discussed in some detail in the previous chapters.

In addition to checking with a calibrator, make an octave-band analysis of some familiar, unchanging steady noise. A small noisy air pump, the discharge vent of the air conditioning system, or almost any broadband noise source that does not change much over a period of time would be very suitable for this purpose. Check the response of the meter in every octave band. If the observed noise levels that have been accurately measured many times in the past suddenly differ from the previously recorded values in one or more frequencies, something may be wrong with the instrument.

Because the sound measuring equipment normally used is battery-powered electronic equipment, the user should constantly be aware of possible faulty operation. Equipment should be checked before it is used to take sound measurements. A suggested series of steps for checking the instrument response is as follows:

1. Turn the power on and allow the instrument to warm up for a few minutes.

2. Check the batteries according to the instructions in the instrument manual.

3. Make an acoustical calibration of the instrument system.

4. Take octave-band measurements of a known convenient broadband steady noise and compare with previously recorded data.

5. If possible, disconnect the microphone and take electrical circuit (instrument noise) measurements.

6. Recheck the acoustical calibration of the instrument system.

If a malfunction of the sound measuring equipment becomes evident, consult the instruction manual supplied with the instrument for necessary adjustments or repairs. Step 4, for example, can reveal a defective octave-band filter. Step 5 will indicate a faulty oscillating amplifier. Step 6 could indicate weak batteries if the indicator dial reading has changed by more than one decibel. If the reading has changed, the batteries should be replaced before making sound level measurements. If the instrument cannot be made to operate properly, consult the manufacturer.

Care of equipment

The sensitivity of a microphone can be expected to change when it is exposed to extremely high or low temperatures. If a microphone is to be used in an unusual temperature environment, calibration information should be obtained at comparable temperatures.

The presence of high humidities can cause condensation to occur in electrical connections. Certain types of condenser microphones malfunction in the presence of high humidity and cause a "popping" or "frying egg" sound to appear at the microphone output. Most of these humidity problems disappear when the equipment is dried out.

At high noise levels, vibration may be present to such an extent that it will affect the instrument readings. It may be necessary to remove the instrument from the area and use an extension cable to the microphone. A test to determine the effect of vibration on the sound level meter is to disconnect the microphone and observe if a high reading is obtained on the meter dial. If a high reading is noted, the meter can be hand held or placed on a rubber pad or rubber-tired cart.

Measurement procedures

Before discussing detailed measurement procedures, several rather important facts should be clearly understood. First of all, with accurate, calibrated sound measuring equipment measuring reasonably steady broadband noise, the accuracy of the instrument readings will probably lie within the range of plus or minus two decibels. In other words, no attempt should be made to read the sound level meter dial to a fraction of a decibel. In fact, unless the noise source is remarkably steady or is a pure tone noise with no nodal patterns, an interpolation of the needle swings will be necessary even when using the slow-response scale. Occasionally a peculiar combination of sound waves from the noise source and reflections from hard walls will produce a noticeable standing wave pattern. The sound level or dial reading will increase or decrease as one moves the microphone a few feet toward or away from the source. If this periodic pattern of rise and fall in readings occurs, the average of the maximum and minimum readings should be recorded.

During a sound level measurement, the dial readings tend to fluctuate due to variations in sound intensity. The range of values for each measurement is observed and recorded. For example, if during a noise level measurement with the instrument switched to slow response the needle fluctuates between 90 and 94 dBA, the range would be reported as 90 to 94 dBA.

If the noise level fluctuates markedly, because of sudden erratic impulse noises, an error of several decibels in readings can result depending on which value is recorded.

Most sound level meters and analyzers are provided with both fast- and slow-response characteristics, either of which can be used by operating a switch. The design of the response characteristics follows specifications of ANSI S1.4-1971.

The slow response or movement is such that the dial needle motion is damped to permit an average to be made of fluctuating noise levels. When a steady noise is measured, the same meter reading should be obtained with either the fast or slow movement.

When a random noise is measured, the first important result that is desired is the long-time average energy level. If the meter needle fluctuations are less than about 2 dB, this average can be easily estimated to a fraction of a decibel. If the fluctuations cover a range of 10 or more decibels, choosing the average is much less certain. The

Fig. 6-5.—Measuring and recording sound level during a grinding operation.

extent of the meter needle fluctuation depends on the meter characteristic. Thus, if the fluctuations exceed 3 or 4 dB for the fast meter position, the slow meter position should be used.

If, for some reason, the sound level meter is not used exactly as recommended in the equipment instruction manual or the inherent limitations of the instruments are not recognized, significant errors in noise measurements may result.

Even in a hearing conservation noise survey, the purely mechanical aspects of making the survey are not as simple as they may seem. The difficulties involved in following a man closely during his workday to account for on-time, off-time periods of noise exposure as well as recording fluctuations in the noise levels to which he is exposed add to the problems involved in performing a realistic sound survey (see Figure 6-5).

There is no substitute for having experience in using an instrument. This chapter cannot provide the familiarity that can be gained just by sitting down with an instrument and its instruction book and becoming familiar with the instrument.

Much valuable information is contained in the equipment instruction book. For example, the instruction manual will specify maximum and minimum sound pressure response levels, humidity limits, microphone directional characteristics, effects of electric and magnetic fields, the effects of the observer, and effects of vibration upon instrument readings.

The accuracy of sound level measurements depends upon the manner in which the instruments are used and their operating condition. The immediate recognition of malfunctioning equipment is essential if accuracy is to be maintained.

The general area in which the sound survey is to be made should be investigated to determine the type and number of sound level measurements that are necessary. The possible effects of shielding of sound waves, particularly at high frequencies, should be kept in mind during this preliminary investigation.

Keep this in mind while making the survey—that unless the noise being measured is steady-state broadband, such as that generated by ventilating systems, blowers, and the like, there will be some variation in instrument readings, depending upon distance and the angle of the noise source in relation to the microphone. The microphone pressure-response diagram appearing in the previous chapter illustrates this point.

Microphone calibration is based on random incidence of the sound waves upon the sensitive element of the microphone. If the microphone is only a few feet from the noise source, or if the noise is coming from only one direction, it is necessary to position the microphone so that the sound path is perpendicular to the side or long axis of the microphone.

Do not hold the sound level meter so that your body reflects sound waves toward the microphone. Discourage people from crowding around to see the instrument because they also can partially reflect or block the incident sound waves. Try to obtain sound level measurements in such a way that the noise source is not directly in line with the longitudinal axis of the microphone but as close to a 90-degree angle as possible. Also, it is important to note that noise level readings at a given distance from a noise source may vary depending on the directionality characteristics of the noise source.

Almost all surfaces will reflect some sound waves. Therefore, measurements taken within most rooms will vary from one point to another. As a general rule, any object will reflect sound waves having wavelengths comparable to or smaller than the dimensions of the object.

Some measurements made for engineering noise control purposes may require that the reverberant levels, which include reflected sound from walls, be measured rather than direct levels which are obtained relatively

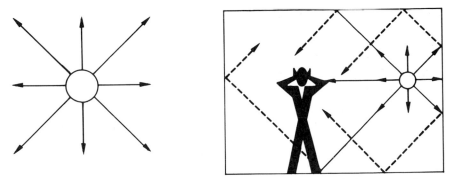

Fig. 6–6.—In free space, a sound radiates uniformly in all directions. In an enclosure, the sound is reflected from the walls.

close to the noise source.

The sound waves at the measuring location may have arrived there in a direct path from the noise source radiating surface and/or an indirect path as reflections from hard ceilings, walls, and other objects (see Figure 6–6). In a small room, or enclosed space, the reverberant sound levels due to reflection will be relatively high. In a very large, high-ceiling room or relatively open space, most of the sound energy measured at the microphone comes directly from the noise source. In an open field (free field) the reduction in measured sound level is about 6 dB each time the distance from the noise source to the microphone is doubled, while a measured reverberant sound level inside a room would be fairly constant throughout the space.

Extremely large noise sources may have sound radiating surfaces that are difficult to localize. In such cases, the microphone may be inserted in a megaphone to assist in pinpointing the direction of the most prominent source of the sound. Listening to headphones plugged into the sound meter will assist in localizing the maximum intensities.

When sound generating equipment is mounted on a wood floor or wall, the floor or wall is likely to emit a considerable portion of the total noise. In most cases, the total sound level (equipment plus background noise) should be measured since it is this condition to which personnel are exposed.

For some noise control purposes, however, the noise of a single machine, operating in a noisy background, often, is desired (see Figure 6–7). In these cases, measurements generally are made relatively close to the machine, and its sound level alone can be calculated by subtracting the background noise by means of a correction (described in the previous chapter).

If it is desired to measure general noise levels including reverberation or reflection effects, then the microphone distance from the source should be relatively far—10 to 20 ft. If only the direct sound from a noise source is to be measured, then a distance such as 1 to 5 ft from the source may be more suitable. If the machine is unusually large and the ceiling and walls are farther away than usual, a greater distance is preferable. Sound level measurements taken outdoors or in large enclosed spaces usually are not affected by reflection from ceiling and walls.

As mentioned previously, rapid air movement past a microphone will give inaccurate readings. If, for example, it is necessary to perform a noise survey in a hot area where man-cooler fans are being used, the microphone should be protected by a wind screen. However, even with a wind screen, there is a limiting air velocity above which errors will occur. Ordinarily, wind screen characteristics are supplied by the manufacturer. Wind screens are available commercially or can be made with simple materials (see Chapter 5).

Before entering the noisy area in which the survey is to be conducted, make a last minute check for the condition of the batteries and the microphone. Repeat the checks during the day if the survey requires several hours.

If the level of noise to which an employee is exposed is to be deter-

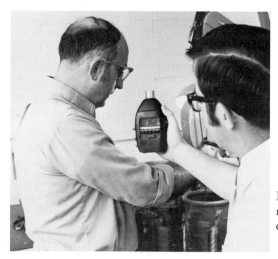

Fig. 6–7.—Taking sound level measurements of a single machine.

Courtesy General Radio Co.

Fig. 6–8.—Sound level measurements should be taken near an employee's work station.

mined, the microphone should be placed in his general working area. A workman often is within 4 or 5 ft from his major noise source and he usually moves around in this general area. Under these conditions, the sound level will be substantially the same for this employee's work area (see Figure 6–8). However, if the workman is closer than several feet from the noise source for long periods of time, then the microphone should be more nearly the same distance to the source as the employee's ear.

An effort should be made to measure the sound level at the employee's work station that is typical of the area in which he works. If the man is doing something like operating a chipping hammer, greater accuracy can be obtained if the microphone is mounted on a tripod and an extension cable used in order to put some distance between the microphone and the instrument case. However, this is not practical in most work situations and, unless a dosimeter is available, the only way to obtain noise level readings is to use the survey meters in the immediate work area.

Measurements should be taken with the instrument held approximately 2 to 3 ft from the person's head to prevent attenuation of the

187

sound by the man's body. Where this is impossible, the measurement should be taken the same distance from the principal noise source as the workman's head. Measurements are made for each distinct operation where employees are exposed to different noise levels.

When hand held during a noise measurement, the sound level meter should be held at comfortable arm's length with the microphone pointed approximately 90 degrees from the noise source. Perhaps the best position is to hold the instrument with the axis of the microphone pointed toward the ceiling.

For a hearing conservation survey, a sufficient number of measurements should be made at the employee's hearing zone to make sure that the sound level reported is typical of the person's exposure. For example, the sound level at a machine tool operator's normal working position should be measured at different times during the operation—several readings at the beginning, several midway through the process, and several at the end of the job. The sound exposure for the entire operation should then be the algebraic summation of these time-weighted levels.

The exposure times can be determined by either a full-shift exposure time for each employee, or calculated based on his partial exposure time and the number of repetitions of the cycle of operations during a full shift.

Statistical sampling techniques

The accuracy of personal noise monitoring, whether using an audio dosimeter or sound level meter, is a function of the statistical sampling of the work environment. D. O. Conn[2] describes a method for determining the number of instruments needed and number of days that sampling is required for valid results to be obtained.

Because noise levels in most work environments may vary considerably from day to day, Conn suggests that these variations lend themselves to a statistical approach. Thus, by calculating the standard deviation of the sound pressure level variations in the environment over a period of time, it is possible to establish limits of measurement precision. Being able to calculate variances in the sampling instrument allows an acceptable statistical estimate of how many people must be sampled over what period of time. The following material adapted from Conn illustrates the calculations for this procedure.

Any monitored work environment will vary in noise level from day to day and the standard deviation of this variation is defined as σ_w.

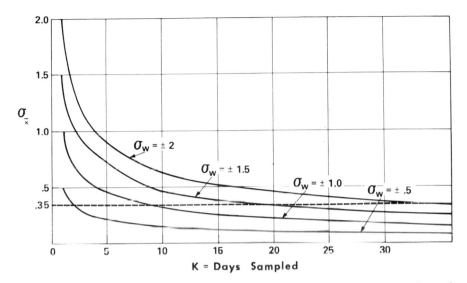

Fig. 6–9.—Plot of precision of measurements in dB against the number of days sampled.

Excluding variances of the instrument, the precision of the measurements ($\sigma_{\bar{x}}$, the standard deviation of the mean of the measurements) is described in Equation (1).

$$\sigma_{\bar{x}} = \frac{\sigma_w}{\sqrt{K}} \tag{1}$$

where: $\sigma_{\bar{x}}$ = precision of measurement in dB

σ_w = standard deviation of work environment in dB

K = number of days sampled

Figure 6–9 is a plot of $\sigma_{\bar{x}}$ versus K, where $\sigma_w = \pm 0.5, 1, 1.5,$ and 2 dB respectively. A tolerance of ± 0.7 dB in the measurement has approximately a ± 10 percent effect on percent OSHA exposure (see Table 6-B). This being the case, to be 95 percent confident of being within ± 10 percent of the OSHA value, $2\sigma_{\bar{x}}$ should equal 0.7 dB. In Figure 6–10, if $\sigma_{\bar{x}} = \pm 0.35$ dB, we find that for a $\sigma_w = \pm 0.5$ dB, two days of sampling are required; for $\sigma_w = \pm 1$ dB, nine days; for $\sigma_w = \pm 1.5$ dB, 19 days; and for $\sigma_w = 2$ dB, 32 days. These represent the minimum number of samples to be taken for a precision of ± 0.35 dB.

To select the number of days to be sampled and the number of instruments to be used for a given work function, Equation (2) takes into account not only variances of the work environment but those of the

TABLE 6-B
THE EFFECT OF dB ERROR
ON PERCENT OSHA EXPOSURE MEASURE

dB Error	Percent OSHA Reading when Expecting 100%	dB Error	Percent OSHA Reading when Expecting 100%
+6.5	246.2	− .1	98.6
+6.0	229.7	− .2	97.3
+5.5	142.4	− .3	95.9
+5.0	200.0	− .4	94.6
+4.5	186.6	− .5	93.3
+4.0	174.1	− .6	92.0
+3.5	162.4	− .7	90.8
+3.0	151.6	− .8	89.5
+2.5	141.4	− .9	88.3
+2.0	132.0	−1.0	87.1
+1.5	123.1	−1.5	81.2
+1.0	114.9	−2.0	75.8
+ .9	113.3	−2.5	70.7
+ .8	111.7	−3.0	66.0
+ .7	110.2	−3.5	61.6
+ .6	108.7	−4.0	57.4
+ .5	107.2	−4.5	53.6
+ .4	105.7	−5.0	50.0
+ .3	104.2	−5.5	46.7
+ .2	102.8	−6.0	43.5
+ .1	101.4	−6.5	40.6
0	100.0		

measuring instruments as well.

$$\sigma_{\bar{x}} = \frac{\sigma_w^2}{K} + \frac{\sigma_i^2}{nK} \qquad (2)$$

Condition: $\dfrac{\sigma_i}{\sqrt{n}} \leqq \sigma_{\bar{x}}$ This condition is necessary for $\sigma_{\bar{x}}$ calculations to be meaningful

where: n = number of instruments

σ_i = standard deviation of measurement instrument

It is important to point out that these calculations are necessary to ensure valid data, regardless of whether dosimetry or conventional measuring techniques are employed.

Example

In order to obtain personal noise exposure data on a group of employees in the same work environment performing similar jobs, how many dosimeters do we use and how many days do we sample?

If there is a 68 percent confidence that the work environment will not change by more than ±0.5 dB from day to day, then $\sigma_w = \pm0.5$ dB. Assuming the sound is equally distributed between 20 and 8000 Hz, $\sigma_i = +0.8$ dB (from Table 6-C). As stated earlier, to be 95 percent confident the answer is within ±10 percent of the actual OSHA value, $\sigma_{\bar{x}}$ should equal ±0.35 dB.

TABLE 6-C

SELECTION OF σ_i* FOR DU PONT DOSIMETER

σ_i	Sound Field Frequency Range
±0.75	100–8000 Hz equal frequency distribution
±0.80	20–8000 Hz equal frequency distribution
±1.5	Predominantly 5000–8000 Hz
±1.0	Predominantly 2500–4000 Hz
±0.4	Predominantly 100–2500 Hz
±0.8	Below 100 Hz

* These are approximations and the σ_i values of any Type 2 sound level meter will not be significantly different.

Sample calculations utilizing equation (2):
where:

$$\sigma_{\bar{x}} = \pm0.35 \text{ dB}$$
$$\sigma_w = \pm0.5 \text{ dB}$$
$$\sigma_i = \pm0.8 \text{ dB (taken from Table 6–C)}$$

condition: $\dfrac{\sigma_i}{\sqrt{n}} = \dfrac{0.8}{\sqrt{n}} \leq \sigma_{\bar{x}} = 0.35$

Solving condition yields $n \geq 5.2$. Rounding to nearest whole number, $n = 5$.

$$\sigma_{\bar{x}} = \sqrt{\frac{\sigma_w^2}{K} + \frac{\sigma_i^2}{nK}}$$

$$\pm0.35 = \sqrt{\frac{(0.5)^2}{K} + \frac{(0.8)^2}{5K}}$$

Solving equation yields K = 3.

Answer: Use five instruments to sample five employees for three consecutive days.

In many practical applications, sufficient information is not available to make accurate estimates of σ_i and σ_w. For these cases, several dosimeters can be used for several days to obtain data which can be easily analyzed for σ_i and σ_w through common statistical procedures. Both the precision of the data already obtained and the sampling procedure required to obtain any other precision can now be calculated using these newly obtained values of σ_i and σ_w.

No matter what the accuracy of the instruments, whether dosimeters or sound level meters, the validity of the measurements is dependent on the use of proper sampling techniques. It is apparent, therefore, that the need for a sampling standard ranks equally with instrument performance standards such as S1.4–1971.

Data and records

In addition to recording noise levels, it is important to record as much meaningful information as possible about the noise source, instrumentation, and the environment being measured—such items as:

1. A brief description of the noise source or sources

2. The number of employees exposed

3. Duration of exposure

4. Characteristics of the environment with respect to its ability to reflect, dissipate, or absorb noise

5. The instruments used to make the measurements

6. Calibration devices

7. The type of measurements, either A-scale or octave-band analysis

8. Methods of sampling in time and space

Measurements should be appropriate to the type of sound being analyzed. Impulse noise requires special measurement techniques, while the so-called steady noises can be measured with sound level meters.

In your notes, describe the exact conditions while making the measurements so that they can be repeated if necessary.

The data recorded will depend on the purpose of the survey—that is, whether it is a hearing loss, engineering noise control, speech interference, or annoyance problem. Because, to a certain extent, each noise survey is unique, the report must fit the requirements of the particular situation. There are, however, certain essential data that must be recorded in almost all cases.

Use data sheets designed specifically for noise problems to make sure that the desired data will be taken and recorded. Here are the important items that will be helpful in preparing data sheets:

1. The time and date of the survey and name of the person conducting the study.

2. Description of equipment used for sound measurements and type of serial numbers of all microphones, sound level meters, and analyzers.

3. Results of acoustical calibration of sound-measuring equipment.

4. The meter speed—slow or fast response, and weighting network used.

5. Description of space in which the measurements are made, such as the dimensions and nature of floor, walls, and ceiling, and location of doors and windows.

6. Description of the (primary) noise source under study; include the main characteristics of the noise to be measured (steady noise, transient noise, impulse noise, broadband noise, pure tones, etc.). Note the most prominent characteristics observed in this way. Include a clear description of the machine (size, nameplate data, speed and power rating, type of operation and operating conditions), the number of machines in operation, the location of the machines, and the type of mountings.

7. Description of secondary noise sources; include location and type of operation.

8. Background, overall, and band levels at each microphone position. (Device under test should not be operating.)

9. Measured overall and octave-band levels at each microphone position with machine in question in operation and the extent of meter pointer fluctuations.

10. Position of the microphone and the direction of arrival of the sound wave with response to the long axis of the microphone. Notation as to tests for standing wave patterns and for the decay of sound level with distance.

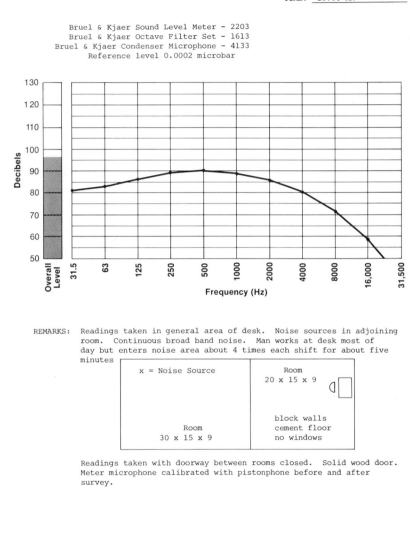

Fɪɢ. 6–10.—An example of the results of a noise survey plotted on a graph. Note diagram of location in relation to work area and employee work station.

194

11. Temperature at microphone location (not critical unless extreme cold or heat).

12. Record the position of the observer with respect to the noise source and the sound measuring equipment.

13. Time pattern of the noise—that is, whether continuous or intermittent and impact and duration of the exposure.

14. Personnel exposed, directly and indirectly. Identify their location with respect to the principal noise sources.

15. Noise control measures in use, including the types of ear protectors and whether they are properly worn.

Experience is necessary to determine those items which are most valuable in a noise control program. The degree of detail is usually dictated by the use which will be made of the data collected. If it is for legal records, there cannot be too much detail. If it is for engineering control purposes, different data may be necessary. It is well to keep in mind that the simpler the form, the easier it is to gather the data.

The kind of data sheet used to record noise measurements varies according to the preferences of the individual making the survey and the kind of information he wishes to record. The data may also be presented graphically as illustrated in Figure 6–10. Special graph paper can be used to plot readings as the survey is being conducted. Figures 6–11 and 6–12 are examples of data sheets that can be used for recording data. As indicated in Figure 6–10, it is often desirable to make a rough sketch of the situation. In many cases the environment being measured will be much more complex than the simple example shown and the sketch will be of great help in reporting the data.

If it is necessary to record a large number of noise measurements, the data can be handled as shown in Figure 6–12. In this example, a blueprint of the work area could be used to locate areas where readings are taken. The blueprint could then become a part of the report. Note that in cases where the arithmetic average of the levels in the speech frequencies exceeds 85 dB, the readings are circled to emphasize the problem.

It is necessary to record as accurately as possible the length of exposure of employees to various sound pressure levels. For a dBA survey the data sheet shown in Figure 6–10 is quite helpful. In addition, there are many times when a stopwatch is useful in a noise survey. It is

(*Text continues on page 198.*)

SOUND SURVEY

Page ___ of ___

Date: _____ Time: _____

Wind Velocity: _____ Wind Direction: _____ Temperature: _____

Sound Level Meter: Type _____ Model _____ Serial No. _____

Microphone: Type _____ Cable Length _____

Analyzer: Type _____ Model _____ Serial No. _____

Other Equipment: _____

_____

Location: _____ Sketch

_____

_____

_____

Location or Situation	Over-all Level	Sound Pressure Levels - Re 0.0002 Microbars Octave Band Center Frequencies (Hz)								
		63	125	250	500	1000	2000	4000	8000	8-Octave added

Remarks: _____

_____

_____ Recorded by: _____

From Industrial Noise—A Guide to Its Evaluation and Control, *Public Health Service Publication No. 1572.*

Fig. 6–11.—A sample of a sound survey form.

INSTRUMENT
Bruel & Kjaer Sound Level Meter - 2203
Bruel & Kjaer Octave Filter Set - 1613
Bruel & Kjaer Condenser Microphone - 4133
Reference level 0.0002 microbar
NOISE SOURCE LOCATION NUMBERS

Freq.	NOISE SOURCE LOCATION NUMBERS							
	1	2	3	4	5	6	7	8
Lin	98	93	102	87	89	94		
31.5	84	82	86	78	71	73		
63	85	85	88	74	72	74		
125	88	86	93	76	75	78		
250	92	88	94	78	79	82		
500	92	87	97	81	85	89		
1000	92	84	95	80	82	90		
2000	86	80	90	76	80	85		
4000	83	74	86	69	74	78		
8000	75	70	82	60	68	72		
16,000	65	58	70	52	60	58		
31,500	—	—	—	—	—	—		

1. General area readings just north of office near column 29; Three people working in this area four hours per day. Noise coming from sorting machines.

2. Same noise source as "1 above - readings taken at shipping dock near column 25 one person exposed 8 hrs/day.

3. Readings taken approximately 10 feet due north of sorting machines - no one directly exposed at this point.

4. Readings near foreman's desk - dept. 10, column 40 - typical activity

5. Levels observed in small utility room, extreme west end of building - no one working in this area continuously.

6. Area adjacent to tool crib in dept. 5 - Several people spend 8 hours per day in this general area.

FIG. 6–12.—Typical data sheet used to report results. Circled readings indicate values that exceeded 85 dB in the speech-important frequencies.

obvious that exposure time and dBA readings should both appear on the data sheet if survey results are to have any meaning.

Noise levels can be recorded as numbers, or plotted on a prepared graph as readings are taken. Because it is easy to misread the dials on a sound level meter, plotting the results of an octave-band analysis survey on a graph would immediately indicate obviously incorrect read-

TABLE 6-D

ADDITION OF
SOUND LEVELS

If the Difference in Decibels Between the Two Levels Being Added Is	Add to the Higher Level
0	3.0
1	2.6
2	2.2
3	1.8
4	1.4
5	1.2
6	1.0
7	0.8
8	.6
9	.5
10	.4
13	.2
16	.1

ings. However, if one is working alone, it may be easier to record numbers in a column than to juggle a meter, take readings, and plot data on a graph all at the same time.

To avoid mistakes in an octave-band analysis, review the readings as soon as all octave bands have been measured. Ordinarily, a reading that is out of line by ± 10 dB or more will be obvious and a recheck can be made immediately. Be sure that no individual octave-band level equals or exceeds the overall level because this is an indication that one or the other of these two readings has been recorded incorrectly. Another check is to add readings in the octave bands to determine if the sum approaches the overall reading. Remember, however, this is not a simple arithmetical addition. The readings are logarithmic in nature

TABLE 6-E
EXAMPLE OF ADDING SOUND LEVELS

Step, or sequence of addition	Frequency (Hz)	Octave band noise level reading in decibels	Noise levels to be added	Difference between levels	dB to be added from Table 6-D	Addition	Sum
	Overall Reading	101					
7	31.5	83	100.8 + 83	17.8	0.1	100.8 + 0.1	100.9
6	63	87	100.6 + 87	13.6	0.2	100.6 + 0.2	100.8
4	125	91	99.6 + 91	8.6	0.5	99.6 + 0.6	100.2
2	250	94	96.0 + 94	2	2.2	96.0 + 2.2	98.2
1	500	96	96				
3	1000	94	98.2 + 94	4.2	1.4	98.2 + 1.4	99.6
5	2000	89	100.2 + 89	11.2	0.4	100.2 + 0.4	100.6
8	4000	83	100.9 + 83	17.9	0.1	100.9 + 0.1	101.0
9	8000	74	101.0 + 74	26.9		INSIGNIFICANT	
10	16000	63	101.0 + 63			"	
11	31500	47				"	

NOTE: Start with highest band level (Step 1) and add to second highest band level as shown in Step 2. Then add this result to third highest octave band level as shown in Step 3 and continue until the contributions from each of the bands are summed up. If two octave band levels have the same value, treat as separate readings, as in steps 2 and 3. (Note: Steps 9, 10, and 11 may be disregarded.)

and the rules for addition shown in Table 6-D must be observed.

If the sum differs considerably (= 2 decibels) from the overall level, something is wrong; perhaps there is an error in addition. Also, if the sum is greater than the overall level there is the possibility that high levels of impact noise are present in the noise being measured. Table 6-E illustrates the application of this summation procedure.

A slide rule, called the "Sound Level Calculator," is available from the National Safety Council. This calculator simplifies and speeds up the addition of decibel values.

Summary

A careful study of this chapter should enable the user of sound measurement instruments to perform thorough and meaningful sound measurements in most of the situations encountered in industrial environments.

The selection of instrumentation for a sound survey is a very important task. A basic understanding of the nature of the problem is

COMPANY NOISE SURVEY

PLANT **EAST SIDE**

DATE OF SURVEY **APRIL 19**

PAGE **1**

INSTRUMENT USED: (Reference level
0.0002 microbar)

1. Bruel & Kjaer Sound Level Meter - 2203
2. Bruel & Kjaer Octave Filter Set - 1613
3. Bruel & Kjaer Condenser Microphone - 4131
4. General Radio - 1565A
5. General Radio - 1556B

Item Number	dBA	REMARKS (Location, Exposure time, etc.)	* Hrs./Mins.
1	86	IN AISLE - COLUMN 1220 NEAR SCREW MACHINES (WITH QUIET TUBES)	8
2	88-90	IN GENERAL AREA OF AUTO. SCREW MACHINES - WORK AREA 5 PEOPLE WORKING EIGHT HOURS	8
3	83-85	IN AISLE NEAR COLUMN 1423	8
4	94-96	WORK AREA BETWEEN ACME-GRIDLEYS ONE MAN — WEARING PLUGS EIGHT-HOUR JOB	4
5		GLEASON NEAR COLUMN 1221	
	86	COMPRESSED AIR OFF	
	98	" " ON	
		(NOT A REGULAR WORK STATION)	
6	90	COLUMN 1220 - WORK STATION NEAREST GLEASON (ABOVE ITEM)	8
7	100	RUN-IN TEST AREA - INSIDE ENCLOSURE	2
	88	IN AISLE JUST SOUTH OF ENCLOSURE	—

* Permissible Exposure Limits to This Sound Pressure Level

FIG. 6–13.—Sound survey form shows the sound level and length of exposure.

needed in addition to a fairly good idea of the type of noise or vibration that is involved. The amount of information about the noise that you will need determines the amount of time that you can spend collecting data and then analyzing it, either in the field or in the laboratory.

One should know the limitations and applications of the various types of instrumentation available. If all of the sound measuring instruments mentioned in this chapter are available, then the problem is one of selecting the proper instruments for the job at hand. If, as is usually the case, only a limited amount of noise measuring instrumentation is readily available, then the problem is one of recognizing the limitations of the particular instrument with reference to the noise problem being studied.

References

1. Karplus, H. B., and G. L. Bonvallet, "A Noise Survey of Manufacturing Industries." *American Industrial Hygiene Association Quarterly*, Vol. 14, No. 4.
2. Conn, D. O., III, "The Audio Dosimeter." *National Safety News*, Vol. 106, No. 4.

Bibliography

Beranek, L. L. *Acoustics*. New York, N.Y.: McGraw-Hill Book Company, 1954.

————. *Noise Reduction*. New York, N.Y.: McGraw-Hill Book Company, 1960.

Botsford, J. H. "A New Method for Rating Noise Exposures." *American Industrial Hygiene Association Journal* 28:431–40.

"Guidelines for Noise Exposure Control." *American Industrial Hygiene Association Journal* 28:418.

Harris, C. M. ed. *Handbook of Noise Control*. New York, N.Y.: McGraw-Hill Book Company, 1957.

Hosey, A. D., and Powell, C. H., eds. *Industrial Noise—A Guide to its Evaluation and Control*. (Public Health Service Publication No. 1572, U.S. Dept. HEW). Washington, D.C.: U.S. Government Printing Office, 1967.

Industrial Noise Manual. American Industrial Hygiene Association, Akron, Ohio, 1966.

Kryter, K. D.; Ward, W. D.; Miller, J. D.; and Eldridge, D. H. "Hazardous Exposure to Intermittent and Steady State Noise." *Journal of the Acoustical Society of America* 39:451–64.

Peterson, A. P. G. and Gross, Jr., E. E. *Handbook of Noise Measurement*. West Concord, Mass.: General Radio Co., 1967.

"Sound Level Calculator," Stock No. 129.94–9. Chicago, Ill.: National Safety Council.

Part **3** **Effects of
Noise on Man**

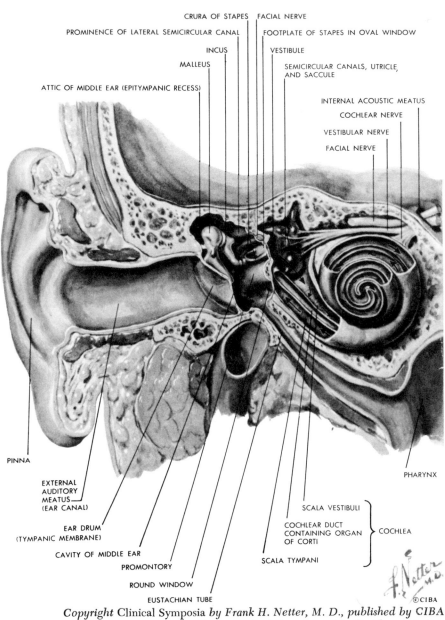

CRURA OF STAPES   FACIAL NERVE

PROMINENCE OF LATERAL SEMICIRCULAR CANAL

FOOTPLATE OF STAPES IN OVAL WINDOW

INCUS

VESTIBULE

MALLEUS

SEMICIRCULAR CANALS, UTRICLE, AND SACCULE

ATTIC OF MIDDLE EAR (EPITYMPANIC RECESS)

INTERNAL ACOUSTIC MEATUS

COCHLEAR NERVE

VESTIBULAR NERVE

FACIAL NERVE

PINNA

PHARYNX

EXTERNAL AUDITORY MEATUS (EAR CANAL)

SCALA VESTIBULI

EAR DRUM (TYMPANIC MEMBRANE)

COCHLEAR DUCT CONTAINING ORGAN OF CORTI

COCHLEA

CAVITY OF MIDDLE EAR

PROMONTORY

SCALA TYMPANI

ROUND WINDOW

EUSTACHIAN TUBE

Copyright Clinical Symposia *by Frank H. Netter, M. D., published by* CIBA *Pharmaceutical Company. Used with permission.*

FIG. 7–1.—An illustration of the outer, middle, and inner ear.

Chapter Seven

Anatomy, Physiology, and Pathology of the Ear

Jack D. Clemis, M.D.

Anatomy of the human auditory mechanism is exceedingly complex. Its minute size and encapsulation in hard, dense bone complicates the study of this delicate mechanism. Thus, there remain many unknown aspects of how the ear functions, particularly the inner ear and the pathways that lead to the brain. Diseases of the ear and disease states of the body that affect the ear or its function is the medical specialization of Otology that requires years of intensive study.

This chapter is intended to provide the reader with an overview of the structure of the ear, how it seems to function, and the more common causes of impairment of this sensitive mechanism.

This subject is presented in a manner that does not require a medical background. References are included at the end of this chapter for those who wish to pursue in greater depth the subjects of anatomy, physiology, and pathology of the auditory mechanism.

The organ of hearing enables an individual to detect sound waves within a range of 20 to 20,000 Hz and convert them into electrical impulses that are transmitted to the brain for interpretation. This structure can be divided into three parts—the outer or external, middle, and inner ear (see Figure 7–1).

This chapter discusses each part in turn and tells how it works and what can damage it.

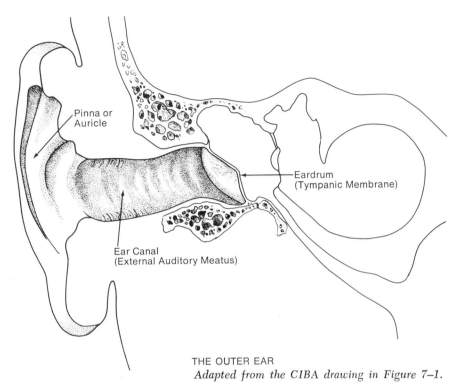

Pinna or
Auricle

Eardrum
(Tympanic Membrane)

Ear Canal
(External Auditory Meatus)

THE OUTER EAR
Adapted from the CIBA drawing in Figure 7–1.

Fɪɢ. 7–2.—The outer ear shown here can be divided into the auricle (or pinna) and the external auditory canal.

External ear

Anatomy

The outer (external) ear is divided into two sections—the portion seen on the outer surface of the head (called the *auricle* or *pinna*) and the *external auditory canal* or ear canal (see Figure 7–2). The pinna is constructed from a convoluted mass of cartilage covered by skin. The external auditory canal is a skin-lined pouch supported in its outer third by the cartilage of the pinna and in its inner two-thirds by bone of the skull. At its innermost end lies the tympanic membrane or eardrum that separates the external from the middle ear.

The small hairs or *vibrissae* and *ceruminal glands* that secrete waxy substance called cerumen are located in the skin of the outer third of the ear canal. The function of the hairs is to filter out particulate matter

206

and other large foreign bits of debris and they are therefore protective. Cerumen, both sticky and bactericidal, prevents smaller particles from entering the ear canal and also keeps a healthy canal free from infection.

Physiology

The function of the outer ear in the hearing process is relatively simple. The external portion of the ear collects sound waves from the air and funnels them into the ear canal where they are transported to the eardrum. The collected sound waves cause the eardrum to move back and forth in a vibrating mechanical motion that is passed on to the bones of the middle ear.

Pathology

Disorders of the auricle include congenital malformations (in which the cartilage is misshaped) and protruding or lop ears, both of which may be surgically corrected. The auricle is the most common site in the ear for malignancies. Dermatitis and infection are common in this area.

The external ear canal is prone to infection because of the increased skin temperature and humidity. In this area, bacterial infections are most common; while fungus (otomycosis)and viral causes are less frequent. Skin disorders (dermatitis) are also frequent canal problems. Generalized skin disorders, usually of the scalp, may extend to the outer ear.

An abnormal narrowing of the ear canal is called a *stenosis* and may be caused by infection or injury. Tumors are more rare in this area, the most common being benign bony masses or osteomas. Malignancies are infrequently seen. A variety of foreign bodies find their way into the external ear canal; these objects should be removed by the physician or the otologist (ear specialist).

Normally, ear canals are self-cleansing but occasionally a failure of this mechanism results in wax impaction. The use of cotton-tipped swabs for cleaning purposes and the use of earplugs tend to pack wax into the ear canal. The external ear canal must be almost totally occluded (blocked) before attenuation of sound occurs. If there is an opening in the ear canal as small as the diameter of one red blood cell, sound can get through to the eardrum and normal hearing can be maintained. Consequently, before any loss in hearing can occur from ear wax, there must be a considerable quantity involved. Wax impaction should preferably be cleaned by an otologist because of the risks of injury and infection inherent in the cleaning process.

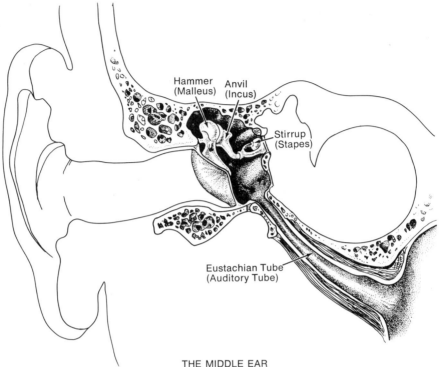

Hammer (Malleus)

Anvil (Incus)

Stirrup (Stapes)

Eustachian Tube (Auditory Tube)

THE MIDDLE EAR
Adapted from the CIBA drawing in Figure 7–1.

Fig. 7–3.—The middle ear shown here is contained within the temporal bone and is made up of the Eustachian tube, the middle ear space, and the mastoid air cell system.

Middle ear

Anatomy

The middle ear consists of the Eustachian tube, the sound conducting mechanism, and the mastoid air cell system (see Figure 7–3).

Eustachian tube. This is a mucous membrane-lined tube extending between the middle ear space and the upper part of the throat behind the soft palate. Like the external canal, the tube is supported in part by bone and in part by cartilage. The tube is normally closed. But contraction of the palate muscles during yawning, chewing, or swallowing opens the tube, establishing ventilation of the middle ear and equalization of middle ear pressure.

Failure of the Eustachian tube to ventilate results in a vacuum in the middle ear space which in turn causes one of two pathological events to occur: (*a*) It pulls fluid into the middle ear, resulting in a condition called nonsuppurative otitis media (noninfectious inflammation of the middle ear); (*b*) it pulls the eardrum inward until the air space has collapsed (atelectasis) or until a little balloon of eardrum is sucked into the middle ear. As the balloon grows, it becomes destructive to the contents of the middle ear space. It is serious, potentially dangerous, and is called a *cholesteatoma*.

The opposite condition, which is uncommon, is a patent Eustachian tube in which the tube constantly remains open. However, if it does occur, the results are an annoying symptom of hearing one's own voice and breath sounds (autophonia) in the involved ear.

Middle ear space. The space or cavity approximately one to two cubic centimeters in volume lying between the eardrum and the bony wall of the inner ear is called the middle ear or *tympanum*. The middle ear is lined with mucous membrane essentially the same as that which lines the mouth. Within the middle ear cavity are located the *ossicles*, the smallest bones in the body, which connect the eardrum to an opening in the wall of the inner ear called the oval window (see Figure 7–4).

The middle ear space is best conceived as a box with six sides or a cube:

1. The outer wall is formed by the eardrum.

2. The inner wall is the bony partition separating the inner ear from the middle ear. The round and oval windows fit into this wall and include the only two movable barriers between the middle and inner ear.

3. Front wall opens into the Eustachian tube.

4. Back wall opens into the mastoid air cells.

5. Roof separates the middle ear from the temporal lobe of the brain.

6. Floor separates the middle ear from the jugular vein and the internal carotid artery which lie high in the neck.

The sound conducting mechanism is housed in the middle ear and includes the eardrum and three ossicles (bones) that are supported by

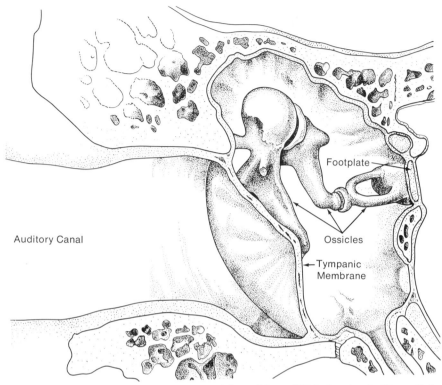

Adapted from CIBA drawing in Figure 7–1.

FIG. 7–4.—The ossicles, located within the middle ear cavity, link the eardrum to an opening in the wall of the inner ear (the oval window).

ligaments and moved by two muscles.

The eardrum. This is normally a transparent membrane that separates the external ear canal from the middle ear. It consists of an inner layer of mucous membrane and a middle layer of fibrous tissue. It is spider web in form, with radial and circular fibers for structural support.

Ossicular chain. The ossicles, which together are called the ossicular chain, are the malleus, incus, and the stapes (see Figure 7–5).

• The *malleus* or *hammer* is fastened to the eardrum by the handle while the head lies in the upper area of the middle ear cavity and is connected to the incus.

210

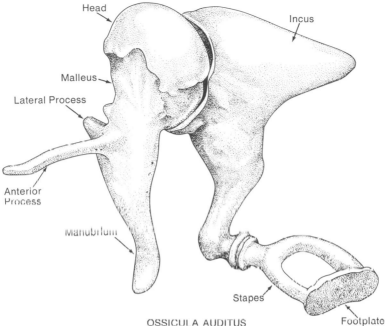

Head

Incus

Malleus

Lateral Process

Anterior
Process

Manubrium

Stapes

Footplate

OSSICULA AUDITUS
Adapted from Stedman's Medical Dictionary, *22d ed., 1972.*

Fig. 7–5.—The ossicular chain consists of the malleus, incus, and stapes.

- The *incus*, also called the *anvil*, is the second ossicle and has a long projection which runs downward and joins the stapes.

- The *stapes*, also called the *stirrup*, lies at almost right angles to the long axis of the incus. The two branches of the stapes are the anterior and the posterior, ending in the footplate which fits into the oval window.

When the handle of the hammer is set into motion by movement of the eardrum, the action is transferred mechanically through the ossicular chain to the oval window.

The *oval window* and the *round window* are located on the inner wall of the middle ear. The round window is covered by a very thin membrane that moves out as the footplate in the oval window moves in. As the action is reversed and the footplate in the oval window is pulled out, the round window membrane moves inward.

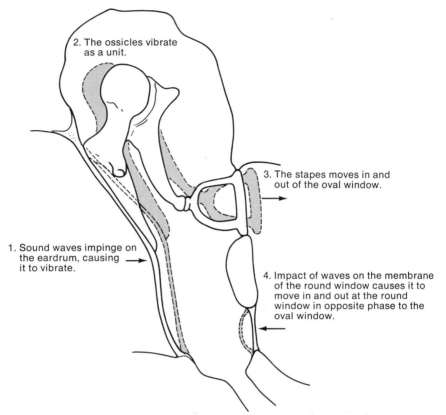

2. The ossicles vibrate as a unit.

3. The stapes moves in and out of the oval window.

1. Sound waves impinge on the eardrum, causing it to vibrate.

4. Impact of waves on the membrane of the round window causes it to move in and out at the round window in opposite phase to the oval window.

Adapted from the CIBA drawing in Figure 7–11.

FIG. 7–6.—The eardrum has about 25 times as much surface area as the oval window. All of the sound energy collected on the eardrum is transmitted through the ossicular chain to the smaller area of the oval window.

Mastoid air cell system. On the back wall of the middle ear space is an opening that extends into the mastoid that resembles a honeycomb system of spaces filled with air.

Physiology

The primary function of the middle ear in the hearing process is to transfer sound energy from the outer to the inner ear. As the eardrum vibrates, it transfers its motion to the attached hammer (malleus). Since the bones of the ossicular chain are connected to one another, the movements of the hammer are passed on to the anvil, and finally to the stirrup which is imbedded in the oval window.

As the stirrup moves back and forth in a rocking motion, it passes the vibrations on into the inner ear through the oval window. Thus, the mechanical motion of the eardrum is effectively transmitted through the middle ear and into the fluid of the inner ear.

Amplification. The sound conducting mechanism also amplifies sound by two main mechanisms. First, the large surface area of the drum as compared to the small surface area of the base of the stapes (footplate) results in a hydraulic effect. The eardrum has about 25 times as much surface area as the oval window. All of the sound pressure collected on the eardrum is transmitted through the ossicular chain and is concentrated on the much smaller area of the oval window. This produces a significant increase in pressure (see Figure 7-6).

The bones of the ossicular chain are arranged in such a way that they act as a series of levers. The long arms are nearest the eardrum, and the shorter arms are toward the oval window. The fulcrums are located where the individual bones meet. A small pressure on the long arm of the lever produces a much stronger pressure on the shorter arm. Since the longer arm is attached to the eardrum and the shorter arm is attached to the oval window, the ossicular chain acts as an amplifier of sound pressure. The magnification effect of the entire sound conducting mechanism is about 22-to-1.

Two tiny muscles attach to the ossicular chain, the *stapedius* to the neck of the stapes bone and the *tensor tympani* to the malleus. Loud sounds cause these muscles to contract, which stiffens and diminishes or damps the movement of the ossicular chain. This presumably offers some protection for the delicate inner ear structure from physical injury. More convincing research is still needed to test the validity of this theory of the function of these muscles.

Pathology

Eardrum. Infections localized to the eardrum are rare and when they do occur, are caused by viruses (myringitis bullosa). The eardrum is, however, frequently involved by infections of the external auditory canal or the middle ear.

Perforations are most frequently caused by infections or by injuries. A blow to the ear compressing air in the external auditory canal can cause sufficient hydraulic force to rupture the eardrum. Sudden pressure changes, such as occur in diving or changing altitude while flying in unpressurized airplanes, cause perforations as does trauma from foreign bodies and unskillful manipulation in the ear canal. Perforations can be repaired by medical treatment or by surgical grafting procedures.

Retractions of the eardrum are caused by poor middle ear ventilation due to Eustachian tube dysfunction. They may be partial retraction pockets or total atelectasis (collapse). The same mechanism of retraction of the eardrum produces cholesteatomas. Treatment must be directed toward the Eustachian tube. Sometimes the Eustachian tube may be bypassed by surgically inserting a small tube in the eardrum. This permits equalization of air pressure on both sides of the drum via the ventilating tube.

The middle ear space is prone to infectious diseases, especially in childhood years. These are predominantly bacterial in origin and called *suppurative otitis media*. Since the middle ear space connects with the mastoid air cell system, infection may easily spread to this area (mastoiditis). Before the days of antibiotics, these were serious often life-threatening problems due to the risk of spread to the brain or major vessels surrounding the ear. While less likely to occur today, these dangers still exist.

Congenital deformities of the middle ear are not too common and are usually associated with structural abnormalities of the sound conducting system. Tumors originating in the middle ear are rare.

Cholesteatomas, the most common destructive problem of the middle ear, are skin-lined pouches. Concentric growth occurs when discharged epithelial debris is compacted into the pocket. As this pocket gets concentrically larger, it destroys all adjacent tissues including bone. Cholesteatomas are destructive to the sound conducting system and to hearing. Extension of the cholesteatoma into the brain or inner ear is a disastrous complication.

Ossicles. There are only two ways that disease adversely affects hearing via the ossicular chain: (a) by fixation, so that the chain cannot vibrate or vibrates inefficiently, and (b) by interruption, producing a gap in the chain. Fixation may occur as a result of developmental errors, adhesions or scars from old middle ear infections or bone diseases that affect this area. Otosclerosis is a prime example of the latter and is the most common cause of progressive conductive hearing loss seen in this country. It usually begins in early adult life. Interruptions are

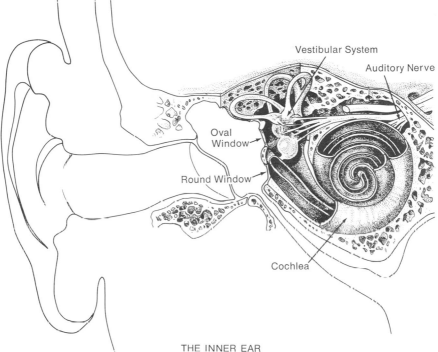

THE INNER EAR
Adapted from the CIBA drawing in Figure 7–1.

FIG. 7–7.—Shown here are the major components of the inner ear including the vestibular system and the cochlea.

usually caused by middle ear infections or cholesteatomas. Head injuries account for a few of them.

It is important to realize that if disease is confined to the outer ear or middle ear or both, the resulting hearing loss will be conductive. Conductive losses are losses for loudness, not for clarity. Most of these disorders causing conductive losses are medically or surgically correctable. Some are dangerous and some are progressive. Therefore, all cases should have the benefit of otologic evaluation and care.

Inner ear

Anatomy

The major components of the inner ear include the vestibular recep-

(*Text continues on page 218.*)

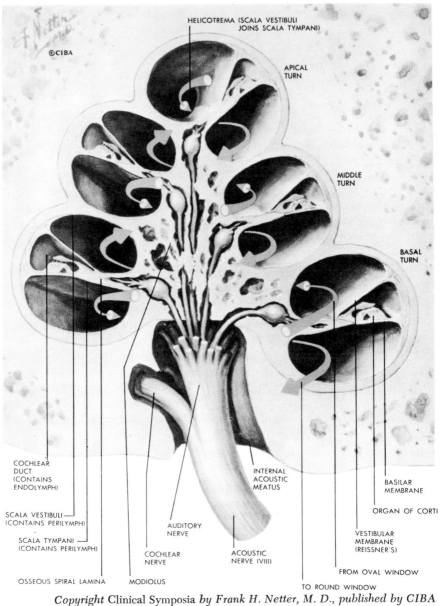

HELICOTREMA (SCALA VESTIBULI
JOINS SCALA TYMPANI)

©CIBA

APICAL
TURN

MIDDLE
TURN

BASAL
TURN

COCHLEAR
DUCT
(CONTAINS
ENDOLYMPH)

INTERNAL
ACOUSTIC
MEATUS

BASILAR
MEMBRANE

SCALA VESTIBULI
(CONTAINS PERILYMPH)

ORGAN OF CORTI

SCALA TYMPANI
(CONTAINS PERILYMPH)

AUDITORY
NERVE

VESTIBULAR
MEMBRANE
(REISSNER'S)

OSSEOUS SPIRAL LAMINA

COCHLEAR
NERVE

ACOUSTIC
NERVE (VIII)

FROM OVAL WINDOW

MODIOLUS

TO ROUND WINDOW

Copyright Clinical Symposia *by Frank H. Netter, M. D., published by CIBA
Pharmaceutical Company. Used with permission.*

Fig. 7–8.—Shown here is a cross-section view of the cochlea.

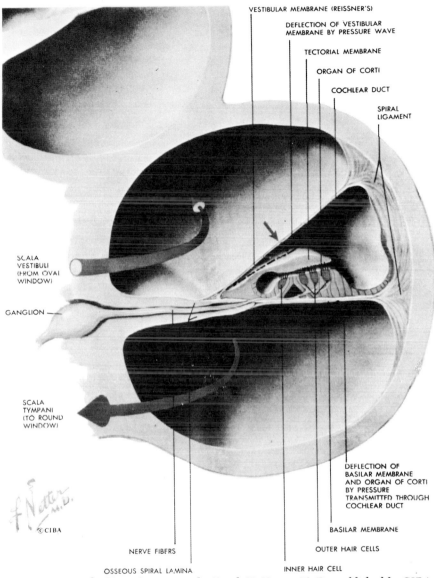

VESTIBULAR MEMBRANE (REISSNER'S)

DEFLECTION OF VESTIBULAR
MEMBRANE BY PRESSURE WAVE

TECTORIAL MEMBRANE

ORGAN OF CORTI

COCHLEAR DUCT

SPIRAL
LIGAMENT

SCALA
VESTIBULI
(FROM OVAL
WINDOW)

GANGLION

SCALA
TYMPANI
(TO ROUND
WINDOW)

DEFLECTION OF
BASILAR MEMBRANE
AND ORGAN OF CORTI
BY PRESSURE
TRANSMITTED THROUGH
COCHLEAR DUCT

BASILAR MEMBRANE

OUTER HAIR CELLS

NERVE FIBERS

OSSEOUS SPIRAL LAMINA

INNER HAIR CELL

Fig. 7–9.—This schematic diagram depicts the transmission of sound across the cochlear duct, stimulating the hair cells.

217

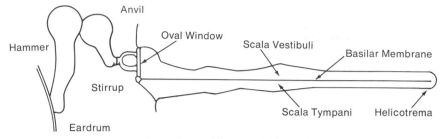

Fɪɢ. 7–10.—Here is how the cochlea would look if it were uncoiled.

tive system and the cochlea, housed within the compact temporal bone called the osseous labyrinth. It is filled with fluid (perilymph) in which a tubular membrane (membranous labyrinth) floats. The membranous labyrinth is filled with a fluid of a slightly different chemical composition called endolymph. Endolymph bathes the balance receptors and the hearing organ (organ of Corti) located within the membranous labyrinth.

The back portion of the inner ear consists of three semicircular canals positioned at right angles to one another, each containing a single balance receptor. The front section (cochlea) is shaped like a snail shell, coiled two and one-half times around its own axis; it houses the organ of Corti. These two regions are separated by the vestibule that contains two additional receptors for balance. This chapter is concerned only with the front section and the organ of hearing (see Figure 7–7).

The cochlea. If the cochlea were unwound two and one-half turns it would resemble an elongated tube. What would be seen if one looked into the open end of the tube is illustrated in Figure 7–8 (cross section of cochlea). Note that the membranous labyrinth has become triangular in shape, attached to one wall of the bony labyrinth by a large surface area and to the opposite wall at a point. The space enclosed by this triangle is the scala media that is filled with endolymph. The hearing organ (organ of Corti) is housed within the scala media. The space above and below the scala media is filled with perilymph and called the scala vestibuli and scala tympani, respectively. They are continuous with each other through a tiny opening (helicotrema) in the apical end of the cochlea (see Figure 7–9).

Follow the path of the perilymphatic space through the cochlea. The scala vestibuli begins beneath the footplate of the stapes and is separated from the scala media below it by Reissner's membrane. The scala

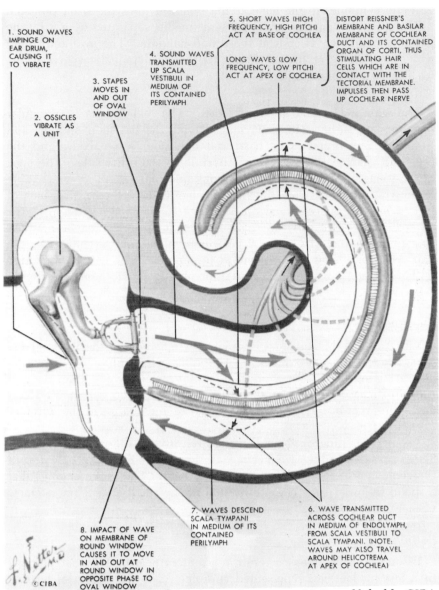

1. SOUND WAVES IMPINGE ON EAR DRUM, CAUSING IT TO VIBRATE

2. OSSICLES VIBRATE AS A UNIT

3. STAPES MOVES IN AND OUT OF OVAL WINDOW

4. SOUND WAVES TRANSMITTED UP SCALA VESTIBULI IN MEDIUM OF ITS CONTAINED PERILYMPH

5. SHORT WAVES (HIGH FREQUENCY, HIGH PITCH) ACT AT BASE OF COCHLEA

LONG WAVES (LOW FREQUENCY, LOW PITCH) ACT AT APEX OF COCHLEA

DISTORT REISSNER'S MEMBRANE AND BASILAR MEMBRANE OF COCHLEAR DUCT AND ITS CONTAINED ORGAN OF CORTI, THUS STIMULATING HAIR CELLS WHICH ARE IN CONTACT WITH THE TECTORIAL MEMBRANE. IMPULSES THEN PASS UP COCHLEAR NERVE

8. IMPACT OF WAVE ON MEMBRANE OF ROUND WINDOW CAUSES IT TO MOVE IN AND OUT AT ROUND WINDOW IN OPPOSITE PHASE TO OVAL WINDOW

7. WAVES DESCEND SCALA TYMPANI IN MEDIUM OF ITS CONTAINED PERILYMPH

6. WAVE TRANSMITTED ACROSS COCHLEAR DUCT IN MEDIUM OF ENDOLYMPH, FROM SCALA VESTIBULI TO SCALA TYMPANI. (NOTE: WAVES MAY ALSO TRAVEL AROUND HELICOTREMA AT APEX OF COCHLEA)

FIG. 7–11.—The mechanism for transmission of sound vibrations from the eardrum through the cochlea is shown in this figure.

vestibuli continues to the helicotrema where it joins the scala tympani.

The scala tympani lies below the scala media, separated from it by the basilar membrane. It ends at the round window. The oval and round windows then lie at opposite ends of the perilymphatic space of the cochlea (see Figure 7–10).

The scala media and organ of Corti. The scala media is triangular in shape and 1¼ in. (31.5 mm) long, having the basilar membrane and Reissner's membrane as its base and one side, respectively, and the broad attachment to the osseous labyrinth as the other side. The apex of the triangle is attached to the osseous spiral lamina.

Growing upwards from the basilar membrane are cells that have a tuft of hair on one end and are attached to the hearing nerve at the other. These hair cells are classified as inner or outer in relationship to the tunnel of Corti. A gelatinous membrane extends over the hair cells from its own attachment at the limbus. The hair is embedded within the substance of this tectorial membrane above them (see Figure 7–9).

Physiology

Vibrations of the stapedial footplate set into motion the fluids of the inner ear and the organ of Corti. As the basilar membrane is displaced, a shearing movement occurs on the tectorial surface which drags the hair cells attached to the nerve endings. This sets up electrical impulses that are appropriately coded and transmitted to the brain via the auditory (cochlear) nerve.

The nerve endings in the cochlea are sensitive to different frequencies. Those sensitive to high frequencies are located at the large base end of the cochlea near the oval and round windows. The nerve endings that respond to low frequencies are located at the small end of the cochlea. (See Figure 7–11.)

Pathology

Disorders of the inner ear result in a sensory hearing loss characterized by a loss for loudness represented audiometrically by abnormal hearing for bone conduction. Inner ear hearing losses may or may not have associated losses for clarity represented audiometrically by poorer speech discrimination scores.

Congenital sensory hearing losses (inner ear hearing losses present at the time of birth) may be of many different types. Some of these are inherited and called familial sensory hearing losses.

Inflammatory disorders include: suppurative labyrinthitis, which is usually an extension of suppurative otitis media and usually terminates with impaired hearing in the affected ear; viral labyrinthitis caused by such organisms as mumps and herpes viruses; and toxic labyrinthitis which relates to inner ear dysfunction, caused by hormonal (hypothyroidism) and allergic factors, and to drugs that have an adverse effect on the hearing mechanism. Trauma to the ear may be physical as in temporal bone fracture, or caused by noise, and termed *acoustic trauma.*

Concussions of the inner ear may also occur and usually cause an incomplete sensory loss with some, if not complete, recovery of hearing. Normal aging causes a sensory hearing loss called presbycusis. Tumors affecting the inner ear are rare and usually do not originate within the cochlea but affect the inner ear by way of extension to this structure. Some types of bone disease affecting the inner ear include Paget's disease and fibrous dysplasia.

Meniere's disease affects both parts of the inner ear (hearing and balance) and its cause is unknown. It is characterized by episodic dizziness, often severe and associated with nausea and vomiting, fluctuant sensory hearing loss that is generally progressive, noise in the ear or tinnitus, and a peculiar sensation of fullness in the involved ear. Usually medical management can control this problem but occasionally surgery is necessary.

Further considerations in noise and hearing loss

Temporary noise induced hearing loss

When a person is first exposed to hazardous noise, the initial change usually observed is a loss of hearing in the higher frequency range, usually a dip or a notch at about 4000 Hz. After a rest period away from the noise, the hearing usually returns to its former level. For practical purposes, 14 hours or so away from the noise is adequate to return the threshold to previous levels.

Permanent damage from noise is generally classified as *noise induced hearing loss* or *acoustic trauma,* depending upon the nature of exposure. The long-term cumulative effects of repeated and prolonged hazardous noise exposure result in permanent pathologic changes in the cochlea and irreversible threshold shifts in the hearing acuity. This is referred to as noise induced hearing loss. It is usually represented audiometrically by a notch at 4000 Hz. Because the hearing loss, however, does not necessarily stop here, further exposure may result in a deepening and widening of the notch. When the hearing loss involves the speech frequency range, considerable difficulty in hearing conversational speech develops.

The effect of noise on hearing depends on the amount and characteristics of the noise as well as the duration of exposure. In some instances, employment for a few hours or days in a noisy industrial environment or exposure to a single sound of damaging intensity may suffice to produce a permanent hearing loss. This is often referred to as acoustic trauma. Yet, others working in the same occupational noise atmosphere for many years retain normal hearing acuity and are unaffected. The major deterioration of hearing, however, occurs during the initial 5 to 10 years of employment in a noise risk environment.

Factors concerned with the causation of permanent threshold shifts include: (a) intensity of noise, (b) frequency spectrum of noise, (c) duration of exposure, (d) character of the noise whether continous, explosive, impulsive, or intermittent, (e) time interval between exposures, and (f) individual susceptibility.

Although susceptibility to noise induced hearing loss varies greatly from one individual to another, noise induced hearing loss is a trait which follows a normal distribution curve. Beyond certain levels of extremely high intensity, it is generally agreed that all individuals are susceptible—provided the exposure is long enough. Generally speaking, tests of temporary threshold shift throw no specific light on the susceptibility of individuals to permanent threshold damage.

Some factors of consideration are sex, age, and associated aural pathology. Some investigators have concluded that women are more resistant to noise than men; others, that there is no sex difference. A committee of the American Standards Association reporting in 1954 on the relation of hearing loss to noise exposure were of the opinion that losses due to presbycusis and to noise summate without influencing each other. They considered there were no grounds for the belief that presbycusis may either protect hearing from noise trauma or make hearing more sensitive to it.

In relation to associated ear disease, it is generally considered that (except for Meniere's disease) people with middle or inner ear hearing loss have no specific predisposition to increased susceptibility to noise induced hearing losses. Obviously patients with conductive hearing losses have a built-in protective mechanism such as one would expect with the use of earplugs.

It has been generally accepted, after the work of Davis,[*] that the term *acoustic trauma* is reserved for permanent hearing loss produced by a very brief exposure to a very loud noise such as gunfire or explosions.

The acoustic reflex is a contraction of the muscles of the middle ear, probably only the stapedius muscle, in response to loud sound that restrains ossicular movement and attenuates loud sound before reaching the inner ear. This protective mechanism was discussed earlier.

One must be cautious about recreational and social hearing losses apart from those that may occur in industry. Such hazards would include skeet or trap shooting, motorcycle racing, sports car racing, use of hover crafts, helicopters and some open cockpit airplanes, boats, snowmobiles, etc. The risk of "pop music" is still open to debate but may have a greater effect on the musicians than on the listeners.

With this basic orientation to the structure of the ear, how it works, and what damage it, the reader should be better prepared to deal with the rest of this book. Furthermore, many of the chapters, especially the following two, will necessitate referring to the diagrams in this chapter.

Bibliography

Bast, Theodore H., and Anson, Barry J. *The Temporal Bone and the Ear.* Springfield, Ill.: Charles C Thomas, 1949.

Brown, Scott. *Diseases of the Ear, Nose and Throat,* vol. 2, *The Ear,* 3rd ed. Philadelphia. Pa.: J. B. Lippincott Co., 1971.

Davis, Hallowell, and Silverman, S. Richard, eds. *Hearing and Deafness.* New York, N.Y.: Holt, Rinehart and Winston, 1970.

Glorig, Aram, ed. *Audiometry: Principles and Practices.* Baltimore, Md.: Williams & Wilkins, 1965.

Paparella, Michael, and Schumrick, Donald A., eds. *Otolaryngology;* vol. 2, *The Ear.* Philadelphia, Pa.: W. B. Saunders Co., 1973.

Sataloff, Joseph. *Hearing Loss.* Philadelphia, Pa.: J. B. Lippincott Co., 1971.

Schuknecht, Harold F. *Pathology of the Ear.* Cambridge, Mass.: Harvard University Press, 1974.

Shambaugh, George E., Jr. *Surgery of the Ear.* Philadelphia, Pa.: W. B. Saunders Co., 1967.

Stevens, S. S., Warshofski, Fred, and Editors of *Life. Sound and Hearing.* New York, N.Y.: Life Science Library, Time Inc., 1967.

-----

[*] H. Davis, "Effects of High Intensity Noise on Navy Personnel." *U.S. Armed Forces Medical Journal,* 9:1027–47 (1958).

Chapter Eight Medical Aspects of Hearing Conservation

By M. S. Fox, M.D.

Safety professionals involved in establishing and maintaining industrial hearing conservation programs should know how the ear functions and what can be done to protect it against excessive noise exposure. It is important to know where to get help and when to refer employees with hearing problems to a physician.

The physician—particularly the otolaryngologist—is frequently consulted for guidance and supervision in carrying out hearing-conservation programs in industry. Because the ear, nose, and throat physician knows the many facets of occupational hearing loss, he can play a useful role in minimizing problems arising out of industrial noise exposure. His special training and knowledge of the anatomy and function of the ear and his practical experience in the daily diagnosis and treatment of various hearing problems enables him to provide guidance for establishing practical hearing conservation programs.

Anatomy of the ear

For purposes of orientation, it would be well to briefly review the anatomy and function of the ear (see Figure 8–1). The ear is divided into the external or outer ear, the middle ear, and the inner ear, as described in the previous chapter.

Sound waves are conducted into the external ear as acoustic energy and are changed into mechanical energy by the eardrum and ossicles of the middle ear. This energy in turn is changed into a fluid movement

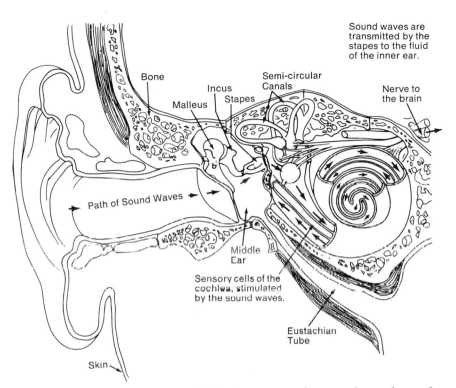

Sound waves are transmitted by the stapes to the fluid of the inner ear.

Bone

Incus

Semi-circular Canals

Malleus

Stapes

Nerve to the brain

Path of Sound Waves

Middle Ear

Sensory cells of the cochlea, stimulated by the sound waves.

Eustachian Tube

Skin

Fig. 8–1.—Schematic diagram of the human ear showing the pathway for sound waves to be transmitted to the inner ear.

within the inner ear which excites the hair cells of the cochlea. Electrical impulses are set up in the cochlea (inner ear) which are conveyed via the auditory nerve to the brain where they are interpreted as meaningful information, based on past experience, training, knowledge of language, and the like.

Types of hearing impairment

Any physical defect or condition of the outer or middle ear which interferes with the passage of sound may result in what is referred to as a *conductive hearing impairment*. Injury or disease of the inner ear can result in a *sensori-neural type of hearing impairment*. A combination of both inner ear and conductive hearing impairment is often seen clinically and is referred to as a *mixed-type hearing loss*. (Table 8-A lists causes of acquired hearing losses.)

225

TABLE 8-A. CAUSES OF ACQUIRED HEARING LOSS

Etiology and Principal Findings	Type of Loss Conductive	Type of Loss Sensori-neural	Onset of Loss
Physical Agents			
Impacted cerumen (wax)	x		Sudden
Foreign body impaction	x	x	Sudden
Trauma, accidental		x	Sudden
Noise-exposure		x	Gradual
Barotrauma	x	x	Sudden
Excessive growth of lymphoid tissue in nasopharynx	x		Gradual
Surgical interference	x	x	Sudden
Toxic Agents			
Quinine		x	
Nicotine (tobacco)		x	
Aspirin (salicylates)		x	
Streptomycin		x	Gradual or
Dihydrostreptomycin		x	sudden, depending on dose
Hydroxystreptomycin		x	
Neomycin		x	
General Infectious Diseases			
Scarlet fever	x	x	Gradual
Measles		x	Gradual
Mumps		x	Gradual
Pertussis		x	Gradual
Varicella		x	Gradual
Influenza		x	Gradual
Pneumonia, virus and pneumococcic		x	Gradual
Typhoid fever		x	Gradual
Diphtheria		x	Gradual
Syphilis		x	Gradual
Common cold	x	x	Gradual
Any disease causing high fever		x	Gradual or sudden
Infections of the Ear			
External otitis	x		Gradual
Otitis media, acute and chronic			
Non-suppurative	x	x	Gradual
Suppurative	x	x	Gradual
Serous	x		Gradual
Mastoiditis, acute and chronic	x		Gradual
Brain Conditions			
Meningitis		x	Sudden
Encephalitis		x	Gradual
Tumors, circulatory diseases		x	Gradual
Concussion		x	Sudden
Fracture of the temporal bone		x	Sudden
Miscellaneous			
Functional			
Psychogenic			
Hysteria			
Malingering			
Advancing age (presbycusis)		x	Gradual

Adapted from Chapter on "Otology" by Aram Glorig, M.D., in Courtroom Medicine, Edited by Marshall Houts. Charles C Thomas, Publisher, Springfield, Ill., 1958.

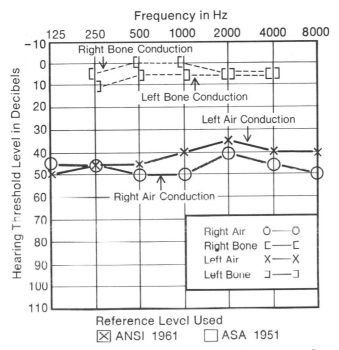

Frequency in Hz

FIG. 8–2.—An audiogram indicating a conductive loss. Note that there are standard symbols to record thresholds on an audiogram. Circles are used to record air conduction thresholds for the right ear; "X's" are used for the left ear. Brackets opening to the right indicate bone conduction thresholds for the right ear, and brackets opening to the left indicate thresholds for the left ear.

Conductive hearing loss

Anything that obstructs the passage of sound waves through the ear canal may cause a conductive hearing loss; for example, a foreign body, impacted cerumen (ear wax), or external otitis. Otitis means "inflammation of the ear." Any inflammation of the middle ear (otitis media) can also cause a conductive hearing loss. A conductive hearing loss may also result from fluid in the middle ear.

Figure 8–2 is an audiogram of a patient with conductive hearing loss. The test for conductive hearing loss is performed by applying sound energy directly to the bones of the skull, where it is transmitted to the cochlea, bypassing the structures in the middle ear. The dotted line in

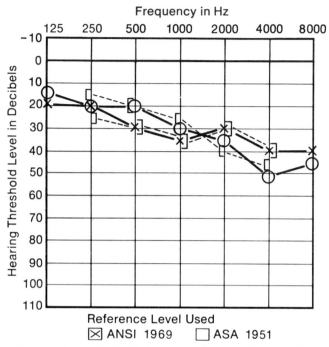

FIG. 8–3.—The results typical of those found on audiograms for individuals having a mild sensori-neural type hearing loss. This person hears just as poorly by bone conduction as he does by air conduction.

Figure 8–2 shows the bone-conduction level or how much the patient could actually hear if the obstruction or condition causing the hearing loss were alleviated.

The solid line shows the amount that the patient hears through air conduction. The difference is called an *air-bone gap*. This is an example of the kind of audiogram that might be seen from a person with otosclerosis. Conductive hearing losses, in some cases, may respond to medical or surgical treatment, and therefore should be referred to an otologist (an ear specialist).

A bone-conduction audiometric test is time consuming and difficult because it involves masking and other technical problems. Although industrial audiometric technicians do not perform bone-conduction audiometry, they should know when it is indicated.

Cause of Hearing Loss	Location of Damage to Ear	Type of Hearing Loss	Medical Name
Continuous noise exposure	Inner ear structures	Inner ear loss	Noise induced hearing loss, occupational hearing loss, or acoustic trauma
Sudden explosive blasts	Inner ear and or middle ear structures	Conductive, inner ear, or mixed loss	Acoustic trauma
Mechanical blows to the head and or ear	Inner ear and or middle ear, or external canal	Conductive, inner ear, or mixed loss	Acoustic trauma
Barometric pressure changes	Middle ear, and or inner ear	Conductive, inner ear, or mixed loss	Aerotitis barotrauma or caisson disease
Burns (Heat as opposed to chemicals)	Middle ear and or inner ear	Conductive, inner ear, or mixed loss	

FIG. 8–4.—Various causes of occupational hearing loss.

Sensori-neural hearing loss

Sensori-neural hearing losses may be caused by drug toxicity, disease, aging, excessive noise, head and ear trauma, blasts, explosions, and other factors. An audiogram indicating a mild sensori-neural hearing loss is shown in Figure 8–3.

The sensori-neural hearing loss resulting from noise exposure is similar to the hearing loss found in individuals who have never been exposed to high noise levels. A baseline preemployment audiogram can be used to establish a record of the employee's hearing at the time of employment.

Definitions

Occupational hearing loss can be defined as a hearing impairment of one or both ears, partial or complete, arising in, during the course of, and as the result of one's employment. It includes acoustic trauma as well as noise induced hearing loss. The term *acoustic trauma* is often misused to denote noise induced hearing loss as well as the sudden hearing loss resulting from intense blasts, explosions, gunfire, and direct trauma to the head or ears. These conditions, however, should be described as separate terms in order that one may logically distinguish between them (see Figure 8–4).

Acoustic trauma should be reserved for the immediate hearing injury produced by one or a few exposures to sudden intense acoustic forms of energy, resulting from blasts and explosions or from direct trauma to the head or ear. Acoustic trauma should be thought of as one single incident to which the worker relates the onset of his hearing loss.

Noise induced hearing loss, on the other hand, is used to describe the cumulative permanent loss of hearing, always of the sensori-neural type, that develops over months or years of hazardous noise exposure. The hearing loss usually affects both ears equally in extent and degree. It should also be pointed out that the onset of the hearing loss, its progression, and its permanency, as well as the characteristic audiograms, vary —depending on whether it is acoustic trauma or noise induced hearing loss.

Noise has been defined as unwanted sound, sound without value to the listener and sound of random intensity and frequency. Sound as ordinarily encountered is a wave motion of air molecules and is produced by vibrating bodies. The wave motion is made up of pressure fluctuations that move outward from the source of vibration at a speed of approximately 1100 feet per second. (See Chapter 2, Physical Characteristics of Sound.)

There are two aspects of the pressure wave which are important: (*a*) the frequency, which is the number of fluctuations per second; and (*b*) the sound pressure level, which indicates the magnitude of the fluctuations.

Let us review some definitions. The frequency of sound largely determines its pitch. Frequency is measured in vibrations per second and designated as hertz (Hz). The young adult with normal hearing can hear sound ranging from as low as 20 Hz to as high as 20,000 Hz (see Figure 8–5). Higher frequencies are generally inaudible and are known as ultrasonic. For further details, see Chapter 3, Subjective Aspects of Sound.

The most important portions of the audible frequency range lie between 500 and 2000 Hz and are referred to as the *speech zone frequencies.* Hearing loss occurring in this frequency region is particularly significant because it handicaps a person carrying by interfering with

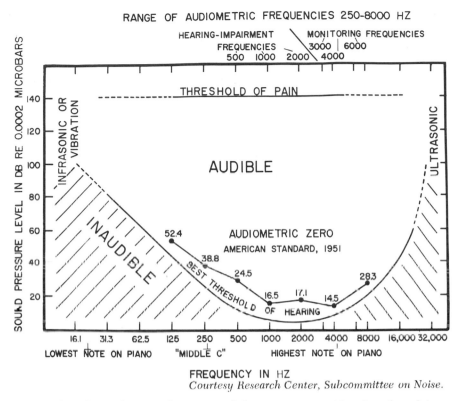

Courtesy Research Center, Subcommittee on Noise.

FIG. 8–5.—Shown here is the range of frequencies considered to be of importance in audiometric measurements. Note: the audible, inaudible, and ultrasonic regions.

conversational speech.

The sound level of a noise is expressed in decibels. A decibel is a dimensionless unit based on the logarithm of the ratio of the sound pressure of the noise and a reference sound pressure. The sound level of a noise can be measured by a sound level meter using the A-scale reading. More detailed analysis of the noise can be made by means of a combination of a sound level meter and an octave-band analyzer. For further details, see Chapters 2, 4, 5, and 6.

Effects of noise exposure

The known effects of noise exposure are of two types:

- *Nonauditory effects*, which influence human behavior

- *Auditory effects*, which consist of temporary or permanent hearing loss

The ability of noise to interfere with speech communication is well known. In industry, the ability to communicate using speech sounds is important. It is essential in employee training, in giving and understanding orders, in hearing warning signals, and other phases of plant operation. In addition, the inability to hear can be very annoying to those concerned. It should be pointed out that many noises which are not intense enough to cause hearing loss can interfere seriously with communication by speech. (For further details, see Chapter 3.)

The annoyance caused by noise is an individual psychologic reaction and depends a great deal upon the physical status, mental attitude, and personal motivation of the person or persons being annoyed. One person may enjoy sounds that are objectionable to another person. These factors are difficult to predict and they defy objective measurements.

In general, sounds of high intensity, high frequency, and intermittency tend to be more annoying, but there are substantial individual differences. The fact that a noise is annoying does not establish it as a damage risk. A good example is a dripping faucet or the neighbor's violin-playing daughter, either of which can be very annoying particularly when one is trying to sleep. For further details, see Chapter 3.

Tests made on the production efficiency of workers performing mental and motor tasks under various conditions of noise exposure reveal that sudden and unfamiliar noise temporarily reduces efficiency, but when employees become accustomed to the noise, the production returns to previous non-noise performance levels. For further details, see Chapter 10.

It has been claimed that noise is a hazard to health and that certain changes occur in the human body as a result of noise exposure. Statements have been made that noise will cause high blood pressure, vascular and nervous disease, ulcers, stress reactions, and other illnesses. Convincing evidence for these claims, however, is lacking. As far as health is concerned, the only definite relationship that can be shown is that noise can cause a hearing loss.

Temporary threshold shift

When a normal ear is exposed to noise at damaging intensities for sufficiently long periods of time, a temporary depression of hearing results. This temporary hearing loss is a physiological phenomenon referred to as temporary threshold shift (TTS). It is believed to occur in the hair cells of the organ of Corti of the inner ear as the result of overstimulation and may be associated with metabolic changes in the hair cells or chemical changes in the inner ear fluids. A temporary threshold shift or hearing loss may be experienced by employees at the end of their work shift.

When the noise exposure is continued and is longer or at a higher level or both, a stage is reached where the hearing threshold level does not recover; that is, the hearing no longer returns to its original threshold. This hearing loss is called *noise induced permanent threshold shift* (NIPTS) and is designated as *noise induced hearing loss*. Studies by Aram Glorig and his staff[1] have revealed the following important facts regarding temporary and permanent threshold shift:

• The temporary threshold shift resulting from one day's exposure to noise levels of 100 dB or more may vary from no shift to one of 40 dB.

• Typical industrial noise exposure produces the largest temporary threshold shifts at 4000 and 6000 Hz.

• Most of the temporary threshold shift occurs during the first two hours of exposure.

• The amount of temporary threshold shift and its location on an audiogram vary with the amount and location of the permanent shift—in other words, the greater the permanent change of hearing level at any frequency, the less the temporary shift at that frequency.

• Recovery of hearing level after temporary threshold shift occurs mostly within the first hour or two after the noise exposure has ended.

• There appears to be a distinct relationship between temporary and permanent threshold shifts.

1. Exposure to noise levels that do not produce a temporary hearing loss, upon continued exposure will not produce a permanent hearing loss.

2. The configuration of the audiogram produced as a result of a temporary threshold shift will resemble the audiogram that will be found when the permanent hearing loss is measured.

Experimental studies on the causes of a temporary threshold shift have demonstrated that continuous noise exposure produces different effects in the ear than interrupted noise exposure. The studies show that, if employees are allowed relatively quiet rest periods away from noise exposure at intervals throughout the day, then long-term changes in their hearing thresholds will be less than changes produced by continuous exposure with no rest periods.

Factors involved

There are many factors that affect the degree and extent of hearing loss.

- Intensity or loudness of the noise (sound pressure level)
- Type of noise (frequency spectrum, continuous, or intermittent)
- Time period or exposure time each day (duty cycle per day)
- Total work duration (years of employment)
- Individual susceptibility
- Age of the worker
- Coexisting hearing loss and ear disease
- Character of surroundings in which the noise is produced
- Distance of ears from the noise source
- Position of ear with respect to the noise source

The first four factors are the most important and are referred to as noise exposure. Thus, it is necessary to know not only how much noise exists, but also what type of noise and how long each day an employee is exposed to this noise.

Continuous vs. intermittent exposure

It has been observed that brief, momentary exposures of most employees to high noise levels will usually not produce any significant hearing loss, while prolonged exposure to continuous noise levels above 90 decibels, as measured on the A-scale of a sound level meter, may result in a significant hearing loss over the years to some employees. While a great deal has been learned about noise exposure and hearing loss as applied to steady-state noises, our present knowledge about the hazard of impact noises—such as blasts, rifle fire, and drop hammers—is still limited and further research is needed in this area.

Individual susceptibility

Certain individuals are more susceptible to the effects of exposure to noise than others. The problem lies in finding this susceptible individual. There are no practical tests currently available that can determine in advance who is more susceptible to noise.

Surroundings

The acoustical character of the surroundings in which the noise source is located influences the noise levels produced. The degree of hearing loss produced in an individual depends, in part, on the distance of the person from the noise source. The further away he is from a loud noise source, the less sound energy a person is exposed to.

Position of ear

The position of each ear with respect to the source of sound waves is also a factor. Some discrete sources of sound energy aimed at an individual can produce a greater effect on one ear with only a minimal effect on the other ear. Visualize a situation where somebody is working outdoors next to a loud noise source, always on his left, through an eight hour day; obviously, his left ear is going to be more exposed than his right ear.

Permanent hearing loss

The early stage of noise induced hearing loss is characterized by a sharply localized dip in the hearing threshold curve at the frequencies between 3000 and 6000 Hz, usually appearing at 4000 Hz (see Figure 8–6) In the early stages of noise induced hearing loss, a person may complain of tinnitus (ringing in the ear), muffling of sound, discomfort of the ears, or a temporary depression of hearing that is noted at work and upon leaving work, but which clears after several hours away from the noise exposure. Pain and vertigo are seldom mentioned by the employee.

As the noise exposure continues, the hearing threshold dip broadens and spreads in both directions, but there is still little noticeable subjective effect on the employee. Impairment of hearing is usually not noticed by the employee until the hearing threshold levels (see Figure 8–7) of the important speech frequencies (500, 1000, and 2000 Hz) average more than 25 dB ANSI (or 15 dB ASA).[2] (These refer to the scales of hearing loss on an audiogram; the −10 dB of ASA is the 0 dB of ANSI.) Substantial losses may occur at frequencies from 3000 to 8000 Hz, with little subjective awareness by the individual of a change in hearing. The noise induced hearing loss is of the sensori-neural type, quite similar in extent and degree in both ears. The eardrums appear

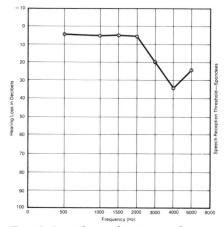

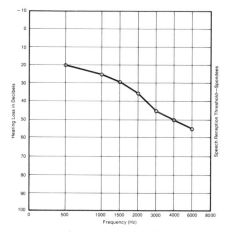

FIG. 8–6.—The early stages of noise induced hearing loss is characterized by a sharp dip in the hearing threshold curve at the frequencies between 3000 and 6000 Hz, usually appearing at 4000 Hz.

FIG. 8–7.—Impairment of hearing is usually not noticed by the individual until the important speech frequencies are affected.

normal. The onset and progress of noise induced hearing loss is slow and insidious, and the worker is likely to be unaware of its existence or disregard it. While the hearing loss from noise exposure may be profound, total hearing loss has not been observed.

Hearing loss due to age

Presbycusis is loss of hearing that accompanies advancing age. This is a sensori-neural hearing loss in older people resulting from atrophy, vascular and neural degeneration, as well as other structural changes in the inner ear. It is part of the general aging process of the body. The term "sociocusis" has been used by Glorig[1] to describe the hearing loss resulting from the presbycusis as well as the cumulative effect of daily nonindustrial noise exposure encountered in everyday life. Clinically, it is usually difficult to distinguish between presbycusis, sociocusis, and noise induced hearing loss on the basis of the audiogram alone.

Hearing conservation

As the name implies, noise induced permanent threshold shift, which results from noise exposure, is permanent and there is no medical or surgical treatment that will restore hearing once it is lost. However, noise induced hearing loss can be prevented or greatly minimized by means of hearing-conservation programs. Through the combined efforts of the Committee on Conservation of Hearing of the American Academy of Ophthalmology and Otolaryngology, the American Industrial Hygiene Association, the Intersociety Committee for Guide Lines on Noise Exposure, and other interested groups, guidelines for permissible noise exposure and hearing-conservation programs have been established.[3]

Essentially a hearing-conservation program consists of three parts:

1. Analysis of noise exposure

2. Control of noise exposure

3. Measurement of hearing

A complete description of how to carry on each phase of a hearing-conservation program is described in detail in Chapter 33 and also the "Guide for Conservation of Hearing in Noise," published by the Committee on Conservation of Hearing of the American Academy of Ophthalmology and Otolaryngology.[3]

A useful rule of thumb for assessing the need for hearing conservation programs is when workers experience the following:

• Difficulty in communicating by speech, at arm's length, when in a noisy environment

• Tinnitus after working in the noise area for several hours or upon leaving work

• A temporary loss of hearing that seems to muffle speech upon completion of work, but which disappears after several hours of rest

These symptoms would be indications for performing more detailed noise analysis. The Intersociety Committee for Guide Lines on Noise Exposure (see Appendix A) proposes that continuous exposure to 90 dB on the A-scale be used as an indication for a more detailed noise analysis and a hearing-conservation program. It is important that a realistic view be taken in selecting the noise exposure level at which hearing-conservation programs would be indicated.

All hearing-conservation programs have certain limits. While protection of hearing of all employees at all audible frequencies is desirable, such a goal is not always attainable and therefore the program must be designed to protect the hearing of the majority of workers for those frequencies necessary for speech communication (500, 1000, and 2000 Hz). If a 90 dBA reading is used as an indication for hearing-conservation programs, it is estimated that approximately 90 percent of the workers would be afforded protection for the speech frequencies. If a 95 dBA value is chosen, about 80 percent of the workers would be protected over a lifetime of work.

Control measures

Control of noise exposure is a problem primarily for the qualified acoustical engineer and should not be attempted without proper consultation. Noise reduction may be accomplished by various means, such as the installation of quieter machinery, change in machine design, mounting of machinery, isolation of the machines from workers, and other engineering controls.

It has also been advocated that operational procedures be revised, so that the workers are exposed to noise for shorter periods of time during the working day. For many industrial operations, environmental control or machinery design has proven to be too expensive and impractical from an operational standpoint, and in these cases personal hearing protection devices, such as properly fitted ear plugs or muffs or a combination of both, have been advocated. For further details, see Chapter 20.

Personal protection

The use of personal protection is primarily a medical problem and should not be attempted without medical guidance and advice. Ear plugs, which are used as inserts into the ears in order to reduce the amount of noise reaching the inner ear, are made in various shapes, colors, and materials. Ordinary absorbent cotton, sometimes used by workers, has proven to have very little attenuation value. Waxed cotton and "cotton down," when tightly packed in the ear canal, have proven to be efficient. These substances are generally one-time affairs and have to be replaced daily. Earplugs made of plastic, neoprene, or other soft materials have proven useful and in most cases when properly worn can offer about the same degree of effective attenuation as earmuffs. The important factor is that an ear defender is worn.

The decision whether to wear earplugs or muffs depends a great deal upon the choice of the worker, the type of work, the environment he is in, etc. Actually there is very little difference in the attenuation afforded by the various earplugs or muffs, if they are properly worn. Unlike other safety equipment, such as safety hats, goggles, safety shoes, respirators, etc., it is difficult to determine whether or not the worker is wearing his earplugs snugly in the ear canal. Educational programs are necessary in order to keep the worker informed as to why he should wear hearing protection.

Hearing measurements

Measurement of hearing is one of the most important parts of the hearing-conservation program. The hearing threshold of the worker should be measured prior to his employment, and the results of these tests will provide a reference base-line audiogram from which subsequent changes in hearing may be determined. These tests may also serve for job placement purposes. The hearing tests performed are pure tone, air conduction audiometric threshold determinations. They must be made under suitable testing conditions, usually in a sound treated room. Followup tests are made as indicated depending upon the noise exposure and work situation. For further details, see Chapters 21 through 28.

Audiometric records

If manual audiometry is used, tabular numerical records are preferred to graphic records. Where single graphic audiogram records are kept on the same employee, the files become bulky and it becomes difficult to compare audiometric tests performed on different dates. A serial audiogram sheet in which each succeeding audiometric test for the same employee is recorded below the previous entry on the same form would be a more practical way of storing audiometric data. Examples of both types of audiometric records are shown in Figures 8-8 and 8-9, on the next page.

Referral for medical evaluation

Employees whose hearing ability is to be tested may present numerous problems to the industrial audiometric technician. For example, an employee may experience a sudden hearing loss in one ear. This may or may not occur as a result of acoustic trauma. Sudden hearing loss does not result from exposure to typical occupational noise levels. Medical conditions such as a viral infection or a vascular occlusion of a blood vessel in the inner ear can result in a sudden hearing loss. This is usually a serious condition and requires prompt medical referral.

Employees who show a marked unilateral or bilateral hearing loss should be referred to specialists for complete otological and audiological

239

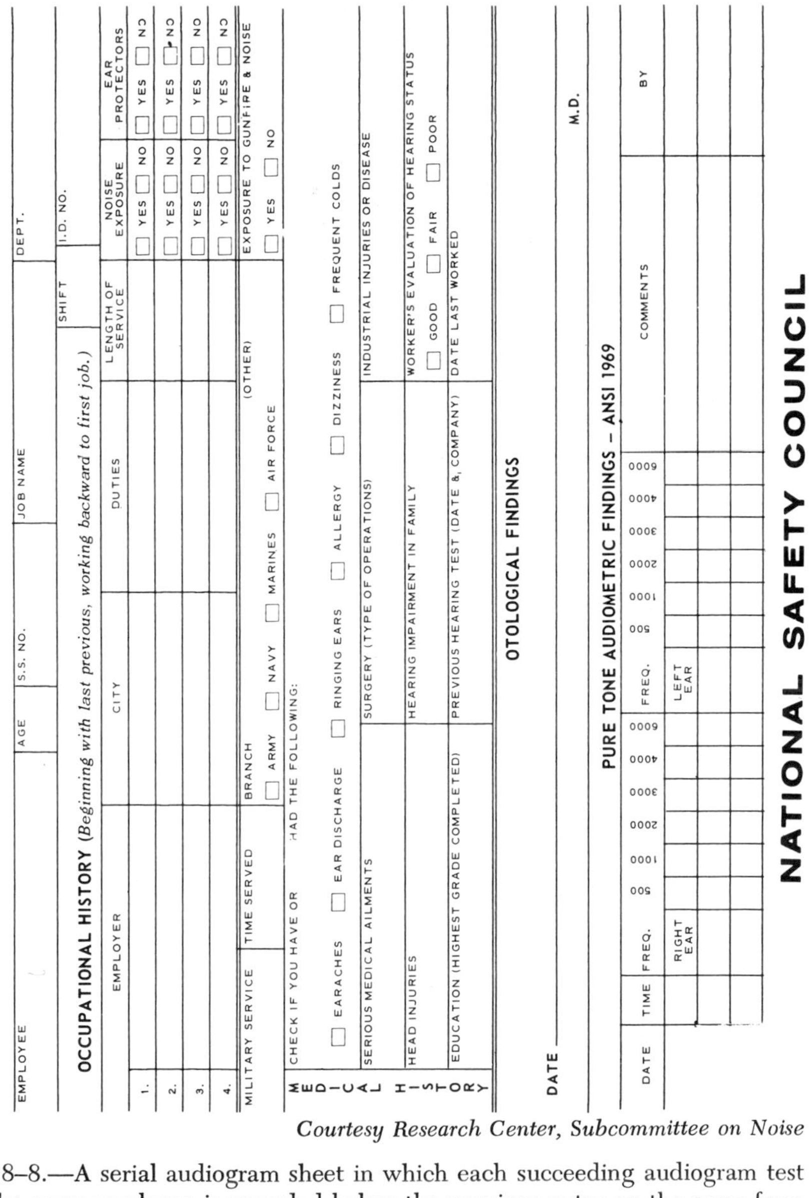

Courtesy Research Center, Subcommittee on Noise

FIG. 8–8.—A serial audiogram sheet in which each succeeding audiogram test for the same employee is recorded below the previous entry on the same form would be a practical way of storing audiometric data.

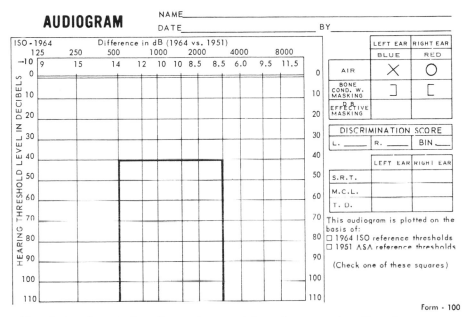

FIG. 8–9.—A typical audiometric record form for manual audiometry.

examinations to determine the cause of the hearing loss as well as possible indications for medical or surgical treatment.

Industrial audiometric technicians should be concerned only with the establishment of pure tone hearing thresholds through air conduction tests. If the results of the audiometric tests performed by the technician show that the worker has a significant hearing loss, further evaluation should be undertaken. It may be necessary to find out whether excessive noise exposure at the work location is present, whether or not hearing protection is being properly worn, etc. Reassigning this employee to a nonnoisy working area may be necessary. Employees with an active visible ear infection should be sent to a physician for treatment prior to performing hearing tests.

In summary, therefore, some indications for medical referral to a physician are one or more of the following:

- An active infection, such as a draining ear
- Obstruction of the external ear

241

- Complaints of severe tinnitus

- Complaints of recurring pain in the ear

- Dizzy spells and problems keeping balance

- A sudden or pronounced unilateral or bilateral hearing loss

- A hearing level threshold which has changed markedly in comparison with the previous audiogram

A complete description of a hearing-conservation program is offered in Chapter 29 and the "Guide for Conservation of Hearing in Noise."[3] The hearing-conservation program, in order to be meaningful, valid, and effective, should have medical supervision and guidance. As stated in the "Guide for Conservation of Hearing in Noise":

"Prevention, diagnosis, and treatment of hearing loss; validation and approval of audiometric records; and the final assessment of measurements of hearing are medical responsibilities. Any hearing conservation program without medical supervision must be considered inadequate.

"Direct medical supervision of a hearing-conservation program is highly desirable. The physician is responsible for the organization and administration of the testing program, as well as for checking and evaluating audiometric records. The physician himself does not perform all the operations necessary to the conduct of the program; he delegates responsibility for many of the technical activities to members of his staff, setting up standards or limits within which they can operate self-autonomously. Whenever medical records show the control of noise exposure may be inadequate, the physician in charge so reports. The responsibility for control of environmental noise exposure then devolves on the industrial hygienist, members of the engineering or safety departments, or other persons assigned to the task. Although the actual operations of measurement and protection are performed by both medical and nonmedical personnel, the physician ultimately is responsible for the health of the employee."

During the past decade, there have been great technological developments and achievements. During this same period, there have also been frequent frustrations in dealing with human, domestic, and international problems. This, too, has been the experience in dealing with the industrial noise problem, which involves many technical and professional fields. It also poses a challenge in human relationships. With an understanding of the problems created by industrial noise, with the realization that it makes good sense to protect human hearing, and with the cooperation of workers, management, unions, and medical and allied professions, the conservation of human hearing in industry can be accomplished.

References

1. Glorig, A., "The Effects of Noise on Hearing," *Journal of Laryngology and Otology,* 75:447-478 (May 1961).
2. "Guide to the Evaluation of Permanent Impairment—Ear, Nose, Throat and Related Structures," *Journal of the American Medical Association,* 535 N. Dearborn, Chicago 60610. 177:489–501 (Aug. 19, 1961).
3. "Guide for Conservation of Hearing in Noise," (a supplement to the *Transactions of the American Academy of Ophthalmology and Otolaryngology*) 1957, Revised, 1970. AAOO, 1100 17th St., N W., Washington, DC 20036.

Handicapping Effects of Hearing Impairment

Joseph Sataloff, M.D.

Poor speech reception is the most obvious handicapping effect of hearing impairment. In addition, there is a damaging effect on the individual's confidence in his ability to function effectively in his social and vocational life. The afflicted individual's natural optimism and belief in his personal competence to deal with his fellow man in a successful manner are undermined. The individual finds himself unsure, apprehensive, and resentful.

The Committee on Hearing of the American Academy of Ophthalmology and Otolaryngology on Hearing Handicap has provided the following definitions relating to hearing loss:

Hearing impairment. A deviation or change for the worse in either structure or function, usually outside the normal hearing range.

Hearing handicap. The disadvantage imposed by an impairment sufficient to affect one's efficiency in the situation of everyday living.

Hearing disability. Actual or presumed inability to remain employed at full wages.

Need for counseling

Most patients go to their physician for the relief of pain or discomfort or to allay their fears of deformity or serious disease. In contrast, a

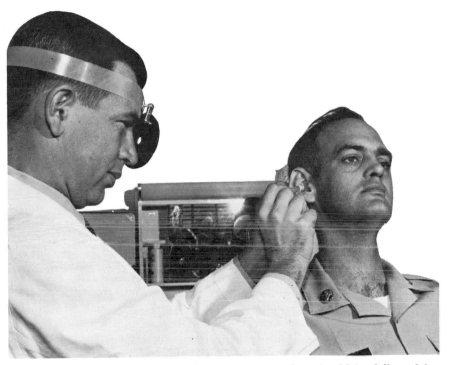

FIG. 9-1.—A hearing test that indicates a hearing loss should be followed by an examination for an accurate diagnosis of the cause.

hearing impairment is painless and rarely a direct threat to life or health. However, the serious social and vocational connotations of impaired hearing are the driving forces that impel the adult patient to see his physician.

Physicians cannot actually cure the hearing loss in all cases. Certainly they are incapable of restoring hearing completely in cases of sensorineural deafness. However, the physician can tell the patient whether medicine or surgery will help, and he can mitigate greatly the social and the emotional consequences of hearing impairment.

Unlike those suffering many other handicaps, the hard-of-hearing individual requires the integrated attention of a team of specialists. In many cases, one specialist alone cannot possibly diagnose, treat, rehabilitate and guide the individual in his efforts to overcome the communicative handicap imposed by his hearing impairment. Therefore, a team, frequently including an otologist, audiologist, and a speech therapist, as well as others, is called upon.

In brief, the physician who suspects the presence of a hearing loss should refer the individual for quantitative hearing tests as a basis for an accurate diagnosis of the cause of the hearing impairment (see Figure 9–1). The physician must then determine whether or not the hearing loss is curable and what measures are necessary to cure it as soon as possible. If a hearing aid can help, a suitable one should be selected and training in its use by an audiologist should be advised. If the patient has a psychological problem, he should be referred for an evaluation and possible counseling.

Consider for a moment what motivates a patient to visit a doctor and complain of hearing difficulties. More than likely it is the embarrassing situations that arise with increasing frequency in his everyday living. The patient may fail to hear or to understand his employer's verbal instructions, especially when there is much noise present. Or, a secretary with a hearing problem may fail to take dictation correctly, and the resulting mistakes can cause a great deal of tension in the office.

These situations are typical of the embarrassing circumstances that produce feelings of inadequacy and insecurity. Yet the patient is unable forthrightly to face the hearing problem that eventually leads him to seek the help of a physician. Furthermore, even when he does go to a physician to complain of hearing problems, he rarely goes of his own free will. The employee usually is nagged into going by his spouse, friends, or boss; sometimes they have been trying for months or years to get him to do something about his hearing difficulty.

Unfortunately, most patients do not really notice any hearing impairment until their hearing level has deteriorated rather markedly. In the early stages of a measurable hearing loss due to excessive noise exposure there are actually very few symptoms. For example, the patient may say that he cannot hear his watch tick in one ear as well as he can in the other. If he is of this observant turn of mind, he may seek help for not being able to hear his watch tick.

The serious effect on the individual's emotional state caused by deficient hearing is the major reason for the categorical statement that every patient with a hearing loss who visits his physician can be helped in some way. Although, it may not be possible to restore hearing to normal or to improve it to a nonhandicapping level, the patient can be taught to hear more effectively with what hearing he has left. A patient's pessimistic and antagonistic attitude toward his problem can be corrected. In many ways, he can be instructed to communicate better.

Effect on the personality

No other physical handicap has so many serious repercussions on the personality of a person as hearing loss has. Interestingly enough, some of the worst effects are associated with hearing losses that are comparatively mild in degree and classified as almost negligible on the basis of audiometric measurements. Conversely, many profound hearing losses may result in severe disruption of communication without seriously affecting the personality. The reasons for such differences are found in the strength or character of the individual and in his mental, spiritual, and economic resources to triumph over adversity and to make the most of his ability to find self-fulfillment and economic security, and participate in the joy of living.

Hearing loss cannot be restricted to the ear alone. It is not possible to divorce the ears from the brain that lies between them. Hearing is a phenomenon that utilizes the pathways between the ears and the brain and is an essential part of the mechanism of human response. In any discussion of hearing, communication, deafness, and handicap, it is necessary to think of the person as a whole and not merely as a pair of ears. For this reason, hearing loss concerns not only the general practitioner but also the otologist, the audiologist, the psychologist, the nurse, the psychiatrist and many others, including, more recently, members of the legal profession.

Reactions to hearing impairment

People react to hearing impairment differently. Some try to minimize or hide their hearing deficiency. Such a person, to keep up with a conversation, makes strenuous listening efforts and fills in hearing gaps by guessing, while he carefully conceals his frustration by acting particularly pleasant and affable. His effort to "save face" leads to numerous embarrassing situations. The effort becomes fatiguing and leads to nervousness, irritability and instability. He sits on the edge of his chair and leans forward to hear better. From the strain of listening, his brow becomes wrinkled, and his face serious and strained. Eventually, the individual becomes worn out from the effort to hide and deny his handicap.

Some people react to hearing loss, particularly that of slow and insidious onset, by becoming withdrawn and losing interest in their environment. This, the most common type of reaction, is reflected in an avoidance of social contacts and in a preoccupation with the subject's own misfortunes. He shuns his friends and makes excuses to avoid social contacts that might cause his handicap to become more apparent to his friends and to himself.

The hearing handicapped executive who must attend group meetings (such as planning or training sessions) soon realizes that he is missing some of the conversation and cannot keep up with what is going on. Rather than tolerate the criticism and caustic remarks reflecting on his alertness, he may resign and step down to a position less worthy of his potential.

When a salesman becomes hard of hearing, his business usually suffers, and his ambitions frequently are suppressed or completely surrendered.

A hearing impairment may not be a handicap, however, to a chipper or a riveter while he is at work. His deafness may even seem to be an advantage, since the noise at his work station is not as loud to him as it is to his fellow workers who have normal hearing. Because there is little or no verbal communication during most jobs that produce intense noise, a hearing loss will not be readily apparent to the exposed individuals.

However, when the employee returns to his family in the evening, the situation assumes a completely different perspective. He may have trouble understanding what his wife is saying, especially if he is in another room reading the paper, and his wife is in the kitchen. This kind of situation frequently leads to a mild dispute at first and later to serious family tension. The wife accuses the husband of inattention, which he denies, while he complains in rebuttal that she mumbles. Actually, the husband becomes inattentive rather than to try to cope with the frustrating and fatiguing strain of trying to decipher the muffled sounds.

When the hearing handicapped individual attends social meetings or visits with friends, he finds he cannot hear what is going on or is laughed at for giving an answer unrelated to the subject under discussion. He soon, but very reluctantly, realizes that something is wrong with him. He stops going to places where he feels pilloried by his hearing handicap. He stops going to the movies, the theater or concerts, for the voices and the music not only sound weak and far away but frequently distorted.

Borderline cases

Strangely enough, some of the most profound personality changes and communication handicaps occur in people who have a comparatively mild hearing loss. This effect is particularly evident in those individuals with otosclerosis (see Chapter 7). Personality changes are seen also in those who have an occupational hearing loss of moderate degree and in cases where the hearing loss is accompanied by distortion and tinnitus. Often, such an individual is constantly disturbed not only by the noises in his ears, the artificial tinny sound of music, the amplified sound of certain voices, but even more by the haunting fear that he will not be able to understand what people say to him. Sometimes he may hear very well and at other times he cannot understand what is being said.

The feeling of uncertainty and insecurity that results from so-called borderline hearing loss may lay the foundation for profound changes in a subject's personality structure that include irritability and suspicion. Frequently, catching only stray words of a nearby conversation, a person comes to feel that others are talking about him in a critical and derogatory manner. This makes him feel resentful and tends to bring out latent neurotic tendencies which might have been otherwise compensated for successfully.

High- and low-tone hearing loss

The employee who develops a noise induced hearing loss has, in the early stages, a high-tone loss (in other words, a loss without involvement of the frequencies below 2000 Hz). Consequently, there is little difficulty in hearing ordinary conversation, but on occasion some difficulty in understanding poorly enunciated speech or whispers. He also may have difficulty hearing speech in the presence of loud noise, because the noise masks out some of the discriminating characteristics of the consonants. It may be difficult to understand certain speech sounds on the telephone, television, records, movies or even radio, because the reproduction of voices through electronic amplification systems often reduces their clarity.

The word "deafness" commonly has been used to refer to industrial or occupational hearing loss, although it may be somewhat misleading. "Deafness" implies obvious difficulty in hearing speech. Actually, the difficulty lies not so much in hearing speech as in understanding it.

An example of this difficulty is reflected in the story of a man who was telling his friend about how good his new car was. "What *kind* is it?" his friend inquired. "About two o'clock," the man with the new car replied.

Hence, unless the physician specifically looks for this lack of consonant discrimination, he is likely to miss the diagnosis of industrial deafness in its early stages, the more so because the employee usually tries hard to compensate for his handicap.

Because of these factors, many representatives of management deny

FIG. 9–2.—An explanation should be given to the individual with a hearing problem so that he may understand why he has difficulty in hearing and discriminating speech sounds.

the presence of occupational hearing loss in their employees; they can walk repeatedly through their noisy shops and converse satisfactorily with all the employees. Too frequently, management fails to recognize that conversing in a noisy environment requires one to speak above the level of the noise, and that the vocabulary used in these brief conversations is such that an employee might be able to carry on that kind of limited and stereotyped conversation even though he has a very serious hearing loss.

Hearing loss in the lower frequencies

It is not until the noise induced hearing loss extends into the lower frequencies and becomes more profound in the higher ones that poor hearing and poor understanding of speech become a prominent symptom. By then, the damage to hearing is severe, irreversible, and often seriously handicapping. Furthermore, it assumes medicolegal importance as grounds for compensation (assuming that it is work connected). The major handicap in the severe cases is in communication, and because this enters into every phase of personal and business life, the "psychic fallout" very easily can cause personality problems.

A factor often present in industrial deafness is a reduction of a person's ability to understand speech because of distortion. For example, the reverberation of the announcer's voice at an airline terminal that makes the message difficult to understand for those with normal hearing becomes a completely unintelligible blur to the person with sensori-neural deafness. The louder the sounds become, the less he understands. Like letters written in ink on blotting paper, the sounds "Flight 231 leaving at 2:30 from Gate 9" dissolve and diffuse to a throbbing roll of thunderous sounds signifying nothing.

Recruitment

Persons with recruitment often will ask a person to speak a little louder, but when the speaker raises his voice a trifle more than necessary, the listener may think the speaker is shouting, but actually understand even less. Instead of helping the listener to understand better, increasing the loudness level for such individuals may actually reduce speech discrimination. This symptom of distortion is in itself one of the major causes for annoyance. It is natural that people who do not hear clearly should become frustrated and irritable.

Profound loss

In general, people with a profound hearing loss are somewhat easier to help than those who are considered to be borderline cases. Individuals with a profound loss are under greater compulsion to admit that they have a hearing handicap. People with a borderline loss tend to hide their handicap and deny it even to themselves. They conceal their impairment just as they try to conceal their hearing aids when they can be persuaded to use them. Of course, the major handicap to those with a profound hearing loss occurs in communication. Often these individuals cannot hear signals, such as horns or a ringing bell. This problem may be of particular importance to mechanics working with motorized equipment who are required to use their hearing to detect abnormal sounds arising from the operation of this equipment.

Another significant handicap to those having a more severe loss is the inability to determine the direction from which a sound is coming. This difficulty is particularly pronounced when the handicap exists in only one ear, or when the hearing loss in one ear is much worse than in the other, because two ears with reasonably normal function are needed to localize the source and the direction of sound.

Another interesting aspect of profound hearing loss is that after a time the person so handicapped tends to speak less clearly. His speech deteriorates, he begins to slur certain speech sounds, and his voice becomes rigid and somewhat monotonous. This frequently happens when a person can no longer hear his own voice clearly. At first he raises his voice often to the point of shouting. After a while, he may find this still unsatisfactory, and then he loses interest in his ability to speak clearly; often he does not even realize that his speech is deteriorating.

The reason is that one of the important functions of the ear is to serve as a monitor for speech. Hearing his own voice tells a person not only how loud he is talking but also whether he is modulating and pronouncing the words correctly. With the loss of this important monitoring system various speech and voice changes often occur.

A hearing loss in older persons, whether from causes associated with aging or owing to other sensori-neural causes, is often quite profound. All too often, the unfortunate oldster begins to believe that his inability to hear and to understand a conversation, particularly when several people are talking, is due to deterioration of his brain. This belief generally is forced on him by his family and friends, who disregard him in group conversation and assume the attitude that he does not know or care about what is going on anyhow, so why include him in the conversation.

Occasionally, the older individual will overhear a remark or notice a gesture signifying that he is getting old and slowing down. Such talk and such attitudes further undermine the old person's already weakened self-confidence and hasten the personality changes commonly associated with deafness and more particularly in the aged deaf. What a great injustice is done to all concerned!

Effect on social contacts

Unlike the blind and the crippled, hearing-impaired individuals give no outward indication of disability, and strangers are likely to confuse imperfect hearing with imperfect intelligence. This hurts the hard-of-hearing person's feelings; this and similar attitudes make for a strained relationship between a speaker and a listener. As a result, the hard-of-hearing person frequently limits his social contacts, which often leads to moods of frustration, insecurity, and even aggression.

The hard-of-hearing person misses "small talk." He does not get the flavor of a conversation so much enriched by side-remarks and innuendoes. This eventually makes him feel shut off from the normal-hearing world around him and makes him a prey to discouragement and hopelessness. Until a person loses some hearing, he can hardly realize how important it is to hear the small background sounds around him, and how much these sounds help him to feel alive, and how their absence makes life seem rather dull. Imagine not being able to hear the sounds of rustling leaves, footsteps, keys in doors, motors running, and the thousands of other little sounds that make human beings feel that they "belong."

When people become aware, if only to a small degree, of the possible effects of hearing loss, they wonder how a money value can be placed on such a handicap. Although the worker's compensation aspects of occupational hearing disability demand a standardized monetary value, a hearing loss is a personal tragedy to each individual. Furthermore, the handicapping effects of hearing loss are dynamic. They are changing even as this chapter is being written. With the development of new media for audiovisual communication, an individual's hearing comes to assume even greater importance.

For example, a hearing loss today is far more handicapping than it was before television, radio, and the telephone began to play such a major role in education, leisure, and the business world. Today the inability to understand speech on a telephone would be a major handicap for the vast majority of people. The loss of the ability to perceive high tones to a professional or an amateur musician or even to an audio high-fidelity fan also is handicapping. The hearing loss of tomorrow will have a different handicapping effect from the hearing loss of today.

Auditory rehabilitation

Although it is true that there is no medical or surgical cure for sensori neural deafness, a great deal can be done to help the individual to compensate for his hearing handicap and to lead as normal a life as possible with a minimal effect on his personality or his social and economic status. This is all done through a method described as "auditory rehabilitation." Thousands of servicemen with severe hearing loss were successfully rehabilitated through hearing centers established by the Army, Navy, and the Veterans Administration. Many otologists and audiologists also can provide some degree of rehabilitation for persons with handicapping hearing losses that cannot be corrected medically or surgically.

Despite the serious limitation of modern knowledge and therapy of so-called "nerve deafness," it can be stated truthfully that practically everyone with a handicapping hearing loss can be helped greatly by effective auditory rehabilitation. The principal objectives of such a program are to provide the following:

FIG. 9–3.—The person who suffers from a hearing disability frequently needs psychological counseling to help him accept the fact of his hearing loss.

- A clear explanation to the patient of his hearing problem

- Psychological counseling to help the patient and family when necessary

- Fitting of the proper type of hearing aid when indicated

- Auditory training to teach the patient to effectively use his residual hearing

- Training in speech reading when required

Explaining the problem

The handicapped individual should be given a clear understanding of his hearing problem and an explanation of why he has trouble hearing or understanding speech. This requires the otologist or audiologist to demonstrate to the patient, on a large diagram of the ear, just how the hearing mechanism works and where the patient's pathology lies (see Figure 9–2). The patient should also be given a clear understanding of the difference between having trouble hearing and having trouble in understanding what is heard. He should be told that the difficulty lies in the ears and not in the brain. It also should be explained to the patient why he has more difficulty in understanding speech when there is much surrounding noise, or when several people are speaking simultaneously.

Psychological counseling

A person with a hearing handicap needs to gain a more penetrating insight into the personality problems that may be already in evidence or likely to develop as a result of his hearing loss (see Figure 9–3). The individual must be treated in relation to his job, family, friends, and way of life. This is not a generalized technique but one that must be specifically designed to meet the needs of the individual whose hearing is impaired. The patient must accept his hearing impairment as a permanent situation and not sit idly by, waiting for a miraculous medical or surgical cure. Above all, confidence and self-assurance must be instilled in the patient. He must be encouraged to associate with his friends and not to isolate himself because of difficulties in communication. It must be impressed on him that by using effectively the hearing he has left, he can compensate and carry on with appropriate modifications.

Hearing aids

The fitting of the proper type of hearing aid, when it is indicated, is a vital part of the auditory rehabilitation program. However, before a patient can be expected to accept a hearing aid, he must be psychologically prepared for it (see Figure 9–4). Many people are reluctant to use hearing aids, and many who have purchased hearing aids never use them or use them ineffectively. Before recommending a hearing aid, it is necessary to determine whether the patient will be helped by it enough to justify his purchasing one. This is particularly important in sensorineural hearing loss in which the problem is more one of discrimination than of amplification.

One of the most important benefits that a hearing aid provides, in those types of hearing loss commonly found as a result of noise exposure, is to permit the individual to hear what he already hears but with greater ease. It minimizes the severe stress of listening. Although the individual may not be able to understand more with an aid than without one, he nevertheless may receive great benefit from the device, because it relieves him of the tension, the fatigue and some of the complications of hearing impairment.

The patient who seeks early medical attention for his hearing loss is wise in many respects. If his condition can be benefited by medical or surgical means, he has a better chance of being helped. If a hearing aid is necessary, the sooner the patient acquires it, the less severe will be the shock to his "nerves" when the hearing aid obliges him to listen to environmental noises that the patient may have been shielded from too long, such as the barking of dogs and the crying of a baby.

Thousands of hearing aids, bought and paid for and given too brief and half-hearted a trial, are relegated to a desk drawer. Overlong postponement in acquiring the aid is sometimes a factor. Often, too, the patient expects to hear normally with a hearing aid, when the condition of his hearing organs makes such a result impossible. The physician, audiologist, and the hearing aid dealer should make it clear to the patient that a hearing aid is not a substitute for a normal ear, especially in a patient with sensori-neural deafness. Other common causes of disappointment with a hearing aid are the patient's preoccupation with an "invisible" or inconspicuous aid, when what he should look for is an aid that will enable him to understand conversation with maximum effectiveness.

Auditory training

Auditory training should be given to teach the patient the most effective way to use his residual hearing with and without a hearing aid. If the patient can be aided with a hearing aid, he also should be made to realize that merely putting on the aid will not solve all of his hearing and psychological problems. The patient has to learn to use the hearing aid with maximum efficiency in such situations as one on one person-to-person conversation, listening to people in groups and at meetings, and when using the telephone. Above all, the patient must recognize the limitations of an aid, so that he will use it when it can be helpful and not use it or turn down the volume when it is more of a detriment than a help, as in certain noisy situations.

Speech reading

Speech reading involves a broader concept than just lipreading. If the individual with a hearing problem can be taught to use his residual hearing more effectively by looking more purposefully at the

FIG. 9–4.—If hearing aids are indicated, the physician should prepare the patient for accepting them and wearing them.

speaker's face and to develop an intuitive grasp of conversational trends so that he can fill in the gaps better than the average person. This is particularly important in patients who have profound hearing losses. It teaches the patient to obtain information from the speaker's face that cannot be obtained by voice communication. Most individuals do a large amount of speech reading naturally. By intensive training, a person can develop this faculty extremely well, though some individuals have a greater aptitude for speech reading than others.

A carefully planned and competently presented rehabilitation program can be of assistance to practically all people with handicapping hearing losses. Not only can it assist individuals to hear better, but, more important, they can be helped to overcome the many personality problems and psychological difficulties that result from hearing loss.

One factor that often complicates the problem of helping the hard of hearing is that they delay so long in obtaining medical attention. It is difficult to get some obviously hearing-impaired people to admit that they have a hearing problem at all, and even more difficult to convince them that they should see a physician about it. This is perhaps one of the reasons that physicians often do not see patients until their hearing (loss) has created marked social and communicative problems both to the patient and to his family and friends.

Summary

The emphasis in this chapter on rehabilitative measures short of medical or surgical cure reflects the fact that a total cure—especially in adults—is not possible in the great majority of cases of occupational hearing loss. The often dramatically successful middle ear surgery is limited mainly to the treatment of otosclerosis that has a reasonably good spread between air and bone conduction (air-bone gap). For such surgery to have a chance of success, the patient must have a cochlea in at least fair working order and a functioning auditory nerve. Although, these requirements are not met by a majority of hard-of-hearing patients, rehabilitation alone often can do a great deal.

The informed specialist can play an important part in helping patients over their psychological hurdles. A hearing aid, speech reading, and similar measures can only do so much and the patient still may need help in adjusting socially, economically, and emotionally to his continuing handicap. The hard-of-hearing patient must learn how to live with a hearing handicap. The informed specialist is the ideal person to share with his patients his knowledge of the causes of their hearing handicap, and to help them to overcome the psychological hurdles that loom so much larger in the average patient's mind than the hearing loss itself.

Bibliography

Sataloff, J. *Hearing Loss,* Philadelphia, Pa., J. B. Lippincott Co., 1966.
Sataloff, J., and Michael, P. *Hearing Conservation,* Springfield, Ill., Charles C Thomas, 1973.

Chapter Ten   Extra-Auditory Effects of Noise

Alexander Cohcn, Ph.D.

Noise is defined as "unwanted sound" because it may adversely affect man in various ways. Hearing impairment and interference with the reception of desired sounds are the two major problems posed by excessive noise exposure.[1-5] Both can be termed auditory (or aural) problems because they are the direct outcome of acoustic insults to normal ear functions or processes of sound transmission and perception.

In contrast, extra-auditory effects of noise constitute those noise-induced problems that extend beyond or are apart from strictly hearing or communicative disorders. These may include disturbances to physical and mental health, behavioral disruption and performance loss, interference with desires or needs for comfort, and the sense of well-being.

This chapter focuses on current thinking about extra-auditory problems in the interest of broadening concern for the full range of real or alleged effects of noise, primarily in the context of workplace noise conditions.

Mechanized job operations can be ranked highest in measured noise levels,[6] with many work routines necessitating frequent and prolonged exposures. Because of this, one could expect that the full array of adverse noise effects on man would be most clearly seen in the industrial setting.

Disturbances to physical and mental health

Disturbances to physical and mental health fall into four categories:

general stress reactions, physical disorders, mental and emotional difficulties, and special problems.

General stress reactions

Various effects of loud noise or sound upon different physiological functions have been reported.[2,3,5,7-15] Distinctive functional changes can be summarized as follows:

• There is vasoconstriction of the small blood vessels of the extremities that becomes progressively stronger with increasing sound intensity.[10] Coupled with this constriction appear to be changes in arterial blood pressure.[14,16] The latter may represent compensating heart action to overcome the constrictive effects of noise. Along these lines, decreased systolic and increased diastolic blood pressures have been observed in workers exposed to industrial noise during the course of their workday.[9,15,16]

• There is evidence of increased corticosteroids in the blood and urine signifying the release of hormones from certain glands.[8,9,13] This glandular response signals the preparation of body defenses to meet and cope with a challenge or threat.

• Respiration rate is decreased with reduction in salivary and gastric secretions resulting in a slowing of digestive processes.[7,9,17]

• The muscle reflexes tend to become more active, and there can be a general tightening of the body musculature, reflecting increased tension or readiness to act.[7-9] As will be noted later in this chapter, the occurrences of loud, unsuspected sounds can trigger reflex-like movements referred to as a *startle pattern*.[18]

Taken together, these physiological reactions to noise constitute a generalized response of the body to stress. Associated with these changes in physiological functions can be a variety of subjective responses also conveying apparent distress due to noise.[19] Depending upon the nature of the noise and the situation or circumstances in which it occurs, the listener may express any one or more of the following—fear, alarm, anxiousness, excitability, irritation, and displeasure.

Physical disorders

Granted that certain noises can induce physiological as well as subjective effects indicative of stress, can repeated occurrences of this type lead to health problems? Glorig,[20] Kryter,[1,3] and other leading noise

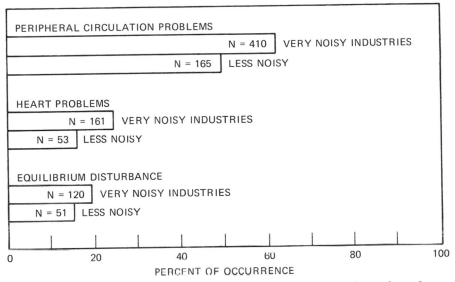

PERIPHERAL CIRCULATION PROBLEMS

N = 410   VERY NOISY INDUSTRIES

N = 165   LESS NOISY

HEART PROBLEMS

N = 161   VERY NOISY INDUSTRIES

N = 53   LESS NOISY

EQUILIBRIUM DISTURBANCE

N = 120   VERY NOISY INDUSTRIES

N = 51   LESS NOISY

0   20   40   60   80   100

PERCENT OF OCCURRENCE

FIG. 10–1.—These data were taken from 1005 German industrial workers showing a correlation between the noise level worked in and occurrence of physiological problems.[3] Peripheral circulation problems include pale and taut skin, mouth, and pharynx symptoms, abnormal sensations in extremities, paleness of mucous membranes, and other vascular disturbances.

experts[2] in the United States uphold the view that man's tolerance to noise is quite high, and that one can adapt to the noise conditions encountered in most present-day environments without harmful effects. Many of the aforementioned physiological responses to excessive noise have been observed to subside with recurrent exposures, suggesting accommodation and, presumably, no ill effects.

Other experts[8] argue that such observations are not taken over a sufficiently long period to judge the possible long-term costs of this adaptive process to the health of the organism. In elaborating on this point, there is evidence that prolonged exposure to high-level sounds can lead to physiological disorders. For example, guinea pigs subjected to intense siren-type sounds for fairly long periods of time eventually revealed depletion and imbalance in endocrine secretions and metabolic deficits, thereby reducing the animal's ability to cope with the sound stress.[11] Additional siren exposure caused gastrointestinal ailments, cardiovascular disease, and tissue damage to kidney and liver tracts. Reproductive dysfunction[12] and reduced resistance to infectious disease[8] have also been

261

reported in laboratory animals exposed to recurrent and prolonged high noise levels.

Results of these studies have not been without criticism. Miller[2] has noted that in some instances the experiments lacked certain controls, such as matched handling of test animals in both the noise and non-noise exposure groups. Also, rodents are known to have special susceptibility to the effects of certain sounds such than any effects realized here may not be generalizable to humans. Further, the sound levels generated in many of these experiments have usually been well above those normally encountered in man's present environment.

With regard to human exposures, there have been reports that couple intense chronic exposures to occupational noise with increased prevalence of circulatory and neurological irregularities in workers.[9,10,14] Jansen[10,22] reported additional cases of abnormal heart rhythms in steel workers exposed to high-level noise in their jobs as compared with those working in quieter areas. Also evident in the high-noise group were more frequent problems of equilibrium disturbance and peripheral circulatory ailments. Figure 10–1 summarizes Jansen's findings. Shetalov and Glotova[14] noted signs of bradycardia (slowed heart beat and arterial blood flow) in significant numbers of workers in noisy ball bearing and steel plants in Russia.

A more recent NIOSH-supported study[21] compared medical problems, accidents, and absence entries in the records of approximately 500 workers located in noisy work areas (measured sound levels were greater than 95 dBA) with a similar size group, nominally matched in age and length of plant experience, but who worked in quieter workplaces (less than 80 dBA). The record entries were collated for a five year period, 1966–1970. Table 10-A summarizes such data for the larger and also noisier of two plants comprising the record sources for this study. Workers subjected to the higher workplace noise levels tended to have a greater number of symptomatic complaints and diagnosed medical problems. Increased record entries here fell primarily into respiratory, allergenic, musculoskeletal, and cardiovascular categories. Respiratory problems common to workers in noisy jobs were hoarseness and laryngitis, apparently from shouting to communicate in high-noise levels. Symptoms and diagnostic signs belonging to the other problem categories were less specific in nature or origin relative to noise.

The fact that those who work in high-noise levels have more frequent medical problems than those who work under quiet conditions may still not offer conclusive evidence that noise is the crucial or main causal

TABLE 10-A

COMPARISON OF TYPICAL FREQUENCIES OF PROBLEMS

Medical Problems, Discrete Absences, and Accidents per Worker in High Noise Level and Low Noise Level Areas of a Boiler Fabrication Plant for a Five-year Period, 1966–1970.

Group	Noise* Level	Sample Tested	Typical Frequency of Occurrence Per Worker			
			Diagnosed Medical Problem	Sympto- matic Complaints	Discrete Absences**	No. of Accidents
Total Plant Group	High	456	3.9	1.7	30.3	9.0
	Low	449	0.4	0.3	4.2	0.4
Foremen	High	34	2.0	2.0	9.1	4.6
	Low	138	3.7	3.4	3.1	4.8
Test & Inspects	High	48	3.9	2.1	8.6	5.7
	Low	38	3.6	1.2	5.7	3.4
Plant Support	High	3	3.6	4.3	60.7	3.0
	Low	10	3.4	1.1	4.1	0.1
Administrative	High	10	4.5	3.3	74.7	7.6
	Low	45	0.8	0.2	4.7	0.3

* High = measured intermittent or steady noise level in work area, 95 dBA or higher
  Low = measured intermittent or steady noise level in work area, 80 dBA or lower
** Discrete absences = defined as a number of separate absence occurrences lasting one
  day or more in length

factor. In each case, it is possible that high-noise conditions may be specific to certain job tasks or work areas that entail greater health risks owing to increased workload or the presence of other noxious contaminants, such as toxic gases, dust, or excessive heat and humidity.

In this regard, comparisons of the frequency of diagnosed medical problems by specific job classifications in both high- and low-noise areas yielded fewer and less clear-cut differences than that shown in the total group data (see Table 10-A). In-plant staff with administrative and support responsibilities show the highest average frequencies of medical problems and workers situated in high-level noise had more symptomatic complaints. However, the relatively small sizes of these two groups of workers may make these results nonrepresentative. These same staff people also show the greatest number of discrete absences, which may be a direct function of their apparent recurrent health problems.

Mental and emotional difficulties

In addition to indications of increased physiological irregularities among industrial workers exposed to high-level noise are also frequent complaints of irritability, fatigue, and maladjustment. Jansen[22] has reported social conflicts both at home and in the plant for a significant number of workers engaged in noisy jobs. On the other hand, psychiatric interviews and psychological tests given to crew members on a U.S. aircraft carrier uncovered no unusual mental problems even though the men worked and lived under continual intense noise conditions.[23] The fact that aircraft carrier crews are a select group, hardened for combat operations, as well as the shortness of their experience as compared with the working lifetime of a civilian employee, limits the applicability of this finding.

Evidence for mental disorders based on noise experience outside the workplace is both sketchy and contradictory.[4,24]

Special problems

The physiological and psychological effects of noise already mentioned are a result of exposures to common sound sources emitting acoustic energy largely contained in the frequency range of 20 to 10,000 Hz. Potential health effects specific to exposure to significant sound energies at ultrasonic or infrasonic frequencies are in need of greater attention. To date, observations concerning infrasonic noise having strong energy below 20 Hz have been restricted to short-term, high-level exposures such as those encountered in launching rocket boosters or in certain phases of rocket acceleration.[25] The levels of 140 dB or more (sound pressure level re 0.0002 microbar), observed in these instances, mechanically drive the chest inward, causing complaints of substernal pressure, coughing, and gagging sensations.

These reactions are further accentuated if there are also present strong audible sounds of low frequency, which can excite the mechanical resonances of the chest. No conclusive information is available in the literature regarding long-term exposure problems or possible adaptation. Also, possible problems of physical fatigue following infrasonic exposure have not been seriously investigated.

Workers using ultrasonic cleaners and drills have complained of subjective effects, including fatigue, headaches, nausea, and malaise.[26,27,28]

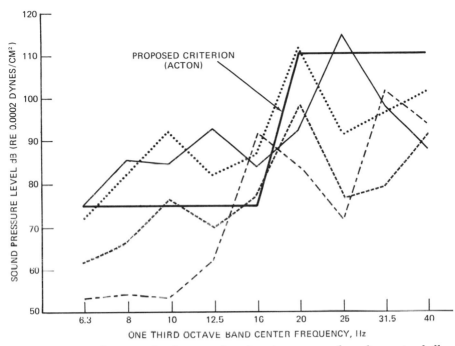

Fɪɢ. 10-2.—Select one-third octave band noise spectra for ultrasonic drills and cleaning equipment yielding subjective effects as compared with proposed criteria (taken, in part, from Acton[28]).

The occurrence of such effects may depend more on the presence of sound energy at the upper end of the audible frequency range rather than on that in the ultrasonic region, that is, above 20,000 Hz (see Figure 10-2).

Energy around 16,000 Hz seems particularly critical in this regard, showing covariation with the symptoms noted. In fact workers who cannot hear these high-frequency sounds do not seem to display the same ill effects as those who have better hearing.[27,28] At this time no information is available on the effects of long-term exposure and possible adaptation to high-frequency sounds.

There is also concern whether excessive noise poses any added hazards to persons with pre-existing health problems. The clinical literature cites individual cases where noises have provoked seizures in persons with epilepsy or caused headaches in those suffering from migraine problems.[29,30] The generality of these findings remains to be demonstrated.

In fact little systematic information is available describing the noise stress tolerance of persons with chronic cardiovascular, neurologic, and gastro-intestinal ailments. Presumably it would be lower than that for persons in normal health. There is also the great likelihood that those unduly distressed by noise (or other stress-producing conditions) would remove themselves from the sources or places of such disturbances.

Some evaluative comments

It is evident from the reports of investigations presented here that no conclusive statements can be offered about the occurrences of physical or mental health problems due to excessive noise exposures. The results to date are suggestive, however, and should invite much more study aimed at clarifying the effects of noise in terms of extra-auditory factors. It is truly surprising that so little work has been done in the United States in this problem area.

Epidemiological surveys concerned with the incidence of acute and chronic ailments in different work groups exposed to high-level noise have never been undertaken in this country and are greatly needed. Indeed, U.S. noise experts show skepticism as to the findings of European research reports acknowledging noise induced health problems, but remain reluctant to become involved themselves in the research basic to validating or disproving such findings.

Disruption of behavior

Aside from possible disturbances to health, noise may disrupt behavior or induce alterations in one's behavioral capacities. Behavioral disruption is most clearly seen in exposures to sudden, unexpected, and loud explosive-like sounds that provoke a startle response.[18,31,32] High-speed photography of human subjects when they are presented with such sounds has served to isolate some of the characteristic elements of the startle pattern.[18] These include flexing of the arms, arching of the torso, widening of the mouth, and blinking the eyes (see Figure 10–3). These reactions in themselves may have little significance, but a sudden noise in certain industrial or other added risk situations could startle a person into injuring himself or others.

Questions remain as to the dependency of startle response on different features of the acoustic stimulus and whether it will habituate with repeated exposures.[33] Some components of the startle reaction do diminish with recurrence of the insults.[31,32] Other elements, particularly blinking the eyes, seem to resist habituation.[18] Shooters, for example, have an eyeblink response in firing their weapons after years of such activity. This finding is even more remarkable because the shooter expects the sound and the element of surprise is eliminated.

Research on startle reactions of animals in response to loud impulsive sound has shown that regularly preceding the stimulus with one of lesser intensity serves to reduce materially the amount of startle re-

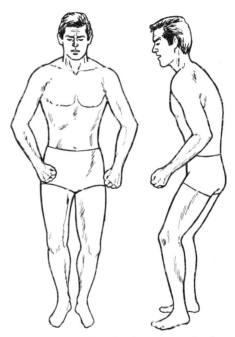

Fig. 10–3.—A sketch of a person displaying the startle response pattern.[18]

sponse.[32] On the other hand, startle reactions to gunshot-like sounds are less intense and more readily habituated in a quiet background than under moderate steady-state background noise conditions. This result seems counter to the widely held notion that the presence of significant background noise helps to mitigate the disturbing effects of intruding sounds of high level.

Steady-state noise may also induce convulsive reactions in rodents. This phenomenon, known as *audiogenic seizure,* is uncommon in man.

267

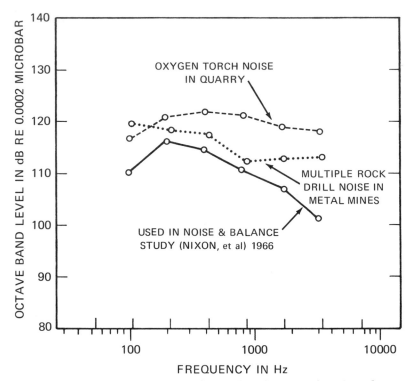

Fɪɢ. 10–4.—Octave-band levels (in dB) for noise found at the ear of a quarry worker with an oxygen torch, a miner operating a multiple rock drill (jumbo), and noise levels used in a study by Nixon, *et al.*

Intersensory effects

Steady-state noise of fairly high level may have certain effects on non-auditory sensory functions that deserve mention especially in the context of safety. For example, Stevens[34] found that exposure to high-level noise (120 dB) reduced the speed by which the eyes moved through certain angles to focus clearly on objects. This result was believed due to noise affecting the ciliary muscles, which control the curvature of the lens of the eye. In another experiment, Benko[35] examined workers exposed to 100 to 125 dB noise and reported a narrowing of their visual field as well as some modification of color perception. The latter suggested a partial deficiency for perceiving red, termed *protonomalia*. Complaints of *nystagmus* and *vertigo* have also been noted under noise conditions

Courtesy Canadian Mining Journal.

FIG. 10–5.—A tread-mounted jumbo rock drill being used in a mining operation. Workers should wear hearing protection whenever operating such drilling equipment. Consideration should also be given to the use of special mufflers on exhausts to reduce the noise level.

TABLE 10-B
FACTORS IN EVALUATING
NOISE PERFORMANCE EFFECTS

Independent Variables

Acoustic-Environmental
Sound level: high vs. low
Exposure condition: continuous vs. intermittent
Type of noise: impulse vs. steady-state; familiar vs. unfamiliar; high pitched vs. low pitched
Noise combined with other physical stressors

Task
Cognitive vs. mental vs. perceptual requirements
Added features: simple vs. complex; paced vs. unpaced; workload; practiced vs. unpracticed

Individual Differences
Attitude: acceptance vs. rejection; ability to control Personality aspects
Psycho-physiological aspects

Dependent Variables

Gross
Total output
Total errors
Finite
Variations in error rates
Variations in performance rates

utilized in the laboratory and found in the field. Levels needed to cause these latter effects are quite high, perhaps 130 dB or more.[5,36] Less intense noise conditions (about 125 dB) can upset one's balance especially if the noise stimulation is unequal at the two ears.[37] This was most clearly shown in a laboratory study where subjects were required to balance themselves and walk on rails of different widths. These disturbances to equilibrium are believed due to noise directly stimulating the vestibular organ, whose receptors are part of the inner ear structure.[38]

There are references to sound affecting still other nonauditory sensory functions. Workers with histories of long-term noise exposure indicate loss of tactile sensitivity in some instances.[5,22] Moderately high-level acoustic stimulation also seems to have analgesic (pain suppressing) properties, and it has been used in place of usual anaesthetics in minor surgery.[39] While its use as an anaesthetic is beneficial, loss in touch and pain sensibilities due to noise could also rob the perceiver of cues of-

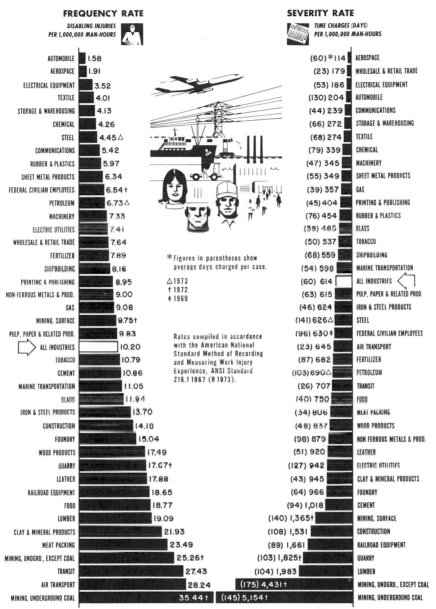

FREQUENCY RATE

DISABLING INJURIES
PER 1,000,000 MAN-HOURS

AUTOMOBILE	1.58
AEROSPACE	1.91
ELECTRICAL EQUIPMENT	3.52
TEXTILE	4.01
STORAGE & WAREHOUSING	4.13
CHEMICAL	4.26
STEEL	4.45△
COMMUNICATIONS	5.42
RUBBER & PLASTICS	5.97
SHEET METAL PRODUCTS	6.34
FEDERAL CIVILIAN EMPLOYEES	6.54†
PETROLEUM	6.73△
MACHINERY	7.33
ELECTRIC UTILITIES	7.41
WHOLESALE & RETAIL TRADE	7.64
FERTILIZER	7.89
SHIPBUILDING	8.16
PRINTING & PUBLISHING	8.95
NON-FERROUS METALS & PROD.	9.00
GAS	9.08
MINING, SURFACE	9.75†
PULP, PAPER & RELATED PROD.	9.83
ALL INDUSTRIES	10.20
TOBACCO	10.79
CEMENT	10.86
MARINE TRANSPORTATION	11.05
GLASS	11.94
IRON & STEEL PRODUCTS	13.70
CONSTRUCTION	14.18
FOUNDRY	15.04
WOOD PRODUCTS	17.49
QUARRY	17.67†
LEATHER	17.88
RAILROAD EQUIPMENT	18.65
FOOD	18.77
LUMBER	19.09
CLAY & MINERAL PRODUCTS	21.93
MEAT PACKING	23.49
MINING, UNDGRD., EXCEPT COAL	25.26†
TRANSIT	27.43
AIR TRANSPORT	28.24
MINING, UNDERGROUND COAL	35.44†

SEVERITY RATE

TIME CHARGES (DAYS)
PER 1,000,000 MAN-HOURS

(60)	*114	AEROSPACE
(23)	179	WHOLESALE & RETAIL TRADE
(53)	186	ELECTRICAL EQUIPMENT
(130)	204	AUTOMOBILE
(44)	239	COMMUNICATIONS
(66)	272	STORAGE & WAREHOUSING
(68)	274	TEXTILE
(79)	339	CHEMICAL
(47)	345	MACHINERY
(55)	349	SHEET METAL PRODUCTS
(39)	357	GAS
(45)	404	PRINTING & PUBLISHING
(76)	454	RUBBER & PLASTICS
(39)	485	GLASS
(50)	537	TOBACCO
(68)	559	SHIPBUILDING
(54)	598	MARINE TRANSPORTATION
(60)	614	ALL INDUSTRIES
(63)	615	PULP, PAPER & RELATED PROD.
(46)	624	IRON & STEEL PRODUCTS
(141)	626△	STEEL
(96)	630‡	FEDERAL CIVILIAN EMPLOYEES
(23)	645	AIR TRANSPORT
(87)	682	FERTILIZER
(103)	690△	PETROLEUM
(26)	707	TRANSIT
(40)	750	FOOD
(34)	806	MEAT PACKING
(48)	837	WOOD PRODUCTS
(98)	879	NON FERROUS METALS & PROD.
(51)	920	LEATHER
(127)	942	ELECTRIC UTILITIES
(43)	945	CLAY & MINERAL PRODUCTS
(64)	966	FOUNDRY
(94)	1,018	CEMENT
(140)	1,365†	MINING, SURFACE
(108)	1,531	CONSTRUCTION
(89)	1,661	RAILROAD EQUIPMENT
(103)	1,825†	QUARRY
(104)	1,983	LUMBER
(175)	4,431†	MINING, UNDGRD., EXCEPT COAL
(145)	5,154†	MINING, UNDERGROUND COAL

*Figures in parentheses show
average days charged per case.

△1973
†1972
‡1969

Rates compiled in accordance
with the American National
Standard Method of Recording
and Measuring Work Injury
Experience, ANSI Standard
Z16.1 1967 (R 1973).

Courtesy Accident Facts 1975.

CHART I.—Injury Rates in Different Industries as Reported to the National Safety Council.

271

fering early warnings of impending danger.

The sensory alterations associated with noise might be dismissed by some individuals as having little importance in rating noise hazards at the workplace. That is to say, truly high noise levels are required to evoke these changes in most cases, and the magnitudes of the resultant sensory alterations can be rather small. Yet, noise levels of industrial equipment and operations can equal or exceed those necessary for inducing nonauditory sensory effects. For example, the octave-band levels used in the test of rail balancing under noise are slightly lower than similar spectral values of noise measured at the ears of metal miners operating jumbo rock drilling equipment and quarry workers with oxygen torches (see Figures 10–4 and 10–5). While perhaps slight, the changes in sensory functions caused by noise could decrease needed safety margins in risky jobs. In this regard, many of the noisiest jobs are also those having the highest accident and injury rates per manhours of work (see Chart I).

Effects on performance

Most investigations of noise effects on behavior have been conducted in the context of performance evaluations. Attempts at demonstrating well-defined effects of noise have generally proven disappointing. While anecdotal references suggest that a person's ability to cope with job demands might be impaired by noise, research studies do not yield consistent data to support this contention. Studies of performance under noise sometimes show losses; sometimes improvement, and, in many instances, no significant change when compared with performance under non-noise or quiet conditions.[3,19,41,42] Under certain test conditions, performance has been shown to improve as well as to deteriorate in the course of the same noise exposure. Clearly, a simple, concise description of noise effects on work performance is quite elusive.

The material presented here will elaborate on the complex nature of noise effects on performance. Specifically, it will enumerate factors which singly, or in combination, influence such effects. These factors, shown in Table 10-B, are grouped under headings of acoustic (or other environmental) variables, task performance variables, and human subject variables.

Acoustic variables

Identified under this heading are noise levels, types of exposure, and other such factors that have been evaluated for their effects on assorted

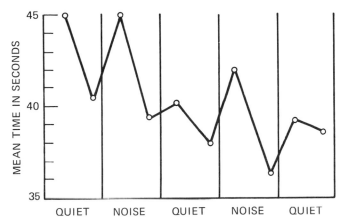

Fig. 10–6.—Results of a study that determined the time needed to code letters of different colors into a prescribed digital form under intermittent noise conditions. The initial slowing down in both the noise and quiet intervals was soon replaced by improved performance.[41]

performance tasks. Even the simplest task performed under quiet and increased noise conditions has varied consequences. For example, with repetitive, simple tasks, a moderately intense noise background might facilitate performance as compared with quiet conditions. This beneficial effect is explained by noting that simple tasks do not demand undivided attention and, thus, are more prone to distractions caused by stray low-level sounds that may occur in the work situation. A moderately intense noise, by masking these kinds of sounds, eliminates such sources of distraction.

The presence of an intense noise is also believed to supply added stimulation to a simple performance situation to keep the subject alert in a task that is otherwise monotonous or boring.[43] On the other hand, this same noise may disturb those individuals in a work situation that requires complete and constant attention to detail.

Noise performance studies involving manipulation of noise levels are further complicated by the phenomenon of loudness adaptation.[44] This refers to a fairly rapid decrease in the perceived loudness and the sensitivity to a sound with continued exposure. Loudness adaptation is complete for most sounds in 10 minutes, with sounds of greater intensity showing more sizable adaptation effects. After 10 minutes of exposure, for example, a 100 dB sound would have dropped the equivalent of 20 dB in apparent loudness as compared with a 70 dB sound, which would show little adaptive change. Such differences in loudness adaptation can nullify the physical differences between high- and low-level noise test conditions and thus suppress possible performance effects. This type of adaptation, in addition to other forms of adaptive psychological and physiological processes, might explain why performance losses measured at the outset of a noise test situation may dissipate later on in the testing.[45]

Adaptation can be offset or delayed by using intermittent background noise. Such exposures may yield poorer performance than that found under continuous noise conditions, especially for complex tasks.

An interesting point about the effects of intermittent noise effects on performance is that losses occur both when the noise is turned on and also when it is turned off. That is, the losses in performance here are not peculiar to the presence of noise, but rather they seem to be a function of changes in the stimulus conditions. Figure 10–6 describes the results of a study made under intermittent noise conditions. Note that there is an increase in mean coding time (a decrease in performance) just after the noise is turned on and just after it is turned off. The initial slowing down of coding time in both the noise and quiet intervals is soon replaced by improved performance.

High-pitch or unfamiliar sounds, because of their greater annoyance or attention-getting value, might cause greater losses in performance than low-pitch or familiar sounds. Here too, however, there are exceptions. A background of near-intelligible speech sounds can be a potent distraction in a work situation and cause poor performance.

Impulsive sounds presented at random intervals interfere more with task performance[46] than does steady-state noise. The performance losses for impulse sounds are confined, however, to the brief periods during or immediately following their occurrence.

Background music. Rhythmic or throbbing types of noise may actually pace performance and accentuate movements or other activities that are in phase with the noise. These so-called "dynamogenic" effects can serve to sustain performance on jobs that might be physically tiring. Related to this point, music programmed with increasingly livelier tempo was found to maintain, if not facilitate, reaction-type measures taken throughout a task session involving visual detection.[47] The musical program, when presented in the reverse order (decreasing liveliness), resulted in relatively slower response times for a matched test group in an otherwise identical task session. These findings suggest that more stimulating

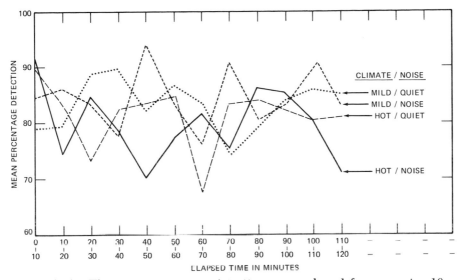

FIG. 10–7.—The mean percentage detection scores plotted for successive 10-minute intervals observed for four ambient conditions, namely, hot-noise, hot-quiet, mild-noise, and mild-quiet.[42]

music should be played toward the latter part of the workday or shift when there may be a tendency toward decreased work output.

Music in industry, generally speaking, seems to create positive attitudes in many employees toward their work whether or not it has any significant effects on their productivity.[41,48]

Environmental stress. Several studies have evaluated performance changes due to noise combined with other types of environmental stress with the results not always being predictable.[42,49,50] One such investigation involved combinations of noise and heat stress conditions.[42] The task was to detect infrequent signal deflections on a 10-dial panel during a two-hour observation period. Plotted in Figure 10–7 are mean percentage detection scores for successive 10 minute intervals as obtained under four ambient conditions, namely, hot noise, hot quiet, mild noise, and mild quiet.

The "hot" thermal conditions represented a 92 F (33 C) dry-bulb, 82 F (28 C) wet-bulb state (about 65 percent relative humidity (RH) and very uncomfortable); the "mild" thermal condition was 70 F (21 C)

275

dry-bulb, 50 F (10 C) wet-bulb (about 20 percent RH). The "noise" condition was a 95 dB broadband sound presented continuously; "quiet" represented a background sound level of 60 dB.

While performance vacillated for all environmental conditions, the hot quiet condition resulted in poorer performance than the mild quiet condition for many of the time intervals. Combining the noise exposure with heat stress, however, did not materially worsen performance. Rather it tended to shift the time course of the losses noted for just the heat stress condition; what happened was that the addition of noise to heat stress advanced the dips in performance to relatively earlier intervals in the test session.

Under mild thermal conditions, there was a significant improvement in detection performance at certain points in the observation period relative to the mild quiet condition. This latter effect might reflect the stimulating effects of noise that serve to maintain the alertness of the subject. Whether these facilitating effects of noise would be sustained for test conditions of longer duration is not known. In fact, truly long-term evaluations of noise effects on performance have not been undertaken.

Renshaw[50] has recently studied performance on a multiple-choice serial reaction test under variable noise and heat conditions presented singly and together. Significantly greater tendencies toward blocking or gaps in responding to the task elements were noted under the combined noise and heat conditions as compared with single exposure to either agent. Harris and Sommer[49] have also reported combined noise and vibration conditions cause greater impairments in performance on short-term memory and subtraction types of mental tasks than individual exposures to either stressor.

Task variables

Some attention has been directed toward identifying the makeup of performance situations that are peculiarly disrupted by noise. Obviously, a task that requires reliable voice communication will be hindered by noise because of the masking effects of noise on speech. At issue here, however, is whether certain kinds of task performance, not dependent on voice communication or other sound signals, can also be degraded by noise.

Efforts designed to show that noise has differential effects on perceptual, psychomotor, and cognitive types of tasks have not been very rewarding.[3,19,42]

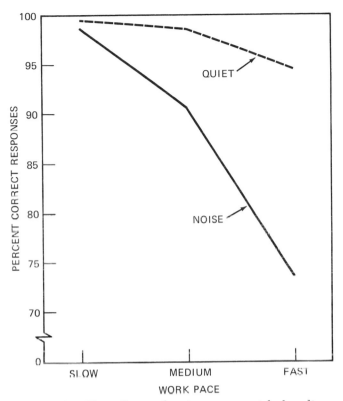

Fig. 10–8.—The effects of noise on a serial decoding operation involving multiple choice responses to light signals dependent on a programmed code. The signals were presented at three different rates: 30, 40, and 50 signals per minute under 50 dBA (quiet) and 100 dBA aperiodic noise conditions.[51]

Perceptual and psychomotor tasks embodying reaction-time measurements, response to faintly illuminated targets, and tracking a moving target via various control movements reveal little or no noticeable changes due to excessive noise. Similarly, mental tasks such as addition or multiplication, word puzzle solving, letter and number sorting, and coding show only insignificant differences between noise and quiet test conditions.

The adverse noise effects on performance seem more dependent on the apparent difficulty or demand imposed by the task than the particular

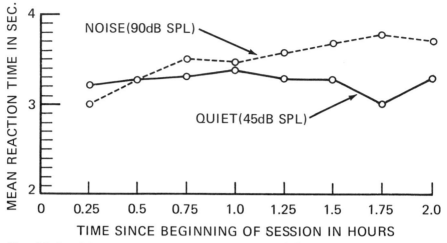

Fɪɢ. 10–9.—Mean reaction time to an occasional faint visual signal in two levels of noise.[41]

behavioral skill being tapped. For example, unpracticed or unfamiliar tasks of all types are more subject to noise induced impairments. The responses required by such tasks, lacking in habit strength, are simply less resistant to the interference or distractive effects of the noise intrusion. In this regard, overlearning a job or work routine can preserve the work performance record under the most stressful situations.

Machine-paced or programmed types of work are also more likely to be affected by noise than tasks that allow a person to work at his own rate. A worker whose performance is momentarily lowered by the presence of noise could offset this loss by working harder in the remaining time on the task. Such opportunities for compensation are not provided in machine-paced operations. This result would hold implications for inspection and assembly line workers who are permitted only limited time to make a judgment or to perform certain operations on a flow of production items.

By increasing the pace of required work, performance losses caused by noise can be amplified. A recent study varied the presentation rates of light signals requiring differential response under 50 dBA (quiet) and 100 dBA periodic noise conditions. The results for these conditions of pacing are shown in Figure 10–8.

Perhaps the most reliable indications of adverse noise effects on performance are found in task situations requiring complete and unremitting attention to complex displays or events. Vigilance tasks are of this nature.[41,52] In general, they require the subject to maintain a watch over numerous dials or indicators, any one of which may show a faint deflection at any time. The number of deflections correctly detected and the speed of detection constitute measures of performance and both display losses under high-level noise (see Figure 10–9).

Noise effects on vigilance are believed especially important because of their relevance to job situations involving monitoring. One field situation has in fact found performance to improve with reduction in noise level.[52] In this instance, cotton weavers, supervising the operations of a number of largely automatic machines, had to be constantly alert for stoppages caused by thread breakage. Such stoppages required quick detection and repair in order to get the machine back in operation without undue delay. Reducing noise levels through the use of ear plugs caused an estimated increase of 12 percent in individual weaver efficiency.

Secondary tasks have sometimes been added to primary ones so as to increase subject burden in noise-performance testing as a means of dramatizing the disruptive effects of noise. For example, subjects had to perform a choice reaction time task at two levels of complexity, and at the same time monitor auditory signals in a secondary task. These simultaneous tests were given in both quiet and in repeated bursts of sound (not coincident with the auditory signals). The experiments showed that secondary-task errors were greater in noisy environments when subjects were tested on the more complex reaction time task. No significant noise effects on the primary reaction time task were found, irrespective of complexity level.[53]

These results suggest an explanation for the failure of noise to have more disturbing effects on simple performance tasks as compared with more demanding ones. Briefly stated, simple tasks demand less than the total performance capacity of the individual. Thus, noise disturbances can be overcome by drawing upon more of this performance reserve. A more demanding, complex task, however, preempts much of this reserve and leaves little, if any, capacity to withstand an added burden such as noise.

The results of a dual task study[54] suggested a variation on this idea in appreciating noise effects in complex task situations. Subject performance was measured under 100 dBA noise on a primary tracking task, which involved following the movements of a pointer in the central visual field while at the same time monitoring flashes of light located at the periphery. Relative to testing under control conditions (65 dBA noise), high-level noise was found to facilitate performance on the primary task (tracking) and to impair it on the secondary one (peripheral monitoring).[54]

It was concluded that the effect of noise in this situation makes attention more selective, especially enhancing concern for the high-priority aspects of a task and neglecting the low-priority elements.

Human subject variables

Individual differences are quite commonly found in the investigation of noise effects on performance. An individual's attitude about noise may be a basic factor in this variance. In highlighting this factor, one study[55] employed three groups of subjects: one group being told that noise would improve performance; another group, that it would cause poorer performance; the remaining one, that there would be some initial improvement followed by a loss.[55] Upon performing the task in a noisy environment, the results of each group differed from one another in directions agreeing with the preliminary briefing. Poorer performance in noise can result from a belief that noise impairs efficiency.

In a more recent study,[56] subjects believing that they had no control over randomly occurring noise intrusions performed much poorer than those who felt that they could terminate such sounds.

Personality factors also seem to influence performance variations. Anxious, tense persons (as defined by personality inventories and certain physiologic indicators) seem less able to cope with noise performance tasks.[57]

The importance of individual-difference factors in noise performance effects becomes even more evident in field investigations. For example, morale and ego involvement in a job can override stresses imposed by noise.[58] Other employees dissatisfied with their work can use the same noise condition as an excuse for poor performance.

It should be mentioned too that, through a process of self-selection, only the more noise tolerant people probably stay in noisy jobs. The more noise sensitive individuals would remove themselves from these situations. Measures of absenteeism or labor turnover should reflect this occurrence but may not always be causally related to noise.

The previously mentioned NIOSH evaluation record of workers in noisy versus quiet jobs is of relevance here because absence entries were compared between groups.[21] Table 10–A shows the typical number of discrete absences per worker in noisy jobs to be almost six times greater than for those in quiet jobs. When equated for job types, differences

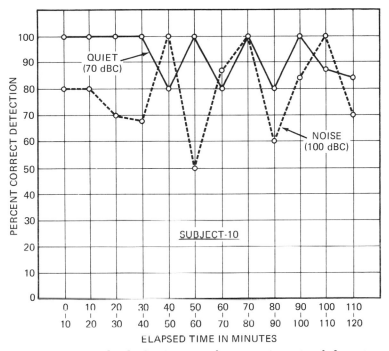

Fig. 10–10.—Individual subject performance in a signal detection task under two levels of noise. Note the wider performance variations under the high level noise as compared to the quieter condition.[19]

in absenteeism were somewhat less between workers in noisy and quiet areas two groups (foremen and test inspectors), but differences were greater in two other groups (plant support and administrative staff).

Also recorded in Table 10–A are more frequent accidents per worker in noisy versus quiet areas, based on total sample data. This result could be due to the higher accident rate peculiar to jobs typically performed in noisy areas relative to those conducted in quieter ones. Consideration of accident frequency rates in specific jobs again shows smaller differences relative to the overall result. As a followup to this study, the Public Health Service plans to evaluate medical, accident, and absence statistics subsequent to the establishment of a hearing protection program in the plants originally studied. Reduction of noise levels through the use of hearing protectors should also diminish the frequency of occurrence of the various problems noted if, in fact, excess noise was the significant causal factor.

In reflecting on the complexity of noise effects on performance, it is important to realize that noise may have greater influence on performance measures other than gross output or total number of errors. For example, performance under noise may be subject to marked fluctuations with periods of poorer work being interwoven with periods of heightened efficiency. These performance swings, when summed over the total work session, may reveal little or no overall performance change compared with a quiet work condition. In short, noise may have more effect on the quality or the consistency of performance than on performance output (see Figure 10–10).

The array of factors influencing noise performance changes makes any conclusions or predictions highly qualified and very conservative. It would seem that intense sounds, impulsive or irregular in occurrence, combined with a taxing task performed by a tense, nervous person offer the combination of conditions most likely to display a noise induced decrease in performance. Whether these performance changes could lead to accidents is speculative, and would depend on the particular circumstances that are present.

Aspects of annoyance

Although annoyance from noise stems in large measure from the interruption or interference with desired or purposeful activities, some noise features are perceived as being inherently more objectionable or bothersome than others. For example, the sound of chalk scraping on a blackboard can cause discomforting chill-like feelings in the listener and avoidance actions.

Annoyance ratings and paired comparison procedures, in which observers are asked to judge the acceptability of a variety of test sounds against a reference sound, have been used to learn how different sound features might affect perceived annoyance.[3,59] The results of such studies suggest the following, holding all other factors constant:

• Single-frequency sounds or pure tones are more annoying than sound composed of bands of frequencies.

• Annoyance is greater for sounds of higher frequency (and thus higher pitch) than for those of lower frequency.

• Annoyance increases with increasing intensity level of a sound; a 10 dB increase corresponds to a doubling in magnitude of annoyance response.

• Sounds that are variable in nature (namely those that occur randomly in time or vary in level and point of origin) are judged more annoying than those that are unchanging.

In some instances, office noise control treatment has taken the form of generating a sound, deemed qualitatively more acceptable in nature, to mask another that has more annoying qualities as just described. It should be mentioned that generated sound levels here are quite moderate and pose no interference to speech communication or other office functions.

Apart from their acoustic composition, certain sounds may evoke annoyance because they convey distress or alarm or have other unpleasant associations to the listener.

A survey of noise conditions in hospitals found that a prevalent source of annoyance was staff conversation in the halls.[60] Patients said these sounds were objectionable, not because of their loudness, but because of the information communicated—such as descriptions of other patients' conditions, operations, and symptoms. Another frequent source of noise annoyance in hospitals is the sound of other patients in distress, such as moaning, or calling for a nurse.

Another example is the sound of approaching aircraft. Because of the possibility of an airplane crash, the sound can elicit fear. This fear appears to be a factor motivating complaints of annoyance in neighborhoods near airports.[61] Similarly, the screaming siren of a patrol car or the clanging bells of a fire engine, because of their significance, can engender annoyance out of fear.

Unpleasant feelings or attitudes about certain sounds based on their information content can lead to annoyance that might be altogether out of proportion to the physical characteristics of the sound. These associations and still others, such as the necessity or advantage identified with different noise sources, hopelessly complicate acoustic scales or schemes for quantifying noise annoyance.

Theoretical explanation of extra-auditory effects of noise

A suggested theoretical basis accounting for the extra-auditory noise effects described in the preceding discussion is presented in this section. Most emphasis will be given to noise performance changes.

It should be noted first that the auditory pathways of man's nervous system are actually organized into two distinct systems (see Figure 10–11).[62] One is a specific projection system whose primary function is to transmit neural impulses underlying sound or noise stimuli from the ear receptors to the higher brain centers for perception and interpretation. The major auditory problems caused by noise—namely, hearing loss and speech interference—involve impairment or interference with this primary system.

The other is a nonspecific projection system that branches off the main

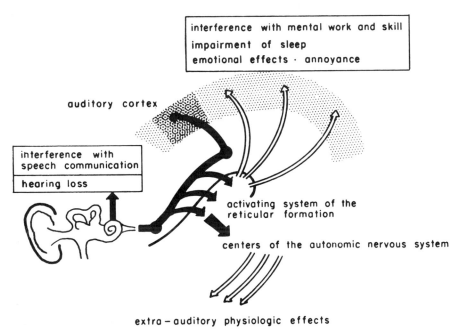

interference with mental work and skill
impairment of sleep
emotional effects · annoyance

auditory cortex

interference with speech communication

hearing loss

activating system of the reticular formation

centers of the autonomic nervous system

extra – auditory physiologic effects

Courtesy of American Industrial Hygiene Association Journal.

FIG. 10–11.—The physiological organization of the auditory pathways in the brain and their relation to noise effects on man (modified after Grandjean[62]).

auditory pathway into an activating-regulating center, called the *reticular formation,* and then spreads diffusely into different functional areas of the brain as well as affecting the peripheral autonomic system. This nonspecific projection system is concerned with the arousal and regulation of different sensory, motor, and autonomic activities, and it is believed to be responsible for many of the observed extra-aural effects of noise or sound stimulation.

Effects of stimulation

Bombarding the reticular formation with high-level noise, or, for that matter, with any other source of physical stimuli, can create conditions of over-arousal and cause individuals to show signs of undue excitability and nervousness. As such, overstimulation of the reticular formation might be the basis for the hyperactive state of muscles, and subjective feelings of tension, excitability, and anxiousness.

On the other hand, more limited stimulation establishes an arousal state that is essential for organized cortical activity. That is, the brain is believed to require a certain level of nonspecific stimulation to make optimum use of the incoming sensory information upon which efficient behavior depends. The arousal function, in turn, is assumed to depend on the variability of the stimulus situation together with factors such as motivation and interest, which are considered as internal inputs.

Arousal

Arousal is considered to vary along a dimension ranging from a sleep-like state at one end to extreme agitation at the other.[63] Scaling of arousal is indicated by the horizontal axis in Figure 10–12. The vertical axis represents a person's effectiveness in doing a job in some arbitrary units. For all jobs there is some value of arousal at which the job is done best by an individual. Noise is an irrelevant sensory input, which adds an increment to arousal.

Thus, noise will reduce the effectiveness of a person who is already at or beyond the high point on the arousal curve (point C). On the other hand, noise will improve a person's performance if arousal is low as shown in point B (which is then shifted to B' by the noise increment). Noise may also shift arousal from a front position to a matching position on the back part of the arousal curve with little effect on efficiency as depicted by D and D.

This theory makes sense of some of the contradictory effects of noise on performance. If a job is routine or easy, then arousal is likely to be less than optimum, and noise may improve performance. As already noted, simple and repetitive task situations show beneficial performance effects from noise. On the other hand, in complex, challenging, and absorbing work situations that would be considered as highly arousing, the addition of noise might shift the arousal level beyond the point of optimum efficiency, with consequent losses in performance.

Consistent with this expectation is the finding that vigilance losses in

285

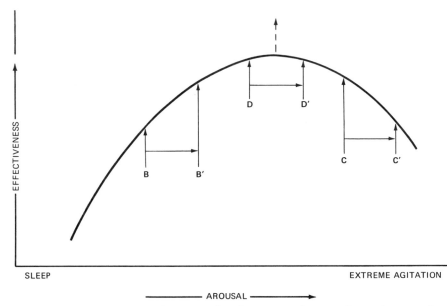

Fɪɢ. 10–12.—How arousal is considered to vary along a dimension ranging from a sleeplike state at one end to extreme agitation at the other. This is shown by the horizontal axis in this figure. The vertical axis represents the effectiveness in doing a job in arbitrary units (adapted from Carpenter[63]).

noisy environments occur only in complex watch-keeping situations where many dials are being monitored, and, consequently, arousal level is high. Single-dial monitoring performance shows little if any effect due to noise. Fast-paced tasks seemingly would maintain a high level of arousal. Thus, the presence of noise, by adding still more stimulation, causes overarousal and thus deterioration in task performance.

How over-arousal upsets performance is not clear. It may be that overarousal creates undue narrowing or chaotic shifting of attention or loss in behavioral control.

Summary

This discussion has treated in separate fashion different adverse effects of noise that are extra-auditory in nature. Referenced here have been noise disturbances to physical and mental health, aspects of noise in-duced behavioral changes and disruption of work performance, and general reactions of noise annoyance. Available data relating noise ex-

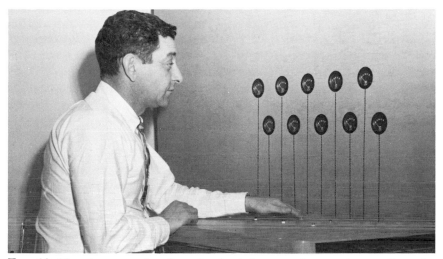

Fɪɢ. 10–13.—An example of a visual vigilance task situation used in the laboratory. The subject is shown responding to a signal deflection on the dial panel. Such deflections can occur on any of the dials according to a random schedule.

posure to physical and mental health problems are suggestive, but are not sufficient to permit definite conclusions. Indeed, systematic and well-controlled studies are needed to verify assorted observations in this area.

Noise induced changes in certain nonauditory sensory functions are noted as are characteristics of the startle pattern evoked by sudden, unexpected sound bursts. The implications of these changes for safety are not certain but inferences are possible.

Studies of noise and task performance have been extensive with the outcomes variable, depending upon the nature of the acoustic (or other environmental conditions present), features of the task being performed, and the attitude or psychological makeup of the performing persons. Transient, irregular noise occurrences in combination with a complex, fast-paced task offer the test conditions most likely to reveal noise degradation, especially with tense or anxious subjects.

Annoyance response to sound stimulation tends to vary with differences in intensity, frequency composition, temporal, and localization factors. The information conveyed by sounds can contribute to their acceptability or annoyance potential.

References

1. Kryter, K. D. "The Effects of Noise on Man" (Monograph Supplement 1). *Journal of Speech and Hearing Disorders*, 1950.
2. Miller, J. D. *Effects of Noise on People* (Reprint No. NTID 300.7). Washington, D.C.: Environmental Protection Agency, 1971.
3. Kryter, K. D. *Effects of Noise on Man.* New York, N.Y.: Academic Press, 1970.
4. Wilson, A. *Noise.* London: Her Majesty's Stationery Office, 1963.
5. Bell, A. "Noise, an Occupational Hazard and Public Nuisance" (Public Health Paper No. 30). Geneva: World Health Organization, 1966.
6. Cohen, A. J.; Anticaglia, J. R.; and Jones, H. H. "Sociocusis—Hearing Loss from Non-occupational Noise Exposure." *Sound and Vibration* 4:12–20, 1970.
7. Department of the Army, Army Research Office. *A Review of Adverse Biomedical Effects of Sound in Military Environment* (Contract DAH 19-71C-001). Washington, D.C. 1971.
8. Welch, B. L. and Welch, A. S. eds. *Physiological Effects of Noise.* New York, N.Y.: Plenum Press, 1970.
9. Anticaglia, J. R. and Cohen, A. "Extra-auditory Effects of Noise as a Health Hazard." *American Industrial Hygiene Association Journal* 31:277–281.
10. Jansen, G. "Effects of Noise on Physiological State." *Proceedings of National Conference on Noise as a Public Health Hazard* (Report No. 4). Washington, D.C.: American Speech and Hearing Association, 1969.
11. Anthony, A. and Ackerman, E. *Stress Effects of Noise* (WADC Technical Report 58–622). Wright-Patterson AFB, Ohio, 1959.
12. Zondek, B. and Tamari, I. "Effects of Audiogenic Stimulation on Genital Function and Reproduction." *American Journal of Obstetrics & Gynecology* 80:1041–1048.
13. Sackler, A. M.; Weltman, A.; Bradshaw, M.; and Justshuk, P. "Endocrine Changes Due to Auditory Stress." *Acta Endocrinologia* (Copenhagen) 31:405–418.
14. Shatalov, N. M.; Saitanov, A.; and Glotova, K. V. "On the State of the Cardiovascular System Under Conditions of Noise Exposure." *Gigiyena Truda I Professional'nye Zabolevanija* (*Labor Hygiene and Occupational Disease*) (Moscow), No. 7 (1962) pp. 10–14.
15. Ponomarenko, I. "The Effect of Constant High Frequency Industrial Noise on Certain Physiological Functions in Adolescents." *Giglyena i Sanitariya* (Moscow), No. 2 (1966), pp. 188–193.
16. Taccola, A.; Straneo, G.; and Bobbio, G. "Modificazione della dinamica Cardiaco Indotta dal Rumore." *Lavoro Umano* (Naples), No. 12 (1963) pp. 571–579.
17. Mangeri, S. "Respiratory Effects of Industrial Noise." *Lavoro Umano* (Naples) No. 7 (1965) pp. 331–338.
18. Landis, C. and Hunt, W. A. *The Startle Pattern.* New York, N.Y.: Farrar Book Co., 1939.
19. Cohen, A. "Effects of Noise on Psychological State." *Proceedings of National Conference on Noise as a Public Health Hazard* (Report No. 4). Washing-

ton, D.C.: American Speech and Hearing Association, 1969.

20. Glorig, A. "Medical Aspects of the Noise Problem." Paper presented to 6th AMA Congress on Environmental Health. April 1969, Chicago.

21. National Institute for Occupational Safety and Health. *Industrial Noise and Worker Medical, Absence and Accident Records* (Contract Report HSM 099–71–6). 1972.

22. Jansen, G. "Adverse Effects of Noise in Iron and Steel Workers." *Stahl und Eisen* (Dusseldorf) 81:217–220.

23. Broadbent, D. "Non-auditory Effects of Noise." *Advancement of Science* (London) 1961.

24. Cohen, A. "Airport Noise, Sonic Booms and Public Health." *Society for Automotive Engineering Journal* 1972, 1315–1328.

25. Mohr, G. C.; Cole, J.; Guild, E.; and von Gierke, H. E. "Effects of Low Frequency and Infrasonic Noise on Man." *Aerospace Medicine* 36:817–824.

26. Skillern, C. P. "Human Response to Measured Sound Pressure Levels from Ultrasonic Devices." *American Industrial Hygiene Association Journal* 26:132–136.

27. Acton, W. I. and Carson, M. B. "Auditory and Subjective Effects of Airborne Noise from Industrial Ultrasonic Sources." *British Journal of Industrial Medicine* 24:297–304.

28. Acton, S. I. "A Criterion for Production of Auditory and Subjective Effects due to Airborne Noise from Ultrasonic Sources." *Annals of Occupational Hygiene* (England) 11:227–234.

29. Merritt, H. H. *Textbook of Neurology.* 4th rev. ed. Philadelphia, Pa.: Lea & Febiger, 1967.

30. Graham, J. R. "Migraine: Clinical Aspects in Headaches and Cranial Neuralgia," in *Handbook of Clinical Neurology.* eds. P. J. Vinken and G. W. Bruyn. Amsterdam (Holland): North Holland Publication Co., 1968.

31. Hoffman, H. S. and Searle, J. L. "Acoustic and Temporal Factors in the Evocation of Startle." *Journal of the Acoustical Society of America* 43:269–282.

32. Hoffman, H. S. and Fleshler, M. "Startle Reactions—Modification by Background Noise Stimulation." *Science* 141:928–930.

33. U.S. Department of Transportation. *Human Response to Sonic Booms* (FAA Report 70–2). Washington, D.C. 1970.

34. Stevens, S. S. *The Effects of Noise and Vibration on Psychomotor Efficiency.* Cambridge, Mass.: Harvard University Press, 1941.

35. Benko, E. "Further Information about Narrowing of the Visual Fields Caused by Noise Damage." *Ophthalmologica* (Switzerland), No. 1 (1962) pp. 76–80.

36. U.S. Department of the Navy, Navy Research Laboratory. *Nystagmus Elicited by High Intensity Sound* (Research Project NM 13–01–99, Subtask 2, Report 6). Pensacola, Fla. 1957.

37. U.S. Air Force. *Rail Test to Evaluate Equilibrium in Broadband Noise* (AMRL Technical Report 66–85). Wright-Patterson AFB, Ohio. 1966.

38. McCabe, B. F. and Lawrence, M. "Effects of Intense Sound on Non-auditory Labyrinth." *Acta Otolaryngologica* (Stockholm, Sweden) 49:147–157.

289

39. Gardner, W. and Licklider, J. C. R. "Auditory Analgesia in Dental Operations." *Journal of American Dental Association* 59:1144–1149.
40. Recht, J. L., ed. *Accident Facts.* Chicago, Ill.: National Safety Council, 1973.
41. Broadbent, D. E. "Effects of Noise on Behavior," in *Handbook of Noise Control,* edited by C. M. Harris. New York, N.Y.: McGraw-Hill Book Co., 1957.
42. Cohen, A. "Effects of Noise on Work Performance." *Proceedings of Research Conference on Applied Work Physiology,* New York, N.Y.: New York University of Rehabilitative Medicine, 1968.
43. Kirk, R. E. and Hecht, E. "Maintenance of Vigilance by Programmed Noise." *Perceptual and Motor Skills* 16:558–560.
44. Small, A. M., Jr. in *Modern Developments in Audiology,* edited by J. Jerger. New York, N.Y.: Academic Press, 1963.
45. Teichner, W. H.; Arees, E.; Reilly, R. "Noise and Human Performance—A Psychophysiological Approach." *Ergonomics* 6:83–97.
46. Vlasek, M. "The Effects of Startle Stimuli on Performance." *Aerospace Medicine* 40:124–128.
47. Wokoun, W. *Effect of Music on Work Performance* (Technical Memo 1–68). Aberdeen Proving Ground: U.S. Army Human Engineering Laboratories, 1968.
48. Uhrbrock, R. S. "Music on the Job—Its Influence on Worker Morale and Production." *Personnel Psychology* 4:9–38.
49. Harris, C. S. and Sommer, H. C. *Combined Effects of Noise and Vibration on Mental Performance* (AMRL Technical Report 70–21). Wright-Patterson AFB, Ohio. 1971.
50. Renshaw, F. "The Combined Effects of Heat and Noise on Work Performance." Ph.D. Dissertation, University of Cincinnati, 1972.
51. Cohen, H. H. "Working Efficiency as a Function of Noise Level, Work Pace, and Time at Work." Ph.D. Dissertation, North Carolina State University, 1972.
52. Carpenter, A. "Effects of Noise on Performance and Productivity," in *Control of Noise,* National Physical Laboratories. London: Her Majesty's Stationery Office, 1962.
53. Boggs, D. H. and Simon, J. R. "Differential Effects of Noise on Tasks of Varying Complexity." *Journal of Applied Psychology* 53:148–153.
54. Hockey, G. R. J. "Effects of Noise on Human Efficiency and Some Individual Differences." *Sound & Vibration,* March 1972, p. 299.
55. Mech, E. V. "Performance on a Verbal Addition Task Related to Experimental 'Set' and Verbal Noise." *Journal of Experimental Education* 22:1–17.
56. Friedman, L. N. "Psychic Cost of Adaptation to an Environmental Stressor." *Journal of Personal and Social Psychology* 12:200–210.
57. Cohen, A., et al. *Effects of Noise on Task Performance* (Technical Report No. RR4). Cincinnati, Ohio: National Institute for Occupational Safety and Health, 1966.
58. Felton, J. S. and Spencer, C. "Morale of Workers Exposed to High Levels of Occupational Noise." *American Industrial Hygiene Association Journal* 22:136–147.
59. Kryter, K. D. "Concepts of Perceived Noisiness—Their Implementation and

Application." *Journal of the Acoustical Society of America* 43:344–361.

60. Goodfriend, L. S. and Cordinell, R. L. *Noise in Hospitals* (Public Health Service Report No. 930-D-11). Washington, D.C.: Government Printing Office, 1963.

61. McKennel, A. C. "Noise Complaints and Community Action." Seattle, Wash.: University of Washington Press, 1970.

62. Grandjean, E. "Summary to Panel I—Effects of Noise on Man." *Proceedings of Conference on Noise as a Public Health Hazard* (Report No. 4). Washington, D.C.: American Speech and Hearing Association, 1969.

63. Carpenter, A. "How Does Noise Affect the Individual." *Impulse* (India), No. 24 (1964).

Chapter Eleven

Standards and Threshold Limit Values for Noise

By Herbert H. Jones, P.E.

Developing good damage-risk noise criteria for people who must live and work in a highly mechanized and consequently noisy civilization is a very complex procedure. Not only does this require sound technical knowledge, but also an understanding of the needs of the total community. Technical, social, economic, and political considerations, to name a few, must also be examined. And even after all this, the decisions made will be arbitrary.

This chapter summarizes the findings of many investigators and examines critically the efforts to designate noise exposure levels that can serve as standards to differentiate between noise levels that are safe or potentially hazardous. The general criteria involved in determining the risk of hearing loss as a result of excessive noise exposure will be presented here.

It has long been acknowledged that sustained exposure to high noise levels will produce a sensori-neural hearing loss in some individuals. Even noise levels below those capable of causing pain or acoustic trauma can also produce hearing loss, if the noise exposure is of sufficient intensity and duration. Because of this, there have been intensive studies by many investigators of the relationship of hearing loss to noise exposure. These studies are basic to the setting of valid noise exposure standards and threshold limit values.

As discussed in the previous chapter, the effects of noise on individuals may be physiological, psychological, or both. The specific response varies

292

TABLE 11-A
DAMAGE RISK CRITERIA PRIOR TO 1950

Author	Overall Sound Pressure Level		
	Safe	*Borderline*	*Harmful*
McCord (1938)			90
Rosenblith (1942)	75–80		
Bunch (1942)		80–90	
McCoy (1944)	80–85	90–100	110–130
Davis (1945)		100	115–120
Goldner (1945)			80
Schweishmer (1945)		80–90	
MacLaren (1945)		100	
Fowler (1947)		100	
Canfield (1949)	80		100–110
Grave (1949)	90		
Guild (1950)	<90 dB above hearing threshold		

with the individual and is unpredictable—a noise exposure that may produce a considerable loss of hearing in one individual may produce little or no hearing loss in another. Although an individual's reaction to a given noise exposure cannot be predicted with certainty, it is possible to determine how much noise, of a particular kind, an *average person* in a given environment can tolerate without complaint or damage to his hearing.

General criteria as to the risk of loss of hearing

Loss of hearing due to exposure to industrial noise was recognized as a possibility as early as 1831 by J. Fasbroke.[1] Since that time, numerous surveys of the hearing of industrial workers have been made in both Europe and America. An examination of the published literature reveals that, until 1940, the majority of the investigators concerned with establishing standards or criteria were of the opinion that a single value for the noise level at all frequencies would be adequate for defining a safe level. After 1940, there was a shift from the single-value designation to multiple values representing the contributions of individual octave bands.

Criteria prior to 1950

Representative examples of single values for overall broadband noise (published between 1938 and 1950) that differentiate between harmful, borderline, and safe levels are given in Table 11–A. No attempt is made to assess the reliability of these values, but brief quotations from some of the original articles will help to establish the significance that the respective authors placed on these values.

• "Experience indicates that a noise level of 90 dB or higher is definitely harmful to the human ear. Exposure to prolonged noises of lower level, which is the case with many occupational noises, is also harmful, but the extent of harm done is not well known and cannot be fairly estimated at this time." McCord (1938)[2]

• "It is difficult [to state] the minimum noise level that would be harmful. It appears that levels as low as 75 or 80 dB, if sufficiently prolonged, suffice to bring about premature aging of the ears." Rosenblith (1942)[3]

• "The boundary line between noises which cause hearing losses and those which do not, lies between 80 and 90 dB." Bunch (1942)[4]

• "It is probably true that a noise of 80 to 85 dB will cause some defects of hearing in the high-frequency zones after a period of years." McCoy (1944)[5]

• "We do not have rigid proof of permanent impairment of hearing by noise of less than 115 or 120 db . . . and concern about exposure to noises below 100 dB . . . is probably unwarranted." Davis (1945)[6]

• "A consensus seems to indicate that the minimum safe level is in the neighborhood of 80 dB, and that an exposure of at least two years is necessary to produce changes at this level." Goldner (1945)[7]

• "The boundary line of noise levels which will cause hearing loss seems to lie between 80 and 90 dB." Schweisheimer (1945)[8]

• "The borderline between relatively harmless noise and that which produces steady deterioration lies in the region of 100 dB." MacLaren (1945)[9]

• "Noise of less than 100 dB may reasonably be considered quite safe, except perhaps for a few unusually susceptible individuals." Fowler (1947)[10]

• "The consensus is that noise levels not exceeding 80 dB above the normal human threshold cannot be harmful to the human ear. . . . Exposure of human beings for periods of weeks to noise levels about 100 or 110 dB will definitely result (in damage)." Canfield (1949)[11]

- "For its effect on the hearing function one can, however, safely disregard noise of less than an 80 dB level." Grave (1949)[12]

- "The guess that is usually made . . . is that the upper limit of safety, for factory conditions, is probably about 90 dB above the threshold of hearing." Guild (1950)[13]

The limitations of the single overall level are noticeable from the divergent views expressed here. It was clear by the early 1950's that proposed limits must consider other physical characteristics of noise in addition to its intensity.

Criteria since 1950

K. D. Kryter. In 1950, Kryter's monograph on the *Effects of Noise on Man* was published.[14] It was a comprehensive review of all of the literature on the subject up to that date. Some of the conclusions from this review were:

- "Exposure to intense noise from machinery may cause partial and temporary hearing losses that persist for a few minutes to several hours."

- "Repetitive exposure over extended periods (years) may result in a permanent partial deafness."

- "The more intense the sound the greater the deafening effect."

- "The maximum intensity at which no deafening effect will occur is probably in the neighborhood of 80 dB (above 0.0002 dynes/cm^2) for critical bands of noise."

- "The overall intensity of a noise, as expressed in dB, will not, as a rule, indicate its deafening effect."

In evaluating the hazard of a given exposure to noise, Kryter recognized a need to consider the component frequencies and the bandwidth of frequencies that have common effects. Nearly all of the criteria advanced since the appearance of this monograph have reflected Kryter's concepts. The limits proposed by Kryter and other investigators are shown in Table 11–B.

H. C. Hardy in a paper published in 1952 concluded ". . . the spectrum level which exceeds 100 sones per octave is very probably damaging for long-time daily exposure. For occasional exposures, spectrum levels of perhaps 200 sones or less are probably safe. All evidence indicates that,

TABLE 11-B
CRITERIA OF RISK OF AUDITORY DAMAGE, AS PROPOSED FROM 1950 TO 1963

Author	Year	Level of Sound Pressure in Octave Bands							
		20–75	75–150	150–300	300–600	600–1200	1200–2400	2400–4800	4800–9600
Kryter	1950	—	120	110	96	95	95	88	100
Hardy, 100 Sones	1952	115	112	108	106	104	95	91	102
Hardy, 50 Sones	1952	104	100	97	95	92	87	85	95
Rosenblith and Stevens—Broadband	1953	110	102	97	95	95	95	95	95
Rosenblith and Stevens—Narrowband	1953	100	92	87	85	85	85	85	85
AFR 160-3	1956	—	—	—	85	85	85	85	—
AAOO	1957	—	—	—	85	85	—	—	—
ISO	1961	103	96	91	88	85	83	81	80
Kryter—Broadband	1963	—	98	92	89	86	85	85	86
Kryter—Narrowband	1963	—	93	87	84	81	80	80	81
AAOO	1964	—	—	—	85	85	85	—	—

when no octave-band level exceeds 50 sones, there is very little likelihood that hearing damage will occur, even when the daily exposure persists over many years."[15] This was the first time that a hazardous zone of noise instead of a single level had been formally presented in the literature. Within this zone there was a very slight hazard at the lower limit and an almost certain hazard at the upper limit.

Rosenblith and Stevens. In 1953, W. A. Rosenblith and K. N. Stevens pointed out some of the difficulties in determining a safe level for exposure to noise.[16] About that time, many groups were in the process of attempting to establish standards of safety for various levels of noise exposure. This required establishing target goals for the hearing threshold levels, as measured with a pure tone audiometer. The measured hearing thresholds should not differ significantly in an exposed population from those of a matched control population. Because the concepts of monitoring audiometry had not been fully developed when this criterion was written, most investigators were striving for limits that would provide complete protection of hearing.

Rosenblith and Stevens recognized that it would be difficult to know when complete protection had been achieved, because the threshold shifts associated with the normal aging process (presbycusis) are not always distinguishable from the threshold shifts produced by exposure to noise. Furthermore, among persons not exposed to noise, presbycusis varies greatly, and the threshold shifts due to age and many other causes may be as great as those caused by exposure to high levels of noise.

Rosenblith and Stevens decided further that levels of noise should be measured in octave bands as long as the spectrum of the noise was relatively uniform. The following assumptions made by Rosenblith and Stevens are considered essential and complementary to their proposed limits of exposure to noise.

• The values are not to be taken too literally, because deviations of the order of one or two dB in either direction can probably be disregarded. The values should be used as zones because of the uncertainties involved in assessing exposure and the differences in individual susceptibility to damage from exposure to noise. Rosenblith and Stevens believed, however, that contours 10 dB lower would involve negligible risks, whereas contours 10 dB higher would result in significant increases in hearing loss.

• The standard levels were to be considered safe for occupational exposure, under the usual conditions, throughout the occupational lifetime.

• The standard levels were to apply to exposure to noise characterized by a reasonably uniform occurrence in time.

• For broadband sound, octave-band levels should be used. For pure tones, or for noise in which the major portion of the energy is concentrated in a band narrower than the critical band, the values designated for critical bands were to be used.

• The standard should be considered as tentative only, and subject to further revision as new laboratory and field data are developed and reported.

One of the most important contributions made by Rosenblith and Stevens was their statement describing the statistical nature of the relationship between exposure to noise and loss of hearing. They also introduced a clearer concept of specific goals for noise damage risk criteria.

Air Force Regulations. The United States Air Force instituted a hearing conservation program in 1956. The mechanism for carrying this out was defined in Air Force Regulation (AFR) 160–3, *Hazardous Noise Exposure.*[17] The USAF program goal was to preserve hearing so that no individual would sustain a hearing loss of more than 15 dB for pure tones at test frequencies of 500, 1000, and 2000 Hz. The Air Force determined that, except on rare occasions, there would be a negligible hazard, if the established USAF standards were not exceeded in any of the four octave bands between 300 and 4800 Hz. The regulation recommended the application of *auditory protection* when the sound pressure level exceeded 85 dB in any one of the four octave bands and made such protection mandatory when the sound pressure in any one of the octave bands exceeded 95 dB. These limits were applied to continuous exposure, eight hours in duration.

Limits were also established for shorter periods of exposure. These limits were based on the assumption that the loss of hearing incurred by an individual was directly proportional to the total acoustical energy received by the individual, regardless of how this energy was distributed in time. Because, on the decibel scale, a doubling of the sound intensity or energy results in an increase of 3 dB, the allowable sound pressure level could be increased by 3 dB, if the duration of exposure were cut in half. One advantage of the equal energy rule was the elimination from the hazardous category of many types of noise and occasions of brief

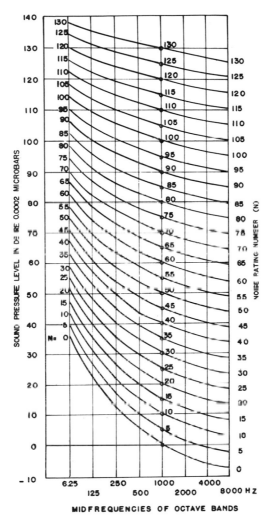

Fig. 11–1.—The International Organization for Standardization (ISO) noise rating curves. In use, the measured noise spectrum is plotted to find that point of the noise spectrum that penetrates the highest noise rating number curve. The corresponding value of the noise rating number curve is the noise rating number (N) of the measured noise. For continuous exposures (five or more hours per day), the noise rating curve of N85 was suggested by ISO as a limit for conservation of hearing. The numbers (−10 to 140 dB) shown along side the left margin represent the measured octave-band levels.

exposure to noise of high intensity.

Air Force Regulation 160–3 employed the more important features of the Rosenblith and Stevens criteria, and, in addition, it specified the use of monitoring audiometry to determine whether those individuals who were exposed to noise were developing actual hearing losses. The regulation also promulgated limits for exposure to noise for periods of less than eight hours per day.

AAOO Subcommittee on Noise (1957). The Subcommittee on Noise of the Committee on Conservation of Hearing of the American Academy of Ophthalmology and Otolaryngology issued a *Guide for Conservation of Hearing in Noise* in 1957.[18] The subcommittee shared the implied goal of the parent committee on conservation of hearing and wrote a limited damage-risk noise criterion that emphasized safety, rather than the acceptance of potential hazards. Their statement read, "If the sound energy of the noise is distributed more or less evenly throughout the eight octave bands, and if a person is to be exposed to this noise regularly for many hours a day, five days a week for many years, then: if the noise level in either the 300 to 600 Hz band or the 600 to 1200 Hz band is 85 dB, the initiation of noise-exposure control and tests of hearing is advisable. The more the octave-band levels exceed 85 dB, the more urgent is the need for hearing conservation." Periods of short exposure and of exposure to impact noises, or those composed of narrow bands, were specifically excluded from this criterion.

ISO/TC 43. In 1961, Technical Committee 43 on Acoustics of the International Organization for Standardization (ISO) proposed a standard specifying *Noise Rating Numbers with Respect to Conservation of Hearing, Speech Communication, and Annoyance.*[19] In this standard, a system of noise rating numbers (N) was proposed. This family of curves is shown in Figure 11–1.

The ISO/TC 43 Committee stated that, when exposure to broadband noise is habitual and the noise is continuous during the working day (five or more hours), the N85 curve shown in Figure 11–1 is proposed as a limit for conservation of hearing because habitual exposure to such a noise for 10 years may be expected to result in a negligible loss in hearing for speech by the average individual. This N85 noise rating number, from the point of view of conservation of hearing in industrial situations, is often considered to be just allowable. (A more rigorous criterion may be appropriate in other situations or jurisdictions.)

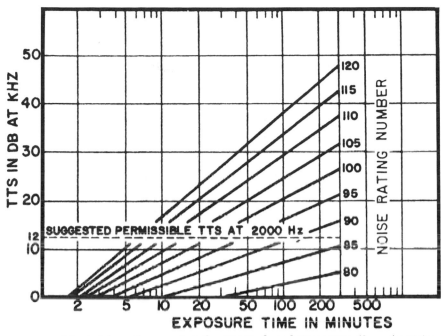

Adapted from the report of the ISO Technical Committee 43 on Acoustics.

FIG. 11–2.—This chart can be used to determine the amount of temporary threshold shift at 2000 Hz due to exposure to 600 to 1200 Hz as a function of level. Noise rating numbers may be determined, if the noise exposure time is known. For example, if it is known that an exposure will be of 100 minutes duration, the noise rating number is 91, because 100 minutes and the N91 curve intercept at the 12 dB temporary threshold shift (TTS) level. It is suggested that the 12 dB TTS level at 2000 Hz be used as a criterion to determine the noise rating number.

The noise rating number of the noise in question is obtained by comparing the measured octave-band levels with the curves given in Figure 11–1. The highest curve that is just exceeded by any of these octave-band levels gives the noise rating number (N).

This proposed standard deviates from the equal energy concept when exposure levels are increased. As the exposure time is reduced by one half, the exposure levels increase from three dB at 90 dB to seven dB at 115 dB as shown in Figure 11–2.

In addition, the ISO/TC 43 Committee proposed a criterion for intermittent exposure as follows:

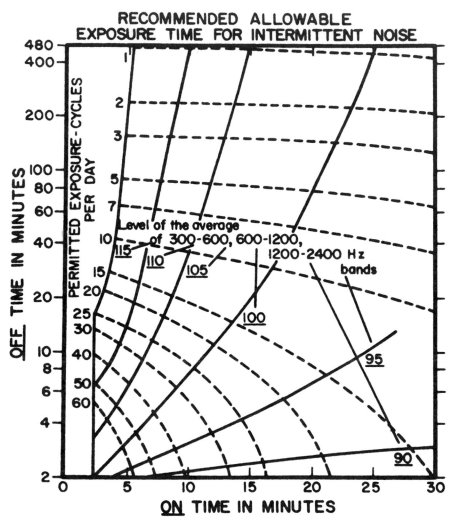

RECOMMENDED ALLOWABLE EXPOSURE TIME FOR INTERMITTENT NOISE

Adapted from AAOO Guide for Conservation of Hearing in Noise, 1964.

Fig. 11–3.—This series of curves is useful for determining "on-time" and "off-time" combinations of intermittent noise exposure to comply with the AAOO recommended hearing conservation criteria. For example, if the average level at the 300–600, 600–1200, and 1200–2400 Hz bands is 105 dB and the noise is habitually on ("on-time") for 10 minutes, the (recommended allowable) "off-time" between noise bursts is the ordinate value of the intersection of the 10 minutes and the 105 dB contour. In this case, it is just under 40 minutes or approximately 10 exposures-cycles per day.

"When habitual exposure is to broadband noise that is intermittently on during the working day, the permissible noise rating number of the noise in question may be obtained by use of Figure 11–3. This figure shows the relationship between the duration of noise bursts (abscissa), the 'off-time' between noise bursts (ordinate), the number of such cycles of noise and relative quiet (broken contours), and the noise rating number (solid contours) that the noise would have if it were on continuously."

Each of these contours is the locus of combinations of on-time and off-time, and, for the noise it represents, it gives a limit of noise exposure, equivalent to that prescribed for a continuous noise with a noise rating number of 85. This criterion may be used as follows: Assume that a noise at a level of N100, as shown in Figure 11–3, persists for 10 minutes after which the noise is discontinued for eight minutes. Enter the abscissa of Figure 11–3 at 10 minutes of duration and the ordinate at 8 minutes of discontinuance. It turns out that this exposure can be repeated on 25 occasions within an eight hour day.

This proposed ISO/TC 43 standard for intermittent exposure again does not follow the equal-energy concept. In the example, an exposure of N100 would be permitted for 250 minutes, whereas, if the equal-energy rule applied, exposure would be limited to N88. This criterion follows that proposed by A. Glorig, W. D. Ward, and J. Nixon in 1961.[20]

K. D. Kryter proposed a criterion in 1963 for exposure to noise in terms of the impairment of speech communication, as the result of noise induced hearing loss.[21] This criterion permits any exposure to noise that, on the average, produces a temporary shift in threshold of a set number of decibels at three frequencies, when measured two minutes after the termination of exposure. The permissible shifts are no more than 10 dB at 1000 Hz or below, 15 dB at 2000 Hz, 20 dB at 3000 Hz or above. It is thought that daily exposure to such noise for about 10 years will produce permanent hearing losses of similiar magnitude. Kryter's criteria for continuous noise exposure for eight hours or less per day, as shown in Figure 11–4, are the same as those contained in the draft of a Technical Report of the Armed Forces, National Research Council Committee on Hearing and Bio-Acoustics, prepared for the Surgeon General of the U.S. Army.

AAOO Subcommittee on Noise (1964). In 1964, the Subcommittee

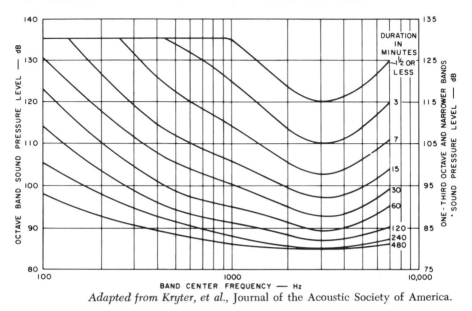

Adapted from Kryter, et al., Journal of the Acoustic Society of America.

FIG. 11–4.—Damage-risk contours for one exposure per day to full octave (left-hand ordinate) and ⅓ octave or narrower (right-hand ordinate) bands of noise. This graph can be applied to the individual band levels present in broadband noise.

on Noise of the Committee on Conservation of Hearing of the American Academy of Ophthalmology and Otolaryngology revised its publication, *Guide for Conservation of Hearing in Noise,* described earlier. The following limits for exposure to noise were given:

"(1) When the exposure to broadband noise is habitual and the noise is continuous during the working day (5 or more hours), the average of the levels at 300–600, 600–1200, and 1200–2400 Hz should not exceed 85 dB. If this average exceeds 85 dB, hearing conservation measures should be initiated.

"(2) When the exposure to broadband noise is continuous for less than 5 hours per day, Table 11–C should be consulted for recommended allowable exposures."

For intermittent exposure to noise, the AAOO Subcommittee also recommended the application of the criteria proposed by Technical Committee 43 of the ISO (shown in Figure 11–3).

TABLE 11-C

RECOMMENDED ALLOWABLE LIMITS TO EXPOSURE TO BROADBAND NOISE

Average Decibel Level of 300–600, 600–1200, and 1200–2400 Octave Bands	On-time Per Day (minutes)
85	less than 300
90	less than 120
95	less than 50
100	less than 25
105	less than 16
110	less than 12
115	less than 8
120	less than 5

Intersociety Committee. In 1967, the Intersociety Committee, consisting of two representatives from the American Academy of Occupational Medicine, American Academy of Ophthalmology and Otolaryngology, American Conference of Governmental Industrial Hygienists, American Industrial Hygiene Association, and the American Industrial Medical Association, published its first report, *Guidelines for Noise Exposure Control.*[23] In this report the use of A-weighted sound level was proposed as a measure of hazard to hearing. The report also presented data on the risk of hearing impairment at various levels of noise exposure, as shown in Figure 11-5.

Botsford. In a study of 580 industrial noises, Botsford[24] showed that the A-weighted sound level indicated hazard to hearing as accurately as did limits expressed as octave-band sound pressure levels in 80 percent of the cases and was slightly more conservative than octave bands in 16 percent of the noises.

Passchier-Vermeer. In an evaluation of the major published studies of the relationship between noise exposure and hearing loss up to 1967, Passchier-Vermeer[25] found that sound level in dBA was as accurate as noise rating numbers in estimating noise-induced hearing loss except for one noise condition.

Robinson. In a study of hearing loss in 759 carefully selected subjects with well-documented noise exposures, Robinson[26] concluded that the

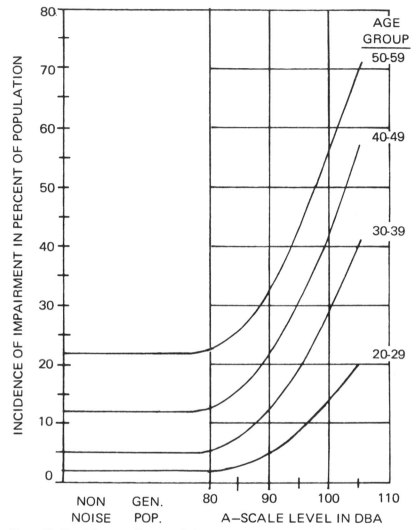

F_{IG}. 11–5.—The incidence of hearing impairment (average hearing threshold level) in excess of 25 dB ANSI at 500, 1000, and 2000 Hz in the general population and in selected population by age groups and by occupational noise exposure as reported by the Intersociety Committee.[23]

error incurred from using dBA in predicting hearing loss was within ±2 dB, even for noises ranging in slope from +4 dB per octave to −5 dB per octave.

Cohen, Anticaglia, Carpenter.   Cohen, Anticaglia, and Carpenter[27] found in a laboratory study that, although the sound level in dBA perhaps discounted too much low-frequency energy, in all cases but one it predicted the temporary threshold shift two minutes after exposure to noises of different spectra (slopes of −6 dB, zero dB, and 6 dB per octave) as well as or better than other noise rating schemes that employed spectral measurements in octave bands.

ACGIH.   As a result of its simplicity and accuracy in rating hazard to hearing, the A-weighted sound level was adopted as the measure for assessing noise exposure by the American Conference of Governmental Industrial Hygienists.

ACGIH committee on physical agents

In May 1967, the American Conference of Governmental Industrial Hygienists established a Committee on Physical Agents.   This committee was directed to review the existing data on exposures of individuals to various physical agents including noise and to recommend to the conference safe limits of exposure.

In establishing the limit of exposure, many factors had to be considered by the committee.   These factors were the types of data available and the validity of this data, methods of control of exposure and their feasibility, and of primary importance, the percentage of the group that would be protected by the established limits.

After considering the information available in this area, the ACGIH Committee recommended a limit of 90 dBA for an eight hour day, 40 hour week exposure.   Data indicated that this would protect about 90 percent of the people exposed to this level for a normal working lifetime. The ACGIH Committee stated that, as more exposure data become available and the cost of engineering controls is reduced, it would be desirable to revise the limit, if necessary, to protect a larger percentage of the exposed population.

For a number of years, it was assumed that equal energy would produce equal damage to the ear.   If this assumption were true, then each time the sound level is increased three decibels the exposure time should be reduced one half.   Laboratory data on temporary threshold shift and field data indicate that for the shorter exposure times, the ear can tolerate more acoustical energy briefly than it can for a continuous eight hour exposure.

Also laboratory data and very limited field data indicate that, if the exposure is intermittent in nature (rest periods between exposures), the ear can tolerate considerably more acoustical energy than for a single exposure to continuous noise. Considering these two factors, the ACGIH Physical Agents Committee increased the limit 5 dB for each halving of

TABLE 11-D
PERMISSIBLE NOISE EXPOSURES

Duration per day Hours	Sound Level dBA*
8	90
6	92
4	95
3	97
2	100
1½	102
1	105
¾	107
½	110
¼	115-C**

* Sound level in decibels as measured on a standard level meter operating on the A-weighting network with slow meter response.
** Ceiling Value.

the exposure time. Thus, these threshold limits are a compromise between the more conservative equal-energy concept and the more liberal intermittent exposure concept.

At one time it was thought by many informed individuals that the limits of exposure for narrow bands of noise or pure tones should be 10 dB lower than for broadband noise. This level was then revised to only 5 dB, but some of the available data indicated that even 5 dB was too conservative. No correction was recommended by the ACGIH Physical Agents Committee for pure tones or for narrow bands of noise.

The ACGIH Physical Agents Committee stated that there is still very little data available on the effects of exposure to impact or impulsive noise. Many factors possibly influence the effects of this type of noise exposure, among which are peak sound pressure level, rise time, decay time, repetition rate, time interval between impacts or impulses, number per day, and background sound pressure levels. It is known that exposure to a small number of 140 dB impulsive noises of short duration will produce a temporary threshold shift. Until additional facts are available, a limit of 140 dB peak sound presure level was recommended by the ACGIH Physical Agents Committee.

Federal noise regulations

The noise exposure limits adopted by the American Conference of Governmental Industrial Hygienists were incorporated in the "Health and Safety Regulations" of the Public Contracts (Walsh-Healey) Act, on May 20, 1969. This was the first federal regulation limiting noise exposure for the prevention of hearing loss. (See Appendix A-1.)

OSHAdministration and NIOSH

The Occupational Safety and Health Act was passed in December 1970, and on May 29, 1971, the Walsh-Healey Safety and Health Regulations were adopted as temporary safety and health regulations under this act.

In 1973, the National Institute for Occupational Safety and Health (NIOSH) of the Department of Health, Education, and Welfare reviewed the published data available on noise induced hearing loss and data they had available from their research studies and made recommendations to the Occupational Safety and Health Administration (OSHA) for a noise health standard.[31] (See Appendix A-3.)

After receiving the recommendations from NIOSH, the Assistant Secretary, Department of Labor, appointed a standards advisory committee on noise to provide evaluation and additional recommendations directly from labor, management, and independent experts. The advisory committee spent about one year evaluating the recommendations from NIOSH and submitted a report to the Department of Labor. The Department has evaluated the committee's recommendations, and determined that the standard should include the elements suggested by the committee. (See Appendix A-2b.)

One of the principal changes recommended by the NIOSH criteria package was a lowering of the basic standard from 90 dBA to 85 dBA. The OSHA advisory committee recommended, and the Department of

Labor agreed, that the 90 dBA limit should be continued. The other major items for consideration were: (*a*) The necessity for audiometric testing as a part of a hearing-conservation program, (*b*) the requirements for administration of such a testing program if it were required, and (*c*) the requirements for personal hearing protectors if they were required for a hearing-conservation program.

The OSHA advisory committee found three factors significant in evaluating these matters: (*a*) The variations in individual susceptibility to hearing damage from the noise, (*b*) the uncertainty inherent in all of the data upon which the standard is based, and (*c*) the variations in existing audiometric programs.

The committee therefore has recommended the institution of audiometric testing programs when the noise exposure of an employee is one half or more of the limiting value (that is an exposure equivalent to 85 dBA for 8 hours). The committee did provide rather detailed requirements for the implementation of such programs.

The objective of this sort of an audiometric testing program is to detect changes in hearing level before they become sufficiently large to be handicapping. It is, therefore, essential that the audiometric environment and technique be well standardized and be stable over a number of years (representing a significant fraction of the employee's working lifetime). It is also essential that these factors be reasonably identical from one employment to another. For these reasons standards are proposed for the audiometric environment, the calibration of audiometers, and the training of audiometric technicians.

Methods are also proposed for determining the adequacy of hearing protectors in instances where these devices are used as part of a hearing-conservation program.

In its communication to OSHA, NIOSH recommended that occupational noise exposure should be controlled, so that no employee would be exposed in excess of a level defined by the formula:

$$T = \frac{16}{2^{(L-85)/5}}$$

Also they recommended that *new* installations should be designed to include noise controls, so that the noise exposure does not exceed the limits defined by the formula $T = 16 \div 2^{(L-80)/5}$. These limits are shown in Figure 11–6. (Appendix A covers this in more detail.)

NIOSH further recommended that the noise limit described by the formula $T = 16 \div 2^{(L-80)/5}$ should become effective for all places of em-

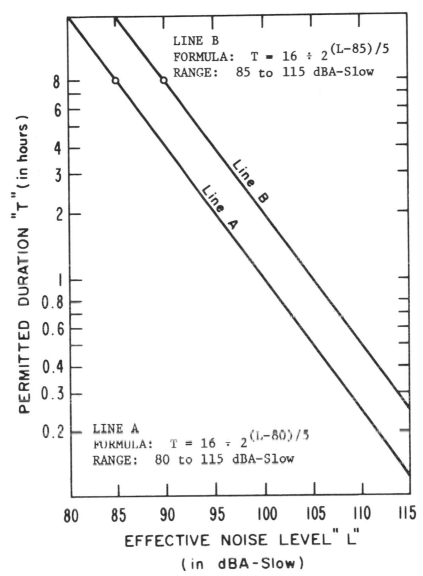

FIG. 11–6.—The permitted duration limits *vs.* noise level as recommended by the NIOSH criteria document. For example, the continuous noise limits, which correspond to line A, would be 85 dBA for eight hours, 90 dBA for four hours, and 95 dBA for two hours of noise exposure. The corresponding levels for line B would be 90 dBA for eight hours, 95 dBA for four hours, and 100 dBA for two hours of noise exposure.

ployment after a time period determined by the Secretary of Labor in consultation with the Secretary of Health, Education, and Welfare. This delay in effective date for all places of employment was believed necessary to permit the Department of Labor to conduct an extensive feasibility study.

NIOSH recommendations for noise limits differ slightly from the 1970 OSHA regulations. The formula $T = 16 \div 2^{(L-85)/5}$ does establish a limit of 90 dBA for an eight hour exposure, which agrees with the current OSHA (1970) regulations, but in computations of Daily Noise Dose it takes into account the exposure to noise in the 85 to 90 dBA range. The 1970 OSHA regulations ignore all exposures below 90 dBA in computations of Daily Noise Dose. The formula $T = 16 \div 2^{(L-80)/5}$ lowers the exposure limit by 5 dBA and also takes into account the exposure from 80 to 85 dBA.

The 90 dBA criterion is primarily aimed at protecting hearing in a restricted range of frequencies, namely 500 to 2000 Hz, which is believed critical to understanding everyday speech. Hearing for frequencies above this range, though even more noise sensitive, receives only indirect and limited protection.

Industrial noise standards do not protect the entire worker population from developing hearing losses even for the critical speech frequencies. Some investigators feel that the 90 dBA limit does permit about 20 percent of the workers exposed to this level to suffer hearing impairment for speech sounds after 30 years of noise exposure.

Off-the-job exposures

The amount of protection offered by this and other comparable standards may be viewed as acceptable in industry, because the worker receives a paycheck for assuming a certain job risk and can be compensated should an occupational-related disorder occur, such as noise induced hearing loss. These considerations, however, may not apply to off-the-job situations where persons may be subjected to possibly harmful noise conditions without receiving any direct compensation as a result of the noise exposure. Much more justifiable in this case would be a noise standard that prevents any noticeable losses in hearing for nearly everyone exposed.

Accordingly, the reference levels used here in rating off-the-job noise conditions might be set 15 dB below those described by ACGIH for exposure durations of eight hours per day or less.[32] These limits are shown in Table 11-E and have been extended on through to the 16 to 24 hour

TABLE 11-E
NOISE LIMITS FOR NONOCCUPATIONAL VS. OCCUPATIONAL NOISE EXPOSURES

Limiting Daily Exposure Times for Nonoccupational Noise Conditions	Sound Level in dBA	Limiting Daily Exposure Times for Occupational Conditions (ACGIH-TLV)
less than 2 minutes	115	15 minutes or less
less than 4 minutes	110	30 minutes
less than 8 minutes	105	1 hour
15 minutes	100	2 hours
30 minutes	95	4 hours
1 hour	90	8 hours
2 hours	85	
4 hours	80	
8 hours	75	
16–24 hours	70	

day with a lower limit of 70 dBA.

Summary

Adverse effects of noise on man include temporary and permanent hearing loss, speech interference, loss in performance capacity, and annoyance. Factors believed critical in evaluating a potential noise hazard to hearing are the overall level, the spectrum of the noise, total exposure duration, time and frequency distribution of short term exposure periods, and susceptibility of an individual's ears to noise induced hearing loss. Specifications for valid damage-risk criteria for noise exposure must take into account all of the factors mentioned here.

A survey of the available literature has been presented in this chapter to assist the reader in search of criteria for the appraisal of noise in various situations. Definite and universally acceptable limits are available for only a few situations, but guidelines have been proposed for the solution of the majority of the problems concerned with noise that will be encountered in the industrial and general environment.

References

1. Fasbroke, J., "Practical Observations on the Pathology and Treatment of Deafness," *Lancet*, London, England. Vol. 19, No. 675 (1931).
2. McCord, C. P.; Teal, E. E.; and Witheridge, W. N. "Noise and Its Effect on Human Beings," *Journal of the American Medical Association* Vol. 110, No. 1553 (1938).

3. Rosenblith, W. A., "Industrial Noise and Industrial Deafness." *Journal of the Acoustical Society of America*, Vol. 13, No. 220 (1942).
4. Bunch, C. C., "Conservation of Hearing in Industry." *Journal of the American Medical Association*, Vol. 118, No. 588 (1942).
5. McCoy, D. A., "Industrial Noise—Its Analysis and Interpretation for Preventive Treatment." *Journal of Industrial Hygiene and Toxicology*, Vol. 26, No. 120 (1944).
6. Davis, H. A., "Protection of Workers Against Noise." *Journal of Industrial Hygiene and Toxicology*, Vol. 27, No. 56 (1945).
7. Goldner, A., "Occupational Deafness." *Archives of Otolaryngology*, Vol. 42, No. 407 (1945).
8. Schweisheimer, W., "Effects of Noise in the Textile Industry." *Rayon Textile Monthly*, Vol. 26, No. 593 (1945).
9. MacLaren, W. R., and Chaney, A. L., "An Evaluation of Some Factors in the Development of Occupational Deafness." *Industrial Medicine and Surgery*, Vol. 16, No. 109 (1947).
10. Fowler, Jr., E. P., "Hearing and Deafness," in *Medical Aspects of Hearing Loss*. New York, N.Y.: Holt, Rinehart, & Winston, 1947.
11. Canfield, N., "Trauma from Noise in Industry." *Connecticut State Medical Journal*, Vol. 13, No. 21 (1949).
12. Grave, W. E., "Will Noise Damage One's Hearing?" *Journal of the American Medical Association*, Vol. 140, No. 674 (1949).
13. Guild, S. R., "Industrial Noise and Deafness." *Journal of Insurance Medicine*, Vol. 5 (April 1950).
14. Kryter, K. D., "Effects of Noise on Man" (Monograph Supplement 1). *Journal of Speech and Hearing Disorders*, 1950.
15. Hardy, H. C., "Tentative Estimate of a Hearing Damage Risk Criterion for Steady-State Noise." *Journal of the Acoustical Society of America*, Vol. 24, No. 756 (1952).
16. Rosenblith, W. A., and Stevens, K. N., *Handbook of Acoustic Noise Control, Vol. II. Noise and Man* (Tech. Rept. No. 53–204). USAF, WADC, National Technical Information Service, Springfield, Va. 1953.
17. Department of the Air Force, Aerospace Medicine, "Hazardous Noise Exposure," AF Regulation 161–35, July 1973.
18. *Guide for Conservation of Hearing in Noise*, rev. Rochester, Minn.: American Academy of Ophthalmology and Otolaryngology, 1964.
19. *Draft Proposal for Noise Rating Numbers with Respect to Conservation of Hearing, Speech Communication, and Annoyance*. International Organization for Standardization, Helsinki, Finland, 1961.
20. Glorig, A.; Ward, W. D.; and Nixon, J., "Damage Risk Criteria and Noise Induced Hearing Loss." *Archives of Otolaryngology*, Vol. 74, No. 413 (1961).
21. Kryter, K. D., "Exposure to Steady-State Noise and Impairment of Hearing." *Journal of the Acoustical Society of America*, Vol. 35, No. 1515 (1963).
22. *Some Damage Risk Criteria for Exposure to Sound* (Draft of Technical Report CHABA Working Group 46) St. Louis, Mo. 1956.
23. "Guidelines for Noise Exposure Control." *American Industrial Hygiene Association Journal*, Akron, Ohio. Vol. 28, p. 418, 1967.

24. Botsford, J. H., "Simple Method for Identifying Acceptable Noise Exposures." *Journal of the Acoustical Society of America*, Vol. 42, p. 810 (1967).

25. Passchier-Vermeer, W., *Hearing Loss Due to Exposure to Steady-State Broadband Noise* (Report No. 35). Leiden, Netherlands: Institute for Public Health Eng., 1968.

26. Robinson, D. W., *Relations Between Hearing Loss and Noise Exposures*. Teddington, England: National Physical Laboratory, 1968.

27. Cohen, A.; Anticaglia, J. R.; and Carpenter, P., "Temporary Threshold Shift in Hearing from Exposure to Different Noise Spectra at Equal dBA Level." *Journal of the Acoustical Society of America*, Vol. 51, p. 503 (1972).

28. *Threshold Limit Values of Physical Agents*. American Conference of Governmental Industrial Hygienists, Cincinnati, Ohio, 1971.

29. Walsh-Healey Public Contracts Act, Title 41, *C.F.R.* Chapter 50. Washington, D.C.: Superintendent of Documents, U.S. Government Printing Office.

30. Occupational Safety and Health Act Regulations, Title 29, *C.F.R.* Chapter XVII. Washington, D.C.: Superintendent of Documents, U.S. Government Printing Office.

31. Department of Health, Education, and Welfare. *Criteria for a Recommended Standard for Occupational Exposure to Noise*. 1972.

Chapter Twelve Federal Regulations and Guidelines

By Floyd A. Van Atta, Ph.D.

The fundamental mission of the U.S. Department of Labor (DOL) is to promote the welfare of working people. The DOL regulations were written with the concept of protection of people. Writing a regulation, however, always represents some sort of a compromise between what is "ideal" and what is feasible. A regulation is also likely to show some evolution over the years with the growth of understanding of the problem on the part of the enforcing agency and of its constituency.

The objective of the Department of Labor in controlling occupational noise exposures is to prevent hearing impairment to employees in the working population. This goal is being approached through regulations requiring the measurement of environmental noise, the use of engineering noise reduction or administrative control measures, and, where necessary, the use of personal hearing protective equipment as a final resort.

Walsh-Healey public contracts act

Regulations governing occupational noise exposure were first issued in 1936 under the authority of the Walsh-Healey Public Contracts Act which stated:

". . . no part of such contract will be performed nor will any of the materials, supplies, articles, or equipment to be manufactured or furnished under said contract be manufactured or fabricated in any plants, factories, buildings, or surroundings or under work conditions which are unsanitary or hazardous or dangerous to the health and safety of employees engaged in the performance of said contract. Compliance with the safety, sanitary, and factory inspection laws of the state in which the work or part thereof is to be performed shall be prima facie evidence of compliance with this subsection." (Prima facie does not mean "for certain"—it means just what it says, "at first glance.")

In 1960, the Department of Labor (DOL) decided to take a second glance at the subsection on prima facie evidence of compliance. The result was the Walsh-Healey regulations of December 1960. Those regulations stated, in essence, that the Secretary of Labor had taken a second look and had decided that the minimum level of environment, which would be considered administratively to be in compliance with the Walsh-Healey Public Contracts Act, would be that stated in the regulations.

The important thing to remember about the Walsh-Healey Regulations is that failure to comply with them is not necessarily a violation of the act. The violation is to provide an environment that is, in fact, unsafe, unhealthful or unwholesome for the employees working on the contract. The Department of Labor had to prove this as a matter of fact in every instance when it cited a firm for a hearing under the Walsh-Healey Regulations.

What it really amounts to is that if the working conditions in a particular firm are in compliance with the Walsh Healey Regulations, the Department of Labor will not cite the employer. If, however, the employer is cited to a hearing and can convince a hearing examiner or a judge that the working conditions are, in fact, satisfactory, safe, healthful, and sanitary, the employer is not obligated to follow the Walsh-Healey regulations. The contract says only that the employer will provide a safe and healthful environment for its employees.

Occupational safety and health act

Under the Williams-Steiger Occupational Safety and Health Act, the situation is quite different—although the words of the regulation are identical. The OSHAct provides specifically that compliance with the regulations promulgated by the Secretary of Labor is mandatory, so violation of the regulation is equivalent to violation of the act.

After a citation is issued, there can be no defense by the employer

317

that the working conditions provided for the employees are as safe and healthful as would be provided by compliance with the OSHAct regulations. An employer can, however, request a variance from the OSHAct regulations, prior to the issuance of a citation, based upon the provision of equivalent safety by a different method.

Under the Williams-Steiger Occupational Safety and Health Act of 1970, the Bureau of Labor Standards was replaced by the Occupational Safety and Health Administration (OSHA). Nonetheless, the Walsh-Healey Public Contracts Act, the McNamara-O'Hara Service Contracts Act, and the Construction Safety Act are still in existence and fully effective. Under OSHA, the regulations under all of these acts have been unified and have been identical with those under the OSHAct of 1970, which has the broadest coverage.

Historical development

The federal regulation of occupational noise exposure began with the rules issued under the authority of the Walsh-Healey Public Contracts Act. Those regulations required that occupational noise exposure must be reasonably controlled to minimize fatigue and the probability of accidents. When the Bureau of Labor Standards was assigned the responsibility for the safety and health aspects of the Walsh-Healey Act in 1964, it immediately started thinking about making the noise regulations both more definitive and more useful.

The Bureau had fairly definite ideas about what noise regulation was needed. The overriding requirement was that it should prevent significant loss of hearing to employees in a large majority of the exposed population. Other requirements were (a) that the criterion chosen should be a good index of the hazard to hearing, and (b) that safe noise levels could be easily determined with a minimum of equipment and special training.

The Department of Labor wanted to follow the guidance of the safety community, insofar as possible, by using voluntary consensus standards as the basis for regulations. However, at that time there were no consensus standards in existence for noise exposure. In 1964 the only standards available were some company standards in a few of the largest companies, the Air Force Regulation 160–3 of 1956 and 160-3A of 1960, and a number of private proposals. (These were discussed in Chapter 11.)

The Bureau of Labor Standards decided that the noise regulations should be concerned only with loss of hearing. There were many claims being made about the physiological effects other than hearing loss, but the Bureau could find no conclusive evidence that such physiological effects occur at noise levels below those necessary to produce hearing loss. If the physiological effects do indeed occur at noise levels that also produce hearing loss, the physiological effects will be controlled incidentally by controlling the noise levels that produce hearing loss. (Physiological effects are discussed in detail in Chapters 9 and 10.)

Nature of occupational deafness

There was no doubt about the real loss of hearing among people in certain crafts—this has been common knowledge for generations. The report of the Z24-X-2 Committee of the Acoustical Society of America in 1954 implied, although it was not stated as such in the written conclusions, that those individuals who were exposed to noise levels greater than 80 dB in the audible octave bands showed some loss of hearing.

The Z24-X-2 Committee did not write a standard in 1954 because it could not pinpoint the amount of hearing loss that could be considered to constitute an impairment and the amount of hearing loss that should properly be attributed to aging. This Z24-X-2 Committee was also searching for a magic number that would make it possible to express the hazard potential of noise in terms of a single, easily determined quantity. They could not find it at the time.

When the Bureau of Labor Standards was ready to start writing a regulation, some of these problems had been fairly well solved. The medical profession had defined hearing impairment as a function of hearing loss. An impairment is a hearing loss that begins to interfere with the understanding of spoken sentences (see Chapter 13 and Part V —Audiometry.

Notice that the definition does *not* imply the ability to recognize every syllable or even every word. It means that impairment is the beginning of difficulty in understanding or getting the sense of a sentence as a whole. The working definition is that impairment exists if the hearing threshold level is in excess of an average of 25 dB (ANSI 1969) for the three audiometric test frequencies of 500, 1000, and 2000 Hz, which are the most important ones for the understanding of speech.

The Bureau was also aware that the hearing loss caused by noise exposure is sensori-neural (or the permanent type) and cannot be corrected by any known medical treatment. The problem caused by this type of hearing loss is difficulty in distinguishing between many consonant speech sounds.

Studies of the amount of hearing loss to be found in an aging population showed the hearing loss due to noise adds to the loss due to aging; therefore, both factors must be taken into consideration. Up to the age of 65 in the range of 500 to 2000 Hz, there is no effect of aging on hearing acuity under the medical definition of a handicapping hearing loss.

There also was, by the time that the Bureau was ready to work on its regulation, a considerable body of data that indicated that the "magic number" should be the A-weighted sound level. It is, perhaps, not too surprising that A-weighting, which was devised to give an indication of the loudness of sounds of moderate intensity, should also give an indication of their relative ability to damage the hearing mechanism.

Fraction of population protected

The Bureau also had to decide what part of the working population should be protected. Because of the normal variation between individuals, it is not possible scientifically to set a realistic standard for exposure to materials or to energy that will protect everyone who is exposed. Generally, threshold limits are set with the intention of protecting 90 percent or more of an exposed population.

The Bureau had, by late 1967, two estimates of that part of the population that should be protected.

• The first was that of the Intersociety Committee on Guidelines for Noise Exposure Control, representing the five scientific organizations most closely concerned with the problem. (See Chapter 11 for details about the Committee.) This Committee reviewed the major investigations on the effects of noise exposure up to late 1966, and estimated that at 85 dBA about 3 percent of a population exposed throughout a working lifetime would have some hearing impairment as a result of the noise exposure (see Figure 12–1). This is probably not significant in view of the scatter of the data points, as the committee pointed out. At 90 dBA, for a working lifetime, the data showed an increase of about 8 percent in the number of people with hearing impairment, which probably is significant.

• The second set of data were adapted from some work that Dr. William Baughn of General Motors presented at the International Congress on Audiology in Mexico City in 1967. These data represent, perhaps, the most carefully studied large group of employees in the world.

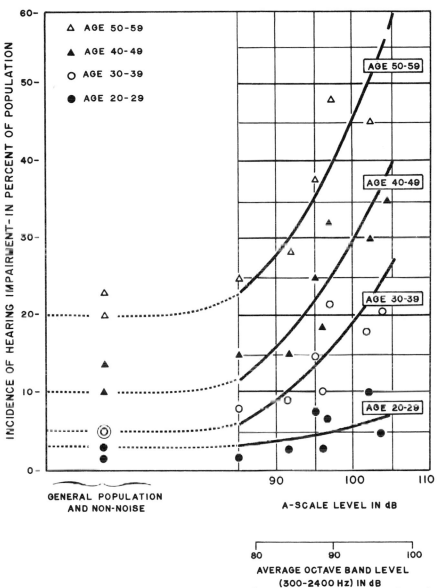

Courtesy American Industrial Hygiene Association Journal.

Fig. 12–1.—The incidence of hearing impairment in the general population and in selected populations by age groups and by occupational noise exposure.

The problems with these data are that no attempt had been made to eliminate from the totals those employees who might have incurred hearing losses from exposures not related to their jobs and that the data may be contaminated by some temporary threshold shifts. Some of these employees had almost certainly been exposed to substantial amounts of gunfire, either as a result of military service or of hunting, and quite possibly to other significant nonoccupational noise exposures away from their jobs. Dr. Baughn recognized this and pointed it out as a problem when drawing conclusions from his data. His data show a significant hearing impairment in about 8 percent of the population exposed for a working lifetime at 85 dBA and in about 17 or 18 percent of those exposed at 90 dBA.

Using both sets of data as a basis, the Bureau of Labor Standards proposed, in September 1968, a basic limit of 85 dBA for habitual exposure to steady state, broadband noise with some adjustments for less than full-shift exposure, and a provision for the use of personal protective devices in situations where the noise could not be controlled by engineering or administrative means. After a rather extensive public hearing, the Bureau of Labor Standards adopted on May 20, 1969, the noise regulation (shown in Appendix A-1).

This regulation provides for a basic level of 90 dBA for continuous noise exposure with a tradeoff of 5 dBA for each halving of the noise exposure time. This includes a rough allowance for interruptions of noise exposure. Impact noises must be limited to less than 140 dB peak sound pressure level.

It is important to note that this is a double requirement. For example, in a power-press department, it is necessary that both the instantaneous peaks of the impact noises arising out of the operation of large presses be below 140 dB and that the continuous ambient background noise be below 90 dBA.

In a forge shop, the continuous background noise (predominately combustion noise from furnace operations) is normally well in excess of 100 dBA, in addition to the impact noises from the hammers. The furnace noise can, and must be, controlled. The impact noise from forging operations is extremely difficult to control, but it may well turn out to be a minor problem due to the intermittent nature of the operation.

When the noise cannot be controlled by administrative or engineering means, it is required that a hearing protection program be administered. This very plainly means that personal protective equipment—ear plugs or ear muffs—must be provided and used.

The federal occupational noise exposure regulations were originally written to apply to contractors under the Walsh-Healey Public Contracts Act and the McNamara-O'Hara Service Contracts Act. These contractors amount to probably about 75,000 establishments at any one time. Having been adopted as the beginning regulations under the Williams-Steiger OSHAct, they now apply to about 2,400,000 industrial and commercial establishments across the country. Since the OSHA rules were adopted as temporary beginning regulations, OSHA must now consider the adoption of a permanent, or at least more permanent occupational noise exposure regulation.

Proposed rules and regulations

The National Institute for Occupational Safety and Health (NIOSH) was established within the Department of Health, Education and Welfare by the Occupational Safety and Health Act of 1970 to conduct research and to recommend new occupational safety and health standards. In 1972, NIOSH issued a 154-page report entitled "Criteria for a Recommended Standard for Occupational Exposure to Noise" (see Appendix A). The Department of Labor then appointed an Advisory Committee on Noise to study the NIOSH document and to develop a proposal for noise regulations for publication in the *Federal Register.*

In order to understand better the evolution of the proposed occupational noise exposure regulations, the following paragraphs were taken from the *Federal Register,* Vol. 39, No. 207, dated October 24, 1974.

Background. On August 14, 1972, the National Institute for Occupational Safety and Health (NIOSH), Department of Health, Education, and Welfare, provided to the Department of Labor a criteria package, "Occupational Exposure to Noise" in accordance with the Occupational Safety and Health Act of 1970. Thereupon, the Assistant Secretary of Labor for Occupational Safety and Health appointed a Standards Advisory Committee on Noise. The purpose of this Committee was to obtain and evaluate additional recommendations from labor, management, government, and independent experts. The Committee in its deliberations considered 138 written comments directed to it by interested parties, as well as numerous oral presentations. It transmitted recommendations for a revised standard to the Occupational Safety and Health Administration (OSHA) on December 20, 1973.

Exposure limits. The 1970 standard on occupational noise exposure limits an employee's exposure to 90 dBA, as an 8 hour time-weighted average. NIOSH indicated a need for reducing this 8 hour exposure level to 85 dBA, but was unable to recommend a specific time period after which the 85 dBA, noise level should become effective for all industry, due to the unavailability of sufficient data relating to technological feasibility of this level. Therefore, NIOSH reluctantly concurs with the generally acceptable 90 dBA occupational exposure level for an 8 hour day. The Advisory Committee's final report recommended retaining the 90 dBA permissible limit for an 8 hour day.

The Environmental Protection Agency (EPA), basing its recommendations on a review of hearing impairment risk, has recommended that OSHA reduce the limit at least to 85 dBA. EPA, which reviewed the draft standard proposal under the authority of the Noise Control Act of 1972, further recommended that additional studies be undertaken to explore reducing the permissible level still further at some future date. (See Appendix B.)

With regard to the risk of hearing loss, OSHA recognizes that comparatively more workers will be at lower risk at 85 dBA than at 90 dBA. However, it also recognizes the technical feasibility problems and the economic impact associated with an 85 dBA requirement as reflected in the Bolt, Beranek, and Newman study and in the draft Environmental Impact Statement. Therefore, OSHA proposes to keep the level at 90 dBA until further empirical data and information on the health risk, feasibility, and economic impact indicate the practicality and necessity of an 85 dBA requirement. We feel that the present level, when coupled with a stringent hearing conservation program beginning at 85 dBA, will reduce the risk to an acceptable level during the period of further consideration.

OSHA is aware of several studies currently under way that may provide additional information for this determination. In addition, the audiometric testing program required by this proposal should provide even more information on this issue. OSHA also requests any available information that can help clarify the question of permissible level.

Another consideration in determining the permissible exposure level is the appropriate doubling rate. This rate is the amount by which exposure intensity may be increased when exposure time is decreased. For example, the 5 dBA for doubling rate incorporated in the present standard and continued in the proposal would allow an exposure of 85 dBA for 16 hours; 90 dBA for 8 hours; and 95 dBA for 4 hours.

EPA recommended a doubling rate of 3 dB. While the 3 dB doubling rate is hypothetically correct for uninterrupted noise exposure, noise exposure in industry is normally interrupted since there are several breaks in the day's work. OSHA agrees with the Advisory Committee that the doubling rate should be adjusted to take into account the various breaks which occur in a workday. Therefore, OSHA believes that a doubling rate of 5 dB is more appropriate than 3 dB.

The present OSHA standard recommends that impact or impulse sounds not exceed a peak sound pressure level of 140 dB. The Advisory Committee suggested that this limit be made mandatory. OSHA has made an addition to the Advisory Committee's recommendation with respect to impulse noise exposure, because the actual exposure is a summation of the peak sound levels of the impulses and the number of impulses. OSHA proposes to limit exposure to impulses at 140 dB to 100 per day and to permit a tenfold increase in the number of impulses for each 10 dB decrease in the peak pressure of the impulse. For example, the number of impulses allowed at 130 dB would be 1000 per day and the number of impulses allowed at 120 dB would be 10,000 per day.

Controls. The 1970 standard states that feasible engineering and administrative controls shall be used to reduce noise exposure to within permissible levels. If such controls are not feasible or cannot reduce the sound levels to within permissible limits, then personal protective equipment shall be used to achieve compliance with the standard. This proposal continues the requirement that engineering and administrative controls be applied first to reduce noise to within permissible levels. The proposal also makes it clear that if engineering and administrative controls are not sufficient to reduce noise exposure to within permissible levels, such controls must nevertheless be used to reduce exposure to the greatest extent feasible and must be supplemented by personal protective equipment to achieve compliance. The proposal requires that all engineering and administrative controls be implemented, except where they are not feasible. (See Appendix A.)

It is the opinion of OSHA that general use of hearing-protective devices as a primary means of controlling noise exposure is not good industrial hygiene practice. It is not a satisfactory method of reducing noise exposure due to administrative difficulties commonly associated with the use of hearing-protective devices. Thus, while hearing-protective devices might technically afford the same protection as engineering controls, experience has shown that the protection afforded is diminished by the difficulty of management in requiring their use, workers' resistance to using them, and improper use and improper maintenance of such equipment. Accordingly, except in certain limited circumstances, this proposal relies primarily on engineering and administrative controls to reduce employee noise exposure. OSHA has, however, received numerous recommendations to permit the use of hearing-protective devices in lieu of expensive engineering controls to reduce the work place noise level. Some claim that hearing-protective devices are as effective as engineering controls in reducing noise exposure and may even have the added benefit of reducing worker exposure far below 85 dBA. Therefore, it has been suggested that the employer be given the choice of which method is utilized to achieve compliance with the standard. In view of the controversy surrounding the desirability of using protective devices to reduce employee noise exposure, OSHA will welcome and consider any submissions concerning the effective use of these devices.

Hearing-conservation program. Due to the wide variation in susceptibility to hearing loss from noise, it is not possible to determine a sound level exposure which will prevent all hearing loss in all members of an exposed population. The proposed requirement for a hearing-conservation program that includes audiometric testing will prevent, or at least minimize, permanent noise induced hearing loss by identifying those workers especially susceptible to noise. Audiometric testing will be initiated for all employees with 8 hour noise exposures of 85 dBA or higher. The audiometric testing program will also be required for those employees who wear personal protective equipment to reduce their noise exposures. Testing is necessary for these employees to assure that their hearing protectors are being used effectively.

OSHA believes that the audiometric testing program will detect any changes in hearing level in these workers so that the employer can adopt corrective action and inform employees before the changes become significant. In order for the results of such a testing program to be valid and meaningful, the audiometric environment and technique must be well standardized and stable over a sufficient number of years to represent a significant fraction of the employee's working life. It is also essential that these factors be reasonably identical from one employment to another. For these reasons, mandatory requirements are proposed for audiometric test rooms and the calibration of audiometers.

Monitoring. Although the current standard requires exposures to be controlled within specified limits, it does not explicitly require monitoring of the sound level of the employee's surroundings nor measurement of the individual employee's resulting exposure. This proposal makes monitoring and measuring requirements explicit. It requires the employer to determine if any employee is exposed to an 8 hour time-weighted average of 85 dBA or above. If any employees are so exposed, the employer must identify such employees and measure their exposure. The proposal also prescribes the minimum acceptable accuracy for monitoring instruments.

Records. The proposal requires the maintenance of records of the results of required measuring, monitoring, and the calibration of the instruments used therein. Employers are required to retain employee audiograms for the duration of employment plus five years; retention of other records is required for five years. The proposed period of retention of these records reflects OSHA's evaluation of their future usefulness to employers, employees, and the government.

In addition, the proposal implements the requirements of the Act which concerns, among other things, employee access to records of noise exposure. In addition, the proposal would require that prompt written notification be given to any employee who has been exposed to noise in excess of the permissible limits. This notification must be accompanied by a statement of the corrective action being taken.

Bibliography

Walsh-Healcy *Public Contracts Act,* Title 41, C.F.R. Chapter 50. Washington, D.C.: Superintendent of Documents, U.S. Government Printing Office.

Occupational Safety and Health Act Regulations, Title 29, C.F.R. Chapter 17. Washington, D.C.: Superintendent of Documents, U.S. Government Printing Office.

Department of the Air Force, Aerospace Medicine, "Hazardous Noise Exposure," AF Regulation 160–3 (1956) and 160–3A (1960)

Department of Health, Education, and Welfare. *Criteria for a Recommended Standard for Occupational Exposure to Noise* (HSM 73–11001), National Institute for Occupational Safety and Health, 1972.

Bolt, Beranek, and Newman, Inc. *Impact of Noise Control at the Workplace,* vol. 1 (report No. 2671). Cambridge, Mass.: Bolt, Beranek, and Newman, 1973.

Chapter Thirteen

Workers' Compensation and Medicolegal Factors

By Meyer S. Fox, M.D.

Noise induced hearing loss in industry has assumed increasing importance during the past 25 years. Although its existence has been known for over a century, very few compensation claims due to noise exposure arose prior to 1947. About that time, however, a large number of workmen's compensation claims for noise induced hearing loss were filed in New York, Wisconsin, and California. The medicolegal and worker's compensation aspects of the occupational noise problem that developed in Wisconsin are described in this chapter.

Workers' compensation laws

The nation's first effective state workers' compensation law was passed by Wisconsin in 1911. Since that time, Wisconsin has demonstrated its leadership in dealing with worker's compensation problems as they have arisen, and this is particularly true as far as the industrial noise problem is concerned. Prior to 1911, if a worker were injured in the course of his employment, he had to sue his employer or fellow employee in order for him to be compensated for his injury, loss of time, or wages. The situation then was very similar to the procedures now being used to deal with personal injury cases resulting from automobile accidents. The employee had to prove negligence on the part of the one causing the injury. The employee had to prove that he suffered either time or wage loss and it was necessary to pursue his claim in a

328

court of law. Needless to say this often caused bitterness and resentment on the part of the individual being sued.

In order to eliminate these factors and provide a more equitable basis for the settlement of claims, the Workmen's Compensation Acts were established. These Workmen's Compensation Acts substituted wage loss and loss of earning capacity in place of negligence or fault as a basis for compensation to the injured employee. Under the Workmen's Compensation Act, the maximum financial liability of the employer was established, and the worker was usually compensated on the basis of a fixed schedule of dollars or a stated percent of his weekly wage for the loss of an arm, eye, loss of hearing, loss of earning capacity, etc. The question of negligence was to be considered only in relationship to penalties for violations of state health and safety regulations.

The original Wisconsin compensation laws covered only liabilities for wage loss or earning capacity resulting from disabling accidents at the work place. Later these statutes were amended to provide benefits for nondisabling accidents and occupational diseases such as lead poisoning, silicosis, radium poisoning, and noise induced hearing loss (which is now considered to be an occupational disease in Wisconsin).

Noise induced hearing loss

A problem is new to the man who has discovered it only recently and this is particularly true in the case of noise induced hearing loss. With the enactment of the Walsh-Healey Public Contracts Act and the Occupational Safety and Health Acts, and the noise regulations promulgated under their authorization,[1] industry has suddenly found that it has a noise problem on its hands. Heretofore many industries, unless they were confronted with hearing loss claims, preferred "to let the sleeping dog lie" and considered it the other man's problem.

Loss of hearing occurring among employees in certain trades (such as blacksmiths, weavers, boilermakers, and the military) has long been known. It was taken for granted that some workers in these trades would sustain noise induced hearing loss. Hearing loss was accepted as a normal hazard by the worker, and the employer considered hearing loss as evidence that the employee had done his work well. In 1937 Dr. C. C. Bunche[2] published a monograph on acoustic trauma in which he discussed numerous audiograms typical of "acoustic trauma." He also predicted the medicolegal problems that would arise as a result of excessive exposure to industrial noise.

AAOO Subcommittee on noise

Any discussion of the medical and medicolegal aspects of noise induced hearing loss must take into consideration the activities and contributions of the Subcommittee on Noise of the Committee on Conservation of Hearing of the American Academy of Ophthalmology and Otolaryngology (AAOO). This subcommittee was formed in 1947 and was charged with the task of studying and investigating hearing loss as a result of noise exposure.

The AAOO Subcommittee, through its Research Center, carried out studies in numerous industries and laboratories. This subcommittee conducted a mass hearing survey at the Wisconsin State Fair in 1954 and 1955. Out of all this work came meaningful, practical information and guidelines which has been of use to industry, the medical and legal professions, compensation boards and those charged with the health and safety of workers.

The AAOO Committee on Conservation of Hearing has been instrumental in alerting industry to the need for establishing hearing conservation programs in order to minimize industrial hearing loss. Two of the most important documents published by this AAOO Committee were the "Guide for Conservation of Hearing in Noise"[3] which contains all the necessary information for establishing hearing conservation programs and "The Guide for the Evaluation of Hearing Impairment."[4]

The AAOO Committee on Conservation of Hearing has been frequently called upon by physicians, state industrial commissions, courts, and the legal profession for medical opinions regarding the effects of noise exposure.

The causal relationship between exposure to excessive levels of noise and the resulting hearing loss requires consideration of many complex, acoustical, medical, and legal factors. Therefore, the physician who examines and evaluates industrial hearing-loss cases must be properly informed, not only of the otological factors, but also of the acoustic and legal aspects of the noise problem.

Physician's medicolegal role

The physician, usually an otologist, who participates as a medical expert in these cases must be an impartial and disinterested expert, who will assist the courts and commissions in resolving the medical problem at hand. He must be objective and scientific and should not be influenced by the social and economic aspects of the problem. Decisions as to when, how much, or under what condition compensation is to be paid for loss of hearing are not matters for the medical expert to decide. These decisions are the functions of the legislatures, commissions, and courts.

The statements of the otologist or medical expert during a hearing should be based upon reasonable accepted otolaryngologic practice and not upon mere speculation. Some of the words that throw medical testimony into the category of speculation include the following: "may," "possible," "could have been," "might," and "might have been." These are all terms of conjecture.

As a rule, the opinions of the medical expert, when appearing before an industrial commission, need be based not upon medical certainty but rather upon reasonable medical probability. Probability is applied to that which is so reasonable or so well evidenced that it induces belief. In other words, the testimony should be based upon what the medical expert would ordinarily expect to find in the usual circumstances.

Diagnosis of noise induced hearing loss

The diagnosis of a noise induced hearing loss should be established and based upon the following facts:

1. History of the patient's hearing loss—its onset and progress

2. Occupational history—type of work, years of employment, and other details

3. Evaluation of noise studies at the employee's work location that reveals noises of a type and of sufficient intensity and duration to cause noise induced hearing loss

4. The result of the otological examination

5. Results of audiological and hearing studies including preemployment, periodic, and termination audiograms

6. Ruling out nonindustrial causes of hearing loss

Too often the worker's medical history and the results of hearing tests are inadequately described and insufficient to permit the differential diagnosis of noise induced hearing loss. Testimony has been presented before state industrial commissions on workers employed in noisy industries for many years in whom the hearing loss was attributed to medical conditions such as high blood pressure, infections, heart disease, diabetes, or advancing age simply because the functional hearing tests disclosed a high-frequency hearing loss.

The otologist or medical expert is not justified in ascribing the hearing loss to medical origin and, at the same time, completely ignoring the fact that the worker has been continuously employed to operate a noisy punch press, riveting machine, or drop hammer for the past 15 or 20 years. Nor is the otologist or medical expert justified in attributing the hearing loss to be occupational in origin, when the history, otological examination, and functional hearing tests reveal findings consistent with middle and inner ear diseases *of medical origin.*

Occupational hearing loss

Occupational hearing loss is the loss of hearing in one or both ears as the result of one's employment. This last phrase is very important; the loss of hearing must be "as the result of one's employment." The loss may be the result of hazardous noise exposure or due to acoustic trauma.

A clear distinction between the terms acoustic trauma and noise induced hearing loss should be made. Many references in the literature on acoustic trauma and noise induced hearing loss use these terms interchangeably, intending them to mean one and the same thing. A distinction should be made between these two terms for clarification.

• *Noise induced hearing loss* is a gradual loss of hearing of the sensorineural type resulting from exposure to hazardous noise levels over long periods of time, usually years.

• *Acoustic trauma* refers to hearing loss of sudden origin which occurs as the result of a single incident such as an explosion, blast, a blow to the ear or head, hot sparks or material entering the ear canal, a fractured skull involving the ear, etc.

The hearing loss in acoustic trauma can be a conductive type, a sensorineural type, or a mixed type of hearing loss. The audiogram of an individual who has incurred acoustic trauma may be different from the audiogram due to noise induced hearing loss. The progress of each type of hearing loss is different and also the recovery is different. Therefore, the physician should clearly distinguish between acoustic trauma and noise induced hearing loss.

Early claims

Claims for occupational hearing loss first occurred in the State of New York, Slawinski v. J. B. Williams & Co.[5] and Rosati v. Dispatch

Shop.[6] About this same time, some 232 claims for loss of hearing were filed against the Bethlehem Steel Company on the basis that the employer had failed to provide a safe place of employment for the workers. These 232 hearing-loss claims were filed under common law but were subsequently directed back to the Workmen's Compensation Board and settled. Shortly thereafter the test case of Albert Wojcik v. Green Bay Drop Forge Company[7] occurred in Wisconsin.

While loss of hearing resulting from sudden accidental acoustic trauma with a specific date of injury had previously been recognized and compensated, hearing loss claims arising from noise exposure were met by surprise and resistance on the part of industries and their insurance carriers. Alarmed and concerned, they felt that it was not the intent and purpose of worker's compensation legislation to compensate for a hearing impairment which did not result in either wage or time loss to the worker. These cases were, therefore, appealed to the higher courts of the various states.

Decisions of the higher courts have practically always upheld the right of the worker to claim compensation for hearing loss resulting from industrial noise exposure. The courts have held that the hearing loss was a scheduled award and often considered it to be an anatomical and social loss to the worker. Some states (Georgia, Oklahoma, and Arizona) have considered the hearing loss to be the result of a series of multiple traumatic incidents and therefore compensable.

Following the Wojcik decision, the Wisconsin Industrial Commission felt that because of the medical and legal uncertainties and possible serious economic effects upon industry, further study of the various problems associated with noise induced hearing loss was needed.

In Wisconsin, the Industrial Commission has maintained an Advisory Committee on Workmen's Compensation and Legislation. This Committee is composed of an equal number of representatives from industry and labor with a few consulting members who do not have voting privileges. When unanimous agreement is reached by members of this Committee on Workmen's Compensation and Legislation, their recommendations are usually adopted by the legislature.

In 1953 the Wisconsin Industrial Commission appointed a medical advisory committee, which consisted of four otologists and an industrial physician. The function of this medical advisory committee was to inform the Industrial Commission and its advisory group on the medical aspects of the noise problem. The original report of the medical advisory committee has recently been reviewed and updated.[16]

Medicolegal problems

Numerous thorny medicolegal problems and questions accompanied the early claims of workers for alleged loss of hearing due to noise. Some of the more important medicolegal questions were as follows:

1. How can the hearing loss resulting from industrial noise exposure be distinguished from the hearing loss due to other causes?
2. What level of noise intensity and over what period or length of exposure time does damage to hearing take place?
3. How can the worker's hearing status be determined at the beginning of his employment?
4. What procedures should be used to determine the hearing ability of workers?
5. What formula should be used to compute the hearing handicap?
6. When can the hearing loss be considered to be permanent?
7. What consideration should be given to other symptoms such as tinnitus which accompanies a hearing loss?
8. What consideration should be given to presbycusis?
9. What consideration should be given to improvement of hearing ability with the use of a hearing aid?
10. Who is responsible for the hearing loss where the employee has worked for several employers, each of whom has contributed to the final hearing loss?

Over the years, many of these questions and problems have been settled, either by industrial and clinical studies, laboratory investigations, or in some instances by compromise.

Damage risk criteria

There was an early demand on the part of industry, industrial commissions, courts, safety directors, and physicians for damage risk criteria. The original estimates of dangerous noise levels attempted to establish a single number—the decibel level—above which hearing loss would occur and below which the noise level would not be injurious to the hearing. The original level used was 90 decibels as measured on the C-scale of a sound level meter. More attention was paid to the level than to the actual time of exposure.

However, subsequent clinical and industrial investigations revealed that noise induced hearing loss depended upon other factors in addition to the overall noise level. It was necessary to know something about the kind of noise (such as impact, continuous, or frequency spectrum) as well as the duration of the noise exposure. Subsequent damage-risk noise criteria, therefore, was based upon exposure to specified octave-band levels for defined time periods.

With the introduction of the Walsh-Healey and OSHAct criteria for noise exposure, the 90 dBA level for an 8 hour exposure is being used to indicate when steady-state noise becomes dangerous and hearing-conservation programs are indicated. (See Table 14-B.) The 90 dBA scale for an 8 hour daily exposure was also recommended by the Intersociety Committee on Guidelines for Noise Exposure.[8]

While it is desirable to reduce industrial noise to levels which protect the hearing of all workers, such a goal is not always feasible from a practical and operational standpoint. As discussed in previous chapters, all hearing-conservation programs have certain limitations. The 90 dBA level was a compromise designed to protect the hearing of the majority of the workers for those frequencies necessary for understanding speech (500, 1000, and 2000 Hz). It is estimated that the 90 dBA level will protect about 90 percent of the workers during their working lifetime. The criteria recommended by the National Institute for Occupational Safety and Health (NIOSH) indicated that an 85 dBA level for an 8 hour noise exposure would be more desirable.[9]

Many of the state enforcement agencies through their Hygiene or Safety Departments are adopting the 1971 OSHA safety codes using the 90 dBA criteria. The State of Wisconsin adopted a noise code under the Safety Division (Wisconsin Administrative Safety Code No. 11).[10]

Baseline audiograms

The sensori-neural hearing loss that results from noise exposure is similar to losses caused by other, less apparent factors. It is, therefore, difficult in any particular case without a baseline preemployment audiogram to state what part of the employee's hearing loss can be attributed to conditions at his present place of employment, and what part existed prior to being hired by his current employer.

The same problem presents itself where the employee has worked for numerous employers where each place of employment might have contributed to the final hearing loss. Which employer then is to be held responsible for the hearing loss found at the time of the claim? Several legislatures (including Wisconsin and New York) have settled this problem by enacting statutes which hold the last employer responsible for the entire hearing loss unless by competent audiometric evidence it can be established what the employee's baseline hearing level was at the beginning of employment.

Rules and statutes have alerted industry to the need for preemployment hearing tests to establish baseline audiograms. Additional audiograms should be taken at periodic intervals on workers exposed to hazardous noise, in order to be able to document the employees' hearing ability through their years of employment.

Evaluation and determination of hearing impairment

An important medicolegal problem pertains to the various testing procedures used by physicians to determine the hearing ability of the claimant and the methods used to convert these audiometric readings into percentages of hearing impairment. After several years of studying the methods and procedures used, the Subcommittee on Noise of the Committee on Conservation of Hearing of the American Academy of Ophthalmology and Otolaryngology published its "Guide for the Evaluation of Hearing Impairment"[4] in 1959.

The principles and recommendations therein were subsequently approved and incorporated into the American Medical Association *Guides to the Evaluation of Permanent Impairment,* Chapter VIII, "The Ear, Nose, Throat, and Related Structures."[11] These guides were the first officially approved formulas for determining the percentage of hearing impairment. They provided a definitive method which could be easily applied and understood by judge, examiner, or jury, and that makes provision for the establishment of the degree of hearing impairment in one or both ears.

It should be pointed out that the evaluation or rating of permanent hearing impairment is a function of the physician alone. However, the evaluation or rating of permanent disability is an administrative and not a medical responsibility or function.

Audiometry. Ideally, hearing impairment should be evaluated in terms of ability to hear everyday speech under everyday conditions; the term "impairment" is used in this sense only. The ability to hear sentences and repeat them correctly in a quiet environment is taken as satisfactory evidence of hearing for everyday speech. The hearing level

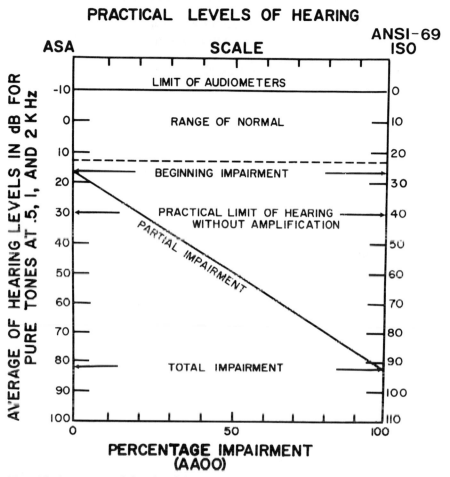

FIG. 13-1.—Practical levels of hearing.

for speech should be established from measurements made with a pure tone audiometer, calibrated to the specifications of ANSI S3.6–1969, *Specifications for Audiometers.*[12] For this estimate it is necessary to obtain the simple average of the hearing threshold level (HTL) at the three frequencies 500, 1000, and 2000 Hz (see Figure 13–1).

Procedure. The following procedure is used to convert hearing threshold levels into percentages of hearing impairment. If the average monaural hearing level at 500, 1000, and 2000 Hz is 25 dB or less, usually

TABLE 13-A
DETERMINATION OF HEARING IMPAIRMENT USING AMA GUIDE
Audiometric Test Results
Hearing Threshold Level (dB)

Frequency (Hz)	500	1000	2000	3000	4000	6000
EAR						
Right	35	40	60	65	70	65
Left	40	50	60	70	75	70

1. Add the HTL for the frequencies most important to speech (500, 1000, and 2000 Hz).

   RIGHT—35 + 40 + 60 = 135 dB
   LEFT—40 + 50 + 60 = 150 dB

2. Divide by three to find the Average HTL for each ear.

   RIGHT—135 ÷ 3 = 45 dB
   LEFT—150 ÷ 3 = 50 dB

3. To convert the Average HTL for the speech frequencies to percent hearing impairment, subtract the Low Fence (according to the AMA formula, it is that point below which no impairment is presumed to exist) from the Average HTL for speech.

   RIGHT—45 dB (Avg. HTL) − 25 dB (Low Fence) = 20 dB
   LEFT—50 dB (Avg. HTL) − 25 dB (Low Fence) = 25 dB

   Then multiply the result by 1.5% for each decibel the Average HTL exceeds 25 dB.

   RIGHT—20 dB × 1.5% = 30% impairment
   LEFT—25 dB × 1.5% = 37.5% impairment

4. To convert the monaural impairment to percent of binaural impairment, proceed as follows:

   a. Percent of hearing impairment in better ear times 5

      $$30\% \times 5 = 150\%$$

   b. Add percent of hearing impairment of worse ear.

      $$\frac{37.5\%}{187.5\%}$$

      Total

   c. Divide the total by 6. 187.5% ÷ 6 = 31.25%
      The binaural loss is 31.25%

no impairment exists in the ability to hear everyday speech under everyday conditions. This is called the *low fence*. At the other extreme, however, if the average of the hearing level at 500, 1000, and 2000 Hz is over 92 dB ANSI, the impairment for hearing everyday speech should be considered total.

The AAOO Subcommittee on Noise recommended the following formula:

TABLE 13-B
CONFLICT BETWEEN STATE SCHEDULES AND AMA FORMULA

MAINE: Schedule for Occupational hearing loss, Effective on November 30, 1967

Compensation for: One ear = 50 weeks
Bilateral = 100 weeks

EXAMPLE:

Suppose an employee has a hearing impairment or loss in the right ear of 60 percent, and a loss in the left ear of 20 percent:

Apply AMA formula

5 times lesser loss = 5 × 20% = 100%
Add percent of greater loss   60%
Total   160%

Divide by 6 = Bilateral loss = 26.67%

26.67 percent of 100 weeks = 26.67 weeks, for bilateral loss.

Loss for right ear alone is 60 percent of 50 weeks or 30 weeks.

For every decibel that the estimated monaural hearing level for speech exceeds 25 dB, allow 1.5 percent up to the maximum of 100 percent, which is reached at 92 dB ANSI.

Binaural. The American Medical Association Guides[11] recommend that any method for the evaluation of impairment include an appropriate formula for binaural hearing that is based on the hearing levels in each ear tested separately. The following formula was recommended:

The percentage of impairment in the better ear is multiplied by five. The resulting product is added to the percentage of impairment in the poorer ear, and the sum is divided by six. The final figure represents the binaural evaluation of hearing impairment not the percentage of hearing loss. (Table 13-A illustrates the use of the AMA recommendations.)

As stated previously, the American Medical Association Guide rates hearing loss of one ear to binaural loss at the ratio of 1-to-5. While many of the state agencies have adopted the use of the American Medical Association formula for determining hearing impairment, they have failed to recognize that in order to determine binaural impairment they would have to change the ratio in their schedules as they apply to

339

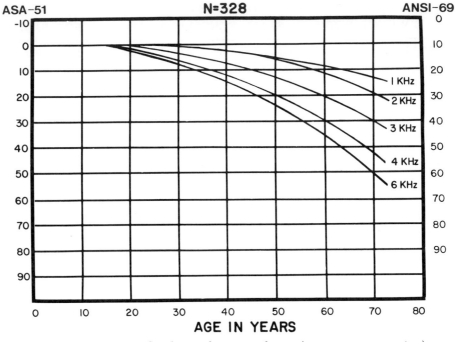

FIG. 13–2.—Hearing level as a function of age (non-noise occupation).

loss of hearing in one ear compared to both ears. This problem is illustrated in Table 13-B.

Presbycusis

The role of presbycusis (hearing loss due to the aging process) in hearing loss claims has received considerable attention. Several state compensation boards have an arbitrary rule requiring a deduction for the hearing loss that accompanies advancing age.

The American Academy of Ophthalmology and Otolaryngology's Committee on Conservation of Hearing[13] has stated that if the AMA formula is used for determining the hearing impairment, no impairment is considered to exist if the average hearing level for the three speech frequencies (500, 1000, and 2000 Hz) is 25 dB or less than the low fence. The average presbycusis curves at age 65 show that this low fence is not reached (see Figure 13-2). Therefore, no deduction for presbycusis should be made.[13] The presbycusis factor is included in setting the 25 dB ANSI low fence.

TABLE 13-C
DEDUCTION FOR PRESBYCUSIS
Results of Audiometric Examination

Frequency (Hz)	Hearing Threshold Level (dB)	
	Right Ear	Left Ear
500	30	30
1000	35	35
2000	45	45
4000	60	60
6000	65	65

1. Find the Average HTL using the speech frequencies only.

500	30
1000	35
2000	45
	110 dB

Divide the sum by 3. 110 dB ÷ 3 = 36.6 Average HTL

2. Deduction for Presbycusis (0.5 dB for each year after age 40).

   Age of employee (65) − 40 = 25 × 0.5 = 12.5 dB Deduction

3. Total deduction equals the Presbycusis Deduction (12.5 dB, in this case) plus the low fence (25 dB using AMA formula) or 12.5 dB plus 25 dB = 37.5 dB.

4. Subtracting the Average HTL (36.6 dB) from the total deduction (37.5 dB) leaves the employee short 0.8 dB of being considered as having a hearing impairment.

Presbycusis deduction. Some states make a mandatory deduction of 0.5 dB per year beginning at the age of 40 (Maine, Maryland, Missouri, and others). This deduction is not justified if the AMA guide is used to determine the degree of hearing impairment.

This problem is best illustrated in Table 13-C where, for example, a retired employee 65 years old, having a hearing loss of 37 dB, would not be entitled to any compensation because of the 25 dB low fence deduction and a 12.5 dB deduction for presbycusis. In other words, this employee would not be considered to have any hearing impairment until his average loss for the 3 frequencies 500, 1000, and 2000 Hz exceeds 37.5 dB.

Waiting period. An accurate evaluation of the employee's permanent hearing loss may be difficult to establish if the claimant is presently working in high-noise areas at the time he presents himself for examination of his hearing ability. The physician must make sure that the worker has been away from hazardous noise for sufficient time to eliminate any temporary threshold shift. Several states, for economic and administrative reasons, require a 6 month removal or waiting period by the employee after termination of noise exposure before a claim can be filed.

When a worker terminates his employment by retirement, his hearing status can be evaluated and the case closed. Where no waiting period is required, there may be a tendency to file claims in an irresponsible fashion. Many of these claims can be nuisance claims where the administrative costs of making the legal and medical examinations exceed the awards. While this is not a medical problem, the 6 month waiting period does have merit.

Hearing aids. What, if any, consideration should be given to improvement with the use of a hearing aid? A few of the state agencies have indicated that a corrective factor is in order where there is improvement in the employee's hearing ability with the use of a hearing aid. The American Medical Association Guide emphasizes that impairment of hearing should be determined without benefit of the hearing aid. This recommendation is consistent with the general policy of determining impairment of body function without the use of any prosthetic device. Any benefit that results from the use of such a device should not be a mitigating factor in determining the hearing impairment.

Hearing loss statutes

While worker's compensation acts in the United States and Canada have some common features, the provisions covering noise induced hearing loss vary greatly in the different states and the Canadian Provinces. A survey of the existing occupational hearing loss compensation statutes and rules in the various states and Canadian Provinces was conducted in 1963 and published in the *Archives of Otolaryngology* in March 1965.[14] This information was subsequently updated in 1969 and in 1972.[15] (See Tables 13-D and 13-E.) In a survey of this type, changes will occur rather frequently; therefore, *it is recommended that you obtain the latest information available in your jurisdiction to check for any legislative or administrative changes that may have occurred.*

(*Text continues on page 349.*)

TABLE 13-D. COMPILATION OF HEARING LOSS STATUTES IN EFFECT

JURISDICTION	Is Occupational Hearing Loss Due to Continuous Noise Exposure Compensable	Basis of Compensation (Per Week—One Ear) Primarily Traumatic	Basic Compensation (Per Week—Both Ears)	Maximum Compensation—One Ear	Maximum Compensation—Both Ears	Must Employee Leave Noisy Work for Any Period Before Filing Claim?	Method or Formula Used for Determining Hearing Impairment (Medical Evidence °)	Do You Deduct for Presbycusis?	Is Any Award Made for Tinnitus?	Do You Compensate for Non Organic Hearing Loss?	Is Provision Made for Hearing Aid?	Is Credit Given for Improvement with Hearing Aid?
1 ALABAMA	No	53 Weeks	163 Weeks	$3,180	$ 9,780	No	Medical Evidence °	°	°	°	°	No
2 ALASKA	Yes	52 Weeks	200 Weeks	$3,600	$ 4,300	No	AMA	Possibly	Possibly	Possibly	Yes	No
3 ARIZONA	Yes°	20 Months	60 Months	$11,000	$33,300	No	AMA	No	Possibly	Possibly	Yes	Not Decided
4 ARKANSAS	Yes	40 Weeks	50 Weeks	$2,520	$ 9,450	No	AMA	Possibly	Yes	Yes	Yes	Yes
5 CALIFORNIA	Yes	150 Weeks	311 Weeks	$2,110	$10,500	No	AMA Plus 3000 +z Frequency	Possibly	Yes	Possibly	Yes	Yes
6 COLORADO	No	35 Weeks	139 Weeks	$2,266	$ 9,000	—	ME	Possibly	—	Law Uncertain	Yes	—
7 CONNECTICUT	Yes	52 Weeks	155 Weeks	$5,824	$17,472	No	ME°	No	No	No	Yes	No
8 DELAWARE	Yes	75 Weeks	175 Weeks	$5,625	$13,125	No	ME	—	—	No	Yes	No
9 DIST. OF COLUMBIA	Yes	52 Weeks	200 Weeks	$10,945	$42,108	No	AMA	Possibly	Possibly	Possibly	Yes	—
10 FLORIDA	Yes	40 Weeks	153 Weeks	$3,200	$12,000	No	AMA-ME	Yes	No	Yes	No	—
11 GEORGIA	Yes°	60 Weeks	150 Weeks	$3,900	$ 9,750	No	ME°	Possibly	No	Possibly	No	No
12 HAWAII	Yes	52 Weeks	200 Weeks	$5,850	$22,500	No	AMA	No	Yes	Yes	Yes	—
13 IDAHO	Yes°	35 Weeks	175 Weeks	$2,600	$11,157	No	ME	Possibly°	Possibly°	Possibly°	Yes	No
14 ILLINOIS	Yes	50 Weeks	125 Weeks	$4,840	$12,112	No	AMA	No	No	Possibly	No	—
15 INDIANA	No	75 Weeks	200 Weeks	$4,500	$12,000	No	ME	No	No	No	No	—
16 IOWA	No	50 Weeks	175 Weeks	$4,200	$14,700	No	AMA-ME	Yes°	—	—	Yes	Yes
17 KANSAS	Yes	30 Weeks	110 Weeks	$1,680	$ 6,160	No	AMA°	—	—	—	—	—
18 KENTUCKY	Yes	75 Weeks	153 Weeks	$6,225	$12,548	No	ME	No	No	No	Possibly	—

TABLE 13-D (continued)

	JURISDICTION	Is Occupational Hearing Loss Due to Continuous Noise Exposure Compensable	Basis of Compensation (Per Week—One Ear) Primarily Traumatic	Basic Compensation (Per Week—Both Ears)	Maximum Compensation One Ear	Maximum Compensation Both Ears	Must Employee Leave Noisy Work for Any Period Before Filing Claim?	Method or Formula Used for Determining Hearing Impairment	Do You Deduct for Presbycusis?	Is Any Award Made for Tinnitus?	Do You Compensate for Non Organic Hearing Loss?	Is Provision Made for Hearing Aid?	Is Credit Given for Improvement with Hearing Aid?
19	LOUISIANA	No°	—	—	$3,675	$7,350	Yes	ME		—	—	no	—
20	MAINE	Yes	50 Weeks	100 Weeks	$4,150	$16,600	6 Months	AMA	Yes ½ dB After Age 40	Yes	No	No	—
21	MARYLAND	Yes*	125 Weeks	250 Weeks	$7,275	$22,601	6 Mpnths	AMA	Yes ½ dB After Age 40	Possibly	—	Yes	No
22	MASSACHUSETTS	Yes*	150 Weeks	400 Weeks	$4,500	$12,000	No	AMA-ME	Possibly	Possibly	Possibly	Yes	No
23	MICHIGAN	Yes°	*	°	°	*	—	ME	No	Yes	—	Yes	—
24	MINNESOTA	Yes	85 Weeks	170 Weeks	$5,500	$17,000	No	AMA-ME	No	Possibly	Possibly	Yes	No
25	MISSISSIPPI	Yes	40 Weeks	150 Weeks	$2,240	$8,400	No	ME	No	Possibly	—	Yes	No
26	MISSOURI	Yes	40 (Noise) 44 (Traumatic)	148 (N) 168 (T)	$2,540 (N) $2,860 (T)	$9,398 (N) $10,920 (T)	6 Months	AMA	Yes ½ dB After Age 40	Yes	Yes	No	—
27	MONTANA	Yes*	40 Weeks	200 Weeks	$2,400	$12,000	6 Months	AMA	Yes ½ dB After Age 40	°	*	*	No
28	NEBRASKA	No	50 Weeks	100 Weeks	$4,000	$8,000	No	AMA	No	Yes	Yes	Yes	No
29	NEVADA	Yes	20 Months	60 Months	$5,655	$16,900	No	AMA	No	No	No	Yes	No
30	NEW HAMPSHIRE	Yes	52 Weeks	214 Weeks	$5,980	$24,610	No	AMA	Possibly	—	—	Yes	No
31	NEW JERSEY	Yes	60 Weeks	200 Weeks	$2,400	$8,000	No	AMA	No	Yes	Yes	Yes	No
32	NEW MEXICO	No	40 Weeks	150 Weeks	$2,600	$9,750	—	ME	—	—	—	Yes	—
33	NEW YORK	Yes	60 Weeks	150 Weeks	$4,800	$12,000	Yes 6 Months	AMA	No	Yes	Yes	Yes	No
34	NORTH CAROLINA	Yes	70 Weeks	150 Weeks	$5,600	$12,000	6 Months	AMA-ME	½ dB for Each Year After Age 38	No	Yes	No	No
35	NORTH DAKOTA	Yes	50 Weeks	200 Weeks	$2,000	$8,000	No	AMA-ME	Possibly	No	No	No	No
36	OHIO	Yes°	25 Weeks	125 Weeks	$1,802	$9,012	No	ME	Possibly	No	No	No	No

	State		100 Weeks	200 Weeks	$5,000	$10,000		ME°					
37	OKLAHOMA	Yes°	100 Weeks	200 Weeks	$5,000	$10,000	No	ME°	No	Yes	Yes	Yes	No
38	OREGON	Yes	60 °	192 °	$4,200	$13,440	No	AMA	Possibly	No	No	Yes	No
39	PENNSYLVANIA	Yes	—	130 Weeks	$600	$6,000	No	ME	No	No	No	No	—
40	RHODE ISLAND	Yes	17 Weeks	100 Weeks	$2.70	$9,000	6 Months	AMA	Yes ½ dB for Each Year After Age 40	No	No	Yes	No
41	SOUTH CAROLINA	Yes	70 Weeks	140 Weeks	$4.35	$10,395	No	ME	No	Possibly	Possibly	Yes	No
42	SOUTH DAKOTA	Yes	50 Weeks	140 Weeks	$3.30	$9,000	—	ME	Possibly	Yes	No	Yes	Yes
43	TENNESSEE	Yes	75 Weeks	150 Weeks	$3.52	$9,330	No	ME	No	No	No	Yes	No
44	TEXAS	Yes	—	150 Weeks	$4.72	$9,450	No	AMA	Possibly	Yes	Yes		No
45	UTAH	fes	—	100 Weeks	$4.35	$8,900	6 Months	AMA	½ dB After Age 40	No	No	Yes	No
46	VERMONT	No	52 Weeks	2.5 Weeks	$4.31	$7,415	No	ME	No	No	No	Yes	No
47	VIRGINIA	fes	50 Weeks	160 Weeks	$4.40	$8,000	No	AMA-ME	No	Yes	Yes	Yes	No
48	WASHINGTON	Yes	40 Weeks	200 Weeks	$2.40	$4,400	6 Months	AMA	No	Yes	Yes	Yes	No
49	WEST VIRGINIA	°es	60 Weeks	180 Weeks	$5.52	$16,560	No	AMA	Possibly	Yes	No	Yes	No
50	WISCONSIN	°es	36 Weeks (N) / 55 Weeks (T)	216 Weeks (N) / 330 Weeks (T)	$1,800 (N) / $2,913 (T)	$10,800 (N) / $7,490 (T)	6 Months	AMA	No	Yes°	Yes°	Yes	No
51	WYOMING	No	*	*	$3,000	$6,000	No	ME	No	No	No	Yes	No
52	PUERTO RICO	Yes	50 Weeks	200 Weeks	$2,250	$3,000	No	AMA	No	Possibly	Possibly	Yes	No
53	CANADIAN PROVINCES	Yes	*	*	*	*	Yes°	*	*	*	*	Yes *	°

°See "Comments" far right column   AMA—American Medical Association "Guide to the Evaluation of Permanent Impairment"   ME—Medical Evidence

TABLE 13-E. ADDITIONAL COMMENTS RECEIVED IN REPLY TO QUESTIONNAIRE

Jurisdiction	Comments
Alabama	*Worker's Compensation laws administered by courts. May use any criteria appropriate to determine issue.
Alaska	Awards for tinnitus and nonorganic loss if it affects hearing and is due to work.
Arizona	Noise induced hearing loss recognized under gradual theory—series of traumatic incidents. Compensation for tinnitus; nonorganic loss based on medical evidence of impairment of function and inability to work.
Arkansas	Might provide hearing aid and credit for improvement as done with glasses.
California	Rating values based on age, occupation, etc. of worker. Formula based on AMA guide but includes 3000 Hz. Awards for tinnitus and non-organic hearing loss on medical evidence. Has experienced frequent claims.
Colorado	At present time only traumatic hearing loss is compensable. Consideration being given to include noise induced hearing loss.
Connecticut	Provision for hearing aid undecided.
Delaware	All Worker's Compensation acts are being reviewed by the assembly.
Dist. of Columbia	Award for tinnitus and nonorganic hearing loss if associated with hearing loss.
Florida	Compensation for nonorganic hearing loss possible on medical evidence if related to employment.
Georgia	Compensation for noise induced hearing loss decided by court decisions. Treated as accidental (traumatic) loss. Provision for temporary as well as scheduled loss depends upon medical opinion for various criteria used.
Hawaii	Compensation for tinnitus and nonorganic loss on medical, psychiatric basis.
Idaho	Consideration of presbycusis, tinnitus, and nonorganic loss based on medical evidence. New law effective in 1972.
Illinois	Hearing loss must be total in one or both ears. New legislation for partial hearing loss being considered.
Indiana	
Iowa	Credit for improvement with hearing aid. Mean between improved and unimproved hearing.
Kansas	Uses modified weighted AMA formula of 1942.
Kentucky	Provision for hearing aid if medically indicated.

Louisiana	Compensation based upon 65 per cent of wages during disability not beyond 300 weeks if temporary, or 400 weeks if total, and must leave work to prove disability. Cases handled by state courts.
Maine	Deduction for presbycusis—½ dB each year after age 40.
Maryland	Occupational deafness Act 101.25A effective 1971.
Massachusetts	*Must be total loss of hearing for all practical purposes to be compensable.
Michigan	Hearing loss must be total in one or both ears to be compensable. Compensation based on wage loss and inability to work.
Minnesota	Award for tinnitus and nonorganic hearing loss on basis of wage loss. Award on one ear up to 85 weeks—50 dBA used.
Mississippi	Award for tinnitus and nonorganic hearing loss if job related.
Missouri	Deduction for presbycusis—½ dB for each year after 40. Separate schedule for loss resulting from noise and trauma.
Montana	Legislature passed law covering occupational hearing loss effective Jan. 1, 1972. Have increase in number of claims. Deduction for presbycusis ½ dB after age 40.
Nebraska	Compensate for tinnitus if associated with hearing loss. Nonorganic on neurosis basis. Has frequent claims.
Nevada	Has general safety rule applying to noise exposure.
New Hampshire	*Anticipate adoption of federal noise exposure standards.
New Jersey	Large number of cases—Large number of claims.
New Mexico	District courts administer the compensation act.
New York	Compensation for tinnitus and nonorganic loss if causally related.
North Carolina	Legislation enacted to compensate for noise induced hearing loss. Effective Oct. 1, 1970.
North Dakota	
Ohio	Loss must be total in one or both ears to be compensable.
Oklahoma	Considered as series of traumatic incidents. Permit use of live voice tests at measured distances.
Oregon	Schedule is based on degrees rather than weeks. Each degree is worth $55.
Pennsylvania	Must have complete loss of hearing in both ears. Receives compensation at rate of 66 per cent of weekly wages for 180 weeks (180 × $60 = $10,800). Legislation suggested but has not been enacted.
Rhode Island	Legislation effective Sept. 1, 1969. Deduction for presbycusis, ½ dB each year after 40.
South Carolina	

347

TABLE 13-E (continued)

South Dakota	Occupational noise compensable as of July 1, 1971. Credit for improvement with hearing aid.
Tennessee	
Texas	Texas law effective June 9, 1971 provides compensation of noise induced hearing loss. Compensation for noise induced hearing loss is based on both ears only.
Utah	Impairment computed on binaural hearing only.
Vermont	May pass legislation in 1972.
Virginia	Hearing table slightly different than AMA. Impairment begins at 17 dB and ends at 80 dB.
Washington	Award for tinnitus and nonorganic hearing loss on medical evidence. Increase in number of claims.
West Virginia	Presbycusis and tinnitus given consideration based on medical evidence.
Wisconsin	Award for tinnitus and nonorganic hearing loss if it is disabling and related to employment. Has separate schedule for noise and traumatic hearing loss cases.
Wyoming	Rated as percentage loss similar to rating for impairment of vision.
Puerto Rico	

NOTE: Numerous states have indicated formulation of state safety codes to govern hazardous noise. Since the OSHA guidelines apply to hazardous exposure in all states, the state code, therefore, must be equal to or exceed the OSHA regulations.

Canadian Provinces	The status and provisions in the Canadian Provinces show slight variations. In general they compensate for noise-induced hearing loss and have specific guidelines (Walsh Healey) for noise exposure and for the use of personal hearing protection devices. Compensation is based upon the percentage of the body as a whole, and varies from three to five per cent for total loss in one ear to thirty per cent for both ears. They use a modification of the AMA formula as follows: No award for average loss less than 25 dB, ASA scale or 35 dB, ISO scale. Complete loss if average is 70 dB on ASA scale or 80 dB on ISO scale. The determination of binaural hearing loss varies from one-to-five to one-to-nine ratio. Most provinces require the applicant to remove himself from noise exposure to collect an award (pension). Some provinces have a presbycusis deduction of ½ dB for each year over 50; some start at 60, and some provinces have no presbycusis deduction. There is provision for furnishing a hearing aid where medically indicated, but no credit is allowed for improvement by use of a hearing aid.

An examination of Tables 13-D and 13-E reveals a definite trend on the part of worker's compensation commissions and agencies to recognize and provide greater coverage for occupational hearing loss. However, significant variations exist in the regulations and provisions regarding medical procedures for determining and evaluating hearing loss claims.

The survey reported here is based on replies from the 50 states, Puerto Rico, the District of Columbia, and the Canadian Provinces. The questions asked and replies received from the industrial accident boards and commissions are summarized in Table 13-D. (Column numbers refer to the following questions.)

Summary and comments on Table 13-D

Question 1. *Is occupational hearing loss, due to continued noise exposure over a period of time, recognized as compensable in your jurisdiction?*

Yes—44   No 9

In three states—Arizona, Oklahoma, and Georgia—hearing loss is considered as a series of repeated traumatic incidents.

The states of Illinois, Ohio, Pennsylvania, Michigan, Rhode Island, and Massachusetts do not provide compensation for hearing loss, either from noise or trauma, unless the loss in one or both ears is total.

From a medical standpoint, there is considerable difficulty in interpreting just what is meant in the statutes by "total hearing loss." Does it mean the inability to hear any sound at the maximum intensity limits of the audiometer? Or does it mean a hearing loss, of sufficient extent and degree, to interfere with the ability to carry on one's duties?

Question 2. *Does your jurisdiction specify either the level, the type of noise, or the period of exposure that is considered hazardous?*

Yes—17

Those states replying "Yes" had regulations similar to those shown in Table 14-B.

The occupational noise exposure regulations in Table 13-A correspond to those set forth in paragraph 50.204.10 of the Walsh-Healey Public Contracts Act and are essentially the same as those set forth in section 1910.95 of the Occupational Safety and Health Act of 1970. It is expected that eventually all the states will adopt similar rules and regulations.

Questions 3, 4, 5, and 6. *Upon what basis is hearing loss compensable in your state? What is the basis for compensation and the maximum compensation for one and both ears?*

An examination of the replies to the questionnaire reveals:

Schedule for Total Loss on One Ear (Questions 3 and 5)
Median number of weeks—between 50 and 55;
Median possible awards—between $2500 and $3000;
   Ranges from 0 to $11,000;

Both Ears (Questions 4 and 6)
Median number of weeks—220;
Median possible award—$8500–$9000;
   Ranges from $4500 to $33,400
Awards more than $12,000 are possible in 25 percent of jurisdictions.

Question 7. *Must the employee remove himself from the noise for any period of time before filing claims for occupational hearing loss? And if yes, for how long?*

The States of Maine, Maryland, Missouri, Montana, New York, North Carolina, Rhode Island, Utah, Washington, and Wisconsin require the employee must be away from noise for 6 months. Louisiana requires that the employee be away long enough to prove disability. The Canadian Provinces require the employee to be away from noise from 1 to 6 months.

Accurate evaluation of the permanent hearing loss may be difficult to establish if the claimant is working in noise at the time he presents himself for examination. The physician must make sure that the worker has been away from the noise for sufficient time to eliminate any threshold shift. Several states, for economic and administrative reasons, require a 6 month removal or waiting period atfer termination of noise exposure before a claim can be filed.

Question 8. *What frequencies, rating scales, and formulas are used to determine the degree and extent of hearing loss for compensation purposes?*

The AMA Guide is used by 31 states. California includes 3000 Hz in addition to 500, 1000, and 2000 Hz. Canada has variations; New Jersey and Kansas use the 1947 AMA formula. Medical evaluation is used by

27 states. Wyoming rates on the same basis as percentage loss of eye-sight.

Many agencies depend upon medical evaluation for the establishment of percentage of hearing impairment. However, 31 agencies do use the AMA guide. What is of primary importance is that the schedules in the various agencies are in conflict with the AMA guide for the evaluation of binaural hearing impairment.

The AMA guide rates hearing loss in one ear to binaural loss at the ratio of 1-to-5; 12 agencies use the ratio of 1-to-1; or 1-to-1.25. In order to use the AMA guide properly you would have to modify most of the schedules as they now exist.

While numerous jurisdictions stated that they were familiar with and were using the American Medical Association Guides to the Evaluation of Permanent Impairment,[11] many of the agencies reported that they relied primarily upon expert medical testimony in determining the causal relationship as well as the extent of the hearing loss. Rather surprisingly, the survey also revealed that some agencies would compensate for tinnitus and nonorganic hearing loss, and they would provide for the cost of a hearing aid, if medically indicated.

Question 9. *Is any deduction made for presbycusis (hearing loss due to age)? If yes, how is the deduction made?*

These six states deduct 0.5 dB for each year after age 40: Maine, Maryland, Missouri, Montana, Rhode Island, Utah.

The Canadian Provinces deduct 0.5 dB for each year after age 50 or 60.

The role of presbycusis in hearing loss claims has received considerable attention. Several compensation boards have an arbitrary rule requiring a deduction for the hearing loss that accompanies age. (See previous discussion on presbycusis.)

Question 10. *Is compensation awarded for tinnitus (ear noises)?*

Yes	—15
Possibly	—11
No	—17
No experience	— 2
No answer	— 8

(Replies include Puerto Rico and Canada.)

Question 11. *Is nonorganic hearing loss (often referred to as psychogenic hearing loss) recognized as a compensable disability?*

Yes —11
Possibly —13
No —17
No answer —12

Basis for award for tinnitus and nonorganic hearing loss is as follows:

• Associated with hearing loss due to work

• If it prevented the worker from continuing employment in the work he was formerly engaged in

• On a neuropsychiatric basis

Question 12. *Is provision made to provide a hearing aid where medically indicated?*

Yes —38
Possibly — 3
No —10
No answer — 2

Question 13. *Is any credit given for improvement with hearing aid?*

No —33
Yes — 4

Summary

A review of the provisions and statutes as outlined in this survey reveals many variations, discrepancies, and inadequacies in the schedules and various medical procedures for determining and evaluating hearing loss claims. These variances are often the result of legislative decisions and judgments and do not necessarily reflect medical opinions.

The vast majority of state worker's compensation laws have schedules or formulas for determining awards—so many weeks for total loss of hearing of one ear, and so many weeks for total loss of hearing in both ears. There are marked variations among the reporting agencies in the number of weeks, and there are great differences in the ratio between the award for loss in one ear as compared to the award for loss in both ears. Many agencies consider total loss in one ear as being equivalent to half of total hearing loss for both ears. This is not in accordance with recommendations made by the American Medical Association in its *Guides to the Evaluation of Permanent Impairment,* where it is recommended that the ratio between the loss of hearing in one ear to both ears be 1-to-5. This brings about a situation where many states find themselves in a bind, because on the one hand they recommend the use of the AMA formula, but on the other hand the schedule of "one ear to two ears" takes preference.

A national commission on worker's compensation laws was established under provisions of the Occupational Safety and Health Act of 1970. This commission conducted a comprehensive study and an evaluation of worker's compensation laws including the occupational hearing loss problem. In its report of July 31, 1972,[17] the commission indicated that improvement is needed in many areas in order to assure an adequate, prompt and equitable system of compensation as well as a safe place of employment free of physiological and physical hazards.

The occupational hearing loss problem is not a static problem. It is constantly being reappraised. The various relationships of acoustic trauma, noise exposure, hearing impairment, protective measures, and other variables are all being tested and reviewed. In addition to the medical problems, certain changes are taking place because of current social and political demands. On the basis of the increase in hearing loss claims and proposed legislation on the part of state and federal agencies, it is safe to say that the noise problem will be very much alive in the years ahead. Regulations and standards will most likely be established in each state patterned after the noise regulations incorporated in the Walsh-Healey Public Contract Act and Occupational Safety and Health Act of 1970.

References

1. Department of Labor. "Safety and Health Standards," Federal Register, Superintendent of Documents, Washington, D.C. 20000. Vol. 34, No. 96, May 20, 1969, page 7948. U.S. Department of Labor, "Guidelines to the Department of Labor's Occupational Noise Standards for Federal Supply Contracts" (Bulletin 334), Superintendent of Documents, Washington, D.C. 20000.
2. Bunche, C. C., "Traumatic Deafness, Historical and Audiometric Study," *Laryngoscope* 47:615, September 1937.
3. "Guide for Conservation of Hearing in Noise," *Transactions of the American Academy of Ophthalmology and Otolaryngology,* revised 1973. Rochester, Minn.
4. "Guide for the Evaluation of Hearing Impairment," a report of the Committee on Conservation of Hearing, *Transactions of the American Academy of Ophthalmology and Otolaryngology,* Vol. 63, No. 2, March–April, 1959.

American Academy of Ophthalmology and Otolaryngology, Rochester, Minn.

5. Slawinski v. Williams & Co., 298 N.Y. 546.

6. Rosati v. Dispatch Shop, 298 N.Y. 813.

7. Wojcik, A. v. Green Bay Drop Forge, 265 Wis. 38.

8. "Guidelines for Noise Exposure Control," *American Industrial Hygiene Journal,* September–October 28:418, 1967.

9. *Criteria for Occupational Exposure to Noise,* U.S. Department of Health, Education, and Welfare, 1972. Washington, D.C.

10. *Wisconsin Administrative Code No. 11 on Occupational Noise Exposure,* July 1971. Department of Industry, Labor and Human Relations, Madison, Wisc.

11. *Guides to the Evaluation of Permanent Impairment,* 1971. American Medical Association, Chicago, Ill.

12. *Specifications for Audiometers* S3.6–1969 ANSI, New York, N.Y.

13. "Statements on Presbycusis," *Transactions of the American Academy of Ophthalmology and Otolaryngology,* Vol. 68:1, January 1964, page 116. Rochester, Minn.

14. Fox, Meyer S., "Comparative Provisions for Occupational Hearing Loss," *Archives of Otolaryngology,* Vol. 81, March 1965, pp. 257–260. American Medical Association, Chicago, Ill.

15. Fox, Meyer S., M.D., "Hearing Loss Statutes in the United States and Canada," *National Safety News,* February 1972.

16. *Wisconsin Industrial Commission Statutes 1972,* Section 102.55 and Appendix.

17. *Analysis of Workmen's Compensation Laws,* 1974 edition, The Chamber of Commerce of the United States, Washington, D.C.

Chapter Fourteen   Evaluation of Noise Exposure

By Julian B. Olishifski, P.E.

Evaluation of noise exposure may be defined as "the decision-making process resulting in an assessment as to how well a given noise environment conforms to a predetermined set of standards or criteria." This chapter discusses the basic factors to be considered in evaluating environmental noise exposure:

- Purpose or objective

- Criteria or standards of judgment

- Accurate and valid measurements

- Interpretation and analysis

Purpose or objective

The purpose or objective of the noise evaluation is determined by the need: Hearing conservation, community annoyance, speech interference, noise control, or equipment testing. (These subjects are discussed in other chapters of this book.)

Evaluation of industrial noise hazards involves knowledge of the levels and duration of exposure in the work place and the effect of these noise stresses upon the health of the worker.

Evaluation, for example, may refer to the analysis of the noise levels arising out of a process to determine the effectiveness of a given piece of equipment used to control the noise hazards arising out of that process.

Control involves the reduction of environmental noise stress to values that the worker can tolerate without impairment of productivity or his health.

Evaluation of occupational noise exposures requires a teamwork approach similar to that used for evaluating the occupational environment for other hazards. Depending upon the nature and extent of the situation, the services of engineers, physicists, and others may be needed to solve a particular noise problem.

To evaluate occupational noise exposures for hearing conservation purposes, it is essential to recognize the sources of hazardous noise. The investigator should determine the number of workers exposed, types of control or protective measures employed, duration of exposure, and other important factors relating to the noise problem. Valuable time can be saved, by conducting a preliminary survey of the plant. The importance of taking adequate notes during a preliminary survey or study cannot be overemphasized.

Many individuals can detect those areas of a plant where sound levels are high enough to interfere with speech reception. A good rule of thumb in judging the noise level is whether a person must shout in order for a nearby person to hear what is said. Thus, a simple walk-through can serve to identify those areas where there may be a noise problem.

Occupational damage risk criteria

The purpose of damage-risk criteria is to define maximum permissible noise levels for stated time intervals, which, if not exceeded, would result in acceptable small changes in hearing levels of exposed employees over a working lifetime.

The acceptability of a particular noise level is a function of many variables. Practical solutions to noise exposure problems can best be achieved by complying with rules or guidelines which have been proven acceptable historically or which are required by legislation.

Increasing attention is being given by regulatory agencies and industrial and labor groups to the effects of noise exposure upon employees; therefore, a need exists for equitable, reliable, and practical damage-risk noise criteria. According to the dictionary, a criterion is a standard, rule, or test by which a judgment can be formed. A criterion for establishing noise damage risk, however, can be based on one or more standards for judgment.

Once the standards upon which to base a judgment of the effects of occupational noise exposure upon employees are selected, damage risk criteria can be developed in terms of a number of requirements and considerations.

• Damage risk criteria should include specified objectives regarding which standard(s) of judgment are being applied.

• Damage risk criteria should specify what proportion of the employees the criteria are designed to "protect" and to what extent. In view of the large individual differences in susceptibility to the effects of noise, the auditory environment that will be satisfactory to a stated percent of the exposed industrial population should be clearly stated.

• Permissible or allowable industrial noise levels should be expressed in units on a scale that includes all relevant aspects of the particular type of noise exposure.

• The specification of the noise exposure should include the effects of the spectral content of the noise, the temporal character and type of noise, the work cycle, and the duration of the noise as well as the sound pressure level. Ideally, the specification should be suitable for all significant sources of noise in the industrial environment making up the total noise exposure.

Damage

A workable definition of *damage* and *risk* should be agreed upon so that *safe* and *hazardous* can then be defined. Damage can be defined in a physical sense as an injury to the cells and tissues of the ear of the exposed employee or in terms of impairment of the individual's hearing functions. Damage can also be defined subjectively as damage of the ability to hear and understand everyday speech. This ability can be approximated in a practical way by conversing with the individual who is being examined. (For more details, see Part Five—Industrial Audiometry.)

Hearing ability. Direct measures for evaluating hearing ability for speech have been developed. These tests generally fall into two classes: (a) those that measure the hearing threshold or the ability to hear very faint speech sounds, and (b) those that measure discrimination, or the ability to understand speech.

Ideally, damage (or hearing impairment) should be evaluated in terms of an individual's ability to hear everyday speech under everyday conditions. The ability of an individual to hear sentences and to repeat them correctly in a quiet environment is considered to be satisfactory evidence of hearing of everyday speech. Hearing tests, using pure tones, are extensively used to monitor the status of a person's hearing ability and the possible progression of a hearing loss. There is a high correlation between a person's ability for hearing pure tones and his hearing of speech.

A person working in an environment noisy enough to cause a progressive hearing loss should have his hearing checked periodically to determine if the noise is having a detrimental effect on his hearing. The noise induced hearing losses that can be measured by pure tone audiometry are those threshold shifts that constitute a departure from a specified baseline. This baseline or "normal hearing" can be defined as *the average hearing threshold of a group of young people who have no history of previous exposure to intense noise and no otological malfunction.*

Threshold shifts. The line delineating temporary threshold shifts and permanent shifts is difficult to determine. When temporary threshold shifts are induced in the laboratory by means of short exposure, the subject's hearing usually returns to its preexposure threshold after a suitable rest period. In industry, where the noise exposure is repeated over and over again, some part of the temporary shift may eventually become irreversible resulting in a permanent threshold shift.

The threshold shift (TS) at any instant of time will be the arithmetic sum of any prior permanent threshold shift (PTS) and any temporary threshold shift (TTS) due to recent noise exposure.

The permanent threshold shift (PTS) may be defined as the summation of any noise induced permanent threshold shift (NIPTS), presbycusis (the loss of hearing acuity with age), and the effect of any physical or medical disorders or diseases resulting in some permanent damage to the hearing mechanism.

Because of marked differences in individual susceptibility, it is not possible to predict the exact conditions under which a permanent threshold shift or a temporary threshold shift will occur in a particular individual. Statistically, however, it is possible to establish a minimum set of noise exposure conditions below which a stated percentage of the exposed listeners will not incur specified threshold shifts.

AAOO-AMA guide. The Committee on Conservation of Hearing of the American Academy of Ophthalmology and Otolaryngology, after years of studying the various methods and procedures used to determine hearing damage, issued a report that was published by the American Medical Association as the "Guide for the Evaluation of Hearing Impairment" (see Chapter 13). It recommended that the average hearing level for the pure tones at 500 Hz, 1000 Hz, and 2000 Hz be used as an indirect measure of the probable ability to hear everyday speech. If the average monaural hearing level at 500, 1000, 2000 Hz is 25 dB ANSI 1969 (15 dB ASA 1951) or less, the AMA Guide stated that usually no impairment exists in the ability to hear everyday speech under everyday conditions.

NIOSH recommendations. The NIOSH criteria document (see Appendix A-3) for occupational exposure to noise recommended a slightly different definition of hearing impairment. NIOSH stated that hearing impairment for speech communication begins when the average hearing level at 1000, 2000, and 3000 Hz exceeds 25 dB ANSI 1969. The principal reasons given by NIOSH for this definition are as follows:

1. The basis of hearing impairment should not be limited to only the ability to understand speech.

2. The ability to hear sentences and repeat them correctly in quiet is *not* satisfactory evidence of adequate hearing for speech communication under everyday conditions.

3. The ability to understand speech under everyday conditions is best predicted on the basis of the average hearing level at 1000, 2000, and 3000 Hz.

4. The point at which the average hearing level at 1000, 2000, and 3000 Hz begins to have a detrimental effect on the ability to understand speech is 25 dB ANSI 1969.

The NIOSH criteria document stated that the ability to hear speech, measured in terms of the lowest sound intensity at which a listener can barely identify speech materials, provides little information concerning communication difficulties under everyday conditions.

An individual with a severe high-tone sensori-neural loss (as is seen in occupational noise induced hearing loss) will fail to hear certain high-frequency sounds and will not have normal speech discrimination. On the other hand, the same individual may easily hear the low-frequency words and have a normal threshold for speech.

TABLE 14-A
INCIDENCE OF HEARING IMPAIRMENT IN THE GENERAL POPULATION AND IN SELECTED POPULATIONS BY AGE GROUPS AND OCCUPATIONAL NOISE EXPOSURE

A-weighted Sound Level (Continuous Noise)	*Percentage of Population Having Impaired Hearing, by Age Groups*			
dB	20–29	30–39	40–49	50–59
Non-Noise	3	5	10	20
General Population	2	5	14	24
80	1	3	6	19
85	3	7	13	27
90	6	14	21	37
92	3	9	15	28
95	10	22	32	47
96	3	10	19	—
97	7	22	32	48
100	16	36	46	60
102	10	18	30	45
104	5	21	35	57
105	24	50	62	75

This table is based on a number of studies that correlate noise exposure with incidence of hearing impairment as defined by the AMA guide. The first column indicates the steady noise levels to which the various groups were exposed in terms of the A-scale reading. The remaining columns show, for various age groups, the percentage of the groups having impaired hearing. The first line of the table shows the incidence of hearing impairment in a population having no exposure to injurious noise and no other explanation for observed hearing impairments. It is presumed that at least this minimal incidence of impairment will be found in any population and that the other groups may be regarded as exhibiting injurious effects of noise only if they show significantly higher rates of incidence. (Adapted from Intersociety Committee Report (1970), "Guidelines for Noise Exposure Control," *Journal of Occupational Medicine,* July 1970, Vol. 12, No. 7.)

Everyday speech rarely takes the form of complete sentence communications; thus, the number of speech cues available for accurate speech perception under everyday conditions is greatly reduced. Therefore, NIOSH concluded that an appropriate predicting scheme for the determination of hearing impairment must include some consideration for an actual daily communication environment rather than some optimum condition as suggested by the AAOO-AMA Guide.

The NIOSH criteria document stated that in order to assess hearing loss accurately for speech under everyday conditions by means of pure tone hearing tests, a modification in the three-frequency average recommended by the AAOO and the AMA is warranted. Such a modification should include the elimination of 500 Hz from the three-frequency formula for determining hearing impairment and the substitution of 3000 Hz in its place.

Impaired hearing for speech cannot be described completely by threshold shifts alone; also affected are such auditory functions as perception of loudness and pitch, discrimination or recognition of speech, and localization of sound.

Risk

Risk may be defined as the difference between the percentage of individuals with a defined hearing impairment in a noise exposed group as compared to a nonnoise exposed group.

Hearing risk to whom and how many of each group should be specified. For example, if it is assumed that it is management's responsibility to see that every employee's hearing ability must be protected against any change in hearing threshold at any audible frequency, the task of providing protection would literally be impossible because of the many nonoccupational causes of hearing loss (such as age, disease, off-the-job noise exposures, and so on) that management cannot control.

The question of how much of the hearing ability should be protected and the percentage of the working population incurring hearing losses of certain magnitudes has long been an issue of much controversy. The ultimate decision as to how much hearing protection is needed must be based on scientific, political, economic, social, and humane values.

Obviously, the decision must be arbitrary. Consequently, both management and labor must be prepared to accept a certain reasonable amount of risk.

The decision to conserve a person's hearing for speech communication determines, to some extent, the degree of hearing loss considered severe enough to constitute a handicap. After this decision has been made and the appropriate data collected, probability-of-hearing-loss tables can be established (see Table 14-A).

Statistical concepts. The damage-risk criteria are, therefore, based on statistical concepts. The criteria should state that, within biological probability, an acceptable degree of hearing impairment will occur in a normal population that has been exposed to stated noise levels for defined periods of time.

Half of the population at risk will have a hearing level equal to or worse than that population's median hearing level. If an allowable noise exposure is chosen that will produce a median threshold hearing level equal to that for the onset of impairment, then about half the noise exposed population will have some impairment.

Two strategies have been adopted to avoid this problem. The first is to set a hearing-level criterion that is better than the hearing threshold measured at the onset of impairment. This would provide some margin for safety, making the limit of acceptable noise exposure nonhazardous for more than 50 percent of the persons exposed. The second is to recommend a system of monitoring audiometry to detect and remove from noise exposure those persons who show evidence of developing significant permanent threshold shifts.

Duration of noise exposure

It is necessary to know, not only how much noise exists in the hearing zone, but also what kind of noise and how long an employee is exposed.

It has been observed that with most individuals, brief, momentary exposures to high-noise levels will usually not produce any permanent significant hearing loss, while prolonged daily exposure to noise levels above 90 dB (as measured on the A-scale of a sound level meter) will result in a significant hearing loss to some individuals.

The incidence of noise induced hearing losses is directly related to the total exposure time. In addition, it is believed that intermittent exposures are far less damaging to the ear than continuous exposures, even if the sound pressure levels for the intermittent exposures are considerably higher. The rest period between exposures allows the ear to recuperate. At the present time, the exact relation between the harmful effects of noise exposure and the energy content of the noise cannot be determined. Doubling the energy content does not produce twice the hearing loss. In general, the greater the total energy content of the noise, the shorter the time of exposure required to produce the same amount of hearing loss, but the exact relation between time and energy is not known at present.

Published guidelines

Published noise criteria are valid for a high percentage of the population but not necessarily for the specific individuals. Because of the wide variations in susceptibility of the individuals to the effects of noise, some may suffer hearing damage when subjected to noise levels equal to, or even below, those given in published noise standards. Thus, the values given should not be considered absolute limits, but as guidelines for practical noise control. Conversely, exceeding the values does not necessarily mean that hearing damage, hostile community reaction, or poor speech communication conditions will occur; but it is safe to assume that adherence to published guidelines will minimize the likelihood of noise problems.

If the sound level is steady and employees are present for known daily periods of time, exposure can be evaluated directly from the table of exposure limits. If the sound level is not steady or if personnel are present for different periods of time, perhaps in areas of different noise levels, it is necessary to ascertain, for each employee, the cumulative daily time spent at each level. This can be done by acoustical sampling procedures or from known schedules of employee location and/or equipment operation.

Off-level of noise

For a fixed noise level, interruptions in noise exposure (intermittency) reduce the amount of temporary threshold shift as compared with the threshold shift resulting from exposure to continuous noise. The increased tolerance to intermittent noise exposure depends on the sound level present during the quiet intervals as well as during the noise segments. The off-level (that level of noise *between* intermittent louder exposures) implies that noises at or below this level do not of themselves cause any significant temporary or permanent hearing threshold shift. With high levels of intermittent noise, some recovery of hearing acuity may take place between noise bursts. NOISH concluded that a level of approximately 65 dBA meets the requirements of criteria established for a true off-level for intermittent exposure (see Criteria Document in Appendix A).

The number and length of quiet periods relative to the amount of on-time noise also influence the degree of threshold shift.

What should be measured

How should noise levels be specified and exposures measured? Different noises are apparently not equally effective in producing hearing loss, so an agreement must be reached on a standard specification of the spectral and temporal characteristics of the noise.

In order to describe noise exposure one must be willing to consider many numbers and many factors. In any data acquisition effort, the validity of the information obtained is directly related to understanding the test objectives and completeness of the planning preceding actual measurements. To make sure that all factors influencing acoustical measurements are considered, careful preparation should be an integral part of the overall effort. (See Chapters 4, 5, and 6 for more information on this topic.)

Numerous concepts, techniques, and definitions have been devised to describe particular noise situations. Often these have been optimized in connection with a particular aspect of human response or effects, such as hearing impairment.

Noise levels

If a definition is accepted that the noise level alone is dangerous or damaging *only* if it gives rise to irreversible significant loss of hearing, it can be concluded that few industrial processes today give rise to dangerous noise levels, although there are numerous instances of potentially damaging noise exposures.

It is generally agreed that after repeated noise exposure some individuals will incur a hearing loss. There is a considerable difference of opinion among the experts as to the boundary that separates the harmless from the harmful noises. This boundary has been variously called the *critical noise level* or the *maximum safety intensity level,* on the assumption that the noise intensity alone was the sole determining factor in evaluating the hazardous quality.

For many years investigators tried to determine this maximum safe intensity noise level, by evaluating the hazardous effect of an industrial noise solely on the basis of overall noise level. It became obvious that this approach was inadequate, especially as the ear is far more sensitive to certain frequencies than to others. Two different noises may each have approximately the same overall sound pressure level, but they may not necessarily present the same hazard to hearing. The use of A-weighted sound level measurements has to some extent overcome this difficulty. Both simplified approaches are subject to error in overlooking the levels in each frequency band, particularly when there are pure-tone components present.

The A-weighted sound level measurement has become a popular measure for assessing the overall noise hazard. The A-weighted network on a sound level meter is thought to provide a rating of most industrial broadband noises in a reasonably similar manner as would the human ear.

As a result of its simplicity in rating the hazard to hearing, the A-weighted sound level has been adopted as the measure for assessing noise exposure by the American Conference of Governmental Industrial Hygienists (ACGIH). The A-weighted sound level as the preferred unit of measurement was also adopted by the U.S. Department of Labor as part of the Occupational Safety and Health Standards. A-weighted sound levels have also been shown to provide reasonably good assessments of speech interference and community disturbance conditions and have been adopted for these purposes.

A-weighted sound levels as single-number ratings have shown excellent agreement in some cases between the A-level and subjective effects, while in other cases relatively wide discrepancies occur. Those investigations that show wide discrepancies usually include comparisons of high-level, narrow-band noise or pure tones with broadband noises. The most consistent results are found when the noises that are being compared are similar in character.

Octave bands

Although the A-scale on the sound level meter has a good correlation with annoyance and a reasonably good correlation with hearing hazard and is used to determine acceptable limits set by the Occupational Safety and Health Act, a full octave-band analysis can be very helpful in identifying the sources of the noise in the plant.

Annoyance is very strongly dependent on energy content in each octave band. Noise control techniques are also dependent on the energy level in each octave band. The identification of pure tone components, when they are present, is an extremely useful diagnostic tool for finding and quieting the noise source.

Some noise sources have a well-defined frequency content. For example, the hum generated by a fan or blower is usually centered at the blade-passage frequency, which is the product of the number of blades of the fan multiplied by the speed (rpm) of the fan. Electrical current transformers usually hum in frequencies which are multiples of 60 Hz. Positive-displacement pumps will have a sound pressure distribution directly related to the pressure pulses on either the inlet or the outlet of the pump.

Noise from the discharge of steam or air pressure relief valves has a frequency peak which is related to the pressure in the system and the diameter of the restriction preceding the discharge to the atmosphere. A peak energy content in any single octave band would provide information as to the predominant noise source.

Sound analyzers can be obtained that can make measurements in third octaves and tenth octaves. The narrower the band for analysis, the more sharply defined is the data.

If a noisy machine is to be used in a room, the acoustical characteristics of the room as a function of frequency and the radiated sound power level in octave bands must be known in order to estimate the noise level that would be produced by this machine. The A-weighted sound level of the machine would not be adequate.

General classes of noise exposure

There are three general classes into which occupational noise exposures may be grouped.

- Continuous noise

- Intermittent, steady, equal intensity noise

- Intermittent, variable intensity noise

Continuous noise is normally defined as broadband noise of approximately constant level and spectrum, to which an employee is exposed for a period of 8 hours per day, 40 hours per week, 50 weeks per year. Into this class of noise exposure fit a large number of industrial operations. Most damage-risk criteria are written for this type of noise exposure because it is the easiest to define in terms of amplitude, frequency content, and time of duration.

An example of noise criteria for this type of noise exposure is the levels set by the Department of Labor for continuous noise shown in Table 14-B. These figures specify an allowable A-weighted level for noise levels that, in effect, permit a 5 dBA increase for each halving of the exposure time. Note that Table 14-B actually defines permissible noise exposure . . . with "exposure" being a function of *both* sound level and duration (in hours per day). Thus, for example, an employee can be exposed to 90 dBA for a full 8 hours a day. But, if the sound level increases to 95 dBA, maximum permissible exposure time drops to 4 hours per day. Every 5 dB increase in sound level *halves* the allowable exposure time.

When employees are exposed to different sound levels during the day, the "mixed exposure" must be calculated by using the formula:

$$\frac{C_1}{T_1} + \frac{C_2}{T_2} + \frac{C_3}{T_3} + \ldots + \frac{C_n}{T_n} = D$$

where each "C" is the total exposure time at a given noise level and each "T" is the total exposure time permitted at that level. If the sum of the fractions equals or exceeds 1, then "the mixed exposure is considered to exceed the limit value."

TABLE 14-B

PERMISSIBLE NOISE EXPOSURE

Duration per Day Hours	Slow Response dBA
8	90
6	92
4	95
3	97
2	100
1½	102
1	105
½	110
¼ or less	115

NOTE: When the daily noise exposure is composed of two or more periods of noise exposure of different levels, their combined effect should be considered, rather than the individual effect of each. If the sum of the following fractions: $C_1/T_1 + C_2/T_2$ ... C_n/T_n exceeds unity, then, the mixed exposure should be considered to exceed the limit value. C_n indicates the total time of exposure at a specified noise level, and T_n indicates the total time of exposure permitted at that level.

For example, an employee is exposed to the following noise levels during his work day:

$$\begin{array}{lll} 85\ dBA & - & 3.75\ hours \\ 90\ dBA & - & 2\ hours \\ 95\ dBA & - & 2\ hours \\ 110\ dBA & - & 0.25\ hours \end{array}$$

Thus, the sum of the fractions is as follows:

$$\frac{3.75}{No\ limit} \text{ or } 0 + \frac{2}{8} + \frac{2}{4} + \frac{0.25}{0.50} = 1.25$$

Since the sum is greater than 1, the employee received an excessive noise exposure during his work day.

The exposure table assumes continuous noise rather than impulsive or impact noise. By definition, "if the variations in noise level involve maxima at intervals of 1 second or less, it is considered to be continuous."

EXAMPLE 1—A drill runs for 15 seconds and is off for ½ second between operations. This noise is rated at its "on" level for the whole eight-hour day—and would be "safe" only if that level were 90 dB(A) or less.

EXAMPLE 2—In a work area, the noise levels are read as 95 dB(A) for two hours a day, 90 dB(A) for four hours a day, and 80 dB(A) for the remaining two hours a day.

1. The problem is to find the permissible durations for each noise level. Table 14–B indicates that permissible exposures are:

> 95 dB(A) = four hours
> 90 dB(A) = eight hours
> 80 dB(A) = no limit

2. When we add the ratios:

Noise Levels	95	90	80
Measured	2	4	2 hours
Permissible	4	8	any period of time

$$\frac{2}{4}+\frac{4}{8}+0=\frac{1}{2}+\frac{1}{2}=1.$$

Or, the noise exposure is *within limits*.

EXAMPLE 3—Noise in an area measures 90 dB(A) for two hours a day, 97 dB(A) for two hours a day, and, for the remaining four hours, there are alternate noise levels of 95 dB(A) for 10 minutes and of 80 dB(A) for 10 minutes.

1. The equivalent steady value of the 95-dB noise is 10 minutes × 3 (the number of times per hour that the noise is "on") × four hours = two hours at 95 dB(A).

2. Add the ratios:

	90 dB	97 dB	95 dB
Measured	2	2	2 hours
Permissible	8	3	4

$$\frac{2}{8} + \frac{2}{3} + \frac{2}{4} = 1\frac{5}{12}$$

This *exceeds the permissible limit* of 1.

Intermittent steady noise exposure may be defined as exposure to a given broadband sound pressure level several times during a normal working day. An example of a person in this noise environment is the inspector or plant supervisor who periodically makes trips from his relatively quiet office into noisy production areas. Criteria established for this type of noise exposure are shown in Table 14-C.

With steady noises, it is sufficient to record the A-weighted sound level attained by the noise. With noises that are not steady, such as impulsive noises, impact noises, and the like, the temporal character of the noise needs additional specification. Both the short-term and long-term variations of the noise must be described.

Impact-type noise is a sharp burst of sound and sophisticated instrumentation is necessary to determine the peak levels for this type of noise. Noise types other than steady are often encountered. In general, sounds repeated more than once per second may be considered as steady. Impulsive or impact noise, such as hammer blows or explosions, generally is less than one-half second in duration and repeats no oftener than once per second. Employees should not be exposed to impulsive or impact noise in excess of 140 dB peak sound pressure level (also, see Chapter 12).

Intermittent sounds of brief duration at intervals greater than one second should be evaluated in terms of level and total duration for an eight hour day. If noise exhibits peaks at intervals of one second or less, the noise may be considered continuous and the highest value should be used in determining exposure in accordance with Table 14-B.

Individual impulse and impact sounds can be characterized in terms of their rise time, peak sound level, and pulse duration. Studies indicate that hearing tolerance to peak sound pressures is greatly reduced by increasing the rise time or burst duration of the sound. Obviously, the rate and number of such impact sounds constituting an exposure period are also factors in making hazard judgments for these types of sound.

It is not easy to measure accurately the sound levels of rapidly varying staccato noises. Sound level meters do not follow sudden peaks in sound pressure and they may systematically distort and misrepresent the true sound pressure levels reached by that noise. This is particularly true for impact noise—for example, the noise of a drop forge. Noise and hearing loss studies, in industry, dealing with these types of exposure conditions are just beginning. Most hearing-loss data reflecting impulse noise hazards have been based on military studies involving noise exposure to gunfire.

Measurement

Sound measurement falls into two broad categories: source measurement and ambient-noise measurement. Source measurements are those which involve the collection of acoustical data for the purpose of determining the characteristics of noise radiated by a source. The source may be a single piece of equipment or a combination of equipment or systems. For example, an electric motor or an entire plant may be considered a noise source.

Source measurements frequently are made in the presence of noise created by other sources which form the background- or ambient-noise level. Although it is not always possible to make a clear distinction between source and ambient-noise measurement, it is important to understand that source measurements describe the characteristics of a particular sound source and ambient measurements describe the characteristics of a sound field due largely to unspecified or unknown sources.

A uniform, standard, reporting procedure should be established to assure that sufficient data is collected in a proper form for subsequent analysis. To be effective, this standard reporting procedure should specify requirements for sound measuring instruments and their calibration. Detailed descriptions of the techniques and methods should include measurement position, operating conditions, instrument calibration, exposure time, amplitude patterns, and other important variables.

Several forms have been devised to record data obtained during a screening survey. Use of these forms will facilitate the recording of pertinent information which will be extremely useful if more detailed studies are to be conducted later. A more detailed discussion on screening surveys appears in Chapter 6. An employee noise exposure survey

TABLE 14-C
ACCEPTABLE EXPOSURES TO NOISE IN dBA AS A FUNCTION OF THE NUMBER OF OCCURRENCES PER DAY

Daily Duration		Number of Times the Noise Occurs Per Day						
Hours	Min	1	3	7	15	35	75	160 up
8		90	90	90	90	90	90	90
6		91	93	96	98	97	95	94
4		92	95	99	102	104	102	100
2		95	99	102	106	109	114	
1		98	103	107	110	115		
	30	101	106	110	115			
	15	105	110	115				
	8	109	115					
	4	113						

This table summarizes the results of TTS studies that may be used to estimate the effect of intermittency of noise exposures on risk of hearing impairment. The information in the table may be approximated by the simple rule that for each halving of daily exposure time the noise level may be increased by 5 dB without increasing the hazard of hearing impairment. To use the table, select the column headed by the number of times the noise occurs per day, read down to the average sound level of the noise, and locate directly to the left in the first column the total duration of noise permitted for any 24 hour period. It is permissible to interpolate if necessary. Noise levels are in dBA. (Adapted from Intersociety Committee Report (1970) "Guidelines for Noise Exposure Control," *Journal of Occupational Medicine*, July 1970, Vol. 12, No. 7.)

is conducted by measuring noise levels at each work position that an employee occupies through the day or by acquiring a sufficient sampling of data at each work station to evaluate the exposure of an employee while at that work station. Work stations which pose particular noise exposure hazards are readily identifiable from measured sound level contours if these are obtained in adequate detail.

All spot measurements should be made under normal conditions of equipment operation with the acoustical environment in normal workday configuration. The measuring microphone should be placed at normal ear level, approximately 5.5 ft for the standing position and 4 ft for the seated position, at each personnel work station with the employee absent at the time of measurement. Additional nearby measurement positions should be used, as necessary, to make sure that sound levels at each work station are adequately described.

In many industrial situations, however, it is extremely difficult to evaluate the noise exposure of a specific worker accurately. This is due in part to the fact that the noise level to which the stationary worker is exposed throughout the working day fluctuates, making it difficult to evaluate compliance or noncompliance with OSHAct regulations. The problem can be due also to the fact that the worker's job may require that he spend time in plant areas where the noise levels vary widely with location, from very low to very high.

Because of the fluctuating nature of many industrial noise levels, it would not be accurate to use a single sound level meter reading to estimate the time-weighted average noise level. A statistical approach should be taken for evaluation of the noise exposure when the level is not constant.

There are many advantages in using the statistical analysis system for surveying industrial noise levels when more than a simple yes/no decision is to be made as to whether the exposure exceeds OSHAct requirements. The noise dosimeter is adequate for this purpose in that it will calculate the equivalent noise level "L" (either directly or indirectly by evaluating the noise dose D). However, when noise control procedures must be implemented or when noise control decisions must be made, the statistical analysis system provides more valuable information.

Accurate results can be achieved when continuous measurements are made throughout a single cycle of operation (which can vary from a few seconds to a few days, depending on the industrial activity). In the situation where the noise level or industrial activity is nonrepetitive, the sample duration should be sufficiently long to permit a reasonable estimate of the parameters of the noise exposure.

A typical industrial noise exposure in a large factory may contain many different types of relatively constant noise sources resulting in considerable variations in noise levels throughout the work space.

Consider the noise exposure of a worker who continuously moves about a large area in performing his tasks. If his routine is not established enough to define specific locations at which nearly all of his time will be spent, there are two different methods that could be used to obtain the equivalent noise level (L) accurately:

1. The worker could wear a dosimeter.

2. A continuous tape recording could be made of the worker's exposure as he moves about the space and the stored information later analyzed.

Analysis of the information stored on the tape recording could provide the following information:

- An accurate value of L.

- A graphical history of the noise level versus time, showing the highest noise level areas. This information is valuable in ranking major noise problem areas, and it provides guidelines as to where any initial noise control should be concentrated.

- A complete description of the noise level versus exposure time. This provides the maximum and minimum noise levels encountered throughout the exposure period and the amount of time these levels were encountered. This information is valuable in evaluating the actual severity of the noise problem.

If, on the other hand, the purpose of the study is to determine which machine or parts of a machine create the most noise, then measurements are made in the area immediately adjacent to the machine.

The period of time over which measurements should be made depends upon several factors. Among them are the characteristics of the noise and the duration of the operation. If the noise is steady and the operation is continuous, a minimum of three measurements should be made. These should be of sufficient duration to obtain the desired data. Data thus obtained should be representative of the workers' exposures. Conversely, if the machine or operation produces an impulse-type noise, measurements should be made periodically throughout each of about three nonconsecutive days to provide representative data. Finally, if the plant operates more than one shift, it may be necessary to make noise measurements during each shift.

Depending upon the particular plant and type of operations, noise levels and the distribution of energy with frequency could be different on each shift. There is no "cook book" procedure available that covers the selection of sampling sites, or suggests the number of measurements to be made. Decisions on these questions have to be based upon the experience and judgment of the individual conducting the study. The object, however, is to obtain data that are truly representative so that the workers' exposure can be evaluated. Since each survey is unique, the method used as well as the data recorded should closely correspond with the requirements of the situation.

Sound surveys

Uniform plant sound survey procedures have been used to provide an efficient method for determining the extent of noise exposure hazards. The survey procedure is a three-step process.

Step 1. Using a sound level meter set for A-weighting, slow response, the regularly occurring maximum noise level and the regularly occurring minimum level are recorded at the center of each work area. (The size of the work area is limited to 1000 sq ft or smaller.) If the maximum sound level does not exceed 84 dBA, it can be assumed that all employees in this area are working in a satisfactory noise environment. If the minimum sound level measured at center of each work area is 92 dBA or higher, then it is assumed that all the employees in this area are working in an unsatisfactory noise environment. If the range levels measured at the center of the work area fall between 92 and 84 dBA, then more information is needed before individual employee exposure can be determined.

Up to this point, the two types of plant areas have been classified as being satisfactory or unsatisfactory, using a minimum of time and expense. Now it becomes necessary to evaluate the noise exposures for people working in locations where measurements at the center of the weak area range from 84 to 92 dBA. These measurements are too near the permissible limit of 90 dBA to estimate; therefore, readings must be made at the employees' position.

Step 2. Sound level measurements are made at the operator's position. Again, if the level varies on a regular basis, both the maximum and minimum levels are recorded. Work station measurements are much easier to evaluate because if the noise level never goes below 90 dBA, the cumulative exposure must be greater than unity which indicates an unsatisfactory employee noise situation. Conversely, if the noise level is never greater than 90 dBA, the cumulative exposure must be less than unity and the employee station can be recorded as satisfactory.

Step 3. At work stations where the regularly occurring noise varies from 90 dBA to levels below 90 dBA, Step 3 analysis must be used. This simply means that noise exposure factors based on noise level and time must be calculated for this situation. At the conclusion of the Step 3 analysis, the surveyor will be able to determine, for each employee, whether or not the employee is working in a hazardous or nonhazardous noise environment. At this point, the plan of attack can be organized and the job of protecting employees begins.

For employees who have varying work patterns from day to day, it is necessary to ascertain the employee's duration of exposure within each hearing hazard zone or at each sound level. Such information may be obtained by consulting the employee or his supervisor, or by visual monitoring. A briefing/debriefing approach for a particular day's activities may also be used. This is done by requesting each employee to keep a general work area/time log of his daily activities and then debriefing him at the end of the work period to check that sufficient information was logged. In some cases, it may appear desirable for an employee to wear a noise dosimeter which records daily exposure in terms of current regulation requirements.

Once an employee's work time occupancy pattern is known, it is a straightforward procedure to evaluate his cumulative exposure for an 8 hour period. The procedure for determining an employee's daily noise exposure rating is as follows:

1. Identify each hazard zone visited by the employee.

2. Record the time in minutes that the employee spent in each zone.

3. Divide the time spent in each zone by the permissible time allowed in that zone.

4. Add the resulting fractions to obtain the employee's exposure rating.

Interpretation and analysis

After all measurements have been made and other necessary data have been assembled, the final step in the evaluation is to interpret results. In some instances it may be necessary to compare results with noise criteria or standards. This, it should be pointed out, is often no simple task, but guidelines on this subject are given in Chapter 11. In order to use these criteria wisely, it is necessary for the investigator to understand the physics of noise and its physiological and psychological effects. At this point, then, the investigator should consult with others to assist him in the interpretation and application of these standards. It should be emphasized, however, that a great deal of sensible judgment is necessary in this step of the evaluation.

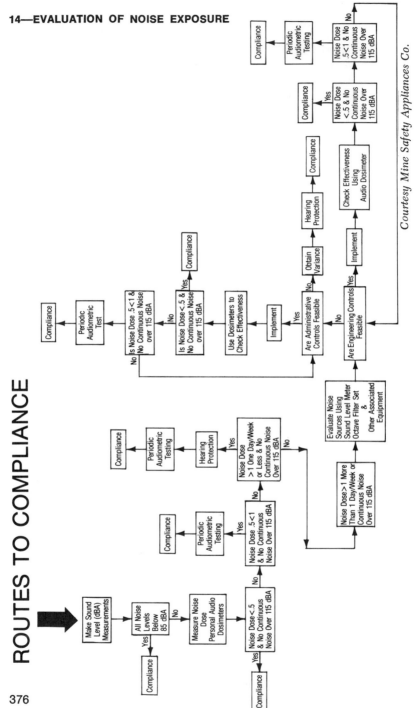

Fig. 14-1.—Summary of individual steps to be taken for evaluation of noise exposure.

Courtesy Mine Safety Appliances Co.

In other cases where regulations or standards have been adopted, it is necessary only to compare noise levels at certain frequency ranges with these standards. If noise levels are less than those permitted, no further action is indicated. On the other hand, if results exceed levels permitted, then a decision must be made on procedures to reduce the workers' exposure. This entails, preferably, the use of engineering control methods to reduce the noise at its source or, if not feasible, the use of hearing-protective devices. A discussion on each of these techniques is given in Chapters 15 and 20. Here again, common sense should be exercised in making recommendations. (See Figure 14–1.)

Finally, the problem at hand may require a decision regarding hearing loss. Information will then be needed in addition to data on noise level and frequency analysis, characteristics of the noise, and duration of exposure. It becomes necessary through audiometry to determine whether or not the hearing of workers is being affected adversely. The team approach, referred to earlier in this discussion, is necessary in evaluating data and rendering decisions in this facet of the problem. A detailed discussion on audiometry appears in Part Five of this book.

Part **4** **Control of Noise**

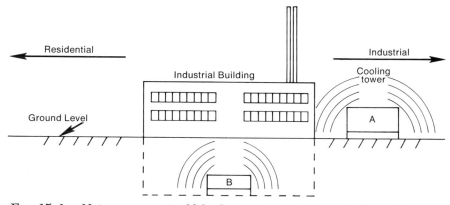

Fɪɢ. 15–1.—Noise sources should be located so they have the least effect on the environment.

Chapter Fifteen

Basic Principles of Noise Control

By Charles L. Cheever

An introduction to the basic engineering principles and typical calculations for solving noise problems are presented in this chapter. Knowledge of these principles is necessary for the most effective and economical application of noise control measures. Noise control efforts should be directed primarily to reduce noise at its source, and secondarily to interfere with its path of transmission.

Another type of noise control involves adding rather than reducing sound energy. For example, in an office where privacy is desired and the noise levels are very low, conversation can be transmitted from one office to another. Increasing the ambient background noise level will tend to mask out the intruding conversation.

Approach to noise control

Avoid the problem, if possible

Try to anticipate noise problems in the initial design stages of a new plant, a building addition, or a new work operation in an existing facility. Eliminating potential noise problems in the "blueprint stage" of a project should minimize subsequent expense for noise control measures. For example, large air-handling and compressor equipment should be located away from critical speech-communication spaces such as conference rooms and private offices. Heavy equipment should be located at grade level to minimize vibration and noise transmission to other areas of

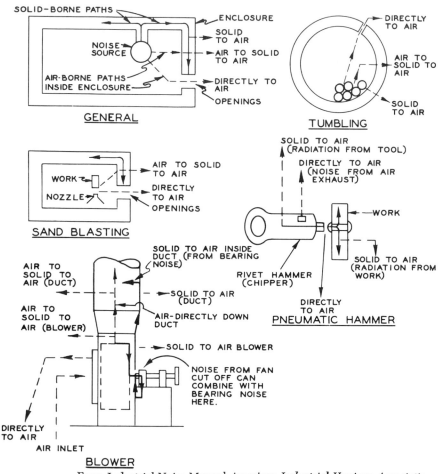

From Industrial Noise Manual *American Industrial Hygiene Association.*

Fɪɢ. 15–2.—Noise flow diagrams showing noise path in typical operations.

the building. Where the advantage of distance and buffer zones between noisy and quiet spaces is not available, special construction and precautions will be required to provide the desired conditions.

Plant layout, construction materials, and equipment selection and installation are factors that should be considered to minimize subsequent noise problems. In planning new facilities, advantage may be taken of distance, isolation, and absorption for the control of noise.

Many sound sources are somewhat directional in nature. It may be practical to locate and orient noise sources so that the direction of emitted sound radiation is least objectionable. In Figure 15–1, an outdoors noise source, such as a cooling tower (A), is shown located away from a residential area where noise levels should be lower than in an industrial area. Also, heavy equipment, such as a reciprocating compressor (B), is located on the service floor grade level to minimize vibration transmission to the building structure.

Architects should consult acoustical engineers during the early stages of building design. In one facility where this was not done, a number of changes were required after the construction project was completed to reduce ventilation noise levels, but the noise problem was never satisfactorily solved. The problem could have been avoided, or at least greatly reduced, if proper attention had been given in the planning stages. The expense involved in making the desired changes after completion of construction can be economically prohibitive. But there are ways to minimize it, even then. These are discussed next.

What to do if you already have a problem

Gather information. The first step for control of existing noise problems is to gather qualitative and quantitative information, which includes making sound pressure level measurements of the desired frequency bands at the locations of interest. (Chapters 4, 5, and 6 discuss proper sound pressure level measurement techniques and instruments.)

Pertinent information concerning the noise problem should be collected, such as the sources of noise, the number of people affected, the nature of the problem, room measurements, and acoustical features of the environment. In many cases, the various sources of noise and paths of noise transmission must be studied in detail, so that a course of action may be plotted. (Figure 15–2 shows various paths of noise transmission from typical sources.) Noise and vibration measurement instruments and frequency analyzers may be used to study the noise flow paths.

Noise criteria. The next step is to compare measured sound levels to the criteria for noise control. In the case of speech interference or annoyance, the noise criteria curves and suggested limits for various applications may be used as guidelines. Recommended noise levels (dBA) and preferred noise criteria ranges for various operations are given in Chapter 11, Standards and Threshold Limit Values for Noise.

The method. After determining the amount of noise reduction needed by comparing sound level measurements with selected criteria, alternate or multiple methods of noise control should be considered.

Reduction of noise at the source, along the transmission path, and at the receiver should be considered from the standpoint of effectiveness, desirability, cost, and maintenance. Often the only way to proceed is to use the experimental or cut-and-try approach to noise control.

Controlling a single noise source at a level far below that of an identical, adjacent source does not reduce the total sound level appreciably. For example, if each of two adjacent machines independently produces sound levels of 80 dB, the combined sound level—the result of adding (logarithmically) the levels of the two machines—is 83 dB when both machines are running simultaneously. A drastic reduction, or even elimination, of sound from one of the machines will produce only a three decibel reduction in the total level, if the other machine continues to produce a noise level of 80 dB.

In many cases, engineering control methods to reduce noise to desired levels may not be feasible economically. Other means of noise attenuation such as personal protection will have to be considered.

After selected noise control measures are implemented or put into effect, the resulting noise levels are measured to determine whether the controls have attained the desired attenuation. Because some noise control measures may deteriorate with time, it is advisable to check noise levels periodically and to make adjustments where necessary.

Noise specifications

Noise problems may often be avoided completely by the use of purchase specifications in the process of selecting equipment. While other factors, such as cost and performance of equipment, weigh heavily in the selection process, noise levels should also be considered. Many equipment manufacturers emphasize the relatively low noise levels of their products as part of their sales promotion efforts. The equipment supplier should supply sound level ratings in octave or one-third octave bands for specified acoustic environment and equipment operating conditions.

Practical considerations must enter into the specification of equipment noise limits. It is impractical, for instance, to specify noise limits that cannot be met. It is also costly to pay a premium for greater noise control than is necessary. For example, specifying a maximum sound level of 60 dBA would be foolish if the equipment is to be located permanently where the existing background noise level is 80 dBA.

Noise control criteria should be also incorporated into architectural or

TABLE 15-A
SUBSTITUTING QUIETER EQUIPMENT OR QUIETER OPERATIONS

Substitute	Previous Noise Source
Belt Drives	Gear Drives
Punch press mechanical parts ejector	Punch press air ejector
Diamond face core drill	Star drill and air hammer
Hydraulic press riveter	Pneumatic hammer riveter
Proper size weld peening tool	Oversized weld peening tool
Pneumatic air cylinders	Electric solenoids
Pneumatic rotary shear	Pneumatic chisel
Welding	Riveting
Grinding, Arcair metal removal or flame gouging	Chipping

Source: Industrial Noise Manual, *American Industrial Hygiene Association*, 1966.

constructional specifications. Acoustical materials can be specified that reduce noise levels. (Further details on this topic are given in Chapter 19, Sound Level Specifications.)

Noise control at the source

The first consideration should be to eliminate or reduce the noise at its source. Sound energy is produced by driving forces which cause vibrations in solids, liquids, or gases, thereby producing sonic vibrations in air. The driving forces may be either repetitive or impact forces. Figure 15-2 shows various paths for transmitting vibrations to the surrounding air. Measures taken to reduce the driving forces or the vibrational response of sound radiating surfaces will result in reduction of sound at the source.

Substitution

In certain cases, it may be feasible to substitute quieter equipment or a quieter operation (see Table 15-A). Substitution with quieter equipment often produces a very dramatic reduction in noise levels. Initial selection of quieter equipment or processes will often be the most economical and, therefore, preferred means for attaining a specified noise reduction.

Reduction of vibration forces

Frequently, noise is produced by forces that cause structural or air vibrations. The vibration forces in equipment are commonly due to reciprocating motion or to imbalance in rotary motion. In air, vibration forces are created by turbulence or pressure pulses. Vibration forces are transferred back and forth between the air and structural materials.

The following measures can reduce troublesome vibration forces:

• Reduction of equipment operating speed

• Reduction of imbalance through the proper alignment and balancing of the rotating equipment (balancing should be done preferably under dynamic load conditions)

• Replacement of worn parts—such as bad bearings

• Provision for proper lubrication to reduce frictional forces

• Reduction of peak forces by extending the force application time (such as the use of stepped punches so that the total work is not done at one instant)

• Reduction of flow velocities of gases and liquids (turbulence and vibration increase as flow rate increases)

• Reduction of turbulence in the flow of gases and liquids by streamlined design (for example, fans with airfoil blades provide reduced turbulence and noise)

• Reduction of impact forces by the use of resilient materials (the use of rubber liners in castings tumblers)

• Tightening of loose parts because increased forces may be produced when motion is not adequately restricted

• Proper assembly of parts (impact or frictional forces may result from improper assembly)

Operation of equipment at high speeds may produce excessive noise levels. For example, pumps operating at 3600 rpm are likely to produce more noise than similar larger pumps operating at 1800 rpm for a given rating. Precision dynamic balancing is one means for reducing the repetitive forces and high sound levels resulting from high-speed operation. Bearings must be properly lubricated and maintained to minimize vibrational forces and noise.

TABLE 15-B
CHANGES IN MACHINERY THAT PRODUCE CHANGES IN VIBRATION LEVELS

• Wear • Erosion • Corrosion • Aging • Inelastic behavior •

• Loosening of fastenings • Broken or damaged parts •

• Incorrect or inadequate lubrication • Foreign matter •

• Environmental changes • Chemical changes in materials •

When maintenance of proper performance or acceptable noise and vibration levels is the goal, symptoms are used as a guide to discover the source of any trouble that may develop and to decide on the remedy. It is helpful to keep in mind the many ways that machine performance is affected by changes that occur with time. A systematic classification of the sources of these changes should serve to point up the many possibilities that exist. (Adapted from *Handbook of Noise Measurement*, General Radio Co., West Concord, Mass.)

Many maintenance groups use periodic vibration measurements or noise stethoscopes in preventive maintenance programs to determine when parts (such as bearings) need to be replaced. The preventive maintenance program, in addition to preventing equipment breakdown, has a desirable by-product—the prevention of increased noise levels that would result from excessive wear of moving parts. Proper lubrication to reduce frictional forces, the tightening of loose parts to avoid rattling, and proper assembly of parts are additional maintenance practices which minimize excessive noise levels. Changes in equipment or machinery that can produce increased vibration and noise levels are listed in Table 15–B. Maintenance programs which effectively minimize these changes will minimize the associated vibration and noise problems.

Reduce vibration response

Reducing the response to vibration forces is a method often used to control many noise problems. This may be brought about through the use of vibration isolators or the use of damping materials. Some examples of various types of vibration isolators are shown in Figure 15–3. Commercially available isolators are, in some cases, a combination of basic types. Isolators are chosen on the basis of the load supported and the deflection needed to reduce transmission of vibration by the desired amount. They are also selected on the basis of cost, resistance to deterioration, expected service life, and damping characteristics.

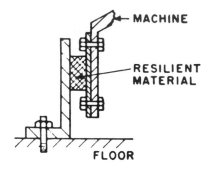

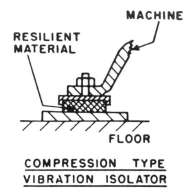

SHEAR TYPE
VIBRATION ISOLATOR

COMPRESSION TYPE
VIBRATION ISOLATOR

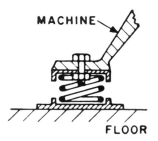

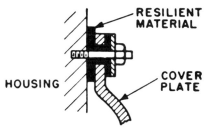

SPRING TYPE
VIBRATION ISOLATOR

ISOLATION IN ATTACHMENT
OF NOISE REDUCING COVER

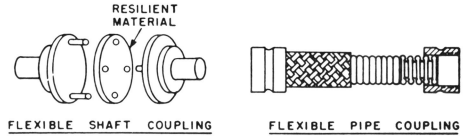

FLEXIBLE SHAFT COUPLING

FLEXIBLE PIPE COUPLING

From Industrial Noise Manual, *American Industrial Hygiene Association.*

FIG. 15–3.—Six types of devices used in vibration isolation.

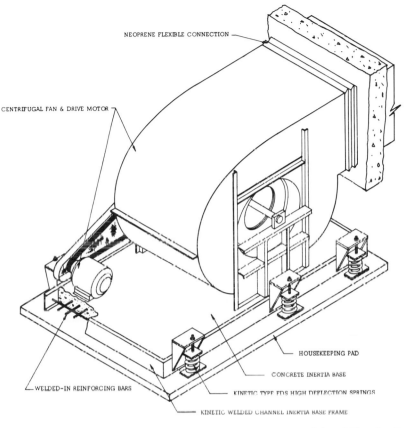

NEOPRENE FLEXIBLE CONNECTION

CENTRIFUGAL FAN & DRIVE MOTOR

HOUSEKEEPING PAD

CONCRETE INERTIA BASE

WELDED-IN REINFORCING BARS

KINETIC TYPE FDS HIGH DEFLECTION SPRINGS

KINETIC WELDED CHANNEL INERTIA BASE FRAME

Courtesy Consolidated Kinetics Corp.

Fig. 15-4.—A centrifugal fan and drive motor mounted on steel spring vibration isolators. Also shown is an inertia concrete block.

Vibration frequency

The natural frequency of a machine, which is supported by vibration isolators, is the frequency at which the system will vibrate if displaced and then released to vibrate freely of its own accord. It is a function of the static deflection of the vibration isolators. As the static deflection is increased, the natural frequency of the system decreases, and in general, the transmission of vibration energy decreases.

The application of periodic forces or impulses can cause a machine or machine part to vibrate with a frequency that may or may not be the natural frequency of the vibrating body. When the period of the

FIG. 15–5.—Instead of mounting a fan directly on a metal bracket, rubber isolators can be placed between it and the mounting bracket.

Courtesy Uniroyal Industrial Products

forced vibration is the same as that of the free vibration, the two effects reinforce each other and large amplitudes of the vibrating body result in a condition that is called resonance. Most mechanical structures resonate at a whole series of frequencies that become more closely spaced at high frequencies.

There are instances where machinery cannot be operated at certain speeds because the frequency of the applied impulse at these speeds corresponds to the natural frequency of some vibrating member. Increased noise levels and even structural failure can occur as a result of resonance.

The natural frequency is also the frequency at which resonance occurs. When the frequency of the driving force coincides with the natural frequency, transmission of vibration force is increased rather than reduced by the isolators. To counteract resonant vibration, damping is added to many isolators to increase energy dissipation. This is done with some sacrifice in the effectiveness of the isolator at other frequencies.

Vibration isolation

Materials commonly used for vibration isolation can be cork, felt, rubber, or steel springs. Cork, felt, or rubber may also be coated with another resilient material to add resistance to deterioration.

Large deflections attained with steel springs can effectively isolate low-frequency vibration of heavy equipment. For large vibration forces, the effectiveness of isolation is improved by mounting the equipment on concrete inertia blocks to increase the weight and inertia of the system (see Figure 15–4). It is important that the isolator supports be located in the plane of the center of gravity of the system. Information regarding selection of isolators for a specific application can be obtained from manufacturers of these devices. It is advisable to consult a specialist in this field for engineering advice on nonroutine applications.

Other examples of vibration isolators are the use of canvas or rubber flexible connectors between ductwork and fans to reduce the transmission of fan vibration. Spring-type hangers are used to support steam lines near pressure-reducing valves to reduce transmission of vibrations from the valve to the building structure.

Vibration isolators can reduce the transmission of vibration from the source to adjacent structural members and thereby reduce the surface areas radiating significant noise (see Figure 15–5). Even with very effective vibration isolators in use, airborne noise can produce sympathetic structural vibrations and transmission of noise to other areas. A combination of measures to reduce structure-borne vibrations may be needed for effective control.

Stiffening

In certain cases, it may be desirable to increase the support points or bracing of a vibrating panel in order to reduce the vibration response. For large panels, the points of contact should be isolated with resilient gaskets or washers made of rubber or plastic. Response of vibrating parts may also be decreased by increasing their mass and stiffness, which shifts the resonances to lower frequencies.

Vibration damping

The term "damping" is used to describe the conversion of resonant vibration energy into heat energy. It is an effective mechanism for noise control—because once converted to heat, vibration energy is no longer available for generation of airborne noise.

The persistence and amplitude of vibration of a resonant part will depend upon its own internal damping. By using materials having high internal damping, or by adding external damping, the displacement amplitude may be decreased. Laminated construction is likely to have high internal damping which is helpful in controlling resonant vibrations.

Vibration damping is used to minimize the amplitude of vibration. Damping is accomplished by internal friction of the material as it is stretched or compressed by the vibratory motion. Damping is usually more effective at high frequencies than at low frequencies because the materials resist the rapid change in motion of high-frequency vibrations.

There is some inherent damping in all materials, but most structural materials—such as metal panels—require damping treatment for noise reduction purposes. In materials such as aluminum or steel, the flexural vibrations persist at various resonant frequencies. Most of the vibration energy in these materials is stored in the bending action and acts as the force to produce repeated vibrations.

Because *mastic materials* resist motion, their vibrational energy is quickly dissipated. Asphalt-base mastic materials with various solid additives have long been used as a damping material. An example is their application to automobile door and body panels to reduce noise from resonant vibrations. Without damping treatment, the sound of an automobile door closing could be described as tinny and noisy.

Since *asphalt-base materials* vary widely in their damping effectiveness, it is necessary to obtain damping ratings for the specific material and conditions of use. Damping ratings may be expressed as vibration-decay rates in decibels per second at 160 Hz at room temperature for the particular treatment applied to a test panel. The thick plate test panel is a 0.25 in. thick, 20 × 20 in. cold-rolled steel plate. Vibration is measured as sound pressure at a microphone placed a few inches away from the center of the panel. In general, damping becomes more effective as the vibration frequency increases.

Asphalt-impregnated felts are also used as damping materials. When a fibrous material such as felt or glass fiber blanket is used with a septum or outer skin attached to it, a very high level of damping occurs. The high level of damping is achieved by the crushing and flexing of the fibrous material between the septum and the vibrating surface. Care must be taken to avoid any solid connection between the septum and the vibrating surface.

Many other damping materials and techniques can be used to reduce resonant vibrations of the source for effective noise control. Prefabricated panels of laminated construction with a layer of viscoelastic material,

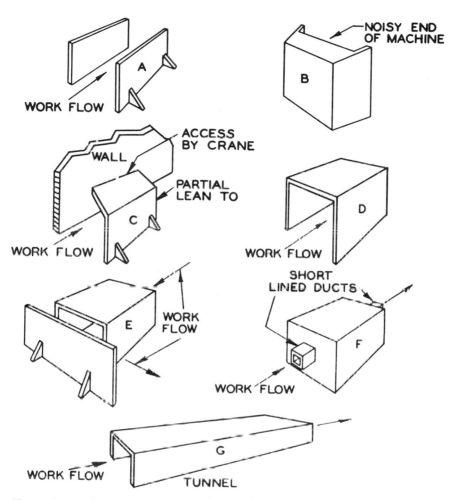

WORK FLOW

A

NOISY END
OF MACHINE

B

ACCESS
BY CRANE

WALL

PARTIAL
LEAN TO

C

WORK FLOW

D

WORK FLOW

SHORT
LINED DUCTS

E

WORK
FLOW

F

WORK FLOW

G

WORK FLOW

TUNNEL

FIG. 15–6.—These enclosures are designed to allow work to flow normally.

From Industrial Noise Manual, *American Industrial Hygiene Association.*

provide an effective approach to damping of resonant vibrations. Since costs are not great, damping treatments can be used in trial applications to evaluate their effect. A technique for determining the presence of resonant vibrations and the need for damping is to vary the speed (rpm) of equipment from well below, to well above the normal speed. Listen for increased noise or changes in pitch, which may indicate that reso-

Courtesy Industrial Acoustics Corp.

Fɪɢ. 15–7.—Enclosure around an electrically operated cut-off saw, driven by a 7½ hp motor, reduced the noise level to an acceptable 84 dBA.

nances are occurring and that damping treatment may be a useful measure for noise reduction. Sophisticated instrument measurement techniques may also be used to determine vibration levels and node patterns (see Chapter 5, Sound Measurement Techniques).

In summary then, the response of a vibrating part to a driving force above the resonant frequency can be reduced by damping the member, improving its support, increasing its mass or stiffness, or otherwise reducing its resonant frequency.

Directionality of source

Many noise sources radiate more noise in a certain direction, so that, if possible, the sources should be oriented so as to direct the noise away from the areas of interest. This is only effective where the areas of interest can be affected by the near fields of noise sources. Access openings in partially enclosed noise sources, intake and exhaust openings, and noisy operating points backed by large reflecting surfaces are examples of directional noise sources where proper orientation and location can be used to advantage.

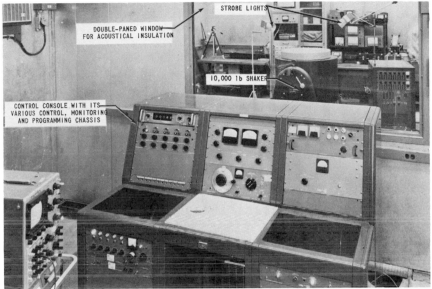

DOUBLE-PANED WINDOW
FOR ACOUSTICAL INSULATION

STROBE LIGHTS

10,000 lb SHAKER

CONTROL CONSOLE WITH ITS
VARIOUS CONTROL, MONITORING
AND PROGRAMMING CHASSIS

Courtesy Argonne National Laboratory.

FIG. 15–8.—A view of a noise barrier which includes a double-paned window for acoustical insulation.

Enclosures

In many cases the practical solution to a noise problem is to enclose the noise source with a sound attenuating barrier. The enclosure must be made of nonporous material to prevent the direct transmission of sound energy. The enclosure walls should have sufficient mass to counteract the driving force of the noise source. Sound absorbing material should be installed inside enclosures to reduce the buildup of reflected sound energy.

Enclosures must be designed to provide convenient access for product flow and for equipment maintenance (see Figure 15–6). Gasketed access doors, removable panels, and acoustically lined tunnel entries are utilized to provide these features. Mechanical ventilation may be required to remove heat from enclosures and the inlet and outlet ducts will require sound traps to attenuate noise from the enclosure.

Special enclosures may be constructed around the noise source or a noisy area may be partitioned off (see Figure 15–7). An example of an enclosure used for noise attenuation is shown in Figure 15–8. One wall

Courtesy Industrial Acoustics Corp.

FIG. 15–9.—Portable soundproof factory office weighs approximately 2000 pounds and can be moved by a forklift.

has a large double-pane window for easy viewing of the shaker apparatus. With the shaker operating at 800 Hz and 90 g (the acceleration due to gravity), the sound pressure level outside the enclosure at the control console is 71 dBA as compared with 95 dBA in the shaker room. The partition allows the shaker operator to work without wearing hearing protectors. A more massive partition would provide a greater noise reduction.

Enclosures may be placed around the receiver rather than around the source of noise (see Figure 15–9). For example, enclosures or partitioned-off work stations are used to separate office or inspection operations from noise production areas. Enclosed air conditioned crane cabs, for example, can also provide substantial noise reduction for the operator.

In cases where transmission loss of more than 50 dB is desired, it may be more economical to use multiple-wall construction than very heavy single walls. The transmission loss from two or more walls separated by air spaces is usually significantly greater than predicted on the basis of mass law attenuation. It is important that the walls should not be solidly tied together in order to gain the advantage of the air space.

Courtesy The Proudfoot Co., Inc.

FIG. 15–10.—Barrier used to block off sound from electrical power transformer.

(More details are given in Chapters 16 and 17.)

Partial enclosures are not nearly as effective as total enclosures for reducing noise levels; however, they can be satisfactory where only a small amount of noise reduction is required. Noise reduction may be improved by lining the partial enclosures with sound absorbing materials. They are most effective in reducing high-frequency noise and least effective at low frequencies.

Barriers

Barrier walls provide a noise shadow effect for the high frequencies but are ineffective for the long wavelength, low-frequency noise. They are most effective when either the noise source or the receiver or both are close to the barrier wall (see Figure 15–10).

Calculations

Transmission loss

Transmission loss (TL) is defined as the reduction in transmission of sound energy, of random incidence, through a partition. It is frequency dependent (see Table 15–C). When a single value is given, it refers to

TABLE 15-C

SOUND TRANSMISSION LOSS (IN dB)
OF VARIOUS MATERIALS AND THICKNESS

Material	Frequency (Hz)				
	256	512	1024	2048	4096
Aluminum sheet, 22 ga.	18	13	18	23	—
Glass, ⅛ in. thick	26	27	31	33	29
Lead, ⅛ in. thick	31	27	37	43	—
Concrete block, 4 in. hollow cinder aggregate	32	37	45	46	48
Concrete, reinforced, 4 in.	36	45	52	60	67
Door, solid oak, 1¾-in. as ordinarily hung	15	20	22	16	—
Walls, 2 in. x 4 in. studs lath and ½ in. plaster both sides	24	28	35	42	59

the average of the transmission losses at 125, 175, 250, 350, 500, 700, 1000, 2000, and 4000 Hz. Figure 15–11 shows the transmission loss according to weight, in pounds per square foot, of various structural materials. Mass theory states that transmission loss should increase by six decibels for each doubling of partition weight. For practical application, however, experimental transmission loss values are used which were obtained from testing full-size partitions. In certain cases, special construction techniques—such as decoupled double-wall construction—can provide increased transmission loss (see Table 15–C). Relatively low transmission losses result when the frequency of sound energy is coincident with resonant partition vibrations.

Sound transmission class

The sound transmission class (STC) rating is favored over the transmission loss (TL) rating as a single-number rating system for subjective

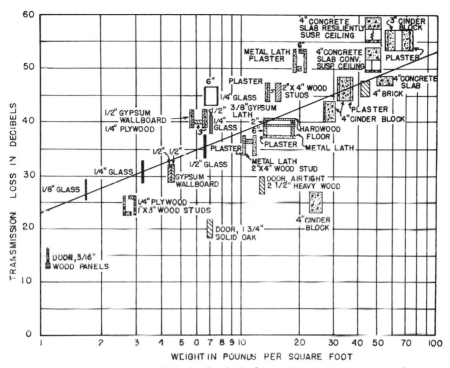

FIG. 15–11.—Transmission loss in decibels for various materials according to their weight in pounds per square foot.

From Industrial Noise—A Guide to Its Evaluation and Control. *Public Health Service Publication* No. 1572.

impressions of sound barrier characteristics. The STC rating takes into account the shape of the transmission loss curve and is more responsive to dips in this curve than is the averaging method employed in the TL ratings (see Figure 15–12). The STC ratings are designed to serve as single-number ratings of the sound insulation provided for speech and music type sounds in offices and dwellings. It is best to use detailed sound transmission loss ratings in conjunction with actual sound spectra measurements where noise sources differ markedly from the characteristics of speech or music.

The STC rating is obtained by fitting a standard contour curve, similar to the A-weighting curve for sound level meters, to the plot of sound transmission losses measured for a barrier at one-third octave from 125 to 4000 Hz. An STC-40 contour curve is shown as an example in Figure

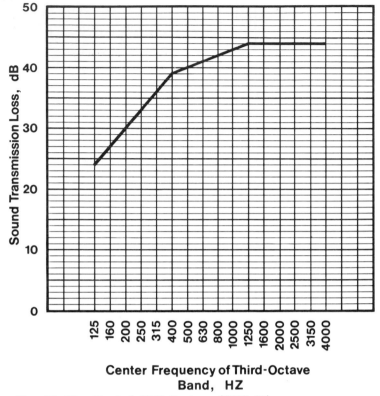

Center Frequency of Third-Octave Band, HZ

FIG. 15–12.—Typical STC Contour (STC-40).

15–12. The STC number is the transmission loss in decibels at the 500 Hz point on the standard contour curve. The determination of the STC ratings is described in detail in ASTM Standard E413-70T, *Tentative Classification for Determination of Sound Transmission Class.*

Sound transmission coefficient

The fraction of incident sound energy transmitted through a partition is called its transmission coefficient (τ). It is related to the transmission loss of a partition by the equation:

$$TL = 10 \log_{10} \frac{1}{\tau} \, dB$$

Most rooms or enclosures are constructed of a number of sections (such

FIG. 15–13.—Using sheet lead as sound barrier.

TABLE 15-D. NOISE INSULATION FACTOR OF A ROOM

	Area Square Feet (S)	T.L. (dB)	τ	τS
Ceiling—four-inch concrete slab one-inch acoustical tile	800	50	0.0000100	0.0030
Walls—four-inch cinder block with plaster (both sides)	1,200	45	0.0000320	0.0384
Floor—four-inch concrete slab plus floor covering	800	50	0.0000100	0.0080
3/16-inch glass windows	60	28	0.001600	0.0960
Two 1½-inch hardwood doors (close fit)	36	20	0.010000	0.3600

A = Total room absorption ft² = 718
    (See Table 16F)

Total transmittance (T) = 0.5104

Noise insulation factor = $10 \log_{10} \dfrac{A}{T}$ dB = $10 \log_{10} \dfrac{718}{0.5104}$ = 31.5 dB

Adapted from Industrial Noise Manual, *American Industrial Hygiene Association.*

as walls, floors, doors, and windows) having different areas of varying transmission coefficients. If each structural element has sound energy of the same level impinging upon it, the average transmission coefficient $\bar{\tau}$ is given by:

$$\bar{\tau} = \frac{\tau_1 S_1 + \tau_2 S_2 + \tau_3 S_3 \cdots \tau_n S_n}{\Sigma S} = \frac{\Sigma T}{\Sigma S}$$

where τ_1, τ_2, τ_3, and τ_n are the transmission coefficients of the different parts of the boundary; S_1, S_2, S_3, and S_n are their corresponding surface areas; and ΣS is the sum of all these areas. ΣT is the total transmittance.

Noise insulation factor

The transmittance and the total number of units of absorption in a room are the principal factors in determining the relative noise insulative properties of its boundaries. This rating is called the *noise insulation factor* (NIF) of a room and is expressed by:

$$NIF = 10 \log_{10} \frac{A}{T} \, dB$$

$$T = \text{transmittance}$$
$$A = \text{total room absorption in sabins}$$

An example of how this formula can be used is shown in Table 15–D which illustrates the importance of eliminating small areas that have relatively large transmission coefficients. For example, double windows having a transmission loss of 40 dB ($\tau = 0.0001$) would increase the noise insulation factor to 32.3 dB. On the other hand, if the windows were open ($\tau = 1.0$), the noise insulation factor would be only 11 dB.

It is obvious from the examples given that increasing the insulation value of the walls has little effect because most of the sound is transmitted through the windows and doors. Also, it should be pointed out that to achieve maximum insulation, openings in enclosures (around doors, services, etc.) should be eliminated. In the example, where double windows were used, an opening of 60 sq in. (approximately 0.42 sq ft) would have permitted the transmission of as much sound energy as the remainder of the room even though the opening is only 0.015 percent of the total room area.

Sound absorbing materials

Sometimes misguided attempts are made to reduce noise levels by wrapping or enclosing noise sources with porous sound absorbing materials such as glass fiber blankets. The results are very disappointing because sound energy passes quite freely through the porous covering. But a material like sheet lead, which has the excellent properties of high density and low stiffness, may be used (see Figure 15–13).

Small openings in sound barriers may greatly reduce their effectiveness. For example, a 1 sq in. hole will transmit slightly more sound energy than the entire surface of a 4 × 12 ft sheet lead barrier rated at 40 dB transmission loss. Door crack openings and ventilation ducts are often noise passages. Commercially available special rubber seals for doors may be used to avoid excessive noise leaks. Where heat buildup in an enclosure is excessive, a fan and lined ducts or package attenuators may be required to provide ventilation without allowing excessive noise to escape from the enclosure.

Porous sound absorbing materials may be used to reduce the reflection or reverberation of sound energy. They cause sound energy to be degraded to heat energy by producing frictional shear forces in the air. Air molecules set in motion by a sound wave are slowed down by friction at surfaces within the absorbing materials. These materials are *not* effective as sound barriers since they transmit air motion and have a low mass per unit of thickness.

Covering of ceiling, wall, and/or floor surfaces with a sound absorbing material will minimize reflection and reverberation of sound energy. Keep in mind that the sound level in the direct field near a noise source will not be affected by sound absorbing materials installed on room surfaces. It is only the reverberant noise field that is reduced by this treatment.

Absorption of sound

Sound absorbing materials are rated by the sound energy ratio of the absorbed to the incident sound. This ratio is expressed as the absorption factor (or coefficient). Even though absorption varies with the angle of incidence of the sound energy, the coefficients are generally related to random incidence by reverberant room testing. Absorption also varies with the frequency of the sound; the absorption coefficient, therefore, is commonly listed at four frequencies—500, 1000, 2000, and 4000 Hz (see Table 15–E). A single "noise reduction coefficient" (NRC) of acoustical materials is the average of the absorption factors at 250, 500, 1000, and 2000 Hz. Using the noise reduction coefficient for calculating noise reduction of acoustical materials is simpler but less exacting than using the coefficients for various frequencies.

TABLE 15-E
SOUND ABSORPTION COEFFICIENTS

Material	Decimal Equivalent of Fraction of Sound Absorbed at indicated Frequency (Hz)			
	500	1000	2000	4000
Concrete, poured	0.02	0.02	0.02	0.03
Plaster, lime sand finish	.06	.08	.09	.06
Glass	.03	.03	.02	.02
Carpet, wool with underpad, ⅝ in.	.35	.40	.50	.75
Drapes, velour, 14 oz/sq yd straight	.13	.22	.32	.35
draped to half area	.49	.75	.70	.60

Absorption coefficients of acoustical materials are affected by the distance from the mounting surface. The most effective conversion of sound energy to heat energy occurs at the point of maximum particle velocity of air molecules in a sound wave. This particle motion is at a minimum at the reflecting or boundary surface (see Figure 15–14). It passes through a maximum, one-quarter wavelength from the point of reflection. For most efficient absorption, the acoustical material needs to be mounted so that the maximum particle velocity occurs within its matrix. In the case of low-frequency noise (125 Hz), a quarter wavelength corresponds to 2.2 ft. That is why suspended ceiling-type mounting of acoustical tile provides better absorption of low-frequency noise than does surface mounting.

The optimum density of an acoustical material should occur at that point where the second energy which penetrates and is reflected back out of the material equals the sound energy reflected from the surface of the material. A denser material reflects too much of the incident energy while a less dense material does not absorb the sound energy as effectively.

The sound absorption of a space is expressed in sabins. One square foot of a completely sound absorptive surface (equivalent to one square foot of open window area) is called a *sabin*. The efficiency of a material to absorb sound energy is termed its *absorption coefficient* at that frequency. The surface area multiplied by the absorption coefficient for this material equals the number of sabins absorption for that area. The

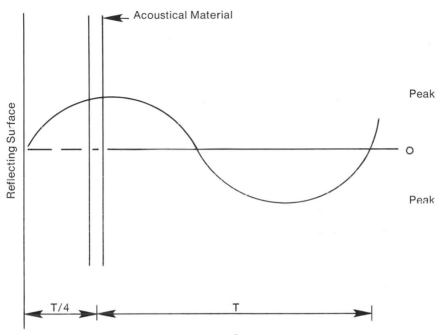

Reflecting Surface

Acoustical Material

Peak

O

Peak

T/4

T

Fɪɢ. 15–14.—Particle velocity in sound wave.

total room absorption in sabins is the sum of sabins of all surface areas. The reverberant noise level in a room decreases three dB for each doubling of the total absorption. In most applications, installation of absorptive materials provides less than 10 dB of noise reduction.

Treatment of 20 to 50 percent of the boundary surface area is a practical approach. Installation of acoustical tile on the ceiling or as a false ceiling is a common practice. It is impractical to treat more than 50 percent of the room surface area because noise reduction gained going beyond this amount is slight.

The average sound pressure level in a room can be calculated from the sound power level of the noise source and the absorption of the room. With a single noise source producing 0.01 watt of sound power in a 10 × 20 × 10 ft room with an average absorption coefficient ($\bar{\alpha}$) of 0.2, the calculation is:

$$\text{Average SPL} = \text{PWL} - 10 \log_{10} S\,\bar{\alpha} + 16.5 \text{ dB}$$
$$\text{SPL} = \text{Sound pressure level, (reference level} = 20\mu \text{ Newtons/m}^2)$$

405

$$\text{PWL} = \text{Sound power level} = 10 \log_{10} \frac{W}{W_{\text{Ref}}} \text{ dB}$$

(reference level $= 10^{-12}$ watts)

$W = $ Acoustic power in watts

$S = $ Boundary surfaces in square feet (walls, ceiling, and floor)

$\bar{a} = $ Average sound absorption coefficient

Therefore:

$$\text{SPL (average)} = 10 \log_{10} \frac{10^{-2}}{10^{-12}} - 10 \log_{10} 1000 \, (0.2)$$

$$+ \, 16.5 = 100 - 23 + 16.5 = 93.5 \, \text{dB}$$

A small amount of sound energy can cause problems as indicated by this hypothetical case.

To evaluate the effectiveness of providing an absorbing surface treatment, calculate the noise reduction that such a treatment would accomplish. The far-field noise reduction, at a given frequency, can be expressed in terms of total room absorption as follows:

$$\text{NR} = 10 \log_{10} \frac{A_2}{A_1}$$

$\text{NR} = $ Noise reduction

$A_1 = $ Total room absorption before treatment in sabins

$A_2 = $ Total room absorption after treatment in sabins

An example of the computations involved in determining noise reduction is as follows:

A room is 20 × 40 × 10 ft and has a plastered ceiling and walls. The sound absorption coefficient (α) of the walls is 0.03 at 1000 Hz. For the floor, $\alpha = 0.04$ at the same frequency. The equipment in the room has an absorption of 50 sabins. The ceiling is to be treated with a material having an absorption coefficient (α) of 0.75 at 1000 Hz. The solution to this problem is given in Table 15–F.

Acoustical materials

There is a wide selection of acoustical materials that can be purchased and used satisfactorily if their capabilities and limitations are understood. Acoustical materials reduce noise in the reverberant sound field but do not alter the noise level coming directly from the source. The sound level at the position of the operator of a machine will likely be unchanged because the operator is in the direct sound field. Other employees, at a distance from the machine, will be exposed to a reduced

TABLE 15-F

NOISE REDUCTION COMPUTATION WITH ACOUSTICAL ABSORPTION MATERIALS

	Surface area (S) In square feet	Absorption Coefficient (α)	Absorption Units (Sabins)
Floor	800	0.04	32
Walls	1,200	0.03	36
Ceiling (before treatment)	800	0.03	24
Equipment			50
Total absorption before treatment			142 (A_1)
Ceiling (after treatment)	800	0.75	600

Total absorption after treatment $(600 + 32 + 36 + 50) = 718 (A_2)$

Therefore: $NR = 10 \log_{10} \dfrac{A_2}{A_1} = 10 \log_{10} \dfrac{718}{142} = 10 \times 0.7 = 7 \text{ dB}$

Adapted from Industrial Noise Manual, *American Industrial Hygiene Association, 1966.*

noise level because they will be in the reverberant or reflected sound field. Selection of acoustical materials should, in addition to absorption properties, take into account fire resistance, cost, esthetic qualities, light reflection, susceptibility to physical damage, and ease of maintenance.

Acoustical tile is available in various sizes, compositions, surface textures, and styles. They are manufactured in standard 0.5, 0.75, or 1 in. thickness. A fissured mineral fiber tile 0.75 in. thick and mounted in accordance with instructions provides the average absorption coefficients as shown in Table 15–G.

The sound absorption is much greater than that for hard reflecting

TABLE 15-G
ABSORPTION COEFFICIENTS OF MOUNTED MINERAL FIBER TILE

Mounting	Frequency (Hz)					
	125	250	500	1000	2000	4000
Cemented directly to ceiling	.10	.26	.79	.93	.86	.80
Mounted on special metal supports	.64	.70	.72	.83	.90	.85

Fig. 15–15.—A package attenuator can be installed as a section of ventilating duct to control noise.

surfaces such as plaster, smooth concrete, metal surfaces, or tile floors. The hard reflecting surfaces generally have absorption coefficients of 0.01 to 0.05. Mineral fiber tile is preferred to cellulose tile because of fire-resistive properties. Properly designed perforated metal facings or thin plastic membrane facings allow transfer of sound energy into acoustical absorbing materials and protect the fibrous sound absorbing materials.

With any of the acoustical materials, it is important to avoid loss of absorption properties due to plugging of porous openings by dirt loading or paint coatings. Painting of acoustical materials should be accomplished such as by light spray painting without bridging over the void spaces.

Sprayed-on mineral fiber acoustical coatings may also be used effectively for noise absorption. They have the advantage of conforming to irregular surfaces. However, they have the disadvantage of being easily damaged and they are difficult to clean.

Space sound absorbers (preformed acoustical units that are hung from the ceiling) have the advantage of providing relatively good low-frequency sound absorption because they are mounted away from the sound reflecting surfaces. Another advantage is their potential for reuse by simply taking them down and rehanging them in another area. They may be in the form of baffles or various geometric shapes. They should be rated in terms of sabins or equivalent square feet of total absorption for a given installation spacing. Their effectiveness is similar to that of the equivalent number of square feet of acoustical tile mounted at the ceiling.

Acoustical materials applied to the interior of equipment enclosures reduce reverberant buildup of noise levels and thereby reduce the amount of noise energy at the exterior surface of the enclosure. Fan noise transmitted through ventilating ducts is frequently reduced by lining the ducts with glass fiber blankets or by installing package attenuators in a section of the ductwork (see Figure 15–15). The package attenuators are convenient to install and provide an appreciable reduction in noise level. They are commercially available, normally in 2 to 8 ft lengths, and are sized for relatively low pressure drops at rated flow. For example, a 5 ft long unit that was installed in a laboratory supply air system was rated at 5000 cfm, 0.25 in. of water pressure drop, and provided 17 to 45 dB attenuation in the octave bands of interest. Anyone selecting such units should be aware that the manufacturer's noise reduction ratings may relate to a static test. Under air flow conditions, some noise may be regenerated due to air turbulence at the discharge.

Summary

More extensive treatment of the principles and applications of engineering control of noise may be found in Chapters 17 through 19 and in the books listed in the bibliography. Reduction of noise at the source through substitution or modification should be the first consideration. When that is not practical or is inadequate, the various means for reducing transmission of noise from the source to the receiver should be considered. Application of the principles discussed in this chapter in a systematic approach is the basis for the successful control of noise.

Bibliography

American Industrial Hygiene Association, *Industrial Noise Manual*, 2nd ed. Akron, Ohio: AIHA, 1966.

American Society for Testing Materials. *Tentative Classification for Determination of Sound Transmission Class*, E413–70T, Philadelphia, Pa.: ASTM.

Beranek, L. L. *Noise and Vibration Control.* New York, N.Y.: McGraw-Hill Book Co., 1971.

Harris, C. M. ed. *Handbook of Noise Control.* New York, N.Y.: McGraw-Hill Book Co., 1957.

Hosey, A. D. and Powell, C. H. eds. *Industrial Noise—A Guide to its Evaluation and Control* (Public Health Service Publication No. 1572, U.S. Dept. HEW). Washington, D.C.: U.S. Government Printing Office, 1967.

Johnson, K. W. "Vibration Control." *Sound and Vibration*, May–June 1962.

Petersen, A. P. G. and Gross, Jr., E. E. *Handbook of Noise Measurement.* West Concord, Mass.: General Radio Co., 1972.

"Sound and Vibration Control" in *ASHRAE Guide and Data Book—Systems and Equipment.* New York, N.Y.: American Society of Heating, Refrigerating, and Air Conditioning Engineers, Inc., 1967.

Yerges, L. F. *Sound, Noise and Vibration Control.* New York, N.Y.: Van Nostrand Reinhold Company, 1969.

Chapter Sixteen Problem-Solving Techniques

By George W. Kamperman

Techniques to solve noise problems include various practical control methods such as vibration isolation, vibration damping, and the use of mufflers, enclosures, or barriers.

Noise control in a machine or device can be done best with an understanding of the application or use of the device and an insight into the noise generating mechanisms. The sound level criteria for a specific facility or noise source should first be established. The next step is to investigate in detail the significant noise sources. To do this may require the use of various sound (and vibration) measurement techniques as described in Chapter 5. The measurements should be aimed at determining both the airborne and structure borne contribution from various individual noise sources. Understanding the nature of the noise transmission paths may require correlation of the results from the structure-borne (vibration) measurements with the airborne measurements. A comparison of these results with the sound level criteria determined previously will lead to the noise reduction requirements. (See Chapter 14.)

Noise generation and transmission

When considering methods of controlling objectionable sound or noise, it is always helpful to consider how that noise is generated or produced (see Chapter 2, Physical Characteristics of Sound). Airborne sound energy or noise, transmitted through the surrounding air space, can be

FIG. 16–1.—The structure-borne path may be the dominant noise transmission path. For example, a high energy noise source located on one floor might easily be heard on other floors in a multistory building.

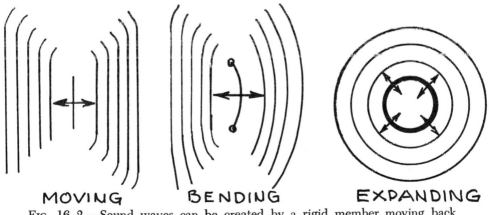

FIG. 16–2.—Sound waves can be created by a rigid member moving back and forth, by a panel constrained at its boundaries and bending in the center, or by a hose or pipe expanding and contracting.

considered to be rapid changes in the ambient air pressure which can affect the ear or a sensitive measuring instrument.

The noise can be generated directly by aerodynamic forces such as the turbulence produced by shearing action of a high-speed airstream. In other cases, the vibration of solid objects or surfaces is transmitted to the air by the housing or frame of a machine and gives rise to excessive sound levels.

Structure-borne transmission

Sound waves may be transmitted from a noise generating source by vibrations of solid structures, by the intervening air path, or by a combination of the two. The noise source itself may not radiate efficiently, but its sound energy may be transmitted by intervening solid paths to other surfaces which may be more efficient radiators. The measured sound levels in air also depend on the characteristics of the space in which the noise source is located.

For example, in a multistory building, a high-energy sound source on one floor can easily be heard on other floors (see Figure 16–1). Vibrations introduced into the structural system of the building cause the sound energy to radiate from the structural columns and other surfaces throughout the building. The noise produced by a jack hammer being used on the first floor can be heard on the twentieth floor. The sound transmission from the jack hammer in this case is not airborne, but is structure borne.

Vibrating surfaces

In addition to the transmission of structure-borne sound, the sound radiated into air by vibrating surfaces should also be taken into consideration. Representative types of vibrating surfaces are shown in Figure 16–2.

• A structural member may move back and forth and radiate sound waves.

• A sheet metal panel may be fixed or constrained at its boundaries but the center may move to radiate sound.

• A flexible pipe or hose from a hydraulic pump may expand and contract at the piston or gear frequency of the pump to radiate sound.

Effect of size. A noise source may radiate sound energy into air which

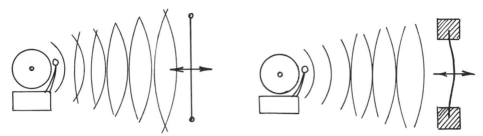

Fig. 16–3.—A noise source may radiate sound waves into the surrounding air. The resulting pressure fluctuation may cause a plate or panel to vibrate and also radiate sound waves.

in turn may cause a plate or sheet metal panel to vibrate which will then reradiate sound energy (see Figure 16–3). The efficiency of a surface to radiate sound energy is determined by its size with respect to the wavelength of sound being radiated.

The *wavelength* of a sound may be defined as the velocity (speed) of sound divided by the sound frequency:

$$\lambda \text{ (wavelength of a sound)} = \frac{v \text{ (velocity)}}{f \text{ (sound frequency)}}$$

The wavelength of a sound is inversely proportional to its frequency. The speed of sound in air at normal temperature and pressure is approximately 1130 feet per second (fps). For instance, at 1000 Hz the wavelength of this sound is approximately one foot. At a higher frequency, the wavelength becomes shorter and at a lower frequency, the wavelength becomes longer. To be more precise, at 200 Hz, a wavelength is 5.65 ft; while at 2000 Hz, a wavelength is 0.565 ft, or $6 \frac{21}{32}$ in. (see Chapter 2, Physical Characteristics of Sound).

When the dimensions of a noise radiating surface are comparable to the wavelength of the sound being emitted, the surface becomes a relatively efficient radiator. If the radiating surface is greater in size than three wavelengths or more, it becomes a very efficient radiator. On the other hand, if the dimensions of the noise radiating surface are reduced in size to one third of a wavelength or less, then this noise radiating surface becomes a relatively inefficient radiator.

A tuning fork vibrating at 256 Hz (middle C) radiates a sound wave having a wavelength of approximately 4 ft (see Figure 16–4). Because the radiating surface of the tuning fork prong is approximately ¼ in. wide or 1/100 of the wavelength, it does not radiate sound very efficiently

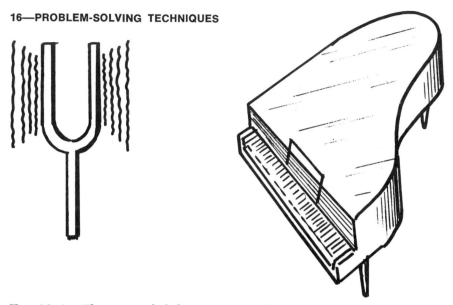

Fig. 16–4.—The tuning fork has a very small radiating surface. The piano sounding board, on the other hand, is an excellent radiator of sound energy.

or effectively. However, if the base of the tuning fork is held in contact with a relatively large surface, such as a table top, the sound level is increased and the tone can be heard—loud and clear. The reason, of course, is that the greater surface area of the table is better able to radiate sound having a four foot wavelength.

Directivity. The directivity of sound waves being radiated by a vibrating surface depends on the size of that surface in relation to the wavelength of the sound being emitted. Surfaces that are small in comparison to the wavelength of the emitted sound radiate the sound relatively uniformly in all directions. Surfaces that are large in comparison to the wavelength are very directive sound sources. A good example of this is the use of a tweeter (a very small loudspeaker which reproduces high-frequency sounds) in a high-fidelity sound system. The bass speaker, called a woofer, is large in comparison to the tweeter and would become very directive if used to reproduce high-frequency sounds.

Since the wavelength of a sound wave is inversely proportional to its frequency, vibrating surfaces become very directive at high frequencies and less directive at low frequencies. When studying a noise problem, it is important to keep in mind the directivity properties of the sound

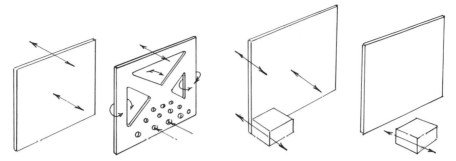

Fig. 16–5.—Noise radiation can be reduced by reducing the radiating area (left), or disconnecting a large noise radiating area from the vibrating part (right).

emitting surface and its relationship to the wavelength of the sound being produced.

Reducing the radiating surface area. A manufacturer may build a machine and then install covers or side panels to dress up the product. He, then, is very surprised that the sound level is higher than before the side panels were placed on the machine. The panels, in this case, not only increased the radiating surface, but because the size of the panels was comparable to the wavelength of the sound being produced, they radiated sound much more efficiently than just the structural members of the basic machine. Sometimes, however, this amplifying effect is desired; for example, the sounding board on a piano (see Figure 16–4) is used to amplify sound.

In cases where a panel or metal cover is radiating sound, the panel may be perforated with holes to reduce the radiating surface. Small holes, less than one-quarter wavelength apart at the highest frequency of interest, will be acoustically transparent (assuming that the panel is not too thick). By reducing the area of the radiating surface, the sound intensity can also be reduced. The sound level may decrease three decibels or more each time the radiating surface is halved (see Figure 16–5). As an example, consider a square or rectangular metal plate where only the edge is functional; in which case, it is a simple matter to cut away the nonfunctional portion and reduce the radiating area. It is easy to see that if a moving machine member has less surface area to compress the air in front of it, creating sound, there will be less mechanical energy converted into sound (see Figure 16–6).

415

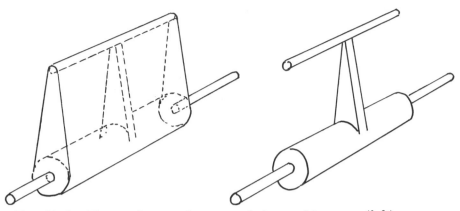

FIG. 16–6.—The nonfunctional portion of the machine part (left) was removed to reduce the radiating area.

Moving air sources

The turbulence of rapid air movements created by a fan, a jet or nozzle, or an obstruction in an airstream can frequently be a major source of noise (see Figure 16–7). In a typical fan, for instance, the rotation of

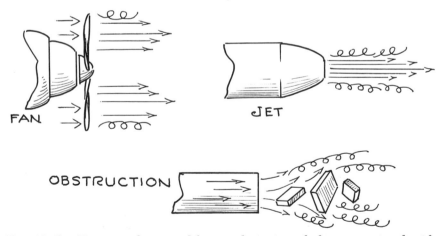

FIG. 16–7.—Noise can be caused by aerodynamic turbulence associated with moving airstreams. Any obstruction in the moving airstream generates additional turbulence with its own characteristic noise spectrum.

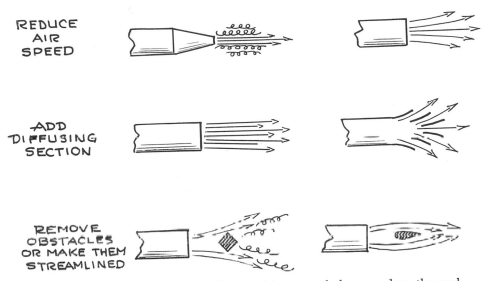

REDUCE
AIR
SPEED

ADD
DIFFUSING
SECTION

REMOVE
OBSTACLES
OR MAKE THEM
STREAMLINED

Fig. 16–8.—To reduce noise levels caused by air turbulence, reduce the peak exit velocity which in turn reduces the shearing effect of ambient air. This may be accomplished by decreasing the pressure and increasing the diameter of the air discharge opening or by adding a diffusing element.

the blades transfers an impulse to the airstream each time a fan blade passes a given point. The result is a sound wave at the fundamental blade-passage frequency. In addition, a fan produces a sound level that varies with the fifth to sixth power of the velocity of fan rotation, so that a two-to-one reduction in fan speed results in a 16 dB reduction in noise level.

The sound energy produced by a high-velocity air jet varies as the seventh to eighth power of the velocity of the airstream, or another way of expressing the same thing is to say that the change in sound level (in decibels) is 70 to 80 times the $\log_{10}$ of the change in velocity of the airstream. Consequently, if the velocity of the airstream is reduced to one half, the noise level will drop more than 20 dB.

The jet airstream created by compressed air ejection systems (on a punch press, for example) generates noise by the turbulence at and just downstream of the air jet exit, and by the high velocity airflow over the sharp edges of the dies or target. Any obstruction in the moving airstream may generate additional air turbulence with its own characteristic noise spectrum.

417

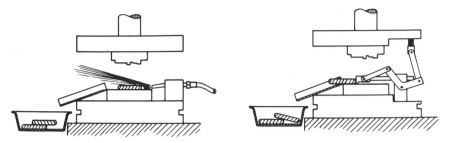

From Industrial Noise Manual, 2nd ed., *American Industrial Hygiene Association.*

FIG. 16–9.—Compressed air is commonly used for part ejection on punch presses, drill presses, and other high-speed production machines. Mechanical ejectors (right) can be readily adapted to many such operations.

To reduce the peak noise level of a high-speed jet airstream, a diffuser section can be placed at the exit so that although the same volume of air is flowing, the maximum exit velocity is reduced (see Figure 16–8). If an air jet is being used as an air-ejection device, the direction of airflow of the air jet becomes important for satisfactory operation with minimum airflow or airstream velocity. Frequently, the peak velocity at the nozzle discharge can be reduced while retaining the necessary thrust, by using multiple openings in one nozzle, or by moving the nozzle closer to the part to be ejected. If there is an obstruction in the airstream, it should either be removed or streamlined to facilitate the passage of air. It may also be possible to substitute a mechanical system and eliminate the air jet altogether (see Figure 16–9).

The velocity and temperature of the moving airstream and the size of the orifice determines the magnitude and spectrum of the sound produced. A small nozzle or orifice produces a hissing sound, while a larger diameter orifice produces a characteristic peak in its noise spectrum at a lower frequency, such as the roar of a jet engine.

For example, at Cape Kennedy during a rocket blast-off, the maximum sound level, below 10 Hz, occurs at ground level while the rocket engine is several hundred to a thousand feet above the ground. The apparent acoustic center of the noise source is far downstream of the vehicle at low frequency and appears to be far behind the vehicle. And yet at 10,000 Hz, the source appears to be within a foot of the combustion chambers. In other words, the high frequencies stay very close to the nozzle, while the lower frequencies peak further downstream.

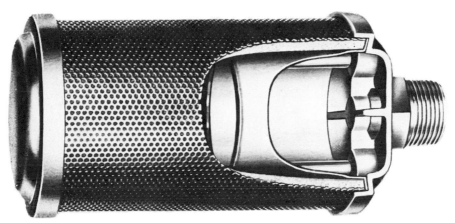

From Industrial Noise Manual, 2nd ed., *American Industrial Hygiene Association.*

Fig. 16–10.—The dispersive-type muffler allows the compressed air to fill the porous container and discharge slowly between cycles.

Mufflers

A muffler is a device which reduces the level of sound traveling in a moving air or gas stream. There are three basic types of mufflers that are available commercially; each uses a different method to accomplish its design purpose to reduce noise.

• DISPERSIVE. The velocity of the air or gas is reduced by spreading the discharge into a large confined area.

• DISSIPATIVE. Sound energy is absorbed by the material lining the duct through which the air or gas stream is flowing.

• REACTIVE. The resonance phenomenon is used to reflect the sound energy back toward the source.

Dispersive mufflers

Dispersive mufflers are usually closed cylindrical containers with porous walls. The velocity of the air or gas flow is reduced by spreading the flow over a large area (see Figure 16–10). The dispersive-type muffler allows waste air or gas to periodically flow into a cavity that is fitted with a porous material having a high flow resistance. Air or other gas fills the cavity quickly and then discharges through the porous walls

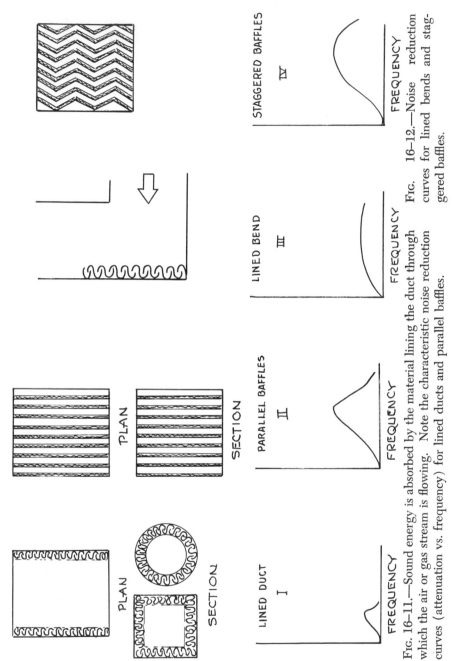

FIG. 16–11.—Sound energy is absorbed by the material lining the duct through which the air or gas stream is flowing. Note the characteristic noise reduction curves (attenuation *vs.* frequency) for lined ducts and parallel baffles.

FIG. 16–12.—Noise reduction curves for lined bends and staggered baffles.

slowly between cycles. By cutting down on the peak velocity and letting the air bleed out slowly, the dispersive muffler is very effective in controlling the noise of compressed air equipment such as pneumatic actuators.

A simple muffler of this type on the air discharge of some pneumatic tools, such as an impact wrench, can result in a 20 to 30 dB reduction in sound levels in the higher frequencies.

Dissipative mufflers

Dissipative mufflers, in contrast, do not reduce air or gas flow velocity but rather absorb the sound energy. For applications such as air and gas intakes and compressor and fan exhausts, the design is usually a straight perforated tube surrounded by sound-absorptive material. This type of muffler is particularly useful for moderate noise reduction over a wide frequency range.

If the problem involves noise in an air duct (such as that used for air conditioning), the duct can be acoustically lined to absorb the sound energy (see Figure 16–11). Depending upon the nature of the problem, a number of design configurations can be used, either individually or in combination. Each configuration has individual characteristics; for instance, parallel baffles provide a more peaked attenuation curve than does a lined duct. If the airflow is small, it is possible to place the parallel baffles close together and achieve more attenuation; because the smaller the spacing or width of the air path, the greater the noise reduction per linear foot.

When designing a system, avoid a line-of-sight noise path from the source. When confronted with a problem involving noise in ventilating ducts that are straight for considerable distances, replace straight sections with one or several lined bends to obtain improved high-frequency attenuation. If, however, it is physically impossible to insert a bend, staggered baffles may be used to achieve the same effect (see Figure 16–12).

Reactive mufflers

Reactive mufflers are usually rectangular or tubular and contain cavities and restrictions or side branches to create a standing wave that will reflect the sound back toward the source. A good example is the automobile muffler which provides maximum attenuation at the low frequencies associated with the engine exhaust noise (see Figure 16–13). Reactive mufflers are very effective in controlling a pulsating noise within a narrow frequency range. The interaction between the various sections, the air-

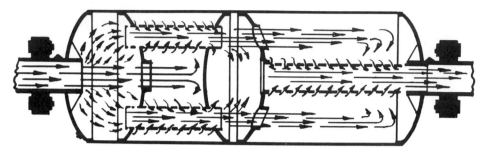

From Industrial Noise Manual, 2nd ed., *American Industrial Hygiene Association.*

Fig. 16–13.—Reactive mufflers are often tubular and contain cavities and restrictions or side branch resonators.

flow, and the sound source, however, can make the reactive muffler design rather complicated.

Because the three types of mufflers have been described separately, it may appear that each is used separately for a specific purpose. Quite frequently, however, a combination of all three types may be used.

Vibration

The sound energy generated by a vibrating surface depends on the degree of movement of the surface as well as the area of the radiating surface. The vibration of machine elements can be a source of excessive sound. In the operation of the machine, sound can arise as a result of impact force, reciprocating action, rotary unbalance, gear tooth forces, or some other form of energy transfer (see Figure 16–14).

When a machine or device is mounted so that only a small portion of the device is in contact with the base, fewer noise problems result. If a small motor is rigidly attached to a structural member of the machine frame, it may produce excessive noise levels. The cause, of course, is not the motor alone, since if it were suspended in the air by elastic bands, there would be very little noise produced.

The electrical induction motor has a rotor that releases a considerable amount of vibration energy as it contacts (at 120 times a second) the thrust bearings on the motor shaft. As soon as this motor is placed or mounted on a supporting member, the increased radiating surface transmits the structure-borne vibrations and may create excessive noise levels.

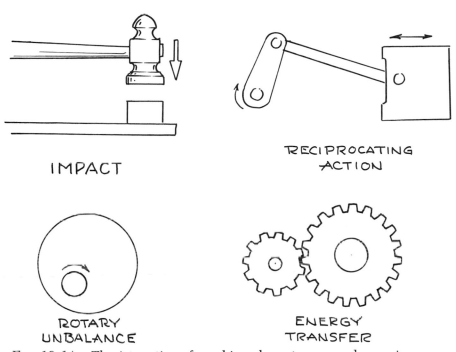

IMPACT

RECIPROCATING ACTION

ROTARY UNBALANCE

ENERGY TRANSFER

Fig. 16–14.—The interaction of machine elements can produce noise as a result of impact, reciprocating action, rotary unbalance, gear forces, or some other form of energy transfer.

Mounts

Frequently, the best solution to a vibration problem is to use resilient mounts to isolate the source of vibration from radiating surfaces, as discussed in the previous chapter. Resilient supports can be interposed between the sound source and the equipment base (frame, cover, or other radiating surface) so that the magnitude of the vibratory force transmitted from the equipment is reduced (see Figure 16–15). However, one must not be misled into assuming that all that is necessary is to place something soft and spongy under a machine to reduce the vibration and resultant noise.

Among important factors to be considered are the mechanical compliance of the device being mounted, the base to which it is to be mounted, and the mount itself. When using a vibration isolator (mount), the problem should not be treated as a simple spring/mass system, but rather as a multiple system or a single complex system. All elements of

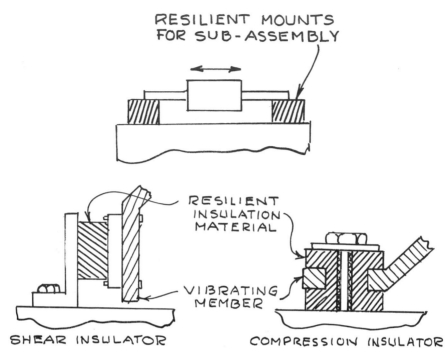

RESILIENT MOUNTS FOR SUB-ASSEMBLY

RESILIENT INSULATION MATERIAL

VIBRATING MEMBER

SHEAR INSULATOR   COMPRESSION INSULATOR

Fig. 16–15.—Transmission of structure-borne vibration may be controlled by reducing the vibration force transmitted to the support or base. Resilient mounts (rubber in shear or steel springs) can effectively isolate structure-borne noise.

the total system must be considered whenever vibration isolators are going to be used.

Vibration isolation refers to the decrease in transmission of vibratory motions or forces from one structure to another. If a force is intermittently applied to a resiliently supported mass, the system will continue to vibrate at a given rate and amplitude. Should the system be vibrating at its natural frequency, it will *resonate*, that is, the amplitude of vibration will increase with each pulse during the cycle.

A case where all the required factors were not considered, with unfortunate results, involved a manufacturer of electric erasers. This company developed a cordless drafting eraser of very fine design with beautifully machined gears and a high-speed motor. Although it was an excellent addition to their product line, the electric eraser had one problem—it was very noisy.

The excessive noise did not hurt sales until another firm marketed a battery-operated eraser which, while not as high in quality, was less expensive and much quieter. The first company then became concerned and called in a noise consultant to help quiet their noisy eraser. During a demonstration of the drafting eraser, the cover was removed, and the noise level dropped dramatically.

A detailed analysis revealed that the noise level was reduced when the cover to the drafting eraser was removed because the radiating area was reduced. The rubber mounts which had been placed under the cover, in an attempt to prevent precisely what happened, were too compressed to act as isolators. In fact, they were compressed so much that it was as if the cover were in direct contact with the case, without the benefit of the mounts.

The idea was correct; the execution was bad. The use of rubber in compression was good, but unless the stiffness and mechanical impedance of the mount is taken into consideration, the vibration isolation measure can fail.

In the example cited, the company had a number of options to correct the eraser noise problem. Without changing the general configuration of the cover, the rubber mounts could be corrugated, waffled, or softer. The rubber mounts require room to move and deform to prevent the transmission of vibrations to the cover.

Another method is to mount the rubber in shear, which in many cases is more effective (see Figure 16–15). When the rubber isolator is used in this fashion, it is usually bonded between two parallel faces. In this position, it is—for all practical purposes—incompressible and yet can change shape without changing volume. Also, the spring constant of the rubber mount can be calculated.

A good example of the application of the rubber-in-shear mount technique is the engine mounts used in automobiles to prevent the motor vibrations from being transmitted to the body or frame.

Resilient materials

In addition to the use of rubber, materials used in vibration isolators include metal springs, cork, plastic, and felt. Other materials, used occasionally, are steel-mesh pads or pneumatic sacks (air bags) depending on the circumstances.

Metal springs are often used because of their versatility. Springs can be used to isolate delicate instruments from external sources of vibration.

Large springs are used to isolate the vibration of huge presses from the building foundation or structure. Other advantages offered by metal springs include interchangeability, resistance to corrosion by oil and water, and adaptation to extremes of temperature.

Felt, another commonly used vibration isolation material, is particularly effective in reducing transmission of vibrations in the audio frequency range. Felt offers a considerable impedance mismatch when in contact with most construction materials.

Cork has been used as a vibration-isolation material for a long time. It is used in pads of varying thicknesses in compression or in a combination of compression and shear. Cork has two distinct disadvantages. As the ambient temperature increases, the effective service life decreases so that at 200 F (93 C) the service life of cork pads is reduced to less than a year. Also, cork is not a very effective isolator in the low-frequency range because great thicknesses are necessary to obtain the large deflections required.

Unless the vibration isolators are selected with care, unsatisfactory levels of noise reduction will result.

Damping

Damping is any mechanism or influence that converts mechanical vibrational energy into a form of heat energy (see Figure 16–16). Once dissipated into heat, the mechanical energy cannot be radiated in the form of noise. Damping should not be confused with resilient isolation where the mechanical energy is accepted, stored, and returned to the vibration system.

Damping can be used to minimize the vibration of panels or other objects. It is important to know what type and method of damping can be used to reduce excessive noise levels.

An accelerometer can be used to determine the vibration frequency of panels and other machine parts. If the panel is vibrating at a much higher amplitude than the machine, damping should be considered. On the other hand, if the panel vibration amplitude is equal to or less than that of the machine, damping material applied to the panel will not help. Damping reduces the vibration level at resonance of the panel only; the driving force is not reduced at all.

Materials. The next problem is to decide what kind of damping ma-

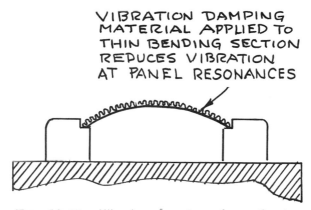

VIBRATION DAMPING
MATERIAL APPLIED TO
THIN BENDING SECTION
REDUCES VIBRATION
AT PANEL RESONANCES

FIG. 16-16.—Effective damping of a vibrating panel can reduce the resonance buildup in the panel. Damping materials should be applied as a layer that is at least equal in thickness to the panel that is being treated.

terial to use. All materials possess the quality of damping to a varying degree. However, some possess it to such a slight degree that additional surface treatment is required to increase those damping capabilities to a level where they are more efficient. This is particularly true of structurally rigid materials (especially metals). There are many commercially available damping materials specifically made for that purpose and many other materials, which, while not intended for that purpose, still can be effectively used for damping if properly applied (see Table 16-A). Damping materials are usually applied to a surface in the form of an unconstrained or constrained damping layer.

The unconstrained damping technique involves applying a layer of the damping material to one or both sides of the vibrating surface. The effectiveness of this damping layer to minimize vibration is proportional to the square of the thickness of the damping material. A rule of thumb used to determine the thickness of the damping layer, for effective sound reduction, is that it must be as thick as the material to which it is applied.

The constrained-layer technique includes an additional rigid stiffening layer over the damping layer, and takes advantage of the resistance to movement or the *shear effect* of the damping layer. The shear effect is

427

TABLE 16-A
DAMPING MATERIALS

Material	Description	Characteristics
Mastic	Asphalt base (cut-back or emulsion type) with varying percentage of solid content.	Semifluid. Effectiveness dependent on temperature. Application should be equal to or twice as thick as plate applied to.
Water-soluble mastic	Nonasphalt base, otherwise the same as the asphalt-base mastic.	Adheres to any clean metal surface, dries quickly and withstands fairly high temperatures. Nonflammable. Can be washed off.
Asphalted felt	Asphalt-impregnated rag, paper, or asbestos felt. Similar to "tar paper" but thicker.	Damping is increased by using multiple construction. Can also be used indented or plain.
Fibrous blankets	Blankets made from glass, cotton, flax, jute, kapok, wood, mineral wool, reclaimed wool, milkweed, etc.	The blanket alone is not very effective in damping vibration. But in conjunction with a loading septum or in constrained layers, it is a good thermal insulator as well as an efficient sound absorbent.
Waterproof crepe paper	Several sheets sewn or spot cemented together.	Is vibration damping and sound absorbent.
Waterproofed pleated paper	Flat sheets of waterproof paper pleated in a number of ways to induce frictional action between pleats during vibration.	Damping capacity dependent on details of manufacturer and application.
Laminated mica	Thin sheets of mica cemented with slow-drying cement.	Effective only while drying (one piece weighing 0.22 lb per square foot produced cay rate of 43 dB per sec. After drying only 3 dB per sec.).
Laminated asbestos paper	Thin sheets of hard asbestos-base paper.	Used by aircraft industry because it is fire resistant.
Loosely felted fiber	Loosely felted reclaimed fibers mostly wool and cotton sewn between sheets of crepe or pleated paper.	Very pliable. Conforms easily to curved surfaces. Damping effectiveness dependent on loaded septum.
Sheet asphalt	Layer of asphalt applied to thick paper and placed against a heated metal surface.	A variation of asphalt-base mastic dampness.

FIG. 16–17.—The silent stock tube for screw machines is a good example of vibration damping and isolation. Mechanical energy is transmitted into the sound absorbent material in the stock tube and converted into heat.

Courtesy Corlett-Turner Co.

the condition or deformation of an elastic body caused by forces that tend to produce an opposite but parallel sliding motion of the body's planes. A relatively thin layer of damping material can be quite effective when used in conjunction with the constrained layer technique.

Another damping technique is the sandwich type with a thin layer of damping material between two identical stiff plates.

Constrained-layer damping is used on aircraft because it conserves weight. Inside aircraft cabins, the main source of sound is the pressure fluctuation against the metal skin of the aircraft caused by external aerodynamic forces. The pressure fluctuation causes the metal skin to vibrate which produces sound inside the aircraft cabin. To damp the vibration of the metal airplane skin, aluminum foil with a viscoelastic layer is applied to the metal skin. When the airplane's metal skin moves, the viscoelastic layer is forced to work in shear, and this shearing action dissipates sound energy.

Isolation

Hydraulic systems can generate high noise levels due to pulsations in the fluid and structure-borne vibration of pumps. A good muffler for a hydraulic system is a flexible hose or an accumulator inserted in the line at the pump discharge. This flexible hose can then expand and contract in response to the hydraulic pulsations. The hose diameter is small compared to the wavelength of the sound being generated by pressure fluctuations in the hydraulic fluid, and the hose, therefore, does not radiate efficiently. To prove this, hold a piece of cardboard or plywood against the hose; more sound will be heard because the vibrations of the hose are amplified by the larger radiating surface.

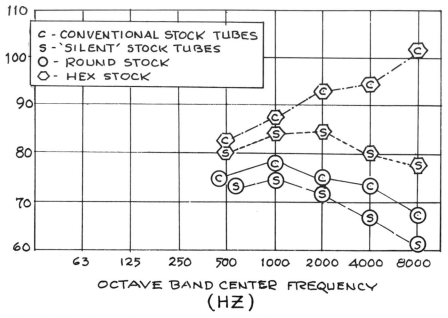

FIG. 16–18.—Comparison of noise levels 5 feet from an automatic screw machine using conventional and silent stock tubes with round and hexagonal bar stock.

Stock tubes

Automatic screw machines use a metal stock tube to keep the metal stock aligned and to prevent it from whipping about uncontrollably. However, as the stock rotates at high speed, a considerable amount of high-frequency noise is created by the metal feed stock hitting the sides of the stock tube.

A commercially available noise control device called a silent stock tube (see Figure 16–17) can reduce the sound level as much as 10 to 20 dBA. The felt inner layer of the silent stock tube attenuates sound energy by:

- Absorption through the open coils of the helical metal strip

- Isolation of the inner metal strip from the outer shell

- Damping of the resonance of the outer shell

Figure 16–18 compares the noise levels of the silent stock tube and the conventional stock tube, using both round and hexagonal stock. Regard-

BARRIERS

BAFFLES GIVE LITTLE SOUND REDUCTION

PARTIAL ENCLOSURES ARE BETTER

AIRTIGHT ENCLOSURES ARE BEST

Fig. 16–19.—Comparison of noise reduction characteristics of baffles, partial enclosures, and total enclosures. A barrier becomes effective when its dimensions are three wavelengths or greater. A partial enclosure may provide 3 to 15 dBA noise reduction. A total airtight enclosure may provide 10 to 50 dBA noise reduction.

less of the type of stock used, the noise levels, particularly at the high frequencies, are considerably less with the silent stock tube.

Enclosures

Sometimes it is not practical or feasible to reduce the sound level at the source. An enclosure or barrier may be used to provide an alternate method of reducing the sound level (see Figure 16–19).

When an enclosure is placed over a sound source, it must be well isolated to minimize transmission of structure-borne sound (vibrations) from the source. The enclosure, being larger, may radiate sound energy more efficiently than the smaller uncovered noise source. It is possible to have an increase in the measured sound level when a machine is en-

431

closed or covered if the structure-borne transmission paths have not been adequately treated.

Acoustic Shields

A wall or barrier can produce a shadow effect. High-frequency sound waves hitting a barrier that is large, compared to the wavelength of that sound, will cast an acoustic shadow, but low-frequency sound waves will flow around the barrier almost as if it were not there.

For instance, a 5 × 7 ft panel would be worthless as a shield for 100 Hz

$\frac{1}{4}"$ X 24" X 48" AUTO SAFETY GLASS

From Industrial Noise Manual, 2nd ed., *American Industrial Hygiene Association.*

FIG. 16–20.—A safety glass shield or rigid plastic barrier between a noise source and the press operator is effective at close range if the noise is predominately high frequency.

sound waves since the wavelength at this frequency is approximately 11 ft. In Figure 16–20, a partial shield or barrier measuring 2 × 4 ft gives a 15 dB noise reduction of the higher frequency sound waves to the operator at the stamping machine. Whether the shield is of safety glass or strong clear plastic is not important. If the shield were made larger, perhaps 4 × 4 ft, it would still not be very effective in controlling the low-frequency noise.

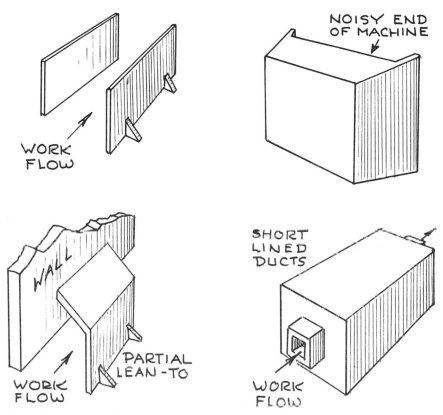

WORK FLOW

NOISY END OF MACHINE

WALL

WORK FLOW

PARTIAL LEAN-TO

SHORT LINED DUCTS

WORK FLOW

FIG. 16–21.—Partial enclosures or barriers with the sound absorptive treatment facing the noise source can provide 10 to 20 dBA noise reduction.

Partial Enclosures

Various partial enclosures (shown in Figure 16–21) can be applied to many noise sources to reduce excessive sound levels. The equipment and process within the enclosure are not visible unless the enclosure is made from a transparent material or viewing ports are placed in the enclosure.

A very noisy machine could be placed inside an enclosure with the work fed in one end and the scrap and the finished product taken out the other side. If necessary to check the process, windows can be placed in the enclosure or if the enclosure is large enough, a door can be used for access. Some enclosures envelop the noise source, while others surround the worker to protect him against excessive noise exposure.

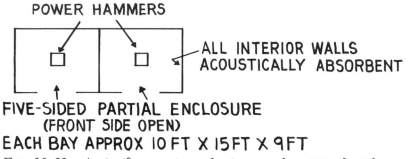

FIG. 16–22.—A significant noise reduction can be attained with an acoustically absorbent lined partial enclosure.

Chipping booths

Many foundries have chipping booths where the scale is removed from castings. An example of two large chipping booths with an open top and partially open side (in order to receive the work) is shown in Figure 16–22. The booths in this case are designed to protect each worker from the sound generated by adjacent work operations. The walls of the chipping booth prevent noise from getting out as well as from getting in. Sound measurements show a 20 dB reduction within the chipping booth due to work operations conducted in adjacent areas. The worker within the booth is able to rest his ears while his neighbor is working on his casting in the adjacent booth. It has been determined that intermittent

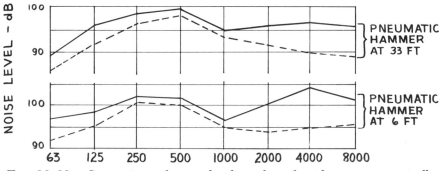

FIG. 16–23.—Comparison of noise levels with and without an acoustically absorbent curtain surrounding a noise source at 6 and 33 feet.

Fig. 16–24.—Vibrations can be transmitted by solid contact between source and floor or enclosure wall.

noise exposure is less hazardous to hearing than is continued exposure. Actually, these booths are of no value in reducing the employee's noise exposure as a result of his own chipping operations within the booth. In fact, unless the booth has highly absorptive walls, the noise level may increase due to reflections.

Caution should be exercised when putting a noise barrier or curtain around a noise source. Figure 16–23 shows measured sound levels of a pneumatic hammer at distances of 6 and 33 ft with and without a curtain. The curtain is not sound absorptive so the sound level is probably higher at the operator's position.

In putting an enclosure around a source and adding absorptive material, a significant degree of noise reduction can be obtained, however,

Fig. 16–25.—Heavy acoustically treated walls may provide great noise reduction; however, a small leak or opening may limit the noise reduction to less than 10 dB.

435

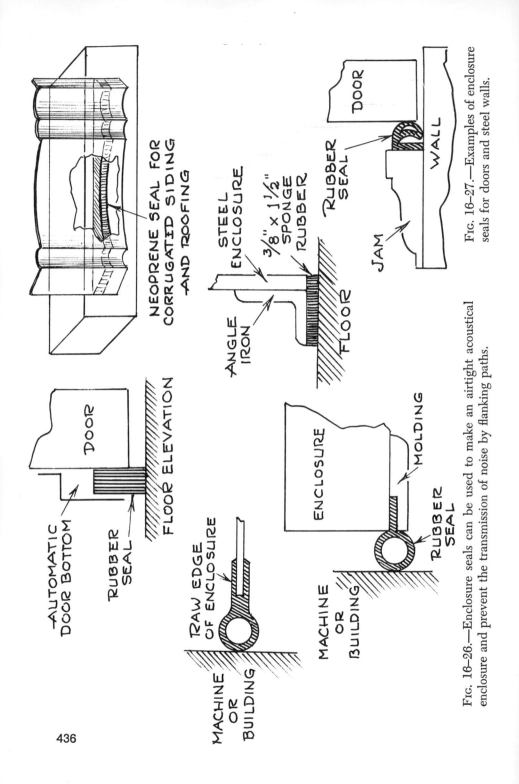

NEOPRENE SEAL FOR CORRUGATED SIDING AND ROOFING

STEEL ENCLOSURE

3/8" x 1 1/2" SPONGE RUBBER

ANGLE IRON

FLOOR

DOOR

RUBBER SEAL

JAM

WALL

Fig. 16-27.—Examples of enclosure seals for doors and steel walls.

DOOR

AUTOMATIC DOOR BOTTOM

RUBBER SEAL

FLOOR ELEVATION

RAW EDGE OF ENCLOSURE

MACHINE OR BUILDING

ENCLOSURE

MOLDING

RUBBER SEAL

MACHINE OR BUILDING

Fig. 16-26.—Enclosure seals can be used to make an airtight acoustical enclosure and prevent the transmission of noise by flanking paths.

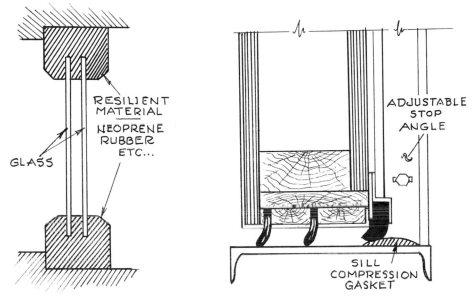

Fig. 16–28.—Enclosure seals for window panes and doors.

only a small degree of noise reduction may occur because of structure-borne transmission to other vibrating surfaces (see Figure 16–24). Proper isolation of the noise source within the enclosure is required to reduce structure-borne noise.

Acoustic leaks

When installing an enclosure, make sure it is acoustically tight so that there are no leaks, since a slight opening in an otherwise acoustically tight enclosure will decrease its effectiveness. A small opening in the enclosure results in the sound pressure level, at the opening, being equal to that inside the enclosure. Even an opening as small as 1/32 in. around a steel door of an enclosure is significant (see Figure 16–25).

Check everywhere for acoustical leaks—floor, ceiling, seams, joints, and any service openings into the enclosure or barrier, such as windows, doors, pipes, or tubing. In most cases, seals and gaskets can be used to close off any openings (see Figures 16–26, 27, and 28). While the choice of the material to be used can vary depending upon the purpose, it is important to remember that most good acoustical absorbents are poor sound barriers.

Fig. 16–29.—Application of the same sound absorptive treatment to the exterior surface of the enclosure instead would be of little value.

Absorption

Sound absorptive material does not make an effective enclosure because it is porous. To be effective, an enclosure must be airtight in order to keep the sound in or out, as the case may be. Not only is the type of the material used for the lining of great importance, but also where and how it is placed.

There have been cases where, in an attempt to attenuate noise, an enclosure was placed around the source and the sound absorptive material

Fig. 16–30.—The sound absorptive material should be placed on the inside of the enclosure to minimize the reverberant build up of sound within the enclosure.

placed on the exterior of the enclosure (see Figure 16–29). Minor noise reduction may occur as a result of the resonances in the panel structures being damped by the sound absorptive treatment. However, to achieve the greatest noise reduction possible, under these circumstances, the sound absorptive material should be placed on the interior of the enclosure so that it will not only damp the resonances in the panels, but also minimize reflections and absorb the reverberant sound buildup in the enclosure as well (see Figure 16–30).

In one example, a tumbler enclosure provided a 10 dB noise reduction by putting glass fiber acoustical insulation on the inside walls of the enclosure. The glass fiber effectively reduced the reverberant field within the enclosure.

Pelletizing operations generally produce high noise levels. A partial enclosure can be effective if a greater portion of the device is enclosed. The plastic feed stock, however, is also a noise radiating surface. The same problem applies to the sheet steel feed used in power press operations. If rubber-lined clamps hold the sheet stock down rigidly to the feed bed, the resonances of the feed material are damped and less noise is radiated.

Walls

Walls of various types and materials are used not only as room dividers and separators, but as a means of containing sound waves—either in or out. The most common wall is the single wall of solid, homogeneous construction, such as sheet metal, plywood, leaded vinyl, brick, concrete, cinder block, or solid plaster. Regardless of the material used, the effectiveness of this wall as a sound barrier is determined by the mass per unit area of the wall. In other words, the heavier the wall material per square foot, the better it acts to prevent sound transmission. There is an increase of about 6 dB in the transmission loss (TL) or noise reduction effectiveness for each doubling of the mass per unit area and/or doubling the frequency.

A one pound-per-square-foot (1 psf) wall (22 gage steel) would produce a 10 dB transmission loss at 125 Hz, and the transmission loss would go up 6 dB per octave so that at 1000 Hz it would be 18 dB more, or a total of 28 dB of noise reduction.

The wall should be impervious to airflow to be effective as a sound barrier. For instance, an unpainted cinder block wall is a poor sound barrier unless the pores are sealed with a rubber-base paint or other similar sealing product. It is obvious that there are limits to increasing

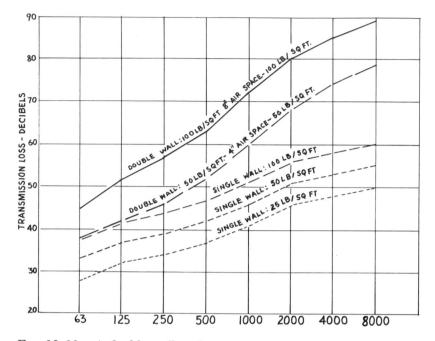

FIG. 16–31.—A double wall enclosure can provide improved noise reduction over a single wall enclosure of the same total weight.

the mass. Not only does the cost increase, but above about 40 dB (TL),
it becomes more economical to have two walls instead of one.

Double walls

A double wall has an air space separating the two walls which increases its effectiveness as a sound barrier. As the air space between the
two walls increases, the effectiveness of the double walls as a sound
barrier also increases. The transmission loss for double-wall construction
increases about 6 dB with each doubling of the air space (see Figure
16–31).

The importance of double walls becomes apparent when comparing
the effectiveness of a single wall with a given mass per square foot, and
two walls each having one-half the mass per square foot but separated
by an air space. The total weight of both the single and double wall is
the same but the transmission loss of the double wall is greater (see
Figure 16–32).

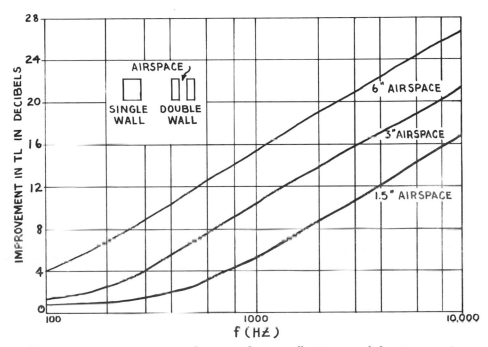

Fig. 16–32.—A comparison of noise reduction effectiveness of the air space in double walls.

Double-wall enclosures start becoming effective approximately one-half octave above the resonant frequency of the double wall system. The resonant frequency (f_o) of the double wall system equals

$$f_o = \frac{170}{\sqrt{m \times d}}$$

m = the mass of one wall times the mass of the other wall divided
 by the sum of the mass of the two walls (mass in lb/sq ft).
d = the dimension of the air space between the two walls in inches.

Double walls are only effective above their double-wall resonant frequency which must be determined for each type of wall construction. If 16 gage steel is used (2.5 lb per sq ft) and there is a 1 in. air space between the plates, the double wall resonant frequency occurs at about 150 Hz.

Sound reflection from walls and ceiling

One of the major areas where noise control measures can be effectively

441

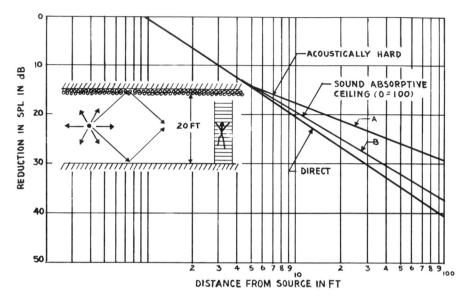

FIG. 16–33.—Noise level between floor and ceiling in a "two dimensional" room as a function of distance; measurements were taken at 1, 20, and 200 ft from point source. Sound absorptive treatment applied to the ceiling in a typical shop space with a ceiling height of 20 ft can reduce the noise 3 dB at 20 ft and 10 dB at 200 ft, respectively.

applied is the transmission path. In working on reducing sound reflections, the direct transmission path of the sound wave from the source to the receiver is all too often disregarded. Attempts are usually made to reduce, substantially, the reflection of sound waves from the walls and ceiling with little or no benefit for the machine operator close to the noise source.

The ceiling and the walls can be treated acoustically to reduce the reflection and increase the absorption in order to lower the overall sound pressure level. Figure 16–33 illustrates the sound pressure level under three conditions:

- An acoustically hard ceiling

- A sound absorptive ceiling

- Direct measurement as if there were no ceiling

442

A glance at the graph (Figure 16–33) shows that at a distance of 20 ft, the difference in the measured sound level between the hard reflective surface and the absorptive ceiling is about 4 dB. Even at 100 ft, there is only about 9 dB difference. Consequently, before spending a large amount of funds on sound absorptive treatment for the ceiling, it is a good idea to study the problem closely.

Noise reduction resulting from acoustically treating the ceiling becomes meaningful when the distance of the receiver from the sound source is at least equal to the distance from the floor to the ceiling, which, in this case is 20 ft. Where would this kind of treatment be used? In any area where there are many identical noise sources grouped together, treating the ceiling with acoustical materials could bring the sound level of the whole room down as much as 5 to 8 dBA.

Baffles

Hanging acoustical baffles to absorb sound energy is another form of ceiling treatment. The baffle area should approximately equal the ceiling area to be as effective as horizontal ceiling treatment. Baffles are often used in food processing plants because they can be wrapped with a thin plastic film and can be scrubbed and washed with a hose. If the baffles are placed closer to the noise source, they are more effective. The baffles should be at least down to the lighting level.

Summary

When confronted with a noise problem, it is well to understand how that sound wave is generated. Only by knowing what sound is and how it behaves can noise control problems be dealt with effectively. Solving noise problems is essentially a matter of understanding the problem and applying known acoustical principles to control the noise problem.

Noise is controlled in one of three ways: (a) at the source, (b) along the transmission path, or (c) at the point of reception (or frequently a combination of any or all). This chapter concerns itself with only the first two.

At the source, the noise producing structure can be altered to produce less noise by reducing the vibrating area, damping the vibrating area, reducing the velocity of flow (air, gas, or liquid), and reducing the vibrating force transmitted by using vibration isolators.

Partial enclosures having minimum dimensions greater than the wavelength of sound of interest and absorptive treatment facing the noise source can be used to provide 10 to 20 dB of noise reduction. Total enclosures can be designed to provide substantial noise reduction.

Absorptive treatment in a large work area can reduce the reverberant noise level in the space but is of no benefit in reducing the direct, line-of-sight noise emanating from the machine to the operator.

Chapter Seventeen

Application of Engineering Noise Control Measures

By William Ihde, P.E.

Excessive industrial noise can be controlled, generally, by engineering changes, administrative measures, or personal hearing-protective devices. Engineering changes for noise control can be applied at the noise source, along the transmission path, and at the recipient. This chapter deals primarily with the application of engineering noise control measures to solve noise problems.

The ultimate goal or purpose for noise control should be determined for each specific case. Is the noise exposure such that the hearing of workers in the area is endangered? Are OSHAct regulations violated? Or, is it merely a noise, which, while not necessarily harmful to hearing, may nonetheless be annoying or interfere with speech communication? More likely than not, the economic justification for instituting noise control measures will be a combination of these reasons.

The following general approach can be used in solving noise problems.

Sound level survey

Recognizing a noise problem is easy. Determining proper corrective action is more difficult because it depends on a detailed sound survey to provide sufficient data upon which to gage the type and extent of the noise problem. The first step, therefore, is to perform a detailed sound level survey.

This survey involves more than just walking around with a sound level meter and recording one's readings; it may at times require a third or

TABLE 17-A
CAUSES OF NOISE AND VIBRATION

- **Machinery Changes**
  Wear
  Erosion
  Corrosion

- **Aging of Equipment Components**
  Crystallization and fatigue
  Loss of adhesion or bonding
  Increased tolerances

- Inelastic Behavior
  Parts stressed out of shape
  Bent parts
  Loosening of fastenings
  Broken or damaged parts
  Incorrect or inadequate lubrication
  Solidifying of grease or packing

- **Foreign Matter in Bearings or Other Critical Areas**
  Dirt, chips, dust, grit
  Contaminants
  Ice accumulation
  Paint and other finishes

- **Environmental Changes**
  Temperature
  Humidity
  Pressure

Adapted from Handbook of Noise Measurement, *General Radio Co., Concord Mass.,* 1972.

tenth octave-band analysis (perhaps in even narrower bands if the problem is difficult to pinpoint). The survey should be as thorough as possible in order to provide the necessary data as a basis for further action. The measured noise levels are compared with acceptable levels and the difference is the amount of noise reduction necessary.

Determine the cause. Pinpoint the principal sound or vibration source. Frequently, the major noise source may not be obvious. Nothing should be taken for granted—make certain that the prime source of the noise has been found. Sometimes, a detailed sound level survey will reveal that the origin of the major noise source is completely different than originally thought.

Engineering studies

After the cause and source of the noise have been determined, engineering studies should be performed to evaluate methods of reducing the existing noise levels. The control method selected usually will (*a*) reduce or eliminate the noise level at the source, (*b*) modify the transmission path of the noise between the source and the employees' work station, or (*c*) lower the noise level at the point of reception in the employee's hearing zone.

Depending, of course, on the magnitude and extent of the noise problem, the control method may be simple or complex. The noise control measure may involve just calling in the millwrights to move the offending piece of noisy machinery to a new location, having carpenters build an acoustically lined box or baffles surrounding the noise source, or perhaps designing and building a sound trap. Whatever method or means is selected, it must do the job properly.

Follow-up sound survey

It should never be assumed that the installation of new noise control measures will "automatically" perform their intended function. A subsequent noise level survey is recommended to test the effectiveness of the control methods. In addition, periodic checks should be performed to make sure that the noise control measures are operating properly. These sound surveys can be used to demonstrate the effectiveness of the noise control measures to others—particularly management who often need graphic proof to show that the money was well spent.

Maintenance

Experience has shown that the application of logical engineering principles (in conjunction with general maintenance) can eliminate or reduce many noises to acceptable levels. Once the noise problem has been identified and analyzed, an orderly process of treatment and reduction can be initiated (see Table 17-A).

Examples of common problems that can be rectified by adequate maintenance are those involving the discharge of compressed air. These noise problems, however, are frequently ignored because they do not normally interfere with production. Reducing noise levels caused by the unnecessary release of compressed air can frequently be accomplished by

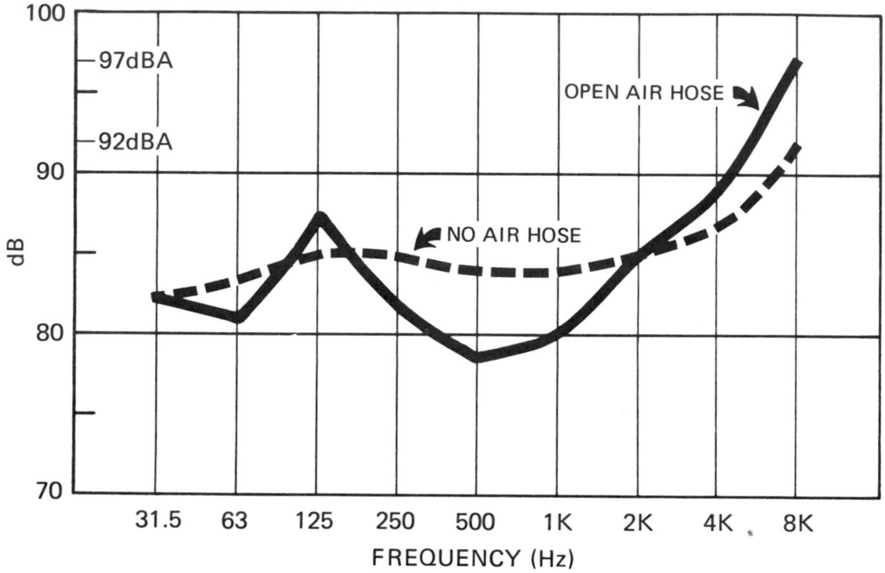

Fig. 17–1.—An octave-band analysis comparing the noise levels at a large press with and without a compressed air hose operating.

capping unused outlets and repairing leaky air valves and hoses.

In addition to preventive maintenance, corrective maintenance is equally important. This includes minor repairs, changes in plant layout, and other maintenance operations. One of the greatest advantages of corrective maintenance is that it can prevent a minor problem from becoming a major one. (In this sense, it is also preventive maintenance.)

The following case histories illustrate the application of maintenance measures in the solution of noise problems.

Combustion noise. A noise survey in a foundry recorded a noise level of 94 dBA at one of the pouring lines located behind some core ovens. The noise was coming from some burners on top of the ovens. Further examination showed that some of the burners were more quiet than others and that when all the burners were turned off, the noise level dropped to 88 dBA. At first, it was assumed that the noise was an inherent characteristic of that type of burner and an acoustic barrier would have to be built in order to protect the workmen at the pouring line.

The noise problem was discussed with the foreman of the department,

448

and it was learned that the burners were purposely set out of adjustment so that the combustion noise could be heard. The burners were operated at a low-pitched roar because the absence of this noise would indicate a flameout.

The problem was solved by installing an electronic flame detector, which not only sounded an alarm but turned the gas off when conditions for a flameout existed. The burners could then be adjusted to operate quietly and the noise level was reduced to acceptable levels.

Compressed-air water traps. A compressed-air hose line, located above and in front of a grinder operator, had a water trap to remove moisture from the factory air supply. During a break period, the sound level was measured at the operator's station and found to be 89 dBA with the water trap bleed-off valve open, but only 76 dBA with the valve closed. Closer checking showed that the vent of the water trap bleed-off valve was partially open. When the operator was asked why the valve was left open, he replied that it was to prevent water from accumulating in the air lines and slowing his grinding tool. The solution was to require the operator to keep the bleed-off valve closed and to bleed off the water at proper intervals. Close supervision was necessary to make certain that the water trap was vented periodically and that the valve was kept closed.

Compressed-air discharge. Another air discharge problem was encountered near a large press where the sound level reached 97 dBA at the operator's station (see Figure 17–1). Close inspection of the area disclosed an open pipe discharging compressed air on the floor at the operator's feet—but concealed behind the foot panel of the press. The operator was asked the purpose of the air blast, and he replied that he thought it was to keep the chips away from his feet. Apparently, the compressed-air discharge vent had been used on a previous operation and was no longer needed. When the press was operated with the compressed air discharge turned off, the noise level fell from 97 dBA to 92 dBA.

Removal of inspection covers. In another case, four floor plates near a return sand elevator in a foundry had been removed to allow for periodic cleaning and maintenance. Although there was a guardrail around the opening to prevent lift trucks from driving in this area, the large floor opening could easily have been stepped into by some person passing by. In addition to the safety hazard, a 103 dBA noise was coming from the open pit. Replacing the plates over the hole brought an immediate 6 dBA reduction to 97 dBA.

Remember, noise control in a complex situation is the control of the various individual noise sources, so that the total noise reduction is the sum of the reduction of the various individual noise sources.

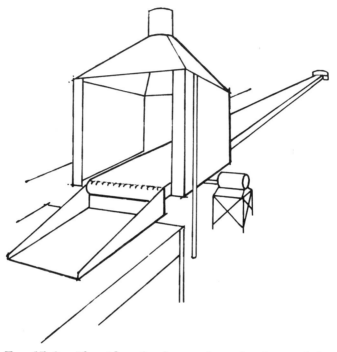

Fig. 17–2.—The sides of a large exhaust hood provided an excellent radiator, particularly at the lower frequencies. After damping, the noise was substantially reduced.

Vibrating hood. In another foundry, a conveyor carrying castings was covered by a galvanized steel hood with an exhaust fan (see Figure 17–2). The noise emanating from that area was broadband, and, when an octave-band analysis was made, it was noted that some of the frequencies were surprisingly low. The reason for the low frequencies was the large size of the hood, the sides of which provided excellent radiating surfaces. The solution to this noise problem was easy to carry out.

Suitable damping material was applied to the hood panels, and the whole hood structure was stiffened to prevent excessive vibration. Previously, the panels (just hanging from the frame) had vibrated like huge free-hanging diaphragms. The point to remember is to prevent noise problems *in the design stage* rather than to create them by providing structures that are efficient noise radiators.

Blowdown valve. A frequent problem in an industrial plant is how to discharge waste compressed air quietly. In one case, an air hoist picked up a flask from one conveyor line, rose about 4 ft, and deposited it on another conveyor. The hoist returned to its former position by gravity after the air ram, which raised the lift, was released by an automatic blowdown valve. The blowdown took approximately one second and the noise level during that time was 122 dBA. Because the air blast was repeated at approximately two-minute intervals for a duration of one second, the noise was considered continuous. (This noise exposure is well in excess of the allowable at 115 dBA.) The solution to this problem was to attach a long hose to the blowdown valve nozzle and release the waste air into a covered hole 6 ft deep and 3 ft in diameter, located in an unoccupied area of the plant.

Noise control at the source

The most effective means of solving a noise problem is to reduce the noise at the source. Consideration should be given to substitution, converting to a different type of operation, or changing the process. Welding or assembly with structural fasteners or adhesive bonding may be used to join parts in place of riveting in metal fabrication operations.

Substituting similar type devices, such as a low-rpm fan with many blades for a high-rpm, two-bladed fan, may help solve fan noise problems. Substituting quieter for noisy equipment may not always be practical, but the need for a quieter machine should be kept in mind when replacing existing production equipment. The permitted noise levels should be specified in the purchase order; many manufacturers now include or make readily available noise data with their equipment specifications.

Modifying the noise source

If practical, modifications should be made directly on the noise source. To reduce the noise level, usually, the most effective effort is to decrease the radiated sound power. This usually involves a reduction in vibration amplitudes of the vibrating parts. Modification of the noise source can be performed by acting on three major areas:

• Decreasing the driving force acting on the vibrating system

• Changing the coupling between the driving force and the radiating system

• Changing or reducing the noise radiating area so that less energy is radiated

In each case, a detailed sound survey would be helpful in determining the major source of noise and the path of transmission. Observing the effects of changes in speed, structure, and mounting, and other variables on the noise spectrum should also help in finding the important noise producing elements.

Decrease in force. The reduction of the driving force that produces

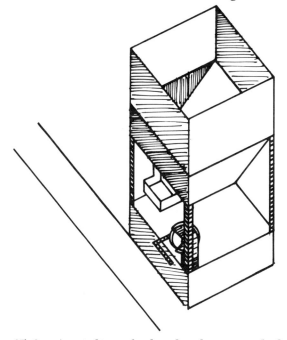

Fig. 17–3.—A grinding wheel pedestal was attached directly to the steel framework which supported a parts hopper. When the grinding wheel was operated, the vibrations of the steel hopper panels radiated noise throughout the whole area.

the vibration is a very important and complex aspect of noise reduction, and is discussed later in this chapter in the section on the Nature of vibration.

Change in coupling. A change in the coupling between the noise source and the radiating system may include the use of vibration isolators. The rigidity of the structural members which transmit the vibration may have to be decreased or increased. Sometimes, a general tightening or fastening of vibrating parts to rigid frame members may be required. Resonant structures are often difficult to control because the resonance may be either in the mechanical structure or in an air chamber. In either case, it is possible to shift the resonance by changing the structure or by adding absorbing material to damp the resonance.

The following example illustrates how a noise problem was solved by changing the coupling between the noise source and the radiating system.

Four grinding stations, where castings were deburred, generated a great deal of noise. The castings were stored in a tote-box hopper built into each work station. The grinding wheel was mounted on a frame, which also served as a stand for the tote boxes (see Figure 17–3). During the grinding operation, a body-press handle was used to force the casting against the wheel as the operator manipulated the casting with his hands. As the casting contacted the grinding wheel, the noise level increased sharply. The tote-box hopper (with high sides and broad panels) radiated excessive amounts of noise into the surrounding work area.

The solution to this noise problem was to mount the grinding wheel frame separately from the tote-box hopper stand. The handle for the body press was isolated from contact with the large frame, which had many radiating surfaces.

Changing the radiating area

An example of changing the radiating area (damping), as a solution to a noise problem is illustrated in the following case.

Residents living adjacent to a paper mill complained of excessive noises emanating from the mill at all hours of the day and night. In response to this community noise problem, management initiated an investigation and a sound survey to locate the offending noises. It was found that during the process of making paper boxes, the trimmed corrugated paper scrap was scooped into floor sweeps discharging into duct conveying systems that sent the scrap material rattling through a 14

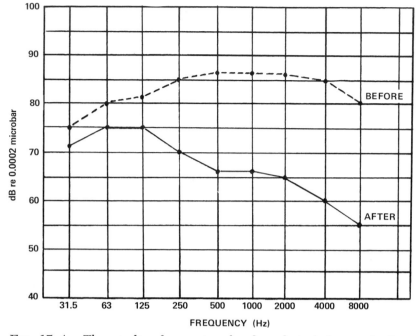

Fɪɢ. 17–4.—The results of an octave-band analysis before and after applying damping material to air-line conveying ducts. Note that a considerable reduction in noise level occurred in the higher frequencies.

in. line for about 200 yards across the plant roof and into a cyclone collector. Initially, it seemed the noise was most likely coming from the cyclone whose blower was humming away at a steady 175 Hz. However, at a meeting with the neighboring residents it was established that the objectionable noise sounded like hammers banging against sheet metal. This noise was then traced to the scrap corrugated paper moving through the air-line conveying system.

To solve this problem, it was theorized that to attenuate the noise, the metal walls of the air-line conveying duct could be allowed to vibrate, but the outside radiating surface must remain immobile in order to prevent propagation of noise from the duct to the surrounding area. To isolate the outside surface, a spacer of inexpensive rock wool building insulation material was used to wrap the lines. Over this insulation material, two layers of tar paper were applied, using a 50 percent overlap. Since high temperatures were not involved, the fire risk was minimal.

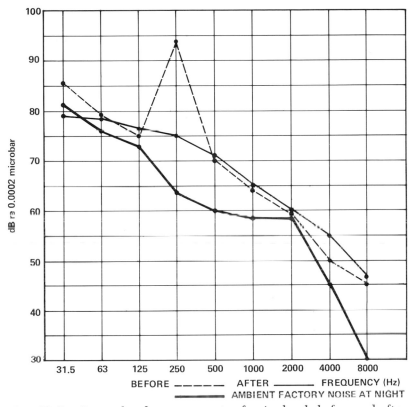

Fig. 17–5.—Octave-band measurements of noise levels before and after a detuning operation by inserting a stub into the air-line conveying duct.

The damping treatment was very satisfactory, resulting in a 15 dB to 25 dB reduction (see Figure 17–4). Prior to the damping treatment of the air-line conveying ducts, the noise of scrap paper rattling through the ducts could be heard at least four blocks away; after the damping material was applied, the noise could hardly be detected when standing next to the wrapped ducts.

Tuning

The following example illustrates tuning stub techniques.

A woodworking factory was using a ducted air pickup system to remove sawdust from its operations. The system had a 12-blade radial

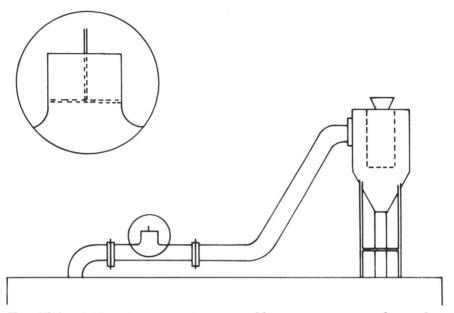

Fɪɢ. 17–6.—A T-section, containing a movable piston, was inserted into the air line. It was adjusted up or down for maximum cancellation of a noise consisting of a single predominant frequency from the cyclone blower.

fan, turning at 1440 rpm, producing a strong noise component at 288 Hz.

The ducts, carrying sawdust, went through and then across the roof to a cyclone air cleaner mounted on a building in which men worked. The noise level in this building was 94 dBA and it was caused primarily by the pure tone noise coming from the cyclone (see Figure 17–5). It was decided to use a piston in a T-section in the piping to reduce the pure tone sound being emitted by the fan. It was necessary, first, to determine whether there was a standing wave in the duct and, if so, to locate its point of maximum amplitude. A 1/10-octave band analyzer was used to survey the sound levels at various points along the duct on the roof. As expected, very definite positions of minimum and maximum sound pressure level were located. A T-section with a movable piston was inserted into the duct that had the highest sound pressure level reading immediately adjacent to the blower (see Figure 17–6).

The piston was inserted or withdrawn so as to cancel the single-frequency noise propagating down the duct. This was indicated by a minimum meter reading on the octave-band analyzer. Only one piston

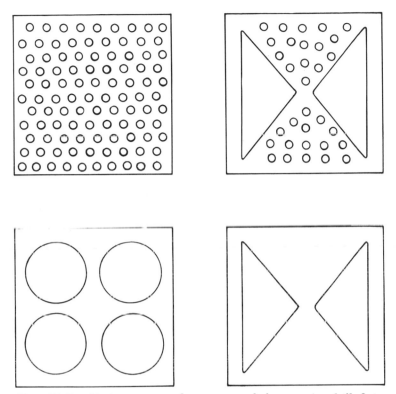

Fig. 17-7.—Various types of cutouts or holes can be drilled in panels to provide less surface area to push against the surrounding air. Stiffening or adding damping material to the back of the panel can also reduce the panel's radiation noise.

was used resulting in a noise reduction of 15 dB, and the level in the building was reduced to about 80 dBA.

Tuning out noise by reflecting waves works best with noise that is predominantly a pure tone or concentrated in a narrow frequency band. It would be impractical to attempt to control noise using tuning stubs when the noise is composed of many different frequencies.

Change in radiating structure

Changing the noise radiating structure to reduce its acoustic radiating efficiency often involves reducing the external surface areas as much as possible. For a device to be an efficient noise radiator, it must have a minimum dimension (length or width) of at least one wavelength at the lowest frequency of interest. If the dimensions are less than one wavelength, the efficiency of that surface to radiate sound is reduced. For instance, if a noise source is a large panel, an octave-band analysis should be made to determine the predominant frequencies present and their corresponding wavelengths. The larger the panel, the more efficiently it will radiate noise.

As a possible corrective procedure, holes can be drilled in the panel, so that, as it vibrates, there will be less surface area to push against the air to generate noise (see Figure 17–7). Stiffening the panel or adding damping material to the back of the panel will frequently lower the noise level. A great variety of damping materials for this purpose are now commercially available.

Directivity patterns

Still another way of modifying the source to decrease noise levels is to change the directivity pattern of the radiated sound waves.

When jet streams of compressed air or other gases are discharged at high pressure from a small-diameter opening, high-frequency sound waves are radiated outward in highly directional patterns. If the direction of air or gas flow is changed by pointing the exit nozzle in a different direction, the noise pattern can be shifted. By changing the nozzle direction in this way, the noise levels in certain directions are considerably reduced.

If the air velocity at the nozzle or discharge end is reduced, the noise generated by the jet discharge will also be reduced. Enlarging the size of the nozzle or aperture reduces the exit air velocity and consequently the noise.

It is important to consider the device or mechanism using the compressed air when investigating the problem. For example, in a machine tool that uses an air-ejection device, every effort should be made to position the air-ejection nozzle properly, so that the finished part will be ejected with the minimum amount of air. Two examples of the high-velocity discharge of compressed air creating a noise problem follow.

1. Defective air valve. A detailed sound survey in an industrial plant revealed a number of noise problems caused by compressed air discharges. An investigation showed that there were a number of punch presses set up to use a cam-operated jet blast of air to eject the finished

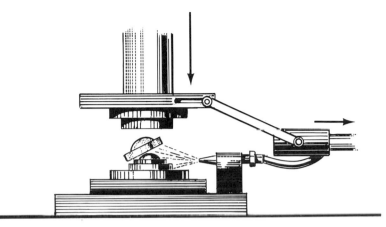

FIG. 17–8.—A cam-operated compressed air valve was used to eject finished parts from the power press. A malfunction kept the air jet on constantly, creating a loud noise.

piece of material from the die (see Figure 17–8). However, the compressed air jet was not turned on and off as needed by the cam-operated air valve. The jet was discharging air continuously.

The solution here was obvious—release the compressed air only when needed—but because production had not been affected, no attempt had been made to reduce the waste discharge of compressed air and the resulting excessive noise levels. When management was made aware of the simple solution to the noise problem, a maintenance crew was sent to make the necessary repairs on the defective cam-operated air valve.

2. Hose coupling leak. During a detailed sound survey, an unattended power press with high noise levels was observed. While there was considerable noise in the area from many power presses, it was noted that a loud shrill sound, due to discharge of compressed air, was coming from the unattended power press. Measurements revealed a noise level in this vicinity of approximately 94 dBA and subsequent investigation showed that an air hose coupling was leaking (see Figure 17–9).

When the hose coupling was properly connected, the noise level in this area was reduced from 94 to 83 dBA. (High velocity air coming from a small aperture has frequencies of 4000 to 8000 Hz.)

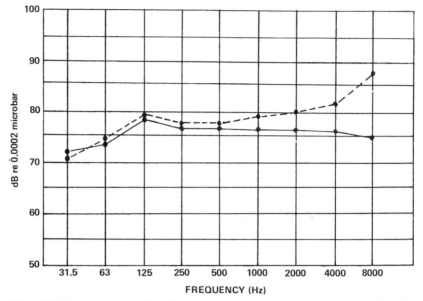

FIG. 17–9.—An octave-band measurement comparing the noise levels before and after a leaking air hose coupling was repaired.

Control of transmission path of sound

Sometimes reducing the noise level at the source is not possible or practical. However, other means can be used to prevent the noise from reaching the ear of the listener by working on the transmission path of the noise at some point between the source and the listener. Attenuation of the noise level may be attained by changing the operator's position relative to the noise source, using sound absorbing materials to change the acoustic environment, or erecting sound barriers or enclosures around the noise source.

Change in position

Reducing the noise level at the listener's ear can be done by increasing the distance between the source and the observer. This can usually be accomplished by moving either the listener or the source (or both), whichever is more practical. If possible, the directionality of the source can be changed to decrease the noise level in a specific area. Or, the listener may change his position with relation to the noise source without necessarily increasing the distance. However, the practicality of these measures depends heavily on the nature of the job of the person being exposed to the noise. Also, the procedures are effective only where approximate free-field conditions exist.

Acoustic materials

Acoustic materials are generally used to counteract sound reflection from hard interior surfaces, and, as such, are of great value in noise reduction. In changing the acoustic environment of a room, a wide variety of sound absorbing material is available. This material can be used to cover the hard, sound reflective surfaces of walls and ceilings. Sometimes, panels of acoustic material are suspended from the ceiling. The principal value of acoustic materials in a room is to reduce the noise level some distance from the source. Consequently, a person working 2 ft away from a loud noise source—as frequently occurs in a factory—gains little benefit from an acoustically treated wall 50 ft away.

Barriers and enclosures

The noise level can be reduced by interposing a structure to interfere with the transmission of sound between the noise source and the observer or point of reception. Earplugs and muffs can be considered to be such attenuating structures. However, of primary concern here are barriers and enclosures. Depending upon the requirements, almost any degree of noise reduction is possible. However, as the degree of required reduction increases, costs and complexity do likewise.

A barrier is a partial enclosure that does not totally contain the noise but reduces its intensity in certain directions. A partial barrier can be used either to reflect the offending noise back towards the source, or to produce an acoustic shadow in the region of interest.

The configuration of partial enclosures may vary almost as much as the acoustic characteristics of each. Naturally, the shape of the enclosure will depend upon the noise source itself.

A partial enclosure, while reflecting a certain amount of noise back within its confines, nonetheless changes the overall pattern of the noise radiation so that higher noise levels may be produced at the open end. However, this effect can be minimized by covering the interior walls of the barrier with acoustic absorption material.

A complete enclosure, obviously, provides the greatest reduction in noise levels outside the enclosure. However, care should be taken that the noise from within the enclosure does not escape through flanking

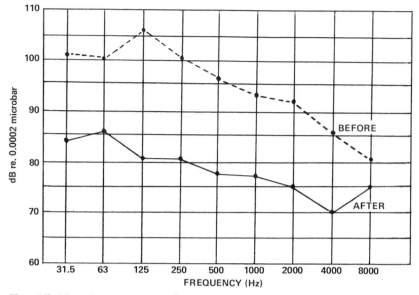

Fig. 17–10.—A comparison of octave-band noise levels before and after the enclosure of the desalinator steam lines.

paths, such as ventilating ducts, or through the floor in the form of vibration. In any enclosure, a door (if used) is the point of greatest weakness as far as transmission loss is concerned. A door should be well insulated and built with air-tight seals at all joints, and it should be installed so that it fits snugly. There can be a 12 dB difference outside a snug-fitting door compared to a loose-fitting one.

A total enclosure should also be lined, at least on part of the inside walls, with absorptive materials, which should then keep the noise buildup inside the enclosure at the lowest practical level.

Less effective than a total enclosure, partial barriers are nonetheless useful in helping to shield high-frequency sound. Attenuation of low-frequency sound is slight unless the barriers are very large. The attenuation of high-frequency sound by the barrier is no more than a few decibels unless the opening that remains is relatively small. The barrier, too, should be covered with absorbing material to avoid sound reflections.

Desalinator. The following example illustrates the application of a

462

total enclosure to solve a noise problem.

A desalinator used on a merchant ship to remove salt from sea water was creating excessive noise levels. The steam that was used in the desalination process could be heard roaring through the lines. During the initial noise survey, attention was centered around the desalinator; however, a detailed analysis revealed that the steam lines feeding the desalinator were the real problem.

In terms of noise control, this problem was rather easy to solve. The process involved wrapping the steam pipes with insulating materials, and then putting a sound transmission barrier (such as a lead vinyl wrapping) around the insulation material. The insulation material in contact with the pipe allowed the passage of noise with very little attenuation. The addition of the solid cover over the insulation acted as a sound barrier. After the installation of both the insulation material and the solid cover over the pipe, the noise level dropped to 70 dB st 4000 Hz (see Figure 17–10).

Most noise problems involving steam center around the noise radiated by the pipe that conducts the steam such as in the example just given. For instance, the lowest noise frequency that a 10 in. diameter pipe will reradiate is on the order of 1200 Hz.

Vibration analysis and control

The majority of industrial noise problems are essentially vibration problems in one form or another. Vibration is a complex phenomenon that calls for detailed analysis and special techniques. (For more details, see Chapter 18.)

A vibration problem may be first noticed while performing a noise level survey of the plant, or it may become apparent from the poor performance of a machine. The first step in solving vibration problems is to locate the point or area where the highest vibration level is observed. A detailed vibration analysis at this point may reveal the true source of the problem. It should be remembered, however, that vibration is transmitted very readily by metal, and the point at which the noise problem is best corrected may be some distance from the point of maximum vibration.

The next step in the search for a solution is a detailed study of the vibration characteristics. A vibration meter that can measure both displacement and acceleration in addition to velocity can be quite useful, because displacement tends to emphasize low-frequency vibrations while acceleration treats all frequencies equally.

FIG. 17–11.—A typical isolator used to prevent the transmission of vibration from the machine to the foundation.

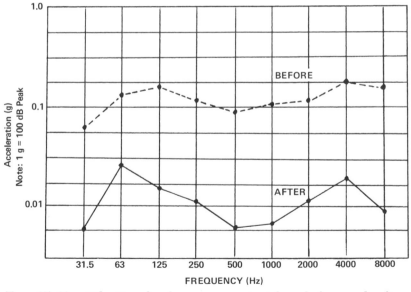

FIG. 17–12.—Vibration level measurements taken before and after placing four vibration isolators under a punch press.

Useful information can be obtained by listening with earphones to the output of a vibration meter to see if there is a dominant frequency, whether it is a random tone such as an impact-type vibration, or a continuous, rough, rushing, or roaring noise. When high-frequency vibration or impact vibration is significant, listening to the character of the vibration signal can often provide an additional clue. For example, worn or bad ball bearings have a characteristic rough tone.

A good example of an impact vibration problem concerned a company that had two 800 ton presses that were used to cut 0.25 in. steel stock. In performing the operation, a high-level impact noise was created. In addition, the presses were pounding themselves into the foundation at a rate of about six inches a year. Also, cracks appeared in the walls of the building as a result of the transmission of vibration to the foundation.

After a thorough investigation, it was decided to isolate the presses by using four vibration isolators weighing 500 pounds each (see Figure 17–11). The entire punch press operation was a completely closed system, so that all movement occurred within the machine framework, and the vibration isolators prevented the transfer of most of the reaction forces to the foundation. After installation of the isolators, vibration measurements were again taken, and a rather impressive drop in decibels was recorded—measurement of the resulting vibration showed a reduction of 15 to 25 dB (see Figure 17–12).

Nature of vibration

Vibratory motion can be placed into three broad categories:

- Low-frequency vibration of the order of shaft or belt speeds

- High-frequency vibration

- Impact vibration and rattles

Table 17–B shows these three vibration categories and a number of conditions that would cause each type of vibratory motion. Use this as a checklist to find the points of highest vibration. In a typical vibration survey, the number of possibilities is narrowed, so that by a process of elimination the major cause of vibration can eventually be found.

Reducing vibration

The reduction of vibratory motion generally can be accomplished by modifying either the driving force or the response to the driving force. To understand how the driving force can be modified, the various

TABLE 17-B. VIBRATION CHARACTERISTICS AND THEIR CAUSES

- **Low-frequency vibration (frequency of order of shaft or belt speeds)**

  Unbalanced rotor (worn, eroded, broken, or corroded parts)
  Misalignment (induces significant axial vibration)
  Eccentric shafts
  Slipping clutches
  Mechanical looseness
  Loose foundation bolts
  Oil whirl or whip (½ or less times shaft speed)
  Worn belts
  Belts and pulleys out of adjustment
  Aerodynamically driven galloping, twisting or reciprocating elements
  that introduce added torsional vibration

- **High-frequency vibration**

  Defective bearings (random or rough vibration)
  Inadequate lubrication
  Poor gears
  Slipping clutches
  Rubbing or binding parts
  Air leaks
  Hydraulic leaks

- **Impact vibration and rattles**

  Parts colliding
  Broken or loose pieces
  Electromagnetically driven loose pieces
  Water hammer
  Surge

TABLE 17-C. VIBRATING FORCES AND COUPLING SYSTEMS

- **Mechanical**

  Unbalanced rotating masses
  Reciprocating masses
  Fluctuating mechanical forces or torques
  Fluctuating loads
  Fluctuating mass or stiffness
  Poorly formed moving components
  Mechanical looseness
  Misalignment

- **Transformation from Another Form of Energy**

  Varying electrical fields
  Varying hydraulic forces
  Aerodynamic forces
  Acoustic excitation
  Varying thermal conditions

ways that a vibrationary force is developed should first be studied. There are two basic processes involved; mechanical energy of some type is coupled to vibratory energy by one or more methods; energy in some other form is transformed into mechanical vibratory energy (see Table 17–C).

Modification of the source or the coupling to the vibrational driving force is accomplished by one or more of the following methods:

- Reduce the vibrations at the source by substitution, or isolation

- Change the character or frequency of the vibration

Reduction of response to the driving force can be accomplished as follows:

- Insert isolating members between the source and other parts of the structure

- Damp vibrating elements

- Detune resonant systems

- Change the mass (increase the mass of stationary elements or reduce the mass of moving elements)

- Change the stiffness or rigidity of machine frame members

- Add auxiliary mass damping or resonant absorbers

Special instruments

Sometimes the source of the vibrationary force is apparent or is well known from experience. In most situations, however, a measuring device is needed to analyze the vibration forces. For example, a stroboscope and a motion picture camera may be used to photograph an operation for subsequent detailed analysis.

A frequency analysis of a vibration can provide information about vibration frequencies that can be related to certain speeds of component members or gear-tooth-meshing frequencies.

Another valuable piece of equipment in vibration analysis is a pair of earphones. Listening to the vibration signal can be helpful in determining the cause of the vibration. For best results, the earphones should have good ear cushions or muffs to keep out extraneous background sound.

Electromagnetic induction

When a noise problem involves devices that are electrically driven, strong vibration forces with frequencies that are multiples of 120 Hz are good indications that the vibration forces are electromagnetically induced. This deduction can be verified, under the proper circumstances, by monitoring the frequency, first during normal operation of the device, and again when the electric power is suddenly removed. Normally, when the electrical power is suddenly removed, the rotating members will coast long enough to avoid a rapid change in mechanical forces although the electrical forces are changed abruptly.

Operating speeds

When machinery is operated at varying speeds, the different speeds can affect the frequencies and amplitudes of the various important vibration components. Measurement during the various portions of the work cycle are an important factor in determining the source of vibrations. This technique is particularly helpful if some rotating parts of the machine can be operated at different speeds or if some sections can be disengaged.

When the vibration analyzer shows an erratic frequency fluctuation over a range of 6 dB or more, the vibration is usually random. The cause of the vibration problem probably may be a rattle, air turbulence, worn ball bearings, fluids in motion, or combustion processes.

This erratic type of vibration can be distinguished from the simpler harmonic motion of rotating unbalanced masses by comparing peak and average readings. The peak value is about 1.5 times, and the peak-to-peak value is about 3 times the average value for simple harmonic motion. For random signals, the ratio is usually much higher, that is, the peak value is 3 to 4 times above the average value (or 6 to 8 times for peak-to-peak readings).

Reduce driving force

Once the source of the vibratory motion is recognized, the next step is to determine the most practical method for reducing the driving force. The possibilities that should be considered include balancing techniques, replacement of worn gears or bearings, application of a better lubricant. Improving the mechanical structure, for example, by reducing the weight of the moving members, increasing the weight of the stationary members, and reducing the velocities of gases or liquids should also be considered.

Reducing response to the driving force

In addition to reducing vibration at the source or modifying the coupling to the driving force, a reduction can be effected by reducing the response to the driving force. Again a careful study of the vibration measurements can be valuable in determining the approach to be used to reduce the response. For example, peak vibration levels provide information concerning:

- Resonant modes of vibration of plates and other structural members

- Where the application of damping materials may be most effective

- Where resonant absorbers can be added

- Where detuning can be used

The following case history is an example of the corrective action necessary to reduce the response to the driving force.

A company had a large power press operation that was used in manufacturing trim for automobiles. These presses were constantly being relocated in various areas of the plant to provide the most efficient manufacturing layout required to produce the product. During the previous year, the firm received many complaints from nearby residents about excessive vibration levels in homes surrounding the factory property. As the company had several very large power presses, it was assumed that they were the vibration sources responsible for the complaints.

Providing vibration isolators for all of these large presses would have been very expensive—so it was decided to study the entire manufacturing operation to see if the particular machine or machines producing the highest vibration levels could be identified. A preliminary survey of the plant indicated that there were five power presses and two large air compressors capable of being major offenders in producing excessive vibration levels.

The engineering department manager stated that one of the air compressors had been relocated recently, and that the complaints seemed to start about this same time. The manager did not feel, however, that the air compressor was the cause of the vibration because most of the complaints came on weekends when the compressor was supposedly shut down. To be safe, however, measurements at the air compressors were included in the survey of vibration levels.

The equipment used for the sound and vibration survey consisted of a recording analyzer, an octave-band analyzer, an impact analyzer, a

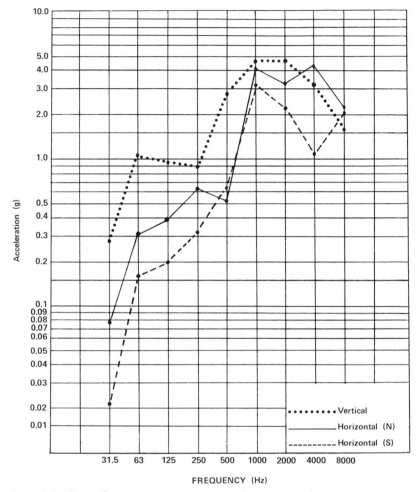

Fig. 17–13.—Vibration measurements taken on a 500-ton press suspected of being the cause for numerous complaints by neighboring residents.

vibration measuring system, a microphone, and the necessary calibration equipment. Because of the impulsive nature of the noise produced by the power press, the impact analyzer was used with an accelerometer. The data presented in Figure 17–13 were the result of the vibration survey on the 500 ton press.

During the noon lunch hour when all of the presses were shut down,

470

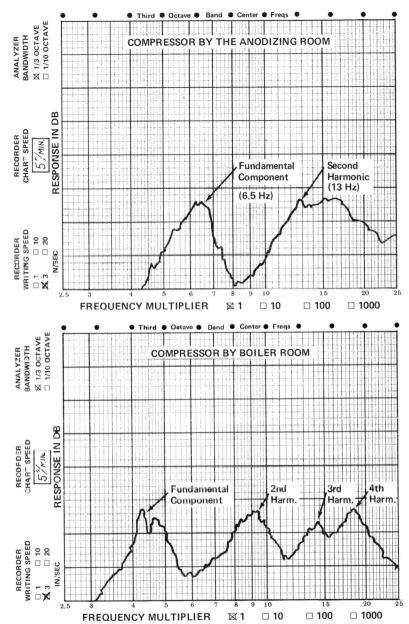

Fig. 17–14.—One-third octave-band vibration measurements taken in the vicinity of a large air compressor located in an anodizing room and one located near a boiler room. These compressors were the real cause for complaint.

471

the vibration spectrum on the two compressors was taken (see Figure 17–14). (Note the large component at the low frequency end of the plot. One compressor has a very large component at 4.25 Hz while the other has a large component at 6.2 Hz. This was measured with the analyzer set to one-third octave.)

Next, the vibration measurement equipment was taken to the home of the person who was doing the most complaining. He demonstrated the presence of objectionable vibrations by placing a glass of water on the window sill of his bedroom. The surface of the water could be seen to ripple. This, however, did not identify the cause of the vibration.

A narrow-band analyzer was set up in the bedroom with the vibration pickup oriented to produce a maximum reading on the instrument dial. It was suspected that the vibration was horizontal rather than vertical, so the vibration pickup was placed on its side with the long axis pointed directly at the center of the plant. The readings on the narrow band analyzer showed a definite maximum indication at 6.2 Hz and none at 4.25 Hz. Very little instrument response at any of the other frequencies tested was detected.

The maintenance department was then called and they were asked to shut down the suspected air compressor. When the air compressor was shut down, the measured vibration at the 6.2 Hz frequency dropped below measurable limits. When the air compressor was again placed in service, the signal at 6.2 Hz reappeared. This was considered a positive indication that this particular air compressor was the source of the offending vibration levels.

A close examination of the foundation for this air compressor revealed a crack in the floor produced by the vibration of this machine.

The solution to this vibration problem was to mount this air compressor on a floating concrete slab, isolated from the concrete floor and building foundation. Connections to the plant compressed-air lines were made through flexible couplings.

A systematic organized approach to this problem eliminated the guesswork, which had been going on and which had corrected nothing. This approach positively identified the vibration source, and avoided the unnecessary investment in vibration isolation on the other machines, which would not have eliminated the problem and complaints.

Resonant vibration

Sympathetic or resonant vibration is a common phenomenon that occurs in many common objects. In fact, every plate or bar has a great number of natural frequencies of vibration, which are related in a rather complex fashion with one another. The frequencies of the resonant vibration modes are affected by the shape, dimensions, stresses, mounting, and material characteristics. In addition, the resonant modes can also be affected when connected to other structures. Except for shapes of the simplest geometry, it is almost impossible to predict the resonant vibration frequencies of a structure or device. Fortunately for equipment designers, the natural modes of vibration for many types of simple structures have been calculated.

The nature of resonance can be easily shown by the movement of a board to which a mass is flexibly mounted. The board is then driven by vibrational forces at a constant amplitude but different frequencies. At certain critical frequencies, the motion of the flexibly mounted mass will be greater than for those frequencies just slightly higher or lower. This critical frequency at which a maximum vibration response occurs is a resonant frequency. If the structure being shaken is complex, many such maxima may be observed. (Minima of motion may also be observed.)

Locating the resonating part

In actual practice, resonant conditions cause excessive noise or vibration at certain speeds. If the speed of a machine or device is slowly varied from below normal operation to a point considerably above, and if definite increases in loudness are noted at certain speeds (assuming no standing wave effects), it may be due to either excitation of different resonant frequencies or to successive excitation of different resonances by different exciting components. When resonant frequencies are so prominent that they can be detected by changes in loudness level, location of the resonating part is easier. However, more often than not, a resonant frequency will be characterized, during a speed change, by the quality of the noise rather than by the loudness (which may remain unaffected). Using a vibration pickup to explore for those points at which the vibration is much greater than other locations can often locate the resonant parts without too much difficulty.

Types of resonance

The resonances may be of the simple type, where the mass is mounted on a flexible support or they may be of the plate mode type where the mass and flexibility of a plate or sheet are in resonance, so that different parts of the plate are moving differently. In the latter case, very complicated motions may result.

Solutions to resonant noise problems

There must be a significant dissipation of energy as the system vibrates, or the resonant amplitude of motion may become very large even with a relatively small driving force. These large amplitudes must be avoided by the proper application of control measures. There are two generally accepted ways of reducing these excessive amplitudes—damping and detuning.

Damping

Many techniques for damping vibration have been developed, including inherently dissipative plastics or metals, frictional rubbing devices, mastic deadeners, sandwich-type dissipative materials, electromagnetic damping, dynamic absorbers, dashpots, and other viscous absorbing systems.

A measurement of the vibration levels at various positions on the device can help determine where damping materials can be applied most effectively. Consequently, if a resonant condition is to be damped, an analyzer, tuned to the resonant frequency, should be used on the output of the vibration pickup. Then during measurements at various points, the vibration component at only the resonant frequency can be singled out to obtain the actual resonant maxima without being obscured by high-amplitude, low-frequency vibrations. Such measurements should be made with a lightweight vibration pickup (in comparison to the mass of the resonant element), so that it does not appreciably detune the resonant system. Whenever possible, make stroboscopic observations; they can be performed without affecting the mass of the system.

Detuning

Detuning (tuning out) an unwanted resonant frequency is not used as often as damping because of the restrictive conditions under which detuning is applicable. Detuning is normally applied if the driving force is at a relatively fixed frequency (pure tone). When that condition exists, it may be relatively easy to adjust the critical resonant frequency out of the operating range by changing the resonant-element mass or stiffness or both.

Resonance can be detected in many cases, when the natural mode frequency of a vibrating piece coincides with, or is very close to, one of the component frequencies of the driving force. The conditions may be such as to make a system self-oscillatory. A reasonably steady driving force, however, is required. The brush squeal on motors and some forms of machine-tool chatter are examples.

Sometimes trial-and-error methods must be used to track down the source of vibration difficulties. Actually, this technique involves nothing more than changing some element of operation, such as the mass, and observing the effects. Good measurements are an essential ingredient in this type of technique.

Balancing rotating machines

One of the principal causes of excessive vibration is an unbalance in a rotating device. The unbalance occurs when the principal axis of inertia of the rotating member does not coincide with its axis of rotation. The addition of a counterweight to the rotating item is the usual remedy. One of the best examples of unbalance in a rotating part is an out-of-balance automobile wheel. Every car owner is familiar with the necessity of getting the wheels on his automobile balanced, and many have watched as the mechanic places the counterweights on the wheel. In industrial application, however, balancing is not usually as simple as balancing automobile wheels.

Summary

An effective noise reduction program requires careful noise level measurements, analysis, and planning. The individual who finds a noise level too high and says, "We'll just slap some acoustical tile here and there; that ought to take care of it," will probably be disappointed at the results. Not that acoustical tile is not a good way to reduce noise—it often is, but, that approach is like shooting a person full of penicillin at the first sign of a cold. The more prudent course is to analyze the noise in detail with a sound analyzer, decide from this analysis what to do about it, do it, and then check the results by further analysis.

Experience is probably the best teacher in solving noise problems, but a few principles are helpful:

• Reduce the noise at its source, if possible

• Reduce the noise level at the ear of the listener by changes in the path from the source

The first approach should always be to look closely at the device generating the noise and see what can be done to quiet the machine.

(*Text continues on page 478.*)

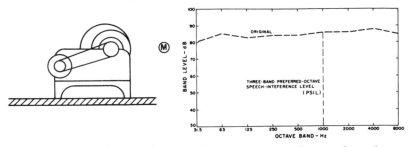

Fig. 17–15.—A machine is shown with no attempt made to reduce the noise. The octave-band analysis shown on the right was obtained by having a microphone positioned where the "M" is shown.

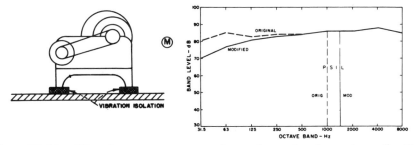

Fig. 17–16.—The first attempt to reduce the noise by using vibration isolators.

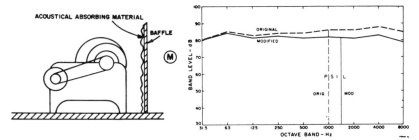

Fig. 17–17.—A partial barrier with the sound absorbent material on the inside.

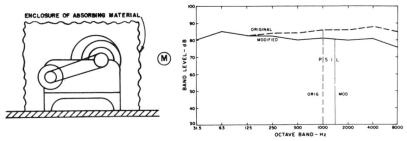

Fig. 17–18.—A total enclosure of sound absorbing material does not appear to have much effect.

476

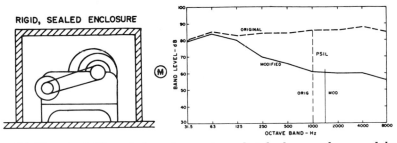

FIG. 17–19.—A rigid, sealed enclosure immediately lowers the sound level; particularly in the speech interference range.

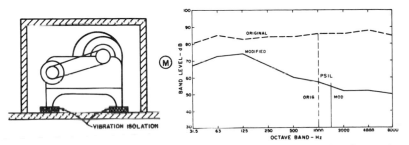

FIG. 17–20.—When isolation mounts are added, the sound level is reduced in the lower frequency range.

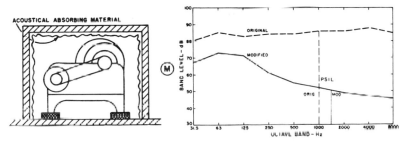

FIG. 17–21.—In adding sound absorbent material to the interior of the enclosure reduces the sound level in the speech interference range.

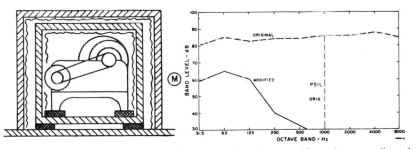

FIG. 17–22.—The ultimate in isolation is the "room within the room" technique.

Illustrations courtesy of General Radio Co.

If the device cannot be quieted, consider a different type of device—one that is quieter—or a reworked unit with a different power-transmission system.

When modification of a noise source is attempted, it usually means a reduction in vibration amplitudes of the source and of the sound generated by vibration panels or other components.

A summary and review of the various noise control measures—such as vibration isolation, barriers, enclosures, and acoustic treatment—are shown in Figure 17–15 through 17–22. A machine setting on the floor with the microphone position marked by "M" in the circle is shown in Figure 17–15. The octave-band analysis of the noise generated by this machine is found alongside each figure. Only the general nature of the effects of various control methods on noise reduction is shown. In an actual case the results may differ materially from those shown because of factors not considered here. But, in general, the degree of noise reduction shown in the figures can be considered typical.

Simple vibration isolation reduces the low-frequency energy and may be an important step in noise control. As shown in Figure 17–16, vibration isolation mounts alone provide only a moderate reduction of the low-frequency noise. The machine tool itself usually radiates high-frequency noise directly to the air, and the amount transmitted to and reradiated by the floor is small. A reduction in the vibration level at the floor alone would not greatly affect the high frequencies. At low frequencies, however, the machine will transmit sound energy to the floor, which may then act as a sounding board to contribute materially to low-frequency sound radiation to the workroom.

The wrong type of vibration isolation mounts, however, may increase the noise level. This may occur when the stiffness of the mounting is such that some vibration mode is exaggerated by resonance. Resonance can be avoided by proper selection and design of vibration mounts for a particular application. In the illustrative example, Figure 17–16, the mounting is sufficiently soft, so that the basic vibration resonance of the machine on the mounting system is below 20 Hz.

The results shown in Figure 17–17 illustrate that a partial barrier affects only the high frequencies, and there it produces only a moderate reduction in noise level.

It is sometimes assumed that the materials used for sound absorption are also effective and can be used for sound isolation. An enclosure constructed solely of lightweight, sound absorptive materials, mounted on a lightweight framework, would typically produce the result shown in Figure 17–18. A noticeable reduction in level would only be found at the high frequencies, and even there it is very small.

A more satisfactory enclosure is built of more massive and rigid constructional materials. Assume that one encloses a machine by a well-sealed, heavy, plasterboard structure. Then the result shown in Figure 17–19 might be observed. Here an appreciable reduction is obtained over the middle- and high-frequency range. The enclosure is not as effective as it might be, however, because two important factors limit the reduction obtained. First, the vibration of the machine is carried by the supports to the floor, and then to the whole enclosure. This vibration then may result in appreciable noise radiation. Secondly, the side walls of the enclosure absorb only a small percentage of the sound energy.

The addition of a suitable vibration isolation mounting will reduce the noise transmitted by solid-borne vibration. This effect is illustrated in Figure 17–20. Here we see a noticeable improvement over most of the audio spectrum.

When the sound absorption within an enclosure is small, the noise energy from the machine produces a high level within the enclosure. Then the attenuation of the enclosure operates from this initial high level. The level within the enclosure can usually be reduced by the addition of some sound absorbing material within the enclosure, with the result that the level outside the enclosure is also reduced. This effect is shown in Figure 17–21, which should be compared with Figure 17–20.

The ultimate in isloation, however, is the so-called "room within a room" technique. If even more noise reduction is required than that obtained by the one enclosure, a second, lined, well-sealed enclosure can be built around the first. The first enclosure is supported within the second on soft vibration mounts. Then a noise reduction of the magnitude shown in Figure 17–22 can be obtained. Here the high frequencies are below the level that can be measured without very special techniques.

An application of some of the techniques as illustrated in Figure 17–15 through 17–22 most likely can solve a large number of the noise problems in industry. Of course, the most important thing to remember is that the place to start noise control is with the noise source. Reducing the noise at the source pays off quickly in the overall picture. It is usually here that simple procedures are most effective. The control of noise by better maintenance of the machinery or by changing material or gears may increase the service life and efficiency of the machine enough to pay for the effort at noise control.

It is possible that a noise problem will defy your best efforts; some problems are like that. A reasonable course then is to obtain a competent acoustical expert. Someone who has studied noise problems for many years can often find a more economical solution than would occur to an inexperienced individual. This chapter should help anyone who has to call in a consultant to better understand what the consultant does and to make better use of his service.

Acknowledgment

Some of the material appearing in this chapter is adapted from the *Handbook of Noise Measurement* (Seventh Edition), Arnold P. G. Peterson and Ervin E. Gross, Jr., General Radio Corp., 300 Baker Ave., Concord, Mass. 01742.

Chapter Eighteen

Vibration Measurement and Control

Julian B. Olishifski, P.E.

Vibratory motion is generally produced whenever machinery or equipment is operated. A brief description of the instruments necessary to measure vibration and some methods of vibration control will be given in this chapter.

Vibration is usually considered to be an oscillatory motion of a system. This motion can be simple harmonic motion or it can be extremely complex. The system might be gaseous, liquid, or solid. When the system involves vibration in the range of 20 to 20,000 hertz, audible sound is produced. The vibratory motion may be periodic or random, steady state or transient, continuous or intermittent.

Undesired vibrations may originate from a great variety of sources, such as unbalance and reciprocating motions in mechanical machinery, improper alignment or operation of equipment, and electromagnetic, hydraulic, and aerodynamic effects. The adverse effect of these vibratory disturbances can range from negligible to catastrophic, depending on the severity of the disturbance and the sensitivity of the equipment or people involved. (See Figure 3–4, page 92.)

Unbalance in rotating machinery members is a frequent cause of vibration. In many cases the vibration caused by a small degree of unbalance may not be too serious; however, in cases involving machinery rotating at high speed, the problem of unbalance becomes critical. Basically two different kinds of unbalance occur in practice: unbalance in one plane, called *static unbalance;* and unbalance in two or more planes, commonly termed *dynamic unbalance.*

481

Static unbalance can be corrected by positioning the rotating machine part with its axis of rotation horizontal, and adding weight on top of the part until it is brought into equilibrium. If the center of gravity of the rotating part does not coincide with the shaft center (center of rotation), the rotating part will vibrate during rotation and cause vibrational forces to be transmitted to the bearings. These vibrations can be eliminated by adding mass to the light side of the machine part opposite to the mass load causing the unbalance.

If the rotating machine part resembles a cylinder rather than a wheel, not only can a center-of-gravity-displacement kind of unbalance be present, but also an unbalanced couple may affect the rotary motion. To correct this condition requires balancing in two planes. For the detection and correction of dynamic unbalance, it is essential that the machine part rotate. An excellent example is the dynamic balancing of auto tires—which are left on the wheel and rotated rapidly by a machine.

Mechanical failure due to material fatigue is by far the most commonly known deteriorating effect of vibration. Failure may also occur as a result of excessive vibration amplitudes, collisions between two vibrating systems, or when a certain vibration amplitude value is exceeded a specified fraction of time.

The motion of a mechanical system acted upon by external forces is commonly termed the *response of the system*. The external forces acting upon the system are termed the *exciting forces*. These terms are general and have to be specified in greater detail when the behavior of a particular system is being investigated.

Moving machine parts vibrate and generate sound and vibration in a unique pattern commonly called its *signature*. This sound wave is made up of many distinct fundamental frequency peaks corresponding to gear meshing, and bearing rotational speeds. Frequency peaks also may occur as the sum, difference, or product of the fundamental frequencies. The sound and vibration pattern or signature changes with variations in rpm and as the moving parts wear and loosen. (See Figure 18–1.)

When the machine or equipment is operating under normal conditions, a vibration pickup mounted on the structure can be used to collect data for analysis. From the sound and vibration pattern or signature, the frequency and magnitude of every important component of the machine's self-generated vibration can be singled out for closer analysis.

The measurement system for a preventive maintenance program based on vibration analysis should have the ability to measure displacement, velocity, and acceleration. The system should also have frequency

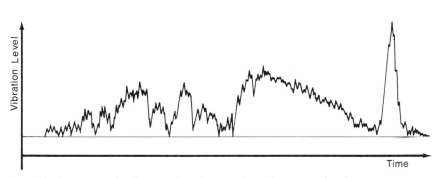

Fig. 18–1.—Typical vibration level recording (signature) of a noise source.

analysis capability from approximately one-quarter of the lowest shaft speed to twice the highest gear mesh frequency. This range is not always possible within a single instrument but many vibration instruments are available with various ranges or degrees of sophistication. Measurement will be discussed in detail later in this chapter.

Narrow-band spectrum vibration analysis provides information on the mechanical condition of an individual machine. Instruments that can provide the vibration amplitude at the predominant frequencies can be a tremendous help in determining the causes of vibration problems.

Any vibration of machine parts may present a potential problem. An analysis of excessive vibration levels should help determine the possible causes, the relative severity, and the necessity for corrective action. The units of measurement at low frequencies (below 30 Hz) should be in mils (0.001 in.) peak-to-peak displacement. At the low frequencies, mechanical interference and rubbing frictions are of interest. Above 30 Hz, velocity measurements should be taken. At very high frequencies, acceleration measurements are usually taken.

Periodic vibration

Vibration is considered periodic if the oscillating motion of a particle around a position of equilibrium repeats itself exactly after some period of time. The simplest form of periodic vibration is called pure harmonic motion which, as a function of time, can be represented by a sinusoidal curve. The motion of any particle can be characterized at any time by (a) a displacement from a position of equilibrium, (b) velocity or rate of change of displacement, or (c) acceleration, or rate of change of velocity.

483

Displacement

Of all the possible vibration measurements, displacement probably is the easiest to understand and is significant in the study of deformation and bending of structures. Frequency-independent vibration measuring equipment can be used to obtain an idea of the overall vibration level at a certain machinery or equipment position. The frequency response of the combined system (pickup + amplifier) is considered to be flat over a certain specified frequency range. However, displacement is not considered to be the most important vibration measurement in many practical problems.

Velocity

The velocity of the vibrating part may be the best single criterion for use in programs of preventive maintenance of rotating machinery. Although peak-to-peak displacement measurement has been widely used in the past for this purpose, its drawback is the necessity to establish a relationship between the limits for displacement at various rotational speeds for each machine part.

Acceleration

In many cases of vibration analysis, the actual forces set up in the vibrating parts are the critical factors. The acceleration of a particular machine member is proportional to the applied force. The moving parts of a vibrating structure exert forces on the total structure that are a function of the masses and accelerations of the vibrating parts.

Other terms, such as *jerk*, defined as the time rate of change of acceleration, are sometimes used in vibration measurements. At low frequencies, jerk is related to riding comfort of automobiles and transportation equipment.

Pure harmonic periodic vibrations only have been mentioned; however, it must be pointed out that most of the vibrations encountered in industry are not pure harmonic motions, even though many of them certainly can be characterized as periodic.

Random vibrations

Random vibrations occur quite frequently in nature and may be defined as motion in which the vibrating parts undergo irregular cycles that never repeat themselves exactly. Obtaining a complete description of the vibrations theoretically requires an infinitely long time to record.

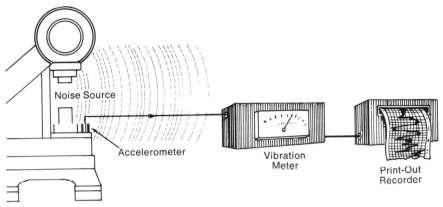

Noise Source

Accelerometer

Vibration Meter

Print-Out Recorder

FIG. 18–2.—Measurement system for narrowband random vibration analysis.

This, of course, is an impossible requirement and finite time intervals are used in practice.

Measurement

A wide variety of component systems, consisting of mechanical or a combination of mechanical, electrical, and optical elements, are available to measure vibration. The most common system uses a vibration pickup to transform the mechanical motion into an electrical signal, an amplifier to enlarge the signal, an analyzer to measure the vibration in specific frequency ranges and a metering device calibrated in vibrational units (see Figure 18–2). Vibration meters are used to measure the displacement, velocity, or acceleration of the vibrating machinery part. If the data is desired in terms of velocity or displacement, electronic integrators are added at the output of an accelerometer (see Figure 18–3).

Vibration pickups

Accelerometers are the most common type of vibration pickup. An accelerometer is an electromechanical transducer which produces an output voltage signal proportional to the acceleration to which it is subject (see Figure 18–4).

Vibration transducers should be small so that the load on the vibrating member that is being measured will be as little as possible. Also, a small transducer can measure the vibration at a point on a structure, rather than on a large area (see Figure 18–5).

485

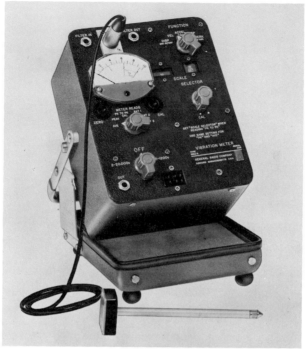

Courtesy General Radio Co.

FIG. 18–3.—Vibration meter and accelerometer pickup.

Two types of accelerometer sensitivities are usually stated—the voltage sensitivity and the charge sensitivity.

Preamplifier

A preamplifier is introduced in the measurement circuit to amplify the weak output signal from the accelerometer.

It is possible to design the preamplifier in two ways: (*a*) the preamplifier output voltage is directly related to the input voltage, or (*b*) the output voltage is proportional to the input charges. The preamplifier is termed a *voltage amplifier* in the former case and a *charge amplifier* in the latter.

Analyzers

The simplest analyzer consists of a linear amplifier and a detection

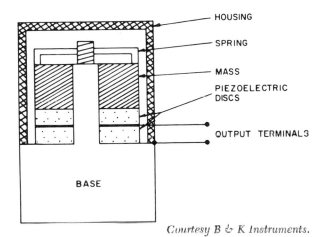

HOUSING

SPRING

MASS

PIEZOELECTRIC
DISCS

OUTPUT TERMINALS

BASE

Courtesy B & K Instruments.

Fɪɢ. 18–4.—The basic construction of a compression type piezoelectric accelerometer (single-ended version).

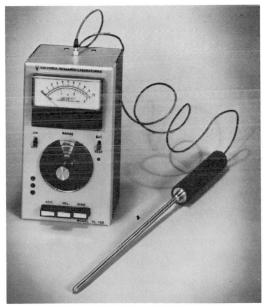

Courtesy Columbia Research Laboratories.

Fɪɢ. 18–5.—A small transducer can measure the vibration level at a point on a structure.

device measuring some characteristic vibration signal value, such as the peak, root mean square, or average value of the acceleration, velocity, or displacement.

In many cases, it will be necessary to determine the frequency composition of the vibration signal. To accomplish this, two types of frequency analyzers are commonly used, *constant bandwidth analyzers* and the *constant percentage bandwidth analyzer.*

If the vibrations are periodic, the constant bandwidth analyzer is preferred because the frequency components are harmonically related. If the signal is not stable and only the first few harmonics are significant, a constant percentage bandwidth analyzer is more suitable. If the vibration signal is random in nature, the type of analyzer that should be used will depend on the use to be made of the data.

The frequency range of interest in vibration measurements has been increasing over the past years. Not long ago, a common upper frequency limit was 50 Hz. Today, many vibration measurements are made at up to 5000 Hz.

Readout or recorder

There are many types of readout or recording devices available that exhibit the analyzer output. These can be divided into three classes: (*a*) the stripchart recorder, which prints the output on preprinted calibrated recording paper, (*b*) a dial face with a needle showing some characteristic value (rms, peak, etc.) on a scale, and (*c*) an oscilloscope, which projects the waveform on a screen for analysis and measurement.

When a strip chart recorder is used, the averaging time is determined by the writing speed of the recording pen, the input range potentiometer, and other internal properties of the recorder instead of observation time.

The oscilloscope presents measurements as a function of time, and makes possible the study of instantaneous values of vibration levels. An oscilloscope with slow sweep rates and a long-persistence screen is recommended.

Accessories

The accessories used with vibration meters are determined by the types of measurements that are needed. However, the accessories needed are usually built into any specified instrument. Integrators to convert displacement, velocity, and acceleration measurements can be specified in the basic circuit design of the measuring device. A series of operating ranges can be made available on the frequency analyzer.

Different recorders can be used in conjunction with any analyzer. For peak or high-speed impact measurements the signal can be recorded on tape and then replayed at varied speeds.

Impulse or shock measurements

The measurement of mechanical impulses and shocks requires particular care in the selection of measuring and analyzing equipment. In some cases, the only quantity to be measured is the maximum acceleration occurring during the shock, and the measurement instrumentation is selected accordingly. However, very often this information is not considered to be sufficient for a relevant description of the shock motion, and quantities such as the acceleration-time integral (total velocity change), and the spectral content of the shock pulse, must be evaluated.

Calibrators

A calibrator is an important accessory for checking the overall operation of a vibration-measuring system. The calibrator is an accelerometer driven by a small, electromechanical oscillator mounted within the mass, and operated at a controlled frequency and acceleration.

Calibration of the vibration measuring system can be made directly from curves and figures supplied by the manufacturer. In addition, special vibration measuring arrangements can be calibrated using a vibration calibrator.

Field measurements

Common sources of error are incorrect mounting, incorrect calibration, connecting cable noises, and thermal effects. Consideration must be given to the location and mounting of the accelerometer in order to measure the original motion and resonant frequencies of the structure. The vibration transducer should load the structural member as little as possible, since any loading effects may invalidate the vibration measurement results.

A number of methods can be used to mount the accelerometer: (a) by means of a steel stud, (b) an isolated stud and mica washer, (c) a permanent magnet, (d) wax, or (e) soft glue (see Figure 18–6). The measurement can also be made with handheld probe. A steel stud provides the best frequency response, approaching the actual calibration curve supplied with the accelerometer.

When electrical isolation is required between the accelerometer and vibrator, a mica washer is used because of its hardness, vibration trans-

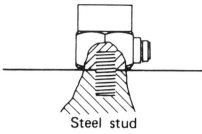

Steel stud

TYPE 1 mounting approaches a condition corresponding to the actual calibration curve supplied with each accelerometer.

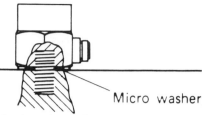

Micro washer

Isolated stud

TYPE 2 mounting. Convenient when electrical isolation between accelerometer and vibrating body is necessary. It employs an isolated stud and thin mica washer. Frequency response is good.

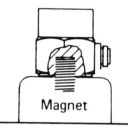

Magnet

TYPE 3 mounting employs the permanent magnet which also gives electrical isolation from the vibrating specimen. A closed magnetic path is used and there is virtually no magnetic field at the accelerometer position.

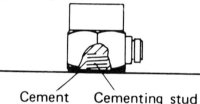

Cement   Cementing stud

TYPE 4 mounting is convenient when a cementing technique is appropriate, with the possibility of removing the accelerometer from time to time.

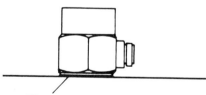

Thin layer of wax

TYPE 5 mounting employs a thin layer of wax for sticking the accelerometer on to the vibrating surface.

Probes

TYPE 6 mounting employs the probe with interchangeable round and pointed tips. The method may be convenient for certain applications, but should not be used for frequencies much higher than 1000 Hz because the natural resonant frequency in this case is very low.

FIG. 18–6.   *Courtesy B & K Instruments.*

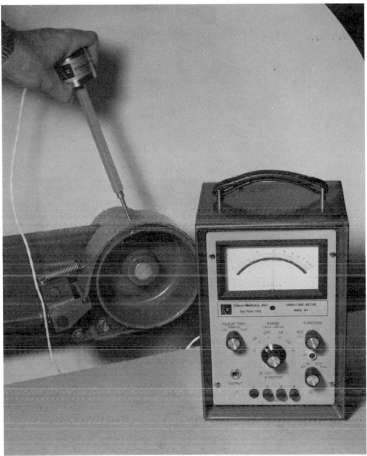

FIG. 18–7.—Handheld probes can be used to conveniently measure vibration level at a particular point.

mission and electrical, characteristics. A permanent magnet also gives electrical isolation, but magnetic, mass, and temperature effects can be introduced.

Handheld probes are convenient but should not be used for frequencies higher than 1000 Hz where mass loading effects begin (see Figure 18–7). Wax, because of its stiffness, gives a good frequency response but cannot be used at high temperature.

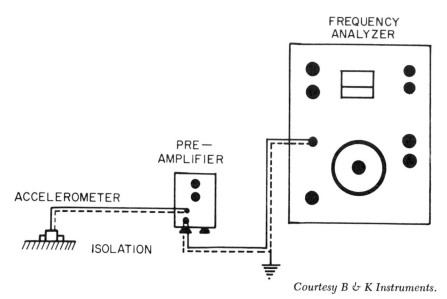

Courtesy B & K Instruments.

Fig. 18–8.—Recommended method of connection so that no ground-loop is formed.

Cable noise

Because piezoelectric accelerometers are high impedance devices, certain problems may also arise from connecting cable noise. These noises can be produced from the mechanical motion of the cable, or from ground loop induced electrical hum noise.

The mechanically caused noises are sometimes called "triboelectric effects," or simply "microphonic noise," and originate from bending or compression and tension of the cable, and may be particularly disturbing at lower frequencies. It is always good policy to clamp the cables as firmly as possible in order to avoid relative movement.

Another type of noise effect is ground loop induced noise. A ground loop induces a current and small voltage drop which adds directly to the weak accelerometer signal. Making sure that grounding of the vibration measuring instruments is made only at one point eliminates ground loop induced noise. (See Figure 18–8.)

General vibration measurement scheme

Important considerations in setting up vibration instruments for field measurements are:

1. Determine placement of the vibration transducer to minimize mass loading effects.

2. Estimate the types and levels of vibrations likely to be present at the mounting point.

3. Select a suitable vibration transducer considering the types of vibrations, temperature, humidity, acoustic, and electric fields present.

4. Determine what type of vibration measurement (displacement, velocity or acceleration) would be most appropriate.

5. Select suitable electronic equipment, considering frequency sensitivity, dynamic range, and convenience.

6. Calibrate the entire vibration measurement system.

7. Select the appropriate mounting method, checking vibration levels, frequency range, electrical insulation, ground loops, and temperatures.

8. Mount the accelerometer and record the results. Record the types and serial numbers of all instruments used.

The apparent vibration level should be recorded by mounting the accelerometer on a nonvibrating part of the equipment that is to be measured. The apparent vibration measurements should be less than one-third of the desired vibration measurements. Or, in other words, the noise "floor" of the equipment should be at least 10 dB below the vibration levels to be measured.

Vibration analysis should include measurements at different points on the machine or equipment to determine the points or areas of greatest vibration. Experience is perhaps the best guide in pinpointing vibration sources in machinery. Since machine vibrations produce nose, reduction of vibration often reduces noise problems as well.

Control of vibration

Ideally all undesirable vibrations should be eliminated at the source although this may not be possible in all cases. If possible, isolate the source by means of shock and vibration isolators, reduce the shock and vibration effects by means of effectively designed vibration absorbers, or consider the use of damping treatments.

Natural vibration sources like aerodynamic turbulence, rough sea movements, and earthquakes cannot be "isolated" in the usual sense of the word. The only way to diminish undesirable vibration effects

Courtesy M/Rad Corporation.

F<small>IG</small>. 18–9.—Inspection bench supported on self-leveling air springs.

originating from these types of sources is to isolate the equipment upon which the vibrations may cause serious damage. (See Figure 18–9.)

The principles involved in shock isolation are very much similar to those involved in vibration isolation, although some differences exist due to the transient nature of a shock. The reduction in shock severity, which may be obtained by the use of isolators, results from the storage of the shock energy within the isolators and its subsequent release in a "smoother" form. Vibration effects may be reduced by:

1. Isolating the source of the disturbance from the radiating surface

2. Reducing the response of the radiating surface

3. Reducing the mechanical disturbance causing the vibration

Isolation

Isolation of structures from the vibrations produced by machinery requires special design considerations. Vibrations from machines can be controlled by using concrete inertia blocks, isolation and absorbing materials (such as cork, felt, timber and rubber), and devices utilizing steel or air springs. Inertia blocks, special materials, and devices are often used in combination to obtain satisfactory results.

Vibration control techniques in the form of shock and vibration isolators provide dynamic protection to all types of equipment. If the equipment is the source of the vibration or shock, the isolator reduces the force transmitted from the equipment to the support structure.

If the support structure is the source, the purpose of the isolator is to reduce the dynamic motion transmitted from the support structure to the equipment. Isolators are used, for instance, in protecting delicate measuring instruments from vibrating floors. In either case, the principle of isolation is the same. The isolator, being a resilient element, stores incoming energy temporarily and releases this energy to the equipment or support structure to be protected at a time interval that reduces the magnitude of motion to the equipment or support structure.

Transmission of vibration to an adjacent structure may be reduced by making connections less rigid. For example, flexible connections for piping and electrical service can be used. When using enclosures, the housing should be anchored to the floor rather than to the machine in order to eliminate transmission of the machine vibration to the enclosures.

Another important consideration is that the vibration isolators be placed correctly with respect to the center of gravity of the machine. In cases of instability, that is, "rocking," the effective center of gravity may be lowered by mounting the machine on a heavy mass and isolating the mass and the machine on a so-called floating floor.

Elastomeric isolators. Many types of elastomers are available to meet specific conditions and applications. Properties important to isolators are damping, bondability, temperature range, bond strength, and resistance to environmental factors. Elastomers can be molded in many sizes or shapes, and stiffness can be varied within relatively wide limits. Elastomers can easily be bonded to metallic inserts for convenient attachment to equipment. (See Figure 18–10, next page.)

Elastomers, such as natural rubber and neoprene, usually possess

(*Text continues on page 498.*)

Courtesy Uniroyal, Inc.

FIG. 18–10a.—Cylindrical mountings can be used in compression or in shear or in a combination of both. The mountings are easy to install because of the threaded studs which are an integral part of each mounting.

FIG. 18–10b.—Glass fiber isolation mount with hold-down bolt and anti-short circuit neoprene grommet.

Courtesy Kinetic Systems, Inc.

Courtesy Vibro/Dynamics Corporation.

FIG. 18–10c.—Vibration mounting pads installed under a machine tool.

FIG. 18–11a.—Air compressor is mounted on steel coil springs.

FIG. 18–11b.—Machinery baseplate rests on coil springs.

enough inherent damping for use on machines that operate momentarily at resonance during starting and stopping. Other elastomers have high degrees of damping and can operate at resonant frequencies for longer periods of time.

Elastomers can be used as shock isolators because of their relatively great energy-storage capacity. Through molding in various shapes, elastomers can attain the desired linearity or nonlinearity required for adequate shock isolation.

When elastomeric isolators are subjected to temperature extremes over long periods of time, their useful life is shortened. On the high end, 400 F (200 C) for some elastomeric materials is attainable. For low-temperature extremes, relatively inexpensive materials are now available.

Elastomeric isolators with a large static deflection may give satisfactory performance temporarily, but may tend to drift or creep excessively over a period of time.

Spring isolators. Coil springs are another form of vibration isolator. Such springs may be loaded in tension, but it is frequently more convenient to load them in compression. The advantages of coil springs are the great range of stiffness characteristics and the freedom from drift or creep. The ratio of vertical to horizontal stiffness can be controlled within wide limits. The stiffness of a spring along its lateral axis has limited applications. Lateral stiffness is a function of the coil diameter, the working height of the spring and the static deflection.

Coil springs are used in vibration isolators primarily for the isolation of vibrating motion having a low forcing frequency. Consequently, coil springs must usually have relatively large static deflection. This introduces the danger of instability, with the possibility that the mounted equipment may fall sideways, unless precautions have been taken to assure stability of the equipment and isolator assembly. Instability is likely to result if the lateral stiffness of the isolator is too small or the static deflection is too large.

Coil springs possess practically no damping. Therefore, transmissibility at resonance is extremely high. Coil springs also allow high-frequency surges to pass through to the equipment being protected. They are also a transmission path for high-frequency vibration resulting in excessive noise levels. (See Figure 18–11.)

Reduction of surface response

Damping. Controlling surface response to vibration is an essential consideration in mechanical system design. Damping, the process of dissipating vibratory energy by converting it into heat, is one of the most important means of achieving this control. Effective operation of mechanical equipment depends on controlling vibratory energy losses.

Complex mechanical systems have many resonant frequencies, and when an exciting frequency or disturbance occurs at a resonant frequency, a potentially dangerous condition results. Uncontrolled resonant response can produce excessive system motion which, in turn, leads to impaired reliability or structural failure.

Structural elements like beams and plates exhibit an infinite number of resonances. When subjected to vibrations of variable frequency, the application of separate dynamic absorbers to structural elements becomes impractical. Since there is usually little inherent damping of resonant vibrations in the structural elements themselves, external arrangements must be made to reduce vibrations.

External damping can be applied in several ways:

1. By means of interface damping (friction)

2. By application of a layer of material with high internal losses over the surface of the vibrating element

3. By designing the critical elements as "sandwich" structures

Interface damping is obtained by letting two surfaces slide on each other under pressure. With no lubricating material the "dry" friction produces the damping effect, although this commonly causes fretting of the two surfaces. When an adhesive separator is used, a sandwich structure results.

Mastic "deadeners" made of an asphalt base are commonly sprayed onto a structural element in layers, to provide external damping. These deadeners are commonly made from high-polymer materials with high internal energy losses over certain frequency and temperature regions. To obtain optimum damping of the structural element, both the internal loss factor of the damping material and the modulus of elasticity (the ratio of stress to strain) must be high.

Damping through the use of sandwich structures refers to a layer of viscoelastic material, placed between two equally thick plates or to a thin metal sheet placed over the viscoelastic material which covers the panel. The damping treatment will be more effective if applied to the area where vibration is greatest, but the actual amount of treatment and the area of coverage are best determined by experiment. As a rule of

Courtesy Consolidated Kinetics Corporation.

FIG. 18–12a.—Flexible noise barrier material can be used as pipe and duct noise jackets. The limp, formable, loaded-vinyl material resists the passage of sound waves and minimizes vibration levels.

Courtesy 3M Company.

FIG. 18–12b.—Pressure-sensitive tapes are used to dampen vibration and sound in fuselage areas of large commercial jet planes.

thumb, the amount (density × thickness) of material applied should be equal to that of the surface to which it is applied. Since this can lead to using large amounts of damping material, the method of covering the damping material with a sheet metal overlay is often used. To a certain extent, the thickness of the damping layer is not the most important factor. Sandwich layers often are preferred due to a greater damping factor than with single layer coatings. (See Figure 18–12.)

Reduction of mechanical disturbance

Mechanical disturbances that produce vibration can be reduced by reducing impacts, sliding or rolling friction, or unbalance.

In all cases, either mechanical energy is coupled into mechanical-vibratory energy, or energy in some other form is transformed into mechanical-vibratory energy. These other forms of energy include varying hydraulic forces, aerodynamic forces, acoustic excitation, and thermal changes.

Often mechanical vibration can be reduced by (a) proper balancing of rotating machinery, (b) reducing response of equipment to a driving force, and (c) proper maintenance of machinery. See Chapters 15–17.

Limitations of vibration control methods

Certain limitations should be noted in the control of vibration:

1. Reduction in transmissibility can only take place by allowing the isolator to deflect. Thus, deflection clearances must be provided around the equipment that is to be isolated.

2. The vibration isolator may actually amplify the vibration characteristics. Select a spring mounting so that the natural frequency of the spring-mass system is considerably (at least one-half) lower than the lowest frequency component in the force system produced by the machine.

3. If the vibration isolator has unexpected nonlinear characteristics, unusual response effects may take place.

All vibration problems should be approached by determining first if a quick, simple, and "common sense" solution is available. If a simple answer is not obvious, the quantitative results of measurements become essential in the analysis and solution of the problem. As various control procedures are tried, vibration measurements can be used to show the progress being made.

References

1. *The Industrial Environment—Its Evaluation and Control.* U.S. Department of Health, Education, and Welfare, National Institute for Occupational Safety and Health, U.S. Government Printing Office, 1973.
2. Beranek, L. L., ed. *Noise and Vibration Control.* New York, N.Y.: McGraw-Hill Book Co., 1971.
3. Beranek, L. L., *Noise Reduction.* New York, N.Y.: McGraw-Hill Book Co., 1960.
4. Blake, M. P., and Mitchell, W. S., eds. *Vibration and Acoustic Measurement Handbook,* New York, N.Y.: Spartan Books, 1972.
5. Brock, S. T. *Application of B. & K. Equipment to Mechanical Vibration and Shock Measurements.* Soborg, Denmark: K. Larsen and Sons, 1972.
6. Harris, C. M. ed. *Handbook of Noise Control.* New York, N.Y.: McGraw-Hill Book Co., 1957.
7. Peterson, A. G., and Gross, E. E. *Handbook of Noise Measurement.* Concord, Mass.: General Radio, 1972.
8. Yerges, L. F. *Sound, Noise and Vibration Control.* New York, N.Y.: Van Nostrand Reinhold Co., 1969.

Chapter Nineteen
Sound Level Specifications

Julian B. Olishifskl, P.E.

Specifications for permissible sound levels for machinery and equipment are an important part of an industrial noise control program. A sound level specification should include the permissible sound level, where and how this level is to be measured, and the type of sound measuring instrument to be used. A typical sound level specification is shown at the end of this chapter.

Objectives

The initial step in planning an acoustically acceptable noise environment is to specify acceptable noise levels for the work place and its surroundings. This involves preparing sound level criteria that set limiting values on noise levels to meet one or more of the following objectives:

• To minimize hearing loss to employees

• To minimize speech interference

• To avoid disturbance and annoyance to the nearby community

Each of these objectives defines a different set of noise control criteria. The end result of an effort to minimize occupational hearing loss may not necessarily satisfy community complaints and vice versa. In each circumstance, it is desirable to make a clear definition of the noise control objective and then to establish sound level criteria for meeting it.

Ultimately, each noise control objective must be expressed as a set of permissible noise levels and then incorporated into the engineering design and performance specifications applying to each piece of equipment or noise source.

Hearing loss criteria

Hearing loss or conservation criteria provide a limit on plant noise levels in terms of maximum acceptable exposure. A good practice to follow for administrative, economic, and safety reasons is to impose an initial noise control design limit for hearing conservation throughout a plant. Allowable noise limit for certain areas may be then lowered for speech communication purposes. The general noise level at the plant boundary lines may then have to be reduced to minimize community annoyance and complaints.

Hearing conservation criteria should apply only to locations where personnel are exposed. Actual locations should be used when known, but assumptions will have to be made for engineering design purposes when the noise exposure to employees is not at a constant level.

Speech interference

In a plant which meets the in-plant hearing conservation criteria, some noise reduction in the speech frequencies may be required in critical locations. Although face-to-face conversation may be possible in high noise areas, communication through fixed speaker systems may be impractical.

Consequently, once the criteria are established, allowable noise levels can be assigned to different parts of a plant to protect the hearing of the workers; for example, all the heavy manufacturing or operating areas can be assigned a maximum 90 dBA noise level and the sound level criteria for light assembly operations can be set at 70 dBA. At the plant's property line, a 55 dBA maximum can be set in the direction of a community, and 70 dBA in directions where only industrial or agricultural land adjoins.

If equipment is to be located near an area where employee communications is a factor, low noise levels are probably justified. However if employees are in the area only briefly, the extra expenditure for quiet machines may not be necessary.

Neighborhood reaction to noisy equipment is another major consideration. A piece of machinery may have a sufficiently low noise level to be acceptable in a manufacturing facility, but not if located in a quiet residential area or if used late at night.

When designing a new plant or operation, a sound level performance specification for machinery and equipment should be written, and the information furnished to the design engineer, design contractor, construction contractor, and equipment suppliers. The use of sound level specifications also provides a procedure for receiving information on noise levels of essential production equipment which exceeds the permissible limits. (See the sample specification on pages 514–524.)

Successful planning for noise control involves (a) the knowledge of the noise characteristics of each machine, process, and acoustical environment, (b) the selection of noise control criteria, and (c) the geographical isolation of noisy operations whose control is not practical.

Plant layout

Plant layout becomes an important factor in controlling employee noise exposure. However, before deciding on a location in the plant for high noise level equipment, answers should be obtained to the following questions:

1. Is the noise produced intermittent or continuous?

2. Is the noise source a single machine or a multiple installation?

3. Is an operator required? At what times?

4. How many employees will be exposed to the noise?

5. Can the equipment be enclosed?

6. What are the ambient noise levels?

The answers to the above questions may be helpful in locating the equipment so as to materially reduce noise exposure.

The proposed plant layout can be used to set the allowable noise level for each machine, or to determine necessary acoustical properties of construction materials. Sometimes, noise levels at certain points are calculated from known or assumed noise characteristics of adjacent equipment. Such information can be used to determine whether the noise level of some of the equipment will have to be lowered.

Equipment operating conditions

A single machine producing intermittent bursts of noise may be located with much more freedom than equipment producing steady noise of the same magnitude. Equipment purchased for intermittent duty quite often comes into full-time operation with a resulting noise problem. In the same manner, a single noisy operation has a way of expanding into a multiple-machine installation that must be handled in a far different manner.

A single machine can usually be isolated acoustically before or after installation. Multiple machine installations require planning if individual enclosures are to be attempted or if machine spacing and room treatment are to be combined to yield minimum noise levels between machines.

Remember, too, that noise from machine tools does not radiate uniformly in all directions. So, just checking at the operator's normal ear level doesn't give a true picture of the sound field around the machine. When installing machines in cluster or groups, the rear of one unit may be close to the ear level of the operator at an adjacent machine.

Normal modes

Permissible noise levels should apply to normal or typical operating modes—startups, standbys, and shutdowns being included as necessary. If there is more than one mode of normal operation, the permissible level should apply to the noisier condition. If emergency condition noise levels are required, these levels should be specifically included. The resulting sound level specification should be very clear as to the conditions assumed. Specifications that cannot be checked for compliance because of unusual operating conditions are difficult to interpret and should be avoided.

When sound level tests show that the specification has been exceeded, some suppliers may claim that the noise is from another source or that it is due to a buildup of reflected noise in that area. When sound level tests are performed correctly, these claims can be easily disproved. If there appears to be a valid exception or if the agreed upon sound level test is unworkable, expert guidance from acoustical specialists should be obtained before agreeing to any variances from the specifications.

Tolerance

Design and performance specifications must have tolerances or built-in margins for unavoidable variations in noise levels. Several factors to be considered in setting noise tolerance levels are limits listed here:

1. Most manufactured equipment in operation will produce sound levels within a specified range, although some units will exceed the noise specification tolerance limit.

2. Noise levels tend to increase with age, wear and tear, lack of main-

tenance, or load conditions of the equipment.

3. The measured equipment noise level may be increased by reflection from nearby walls or objects.

4. Equipment may meet an average noise level limit but may produce directional noise which exceeds the specification when measured in a certain direction.

5. An allowance may be necessary where knowledge of noise levels of individual sources is not precise.

Economics

The most economical noise control design ordinarily involves separate sound level or performance specifications for each type of equipment. Performance specifications should apply to individual noise sources so that the equipment supplier can be held accountable. What equipment is included must be clearly defined; and responsibility must be assigned for each piece of accessory equipment, both when it operates alone and when the complete system is in operation.

Unfortunately many original equipment suppliers may not know how to reduce noise levels and many noise sources do not lend themselves to sound level type specifications. Therefore it is necessary to take a systematic approach which starts early in the equipment design stages.

Engineering specifications for selection of equipment should also incorporate a requirement for noise level data. In most cases, the original equipment manufacturer is in the best position to reduce the noise of the machine at its source with built-in designs. The cost of reducing noise levels after a piece of equipment has been installed may greatly exceed the original factory-installed cost of noise controls for the machine. Noise levels should be one of the initial considerations in the design of new equipment or facilities, or the modernization of existing ones.

Greater horsepower, higher speed, and higher production rates generally mean new equipment has higher noise levels than the older equipment it replaces. Designers have, therefore, an increasingly difficult task to reduce machinery and equipment noise in the face of the user's relentless demand for greater productivity. In a typical purchase negotiation, this dilemma usually forces the buyer to compromise. If he wants prompt delivery and assured maximum production capability, he must often accept noise levels that will require the operator to wear hearing protection or require later modification of the machine or equipment in order to reduce the noise levels.

When an equipment or machinery manufacturer receives a request for a quotation, the manufacturer normally uses his stock equipment as a basis for his bid. If the customer feels that the equipment is too noisy, the manufacturer quotes special designs and acoustic treatment and the purchaser then selects the optimum alternative. If there are a number of suppliers, the purchaser should consider the relative noise outputs when comparing prices and other factors that determine which product will be chosen.

Quieter equipment frequently is more expensive. This extra initial cost is, however in most cases, much less than the cost of the modification necessary to correct a noise problem after installation of the equipment. One should consider how much the application of noise control measures will add to the cost of a less expensive but noisy machine.

If all of the equipment under consideration fails to meet the specified noise level requirements, the purchaser must then decide which of the products are more amenable to noise control.

Acoustical data

Unfortunately acoustical information for design and performance specifications for different types of machines is not always expressed in the same form. In fact, there is an overwhelming diversity of forms. And, while a certain amount of diversity may be necessary, it must be recognized that there are some significant differences in the requirements for acoustical information for different types of applications. It is, therefore, necessary to offer a choice of alternatives to match the requirements of different applications.

Sound level data for comparison and hearing-conservation purposes should be expressed as suitably weighted numbers so that they can be used by people who are not acoustical experts.

Sound power vs. sound pressure

A listener hears sound as a result of variations in the sound pressure level at the point of observation, and this is also what the sound pressure level meter detects. For a given noise source, however, the sound pressure level varies with the acoustical environment. For example, a noise source will sound louder if placed in a hard-walled room than in a so-called "dead" room, which is carpeted and has drapes and upholstered furniture. However, the sound *power* output of the noise source is the same in both rooms. (Also see Chapters 2 and 3.)

Sound pressure describes conditions at the listener location. It is appropriate for characterizing the output of a noise source only if the acoustical transmission path to the listener does not differ much between various environments. Sound power data is usually necessary for taking into account different acoustic environments. There are many cases which require some judgment as to the proper method of expressing sound data.

Sound pressure data, for instance, are useful for large machines which have no clearly defined operator location if the data are averaged over the anticipated operator locations; for example, a path around the machine at ear level and arm's length from the machine.

Sound power data are preferred for machines which are small enough to be tested under laboratory conditions, provided that the directivity sound pattern is not a significant factor.

Suppliers of machinery and equipment, or the purchaser, could elect to have noise data expressed in terms of "best-" or "worst-case" sound pressure levels (SPL) or sound power levels (PWL) at specific frequencies. The decision would depend on the facilities available to the manufacturer when developing the data and the needs of the equipment purchaser.

When free-field sound power levels are listed in the specification for the equipment operating in a particular location, the sound level at a particular distance from the "hearer" can be estimated with reasonably good accuracy. When sound pressure level data are provided they should reflect either the very worst case (maximum noise level for a machine operating and tested in a reverberant environment), or the very best case (an operational free-field situation). Generally, industrial environments lie somewhere between these two limits, and sound pressure levels at a hearer's location can also be expected to be within these extremes.

Single weighted number vs. octave-band analysis

Sound power data, as well as sound pressure data, can be expressed either as frequency spectra or as suitably weighted single numbers.

Engineering design and performance specifications for noise control must provide information as to the sound levels at various frequencies because sound transmission depends on frequency. Octave-band spectra generally are sufficient, but one-third octave-band spectra should be provided when there are significant discrete frequency components.

509

Octave-band data are essential in acoustical design and in engineering noise control, but noise exposure of employees is generally evaluated in terms of A-weighted sound levels.

The A-scale is by far the simplest and, therefore, the most common weighting method. It is often the only practical method for survey measurements. Other weighting methods such as the "perceived noise" method correlate better with subjective judgments of unsteady noises and sounds containing discrete frequency components.

Governmental regulations

With the adoption of the Occupational Safety and Health Act of 1970, the federal government has established maximum level and duration of exposure to noise for employees. The Act does not specify machine or equipment noise levels, only operator noise level and duration of exposure. (See Appendix A-2.)

The noise limits set by the Act are stated as maximum permitted periods of exposure to specified noise levels, measured on the A-weighted network of a sound level meter, and expressed in decibels sound pressure level at the point of hearing. The user or specifier needs to know what the resulting total noise level at the point of hearing will be when all the mechanical equipment is installed and in operation. This will depend on the following:

1. The sound power level of the noise source

2. The distance from the noise source to the point of hearing

3. The size and acoustical characteristics of the room

4. The presence of other noise sources in and around the work space

The sound level specification that is made part of the purchasing agreement must be more than just some general compliance statement, such as, "Must meet the requirements of OSHA." The buyer should advise the machinery or equipment manufacturer in advance (a) the maximum overall dBA level that he will accept, (b) the location, distance, number of noise readings to be taken at "idle" and "operating" conditions, (c) that a detailed sound level report will be required, and (d) what steps to take if the sound level specifications cannot be met.

It is important to realize that the OSHAct sets allowable noise limits in relation to the exposure of the *people* involved. The Act does not relate directly to noise generating equipment. It is not intended to be used as an equipment specification and cannot in practice be used as such.

Despite this fact, a number of machinery and equipment users and specifiers have requested manufacturers to provide "dBA ratings" on equipment, to provide equipment "in accordance with" the Act, or to guarantee a maximum "dBA level" after installation.

The recording and reporting of data is of utmost importance to the sound level specification. This is the data on which an evaluation of the equipment and the decision to purchase must be made. The sound level specification will serve notice that the purchaser is interested in quieter equipment and this factor will influence his decision to purchase. Consequently, suppliers who wish to remain competitive will have to design quieter equipment.

An engineering design and performance specification for noise control should incorporate these items:

1. The responsibility of the manufacturer for noise control in any new equipment

2. Acceptable instrumentation to be used for noise measurements

3. Specific conditions under which noise tests are to be made

4. Specific noise test procedure

5. Method of recording and reporting the data

6. Maximum noise levels that will be accepted and specific proposals for noise control

The acoustical characteristics of the space in which equipment operates can have major effects on the sound pressure levels at a hearer's location. Machinery manufacturers cannot be aware of such localized conditions, and consequently imposing the burden of "complying with OSHA" upon them is inappropriate as a general condition in the "terms of sale." Since OSHA establishes responsibility for compliance with the Act on the employer, the question then arises how the purchaser would negotiate the terms-of-sale considerations as they relate to allowable noise levels.

Acceptable sound level specifications must, therefore, specify whether they apply to average or maximum noise levels, whether there is a tolerance for measurement accuracy, and whether corrections for background noise and reverberant buildup are to be made. Sound level limits for individual sources in a group must be set lower than the acceptable average sound level to assure that the permissible levels are not exceeded. The permissible sound level, as the reader will recall, is a function of the number of noise sources, the intensity, the frequency spectrum of each source, the proximity of the employee, and the immediate acoustic environment.

Guidelines for sound level specifications

A set of guidelines for sound level specifications for purchasing equipment has been developed by the Subcommittee on Noise of the American Iron and Steel Institute's Committee on Industrial Hygiene. Several manufacturing associations have also established standards for measuring the noise levels produced by their particular type of machinery or equipment. (See References.)

A sound level specification should specify the number and type of noise level measurements which are necessary and provide an outline for a uniform method of conducting and recording noise tests. An engineering specification for noise control should be developed to incorporate, as a minimum, the following items:

1. Purpose

2. Scope or application

3. Instrumentation

4. General test requirements

5. Sound level measurement test procedures

6. Specific requirements

7. Permissible sound levels

Purpose. In this section, the vendor should be made aware that the purpose of the specification is to set limiting values of noise produced by equipment considered for purchase, and that the manufacturer is responsible for meeting these values.

Scope or application. The sound level specification should apply to all equipment. In the event that the purchaser is furnished several pieces of equipment that may connect as systems, then the seller will be responsible for his portion of the system as specified.

Instrumentation. A sound level meter and octave-band analyzer are the minimum instruments needed to perform the required noise tests. The instruments must meet the latest revised standards of The American National Standards Institute. The instruments should be calibrated acoustically as recommended by the instrument manufacturer. (See Part Two—Measurement of Sound.)

General test requirements. When noise tests are performed, only the equipment under test is to be observed and recorded. To ensure this, the background noise levels in the test environment must not contribute to the equipment noise test results. The physical dimension and materials of construction of the test room are essential information which the purchaser must have if he is to correctly interpret the reported noise level data. (See Chapter 5.)

Sound level measurement test procedures. In this section, specific test procedures are established to inform the vendor of the manner in which to conduct the noise tests. It is important that certain requirements for testing be met so that reliable and representative data can be reported and that data from several vendors can be adequately compared.

It is always desirable to know the noise level produced by equipment when it is operating under normal conditions. Generally, the purchasers may have specified other performance tests prior to shipping the equipment. This is a good time for the vendor to schedule the required noise tests in conjunction with other performance tests.

Specific requirements. The manufacturer should understand that he is responsible for conducting and recording noise tests according to a uniform standard method. The tests are to be made by the vendor at his plant, as approved by the purchaser. The equipment manufacturer should also be required to discuss his suggestions for controlling the noise output of his product. In doing this, he should provide a cost breakdown for noise control.

Permissible sound levels. The primary purpose for including sound level specifications for noise control of equipment is to inform the vendor of the requirements for acceptable noise levels. Acceptable noise levels for equipment should be such that all personnel exposures are within current guidelines to prevent occupational noise-induced hearing loss.

Sample specification

The following specification is intended only as a guide to make sure that adequate data is recorded. Because of the wide range of differences in noise sources, a standardized data form is not practical for the industry; therefore, the specification should be tailored to meet individual requirements. Because certain aspects of the noise problem lack definition, this specification should be used as a guide and modified to suit the particular test situation. It should be amended to incorporate improved techniques and procedures as they are developed.

SOUND LEVEL SPECIFICATION
FOR MACHINERY AND EQUIPMENT

1. Purpose

1.1. This specification is intended to provide a uniform method for conducting and recording sound level tests to be made on machines and equipment to be purchased by this company.

1.2. The purpose of this specification is to encourage vendors to consider noise abatement in the design of their equipment to control noise at the source.

2. Scope

This specification outlines the methods of testing including operating conditions, measurement techniques, instrumentation, and reporting of data.

2.1 Application

2.1.1. This specification applies to all stationary and mobile equipment and machinery that may produce steady or intermittent noise including all machine tools, material handling devices, pumps, hydraulic power systems, gear drives, washers, collectors, and all other equipment that generates noise.

2.1.2. This specification applies to individual items of equipment furnished separately or to several pieces of equipment connected to form a system and furnished by one vendor who will be responsible for reporting the sound level for the complete system.

---

3. Instrumentation

All instruments used to measure or record sound shall meet the applicable latest revised standard of the American National Standards Institute.

3.1. *Selecting instruments*

Sound pressure level measurements shall be made using a minimum of a sound level meter and octave-band filter set. Other instruments are to be used as deemed appropriate for the type of noise being measured and the data desired.

3.1.1. *Microphones*

All microphones shall meet the requirements of ANSI S1.12 (R1972), *Specification for Laboratory Standard Microphones*.

3.1.1.1. *Condenser*

A condenser microphone or its equivalent in accuracy, stability, and frequency response is recommended.

3.1.1.2. *Ceramic*

Where a cable length of 10 ft or more is required to couple a ceramic microphone to the instrumentation, a preamplifier, located at the microphone, shall be used.

3.1.1.3. *Dynamic*

Dynamic microphones are not recommended because of their sensitivity to magnetic fields.

3.1.1.4. *Frequency response*

A microphone should have a flat response to a noise source located normal to it in a free field, because it reduces the probability of error due to extraneous noise sources.

3.1.2. *Sound level meters*

Sound level meters used to measure steady-state noise shall meet the requirements of ANSI S1.4–1971.

3.1.3. *Octave-band filters*

Octave-band filters shall meet the requirements of ANSI S1.11–1966 (R1971), *Specification for Octave, Half-octave, and Third-octave Band Filter Sets*.

3.1.4. *Recorders*

Two types of recorders are available for use: the tape recorder to preserve the sound signals on tape for later measurement and the graphic level recorder to monitor and furnish a graphic record of the sound over an extended period of time.

3.1.4.1. A tape recorder must be capable of having a frequency response that is flat ±3 dB from 50 Hz to 8000 Hz.

3.1.4.2. A comparison shall be made between the original and the tape recorded noise to determine that clipping of the signal does not occur.

3.1.5. *Calibration of instruments*

Measurements shall be initiated each day by performing the instrument field calibrations recommended by its manufacturer. These shall be verified at the end of each day. If this check indicates a deviation of 1 dB from the established calibration the data acquired that day shall be invalid.

4. General test requirements

Noise measurements shall be made by the vendor in conjunction with other required performance tests prior to shipping the equipment to the purchaser. The purchaser reserves the right to send qualified representatives to the vendor's plant to observe or conduct noise tests described in this standard.

4.1. *Location of tests*

Noise measurements shall be made at the vendor's test location, unless otherwise specified.

4.2. *Test space*

A plot plan shall be included on the data sheet. It shall describe the test space (area), including all major reflecting surfaces (such as walls, cabinets, control panels, etc.), within 30 ft of the machine envelope.

4.2.1. A brief description of the major reflecting surface materials shall be provided.

4.2.2. When tests are not conducted in a room, a sketch showing the position of the equipment and measurement locations should be submitted.

4.3. *Ambient noise levels*

To obtain an accurate measure of the noise produced by a machine, the ambient noise level should meet the following conditions:

4.3.1. The ambient noise level at the frequency band being

measured should be at least 10 dB lower than the noise level generated by the machine.

4.3.2. The ambient level must remain steady for the duration of the test, or if varying, should not exceed a level 10 dB below that of the machine under test.

5. Test procedures

5.1. *General*

5.1.1. Measurements shall be performed at equally spaced positions around the equipment. The first measurement shall be taken at the position of highest sound level with subsequent measurements taken at intervals around the equipment as shown on Figure 19–1, on pages 518–520.

5.1.2. There shall be maximum spacing of 6 ft between measuring positions. The microphone shall be placed at an elevation of 5 ft above the base. The horizontal distance from the equipment surface shall be kept to a minimum but not less than 3 ft (and not greater than 6 ft). For equipment having established operator's positions, additional noise measurements shall be made at each operator's position.

5.1.3. The noise measurement locations selected should include all noise sources.

5.1.4. The rectangular measurement envelope need not include complete structure length if all sound sources are included.

5.1.5. Measurement points may be offset from the established pattern if desired measuring location is inaccessible.

5.2. *Types of noise*

5.2.1. *Steady-state noise*

5.2.1.1. Noises that are continuous or that consist of impulses spaced less than one second apart are considered to be steady state noises.

5.2.1.2. The following measurements should be obtained at each location:

a. Sound pressure level, dBA, with meter on SLOW response.
b. Overall sound pressure level with meter on SLOW response.
c. Octave-band spectral analysis. The octave bands shall be those specified in ANSI S1.11–1966 (R1971) whose center

XYZ COMPANY

NOISE MEASUREMENT DATA

A. MACHINE SPECIFICATIONS

BUILDER _____ BUILDER'S NO. _____ BUYER'S P.O. NO. _____
EQUIPMENT SPECIFICATION: TYPE _____ MODEL _____
SERIAL NO. _____ SIZE _____ CAPACITY _____
SPEED _____ HORSEPOWER _____ AUXILIARIES _____
_____
_____

B. INSTRUMENTATION

INSTRUMENT	MODEL	SERIAL NO.	CALIBRATION DATE
SOUND LEVEL METER			
MICROPHONE			
CALIBRATOR			
TAPE RECORDER			
ANALYZER			
IMPACT METER			
OSCILLOSCOPE			

C. CONCLUSIONS

_____
_____
_____
_____
_____
_____
_____
_____
_____

D. CERTIFICATION

CERTIFIED BY:
NAME: _____
POSITION: _____
COMPANY: _____

Courtesy National Machine Tool Builders Association.

FIG. 19–1.—Page 1 of sample form.

TEST LOCATION_____

TEST SPACE DESCRIPTION_____

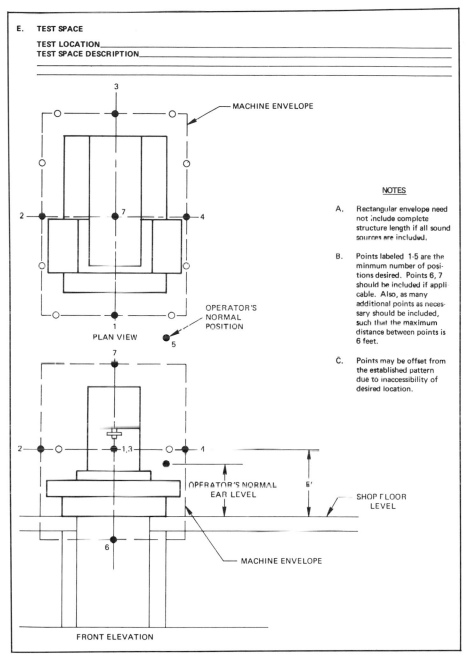

NOTES

A.   Rectangular envelope need not include complete structure length if all sound sources are included.

B.   Points labeled 1-5 are the minimum number of positions desired. Points 6, 7 should be included if applicable. Also, as many additional points as necessary should be included, such that the maximum distance between points is 6 feet.

C.   Points may be offset from the established pattern due to inaccessibility of desired location.

PLAN VIEW

OPERATOR'S NORMAL POSITION

MACHINE ENVELOPE

OPERATOR'S NORMAL EAR LEVEL

SHOP FLOOR LEVEL

MACHINE ENVELOPE

FRONT ELEVATION

Page 2 of sample form.

F.   TEST DATA

OBSERVER: _____ DATE: _____

Test Point	dB(A)	dB(C)	Peak dB of Impulse	Center Frequency - Hertz							
				63	125	250	500	1000	2000	4000	8000
Load Conditions No. 1 (Describe) _____											
_____ Min. Duration Per _____ Min. Cycle _____											
Load Conditions No. 2 (Describe) _____											
_____ Min. Duration Per _____ Min. Cycle _____											

Pages 3 and 4 of sample form.

frequencies range from 63 to 8000 Hz. Where one-third octave filters are used, the octave-band levels shall be determined by adding the appropriate one-third octave-band levels on an energy basis.

d. The average level shall be used when the noise level fluctuates *randomly* over a range of 5 dB or less. Where such fluctuations exceed 5 dB, the maximum and minimum levels shall be reported.

5.2.2. *Cyclical noise*

5.2.2.1. Cyclical noise levels are those that change repetitiously in excess of 5 dB during machine duty cycle. The various levels and their duration shall be recorded along with a listing of their causative events.

5.2.2.2. The following measurements should be obtained at each location:

a. Sound pressure level, dBA, with meter on SLOW response.

b. Overall sound pressure level, dBC, with meter on SLOW response.

c. For engineering evaluations, it may be desirable to use FAST meter response when noise levels with durations less than three seconds are encountered.

d. Octave-band spectral analysis. The octave bands shall be those specified in ANSI S1.11–1966 (R 1971) whose center frequencies range from 63 to 8000 Hz. Where one-third octave filters are used, the octave-band levels shall be determined by adding the appropriate one-third octave-band levels on an energy basis.

5.2.3. *Impulse noise*

Impulse noises shall be considered to be singular noise pulses of less than 1 second duration or repetitive noise pulses occurring at greater than 1 second intervals.

5.2.3.1. The peak levels, durations, and frequency of occurrence of noise impulses shall be measured.

5.2.3.2. Weighting networks or filters shall not be used when measuring impulse peak levels.

5.2.3.3. Duration shall be taken as the time from the initiation of the pulse until the envelope of the positive noise peaks decays 8.7 dB from its maximum.

5.2.3.4. Impulse parameters shall be determined from measurements of at least 10 impulses. The average values and the range of values shall be reported.

5.3. *Operating mode*

Each period where the noise level changes 5 dB or more at any test point shall be considered to be a change in load condition. The description of each load condition must indicate overall machine operating conditions or performance as well as the specific operation or duty cycle events.

5.3.1. *Measurement*

5.3.1.1. Sound levels shall be measured during the simultaneous operation of all components and applicable accessories when running under all anticipated operating conditions.

5.3.1.2. Measurements taken under the conditions defined

under Section 5.2 shall be made as follows:

a. At all probable unloaded modes such as idle or maximum traverse.

b. At the various rated load modes of operation unless otherwise specified.

5.3.2. *Load conditions*

5.3.2.1. The vendor shall state whether the load conditions were actual or simulated; and if simulated, he shall state the means of simulation.

5.3.2.2. The vendor shall be provided parts or materials by the purchaser if these are peculiar to the purchaser, so that actual loaded conditions can be measured.

5.3.3. *Exceptions*

Exceptions shall be only the following devices, when furnished as nonintegral components, which shall be tested under the operating conditions specified by the applicable industry standard.

5.3.3.1. *Nonducted fans*

Sound pressure levels for nonducted exhaust and supply air-moving devices shall be in accordance with the Air Moving and Conditioning Association (AMCA) *"Sound Test Code,"* Bulletin 300–67.

5.3.3.2. *Compressors*

Compressors shall be measured and the results reported in accordance with CAGI-PNEUROP Test Code, latest edition.

5.3.3.3. *Electric motors*

Testing of equipment and reporting of data shall be in accordance with IEEE Publication No. 85, *Test Procedure for Airborne Noise Measurements on Rotating Electric Machinery,* latest edition.

5.3.3.4. *Pneumatic tools*

Air motors, hand grinders, screwdrivers, nutsetters, impact wrenches, etc., shall be measured and reported in accordance with CAGI-PNEUROP Test Code under a "running free" condition.

6. Specific requirements

6.1. *Measurements*

Measurements shall be the responsibility of the vendor. Results shall be certified.

6.2. *Quotation*

The quotation shall include the cost required to meet this sound specification, stated as an alternate and separate item to the cost of the vendor's standard product.

6.2.1. *Data*

6.2.1.1. Vendors providing standard equipment, such as catalogued multipurchased and repeatedly purchased items, shall submit certified data on the Noise Measurement Data Form (see Figure 19–1) with their quotation.

6.2.1.2. Vendors providing special-purpose machines without previously validated sound level measurements shall submit data on the machine as soon as the data becomes available and prior to shipment.

6.2.2. *Test location*

Where it is a common accepted practice or where the vendor deems it impractical to set up and test a complete machine at the vendor's facility, he shall so state in his quotation and request arrangements to perform the test at the purchaser's facility under agreed and negotiated conditions. The vendor shall remain responsible for meeting the sound level specifications relating to the *complete machine* including all subassemblies and components.

6.3. *Specified limitations*

Where the vendor cannot meet the specified sound limitations as described in Section 7 and he can demonstrate that he has fully explored engineering solutions, he shall pursue the following course of action:

6.3.1. Provide the purchaser with a certified octave-band analysis for each test location, for each load condition, together with sound level duration per work cycle in sufficient detail to permit calculations of equivalent employee daily noise exposures.

6.3.2. Provide the purchaser with all information necessary to institute corrective procedures for use after installation.

6.3.3. If the vendor is unable to provide meaningful information, he may request the assistance of the purchaser.

6.3.4. Vendor must request and receive written permission to ship any machine exceeding the specified sound limitation.

6.3.5. Shipping of equipment does not release the vendor from his responsibility to meet the specified sound limitations.

6.3.6. If the vendor has not been specifically relieved in writing of the foregoing requirements he must provide a plan, including a timetable, for corrective action.

6.4. *Special noise problems*

Locations which create special noise problems not covered by these specifications (such as ambient neighborhood noise) require special consideration and can be provided for by including supplementary specifications in respect to the following:

6.4.1. Noise levels as stated in Section 7.

6.4.2. Load or load range during test as stated in Section 5.3.

6.4.3. Microphone positions as stated in Section 3.

6.4.4. Number of readings as stated in paragraph 5.1.2.

7. Permissible sound levels

7.1. This specification is a means of establishing the limiting value of noise (unwanted sound) generated by equipment to be installed in an industrial plant.

7.2. Unless otherwise specified, octave-band analysis will be used to measure equipment sound levels. Maximum levels suitable for the application intended shall be specified by purchaser for each purchase. Sound level specifications for special specific equipment to be installed in environments are to be supplied by purchaser.

7.3. At the point where the nearest listener will be for significant periods of time, all steady-state or cyclical sound levels produced by the machinery shall not exceed 80 dB when measured by a sound level meter meeting ANSI Standard S1.4–1971 *Specification for Sound Level Meters* set to A-weighting and to SLOW response.

8. Special and lower sound limitations

The sound pressure level of _____ dBA shall not be exceeded at any of the specified measurement locations. This limit supersedes the 80 dBA limitation. This option may be imposed and specified herein.

9. Evaluation

Sound level data submitted with quotations on machinery and equipment shall be used in the evaluation of bids. Other factors being equal, the quietest machine or equipment shall be selected.

Bibliography

Air Moving and Conditioning Association, Inc. *Test Code for Sound Rating Air Moving Devices, Standard 300–67.* Arlington Heights, Ill.: AMCA, 1967.

American National Standards Institute, New York, N.Y.: *Methods for the Determinations of Sound Power Levels of Small Sources in Reverberation Rooms,* S1.21–1972.

Method for the Physical Measurement of Sound, S1.2–1962 (R1971).

Specification for General-Purpose Sound Level Meters, S1.4–1961 (R1970).

American Society of Heating, Refrigerating, and Air-Conditioning Engineers, Inc. Chapter 33, "Sound and Vibration," *Guide and Data Book.* New York, N.Y.: ASHRAE, 1970.

Compressed Air and Gas Institute. *CAGI-PNEUROP Test Code for the Measurement of Sound from Pneumatic Equipment.* New York, N.Y.: Compressed Air and Gas Institute, 1969.

Dryden, S. L., and Judd, S. H. "Must Your Plant Be Too Noisy?" *American Industrial Hygiene Journal* 34:241–51.

Fair, E. W. "Shall We Buy That New Equipment?" *Chemical Engineering,* October 1973, pp. 124–126.

General Motors Corporation *Sound Level Specification for Machinery and Equipment.* Detroit, Mich. GMC, 1971.

"Guidelines for Noise Control Specifications for Purchasing Equipment," Subcommittee on Noise of the Committee on Industrial Hygiene, American Iron and Steel Institute, 1969.

"How to Buy Quiet Machine Tools," *Factory,* May 1972, pp. 31–33.

The Institute of Electrical and Electronics Engineers, Inc. *Test Procedure for Airborne Noise Measurements on Rotating Electric Machinery.* New York, N.Y.: IEEE, 1965.

National Electrical Manufacturers Association. *Sound Level Prediction for Installed Rotating Electrical Machines.* New York, N.Y.: NEMA, 1974.

The National Fluid Power Association, Inc. *Method of Measuring Sound Generated by Hydraulic Fluid Power Pumps.* Thiensville, Wis.: NFPA, 1970.

National Machine Tool Builders' Association. *Noise Measurement Techniques.* McLean, Va.: NMTBA, 1970.

"Noise, The Nuisance That Has Become a Menace, Challenges Designers." *Product Engineering,* October 1972, pp. 18–20.

Simpson, C. W. "Use of Purchase Safeguards for Noise Control," *Transactions of National Safety Congress, 1969,* Vol. 15, pp. 19–22.

Chapter Twenty

Personal Hearing-Protective Devices

By Richard L. Swift

Personal hearing-protective devices are acoustical barriers that reduce the amount of sound energy transmitted to receptors in the inner ear. Federal and state occupational safety and health regulations require that whenever employees are exposed to excessive noise levels, feasible administrative or engineering controls should be used to reduce these excessive sound levels. Where these control measures cannot be completely accomplished and/or while such controls are being initiated, personnel must be protected from the effects of excessive noise levels. Such protection can, in most cases, be provided by wearing suitable hearing-protective devices.

The sound attenuation capability of a hearing-protective device at threshold is the difference (in decibels) between the threshold of audibility for an observer with hearing protectors in place (test threshold) and the measured hearing threshold when his ears are open and uncovered (reference threshold).

An important consideration in the design of a hearing-protective device is the attenuation capability or the degree of reduction of sound energy as it passes through the device.

Hearing-protective devices in common use today are generally either insert or muff types. The insert-type protector attenuates noise by plugging the external ear canal, whereas the muff-type protector encloses the auricle of the ear to provide an acoustical seal. The effectiveness of hearing-protective devices depends on several factors that are related to

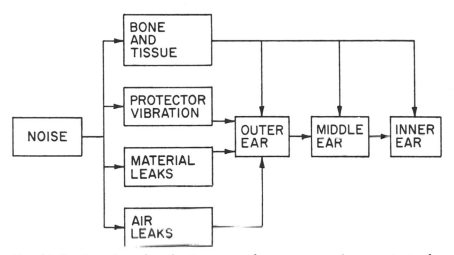

Fig. 20-1.—Sound reaches the inner ear of a person wearing a protector by different methods.

the manner in which the sound energy is transmitted through or around the device. Figure 20-1 shows four pathways by which sound can reach the inner ear when hearing-protective devices are worn: (a) seal leaks, (b) material leaks, (c) hearing-protective device vibration, and (d) conduction through bone and tissue.

Seal leaks. Small air leaks in the seal between the hearing protector and the skin can significantly reduce the low-frequency attenuation or permit a greater proportion of the low-frequency sounds to pass through. As the air leak becomes larger, attenuation becomes less at all frequencies.

Material leaks. A second transmission pathway for sound is directly through the material of the hearing-protective device. That is, the hearing-protective device will attenuate or prevent the passage of most of the sound energy but still allow some to pass through.

Device vibration. A third pathway for sound to be transmitted to the inner ear is when the hearing-protective device itself is set into vibration in response to exposure to external sound energy.

The amount of sound energy transmitted through the protector depends upon the materials of construction, design, and the mass of the device. It is possible, by adding more mass and more attenuating material, to obtain almost any desired degree of attenuation. The amount of attenuation attained is limited only by the cost and the massiveness of the protective device.

Bone conduction. If the ear canal were completely closed so that no sound entered the ear by this path, some sound could reach the inner ear by bone conduction. However, the level of the sound reaching the ear by such means would be about 50 dB below the level of air-conducted sound through the open ear canal. It is obvious, therefore, that no matter how the ear canal is blocked, the protective device will be bypassed by the bone-conduction pathway through the skull. A perfect hearing-protective device worn in or over the ear cannot provide more than 50 dB of sound attenuation.

Hearing-protective devices

Personal hearing-protective equipment can be divided into four classifications:

1. Enclosure

2. Aural insert

3. Superaural

4. Circumaural

Enclosures

The enclosure-type hearing-protective device, as the name implies, is incorporated in equipment which entirely envelopes the head. A typical example is the helmet worn by an astronaut. In such instances, attenuation at the ear is achieved through the acoustical properties of the helmet.

The maximum amount that a hearing protector can reduce the sound reaching the ear is about 35 dB at 250 Hz, and up to about 50 dB at the higher frequencies. By wearing hearing protectors, then adding a helmet which encloses the head, an additional 10 dB reduction of transmission of sound to the ears can be achieved.

The body itself conducts sound through the tissues, so that, unless the entire body is encased in a suitable material, the degree to which the amount of sound reaching the ear is reduced is limited.

Helmets can be used to support earmuffs or earphones and cover the bony portion of the head in an attempt to reduce bone-conducted sound. They are particularly suited for use in extremely high-noise levels and where protection of the head is needed against bumps or missiles. With good design and careful fitting of the seal between the edges of the helmet and the skin of the face and neck, a further 5 to 10 dB of sound attenuation can be obtained in addition to that already provided by the earmuffs or earphones within the helmet. This approach to protection against excess noise is practical only in very special applications. Cost as well as bulk normally preclude the use of helmet-type hearing protection for a general industrial hearing-conservation program.

Aural insert protectors

Aural insert hearing-protective devices are normally referred to as inserts or earplugs. This type is generally inexpensive but the service life is limited, ranging from single-time use to several months. Insert-type protectors or plugs are supplied in many different configurations, and are made from such materials as rubber, plastics, fine glass down, and wax-impregnated cotton. The pliable materials used in these aural inserts are quite soft and there is little danger of injury resulting from accidentally forcing the plug against the tender lining of the ear canal.

It is desirable to have the employee's ears examined by medical personnel before earplugs are fitted. Occasionally the physical shape of the ear canal precludes the use of insert-type protectors. There is also the possibility that the ear canal may be filled with hardened wax. Where wax (cerumen) is the problem, it should be removed by qualified personnel. In some cases, the skin of the ear canal may be sensitive to a particular earplug material and it would be advisable to recommend earplugs that do not cause an allergic response.

Aural insert-type hearing protectors fall into three broad categories or general classifications:

- Formable type

- Custom-molded type

- Molded type

Formable protectors. Formable type of hearing-protective devices can provide good attenuation and fit all ears. Many of the formable types are designed for a one-time use only and then thrown away. Materials from which these disposable plugs are made include very fine

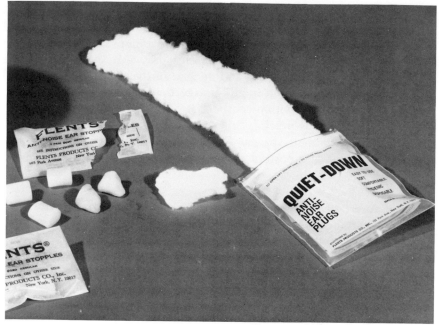

Courtesy of Flents Products Co., Inc.

Fig. 20–2.—Many formable types of hearing protectors are available. Among these are glass fiber and wax-impregnated cotton.

glass fiber (quite often referred to as Swedish wool) and wax-impregnated cotton and expandable plastic (see Figure 20–2).

These materials are generally rolled into a conical shape before being inserted into the ear. However, adequate instruction must be given to emphasize the importance of a snug fit, while at the same time a certain amount of caution must also be given not to push the material into the ear canal so far that it has to be removed by medical personnel.

Another type of material is a plasticlike substance which is similar in consistency to putty. The preparation of this material requires that the individual take a quantity of it and mold or form it so that it can be inserted into the ear canal. The user should be instructed properly so that he knows the correct method of forming the material. In addition, the user must be cautioned to have clean hands when forming the material and placing it in his ear. If the hands are dirty foreign material can get into the ear canal.

Courtesy General Electric Co.

FIG. 20-3.—The custom-molded type of hearing protector must be individually fitted to each person.

The degree to which the formable-type devices reduce the amount of sound reaching the inner ear is dependent upon a snug fit in the ear canal.

• *Wax plugs* can be shaped to fit almost any ear canal. In general, if properly inserted, wax protectors provide adequate attenuation characteristics.

• *Cotton plugs.* Dry cotton plugs can be fitted to almost any shape or size of ear canal; however, they provide very little attenuation. If cotton is mixed with paraffin wax it becomes much more efficient. Wax-impregnated cotton plugs lack elasticity and after repeated changes in shape caused by jaw movements, the relatively inelastic plug is compressed into a shape that no longer fits tightly.

Although the initial cost of some of the wax-impregnated cotton, plastic or glass fiber plugs is lower than that of other kinds of plugs, the total cost over a long period of time will probably be greater because the disposable plugs must be replaced more frequently.

• *Expandable plug.* A variation of the formable plug is supplied in an expandable-type of material. The earplug can be compressed, or formed, prior to placing it in the ear canal. After such insertion, the material

531

attempts to return to its original shape, thus expanding against the ear canal to form a seal. This type of device can be reused a number of times, but it should be thoroughly cleaned after each use.

Custom-molded protectors. Formable hearing-protective devices in this category are, as the name implies, custom molded for the individual user (see Figure 20–3). Generally, in this type, two or more materials (packaged separately) are mixed together to form a compound that when set resembles soft rubber. For use as a hearing-protective device, the mixture is carefully placed into the outer ear with some portion of it in the ear canal, in the manner prescribed by the manufacturer. As the material sets, it takes the shape of the indivdual ear and external ear canal. In some cases, the materials are premixed and come in a tube from which it can be injected into the ear.

Preparation and insertion of the material into the individual's ear canal must be performed only by trained, competent personnel. Because these protectors can be used a number of times, the user must be thoroughly instructed in the proper method of inserting this type of device into the ear canal in order to achieve a maximum degree of sound attenuation. These devices are considerably more expensive than other types of plugs, but they have a longer service life, which can be prolonged by thorough cleaning after each use, in accordance with the manufacturer's instructions.

Premolded aural insert protector. Premolded-type insert protectors are quite often referred to as prefabricated because they are usually made in large quantities in a multiple-cavity mold. The materials of construction range from soft silicone rubber to plastic.

There are two versions of the premolded insert protector. One is known as the *universal fit* type in which the plug is designed to fit a wide variety of ear canal shapes and sizes (see Figure 20–4). The other type of premolded protector is supplied in several different sizes to assure a good fit (see Figure 20–5). The design of the plug is important; for example, the smooth bullet-shaped plug is very comfortable and provides adequate attenuation in straight ear canals; however, its performance falls off sharply in many irregularly shaped canals.

One of the most efficient and widely used earplugs is commonly referred to as the V-51-R (see Figure 20–5). (The V-51-R formulation and design were developed during World War II by the Army, hence its designation. Although many firms use the same basic design, they have improved the original formula and use their own product designation.) The plug consists of a soft plastic bung having a flexible flange that conforms to the shape of the external ear canal and makes as complete a seal as possible. A flanged plug, such as the V-51-R, may provide slightly less attenuation for very straight canals, but it is one of the most versatile designs in that it provides good attenuation for a wide variety of ear canal configurations.

The use of premolded insert-type protectors requires proper fitting by trained personnel (see Figure 20–6). In many instances, the right and left ear canals of a given individual are not the same size. For this reason, properly trained personnel must prescribe the correct protector size for each ear canal. Sizing devices are available to aid in the proper fitting (see Figure 20–7).

The premolded type of earplug has a number of disadvantages which limit its practical acceptability. In order to be effective, it has to fit snugly and, for some users, this is uncomfortable. Quite often because the plug must fit tightly and because of the irregular shape of the ear canals that many persons may have, an incorrect size of plug is selected, or the plug is not inserted far enough, or a good fit cannot be obtained. Further, the plugs must be kept clean to minimize the risk of infection.

Some premolded type insert protectors may shrink and become hard, which is caused primarily by ear wax (present in all ear canals). The wax extracts the plasticizer from some plug materials with a resultant hardening and possible shrinkage of the plug. The degree of hardening and shrinkage of the plug varies from one individual to another depending on such things as temperature, duration of use, and personal hygiene of the user. Regular cleaning of the protectors with mild soap and water prolongs their useful life. To keep the plugs clean and free from contamination most manufacturers provide a carrying case for storing them when not in use.

Some insert protectors are advertised as being able to pass speech sounds, but attenuate sudden loud noises. A small hole through the body of the plug allows low-frequency sounds (below about 1000 Hz) to pass with little attenuation. The lower speech frequencies are, consequently, able to reach the ear more easily than with a comparable solid plug. These characteristics make the plug with a small hole more suitable for use when communication is required during periods of relative quiet. When worn in the presence of a continuous noise, on the other hand, speech reception may be impaired more than with a solid plug.

(*Text continues on page 536.*)

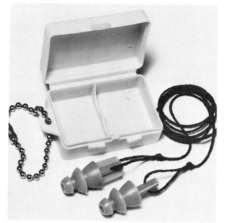

Courtesy Mine Safety Appliances.

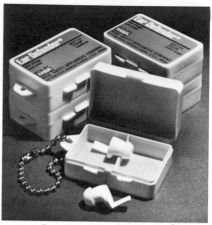

Courtesy Mine Safety Appliances.

FIG. 20–4.—Pre-molded insert protectors are available in the universal-fit type to fit all ears.

FIG. 20–5.—One of the most commonly-used earplugs is the V-51-R, so called from its WW-II designation.

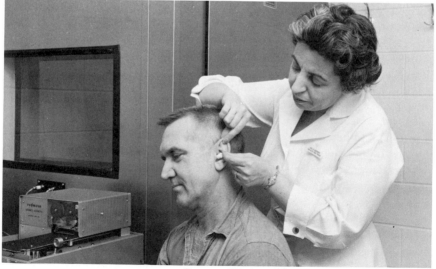

Courtesy Western Electric Co., Hawthorne Works.

FIG. 20–6.—The pre-formed earplug requires proper fitting and instruction in insertion by trained personnel.

534

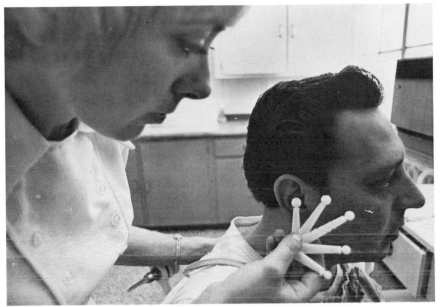

FIG. 20–7.—Rather than determine the size of earplugs by trial and error, a gage such as that shown above can be used.

FIG. 20–8.—Superaural protectors do not plug the ear canal but rather block it at the entrance, such as the type shown above.

FIG. 20–9.—Earmuffs cover the entire ear and rest firmly against the side of the head. The cushion or pad of the cups forms a good seal against intruding noise. Earmuffs are available in a number of styles and shapes.

Courtesy Mine Safety Appliances.

This is because low-frequency masking noise is passed in addition to speech.

Amplitude-sensitive earplugs have a valve in the plug which is momentarily thrust into a closed position by the change in sound pressure caused by the noise. Reports in the literature suggest that amplitude-sensitive earplugs may require extremely high sound pressure levels to actuate the valve. The evident popularity of this earplug in rifle clubs suggests that it may be effective against intermittent impulsive types of noise. It is possible, however, that this may be due to the general sound attenuating properties of the earplug rather than to operation of the valve.

Superaural protectors

Hearing-protective devices in this category depend upon sealing the external opening of the ear canal to achieve sound attenuation (see Figure 20–8). A soft, rubberlike material is used to make the caps normally held in place by a very light band or head suspension. The tension of the band holds the superaural device against the edges of the ear canal.

Circumaural protectors

Circumaural hearing protective devices, usually called earmuffs, consist essentially of two cup- or dome-shaped devices which fit over the entire external ear, including the lobe, and seal against the side of the head with a suitable cushion or pad (see Figure 20–9). In general, the

Fig. 20–10.—Earmuffs arc generally worn with the suspension assembly over the head. However, when protective head gear is worn, it is well to be able to place the suspension brace behind the head or below the chin.

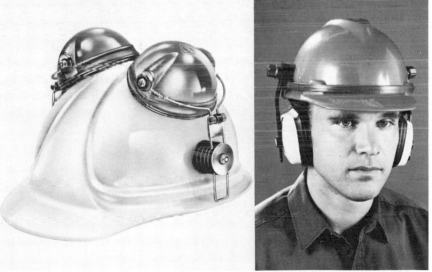

Fig. 20–11.—When it is necessary to wear a hard hat at all times and also be exposed to high noise levels, ear protectors can be attached to the hat.

ear cups are made of a molded rigid plastic and are lined with an open cell-type foam material. The size and shape of the ear cup vary from one manufacturer to another.

The cups are generally held in place by a spring-loaded suspension assembly, or headband. Some earmuff models are designed to be worn only in one specific position, such as with the suspension over the head. Other versions permit placing the suspension assembly in a variety of positions such as: over the head, behind the head (nape type), or under the chin (see Figure 20–10). Where workers must wear head protection at all times but are exposed to intermittent high noise levels, hearing protectors can be attached to a safety hat in such a manner that the muff can be swung up and away from the ears and placed on top of the hat when not needed for protection (see Figure 20–11).

The suspension assembly for over-the-head configuration should be adjustable to allow for differences as great as 4 in. between the entrance to the ear canal and the crown of the head, and each muff must be free to rotate several degrees in both the horizontal and vertical directions to accommodate variations in head shapes. The muff suspension must also provide a force sufficient to provide a good acoustical seal against the head without causing discomfort to the wearer.

The applied force is directly related to the degree of attenuation. The width, circumference, and material of the earmuff cushion resting against the head must also be considered in order to get a proper balance of performance and comfort. The required width of the contact surface to provide a good acoustical seal depends to a large degree upon the material used in the cushion. The cup with the smallest possible circumference that will accommodate the largest ear lobes should be chosen. A slight pressure on the lobe can become painful in time, so it is very important to select a muff dome that is large enough.

The earmuffs currently on the market are supplied with replaceable ear seals or cushions which may be filled with foam, liquid, or air—with the foam-filled type being the most prevalent. The outer covering of these seals is vinyl or a similar thermoplastic material. Human perspiration tends to extract the plasticizer from the seal material which results in an eventual stiffening of the seals. For this reason, the seals require replacement at periodic intervals with the frequency of replacement being dependent upon the conditions of exposure.

Earmuffs have been increasing in popularity for the following reasons:
• There has been a significant decrease in the price of the ear muff-type protector.

- There is a large choice of circumaural models and design which will satisfy a wide variety of hearing-protection requirements.

- Earmuffs generally require no individual fitting, only a simple demonstration at the time of issue is needed to show how the muff is to be worn.

- Because earmuffs are very visible, supervisory personnel can determine more easily that hearing protection is being worn.

Design factors

In the selection of personal hearing-protective devices several design factors must be taken into consideration including:

- Performance

- Comfort

- Communication requirements

- Appearance

- Other Factors

Performance

Reasonably good performance (attenuation) can be achieved by the hearing-protective devices currently being produced in the United States. There are, however, variations from one model to another—some will provide a greater degree of hearing protection at lower frequencies, while others offer better hearing protection in the higher frequency ranges.

The controversy about the relative values of earplugs and earmuffs has gradually died down. Aural inserts are effective only if they are inserted properly in the ear. Preformed molded earplugs do not always provide the necessary protection because some workers insert the plug too loosely. Even if the plugs have been properly inserted, normal jaw movements tend to loosen them and to make them ineffective. The use of preformed earplugs is difficult to police by the supervisor because their small size tends to make them difficult to see at a distance. The custom-molded earplugs which are individually fitted to each ear of an employee have overcome some of the disadvantages of preformed earplugs.

Figure 20–12 shows the low-frequency (125 Hz to 1000 Hz) attenuation capabilities of two very similar earmuff assemblies. Muff No. 1, in this example, has a larger enclosed volume than Muff No. 2, and, in this instance, provides a greater degree of attenuation in the lower frequen-

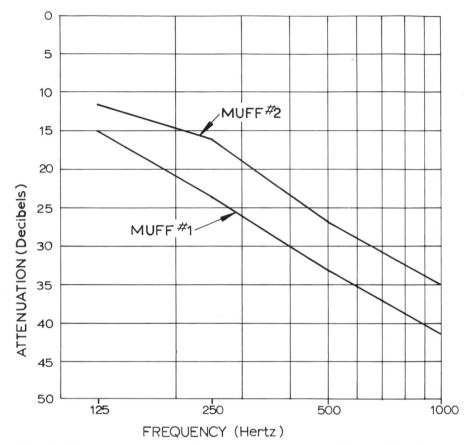

FIG. 20–12.—An attenuation chart comparing two earmuffs: muff No. 1 has a greater enclosed volume and, in this instance, greater attenuation.

cies. On the other hand, the performance efficiency could be reversed in the higher frequencies.

When deciding on the type of protector to be used, look over the manufacturer's attenuation data. Figure 20–13 compares the attenuation capability of a molded-type insert protector and a muff-type protective device. At some lower frequencies, the earplug offers greater attenuation than the muff; however, at higher frequencies, the earmuff generally provides the better attenuation.

If greater protection is required than can be achieved with earmuffs, additional attenuation can be obtained by wearing a combination of plug

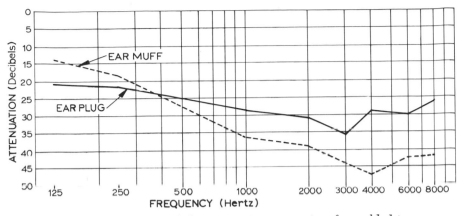

Fig. 20–13.—A comparison of the attenuation properties of a molded type earplug and an earmuff protector. Note that the earplug offers greater attenuation at the lower frequencies while the earmuff is superior at the high frequencies.

and muff. While the reduction afforded by wearing both types of protector simultaneously is considerably less than the sum of the individual attenuations, it is still greater than when either is worn separately.

Comfort

Comfort is one of the most important considerations that must be given to the selection of a hearing-protective device since user acceptance will, to a great extent, depend on this factor. Most protectors are uncomfortable if worn for long periods. Usually earplugs are judged by employees to be less comfortable than earmuffs, despite the latter being heavier, more bulky, and likely to cause perspiration. Aural inserts have to fit snugly in order to obtain maximum attenuation efficiency.

Wearing earmuffs may be uncomfortable in industries where workers are exposed to very high temperatures and humidity, such as in foundries. However, it is possible, even then, to use earmuffs if use is confined to several hours a day at intermittent periods. There are times when a person cannot wear muffs for extended periods of time because they become filled with perspiration. In such instances, special consideration should be given to the type of hearing protection being provided.

Some of the factors which must be considered in the design of an earmuff include the following:

541

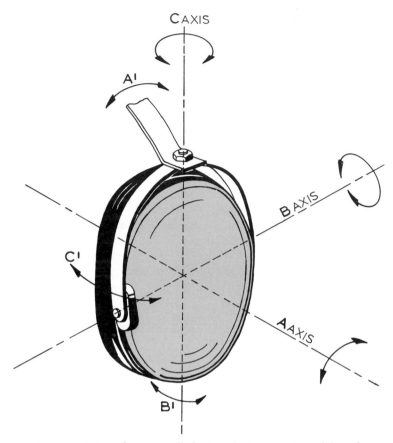

Fig. 20–14.—The earmuff design that is most useful and popular should adjust to a variety of head sizes and shapes.

• There should be no sharp points or edges contacting the wearer's head to make the muff uncomfortable when worn for extended periods of time. Any part of the earmuff that comes into contact with the wearer such as the ear seal, should be made of material which does not absorb perspiration and is easy to clean.

• The ear cups should be adjustable to conform to the various head shapes and sizes (see Figure 20–14). To achieve this they should have the ability to articulate, or turn, to some degree in three dimensions.

Some manufacturers achieve articulation through a yoke design, others use two spring wires with a floating bearing for supporting the cup; while still another uses a ball-and-socket principle.

- The pressure which the two cups exert on the head, commonly known as *head force,* affects both comfort and effectiveness. Generally, as the headband force is increased, the attenuation offered by the device also increases. There are practical limits beyond which the device either provides no better attenuation or becomes too uncomfortbale to wear. The wearer should be cautioned not to deliberately "spring" the suspension beyond its limit to make it more comfortable since this will result in reduced hearing protection.

- Weight is another factor to consider. Although of less importance than the other factors discussed, it certainly must be recognized if the muff is to be worn during the full work period.

Communication requirements

The wearing of a hearing-protective device does not always necessarily reduce the ability to communicate. The hearing acuity of the user, the type of noise, and the type of protective device are all contributing factors. It is important to evaluate the extent of hazard in each situation in order to select the proper hearing-protective device.

When a hearing-protection program is instituted, perhaps the chief complaints of employees are, "we cannot wear hearing protectors because the protectors prevent us from hearing the sounds and voices essential to the operation of our equipment." Hearing protectors in very noisy areas often make it easier—not harder—for employees to understand conversation and to hear the sounds of their machines.

In continuous high-level noise environments, explanation and demonstration will reduce employee resistance to wearing hearing protection. In intermittent noise situations the problem of employee acceptance of hearing protectors is more difficult. Careful consideration must be given to the degree of noise hazard and the risks of impaired communication. The average noise level between bursts may be at a low intensity and be barely audible but the noise peak during the burst may present a hazard to unprotected ears.

In a quiet environment, speech sounds of sufficient intensity will be heard without degradation by a normal-hearing person wearing ear protectors. However, those individuals with a hearing loss will be at a disadvantage when hearing-protective devices are worn.

543

At noise levels of about 85 dB, if the message can be heard at all, then hearing protectors make little difference to its intelligibility. This is because the sound level of both the voice signal and the noise is lowered equally by the hearing protectors, that is, the signal-to-noise ratio is unaltered. At levels above 85 dB, the use of ear protectors may actually be beneficial to communication. The reason is that the ear mechanism is overloaded when the levels of speech and noise are very high, and the hearing protectors simply attenuate both the speech and the noise to a level that the ear can handle more efficiently. This advantage holds true for both insert and muff-type protectors.

A person's impression of the sound level of his own voice when speaking will be altered by the use of hearing protection. In a noisy environment, the speaker must match his voice level against the ambient noise level. The result is that the speaker, wearing hearing protectors, tends to use insufficient vocal effort to overcome the actual noise of his environment.

Appearance

Appearance certainly is an important factor in the selection of a hearing-protective device. The best protector in the world is of no value if it is not worn. It may be helpful to allow employees to select from more than one type of muff or plug to get them to wear hearing-protective devices.

Other factors

Materials used in hearing-protective devices coming in contact with the wearer's skin should be generally nonirritating and exhibit minimum shrinkage, hardening, or cracking during normal use. Materials used should be slow burning or nonflammable. It should be possible to fit and remove the protectors quickly and easily. They should be durable—resistant to perspiration—and nontoxic to the skin.

The cost of a hearing protector should be judged in relation to its expected life and the protection required. A few years ago, the circumaural type protectors cost more than $20 each. Today, however, through better production techniques, quantity purchases can bring this price down to the vicinity of $5 or less. To this, of course, must be added the cost for periodic replacement of the seals.

The longevity of an insert- or plug-type hearing-protective device is limited, so that while the initial cost of the plug is usually considerably less than that of the muff, the frequency of replacement over extended periods of time becomes an important factor when comparing and making the decision between plugs and muffs.

Aural insert-type hearing protectors should not be used when an ear infection is present. The infection may be aggravated particularly when reusable types are not properly cleaned after each use. Earmuffs provide a reasonable alternative in these special cases.

Performance testing

The mean hearing protection and standard deviation of hearing protective devices is determined by the manufacturer in accordance with ANSI Z24.22–1957 or S3.19–1973, *Real Ear Method.* The test method and test facility used by the manufacturer should be clearly stated. When a hearing-protective device is capable of being worn in more than one position, the hearing protection should be indicated in the manufacturers' literature for each position.

Different test methods will show differences in the hearing-protection values for an individual protector. The ANSI S3.19–1973 method, while indicating lower protection at some frequencies, also shows somewhat less data spread.

At the present time, most hearing-protective devices are performance tested in accordance with the American National Standards Institute (ANSI) Standard Z24.22–1957. This method of measuring the attenuation is widely accepted and involves a free field binaural threshold-shift technique. In this, the threshold of hearing for selected pure tones is measured in a free field using both ears of each of a group of subjects with normal hearing. The thresholds are also measured with each subject wearing the selected ear protectors. The average difference between these two thresholds represents the degree of attenuation attained.

In its current form, ANSI standard Z24.22–1957 states that each device will be tested by a jury of ten randomly selected individuals. The base hearing threshold level of each individual is first determined. Then each individual tests the device three different times at each of the following frequencies: 250, 500, 1000, 2000, 4000, 6000, and 8000 Hz. This testing results in 30 values for each test frequency which are totaled and averaged. The averages are then plotted on a graph which becomes the mean *attenuation curve* for that particular hearing-protective device. It represents the difference in sound pressure level or the protection that the hearing-protective device provides (see Figure 20–15). As such, it is a value which is independent of weighting networks. Most manufacturers of hearing-protective devices have this test work done by an independent

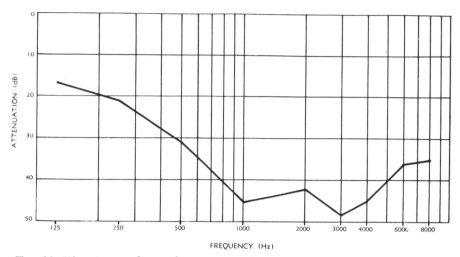

FIG. 20–15.—A typical test of a circumaural hearing protective device showing the attenuation characteristics.

testing agency. Descriptive literature, in general, shows the test results in both chart and tabular form.

Standard deviation

Since the data on hearing protection is an average of the people tested, one-half would realize better than the average degree of protection, while the other half would receive less than that indicated. Such variations from the average are usually referred to as deviations and the term "one standard deviation" refers to a specific variation that would be anticipated among the population being tested.

The standard deviation (σ) is computed using the formula:

$$\sigma = \sqrt{\frac{\Sigma d^2}{N - 1}}$$

where d is the difference between the grand mean and an individual observation, and where N is the number of observations.

Manufacturers' literature normally will tabulate the standard deviation at each of the frequencies tested.

Currently, a number of changes in ANSI Z24.22–1957 are being considered. When evaluating hearing-protective devices, make certain that the reported performance characteristics have been determined in accordance with the latest revision to this standard.

Selection of protector

The attenuation characteristics of a particular hearing protector must be considered before using it for a specific application. As part of a well-planned hearing-conservation program, characteristics of the noise levels for the various areas should be known. From these data and the attenuation information available from manufacturers, it can be determined whether a given device is suitable for the application intended. Consideration must be given to the work area where the individual must use the hearing-protective device. For example, a large-volume earmuff would not be practical for an individual who must work in confined areas with very little head clearance. In such instances, a very small or flat earcup or insert-type protector would be more desirable.

When using muff-type protectors in special-hazard areas (such as power-generating stations where there are electrical hazards), it may be desirable to use nonconductive suspension systems in connection with muff-type protectors. Also, if the wearing of other personal protective equipment, such as safety hats or safety spectacles, must be considered, the degree of hearing protection required must not be compromised. The efficiency of muff-type protectors is reduced when they are worn over the frames of eye-protective devices. The reduction in attenuation will depend upon the type of glasses being worn as well as the head configuration of the individual wearer. When eye protection is required, it is recommended that cable-type temples be used. This type will give the smallest possible opening between the seal and the head.

Other considerations when selecting a hearing-protective device include the frequency of exposure to excess noise (once a day, once a week, or very infrequently). For such cases, possibly an insert or plug device will satisfy the requirement. On the other hand, if the noise exposure is relatively frequent and the employee must wear the protective device for an extended period of time, the muff-type protector might be preferable. If the noise exposures are intermittent, the muff-type protector is probably more desirable since it is somewhat more difficult to remove and reinsert earplugs.

In determining the suitability of a hearing-protective device for a given application, the manufacturer's reported test data must be examined carefully. It is necessary to correlate that information to the specific noise exposure involved. The attenuation characteristics of the individual hearing-protective devices are compared at different frequencies.

A method relating manufacturer's performance test information to specific exposure data is outlined in the proposed standard ANSI Z137.1– 1973.

The OSHA Standards Advisory Committee on noise has recommended that personal hearing-protective equipment be selected on the basis of the pure tone attenuation characteristics of the hearing protector and the specific noise exposure involved.

The NIOSH "Criteria Document on Noise" details one possible approach to the determination of a factor which does correlate hearing protection performance to noise exposure. In this case, it is referred to as an *R-factor*. Regardless of whether the NIOSH approach or some other approach is used, the user of a protective device must be able to relate a specific exposure to the attenuation characteristics of a given hearing-protective device to determine whether that protector will provide an adequate safeguard for the specific noise exposure.

In addition to the NIOSH Criteria Document, a committee of the American National Standards Institute (ANSI) has prepared a standard on hearing-protective devices. The purpose of this standard is to provide a feasible method for determining the requirements for manufacturing and the performance, testing, selection, and use of hearing-protective devices.

Summary

Protecting employees against the hazards of excess occupational noise involves a great deal more than just handing them a set of earplugs or earmuffs. The hazards must definitely be established through systematic noise surveys. If the sound levels cannot be reduced by administrative or engineering means, then personal hearing-protective devices must be supplied and just as importantly, must be used. In connection with such programs, accurate and complete records must be maintained, not only of the noise surveys conducted, but also the condition of the hearing of the individuals employed. This is accomplished through audiometric testing.

Published references, such as the American National Standards Institute standard on hearing-protective devices and the National Institute of Occupational Safety and Health (NIOSH) "Criteria Document on Noise" provide valuable information for selecting the most suitable hearing-protective devices. This selection involves many different considerations, including the ability to sell the idea and the need, to both employee and employer. Cost and performance are particularly important to management; whereas comfort, appearance, and protection of the ability to communicate are of greatest concern to the employee. If, however, through good indoctrination, communication, and training, the basic rules and objectives are understood and followed, a successful hearing-protection program can be implemented.

Bibliography

American Industrial Hygiene Association. *Industrial Noise Manual.* Detroit, Mich.: AIHA, 1966.

American National Standards Institute, New York, N.Y.: *Method for the Measurement of Real-Ear Protection of Hearing Protectors and Physical Attenuation of Earmuffs,* S3.19–1974.
Personal Hearing Protective Devices For Use in Noise Environments (Proposed Draft), Z137.1–1973.

Coles, R. R. A. "Control of Industrial Noise Through Personal Protection." *American Speech and Hearing Association Journal,* October 1969, pp. 10–15.

Hosey, A. D., and Powell, C. H., eds. *Industrial Noise—A Guide to its Evaluation and Control,* Public Health Service Publication No. 1572. Washington, D.C.: U.S. Government Printing Office, 1967.

Michael, P. L. "Ear Protectors—Their Usefulness and Limitations." *Archives of Environmental Health* 10:612–18.

National Institute for Occupational Safety and Health. *Criteria for a Recommended Standard—Occupational Exposure To Noise,* HSM 73–11001. Washington, D.C., 1972.

Part 5 Industrial Audiometry

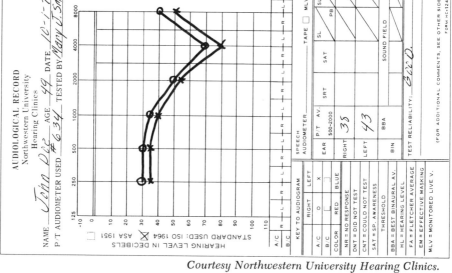

Courtesy Northwestern University Hearing Clinics.

Fig. 21–1.—The first stages of the effects of exposure to excessive noise are shown on this audiogram. Note the decided notch at the 4000 Hz frequency. (Both sides of form are illustrated.)

Chapter Twenty-One

Fundamentals of Air-Conduction Pure Tone Audiometry

By Earl R. Harford, Ph.D.

Pure tone audiometry is widely accepted as the basic test of hearing sensitivity. It requires that a person whose hearing is being tested report when he is able to detect the presence of a specially controlled sound. This is an indirect or inferential test because the subject whose hearing is being tested is depended upon by the tester to give correct information. If the subject does not give correct information for one reason or another, the hearing test results will be wrong—that is, the test results will not truly represent the minimum level of sound that the subject is barely able to detect. Pure tone audiometry requires the cooperation of the person being tested who must understand what he has to do. The technician, also, is required to make accurate judgments about the responses of the person being tested.

Accurate measurement of hearing also depends upon a number of other important factors. The major ones are discussed here. Additional details are included in Chapters 22 through 28.

Purpose of industrial audiometry

There are four major purposes for which hearing tests are used: (*a*) identification of hearing impairment, (*b*) as an aid to diagnosis, (*c*) as a guide to the management of the patient, and (*d*) monitoring the hearing status of an individual.

Identification of a hearing impairment

Pure tone audiometry can detect the slightest deviation from normal

553

hearing over a broad pitch or frequency range. Pure tone audiometry is an accurate method to identify the presence of a subtle abnormality in hearing. Of course, as with most other physical problems, the earlier that a hearing loss is identified, the more effective is the management of the patient. In its early stages, a hearing loss caused by exposure to noise is usually slight; the impairment is not obvious to the noise exposed person or to others who communicate with him (see Figure 21-1).

Aid to diagnosis

The measurement of hearing is used as an aid to diagnosis. All otological (ear) examinations should consist of a history, a physical examination of the ears, nose, and throat, and a hearing test. In a typical industrial hearing-conservation program, the hearing tests conducted in the plant ordinarily are *not* performed for diagnostic purposes. In some situations, diagnostic tests might be performed when there is a fairly elaborate medical department. Regardless, all employees *identified* as having a hearing loss through audiometry performed in the plant should be referred for an ear, nose, and throat examination. The results should appear in the employee's personnel or medical file. Industrial audiometric technicians should compare the results of their tests of the employee's hearing with those done under clinical conditions.

Guide to management of the patient

Hearing tests can be used as a guide for both medical and nonmedical management of the individual with a hearing impairment. Hearing tests are often used to determine if a person needs a hearing aid, or is a possible candidate for medical or surgical management.

Monitoring

Finally, hearing tests are used to monitor the status of a person's hearing and the possible progression of a hearing loss. If a person is suspected of having a progressive hearing loss, his hearing can be checked periodically to determine if his hearing loss is becoming worse. Also, if a person is exposed to an environment noisy enough to cause a progressive hearing loss, then the person's hearing should be checked periodically to see if the noise is in fact having a detrimental effect on his hearing (see Figure 21-2).

Audiometric records are essential to an industrial hearing conservation program by serving as a gage of the effectiveness of noise reduction or engineering control efforts and of the success of a personal hearing-

Courtesy Employers Insurance of Wausau.

Fig. 21–2.—The nurse, using an enlargement of an audiogram, is explaining to the employee what the marks on an audiogram indicate. The nurse can use this opportunity to explain to the employee why personal hearing protection should be worn when excessive noise levels are present.

conservation program. Monitoring audiometirc examinations are, at present, the best way to detect an individual's susceptibility to noise. By comparing an employee's audiogram taken at successive intervals, it can be determined how well the employee's hearing ability is bearing up under the strain of a noisy environment.

Basic minimum requirements

The success of an effective industrial hearing conservation program depends upon accurate audiometry. Accurate audiometry has two basic ingredients: *validity* and *reliability*. The distinction between validity (or a valid measurement) as contrasted with reliability (or a reliable measurement) should be clearly understood.

555

Validity

A valid measurement is one that accurately measures what it purports or claims to measure. For example, when the internal temperature of the human body is desired, the temperature is measured directly with a thermometer. If the thermometer scale indicates 102 F (39 C) when the body temperature is actually a normal 98.6 F (37 C), then the thermometer may be defective and the scale reading would be an invalid measurement of body temperature.

A clue to the validity of pure tone audiometry can be obtained by making a judgment of how well the same person hears speech. There is a high correlation between a person's hearing for pure tones (the sounds used in basic audiometry) and his hearing for speech. If an individual's audiogram indicates a severe hearing loss, the test would have poor validity if the subject were able to carry on a conversation with the technician speaking at a normal level. As a person gradually starts to lose his hearing, he or she relies more and more on visual communication. Many people with a significant hearing loss can carry on a fairly good conversation if they can see the lips of the other person.

When an individual reports for a hearing test, the audiometric technician should note whether the subject appears to have difficulty hearing ordinary conversation. In fact, the technician should ask a question, being careful not to allow the subject to see the speaker's lips. The subject's answer may give the technician a clue as to what to expect on a hearing test. However, it should be considered only as a crosscheck, because it does not work in every case. But even a crosscheck is often better than making no attempt at all to evaluate a person's ability to communicate and to compare his hearing for normal conversational speech with pure tone audiometric results.

The audiologists, the professionals who measure hearing in a clinical environment, use calibrated speech tests and other techniques to establish the validity of their pure tone audiometric tests. The person conducting hearing tests in industry is not expected to get this involved; rather, they should concentrate on good reliability, because it can be controlled by a technician with limited training. As shall be discussed next, good reliability encourages good validity.

Reliability

The second objective for accurate audiometry is *good reliability*. Good reliability means consistent repeatability—that is, the ability to obtain the same information each time it is sought, provided nothing has happened to the person's hearing to cause a real change. Good reliability is very dependent upon controlling the other variables, or factors, that can change hearing test results, in addition to an actual change in the status of the person's hearing.

To illustrate this point using a thermometer reading, if something is wrong with the thermometer so that it *consistently* gives a reading of 100 F (38 C) day in and day out on a person with normal body temperature, it would be a very reliable measurement—but the measurement would be invalid, because the thermometer is not accurate. The instrument used to measure body temperature is simply one variable, just as the operation of an audiometer is a variable in determining the reliability and validity of a hearing test. The objective in audiometry is, therefore, to end up with both a valid and a reliable test. (See Chapters 27 and 28 for additional details.)

Prerequisites for valid and reliable audiometry

Four major factors influence the accuracy of a hearing test.

- A constantly quiet and comfortable test environment

- Stable, accurate test equipment

- A well-instructed, cooperative subject

- Competency of the audiometric technician

A competent, trained, and skilled tester is one who has had systematic professional instruction. This is probably the most important prerequisite, because a competent tester can often compensate for slight deficiencies in the test environment, equipment, and difficult-to-test subjects.

Constantly quiet test environment

A constantly quiet and comfortable test environmental facility can be purchased as a prefabricated unit or it can be a room taken over for testing hearing, provided the ambient noise level does not exceed the allowable established background noise criteria (see Figure 21–3). (Refer to Chapter 27.)

There are ways to check a booth or room to determine if it is quiet enough for audiometry. One method is to accurately measure the sound level using appropriate instruments. Another method is to test the hearing of a person who has very good hearing or test your own hearing to see how easily the faint signals can be heard at all test frequencies. (See

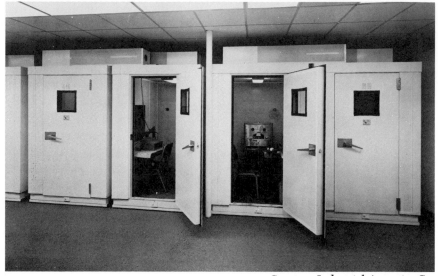

Fig. 21–3.—A constantly quiet and comfortable test environment for audiometric testing is mandatory.

Chapter 27 for a detailed procedure.) To check the sound level within the audiometric test booth or room, an octave-band analyzer should be used—the same kind of meter that is used for the measurement of noise in the plant. Using the octave-band filter, the sound levels are measured in the different bands to determine if the sound levels are within the allowable levels for audiometric test rooms. (See Chapter 27 for more detalis.)

If the sound level is higher than the established criteria in any one or two of the octave bands, then the source of the noise should be found and eliminated. If this is not possible, a suitably constructed audiometric test facility should be secured. Elimination of the noise may involve controlling the noise from the ventilation system, the fluorescent lighting, or carpeting the hallway outside the room.

In most industrial situations, it would be more practical to buy a prefabricated acoustic test environment than to modify a marginal room. Audiometric testing booths or rooms meeting the necessary requirements are readily available. After installation, sound level tests should be made to make certain the booths meet the specifications.

It is important to specify proper ventilation and attenuation of sound

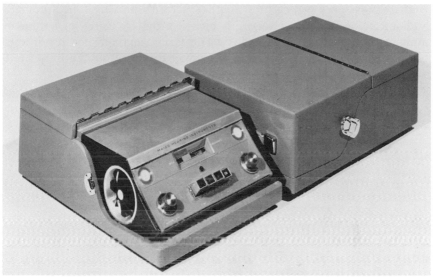

Fig. 21–4.—The audiometer should meet ANSI S3.6–1969 specifications and be calibrated periodically.

to keep outside noises from penetrating the room. Having a good acoustic environment in which to test hearing is just as important as having a well-lit environment in which to test eyesight. Unpredictable, sporadic noise interferes with the accuracy of hearing tests. Hearing tests should not be performed in an area that is conducive to invalid and unreliable results. The consequences could be very costly.

Accurate test equipment

It is important that an audiometer be checked periodically for accuracy. An audiometer can continue to produce test tones just as a watch can continue to tick, yet a watch can be off five minutes a day—or even five minutes an hour. Similarly, an audiometer can become inaccurate. Sometimes it is very difficult to detect a malfunctioning instrument. These subtle flaws in the audiometer can have an effect on its accuracy and lead to invalid and unreliable test results (see Figure 21–4).

It will help to remember that an audiometer is *not* an instrument that *tests* hearing. *There is no instrument that tests hearing. The audiometer offers a controlled stimulus that is used for the measurement of hearing.*

Actually, one person measures another person's hearing. In automatic audiometry, a person measures his own hearing by using an automatic audiometer to control the stimulus. This is a basic point because it must not be thought that the instrument is going to do the job entirely by it-

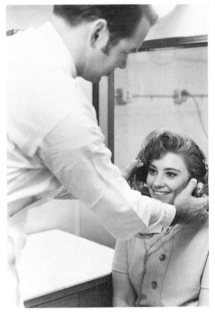

Fig. 21–5.—The technician is shown adjusting the earphones on a subject, letting her get the feel of them. A well-instructed subject is one of the critical aspects of a reliable hearing test.

self. Even with an automatic audiometer, the technician must not have the impression that the audiometer does all the work—the subject measures his hearing by telling the automatic audiometer what to do.

For detailed information on the care, operation, and testing of the audiometer, see Chapters 22, 23, and 27.

Well-instructed subject

Having a well-instructed subject is critical to making a valid and reliable hearing test. Without using technical jargon, the technician should tell the subject just exactly what he needs to know in order to participate in the basic test (see Figure 21–5). Let it go at that, and see how it works out. If the subject does not seem to understand, modify or add to the instructions. (Chapter 24 describes the actual hearing-test technique.)

The listener *must* be well instructed. This is very critical because it influences the end result of either automatic audiometry or manual audiometry. In the latter case, the tester has the opportunity to control the subject's behavior on the test. For this reason, manual audiometry often works where automatic audiometry fails.

Audiometric technician competency

The competency of the industrial audiometric technician is important in the overall determination of hearing thresholds. His (or her) competency can affect the validity and reliability of the hearing test. *Remember that legal decisions involving large sums of money may be based in part on the work of the audiometric technician.* Even more important is the accuracy of medical decisions affecting a person's hearing for the rest of his life.

The tester, by giving a test one way one day and another way the next day, flirts with invalid and unreliable test results, or inaccurate audiometry. Test methodology or technique is just one of the many variables that can affect the outcome of a hearing test. Remember, by maintaining control of those events that can influence the test results, there is a better chance of accurate results. The tester has the greatest influence on the many variables. The tester must have enough information so that when a problem develops, he can recognize that this is a variable that must be controlled.

Evaluating test accuracy

Measurement of hearing is an indirect measurement of a biological function. Heartbeat, breathing rate, blood pressure, temperature, and, in fact, almost all other biological functions are measured directly. About the only other measurement that is not direct is the test of vision, where the tester asks, "Tell me what you see." How does the tester know if the subject being tested is giving him the correct information? What if the subject reads off, "P, R, Z, S"? The tester must compare this response to the letters that were actually presented.

Similarly, the audiometer produces sound going into the subject's ear. The subject, in turn, tells what he hears. *The variable that is injected is the subject. He* must tell what he hears, and the tester must judge and determine the validity of the subject's response (see Figure 21–6).

The tester is constantly being challenged to make decisions—a hear-

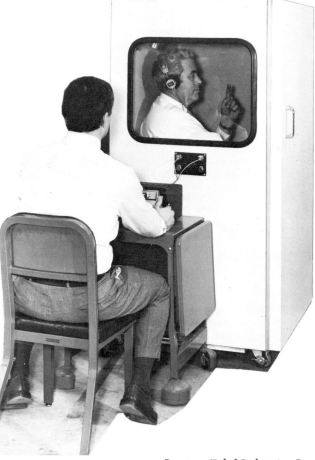

Courtesy Eckel Industries, Inc.

Fig. 21–6.—The raised finger of the subject's hand indicates that he is, in all probability, hearing the test tone. The technician must make a decision as to whether this was a good or an inaccurate response.

ing test requires a series of decisions. Every time a person responds when a pure tone signal is presented in the earphone, every time that he does not respond when the signal is presented—the tester must make a decision about the validity of the response, and he must make it very rapidly. Fortunately, experience improves the speed and accuracy of decision making. The alert tester learns to pick up very subtle clues. A good tester makes a high percentage of accurate decisions.

The industrial audiometric technician can never blandly assume when conducting a hearing test, that when the subject's finger goes up, the subject actually heard the signal; and that when the finger does not go up, the subject did not hear the signal. Such a blind assumption could easily lead to invalid, unreliable audiometry.

Testing hearing, like driving a car, is a skill—some people can do it better than others. Just as some people find that they are not driving a car—that the car drives them—other people find that the audiometer runs them. The tester should not really count on the audiometer to do the work—*he* must do the work, and *he* must be the master of the audiometer. When the tester feels that his hearing is being tested at the same time that the subject's hearing is being tested, then he has "arrived." It is as though he can almost hear the tones that are being presented—he knows how long they last, and how loud they should be sounding to the person wearing the earphones. That is how he learns to "read" the listener's responses, so that he can make an accurate decision as to whether it was a good or inaccurate response.

Each tester, after giving a test, should look at that test result, look at the subject being tested, look at the audiometer, and make a decision about the accuracy of that test. The degree of reliability on the audiogram should be noted and no one can do this better than the tester who made the test.

When a tester is all through with the test and then really does not know what degree of reliability to assign to it, the subject should be scheduled for another test at a later time. No audiogram should be taken just on its face value; it should be taken in light of the tester's assigned judgment of both the validity and reliability of the test. This skill must be developed. No instructor can merely list the qualifications of a good tester or of a poor tester. He must also observe the tester at work. Finally, a good tester must have experience and repeat it time and time again.

The necessity for valid and reliable audiometric data when conducting industrial hearing conservation programs cannot be overemphasized. Only through proper training procedures will the person selected to perform audiometric examinations—the industrial audiometric technician—understand that effective hearing measurement techniques involve more than the mere pressing and turning of audiometer buttons and dials.

A highly motivated, properly trained, and competent industrial audiometric technician is the *key to a successful industrial hearing conservation audiometric program.*

Summary

The four major purposes of industrial audiometry are:

• *Identification* of the hearing loss

• *Aid to diagnosis*—the employee should also have an ear, nose, and throat examination

• *Guide for management of the patient*—determine if a hearing aid or medical or surgical treatment is needed

• *Monitor* the status of a person's hearing and possible progression of hearing loss

Accurate audiometry, a necessity for an effective hearing-conservation program, consists of having valid and reliable measurements. A valid measurement accurately measures what it claims to measure, while reliability means consistent repeatability. Both validity and reliability must be present for an accurate hearing test.

Four major factors influence the validity and reliability of a hearing test:

1. A constantly quiet test environment

2. Accurate test equipment

3. A well-instructed subject

4. The competency of the audiometric technician

Chapter Twenty-Two

The Audiometer— Its Care and Operation

By Noel D. Matkin, Ph.D

An essential component of any industrial hearing-conservation program is the proper care and operation of the audiometer. The accuracy of any measurement of hearing is dependent upon a well-functioning audiometer. This chapter will familiarize the reader with the essential components of an audiometer and discuss the basic procedure for its proper care and maintenance.

An audiometer is a frequency-compensated, audio-signal generator. It produces pure tones at various frequencies and intensities for use in measuring hearing sensitivity. The purpose of such hearing tests is not only to identify existing hearing impairments, but more importantly, to monitor the effectiveness of hearing protection programs. In contrast to a clinical setting, where a variety of audiometric tests are undertaken, hearing tests in industry are limited to pure tone air-conduction audiometry. The audiometer was developed to provide electronically a stimulus similar to that generated by the tuning fork. In one respect the audiometer is superior to the tuning fork, in that, intensities can be controlled much more accurately; therefore, the results can be more carefully quantified.

The **wide-range audiometer** produces signals in the major portion of the human auditory range. It includes a pair of air-conduction earphones, a bone vibrator, a tone switch, and facilities for masking the opposite ear during testing. The wide-range audiometer is intended for use primarily in clinical and diagnostic applications.

565

The **limited-range audiometer** is more restricted than a wide-range audiometer in its ranges of frequency and of sound pressure level. It produces pure tones at 500, 1000, 2000, 3000, 4000, and 6000 Hz, at intensity levels from 10 dB to 70 dB standard reference threshold level. The limited-range audiometer is intended for measuring the hearing threshold levels of adult populations typical of those found in industry.

Vacuum tubes were used in the early audiometers and were continued in exclusive use until the recent development of solid-state circuitry, which is now basic in the newer audiometer design. If fully transistorized, these instruments are lighter, more compact, and require no warm-up period. They may be turned on and off as needed. Transistorized audiometers are more stable and more reliable than vacuum tube instruments.

Most audiometric technicians are well aware of the importance of a properly serviced and calibrated instrument; however, the technician, although trained and qualified to make basic audiometric measurements, may not be aware of the problems with an audiometer that can affect the test results.

An audiometer is an electronic device, and like any other electronic device, it can deteriorate over a period of time. The day-to-day change may go unnoticed unless the user is aware of certain problems that can exist. Some guidelines and procedures to aid the operator in determining when the instrument needs technical service are outlined in this chapter. (Routine and factory calibration is discussed in the next chapter.)

Audiograms

The information on an audiogram will vary according to the need or purpose for which hearing testing is being done. However, it basically consists of a graph with frequency indicated along the top (abscissa) and intensity along the side (ordinate) (see Figure 22–1).

The frequency scale across the top of an audiogram is expressed as either *cycles per second* (cps, the older term) or the newer term *hertz* (Hz). Both mean the same thing, but hertz is the preferred term. The frequencies involved in a hearing test vary with the requirements of the hearing-conservation program and federal and state laws. The basic test frequencies are 500, 1000, 2000, 3000, 4000, and 6000 Hz. Some state regulations or medical departments also call for tests at 250 and 8000 Hz. The left side of the audiogram shows a scale expressed in decibels (dB) of intensity, from zero to as high as 110 dB.

If a test of hearing is desired, obviously an instrument should be designed so that the technician can control the frequency and the intensity of each test stimulus. The lowest intensity at each test frequency at which the listener signals that he hears the pure tone approximately 50 percent of the time is the hearing threshold and is recorded on the listener's audiogram. It is well to remember that the *lower* threshold numbers—expressed in decibels—indicate *better* hear-

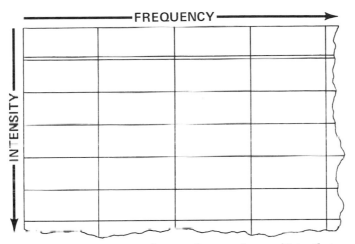

FIG. 22–1.—A section of an audiogram form. Note that frequency increases from left to right and intensity increases from top to bottom.

ing than the *higher* numbers which indicate a worse, or *poorer* threshold of hearing.

Numerical charts facilitate the keeping of records for successive hearing tests on the same individual. Consequently, numerical charts rather than audiograms are often used in an industrial hearing-conservation program to minimize the paperwork in each employee's health records. With such charts, the audiometric technician simply records the intensity noted at each test frequency which represents the subject's threshold for the right and left ears. Test frequencies are noted at the top of vertical columns. Dates of the tests usually appear on the left, and the initials of the audiometric technician appear on the right side of each row of figures (see Figure 22–2).

Other types of audiometric forms have been suggested, but they are

(*Text continues on page 570.*)

DATE	FR.	250	500	1000	1500	2000	3000	4000	6000	8000		250	500	1000	1500	2000	3000	4000	6000	8000	COMMENTS	BY
8-1-54	RT.	5	5	0	5	0	-5	-5	0	5	LF.	10	5	5	0	-5	-5	0	5	5		A. B.
8-1-55	EAR	10	5	5	0	5	5	0	5	5	EAR	10	5	0	-5	-5	0	0	5	5		A. B.
8-1-56		10	5	5	0	0	5	0	5	5		10	5	5	0	0	-5	-5	0	0		A. B.

FIG. 22–2.—Numerical charts can be used to facilitate the storage of information on an employee's hearing level. Date of test and the initials of the audiometric technician appear on the right side of each row of test findings.

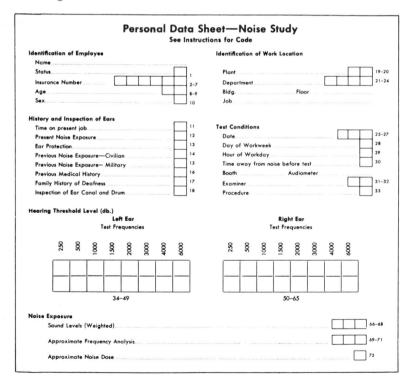

FIG. 22–3.—The audiometric form shown here would be especially useful for storing information that may later be computerized.

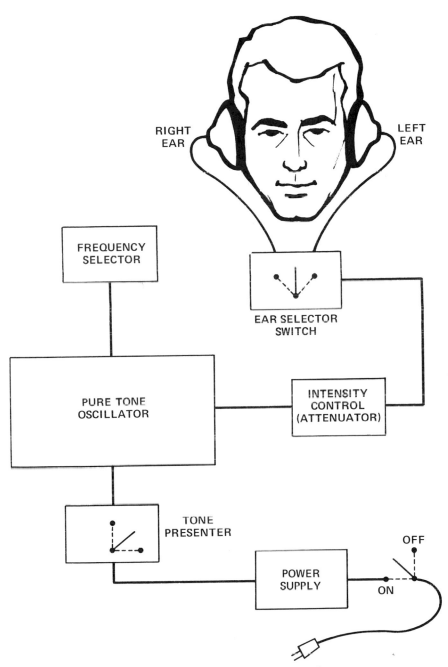

FIG. 22–4.—A schematic diagram of an audiometer. The various components needed are: an on-off switch, tone presenter, pure tone oscillator, intensity control (attenuator), frequency selector, ear selector switch, and earphones.

FIG. 22–5.—A face panel of a manual audiometer to illustrate the on-off switch on the left.

simply different versions of the basic types, adapted for a particular usage. For example, one form is especially adapted for use with computers for research purposes (see Figure 22–3).

Audiometers

It is important that the audiometric technician feel comfortable with the particular type of audiometer used in the hearing-test program. There are two distinct types of audiometers—the conventional or manually operated unit and the automatic which allows the subject to conduct his own hearing test. Both the manual and the automatic require a highly motivated, properly trained, industrial audiometric technician with demonstrated competency. The pros and cons of each class of instrument should be considered carefully before selecting one for a hearing-conservation program. For more details, see Chapter 25, Manual and Automatic Audiometry.

The basic design or block diagram of an audiometer is shown in Figure 22–4. With the aid of this diagram, the trained technician should easily be able to visualize the basic components of the audiometer, and with a few minutes study be able to determine the operating characteristics of any audiometer. The manual audiometer will be described in detail in this chapter; a short description of the automatic audiometer follows that discussion.

Manual audiometer

All manual audiometers are basically the same; manufacturers simply package them differently. Although a variety of styles of control panels are available, the basic function of the audiometer is the same regardless of the type or placement of the controls or the slightly different nomenclature—whether the control is a dial, a pushbutton, or a toggle switch does not really matter.

Power switch. This switch may be labeled simply ON-OFF (see Figure 22–5). Sometimes it is a position on the *earphone selector* switch (on some units called the *output control*). In addition, some panels have an indicator light that glows when the power is on. Still other units have a dial light that illuminates the dials of the unit when the power is turned on. If the dial or power light does not light—after the power has been turned on—check the power connection to the electrical outlet first, then check the audiometer fuse, and finally, check the bulb itself. If none of these solves the difficulty, call the repairman.

Most audiometers use 110 volt, a-c power; however, many portable units operate on battery power. If your instrument uses batteries, always check their condition before beginning a testing program. Here the manufacturers have anticipated the need and most battery-powered audiometers have a visual check; that is, when a button is pressed, a needle reveals the condition of the batteries.

Tone presenter. A tone switch can be used either to interrupt or to present a test signal, depending upon the position of the *tone interrupter* (TI) reverse switch (in the case of the unit shown in Figure 22–6). When the TI reverse switch is in the normal position, a continuous tone is delivered to the listener's ear which can be interrupted by the interrupter switch. The TI reverse switch in the reverse position would serve as a *tone presenter* switch. In other words, a tone is presented to the subject every time the switch is depressed and is terminated when the switch is released.

The tone presenter mode is recommended during pure tone tests in

Courtesy Audiotone.

Fig. 22–6.—The face panel of an audiometer illustrates the Interruption or Presenter switch, shown at the bottom center.

Courtesy Beltone Electronics Corp.

Fig. 22–7.—An intensity or attenuator control dial on the left and a frequency selector on the right.

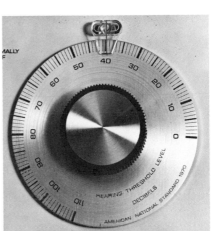

FIG. 22–8.—The attenuator dial, labeled in decibels and scaled in five-decibel steps.

Courtesy Eckstein Brothers, Inc.

FIG. 22–9.—The attenuator or hearing threshold dial scaled in 1 dB steps.

Courtesy Zenith Radio Corp.

order to minimize auditory adaptation. In other words, it is always better, during routine tests, to present a tone out of silence because then the listener's best thresholds are recorded.

There are two parameters of the pure tone that must be controlled—intensity and frequency. Therefore, both an intensity control and a frequency control dial are needed (see Figure 22–7).

Intensity control (attenuator). On an audiometer, the attenuator controls the intensity of the pure tone signal. Frequently, on or near the attenuator, an inscription states the standard by which the audiometer was calibrated—ISO 1964 or ANSI 1969.

The attenuator, which is actually a hearing threshold level dial, may be scaled in one decibel units from zero to 100 dB, or the dial may be scaled in 5 decibel steps (see Figures 22–8, –9, and –10). Either type of attenuator is suitable for hearing threshold measurements in an industrial hearing measurement program.

Frequency selector. The frequency selector dial is used to control the second important parameter of the test signal (see Figure 22–9). On some audiometers the attenuator and frequency selector look very much alike and may be side by side.

The frequencies are specified at octaves of 125, 250, and 500 Hz, and

Courtesy Audiotone.

Fig. 22–10.—The frequency selector
switch set at 1000 Hz.

so on (whatever is shown on the dial). Some of the older audiometers
may have odd numbers instead of rounded numbers. For example, instead of 250, the number on the dial may be 256 Hz. (This is the frequency of middle C on the piano.) If an audiometer has numbers such
as 256 Hz, 512 Hz (and their multiples) on the frequency-selector dial,
the unit is fairly old.

The increments above 500 Hz on a modern unit are shown in Figure
22–10. The first step, 750, is half way to 1000 Hz. Many modern audiometers have the capability to produce tones at half octaves—750,
1500, 3000, and 6000. Why half octaves? If a threshold response is recorded at the 10 dB level at 1000 Hz and zero dB at 500 Hz, can the
subject's hearing level be estimated to be between zero and 10 dB between these two frequencies? Actually in practice, a straight line is
drawn on the audiogram between the two readings to provide an estimate
of what the person hears between the two test frequencies.

It is fairly safe to undertake such straightline interpolations when there
is minimal change in threshold levels between the various octave test
frequencies. However, when there is a significant difference in hearing
levels between two test frequencies, it is not actually known if the hearing thresholds change as a straightline function or whether hearing sen-

sitivity continues and then drops off suddenly. To determine this, a half-octave test frequency is used.

The frequency selector switch supplies additional important information. The maximum intensity output for air conduction is marked on the dial for each frequency. For example, at 1000 Hz it may read 110 dB. At low and very high frequencies, the maximum intensity output will be less than 110 dB because it takes more energy to reach the auditory threshold. For example, at 125 Hz, the maximum output by air conduction is 70 dB (see Figure 22–10). This maximum limit will vary slightly from one audiometer to another.

The normal hearing mechanism is less sensitive for the very low and the very high sound frequencies. Actually the audiometer output across frequencies is similar, but the normal hearing mechanism functions less efficiently at some frequencies.

Some audiometers show the maximum output in decibels for both air and bone conduction. For example, at 125 Hz, the frequency selector dial is marked "70-A, 40-B." A general rule of thumb is that the maximum output for bone conduction will be about 40 dB less than that for air conduction because it takes a good deal of energy to set the bones in the skull into motion before the threshold by bone conduction can be reached.

Ear selector switch. An ear selector switch is also needed to permit the selection of the right or left earphone or bone oscillator. This same switch may also be designed to include an OFF-ON control on some units. The earphones may be color coded to indicate the ear—red or black for right, blue or grey for left. Newer audiometers have R or L printed on each phone.

It is important to know that some audiometers also are provided with a bone conduction device. This type of audiometer has a selector switch that will send the signal either to the earphones or to a vibrator that fits on the mastoid bone.

Transducer. A transducer converts electrical impulses into a sound stimulus, and because industrial audiometry is involved only with air conduction, the transducer is a set of earphones.

Earphones have a stable response when handled properly, and will retain their calibration for long periods of time. But if they are mishandled, for example, if the opening in the earphone is placed against a surface where first a pressure and then a partial vacuum is created, the distance

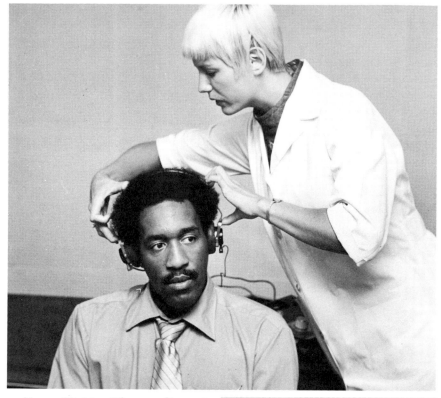

Fig. 22–11.—The audiometric technician should always place the earphones on the subject and remove them after the test—only trained personnel should be allowed to handle the audiometer earphones.

Courtesy General Radio Co.

Fig. 22–12.—Storage of audiometer earphones. Note the cord is wrapped between the earphones.

between the diaphragm and the voice coil may be altered, or the diaphragm itself may be ruptured resulting in a worthless set of earphones.

The characteristics of audiometer circuits are carefully matched to a particular set of earphones, therefore it is necessary to specify the make, model, and serial number of the audiometer and the type and class of the earphones when replacement earphones are requested. The audiometer and its earphones form a matched set that must not be altered.

To prevent damage to the earphones, the operator should always place them on the subject prior to the test and remove them after the test; in other words, only trained personnel should handle the earphones (see Figure 22–11). When the test is complete, the earphones should be placed where they are not likely to fall to the floor. When tests are completed for the day, the earphone cord should be wrapped around and between the earphones so as to provide a space for trapped air to escape (see Figure 22–12).

Automatic audiometer

The automatic audiometer has the same basic components as the manual audiometer. The technician prepares the automatic audiometer prior to the test, gives the subject his instructions, places the earphones on the subject's head, and usually monitors the hearing test. The person being tested responds to a test tone by pressing a button. The button controls a mechanism in the audiometer that moves the stylus and records the threshold on a card. The automatic audiometer may (because of the additional circuitry, moving stylus, gears, and the like) have problems not found in the manual audiometer, and, consequently, require more service calls.

Audiometer checklist

The industrial nurse or audiometric technician is not expected to be a repairman. There are some basic steps that can be taken to keep the audiometer working properly and to prevent delays in the hearing testing program.

There are some common problem areas that should be checked if an audiometer is not functioning properly. First, the power supply plug should be seated properly in the electrical outlet. Next, check the plug/cord junction. Sometimes with improper use, a break in the junction between the plug and the cord can develop. Finally, check the fuse.

Some major checkpoints will be discussed on the following pages.

Earphones

Improperly fitted earphones can result in inaccurate hearing measurements, especially below 500 Hz. Before initiating a hearing test, all obstructions (such as hair covering the ears, eyeglasses, and earrings) should be removed. Adjust the headband so that it rests solidly on the crown of the head and exerts a firm pressure on both ears. Always make the following checks:

• Examine the cushions on the earphones—they should be reasonably resilient and free of cracks; if they are not, they can be replaced quickly if an extra set of cushions is kept on hand. Because earphone cushions often get oily and stiff with continuous use, this can prevent a good fit.

• Take special care in placing the earphones on the subject's ears. If the seal is as tight as it should be, the subject's voice will seem louder to him as he talks; other persons' voices will seem softer to him.

• To obtain an accurate audiogram, the earphones must fit snugly over the ears. With constant use, the headband tends to stretch and lose its tension. Over a period of time it becomes difficult to obtain a good fit to the subject. Properly shaped headbands should be tensioned so that they look like the illustration in Figure 22–13. If an audiometer is going to be used continuously, it would be a good idea to keep a spare set of headbands available.

• Handle earphones with "tender loving care." They are *not* interchangeable, and they *must not* be bumped or exposed to excessive temperature changes.

Earphone cords

Probably the most common problem with earphones is the cords. It would be well to order an extra set of earphone cords when an audiometer is purchased. With constant handling of earphones, the cords eventually get damaged. To check the earphone cords:

1. Set the attenuator control at 50 dB.
2. Set the frequency selector at 1000 Hz.
3. While listening to a continuous tone, flex the cord on each phone at both connections (at earphone and audiometer)—see Figure 22–14.
4. If a scratchy noise is heard or if the tone is intermittent, replace the cord.

(*Text continues on page 582.*)

FIG. 22–13.—Two sets of earphones. Those on the left (cushion approximately ¼ in. apart) are shaped correctly. The headband on the earphones on the right is stretched out of shape, the earphones will not provide a good seal or adequate pressure over the ears.

FIG. 22–14.—The technician, while listening to the tone, should flex the cord at the point of connection of the earphone.

Courtesy Beltone Electronics Corp.

FIG. 22–15.—Two earphones. The one at the left has metal, slotted-head set screws, and the one shown at the right has Allen-head set screws to fasten the earphone cords to the receiver.

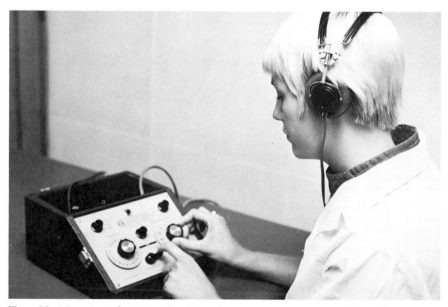

FIG. 22–16.—A technician checking for extraneous noise or hum by setting the attenuator at 50 dB, interrupting the signal, and listening for noise.

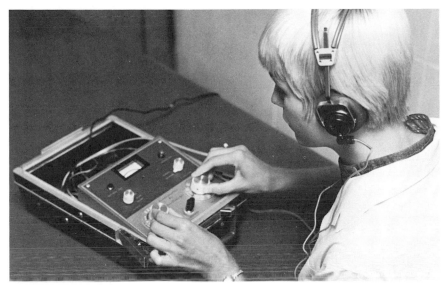

FIG. 22–17.—The technician checking for attenuator noise by setting the frequency dial to 1000 Hz and slowly increasing the attenuator from zero to 60 dB.

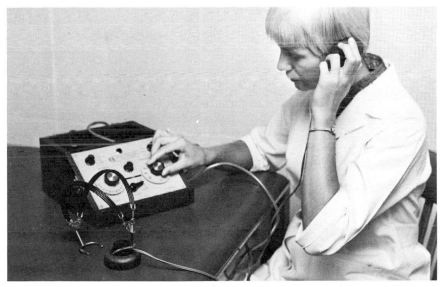

FIG. 22–18.—The technician checking for tone in the quiet or nontest earphone by applying a signal to the opposite phone and listening for tone in the nontest phone.

When a set of cords is defective, the signal may be interrupted intermittently during an actual hearing test due to minor head movements. The person being tested may be responding normally then may turn his head and suddenly stop responding, even though the sound is being presented at greater and greater intensities. The subject may then turn his head again, become suddenly startled, and complain to the operator that the signal is too loud. The most common explanation for such an occurrence is that when the subject turned his head, the defective earphone cord may have shorted out. Many technicians immediately blame the listener for inconsistent responses, but the fault may lie with the equipment and not the subject.

The next common problem area that should be checked, if an intermittent signal is detected, is the two tiny earphone setscrews that fasten the cord to the earphone (see Figure 22–15). These setscrews might be difficult to find, because some manufacturers cover the earphones with a company nameplate. Remove the earphone cushions and take off the plate; the setscrew can then be seen. The screws can be tightened with either a small jeweler's screwdriver or a small allen wrench.

Audiometer Noise

Hum or extraneous noise produced by the audiometer's internal circuitry can affect the test results. To check for this condition:

1. Set the frequencies to 4000 Hz.

2. Set the attenuator to 50 dB (signal continuously on) and listen for extraneous noise or hum in the earphones.

3. Interrupt the signal and again listen for noise (see Figure 22–16).

There should be no hum or extraneous noise at any of the cited settings with the tone on or off. It is possible that some audiometers will have a hum present at higher settings with the tone off. However, this noise should not interfere with hearing tests. If it is audible at 50 dB, the instrument needs technical service.

A second type of extraneous noise encountered in some audiometers is the radiation of the pure tone signal from the unit itself. This problem seems to occur most frequently at 4000 Hz. Often the audiometric technician can hear the test tone while administering a test to a subject. If this problem is present, the audiometer should be checked by an authorized factory representative.

Attenuator noise

The attenuator or hearing level control may become noisy from dirt deposits and lack of lubrication. To check for a noisy condition:

1. Set the frequency selector control to 1000 Hz, and turn the tone on.

2. Slowly increase the attenuator setting from zero to 80 dB, and listen for scratchy noise (see Figure 22–17).

3. If noise is heard, turn the attenuator control back and forth with a rapid motion several times. (If the noise is from dirt deposits, this method is sometimes effective in removing the dirt.) If the scratchy noise does not go away after rotating the dial a few times, call a serviceman.

If the attenuator is a 5 dB step attenuator, a click may be heard as it is turned. This is not a serious problem. Do not confuse this with noise.

Tone in quiet earphone (crosstalk)

An audiometer may produce a tone in the earphone on the ear not being tested (see Figure 22–18). To check this condition:

1. Apply a signal of 80 dB only to the right phone.

2. Disconnect the right phone from the audiometer or from the wall if a two-room installation is used.

3. Rotate the frequency switch through all frequency settings, while listening for tone in the left or nontest earphone.

4. Repeat the same procedure for the left phone.

Any sound in the quiet earphone will interfere with hearing testing, particularly when one ear has depressed hearing and the other is good. The average person will respond when the tone is perceived regardless of which ear is being tested, resulting in invalid audiometry.

Biological check

It is strongly recommended that each day before the audiometer is to be used—not just once a week—the audiometric technician make a hearing test check of the unit's function on his or her own ear.

Briefly, turn on the audiometer and put the headphones on. Now sweep through the various test frequencies with the attenuator dial set at 60 dB and with the tone presenter switch on to make certain that a signal is heard at each frequency. Start with the left ear, and then switch to the right ear, and repeat the procedure. Next, set the frequency selector to 1000 Hz and sweep the intensities on the attenuator switch to check for "dead spots." If the instrument is not used daily, it may be necessary to rotate the attenuator (hearing level dial) rapidly to eliminate scratchiness. Make sure clear signals are heard.

The earphone cords near the receiver, and near the instrument, should be flexed while the hearing level dial is set at 60 dB (or louder) for each earphone to be sure there is no intermittency.

Finally, check for crosstalk at 4000 Hz especially in a new unit or after factory calibration. This means sending the signal to the right earphone but listening with the left earphone and vice versa.

Biological calibration

Depending on how often the audiometer is used, a biological calibration of the instrument should also be made. That is, keep cumulative records on one or two young adults who have not been exposed to noise and who have normal hearing. On a monthly basis, check the hearing of these people.

When biologically checking the calibration of an audiometer, test the same ear of the normal listener with both earphones. If there is a significant difference (10 dB or more) between the two earphones in terms of the intensity level at which the listener responds, the audiometer is probably out of calibration due to a change in the sensitivity of one of the earphones. In contrast, if there is a difference in response levels for both earphones as compared with earlier findings, this would ordinarily indicate a circuit problem in the audiometer. In either case, the calibration of the audiometer should immediately be checked electroacoustically by a factory-authorized service representative.

The tests described in this chapter are not as complete as electroacoustic calibration (see Chapter 23 for a description of electroacoustic calibration), and they should not be expected to replace service calls. The tests described in this chapter are designed to aid the audiometric technician in recognizing when an instrument has developed a problem that may affect the test results.

Annually, return the instrument to the dealer from whom it was purchased for a certified calibration. When the instrument is returned, immediately complete a biological calibration to be sure that nothing has happened in transit to offset the calibration.

Summary

The function, care, and operation of an audiometer have been discussed in this chapter. While many audiometric technicians are well aware of the importance of a properly serviced and calibrated instrument, they may overlook potential trouble spots in the operation of these instruments. An audiometer is an electronic device, and like any other electronic device, can deteriorate over a period of time. The day-to-day change may go unnoticed unless the technician is aware of certain problems that can exist. Some simple tests to aid the audiometric technician in recognizing when an audiometer needs technical service have been outlined.

It is very important that the audiometric technician know his instrument intimately before undertaking audiometric measurements. The various components of an audiometer were described in detail as were the specific function and potential problem areas.

An audiometer checklist was included to outline the steps to be taken to make certain that the audiometer is functioning properly. Potential problem areas discussed were earphones, earphone cords, audiometer hum, attenuator noise, and tone in the quiet earphone and crosstalk.

The procedures involved in the biological calibration of the audiometer were also discussed. A simple daily routine that should be followed was described. Biological calibration procedures, although not as reliable as electroacoustic calibration, are designed to assist the audiometric technician in knowing when an instrument has developed a problem that may adversely affect test results.

Chapter Twenty-Three

Audiometer Calibration

By William Reich

An accurate and reliable audiometric testing program involves more than simply purchasing a good audiometer and having a qualified operator. Periodic calibrations are also an essential part of the audiometric program. Studies have shown that without calibration there is only a 50 percent chance that the audiometer is performing as it should.

A calibrated audiometer (*a*) emits a pure tone test signal at the level and frequency shown on the instrument dials, (*b*) delivers the signal only to the place (that is, a specific earphone) it is directed, and (*c*) produces the signal free from contamination by extraneous noises or unwanted by-products of the test signal. For example, when the audiometer is set to deliver a 1000 Hz pure tone signal at 40 dB hearing threshold level to the right earphone, it should do precisely this; instead of, for example, delivering a pure tone test signal of 1170 Hz at 55 dB to the right earphone with a portion of the same signal to the left phone as well.

There are recognized tolerances for the important physical characteristics of pure tone audiometric signals and these should be taken into account in any critical evaluation of the output of an audiometer.

Audiometer manufacturers are careful to design and produce audiometers which meet the ANSI S3.6–1969 *Specifications for Audiometers.* Unfortunately, once an audiometer is set to these standards of performance, there is no assurance that it will remain in calibration. As a matter of fact, this equipment is sensitive and highly susceptible to unintentional

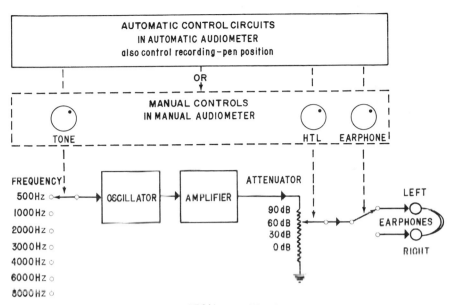

SIMPLIFIED BLOCK DIAGRAM OF TYPICAL AUDIOMETER

Fig. 23–1.—Pure tone audiometers differ widely in details but not in their basic makeup. An oscillator produces a signal (any of the six or seven frequencies) which is amplified and then applied to an attenuator to control its output level. From the attenuator it is applied to either the left or the right earphone.

abuse that can result in faulty operation. OSHA even has specific calibration requirements (see Appendix A).

In the past, a very small percentage of audiometers received periodic calibration; in fact, most calibrations were done only after an audiometer had stopped operating completely. One reason for this neglect may be the result of the inconvenience of packing and shipping the instrument as well as parting with a unit that seems to be operating adequately.

There are numerous reasons for an audiometer to go out of calibration or lose its precision; these include dropping the earphones, overheating (leaving the audiometer turned on after covering it with a dust protector), exposure to excessive dust (transporting an audiometer in the trunk of an automobile over dusty roads), exposure to high humidity and salt air, excessive jarring, and normal aging of the electrical components. Whenever it is known that an audiometer or the earphones have been subjected to abuse, they should be checked to see that they have not gone out of calibration.

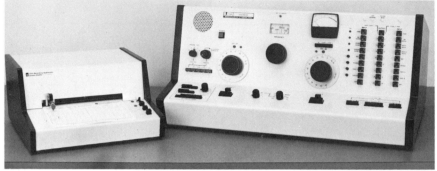

Courtesy of Grason Stadler.

FIG. 23–2.—The speech audiometer shown here delivers spoken materials at known levels to the ear of the listener whose hearing is being tested.

Courtesy of Grason Stadler.

FIG. 23–3.—The research or diagnostic audiometer shown here is a valuable tool in the medical research laboratory or clinic but is not in general use in industry.

Pure tone air conduction audiometers

There are as many ways to calibrate audiometers as there are models of audiometers and any attempt to discuss the procedures involved would span volumes. But very general information is presented in this chapter to assist in performing field checks on the *limited-range, pure tone audiometer,* the one most likely encountered in industrial hearing conservation programs. As its name implies, the limited-range audiometer delivers to the earphone a single frequency sound (or pure tone) (see Figure 23–1).

A second basic type of audiometer is the *speech audiometer* which delivers spoken material to the ear, that is, a complex signal consisting of many frequencies. The speech audiometer is more complex, and is not widely used in industrial applications (see Figure 23–2).

Less common in industry are *research or diagnostic audiometers* (see Figure 23–3) and *screening audiometers*—the former are too slow, too expensive, and too complicated for routine industrial hearing testing, and the latter provide insufficient information.

The following calibration procedures are intended for use with pure tone air conduction audiometers—manual or automatic. Whether these procedures are employed for daily or monthly checks or annual calibrations, they should all be documented at the time the checks are performed. Such records are necessary for legal purposes and as an aid for monitoring the audiometer's long term performance. Repairs, when necessary, should be performed only by skilled technicians.

Daily calibration check

Simple, yet effective, checks on the operation of the audiometer are known variously as operational calibration checks, biologic checks, or subjective-response checks. They require only a few minutes to accomplish, need no test equipment, and are best performed daily prior to the normal audiometric examinations. (For additional details, see Chapter 22, The Audiometer—Its Care and Operation.)

Earphone cord. Set the audiometer at a 50 dB hearing threshold level (HTL), continuous tone at 1000 Hz. Flex the earphone cord for the active earphone, particularly at the ends, and listen for an intermittent tone, abrupt changes in the level, or a scratchy sound when the cord is flexed. Any of these indications is a sign that the earphone cords have broken leads internally or have opened at a junction and, therefore, re-

Fig. 23–4.—The earphone cord is often abused; it should be checked daily.

Courtesy General Radio Co.

Fig. 23–5.—The rasping or edgy quality of any tone can upset the hearing test results. Care must be exercised by the operator when setting the frequency.

Courtesy Grason Stadler.

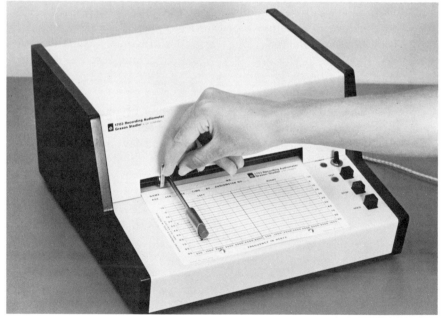

quire repair or replacement (see Figure 23–4).

Hum and noise. Set the frequency selector switch at 1000 Hz and the attenuator (hearing threshold level control) to its minimum output position, and then slowly increase the level to its maximum output position. Any low frequency hum (60 or 120 Hz) or random noise (a hiss or low rushing sound) at settings below 60 dB hearing threshold level (HTL), 1969 ANSI or 1964 ISO, is a sign of internal noise and a cause for repair. Some hum or random noise above 60 dB is permissible.

Any erratic tone or scratchy noise as the attenuator setting is changed is a sign of a dirty or partially open attenuator, and it is also a cause for repair.

Distortion and frequency shift. Reset the attenuator to a comfortably loud (70 to 80 dB HTL) continuous tone at a lower frequency (usually 500 Hz), and work through each frequency to the highest (6000 or 8000 Hz). Any rattling, rasping, or edgy quality to any tone presented is an indication of distortion or frequency shift, and it is a cause for repair (see Figure 23–5).

Monthly electroacoustic check

Procedures and methods for checking the accuracy of an audiometer using a sound level meter are given here. Obviously, precise calibration is not intended, for only sensitive measuring equipment will provide this, but if one follows these suggestions, he should at least be capable of determining when a more extensive laboratory calibration is indicated.

Acoustic output. By far the most prevalent and most serious audiometer deficiency is inaccurate intensity or acoustic output (hearing threshold level).

It is important to know whether the audiometer is producing the intensity output shown on the dial, or if, in fact, it is producing more or less than this reading. For example, if the actual acoustic output of an audiometer is 15 dB less than that shown on the dial, every person tested on this instrument will show a 15 dB hearing threshold shift. Further, a person's hearing loss might be classified as progressive in a case where an audiometer developed a loss of acoustic output between an initial test and followup testing. An audiometer that is too strong (actual output at earphone greater than dial reading), however, can present problems similar to those just described, but in an opposite direction. This problem is critical in industrial audiometry where 10 dB one way or the other (strong or weak) can mean the difference between an effective audiometric program and an almost complete waste of time.

The second problem with intensity is nonlinearity. Figure 23–6 illustrates what happens when an audiometer fails to attenuate (decrease

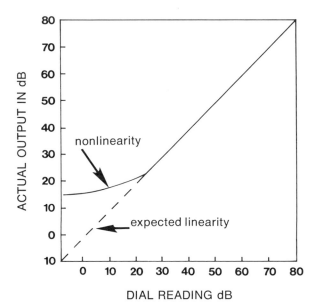

FIG. 23–6.—In this example of a nonlinear auditory test signal, note that as the dial reading is decreased below 25 dB, the actual output of the signal fails to decrease as the dial reading indicates and then remains constant. (A decrease of the dial does not decrease the output of the signal.)

output) uniformly below 25 dB. Anyone with an actual threshold of 10 or 15 dB will report hearing a tone when the dial is set at 0 dB. This may present serious problems in industrial audiometry. Audiometers that are nonlinear below 25 dB should not be used in industrial hearing-conservation programs.

Most of the problems with inaccurate output levels are due to the earphones, especially in the newer audiometers, where stable semicon-

TABLE 23-A
SOUND PRESSURE READINGS
(levels in dB re 0.0002 μbar)

	Frequency (in hertz)						
	500	1000	2000	3000	4000	6000	8000
1969 ANSI 0-dB Hearing Level							
WE705A	+11	+6.5	+ 8.5	+ 7.5	+ 9	+ 8	+ 9.5
TDH-39	+11.5	+7	+ 9	+10	+ 9.5	+15.5	+13
TDH-49	+13.5	+7.5	+11	+ 9.5	+10.5	+13.5	+13
PDR-1	+11	+7	+ 9	+10	+13.5	+ 8.5	+11
PDR-8	+11.5	+6.5	+ 7.5	+ 8	+ 9	+17	+13
PDR-10	+10	+6	+ 6.5	+ 9	+ 9	+18.5	+14
Sound Level Meter Weighting							
20 kHz, all pass or flat	0	0	0	0	0	0	0
C	0	0	0	− 0.5	− 0.8	− 1.9	− 2.8
B	− 0.3	0	0	− 0.5	− 0.8	− 1.9	− 2.8
A	− 3.3	0	+ 1.4	+ 1	+ 1	− 0.1	− 0.9
Typical Ceramic Microphone							
15/16 (or 1 inch)	0	0	0	− 0.5	− 1.8	− 2.6	− 1
1⅛ inch	0	0	+ 1.2	+ 1.7	+ 2.2	− 0.5	− 7
Typical Sound Level Meter Variations							
	0.5	−0.1	− 0.5	+ 0.2	+ 0.9	− 2.4	− 2.9

ductor circuits considerably reduce the amount of drift or change in the acoustic output of the audiometer.

Daily operational calibration checks will detect some errors in the acoustic output, but more precise checks require external test equipment. The equipment traditionally used in the laboratory for these checks is quite expensive, but recent developments have produced relatively inexpensive field test sets that provide essentially the same measurement accuracy for sound pressure level. In addition, the field sets are small, easily operated, and battery powered for complete portability.

Such sets usually consist of three elements—a sound level calibrator, a sound level meter, and an earphone coupler (also known as a standard

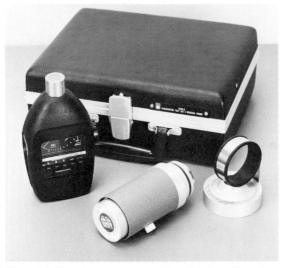

Fig. 23–7.—Accurate monthly checks of audiometer performance require a sound level calibrator, a sound level meter, and an earphone coupler.

Fig. 23–8.—The accuracy of a sound level meter should first be verified by a sound level calibrator.

Fig. 23–9.—To test the earphones, place the sound level meter on a firm, vibration-free table and place the earphone coupler over the microphone.

Fig. 23–10.—Carefully place the earphone (including its cushion if using an NBS Type 9A coupler) on the coupler, and lower the weight into place.

Photographs courtesy General Radio Co.

594

or artificial ear)—see Figure 23–7. One other important inclusion is a chart, or set of charts, that contains calibration data with which to determine the proper readings. In the absence of such a chart, Table 23-A can be used as a close approximation.

There are two steps involved in making the checks: first, be sure of the accuracy of the calibration setup, and second, check the audiometer itself for output and linearity.

Sound level meter check. Set the sound level meter for a C-weighted, fast meter response (C_F). (Some sound level meters have no C-weighting, but usually provide the data for an equivalent A-weighted reading.) Place the sound level calibrator over the microphone of the sound level meter and note the reading on the meter (see Figure 23–8). It should correspond to the correct reading for the calibrator employed. (With the equipment shown in Figure 23–7, the correct reading is 113.9 dB.) If it does not, a simple screwdriver adjustment is all that is required on most sound level meters to achieve the correct indication. After this calibration, the calibrator is removed and turned off. (See Chapter 4 for details on sound level meters and their calibration.)

There are two basic types of earphone couplers, the NBS Type 9A and the ANSI Type 1. Both perform the same function, that of coupling the earphone to the microphone in a manner that simulates the volume of air space of the human ear, and both do it with the same accuracy (see Figure 23–9). The Type 9A is becoming the more popular, however, because it can be used without removal of the ear cushions from the earphones.

Place the earphone coupler over the microphone on the sound level meter. Then remove the earphone to be tested from the headband, lift the earphone coupler weight or spring, gently place the earphone on the coupler, and carefully lower the weight or spring in place on top of the earphone (see Figure 23–10).

Set the output threshold level on the audiometer to 70 dB (1969 ANSI) and the tone to the lower frequency (usually 500 Hz, continuous tone). The indication on the sound level meter should be within ±3 dB of the proper value (see Table 23-A). With the equipment shown in Figure 23–10, for a TDH-39 earphone, the proper reading would be 81.5 dB, and any reading from 78.5 to 84.5 dB would be acceptable.

Perform this test at the 70 dB HTL for each test frequency. The indications should be within ±3 dB of the proper value at the lower frequencies (500 Hz through 3000 Hz), within ±4 dB at 4000 Hz, and within ±5 dB at 6000 and 8000 Hz.

Attenuator linearity. After the 70 dB HTL level at each frequency has been checked, return the tone to 4000 Hz, and decrease the HTL level by 5 dB starting at the maximum output. The sound level meter should indicate a 5 dB ±1.5 dB decrease. This is a check of the attenuator linearity, and it should be performed at each 5 dB interval, down as low as the sound level meter will indicate (about 45 or 50 dB HTL). It is not necessary to do this at more than 4000 Hz.

The lower the HTL setting, the more difficult the measurement becomes due to ambient noise and vibration. Linearity checks at the lower HTL settings require more expensive laboratory equipment with greater sensitivity.

After one earphone has been tested in this manner, perform the 70 dB level checks on the other earphone. The attenuator linearity checks need not be performed on the second earphone, because most audiometers use the same attenuator for both earphones.

Annual factory calibration

Daily and monthly checks will detect the more common defects, but these checks are not a substitute for thorough, annual factory electroacoustic calibrations. Due to the extensive amount of equipment required and the skilled personnel necessary, annual calibrations are usually performed by the audiometer manufacturer or by laboratories specializing in such services.

More and more industrial concerns, however, are equipping themselves with test instruments to perform these calibrations as a part of their overall hearing-conservation program. The test equipment can also be used for other aspects of the noise problem—such as noise level surveys and suitable site selection of the audiometric examination room.

Audiometers for industrial use are manufactured in accordance with a single set of specifications, the American National Standard, *Specifications for Audiometers* S3.6-1969, and most audiometers for industrial audiometric programs conform to the requirements given for limited-range, pure tone audiometers. ANSI S3.6-1969 defines the minimum requirements for the output characteristics of the audiometer, such as its hearing thresholds, test tones, output purity, etc.

In practice, each make of audiometer is calibrated in accordance with a factory procedure that incorporates the provisions of ANSI S3.6-1969 and includes other necessary operating parameters as well, such as power supply voltages, bias levels, and mechanical adjustments. These factory calibration procedures are generally included in the instruction manual for the audiometer, or they are available from the manufacturer on request. When such is not the case, it is often an indication the manufacturer feels that only he or his authorized representative is qualified to perform the calibration.

Courtesy General Radio Co.

Fig. 23-11.—Complete factory calibrations require a full array of laboratory test equipment.

Equipment required (see Figure 23–11). The equipment necessary to perform a complete factory calibration on most makes of limited-range audiometers costs several thousand dollars, and includes the following:

• *Sensitive sound analyzer with calibrated microphones.* Continuous frequency coverage from 500 Hz to 18 kHz with 2 percent accuracy, 1/10-octave bandwidth or narrower, an input range from 20 to 80 dB ($\cong$ 3 μV to 3 mV for most analyzers) with $\pm$2 dB accuracy. Most analyzers require input preamplification for the high sensitivity required.

• *Sound level calibrator*

• *Earphone coupler* (standard ear)

- *Low-frequency oscilloscope*

- *Adjustable autotransformers* 105 to 125 V output with sufficient current capability to power the audiometer.

- *Voltmeter* (a-c and d-c)

- *Frequency counter*

Preliminary conditions. Most characteristics checked in a factory calibration are required by ANSI S3.6-1969 to be within specifications over an input line voltage range of 105 to 125 volts. Therefore the audiometer is connected to an adjustable line-voltage source (an adjustable autotransformer), rather than directly to an a-c receptacle. The initial checks are made at 115 volts, repeated at 105 volts, and at 125 volts.

After initial turn-on and after the line voltage is changed, ANSI specifies a 30 minute wait to allow the electrical components in the audiometer to stabilize before they are checked. For vacuum tube audiometers, this long wait is often desirable; for semiconductor audiometers, a five minute wait is usually satisfactory.

Hearing threshold levels, tone accuracy, and output purity are also required by ANSI to be within the prescribed limits over a temperature range of 60 to 90 F (16 to 32 C). Rarely, if ever, is this verified during a routine factory calibration. Most manufacturers do, however, design their instruments to perform over the required temperature range, and they test periodically to assure conformance.

One other precaution applies to all tests made at the lower HTL outputs from the audiometer. At these low levels, ambient noise or vibration can seriously interfere with the tests, so it is important that they be performed in an area where background noise levels are very low—such as in an audiometric examination room.

Preliminary checks. The following procedure is meant only to show the primary steps involved to assure conformance to the specifications contained in ANSI S3.6-1969. It is not a factory procedure; that is, it makes no attempt to define specific test equipment control settings or connections, to call out particular audiometer controls or adjustments, to describe incidental checks on power supplies, bias networks, or other components.

Before the actual factory calibration process begins, it is good practice to detect gross malfunctions first. Many of these malfunctions can be determined by the first three daily checks (described under Daily Calibration Check). These are followed by seven other checks:

• *Sound analyzer check.* This is the same as the first procedure listed under Monthly Electroacoustic Checks, except that the sound analyzer and preamplifier are used rather than the sound level meter.

• *Acoustic output.* Here again, the procedure is the same as the equivalent step under Monthly Electroacoustic Checks, with the substitution of the sound analyzer for the sound level meter. There is one addition. The acoustic output at each frequency is also checked at 105 and 125 volt line inputs. At each voltage, the output should not vary more than two decibels from its value at the nominal 115 volt line input.

• *Attenuator linearity.* This is the same procedure outlined under Monthly Electroacoustic Check, except that with the more sophisticated analyzer, the attenuator can be checked at the lowest settings rather than simply down to 45 or 50 dB HTL.

• *Tone accuracy or frequency.* Most audiometers have little difficulty meeting the ±3 percent specified tolerance for frequency. For example, on the basis of the current standards, this means that an audiometer can produce a signal from 970 to 1030 Hz when set at 1000 Hz or 3880 to 4120 Hz when set at 4000 Hz. Of course, the lower the frequency, the less variance (that is, 485 to 515 at 500 Hz). Even though these tolerances are permissible, the audiometer should have as accurate a frequency output as possible. The audiogram of a person with a sharp drop in hearing, starting at precisely 1000 Hz, if tested on two audiometers with considerable variance in frequency output, would greatly differ. On the audiometer that produces 950 Hz at the 1000 Hz setting, the threshold could be close to normal, but on another audiometer that produces 1200 Hz at the same 1000 Hz setting, the threshold could conceivably be poorer than normal.

To test for tone accuracy, connect the audiometer to the sound analyzer by means of an earphone coupler and, using a continuous tone at 70 dB HTL, check each frequency for an accuracy of ±3 percent. Do this by setting the input frequency of the analyzer for the maximum reading on its meter, and note the setting of the frequency dial; it should correspond to the values listed in Table 23-B (at line voltages of 105, 115, and 125 volts).

• *Output purity or harmonics.* Even though allegedly pure tones are

TABLE 23-B
SPECIFIED TOLERANCE
FOR FREQUENCY

Audiometer Tone	Analyzer Dial Setting
500 Hz	485 to 515 Hz
1000 Hz	970 to 1030 Hz
2000 Hz	1940 to 2060 Hz
3000 Hz	2910 to 3090 Hz
4000 Hz	3880 to 4120 Hz
6000 Hz	5820 to 6180 Hz
8000 Hz	7760 to 8240 Hz

used in the basic measurement of hearing, it is possible to have harmonics of the fundamental (test frequency) present in the earphones. The ANSI specifications call for the fundamental to be at least 30 dB above the sound pressure level of any harmonic.

A person with a hearing loss in the lower frequencies, with better hearing in the higher frequencies could present optimistic thresholds in the lows if the harmonics are not well below the fundamental. That is, the threshold would be obtained for the harmonic instead of the fundamental.

ANSI S3.6-1969 states, "the maximum sound pressure level allowed for *any* harmonic is 30 dB below the level of the fundamental." Although the specification reads "any" harmonic, only the second and third harmonics need be measured for all practical purposes; there is little likelihood of trouble from higher-order harmonics.

At each tone (and at line voltages of 105, 115, and 125 volts), set the analyzer frequency to the same frequency as that of the audiometer (fundamental), and note the decibel level indicated on the analyzer. Increase the analyzer setting to twice the frequency (second harmonic), and then to three times the frequency (third harmonic) of the audiometer tone. In each case, the level should be 30 dB below that of the fundamental.

• *Tone switching.* Continuous-tone audiometers incorporate a switch to turn the tone on and off manually, and pulse-tone audiometers incorporate circuitry to pulse the tone on and off automatically (usually 200

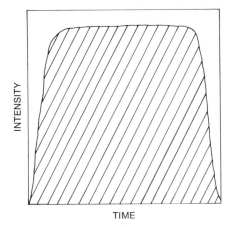

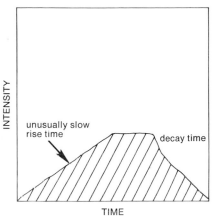

Fig. 22–12.—An illustration of uniform rise (on) and decay (off) times of a pure tone test signal with an absence of undesirable overshoot and uneven plateau.

Fig. 23–13.—An illustration of an unusually long rise time and a decay time which is unequal to the rise time. A desirable rise and decay time is approximately 100 milliseconds.

milliseconds on and 200 milliseconds off). Often, this switching causes undesirable side effects, such as audible transients or extraneous frequencies that interfere with the audiometric examination. For this reason, the switching characteristics are carefully specified by ANSI S3.6-1969.

Figure 23–12 illustrates the upper half of a signal on an oscilloscope when it is turned on and then turned off. According to the specifications, it should rise to its peak within 100 to 500 milliseconds and decay or fall off in the same period of time. Actually, it is undesirable to have a rise-decay time more than 200 milliseconds. The audiometer should not present an unusually long rise and decay time (see Figure 23–13), and there should be an absence of overshoot in the tone (see Figure 23–14). The precision of the signal presentation can be very critical for accurate audiometry.

A slow rise time may fail to elicit maximum excitation of the auditory mechanism and result in a poorer threshold than, in fact, is present. On the other hand, overshoot may result in the establishment of better thresholds than are present. Finally, if the rise time is too brief and overshoot is present (see Figure 23–15), the result could be an audible

601

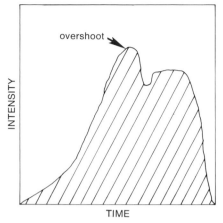

Fɪɢ. 23–14.—An illustration of unwanted overshoot in the presentation of a pure tone test signal.

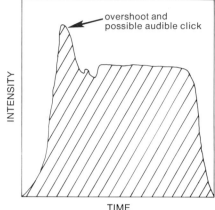

Fɪɢ. 23–15.—An illustration of an undesirably short rise time with overshoot. This situation can lead to an audible click as the tone is presented.

click in the pure tone signal, thus encouraging a threshold for clicks and not for pure tones.

To check for tone switching or tone interruption time, set the audiometer for a 70 dB HTL *pulsed* tone at 1000 Hz, connect the electrical *output* of the analyzer to the input of an oscilloscope, and note the relative magnitudes of the on and off levels; the off level should be at least 50 dB below the on level.

Reduce the audiometer attenuator setting to 30 dB HTL, set the oscilloscope controls so the upper 20 dB of the signal fills at least four major vertical divisions of the oscilloscope display, and check for the characteristics as shown in the diagram (see Figure 23–16). After the test is complete, disconnect the oscilloscope.

• *Noise.* ANSI S3.6-1969 states that "any signal from the earphone not being used for test purposes shall have a sound pressure level at least 10 dB below the hearing threshold reference level, except that it need not be more than 70 dB below the signal from the on earphone." The first part of this specification (par. 4.4.2 ANSI S3.6-1969) is extremely difficult to verify at best and impossible in most cases, even with laboratory

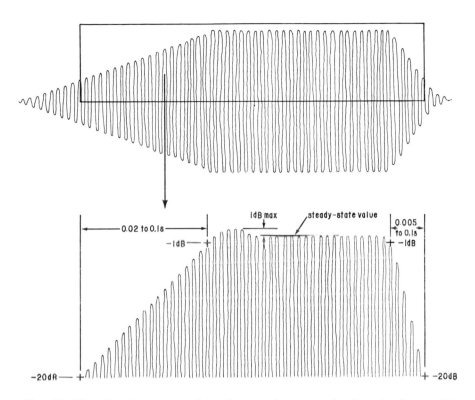

FIG. 23–16.—The time required for the sound pressure level to rise from −20 dB to −1 dB, regarding its final steady value, shall not be less than 20 milliseconds and not more than 100 milliseconds, and the time required for the sound pressure level to decay by 20 dB shall not be less than 5 milliseconds and not more than 100 milliseconds. The operation of the tone switch shall not cause the output voltage level operating the earphone to attain at any time a value more than 1 dB above its steady state.

instrumentation. The second part, although difficult, is possible as follows:

1. Set the attenuator for the highest output level (usually 90 dB HTL), and note the reading on the analyzer. Switch the earphone on the coupler from active to inactive; for example, if the left earphone is on the coupler, switch the audiometer controls so that the tones are presented to the right earphone. The analyzer reading should now be at least 70 dB below the first reading.

603

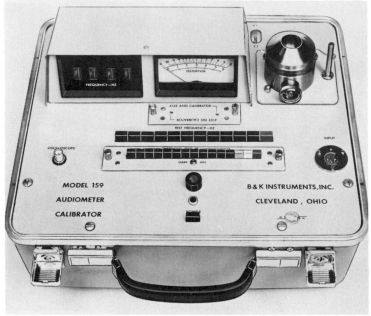

Courtesy B & K Instruments, Inc.

Fig. 23–17.—The audiometer calibrator shown here permits an operator to rapidly test all audiometric frequencies and air conduction hearing level settings. This instrument reads out in dB deviation from the prescribed sound level, digitally indicates frequency of the test tone, and provides a meter indication of harmonic distortion components.

2. Disconnect the earphones from the audiometer and insert a dummy load (a resistive load equal to the impedance of the earphone at 1000 Hz) in their place. Remove the earphone from the coupler, place both earphones back in the headband, and place them over your ears as if you were about to take an audiometric examination. Set the output to 50 dB HTL. No extraneous noise should be heard from the audiometer.

This check is meant to disclose any noise that would furnish a clue, to a person taking the test, that might influence the test results. It should be performed by a person with normal hearing under the conditions normally encountered in the test environment. (In most cases this means the subject would be inside an audiometric examination room, the audiometer would be outside.)

This completes the essential steps of any factory calibration procedure. Attention to the calibration of an audiometer cannot be overemphasized. An audiometer can easily fall short of the physical characteristics required for an accurate assessment of auditory function. Audiometers should receive periodic rigorous maintenance and calibration. (See Figure 23–17.)

Bibliography

American National Standards Institute, Inc. *Specifications for Audiometers* (ANSI S3.6–1969). New York: American National Standards Institute, 1969.

General Radio Co. "Field Calibration of Audiometers." *Noise Measurement*, No. 1 (1967).

————. "New Reference Threshold for Audiometers." *Noise Measurement*, No. 1 (1970).

————. *Simplified Pure Tone Audiometer Calibration with Sound-Measuring Instruments* (Publication IN114). Concord, Mass.: General Radio Co., 1970.

Harford, E. R. "Audiometer Calibration." *Hearing Measurement*, edited by I. M. Ventry, J. B. Chalking, and R. F. Dixon. New York, N.Y.: Appleton-Century-Crofts, 1971.

Sutherland, H. C. and Gasaway, D. C. *On-Site Audiometer Calibration Check.* USAF School of Aerospace Medicine, Brooks Air Force Base, Texas, 1970.

Thomas, W. G.; Prestar, M. J.; Summers, R. R.; and Stewart, J. L. (National Center for Chronic Disease Control). *A Study of the Calibration and Working Condition of Audiometers.* Chapel Hill, N.C.: University of North Carolina Press, 1967.

Chapter Twenty-Four

Determination of Hearing Threshold

By Earl R. Harford, Ph.D.

The basic guidelines for determining hearing thresholds presented in this chapter can provide a starting point for learning the preferred way to determine a hearing threshold. A systematic method for testing hearing is essential for accurate, valid, and reliable audiometry.

The preferred technique for establishing a pure tone threshold described here was researched in 1959 by Carhart and Jerger,[1] and is used by many professional audiologists today. Reading this chapter will not substitute for direct, practical instruction in audiometry. Instead, this chapter simply describes guidelines to follow when measuring hearing ability to assure more accurate industrial audiometry. Adherence by the technician to the procedures described here should help obtain valid and reliable test results a greater percentage of the time.

Audiometry

Units of measurement

Two properties of sound that are controlled when measuring hearing ability with pure tone audiometry are (a), frequency, which is related to the pitch of sound, and (b), intensity, which is related to loudness (see Figure 24–1).

As described in earlier chapters, the unit of measurement for describing frequency is *hertz* (abbreviated Hz), in deference to Rudolf Hertz for his work in acoustics. (The older term, that is still in limited use, is cycles

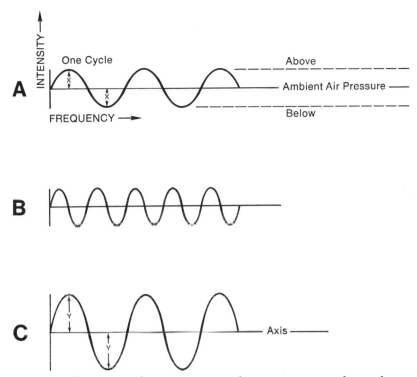

FIG. 24–1.—The curves shown are pictorial representations of sound waves. Frequency is related to pitch, and intensity is related to the loudness of the sound. *Curve B* represents a sound that would be higher in pitch than a sound represented by *Curve A*, because the variations in air pressure, as represented by a point on the curve, cross the axis more frequently. The intensity (loudness) of a sound can be shown by the height of the curve; the sound represented by *Curve C* is louder than the sound represented by *Curve A* (distance *Y* is greater than distance *X*).

per second, or cps.) The *decibel* (abbreviated dB), in deference to Alexander G. Bell, is the unit of measurement for describing the intensity of sound (see Figure 24–1 and refer also to Chapter 2). Thus, when measuring hearing ability, an individual's capacity to hear sound at various frequencies is assessed. In order to do this, the intensity (or loudness) at the earphone is varied until the lowest intensity level is found at which the individual is just able to detect the presence of that sound.

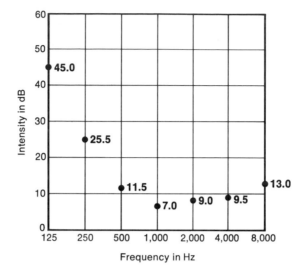

Fɪɢ. 24–2.—The average normal hearing threshold values in decibels above the standard reference level (0.00002 N/m²) for each frequency of interest. These standard reference threshold sound pressure levels apply to the use of the Type TDH-39 earphone and the National Bureau of Standards 9-A coupler—based on ANSI *Specifications for Audiometers* S3.6–1969 (R1973).

Audiometric zero

There are two different decibel (dB) scales that must be differentiated. The sound pressure level (SPL) scale (or the "engineer's scale") has as its base of reference 20 micropascals or 20 micronewtons per square meter, 20 μN/m² (0.00002 N/m²), or 0.0002 microbars (dynes per centimeter squared). This level of sound pressure is so faint that very few persons can hear it.

In contrast, the hearing threshold level (HTL) scale used in audiometry has, as its base of reference, a specific amount of sound pressure that is just audible to the average normal young ear; that is, the level where the average normal-hearing person can just barely detect the presence of that sound. It ranges from 7 to 45 dB above the engineer's reference level, depending upon the frequency (see Figure 24–2).

The human hearing mechanism is most sensitive to sound in the speech frequencies, 500 to 2000 Hz), and less sensitive to sound in the lower frequencies (below 500 Hz). This is why it takes more sound pressure,

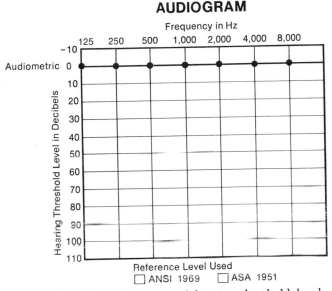

AUDIOGRAM

Frequency in Hz

FIG. 24–3.—The various normal hearing threshold levels at different frequencies, which are shown in Figure 24–2, were arbitrarily assigned a zero value, placed on a chart, and a straight line was drawn to represent zero hearing level for human hearing.

starting at the acoustical engineer's zero, in order to get to that point where the average normal listener is just able to detect the presence of sound for the lower frequencies (see Figure 24–2).

Remember this—the engineer's sound pressure level scale is used to measure the level of sound in our environment. The hearing level scale is applied when measuring human hearing. The two measurement scales are *not* identical. It is important to understand this to avoid making a direct comparison between measurements of environmental noise and the hearing threshold measurements as shown on an audiogram.

The different amounts of minimum sound pressure at the earphone at various frequencies that are just audible to young individuals with average normal hearing are designated as zero. Zero hearing level in audiometry does not mean the absence of sound. It means, rather, "the lowest sound level at a particular frequency at which the average young normal listener is just barely able to hear that sound." The zero hearing levels at different frequencies were arbitrarily assigned a zero value, plotted on a chart and a straight line drawn to represent the standard audiometric reference threshold level, or audiometric zero hearing level, for human hearing (see Figure 24–3).

When examining the chart on which hearing level is plotted, note that it contains a series of straight horizontal lines, and one line is *audiometric zero.* This line represents that level at which the average normal listener is just able to detect the presence of a sound at that frequency. The hearing level dB scale for audiometry is relative to audiometric zero.

Keep in mind that the sound pressure level scale is used by the engineer to measure such things as the environmental sound level. He uses an instrument called a sound level meter. But, when human hearing is measured with an audiometer, the hearing level scale is used, and the audiometric zero reference is louder than the engineer's zero.

The instrument that is used to measure hearing—the audiometer—must not be confused with the sound level meter, even though the unit of measurement used with both instruments is the decibel. This is similar to having two thermometers—one set to measure in degrees Celsius (C) and the other to measure in degrees Fahrenheit (F). Even though the name of the unit of measure (degree) is the same, the reference points are different; that is, 32 F = 0 C.

Pure tone threshold

Definition

The *threshold of hearing* can be defined as "the faintest intensity (or hearing level) of sound (pure tone) that a person is able to detect the majority of the time."

There is one aspect of this definition that should be explained. That is ". . . the majority of the time." Some definitions say "50 percent of the time." That is somewhat unrealistic when applying routine audiometry, because it may not be possible to find the level where a person can just barely hear a sound half the time it is presented.

Most audiometers are designed to change the sound intensity in 5 dB steps. Thus, if a person's "50 percent of the time" point (threshold) is between 10 and 15 dB, for example, it would be virtually impossible to establish this point. Also, to find the 50 percent response level, a sound must be presented at least four times in order to obtain two responses. If the stimulus is presented twice and the response is given once, the sample would be too small, and this procedure could lead to incorrect chance responses. Therefore, most audiologists use a criterion that calls for two responses at the same lowest level when the tone is being increased in increments of 5 dB.

Threshold techniques

There are three techniques that can be used to find that minimum sound level that a person can just detect:

• *Descending technique.* Present a sound at a level that a person can easily hear all the time, then gradually make it fainter. With this technique, seek the point where the subject stops hearing the sound.

• *Ascending technique.* The second approach is to start with a sound level that a person cannot hear and gradually increase it. Then look for that point where he reports that he just begins to hear the sound.

• *The combined method* is a combination of the descending and ascending technique. Start with a sound level that the subject can hear, decrease the level until he cannot hear it, and then increase the sound level to where the subject does hear it. Move back and forth across the person's hearing threshold for confirmation. This is the most common method employed in automatic audiometry.

One may justifiably ask, "Does it make any difference in the final results which technique is used? What if one technician always used the descending technique and another technician always used the ascending technique to find the subject's hearing threshold? Keeping all the other variables the same, would the use of these different techniques lead to different results?"

It has been found that the technique itself is not as important as the consistency of using the same technique each time the tests are given.[1] The same general approach to finding the threshold should be used as much as possible. Also, the technique that is preferred by the technician and the person whose hearing is being tested is an important consideration.

Generally, a person's hearing fluctuates from day to day and from moment to moment. The hearing threshold is not a fixed point, but rather a range; this is true for both normal and abnormal hearing. Measurements can be obtained from the upper or the lower portion of that range, and be valid, but result in a different threshold. Using the same accepted technique day in and day out is likely to yield the most con-

FIG. 24–4.—Seat the person, whose hearing is to be tested, so that you can see his profile. Facial expressions can reveal worthwhile information.

FIG. 24–5.—Do not let the subject look at you or at the dials on the audiometer. The things that you do with the audiometer when presenting a test tone may distract, confuse, or initiate a false response from the subject—especially when he is not sure that he is hearing the tone.

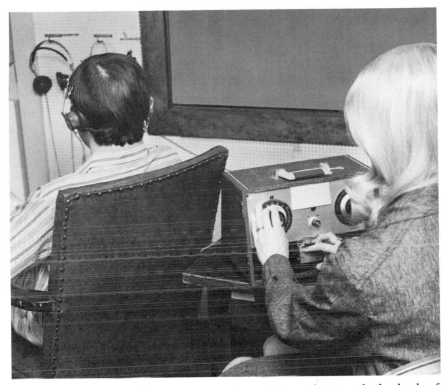

Fig. 24–6.—It is equally important not to seat the subject with the back of his head towards you. You will miss facial expressions and other important clues to the validity of the test.

sistent hearing test results. There is a greater possibility of obtaining inconsistent results when using different techniques; for example, using the descending approach one time and the ascending approach another time. Inconsistent use of these different techniques can lead to discrepant results and cause confusion in the interpretation of the test results.

One of the purposes of special courses for industrial audiometric technicians is to standardize the testing technique used in industrial audiometric programs. This provides the rationale for the "cookbook procedure" for finding hearing thresholds. Of course, this cookbook procedure can and should be modified in those situations that warrant it, but only in those situations where it is necessary to deviate from a standard technique.

Fig. 24–7.—Instructions to a subject should include, "I am going to place these earphones over your ears," and "When I present a tone it will sound like a note from a piano."

Conducting a test

The procedure for finding threshold of hearing should be a standardized, consistent technique. This will encourage consistent hearing test results from place to place and from one time to another.

Seating

The subject whose hearing is being tested is one of the most important variables affecting accurate audiometry. If he is uncomfortable, he is likely to give spurious and inconsistent responses. If the subject is warm, because the audiometric room is poorly ventilated, then he might want to get out quickly—he becomes careless in his responses. If the subject is easily excitable or nervous, he may start "hearing things," which can

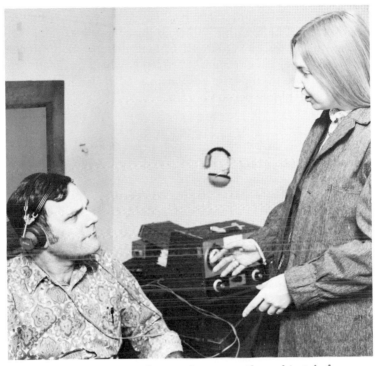

Fig. 24–8.—If you put the earphones on the subject before you complete your instructions, he may have difficulty hearing you. (Earphones introduce a 20 to 30 dB attenuation.)

lead to great frustration on the part of the technician as well as the subject.

Try to put the subject at ease. Assure him that the audiometric test is not threatening; this is time well spent. The subject should sit so that the tester can see his profile, even if the hearing test is conducted in a two-room setup (see Figures 24–4 and 28–1). In this way, sometimes worthwhile information can be obtained from the subject's facial expression; for instance, observe if he starts to fall asleep or becomes nervous or upset during the test.

Under no circumstances should the subject be allowed to see the technician as he operates the audiometer when presenting the test tone. This can distract or confuse the subject, and can lead to spurious responses (see Figures 24–5 and 24–6). This happens unintentionally much

615

more often perhaps than intentionally. The subject should not respond on the basis of what he sees, but rather to the sounds he hears. In short, if the person being tested is allowed to look directly at the technician, false positive and negative responses are likely which can lead to invalid test results. In any event, false responses tend to lengthen the time it ordinarily takes to obtain an audiogram.

Instructions

Good instructions are very important because good audiometry depends on the subject being well instructed. Although the instructions given here will suffice the majority of the time, they may have to be modified once in a while (see Chapter 28, Testing Problems: The Subject and the Technician). These instructions are for the average employee (see Figure 24–7).

Avoid the use of technical jargon, such as "pure tone," "decibels," "audiometric zero," and the like. Here is what can be said.

"I am going to place these earphones over your ears, and you will hear a sound like a note from a piano. Each time you hear one of these notes, no matter how soft it is, raise your finger, and then put it down. That's all you need to do."

Tell the subject to put his finger up and then right down. Do not ask him to indicate how long he hears the tone nor in which ear he hears it because this imposes another task—the task of auditory localization, which some people find very difficult. If the hearing test is a new experience for the subject, it is enough to ask him to do just one thing at a time; that is, "Indicate when you just hear a sound."

One question that should be asked, even if the answer is known from the record of previous hearing tests, is, "Do you hear better in one ear than the other?" The better ear should *always* be tested first. Most adults are aware whether they have a unilateral (one-sided) hearing loss. If the subject states that the hearing in both ears is about the same, then the technician should be consistent as to which ear is tested first. If many audiometric tests are being conducted daily, one after the other, it is considered good practice to start testing the right ear first (unless the hearing in the left ear is better).

Placement of earphones

The earphones should not be placed on the subject until after the instructions have been given. Earphones attenuate external sounds by

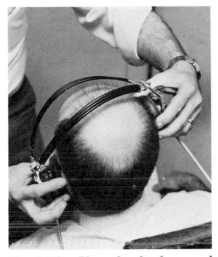

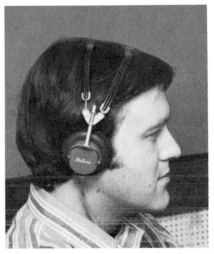

Fig. 24–9.—Place the diaphragm of the phone directly over the ear canal, and run a finger down behind the phone to determine if any part of the auricle of the ear is exposed. If so, adjust the diaphragm to cover it.

Fig. 24–10.—Be certain the headband is adjusted properly before starting the test. On the person shown here, the headband is located in the proper position—immediately over the crown of the subject's head.

20 to 30 dB, and introduce an artificial hearing loss with regard to the spoken instructions (see Figure 24–8). Therefore, place the earphones over the subject's ears *after* the instructions are given. If the subject is wearing glasses or earrings, ask him or her to remove them before putting the earphones in place.

• Open the headband and extend the earphones as far as possible. This will facilitate an easier and more correct placement of the earphones. Occasionally, a subject may have an unusually large head. In that case, be certain the earphone for the ear being tested properly covers that ear. When testing the other ear, reposition the earphone so that it is in the proper location.

• Position the opening to the diaphragm of the earphone directly in line with the ear canal, and run a finger down behind the cushion of the earphone to determine if any part of the auricle is exposed (see Figure 24–9). Occasionally, when a person has a large auricle, a portion of it will be exposed, but this is unavoidable, and it should not influence the test results.

617

FIG. 24–11.—Improper earphone placement—with the headband too far forward or too far to the rear can have a detrimental effect on the test results. The headband could even slip off the back of the subject's head.

FIG. 24–12.—It is suggested that the left earphone be held in the right hand and the right earphone in the left hand and face the subject prior to placing the earphones over the ears of the subject.

• Make certain that the headband is properly positioned on the crown of the subejct's head. Proper placement, initially, will minimize the need for a readjustment of the earphones in the middle of the test. Be certain the headband tension is adjusted snugly before starting the test (see Figures 24–10, 24–11, and 24–12). Improper earphone placement can have a detrimental effect on the test results. If the earphones are placed wrong today and right tomorrow, the subject's hearing will appear to have changed when actually, the difference is due to earphone placement.

• Always observe that the right (or red) earphone is on the subject's right ear, and that the left (or blue) earphone is on his left ear. This seems to be an unnecessary precaution to mention, but it is one of the easiest errors a technician can make.

• After the earphones are placed correctly on the subject's head, he might be asked, "Do they feel okay? Do you want to adjust them a little bit?" But the subject should be watched if he makes an adjustment, because he might move the earphone too far in one direction or the other.

Establishing the threshold

The determination of auditory thresholds should follow a systematic order. The technician should be prepared and know what to do during the test from one event to the next, depending upon the subject's reactions to the presentation of the test signal. There should be little hesitation by the technician, so that the determination of threshold can be established in a minimum of time and with a maximum of confidence in the final results.

Establish a *criterion for a satisfactory auditory threshold*, and once it is met, be satisfied, and go on to the next frequency. Audiometry performed in this manner is the way it should be done; that is, with dispatch —precisely and accurately. It should not be a tedious, long, and drawn out procedure.

The following is a "cookbook procedure" for determination of auditory threshold:

Preparatory phase

1. Set the frequency selector at 1000 Hz.

2. Set the tone reverse switch so that the signal is normally off when the tone presentor is in the *rest* position. (See Chapter 22 for a description of the switches and their function.)

3. Be certain the earphone (or output) selector is set to deliver the first test tone to the correct earphone—the right earphone unless the left is the better ear

4. Adjust the intensity or hearing level control to a dial reading of 50 dB. Some technicians prefer to start with the hearing level dial set at zero dB or 10 dB and, with the tone on, increase the level until the subject responds. Then the tone is interrupted, and the testing begins at (or just a little louder than) this level.

5. Present the pure tone sound for one to two seconds (see Figure 24–13).

6. If a response is obtained from the subject, decrease the intensity control to 40 dB, and present the tone again for one to two seconds.

7. If again a response is obtained, decrease the intensity level another 10 dB, and present the tone. Continue decreasing the level of the tone in steps of 10 dB, until an intensity level is reached where the subject fails to respond when the tone is presented.

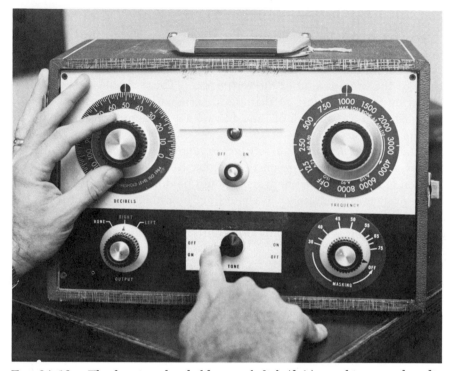

FIG. 24–13.—The hearing threshold control dial (*left*) on this manual audiometer is shown set at 50 dB, and the frequency selector dial (*right*) is set at 1000 Hz while the tone presentor switch (*lower center*) is being depressed.

The initial decreasing or descending portion of a hearing threshold determination might be referred to as the preparatory phase. That is, the technician allows the subject to hear the test tone loudly enough so the subject can get a good idea of what the signal sounds like so that he will be better able to recognize it when it becomes very faint. The preparatory phase also gives the technician a clue as to how good or poor the subject's hearing is at that particular frequency. Of course, if the subject fails to hear the pure tone or sound at 50 dB, when it is first presented, the technician should immediately increase the intensity or hearing level to 70 or 80 dB for the start of the preparatory phase. The aim is to allow the subject to hear the tone easily at the outset of the hearing threshold determination at each frequency and then descend in 10 dB steps until the subject fails to respond.

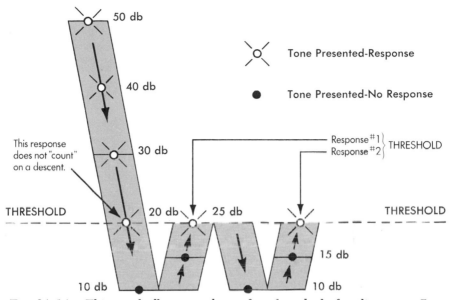

FIG. 24–14.—This graph illustrates the preferred method of audiometry. Begin with a sound that the subject can hear, decrease the intensity in 10 dB steps until he cannot hear it, and then increase the intensity in 5 dB steps to the level the subject hears again. The signal is then decreased by 10 dB (where the subject is not likely to hear it). Then increase the intensity in 5 dB steps until the subject responds. If the response occurs at the same level as the first ascending response, the threshold has been reached for that frequency. If it is not the same level as the first ascent, repeat the ascent in 5 dB steps until two responses are obtained at the same level.

Test phase

Now the subject is ready for the test phase (see Figure 24–14). In this illustrative example, a hearing level of 10 dB was reached where the subject did not respond.

1. Increase the intensity or hearing level of the tone by 5 dB, and present the signal for one to two seconds. In this example (Figure 24–14), the subject did not respond.

2. Increase the hearing level still another 5 dB, present it for one to two seconds, and assume the subject does respond. This counts as the first response while increasing the intensity or hearing level of the test tone.

621

3. Now decrease the hearing level by 10 dB, and present the tone for a second or two. Assume the subject does not respond (and under ordinary circumstances he should not respond).

4. Increase the intensity or hearing level 5 dB and present the tone. Assume the subject does not respond.

5. Increase the hearing level still another 5 dB and present the tone. Assume now the subject does respond. This is the second time he has responded at the same level while increasing the intensity of the tone.

This level consitutes the subject's hearing threshold at that frequency.

Be certain that the tone is *off* each time the intensity or hearing level control is changed. Be certain that the level is decreased in 10 dB steps and is increased in steps of 5 dB. (Some audiometric technicians decrease the level in 5 dB steps during the test phase, but this tends to induce false responses and lengthens the test phase of the threshold determination by introducing an extra step.) Although it is acceptable to decrease the intensity or hearing level of the tone 5 dB during the test phase, 10 dB decreases tend to offer more clear-cut responses.

The important point to remember is that the intensity or hearing level should always be increased in steps of 5 dB once the test phase starts. If greater steps than 5 dB are used, the true hearing threshold may be missed. Another important point to remember is that the responses during the preparatory phase really do not count. Also, remember that you are seeking the level at which the subject first reports hearing the test tone on two ascending approaches. These responses need not be two in consecutive succession, as shown in the example in Figure 24–14. That is, in some cases on the second ascent, the subject may not respond until the intensity is increased to 25 dB, but on the third ascent, he may respond at 20 dB. Thus, there would be a response at 20 dB (first ascent) and another response at 20 dB (third ascent). Therefore 20 dB would be recorded as the subject's hearing threshold at 1000 Hz.

Recommended test frequencies

The recommended order is 1000, 2000, 3000, 4000, 6000, and 8000 Hz. Then, test at the lower frequency of 500 Hz. Although 8000 Hz is not a required test frequency for industry, it is usually essential for establishing the noise induced notch (see Chapter 26 for a description).

Intratest reliability

The subject's hearing threshold at 1000 Hz should be reestablished to check intratest reliability. If the recheck yields a threshold within ±5 dB of the first threshold established at 1000 Hz, then the technician should have increased confidence in the reliability of the thresholds for the intervening frequencies. If more than a ±5 dB difference is obtained for the second threshold determination at 1000 Hz, the tester should repeat the hearing threshold determination at other frequencies to assure good reliability or the need for a complete retest. After this has been done, the technician should switch the output selector to the other ear-phone, and the same test procedure followed for the subject's opposite ear.

Qualification of the test

Somewhere on the audiometric record, write down an opinion of the reliability of that hearing threshold test. If the reliability is good, do not hestiate to put down "good." But if the test is anything less than good, it should be qualified. "Poor" or "fair" reliability must be indicated; then the reason should be stated. If the subject did not seem to be cooperative, then put that down, so that anyone else who might be testing or managing that person will get a clue as to what the problem might have been. It might not always be the subject being tested that causes the validity and/or reliability of the test results to be questioned. Perhaps the audiometer is not working properly. If so, an appropriate notation to this effect should be included on the audiometric record.

If a poor reliability is assigned to a high percentage of audiometric records, there is probably a basic problem that needs evaluation— such as the background noise level in the test room or the testing technique. Keep in mind, however, that an audiometric record that says nothing about reliability is usually assumed to be accurate and reliable by those individuals doing the interpretation. If the audiometric test did not appear to be accurate, then do not hesitate to make this statement on the audiogram. By following this approach, a problem may be prevented that might otherwise occur had no notation been made. Also, you are protecting your own reputation as a careful and competent technician when you qualify each test according to your best judgment.

Summary

In order to achieve valid results that can be accepted from place to place and from one point in time to another, the technique for measuring air conduction, pure tone hearing thresholds should be standardized. Of the three methods—descending, ascending, and combined techniques—

a modified ascending technique is recommended in this chapter.

When performing a manual, air-conduction, pure tone audiometric test, the following sequence is suggested:

1. *Seating.* Because the subject is possibly the most important variable in an audiometric test, make sure he is seated comfortably and is at ease.

2. *Instructions.* Give the subject clear, concise instructions about what is expected of him. Do not get technical; use language that has meaning to him.

3. *Placement of earphones.* Place the earphones correctly on the subject's head. Bad placement of earphones can adversely affect the hearing test results.

4. *Determining threshold.* Determination of the hearing threshold should follow a systematic order. The test should be performed with care, but should not be a long drawn out procedure.

Evaluate the reliability of the test. It is very important that this notation be made on the record. If nothing is stated about the technician's judgment of the accuracy of the test, anyone interpreting the audiogram is expected to assume it is valid and reliable.

Clearly, there are problems and pitfalls associated with audiometry that have not been mentioned in this chapter. However, the subsequent chapters in Part Five deal with such problems and with suggestions on how to cope with them.

References

1. Carhart, R., and Jerger, J. F., "Preferred Method for Clinical Determination of Pure Tone Thresholds." *Journal of Speech and Hearing Disorders* 24:330–345.

Chapter Twenty-Five

Manual and Automatic Audiometry

By William F. Rintelmann, Ph.D.

Automatic audiometry differs considerably from manual audiometry, not only in the instrumentation, but also in the technique or method of determining the hearing threshold. To determine the hearing threshold, the method of adjustment is employed in automatic audiometry; whereas, in manual audiometry, the preferred technique is the modified, ascending method.

Laboratory studies have demonstrated good inter-test reliability when comparing the results of hearing tests made with both types of instruments. However, in industrial hearing conservation programs, hearing threshold differences of 10 dB or greater may be found for a large percentage of employees when comparing automatic to manual audiometric test results. This 10 dB or greater discrepancy can have an important impact on decisions regarding medical or audiological referrals of employees.

The automatic, self-recording audiometer is a complex electronic device, and its proper use requires considerable knowledge and skill. If an industrial hearing conservation program is to be effective, both the individual responsible for the conduct of the program and the industrial audiometric technician must be familiar with the special problems associated with automatic audiometry.

Pure tone audiometry

Purpose

The purpose of administering a pure tone air conduction audiometric

AUDIOGRAM

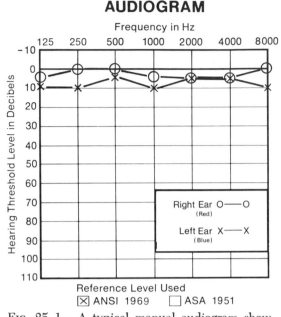

FIG. 25–1.—A typical manual audiogram showing hearing thresholds within the normal range.

test to an employee is essentially two fold: (a) to determine whether or not the employee's hearing for pure tones is within normal limits and (b) to obtain a record of the employee's audiometric configuration over a specified frequency range. (See Figure 25–1). Detailed information concerning the employee's audiometric configuration is helpful in making appropriate medical or audiological referrals. For example, an employee with a noise induced hearing impairment will typically display a high-frequency loss with the greatest amount of hearing threshold deficit somewhere between 3000 and 6000 Hz.

Valid audiograms

There is considerable scientific evidence[1] to demonstrate that after exposure of an individual to intense noise, a temporary threshold shift occurs that is directly related to the sound level and duration of noise exposure. The higher (more intense) the noise level or the longer the exposure time, the greater the temporary threshold shift (see Figure 25–2).

626

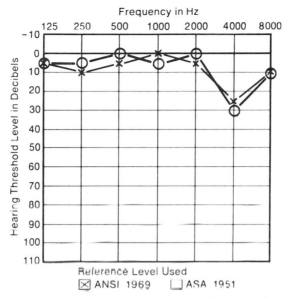

Frequency in Hz

Reference Level Used
☒ ANSI 1969   ☐ ASA 1951

Fɪɢ. 25–2.—A typical manual audiogram that was taken immediately after the employee was exposed to excessive noise. Compare the hearing threshold levels shown here with those plotted on the audiogram shown in Figure 25–1. Note the sharp drop at 4000 Hz.

The temporary decrease in hearing acuity or pure tone threshold sensitivity in an individual resulting from previous noise exposure has been termed temporary threshold shift (TTS), provided that the hearing thresholds return to pre exposure levels after the person has been away from noise for several hours.

Many employees working in typical industrial environments are routinely exposed to noise levels that can produce a TTS. Therefore, the validity of audiometric tests can be seriously affected by the level and duration of previous noise exposure. To avoid the problem of TTS, all audiograms should be obtained when the employee is in a "noise rested" condition.

This can be achieved either by making sure that the worker is away from noise for many hours (at least 14 or more) or by having the employee wear some type of hearing-protection device (plugs or muffs) that will effectively reduce his noise exposure. Because of work scheduling problems, the most feasible method of attaining valid "noise rested" hearing thresholds is the proper use of hearing protection.

627

Fig. 25–3.—A typical discrete frequency, manual, pure-tone, air conduction audiometer used in industrial hearing conservation programs.

Courtesy of Beltone Electronics Corp.

A **baseline audiogram** is a reference audiogram to which subsequent pure tone hearing tests are compared. Therefore, the initial audiogram of an individual's hearing thresholds after he has been hired is a baseline audiogram. Precautions should be taken to minimize the effects of a temporary threshold shift on the hearing test results.

Subsequent pure tone hearing tests, which are administered to employees at intervals of one to two years after the baseline test, are termed *monitoring audiograms.* The results of these tests are then compared to the initial or baseline audiogram (see Chapter 26). Precautions to avoid TTS should also be observed when administering monitoring audiometric tests (additional details are in Chapters 27 and 28).

Audiometers

Manual audiometry

The conventional, discrete frequency, manual, pure tone air-conduction audiometer is manufactured in three basic types: wide range, limited range, and narrow range (see Figure 25–3).

The limited-range, pure tone, air-conduction audiometer is the type of manual instrument best suited for use in industrial hearing conservation programs. The limited-range audiometer has the frequency and inten-

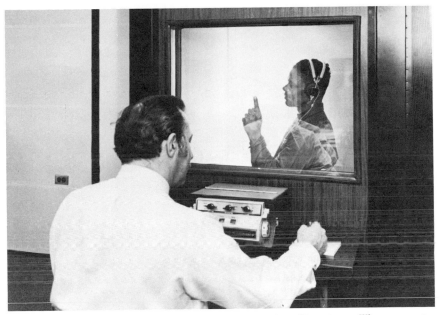

Fig. 25–4.—An audiometric test using a manual audiometer. The presentation of the pure tone signal is under the control of the industrial audiometric technician. The individual whose hearing is being tested responds when he hears a tone in the earphone by raising and lowering his finger.

Courtesy of Industrial Acoustics Co.

sity capability required for air-conduction hearing threshold testing in industrial hearing-conservation programs (see Chapter 22 for additional details).

The audiometric test frequencies incorporated in the instrument should include 500, 1000, 2000, 3000, 4000, and 6000 Hz. The intensity range should be at least from 10 dB to 70 dB hearing threshold level.

The distinguishing characteristic of finding the subject's hearing threshold with a manual audiometer is that the presentation of the pure tone signal is under the control of the audiometric technician—not the person whose hearing is being tested. Hence, when a pure tone signal is presented in the earphone by the audiometric technician, the individual whose hearing is being tested responds when he hears a tone simply by raising his hand (or finger) or by pushing a signal button (see Figure 25–4).

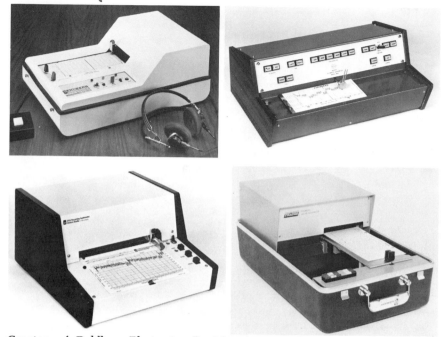

Courtesy of Dahlberg Electronics, Precision Acoustics Corp., Grason Stadler, and Tracor, Inc.

FIG. 25–5.—Typical models of automatic self-recording audiometers used in hearing conservation programs.

Three techniques can be used for obtaining air-conduction, pure tone hearing thresholds. They are the ascending, the descending, and the combined technique. Any of these methods can be used to obtain valid hearing thresholds; however, the important thing to remember is to be consistent and whenever possible always to use the same method or technique. The modified ascending technique has generally been adopted as the preferred method for manual audiometry in industrial hearing conservation programs. (See Chapter 24 for a detailed description of this technique.)

Automatic audiometry

The automatic audiometer is sometimes called the Bekesy type, because it was first described by Georg von Bekesy in 1947.[2] It is also referred

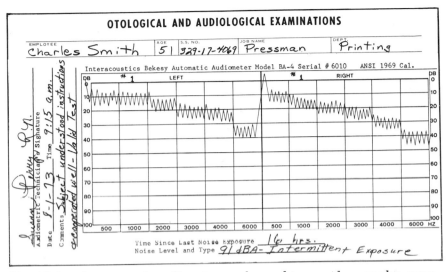

FIG. 25–6.—An automatic audiometer can be used to provide a graphic representation of a subject's hearing threshold. A pen is moved along the audiogram's horizontal axis (as a function of time) for each frequency. The subject presses the handheld pushbutton to make the audiometer tone softer; releasing the pushbutton makes the signal louder. The subject is instructed to keep the audiometer tone in his earphones at a level that he can just hear. "Press the pushbutton when you hear the signal and release the pushbutton when you don't." As long as the subject indicates by pressing a pushbutton that he hears a signal, the attenuator continuously decreases the intensity of the tone while simultaneously moving the recording pen up the vertical axis of the audiogram. The moment he indicates by releasing the pushbutton that he no longer hears the tone, the attenuator, associated with the pushbutton, increases the sound intensity in the earphones while simultaneously moving the recording pen down the vertical axis of the audiogram.

to as a self-recording audiometer (see Figure 25–5). As the name implies, the instrument has certain automatic features that permit the audiometric technician a certain amount of freedom while the hearing thresholds of the person being tested are automatically recorded on an audiometric chart (see Chapter 22 for description).

Depending on the model and complexity of the automatic audiometer, hearing threshold levels can be obtained by either the sweep- or fixed-frequency method.

In the sweep-frequency type of automatic audiometer, the frequency of the pure tone signal being sent to the earphone slowly but constantly changes or sweeps across the frequencies from low to high (for example, 100 to 10,000 Hz).

In the fixed-frequency type, the pure tone signal remains at a constant frequency for a set period of time, and then is automatically changed to the next higher test frequency. For example, a pure tone signal at 500 Hz is delivered to the earphone for 30 seconds, then the frequency is automatically changed to 1000 Hz, the next higher frequency, for 30 seconds. This sequence is repeated for the other test frequencies.

With either method, the person whose hearing is being tested controls the intensity of the pure tone signal being delivered to the earphone by means of a hand-held switch. He is instructed to activate the switch as soon as he hears a sound in the earphone, and to release the switch when the sound is no longer heard. By following these instructions, the pure tone signal being delivered to the earphone repeatedly goes from just above to just below the subject's hearing threshold, while the threshold tracking is automatically recorded by an ink pen on a chart (see Figure 25–6).

The exact threshold (hearing threshold level) at a particular frequency is the midpoint between the peaks and valleys on the chart. For example, in Figure 25–6 the threshold at 500 Hz in the left ear is 10 dB hearing threshold level. A generally accepted requirement for a valid hearing threshold measurement is that there must be at least six full threshold crossings at a given test frequency.

This method of measuring an individual's hearing ability or obtaining a hearing threshold with an automatic audiometer is similar to the *combined* method in manual audiometry (see the previous chapter). In other words, the individual's hearing threshold is traced in a combination of descending and ascending runs, while the intensity of the pure tone signal being delivered to the earphone is being controlled by a hand-held switch, operated by the person whose hearing is being tested. The technical name for this method (provided the intensity of the pure tone signal is under the control of the person receiving the test) is the *method of adjustment*.

The fixed-frequency type of automatic audiometer should provide test tones for at least six frequencies (500, 1000, 2000, 3000, 4000 and 6000 Hz) which can be automatically and sequentially presented for 30 seconds—first to one earphone and then to the other (for instance, left and then right) followed by a retest at 1000 Hz of at least one ear.

The pure tone signal can be presented as either a pulsed tone (200 milliseconds on and 200 milliseconds off) or a continuous tone; however, the pulsed method is recommended for threshold determination. The intensity of the pure tone signal should range from -10 to $+90$ dB hearing threshold level re ANSI S3.6–1969.[3] The intensity level should be controlled by a hand-held switch by the person whose hearing is being tested.

Manual vs. automatic

A question often asked is, "Are automatic audiometers more suitable for use in industrial hearing conservation programs than manual audiometers?"

There is no simple answer to this question because both types of audiometers have certain advantages and limitations. In terms of the time required for testing, neither type of audiometer has a clear-cut advantage. Some employees with hearing problems can be tested faster using the automatic audiometer; however, most young adults with normal hearing can be tested more quickly using the manual-type audiometer.

Manual

When using the manual audiometer, the industrial audiometric technician has greater flexibility regarding test procedures and more control over the test situation than with the automatic type. Employees whose hearing is being tested seem to find the instructions and the task easier to understand and perform when tested using the manual audiometer. As a consequence, less time is spent by the audiometric technician on reinstruction of the subject with the manual than with the automatic audiometer.

In fact, industrial audiometric technicians and audiologists may encounter a few employees who simply cannot perform the required task with the automatic audiometer without considerable reinstruction and practice. However, experience has shown that many of these same individuals will respond appropriately when tested using the manual audiometer.

Automatic

The automatic audiometer has one major advantage over the manual audiometer. The audiometric technician need not devote full time and attention to the task once the technician is certain that the person whose hearing is being tested is responding properly. The industrial audiometric technician is free for a few minutes to do other things, *provided the automatic threshold recording can be continuously visually monitored*

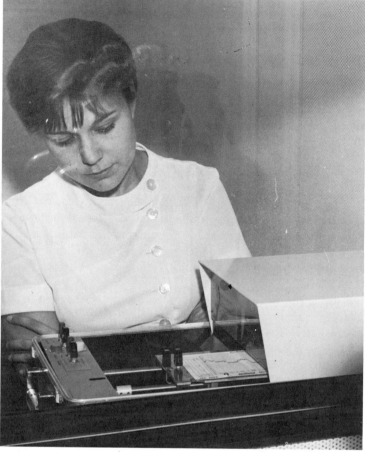

Courtesy Tracor, Inc.

FIG. 25–7.—An industrial audiometric technician visually monitoring the tracings produced by an automatic audiometer.

(see Figure 25-7).

Thus, depending upon the equipment available and the physical arrangement of the testing booth, it is possible for the audiometric technician to administer and visually monitor two or more automatic

634

audiometric tests simultaneously.

It may also be possible for the audiometric technician to administer a manual test to one employee while another employee is tracing his own thresholds with the automatic audiometer. A controversial question, however, remains, "Can the typical industrial audiometric technician effectively monitor more than one hearing test at a time?" This author is against this practice in the industrial setting.

One caution regarding the use of automatic audiometers should be clearly understood. Although the instrument is essentially automated, the industrial audiometric technician must still have the same basic knowledge and training concerning industrial audiometry as would be needed for administering manual pure tone hearing tests. In fact, many aspects of automatic audiometry are considerably more complicated than manual audiometry. Hence, only a properly trained, well-qualified industrial audiometric technician can obtain valid and reliable pure tone, air-conduction hearing thresholds with an automatic audiometer.

The self-recording features of an automatic audiometer minimize the chances of human error by the industrial audiometric technician in recording the subject's hearing threshold. Audiometry, using the manual instrument, requires that the industrial audiometric technician record the subject's hearing threshold; this is done automatically with an automatic audiometer.

Some authorities feel that the use of an automatic audiometer eliminates false recording or misleading results being reported by the industrial audiometric technician. At times, due to the press of other duties or approaching quitting time, the industrial audiometric technician may have a tendency to rush through a manual audiometric test, thereby increasing the possibility of obtaining an invalid audiogram. The automatic audiometric test requires a fixed unchangeable period of time for each test.

A final point that should be considered by the individual in charge of the industrial hearing-conservation program is the communication problem. If the employee, whose hearing is to be tested, does not fully comprehend the instructions or what is required of him, invalid test results will be obtained. This communication gap between the employee and the industrial audiometric technician can cause problems whether the manual or automatic type of instrument is used. The industrial audiometric technician should not proceed with an audiometric test until he is certain that the employee, whose hearing is to be tested, fully understands the required task, including how and when he is to respond.

Relative popularity

Approximately two decades have passed since it was first suggested that automatic audiometry could be usefully employed in the industrial setting.[4,5] Only limited evidence exists, however, concerning the relative popularity of either the manual or the automatic audiometer in present-day industrial hearing-conservation programs.

In a survey of hearing-conservation programs in representative aerospace industries, Rintelmann and Gasaway[6] reported in 1967 that, in a sample of 50 companies, the use of the manual-type audiometer outnumbered the automatic type by approximately four to one. Further, surprisingly, they found no relationship between the size of the work force tested and the type of audiometer (manual *vs.* automatic) employed. Thus, it appears that further research is needed to ascertain the pros and cons of automatic audiometry for industrial application.

Medical or audiological referral

The audiometric test results along with certain case history information form the primary basis for medical or audiological referral. If the hearing-conservation program has been recently established (within the last one or two years), then referrals are usually made on the basis of a single audiogram.

If, however, the hearing-conservation program has been in existence for several years, referrals are usually based on comparing a worker's most recent audiogram to his baseline audiogram. For example, if the employee's pure tone hearing thresholds show an increasing hearing loss between successive audiograms, then a referral for medical or audiological examination is indicated.

Referral criteria

Specific referral criteria are hearing threshold shifts (poorer thresholds) in either ear that equal or exceed a predetermined amount, for example, 10 dB at 500, 1000, 2000, or 3000 Hz, or 15 dB at 4000 or 6000 Hz. This example of referral criteria is from the National Institute for Occupational Safety and Health (NIOSH) Criteria Document *Occupational Exposure to Noise.*[7] The important point is that whatever the amount of threshold

---

At the time of this writing, the most recent definition of a *"significant threshold shift* [is] an average shift of more than 10 dB at frequencies of 2000, 3000, and 4000 Hz relative to the baseline audiogram in either ear." Quotation is from the proposed OSHA rules (*Federal Register,* Vol. 39, No. 207—October 24, 1974).

shift that is ultimately accepted as the threshold change criterion, the individual responsible for the conduct of the hearing-conservation program must decide in each case whether or not the threshold shifts are valid (real) changes in the employee's hearing thresholds.

If the hearing threshold shift, due to some testing error, equals or exceeds the referral criterion (for example, 10 dB at 2000 Hz), then, obviously, large numbers of employees would receive unnecessary medical or audiological referrals. This could result in a considerable waste of the employer's economic resources. On the other hand, if the tests do not disclose real changes in hearing, problem cases that should receive medical attention may go undetected. Thus, it is extremely important that audiometric results are both reliable (repeatable within accepted limits) and valid.

Threshold differences

The preferred method for hearing threshold determination with a manual audiometer is the *modified ascending technique;* whereas with the automatic audiometer, thresholds are obtained using the *combined* technique (method of adjustment). In order to obtain the most reliable audiograms possible, industrial audiometric technicians should consistently use the same method for measuring an employee's hearing thresholds (see Chapter 24).

In an industrial hearing-conservation program where both manual and automatic audiometers are used, an employee's baseline audiogram might be obtained using a manual audiometer, and one or two years later, a followup audiogram might be obtained with an automatic audiometer (see Figure 25–8). If an employee's pure tone hearing thresholds have changed, does this represent a real (valid) change in his hearing or is it due to questionable audiogram reliability? In other words, are the test-retest hearing threshold differences due simply to the fact that two different types of audiometers and two different threshold measurement techniques were used? Unfortunately, a definite answer cannot be given.

It is highly likely, however, that the instrument (audiometer type) and methodological (modified ascending *vs.* adjustment techniques) differences could have an important negative effect on threshold test-retest reliability.

Research by Burns and Hinchcliffe (1957)[8] and Corso (1956)[9] reported that hearing thresholds obtained with an automatic audiometer generally agree with thresholds measured with a manual conventional audiometer

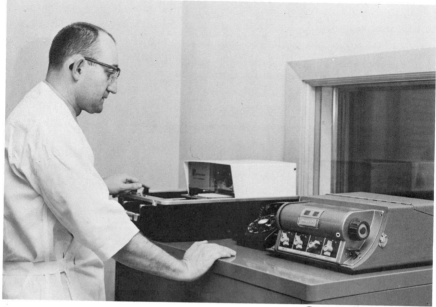

Courtesy Erie Mining Co.

Fig. 25–8.—An automatic audiometer displayed at left and a manual type on the right. Both are used in the industrial audiometric program in this plant.

when using the method of limits (combined method with the technician controlling the signal), provided that the thresholds are measured in 2 dB steps.

Rodda (1956)[10] compared automatic (Bekesy) audiometer test results with those of four different manual audiometers using 5 dB steps, and found that at 250, 6000, and 8000 Hz differences were greater than 10 dB 50 percent of the time. Rodda's results must be interpreted with caution, however, because each audiometer was operated by a different individual who may have used a different technique for measuring hearing threshold.

O'Connell (1957)[11] compared the results of audiometric tests using fixed-frequency (Bekesy) audiometry with manual conventional audiometry at 500, 1000, 2000, 4000, and 8000 Hz. He found that the percentage of agreement (±5 dB) was lowest at 2000 Hz (37.5 percent) and highest at 1000 Hz (87.5 percent).

Price (1963)[12] measured the hearing thresholds of 20 normal young

TABLE 25-A

AUTOMATIC AUDIOMETRY THRESHOLD VALUES (FROM PRICE[12])
COMPARED WITH ANSI S3.6–1969 (R1973)

All data shown are in dB Sound Pressure Level re 0.0002 dynes/cm².

Source	Frequency (in hertz)								
	125	250	500	1000	2000	3000	4000	6000	8000
AD/2*	50.4	28.5	12.6	6.4	6.7	5.8	6.6	21.4	15.5
S Mid**	51.3	27.4	12.1	4.6	5.8	2.8	4.7	9.6	14.8
ANSI***	45	25.5	11.5	7	9	10	9.5	15.5	13

* Average of ascending and descending threshold (AD/2).[12]
** Midpoint average of sweep frequency tracings at 2.5 dB/sec (S Mid).[12]
*** Reference SPL for zero dB Hearing Threshold Level with the Telephonics TDH-39 earphone mounted in MX-41/AR earphone cushions. (ANSI, 1969).

adult listeners with an automatic audiometer. The stimulus or pure tone signal was presented under a variety of conditions. He used a pulsed tone signal at 2.5 interruptions per second, two attenuation rates (2.5 and 5 dB/sec), and obtained both sweep- and fixed-frequency hearing thresholds on the subjects. He also measured thresholds using the ascending and descending technique separately in the fixed-frequency mode and found good agreement (±2 dB or less).

Table 25-A, based on Price's[12] data shows the average of the ascending and descending hearing thresholds and the midpoint average of the slow speed (2.5 dB/sec) sweep-frequency thresholds compared with the 1969 American National Standard. This table shows that with the exception of 125, 3000, and 6000 Hz, automatic audiometry thresholds for both the sweep- and fixed-frequency modes are within ±5 dB of the American National Standard. In fact, at some frequencies (250, 500, 1000, 2000, and 8000 Hz) the differences are ±3 dB or less. The greatest average difference (sweep frequency at 3000 Hz) is 7.2 dB below the American National Standard, indicating that reasonable agreement can be obtained between automatic audiometry thresholds and the present standard.

A few studies comparing results of automatic versus manual audiometry have been reported for young children. However, testing the hearing ability of young children under rigid controls cannot be compared to testing adults in the industrial setting. It has been found that with adequate time and proper pretraining, valid and reliable hearing thresholds can be measured, although automatic audiometry procedures present more problems with children than with adults.[13]

In 1971, Gosztonyi, Vassallo, and Sataloff[14] published the results of their audiometric investigation and evaluation of 100 persons (50 office and 50 shop) employed in a heavy industry. In this study, four modes of testing were investigated:

1. Automatic thresholds obtained by certified industrial audiometric technicians

2. Automatic thresholds obtained by a certified clinical audiologist

3. Manual thresholds obtained by the clinical audiologist

4. Speech Reception Thresholds (SRT) obtained by the clinical audiologist

In addition, the 100 audiograms obtained by the industrial audiometric technicians using the automatic self-recording audiometer were evaluated by Gosztonyi *et al.*, and some were classified as being "unreliable" on the basis of the occurrence of one or more of the following reasons:

1. A bilateral lack of response by the subject when the hearing level is at full intensity

2. Audiometric tracings indicating severe hearing losses at all frequencies

3. Wide or erratic audiometric tracings throughout

4. Lack of superposition of the audiometric tracings (where the industrial audiometric technician had repeated the test on the same card)

The audiograms of all 50 office employees were judged reliable, but only 27 of the 50 shop workers had audiograms that were evaluated as being completely reliable. Further, for the office employees test-retest reliability was good (±5 dB) for automatic audiometry thresholds at 500, 1000, 2000, 3000, 4000, and 8000 Hz.

Among the shop employees, 23 workers required further study because their original automatic audiometry thresholds were judged to be questionable. Of these, 17 were judged to be unreliable. The following specific results were obtained by Gosztonyi *et al.*:

> The difference between company-performed self-recording and audiologist-performed manual thresholds averaged 15 dB, but there was a wide range of differences of −18 to +53 dB. . . . The difference between self-recording and manual thresholds obtained by the audiologist averaged 10 dB. The manual thresholds almost ex-

clusively were better than self-recording thresholds. The range was from a 0 to 44 dB difference. . . . The only real agreement in thresholds was found in manual thresholds and SRT's obtained by the audiologist. The average difference was less than 3 dB with a range of 0 to 27 dB. (p. 116)

Gosztonyi and his coexperimenters concluded that a gross difference existed in the reliability of automatic audiometric examinations of office personnel as compared to shop employees. All of the audiograms under consideration were evaluated and judged to be reliable for office personnel; whereas 34 percent of the shop employees were judged to have unreliable automatic audiograms.

In attempting to determine why there was such a large percentage of unreliable audiometric test results among the shop workers and none among the office workers, the researchers found that a large number of the hourly (shop) workers had filed claims for worker's compensation due to occupational hearing loss, while no such claims had been filed by the salaried (office) employees.

The researchers further discovered that of the 23 shop workers with questionable test results, 15 were later found to have filed compensation claims. The unreliable audiograms were perhaps deliberate falsifications and could be called malingering.[14]

Recommended procedure

Two facts have been stressed thus far. First, automatic audiometry presents certain special problems that are not encountered in using manual audiometry (see Chapters 27 and 28). Second, as a consequence of the different techniques used with automatic and manual audiometers, hearing threshold differences greater than 10 dB can be obtained from a large percentage of industrial employees whose hearing is being tested.

Great care must be taken to make sure that the person whose hearing is being tested with an automatic audiometer understands what he is to do. Just as in manual audiometry, the audiometric technician should be prepared to stop the automatic self-recording test, reinstruct the employee, and allow him enough practice to be sure he makes the proper responses.

If, at the outset of the automatic self-recording hearing test, no response is made by the subject at the maximum output of the audiometer in the low to midfrequency region (500 and 1000 Hz), it is unlikely that a valid audiogram will be obtained.

Another indication of an invalid audiogram is when the threshold swings or tracings are wider (distance between the peaks and the valleys) than 10 dB. If this problem occurs and reinstructing the subject (plus giving him more time to practice) does not improve his responses, the automatic self-recording test should be discontinued, and the subject should be retested with a manual audiometer. *In problem cases, better responses will usually be obtained with a manual audiometer.*

Even though the pure tone signal presentation for each frequency is programmed in automatic audiometry, the industrial audiometric technician should constantly visually monitor the threshold recording so that erratic responses can be spotted immediately.

Psychological factors can have considerable influence on the subject being tested. An individual might be less likely to feign a hearing loss for worker's compensation purposes if he feels his responses are being closely watched.[14]

Because constant monitoring of the audiometric test by the audiometric technician is recommended, the frequently expressed "advantage" by advocates of automatic audiometry—giving the industrial audiometric technician "free time" during the hearing test—becomes questionable.

If both manual and automatic audiometers are routinely used in an industrial hearing-conservation program, it is recommended that, whenever possible, retests or followup audiograms be administered on the same type of audiometer used to obtain the baseline audiogram for that particular employee. This procedure avoids the basic problem of interpretation of threshold differences as a function of audiometer type (manual *vs.* automatic). This recommendation, of course, assumes that the initial automatic self-recording test results have been reviewed and evaluated as valid.

As a final point, discussion of the problem of malingering (faking a hearing loss) by the employee whose hearing is being tested is beyond the scope of this chapter. Whenever this is suspected by an industrial audiometric technician, a medical or audiological referral is in order.

Summary and conclusions

Based on the studies reviewed here, it appears that automatic audiometric thresholds can show good agreement with the 1969 American National Standard and also with manual conventional thresholds, provided that the latter are obtained with the method of limits using a 2 dB-step attenuator.

When the modified ascending method of determining the employee's hearing threshold is used with the more commonly employed 5 dB-step attenuator (as in the clinical or industrial setting), a large percentage of audiometric test results show threshold differences of 10 dB or greater when the results of automatic and manual audiometry are compared. Malingering by employees in the industrial setting, motivated by industrial compensation claims, tends to affect pure tone threshold reliability and validity substantially.

Automatic audiometry in industry can present a substantial problem in terms of employees being sent for medical or audiological referral. If strict audiometric referral criteria are followed, it is very important that valid audiograms be obtained. The purpose of baseline and monitoring audiometry, of course, is to detect possible changes in hearing due to noise exposure and not to refer large numbers of employees as a result of procedural differences in the audiometric testing program.

According to the studies reviewed here, careful attention to proper audiometric testing procedures can result in good agreement between automatic and manual audiograms. However, it appears that in an industrial setting more than 30 percent of the employees might be un necessarily referred, perhaps because successive audiometric tests were administered using first one type of audiometer (such as a manual) and then a year or so later the other type (automatic). A more serious problem in an industrial hearing conservation program is that poor test-retest reliability of successive audiograms could obscure actual hearing threshold changes that might be due to excessive noise exposure.

References

1. Ward, W. D. "Adaptation and Fatigue," in *Modern Developments in Audiology*, edited by J. Jerger. New York: Academic Press, 1973.
2. Bekesy, G. von. "A New Audiometer." *Acta Otolaryngologica* (Stockholm, Sweden) 35:411–422.
3. American National Standards Institute. *Specifications for Audiometers* (ANSI S3.6–1969). New York: American National Standards Institute, Inc., 1970.
4. Carhart, R. "Automatic Audiometry for Industry?" *American Industrial Hygiene Association Quarterly* 17:381–387.
5. McMurray, R. F., and Rudmose, W. "An Automatic Audiometer for Industrial Medicine." *Noise Control* 2:33–36.
6. Rintelmann, W. F., and Gasaway, D. C. "A Survey of Hearing Conservation Programs in Representative Aerospace Industries—Part I: Prevalence of Programs and Monitoring Audiometry." *American Industrial Hygiene Association Journal* 28:372–380.
7. U.S. Department of Health, Education, and Welfare. National Institute for Occupational Safety and Health. *Occupational Exposure to Noise*. Washington, D.C. Government Printing Office, 1972.

8. Burns, W., and Hinchcliffe, R. "Comparison of the Auditory Threshold as Measured by Individual Pure Tone and by Bekesy Audiometry." *Journal of the Acoustical Society of America* 29:1274–1277.

9. Corso, J. F. "Effects of Testing Methods on Hearing Thresholds." *Archives of Otolaryngology* 63:78–91.

10. Rodda, M. "Consistency of Audiometric Testing." *Annals of Otology, Rhinology, and Laryngology* 74:673–681.

11. O'Connell, M. H. *Auditory Acuity Measured by Automatic and Manual Audiometry* (Report No. 58–13) Randolph AFB, Texas: U.S. Air Force School of Aviation Medicine, 1957.

12. Price, L. L. "Threshold Testing with Bekesy Audiometer." *Journal of Speech and Hearing Research* 6:64–69.

13. Fulton, R. T. "Bekesy Audiometry," in *Audiometry for the Retarded with Implications for the Difficult-to-Test,* edited by R. T. Fulton and L. L. Lloyd. Baltimore: Williams and Wilkins Co., 1969.

14. Gosztonyi, R. E., Jr.; Vassallo, L. A.; and Sataloff, J. "Audiometric Reliability in Industry." *Archives of Environmental Health* 22:113–118.

Chapter Twenty-Six

Interpretation of Audiograms: Communication Problems

By Earl R. Harford, Ph.D., and Noel D. Matkin, Ph.D.

This chapter provides information on the interpretation of audiometric test results, the patterns of audiograms, and the distinctive aspects of different types of hearing loss. It also includes a description of communication problems that are caused by occupational hearing losses.

The ear and how it functions

Before discussing audiometric interpretation, it is helpful to review the anatomy of the human auditory mechanism (see also Chapter 7). Hearing losses can be classified as to type of impairment on the basis of audiometric data which in turn reflects the location of the anatomical dysfunction. The human ear can be divided into three major parts—the outer ear, the middle ear, and the inner ear, including the auditory nerve (see Figure 26–1).

The outer ear serves to funnel and conduct sound vibrations to the eardrum through the ear canal. The eardrum vibrates in response to the sound waves that strike it. This vibratory movement, in turn, is transmitted to the chain of three tiny bones in the middle ear. These small bones, the ossicles, conduct the sound vibration across the air-filled middle ear cavity to fluid in the delicate inner ear.

The vibrating action of the ossicles creates waves in the inner ear fluid that stimulate thousands of microscopic hair cells. The stimulation of these hair cells generates nerve impulses, which pass through the auditory nerve and travel to the brain for interpretation.

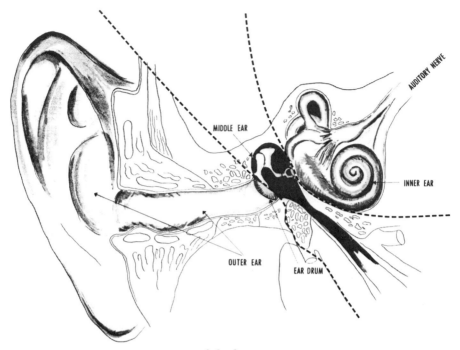

Fɪɢ. 26–1.—A cross-section view of the human ear.

In brief, the outer and middle sections of the ear serve to conduct sound energy to the deeper structures. Therefore, the outer and middle ear, collectively, serve as the conductive hearing mechanism.

In contrast, the deeper structures, including the inner ear and the auditory nerve, are referred to as the sensori-neural mechanism. The terms *conductive* and *sensori-neural* will be used later in this chapter to describe two major types of hearing impairment. (See also Chapter 7.)

As long as the hearing mechanism functions normally, the ear has the ability to detect sounds of minute intensity, and at the same time to tolerate sounds of extremely great intensity. The loudest sound the normal ear can tolerate is more than one hundred million million (10^{14}) times more powerful than the faintest sound the ear can detect.

Further, a young listener with normal hearing can detect sounds across a very wide frequency range—from very low-pitched sounds of 20 Hz to very high-pitched sounds of 20,000 Hz. To hear speech well, a much narrower range of normal hearing is required: 500 to 2000 Hz, commonly referred to as the *critical speech frequencies.*

Although nature has surrounded the delicate ear mechanism with hard, protective bone, any portion of it can become defective. The part of the hearing mechanism affected and the extent of damage has a direct bearing on the type of hearing loss that results. The common types of hearing impairment are reviewed later in this chapter. (See also Chapter 7.)

The audiogram

Let's review. The minimal intensity level at which various sounds produced by an audiometer are just barely heard by the person being tested is usually recorded on a standard chart—the audiogram.

Frequency

Shown on the audiograms in Chapter 25 are several numbers across the top (125 to 8000 Hz). These numbers represent the frequency or pitch of sounds, expressed in hertz (Hz). The lower numbers to the left represent low-pitch sound. For example, a 250 Hz tone sounds like middle C on a piano. The tones become progressively higher in pitch as one moves to the right across the higher numbers. A 4000 Hz tone sounds much like a piccolo hitting a high note.

Intensity

In contrast, the numbers down the side of the audiogram indicate the intensity or loudness of the sound, which is measured in decibels. The smaller the number, the fainter the sound. When measuring a person's hearing, the level is established at each test frequency where the sound can just barely be heard. This level is called the *threshold of hearing sensitivity.*

Note that in audiometry, the further a person's threshold is below the zero line of the audiogram, the greater is the loss of hearing. Common practice is to record the pure tone thresholds by air conduction for the right ear on the audiogram as red circles. In contrast, a blue X is recorded to reflect each threshold for the left ear (see Figure 25–1, page 626).

If a sound must be made louder than 25 dB (ANSI) for a person to detect its presence, his thresholds begin to fall into the range of hearing impairment.

Thus, the more intense, (louder) the sound from the audiometer must be for a person to hear it, the greater is that individual's hearing loss.

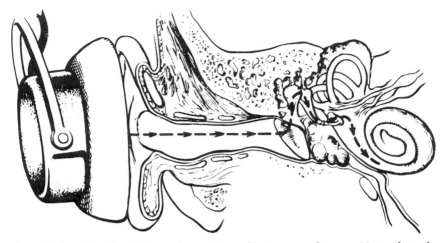

Fig. 26–2.—An illustration of an air-conduction earphone. Note that the phone is placed directly over the external ear canal, and the sound waves are conducted (by air) to the eardrum and through the middle ear to the inner ear.

However, as long as hearing is normal or nearly normal across a frequency range from 500 to 2000 Hz, the person should have little difficulty hearing speech in ordinary listening situations.

The American Academy of Ophthalmology and Otolaryngology has recommended that audiograms should be drawn to a scale like the illustration shown in Figure 24–3, page 609. Note that for every 20 dB interval as measured along one side and for one octave as measured across the top (250 to 500 Hz, for example), there will be a perfect square.

The reason for this recommended scale is that the appearance or pattern of a hearing loss can be altered a good deal by changing the dimensions of an audiogram. If the proportions of the audiogram are different from standard dimensions, a person's hearing loss may look quite different than if it were plotted on the standard audiogram format.

Hearing losses plotted on a chart produce a profile of a person's hearing. A trained person can review an audiogram to determine the type and degree of hearing loss, and can speculate on the degree of the difficulty in communication this loss will cause.

Customarily, audiograms are scaled in 10 dB steps. Obviously, if a person has a threshold of 55 dB, it is plotted on the appropriate frequency

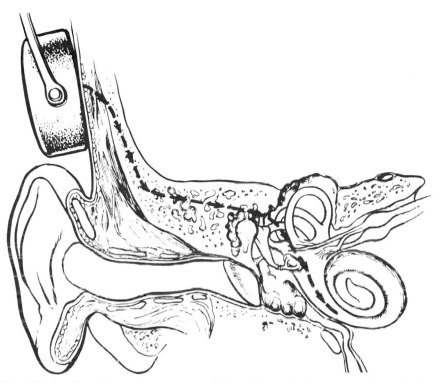

Fig. 2ß–3.—Sound can be transmitted directly to the inner ear through the bones of the skull by means of a bone-conduction vibrator placed on the mastoid bone behind the outer ear. The broken rule (with arrows) shows the path taken by the sound waves through the bony areas of the head to the inner ear.

line at the halfway point between 50 and 60 dB.

Hearing level for a pure tone is the number of decibels that the listener's threshold of hearing is greater than the standard audiometric zero for that frequency. It is the reading on the hearing threshold level (hearing loss) dial of an audiometer that is calibrated according to American National Standard *Specifications for Audiometers* (S3.6–1969). Prior to the adoption of this standard, audiometers were calibrated according to American Standard Z24.5-1951 (ASA-1951).

There are still in use some audiometers calibrated to the older values (ASA-1951). For this reason, each audiogram or record of an audiometric test should include a specific notation as to whether it is based on the current ANSI-1969 or the ASA-1951 reference levels.

Hearing loss

Bone-conduction tests

Until now, only one of the two ways that sound reaches the inner ear has been discussed—that is by air conduction where sound travels from the outer ear through the bones of the middle ear into the fluid of the inner ear (see Figure 26–2). However, there is another way to introduce sound and to measure hearing. This is by *bone conduction,* where sound travels directly to the inner ear via the bones of the skull; a route that by-passes the outer and middle ear. Bone-conduction audiometry is rarely performed by the industrial audiometric technician; however, a basic understanding of bone-conduction tests can be helpful for the interpretation of audiograms.

During such a test, a bone vibrator is placed on the mastoid bone behind the auricle (outer ear), and it is held in place by a headband. It is, in fact, a unit that vibrates the skull (see Figure 26–3).

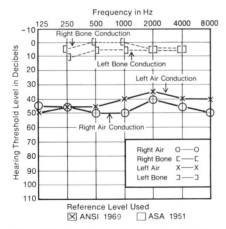

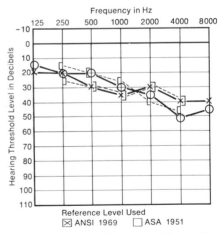

FIG. 26–4.—An audiogram indicating a conductive loss. (This figure and the next were also shown as Figures 8–2 and 8–3, pages 227 and 228.)

FIG. 26–5.—Results typical for individuals having a sensori-neural type hearing loss. The person whose audiogram is shown here hears just as poorly by bone conduction as he does by air conduction.

650

Obviously, bone has more resistance to vibration than the air column in the outer ear canal. As a result, it takes a good deal more intensity for a listener to detect sound from the bone vibrator than from an earphone. This increase in sound output is built into the audiometer when it is manufactured and calibrated at the factory.

Types of hearing loss

When test results show that a person has depressed hearing by air conduction but normal hearing by bone conduction, the presence of a *conductive hearing loss* is indicated (see Figure 26–4). In other words, the conductive mechanism is impaired in some way since the normal bone conduction responses indicate that the deeper structures of the ear are intact.

If, however, the person hears just as poorly by bone conduction as he does by air conduction, then the hearing loss must be due to damage in the deep structures of the ear. (No matter how sound is presented to the sensori-neural mechanism, it is met by an insufficient receiver in the inner ear.) This then would indicate a *sensori-neural loss* (see Figure 26–5).

A third type of hearing loss is a combination of conductive and sensori-neural. This is referred to as a *mixed hearing loss*. If the person has a mixed type of impairment, he will show a hearing loss on both types of tests. However the loss for air conduction will be greater than the loss for bone conduction. A mixed impairment is shown in Figure 26–6.

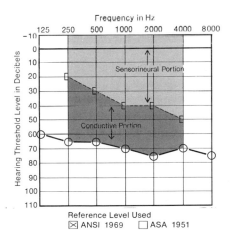

Fig. 26–6.—An audiogram typical of a person suffering a mixed type of hearing impairment. The loss for air conduction is greater than the bone-conduction loss. (Only right ear results are shown.)

In summary, by comparing the relationship of the audiometric test findings from air-conduction and from bone-conduction audiometry, the type of hearing loss can be indicated.

A conductive loss is due simply to some impairment of sound transmission before it reaches the inner ear. A conductive impairment, then, is one that results from some interference with the function of the outer ear or the middle ear.

Any blockage—usually ear wax or infection—of the outer ear that results in a loss of energy being conducted to the middle ear can cause a conductive hearing loss. Similarly, any impairment in the sound transmission system of the middle ear can cause a conductive hearing loss. Of course, such a loss could also be due to malfunction in both the outer and the middle ear.

In contrast, a hearing impairment that involves only the inner ear or the auditory nerve is classified as a *sensori-neural impairment.* (*Sensori* refers to the sense organ in the inner ear; *neural* refers to the nerve fibers.) A sensori-neural loss can involve either an impairment of the cochlea, the auditory nerve, or both.

It is virtually impossible to tell from an audiogram whether the damage is in the inner ear or in the auditory nerve, which is the transmission line to the brain. For this reason the loss is labeled sensori-neural, because the specific area of damage cannot be determined from audiometric findings.

In the past, two different terms were also used to describe sensori-neural losses. These terms were "nerve deafness" and "perceptive deafness." These are losing their popularity, but they may still be encountered occasionally.

Of course, it is possible for a person to have something wrong with both his conductive hearing mechanism and his sensori-neural hearing mechanism. In fact, the damage does not even have to occur at the same time. For example, a person may have a sensori-neural loss and then later develop a middle ear infection, which would then produce a conductive hearing loss. If the person had both types of hearing loss, it would be called mixed hearing loss, as discussed earlier.

Degree of hearing loss

Although the type of hearing loss is important, of equal importance is the degree of the impairment. Because the critical range for hearing speech is 500 through 2000 Hz, most professionals focus on this range of hearing when describing the extent of a hearing loss. An individual's hearing loss for speech can be estimated by taking an arithmetic average

TABLE 26-A
CLASSIFICATION OF DEGREE OF HEARING LOSS

Degree	Average Hearing Level for 500, 1000 and 2000 Hz in the Better Ear*		Ability to Understand Speech
	At Least	Less Than	
None	—	25 dB	No difficulty
Slight	25 dB	40 dB	Difficulty with faint speech
Mild	40 dB	55 db	Frequent difficulty with faint speech
Moderate	55 dB	70 dB	Frequent difficulty with loud speech
Severe	70 dB	90 dB	Can hear only shouted or amplified speech
Profound	90 dB	—	Very limited usable hearing

* Re: 1969 ANSI Reference Level

of the thresholds for pure tones, as seen on the audiogram, for the three test frequencies of 500, 1000, and 2000 Hz.

Averaging method. For example, if a person has an average hearing loss of 60 dB in each ear for the critical speech frequencies, the hearing loss would be classified as moderate in degree. Table 26-A is a guide for classifying the degree of hearing loss. The table includes a general indication of the communication handicap imposed by impairments of different degrees. It should be noted, however, that this classification refers to the level of hearing for speech in the better ear.

Audiologists use the averaging method just described, because of their concern with the effect of hearing loss on communication. In contrast, for medicolegal purposes, the degree of hearing loss is usually expressed as a *percentage of loss*. (See Chapter 13, Workers' Compensation and Medicolegal Factors, for additional information.)

Percentage method. Many states have adopted, by legislative action or administrative ruling, formulas to be used for determining the percentage of hearing disability. Several different formulas can be used to

TABLE 26-B

CONVERSION OF ESTIMATED HEARING LEVEL
FOR SPEECH TO A PERCENTAGE OF
MONAURAL HEARING IMPAIRMENT

Hearing Threshold Measurements in Decibels of One Ear at 500, 1000 and 2000 Hz on Audiometers Calibrated to:				Monaural Hearing Impairment (Percent)
ANSI 1969		ASA 1951		
Decibel Sum	Hearing Level for Speech	Decibel Sum	Hearing Level for Speech	
75 or Less	25 or Less	45 or Less	15 or Less	0
80	26.7	50	16.7	2.5
85	28.3	55	18.3	5.0
90	30.0	60	20.0	7.5
95	31.7	65	21.7	10.0
100	33.3	70	23.3	12.5
105	35.0	75	25.0	15.0
110	36.7	80	26.7	17.5
115	38.3	85	28.3	20.0
120	40.0	90	30.0	22.5
125	41.7	95	31.7	25.0
130	43.3	100	33.3	27.5
135	45.0	105	35.0	30.0
140	46.7	110	36.7	32.5
145	48.3	115	38.3	35.0
150	50.0	120	40.0	37.5
155	51.7	125	41.7	40.0
160	53.3	130	43.3	42.5
165	55.0	135	45.0	45.0
170	56.7	140	46.7	47.5
175	58.3	145	48.3	50.0
180	60.0	150	50.0	52.5
185	61.7	155	51.7	55.0

calculate the percentage. The results may differ slightly, depending on which formula is used. Different states require different formulas; therefore, it is important to become familiar with the regulations in effect in the particular state or jurisdiction of interest. The American Medical Association's "Guide to the Evaluation of Hearing Impairment" states that for every decibel that the hearing level for speech exceeds 25 dB, ANSI-1969 (15 dB, ASA-1951) 1.5 percent of monaural impairment is allowed

Hearing Threshold Measurements in Decibels of One Ear at 500, 1000 and 2000 Hz on Audiometers Calibrated to:				Monaural Hearing Impairment (Percent)
ANSI 1969		ASA 1951		
Decibel Sum	Hearing Level for Speech	Decibel Sum	Hearing Level for Speech	
75 or Less	25 or Less	45 or Less	15 or Less	0
190	63.3	160	53.3	57.5
195	65.0	165	55.0	60.0
200	66.7	170	56.7	62.5
205	68.3	175	58.3	65.0
210	70.0	180	60.0	67.5
215	71.7	185	61.7	70.0
220	73.3	190	63.3	72.5
225	75.0	195	65.0	75.0
230	76.7	200	66.7	77.5
235	78.3	205	68.3	80.0
240	80.0	210	70.0	82.5
245	81.7	215	71.7	85.0
250	83.3	220	73.3	87.5
255	85.0	225	75.0	90.0
260	86.7	230	76.7	92.5
265	88.3	235	78.3	95.0
270	90.0	240	80.0	97.5
275 or more	91.7 or more	245 or more	81.7 or more	100.0

In using this table, note that a value for the hearing level in decibels at 500, 1000, and 2000 Hz is needed. If the loss at a given frequency is beyond the range of the audiometer, the level shall be taken as 100 dB for that frequency. If the hearing level at a given frequency is better than normal, the level shall be taken as zero decibels for that frequency. For every decibel that the estimated hearing level for speech exceeds 25 dB (15 dB, ASA-1951), 1.5 percent of monaural impairment is allowed up to a maximum of 100 percent. This maximum is reached at 91.7 dB (81.7 dB, ASA-1951). This table is adapted from *Guides to the Evaluation of Permanent Impairment,* American Medical Association, 1971.

up to a maximum of 100 percent. This maximum is reached at 91.7 dB, ANSI-1969 (81.7 dB, ASA-1951)—see Table 26-B.

Unfortunately, neither of these methods for establishing the degree of hearing loss consider *the person's ability to understand speech clearly.* This ability is very important, because loss of clarity in hearing for speech can also have a very detrimental effect on a person's ability to communicate.

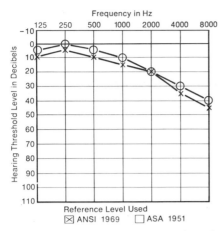

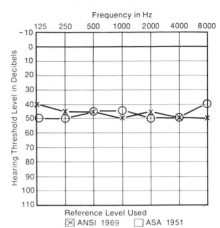

Fig. 26–7.—An audiogram, based upon pure tone air-conduction tests, that has a gradually falling configuration. The gradual slope to the right indicates a greater loss in the high frequencies.

Fig. 26–8.—An audiogram having a flat configuration, indicating that hearing loss is relatively constant across the test frequencies. (The individual is uniformly missing many speech sounds.)

Common audiometric configurations

In addition to considering the type and degree of hearing loss, the pattern of the air-conduction findings on the audiogram can provide valuable information when attempting to understand the effect of the impairment on communication.

The following are common audiometric configurations:

1. *The gradual falling configuration* (Figure 26–7) slopes gradually to a greater loss in the higher frequencies.

2. *The flat configuration* (Figure 26–8) is seen where a hearing loss is relatively constant across the test frequencies; of course, it can fall at any intensity level from a mild to a profound impairment. The person with this type of audiogram has an overall loss of sensitivity for all speech sounds. His hearing loss at 500 Hz is about the same as it is at 1000 and 2000 Hz, so he is uniformly missing many speech sounds.

3. *The marked falling configuration* (Figure 26–9) tapers off abruptly. In such a case, the person may hear low-frequency vowel sounds quite

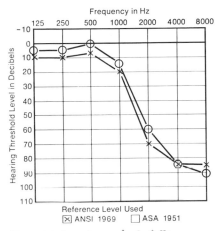

Fig. 26–9.—A marked falling configuration that tapers off abruptly.

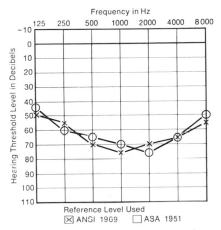

Fig. 26–10.—An audiogram with a trough configuration. The trough shape follows the normal loudness contour.

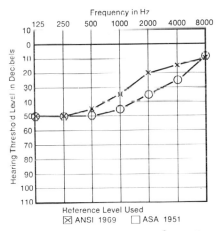

Fig. 26–11.—A rising configuration, indicating a hearing loss for low pitch sounds (vowels).

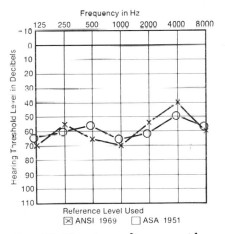

Fig. 26–12.—An audiogram with a jagged configuration, which is quite typical of a conductive hearing loss. This type of loss could occur at other threshold levels.

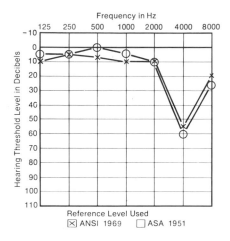

FIG. 26–13.—A configuration called "the high frequency notch" indicates hearing is quite good through much of the speech range; then drops off suddenly, but returns to normal after 4000 Hz.

well, but he will miss many of the high-frequency consonant sounds. As a result, speech seems muffled.

4. *The trough configuration* (Figure 26–10) is less common. In the past, it was believed that it was a typical audiometric pattern of the malingerer. The trough shape follows the normal loudness contour. Therefore, it appears that the malingerer sets a fix or a peg on a particular loudness of sound. He, then, refuses to admit to hearing test sound less loud than his reference. Because sounds in the lower and higher frequency range tend to be heard louder, the audiogram for a malingerer may be trough shaped. However, it has been found that malingerers can come in "different packages." It also has been found that this configuration may be encountered in cases with real hearing loss.

5. *The rising configuration* (Figure 26–11) shows a hearing loss for low sounds, such as vowels, but near-normal hearing for higher pitched sounds such as consonants.

6. *The jagged configuration* (Figure 26–12) is quite typical of a conductive loss. Note that the configuration goes first in one direction and then in another, especially around 2000 Hz.

7. *The fragmentary configuration* is just what the term implies. There are responses in only a limited portion of the frequency range.

8. *High-frequency notch* is one audiometric pattern or configuration

where hearing is usually quite good through much of the speech range. It then drops off very sharply but it usually comes back up again after 4000 Hz. Such a configuration is referred to as a high-frequency notch (Figure 26–13). This high-frequency notch is commonly associated with a permanent threshold shift caused by exposure to intense noise. Some audiograms, however, do not show a return to better hearing at the very high frequencies. One major variable is the person's age— the older a person, the less likely he is to have a pronounced notch. The aging process tends to cause some hearing loss at 8000 Hz, and this obscures the notch at 4000 Hz.

It should be clearly stated that one can never accurately determine the type of hearing loss on the basis of audiometric configuration alone. In fact, it is a dangerous practice to attempt such an interpretation.

Communication problems

The communicative problem of a person with a noise induced hearing loss is very frustrating to him and is easily misunderstood by his family and friends. This problem causes a great deal of inconsistent auditory behavior—the person appears to hear very well at some times and very poorly at other times. Thus, he is often accused of "not paying attention."

It is helpful for the industrial audiometric technician to understand the kind of communicative problem imposed by a hearing loss caused by substantial exposure to noise.

Hearing vs. understanding

There are two important characteristics of normal hearing—the ability to hear sounds as loud as they truly are, and the ability to hear sounds with complete clarity. It is important that the distinction between these two characteristics of hearing be understood.

If Figure 26–14 is held at arm's length, the printing in the center is obvious but almost impossible to read. A close look confirms that it is indeed a word and an even closer look reveals that the word is LOUDNESS. This example illustrates a very common problem associated with hearing loss—the inability to hear soft sounds. If the sounds are made louder, a person with only a loss of hearing sensitivity will have much less difficulty.

Now, if Figure 26–15 is held at arm's length, there is no difficulty in seeing everything within the box. But can the word be read? If not, the reason is that too much of the word is missing, even though it is large enough. No matter how closely you look, it is difficult to read because

Loudness

FIGS. 26–14 and 26–15.—These represent, by analogy, two important dimensions of normal hearing—the ability to hear sounds as loud as they truly are and the ability to hear sounds with complete clarity. In the above illustration, it is almost impossible to read the word, LOUDNESS. This is comparable —visually—to not being able to hear faint sounds. Moving the figure closer makes it easier to read, just as increasing the volume makes it easier to hear. On the other hand, no matter how closely the lower illustration is held, it is difficult to interpret. The word is not clear because some important parts of the letters are missing. This illustrates a hearing difficulty caused by a loss in the ability to distinguish between various sounds.

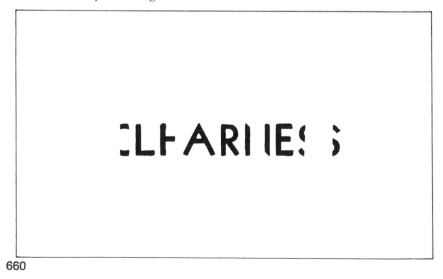

some of the important parts are missing. This illustrates a problem in loss of clarity or an inability to distinguish between the various sounds in the language. (The word, incidentally, is CLEARNESS.)

If damage is sustained to any portion of the outer and middle ear, the primary result is an inability to hear soft speech. Clarity, however, is preserved.

The key to clarity in hearing is held by the inner ear mechanism and the nerve fibers that carry the message to the brain.

If the inner ear or auditory nerve is damaged, not only a loss in the loudness of sounds can be expected, but, in many cases, a loss in clarity as well. In such dual problems associated with sensori-neural hearing loss, speech seems muffled or fuzzy no matter how loud it is. In these cases, the major problem is trouble in understanding what is being said.

The person with a hearing loss from noise exposure often has this kind of problem. The tiny hair cells that respond to specific speech sounds may be so severely damaged that they cannot react when the vibrations from the outside strike them. At the same time, hair cells for other speech sounds may be functioning normally. The person with such a hearing impairment misses parts of words and parts of sentences so that he often misunderstands what is being said. This can be a very subtle problem. In fact, the person with this loss is often not aware of the reason he does not hear everything.

To summarize, then, a person with a high-frequency, noise induced loss can hear persons talking, but may not understand what is being said.

Characteristic speech sounds, shown in Figure 26 16, are for average conversation at 3 to 4 ft distance between the speaker and the listener in the average home or office. These can be related to the two principal parts of speech—vowels and consonants.

The vowel sounds—located in the lower frequencies—are the more powerful speech sounds. Therefore, vowels carry the energy for speech. In contrast, the consonant sounds—located in the higher frequencies— are the keys to distinguishing one word from another, especially if the words sound alike. This is the heart of the communicative problem of the person with a noise induced (high-frequency) hearing loss. He cannot distinguish the difference between similar words, such as "stop" and "shop."

It is quite easy to miss a key sound in a word. In turn, this could change the meaning of a key word in a sentence. As a result, the entire sentence might be misunderstood.

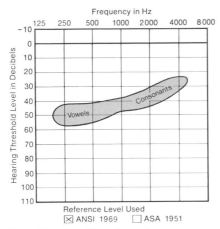

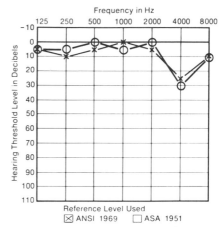

FIG. 26–16.—Vowels and consonants are the principal sounds of speech. This audiogram shows the frequency ranges where these speech sounds are heard. Vowel sounds carry the energy for speech and are located in the lower frequencies. Consonants "shape" the sounds of the vowels and are heard in the higher frequencies. Consonants are the keys to understanding words.

FIG. 26–17.—An audiogram illustrating a drop in hearing sensitivity at the 4000 Hz frequency. Such an audiogram may represent the earliest stage of a hearing loss. It might occur after a month, a year, or even longer exposure to excessive noise. This type of audiogram is referred to as a "minor high-frequency notch."

Persons with a high-frequency hearing loss often get along fairly well in quiet listening situations. But as soon as they are in a place where there is a lot of background noise, such as in traffic or on the job, it becomes difficult to communicate through hearing alone. For example, if a speaker continued talking in the presence of a typical background sound, a listener should hear him quite well. But, if the listener were to develop a hearing loss for all speech sounds above 1000 Hz, he would immediately notice a marked loss in communication. In other words, so long as it is quiet, a person can use his good hearing below the mid-pitch range to his advantage. Unfortunately, most noises around us interfere with the low-pitched speech sounds and cause us to miss many of these sounds as they occur in words and sentences. Thus, the hearing loss caused by the noise, plus the hearing loss resulting from damage to the ear, results in a

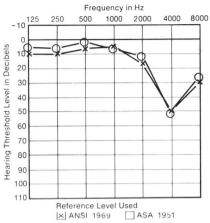

Frequency in Hz

125 250 500 1000 2000 4000 8000

Hearing Threshold Level in Decibels

Reference Level Used
⊠ ANSI 1969   ☐ ASA 1951

FIG. 26–18.—How an individual's audiogram might look after additional exposure to excessive noise. Compare the hearing threshold levels shown here with those plotted on the audiograms shown in Figures 26–17 and 26–19.

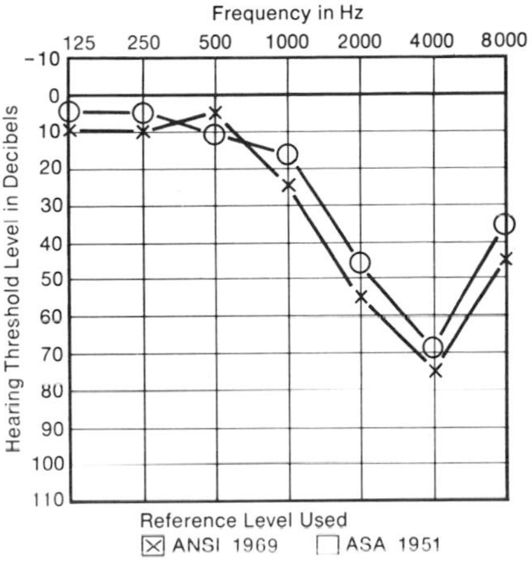

FIG. 26–19.—An audiogram where the high-frequency notch has widened, and the individual's hearing level is more affected at 2000 Hz. This is the state where the individual will begin to note a hearing problem.

greater hearing problem than one would have with just one of these conditions alone.

Increased dependence on vision

The more difficult it is to hear, the more natural it is to rely on vision for communicating. Persons with normal hearing will naturally watch a talker's face and lips more carefully in a noisy place where it is difficult to hear. It is also necessary to concentrate intently on what is being said in order to follow the conversation.

It is natural for the hearing-impaired person to rely on his "helpers" (his eyes and his power of concentration) in order to grasp many conversations. When a person is tired after a hard day's work, or when he is not feeling up to par, it is quite natural for him to feel that his hearing is getting worse.

This same suspicion may arise in a poorly lighted room. The hearing loss is no worse under these circumstances, but communication is indeed more difficult.

It is important that all persons with hearing losses realize how much help they can receive from their vision and power of concentration. Most persons who lose their ability to hear normally will develop some lip-reading skills on their own.

Figures 26–17 through 26–19 show the progression of a noise induced hearing loss. Figure 26–17 represents the earliest stage showing a minor high-frequency notch. This could occur after a month or a year that an employee has worked in the plant.

Figure 26–18 shows the next stage when something really serious starts to happen. The employee's hearing starts to drop at the pivot frequency of 2000 Hz. As a result he begins to have difficulty hearing speech in group situations and in noisy places. How soon this second stage is reached will depend upon the individual's susceptibility to noise, the noise level, the time of exposure, and the nature of the hazardous noise.

Then, as shown in Figure 26–19, the hearing loss for the higher frequencies becomes pronounced. When this stage is reached, the person develops a significant communicative problem. This is the time when the person is going to become quite aware of the problem himself.

A typical communicative problem that such people try to describe is, "I can hear you talking, but I can't understand what you are saying." "I can hear my wife clearly most of the time if we are in the same room, but with a group of people, I can hardly understand anything." Or, "I can hear okay in the office, but when I go into the cafeteria, I have to sit next to a person in order to hear him."

The type of hearing loss resulting from noise exposure is subtle and evasive—both to the person who has it and to the industrial audiometric technician who tries to detect it. The *only* way to identify a noise induced hearing loss in the early stages is by careful audiometric testing. An early noise induced hearing loss cannot be identified with any degree of confidence through conversation. If industrial noise caused a flat hearing loss, as shown in Figure 26–8, it would be possible to determine that the person was having difficulty in hearing by merely talking to him. But exposure to high-intensity industrial noise initially causes the subtle type of hearing loss shown in Figure 26–17.

Summary

This chapter described the various characteristics of hearing loss such as type, degree, and configuration. Particular emphasis is placed upon the stages of noise induced hearing impairment as encountered in industry. An attempt has been made to describe the handicap in listening imposed by such hearing losses. This information is presented with the intention of introducing occupational health and safety personnel to the interpretation of audiograms and to the implication of industrial hearing loss on communication. Also, the information in this chapter should be more meaningful after reading the previous chapters in this section on industrial audiometry.

Chapter Twenty-Seven

Testing Problems: The Test Environment and Equipment

By Earl R. Harford, Ph.D.

An important objective of every effective industrial hearing-conservation program should be accurate audiometry. Accurate audiometry is dependent upon four prerequisites: (*a*) a constantly quiet test environment, (*b*) stable, accurate audiometric test equipment, (*c*) a well-instructed, cooperative subject, and (*d*) a competent, highly motivated, and properly trained audiometrist.

Accurate audiometry yields test results that represent a person's true level of hearing ability, and it can be repeated to yield the same results, provided the person's hearing has not changed. Only by monitoring the hearing of employees can the effectiveness of a hearing-conservation program be evaluated.

Some of the more important problems and pitfalls in audiometry as a result of an inadequate acoustic test environment and malfunctioning or out-of-calibration audiometric test equipment are discussed in this chapter.

Constantly quiet test environment

The first requirement for accurate industrial audiometry is to have the background noise in the test booth or room at or below the appropriate values specified in ANSI S3.1-1960 (R1971) as shown in Table 27–A.

In selecting an adequate audiometric testing area, one must first decide on the required dimensions. If only one person is to be tested at any given time, a relatively small booth can be selected. Such a booth could

TABLE 27-A
MAXIMUM BACKGROUND NOISE IN AUDIOMETER ROOMS

| | Octave-band Center Frequencies (Hz) | | | | | |
	500	1000	2000	4000	6000	8000
Band Sound Level (dB)	40	40	47	57	62	67

TABLE 27–A.—The maximum allowable sound levels at different frequencies for no masking to occur above the zero hearing-loss setting of a standard audiometer, calibrated according to ANSI S3.6–1969. The information in this table is based on American National Standard S3.1–1960 (R1971) as it applies to audiometric test rooms used in industrial hearing-conservation programs.

be located inside a larger room (see Figure 27-1). This would provide space for the audiometrist and the audiometer outside the testing booth and also for storage of pertinent records and other reference materials. If two persons are scheduled to be tested at the same time, two audiometric test booths would be required, along with a correspondingly larger area outside the booths. Ideally, the location for the audiometric test facility should be in close proximity to the medical or personnel department.

Background noise levels

An engineering noise survey of the proposed test area should be made with an octave-band analyzer. The ambient background noise levels present in the area proposed for the location of the booth will determine the acoustical characteristics of the booth necessary to obtain a satisfactory audiometric testing environment (see Figure 27–2).

The maximum allowable noise levels inside the audiometric test booth are listed in ANSI S3.1-1960 (R1971), *Criteria for Background Noise in Audiometer Rooms.* A safety factor of 10 dB should be used when specifying maximum background levels (see Table 27–B). Frequently, the ambient noise levels in the audiometric test area fluctuate; therefore, the use of a safety factor of at least 10 dB is good practice, and its use

(*Text continues on page 670.*)

TABLE 27-B

SAMPLE CALCULATION USING AUDIOMETRIC BOOTH
ATTENUATION DATA

Preferred Center Test Frequencies (Hz)	Octave Bands				
	500	1000	2000	4000	6000
Octave band cutoff frequencies (Hz)	300 to 600	600 to 1200	1200 to 2400	2400 to 4800	4800 to 9600
Maximum allowable band sound levels ANSI S3.1–1960 (dB)	40	40	47	57	62
Attenuation (dB) of audiometric booth (data supplied by manufacturer)	46	53	58	61	63
Maximum allowable ambient background levels (dB) outside booth to comply with ANSI S3.1–1960	86	93	105	118	125
Safety factor (−dB)	−10	−10	−10	−10	−10
Suggested maximum allowable background noise levels (dB) outside this particular audiometric test booth	76	83	95	108	115

TABLE 27–B.—This table shows the suggested maximum background levels that can be present outside this particular audiometric booth or room while a hearing test is being conducted inside. A safety factor of at least 10 dB (row six) should be considered at the design stage to allow for unexpected fluctations in the ambient noise levels. Therefore, it would be wise to place the audiometric booth in this example in an area with ambient noise levels no greater than shown in the last row in this table. Moreover, ambient outside noise levels may increase from time to time. The data in this table is based on ANSI S3.1–1960 (R1971), *Criteria for Background Noise in Audiometer Rooms*.

Courtesy Industrial Acoustics Co.

FIG. 27–1.—Audiometric testing in the typical plant environment usually requires a specially-constructed test booth. If only one person is to be tested at a time, a small booth can be located inside a larger room.

FIG. 27–2.—An engineering noise survey of the proposed audiometric test area should be made with an octave-band analyzer to determine the ambient background noise levels.

Courtesy Xerox.

Courtesy Industrial Acoustics Co.

FIG. 27–3.—An audiometric test booth with double doors and walls provides considerable attenuation. Such a booth can be placed in a fairly noisy location and still provide an acceptable test environment. Audiometric test booths must be carefully designed, constructed, and installed to achieve a satisfactory acoustic environment.

will ensure long-term adequacy of the audiometric testing facility.

It has been suggested that audiometer earphones be mounted inside earmuffs to replace an audiometric testing room. However, the attenuation characteristics of some earmuffs may not be sufficient to meet ANSI standards when used in some areas, and could yield unreliable results. Earphone-to-ear coupling may vary depending on how well the earmuffs fit each individual. In certain special cases in industry, earphone/earmuff combinations may be useful. But in the typical industrial situation, such as in a first aid station or medical department close to a noisy area, earphones inside earmuffs alone do not ordinarily offer enough attenuation to satisfy the requirements for a quiet test environment. This is unfortunate because such a device for providing an acceptable test environment could be extremely beneficial in solving a major problem for many industrial firms that have difficulty justifying the cost of an audiometric test booth.

Individual rooms, or at least sections of a room, should be acoustically and visually separated from each other. The floor should be covered with a carpet or an absorbent material to reduce impact and scraping noises in the room. Walls and ceiling should be acoustically treated to give maximum absorption of sounds arising from inside the room as well as from the outside. Adequate ventilation should be provided, but the ventilation system should not produce background noise exceeding the levels specified in Table 27–A. Make certain there is no ballast hum from fluorescent fixtures. Such hum can be avoided by using incandescent lighting. It may be necessary to install acoustical treatment in ventilating ducts leading to and from the room, and to maintain low face velocities to minimize diffuser noise.

When the walls of an audiometric test room are heavy, and the noise and vibration transmitted through the floor and ceiling is minimal, the major source of the background noise in the room may be acoustic leaks around doors, windows, pipes, and ducts. Thus, audiometric test rooms must be carefully designed, constructed, and installed to avoid such problems (see Figure 27–3). As with most products, there is a difference in the price and quality of audiometric test rooms or booths.

Effect of ambient noise. There are two forms of background noise that can have a detrimental effect on the accuracy of audiometry and the efficiency of the audiometric testing program. First, there can be a constant background noise, which masks a person's hearing (especially if he has good sensitivity) and creates an artificial hearing loss. This situation can be recognized by testing a person with good hearing in a quiet room and then in a room with excessive constant background noise. The audiogram obtained in the quiet room should show good normal hearing. The audiogram obtained in the noisy room will show an apparent hearing loss (see Figure 27–4). The loss ordinarily occurs at the lower frequencies, because common ambient noise is low frequency in nature.

A frequent question from persons who are not familiar with audiometric testing is, "If I am performing audiometric tests in a room that has a constant low-frequency noise from the ventilation system blowers, can I correct for the possible masking effect?" The answer is, "No." The hearing threshold level should be recorded at its measured value. A value cannot be subtracted from the measured hearing threshold,

AUDIOGRAM

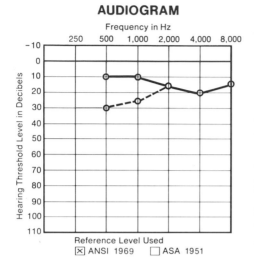

Reference Level Used
☒ ANSI 1969   ☐ ASA 1951

FIG. 27–4.—The results of two hearing tests taken on the same individual. The upper curve, showing the hearing threshold levels connected by solid lines, was taken in a quiet audiometric test booth. The lower curve shows the results of interference of background noise during a hearing test taken on the same individual in a noisy location. Typical industrial background noise will seriously interfere with measuring hearing threshold levels at the lower frequencies.

even if the level of the background noise is known because it is not known whether the measured hearing threshold obtained represents a real hearing loss or is due to a masking effect from noise in the test environment.

The constant noise in the test room will not add to a hearing loss, unless, of course, the noise is very loud and the hearing loss is mild. It is impossible to subtract only the difference between the real hearing loss and the one which was measured, because the level of the real hearing loss is unknown. The only thing to do is get rid of the masking noise. The masking noise might reward some persons and penalize others. This is one of the reasons why it is important that audiograms be done under testing conditions meeting ANSI S3.1-1960 (R1971) specifications.

The other form of background noise that can have a detrimental effect on an audiometric test is a sporadic, unexpected interference, such as from aircraft, persons walking through a hall, ringing telephones, or intermittent use of a noisy machine nearby. The problem this creates is one of distraction. It not only annoys the person being tested, but such interference can also frustrate the audiometrist, who must interrupt the test and wait for the noise to subside.

This form of interference can have a very detrimental effect on the length of time allotted for the typical hearing test and cause a gross waste of manhours. If automatic audiometry is used, the person testing himself may not be aware of the need to stop the test; therefore, the final results could be invalid because of the masking effect of these distractions.

Thus, the company that elects not to invest in a high-quality audiometric test environment because it has a "fairly quiet" room on the premises or nearby, may soon discover it has been "penny wise and dollar foolish."

Ventilation. The absence of acoustic leaks is a basic requirement of a good quality booth for testing hearing. Unwanted sounds can enter a booth through a tiny crack; this means the booth must be airtight. It takes only a matter of a few minutes for the temperature to increase in an occupied small test booth. Practically all booths designed for testing hearing have some form of ventilation, usually a simple circulation system.

It is very important to keep the temperature and humidity in the test booth at comfortable levels. If the area surrounding the booth is not air conditioned, the temperature in the booth could become very uncomfortable during a test. Temperature and humidity control needs to be anticipated at the time of selecting the booth to avoid unnecessary added expense at a later date. Generally, it is more economical to order the booth with an air conditioner than to add it as an accessory at a later date. Also, the overall efficiency of the system is likely to be better with "factory-installed air."

If the person being tested is in a hot, humid, uncomfortable booth, he will often present the audiometrist with problems. The subject may become drowsy and respond inconsistently to the audiometric test tones. He may decide to rush the test and begin responding even when he does not hear the tone. He may find it more difficult to concentrate on the listening task. In short, the more uncomfortable the person being tested is, the more likely the test results will be inaccurate. Background sound attenuation and a humidity- and temperature-controlled atmosphere are the two major aspects of the test environment that cause problems in achieving accurate industrial audiometric test results.

Several manufacturers offer a wide selection of preengineered booths and rooms. Extensive experience clearly dictates that the sensible approach toward achieving a quiet audiometric test environment in noisy industrial locations is to purchase a preengineered room (see Figure 27–5). It is rare for a firm to save money and achieve adequate attenua-

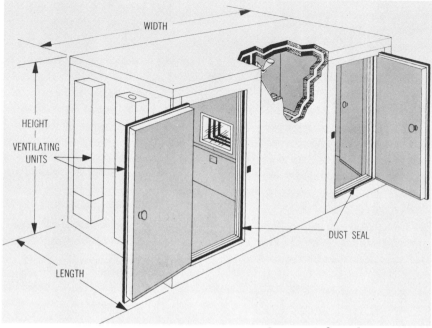

WIDTH

HEIGHT

VENTILATING
UNITS

DUST SEAL

LENGTH

Courtesy Industrial Acoustics Co.

Fig. 27–5.—Details of construction for a two-room audiometric booth setup showing the ventilation ducts. Sound isolation and humidity-temperature control are usually necessary for a suitable audiometric test booth. Rarely does one save money or achieve adequate attenuation by building his own audiometric test room—unless the area surrounding its location is constantly quiet.

tion by building its own room, unless the area surrounding the test room is constantly quiet. Even then, the difference in cost is negligible between a commercial room and a "homemade" room, when all factors are considered.

Properly functioning audiometer

The importance of an accurate, calibrated, stable audiometer cannot be stressed too strongly. Unfortunately, the typical audiometrist may not realize that audiometers are sensitive instruments that warrant constant monitoring to guard against defects (see Chapter 22).

A properly functioning audiometer has the following characteristics:

Fig. 27–6.—Each day, prior to use of the audiometer, the audiometric technician should make a complete auditory and visual inspection. Check the earphone cords for loose connections and worn or cracked insulation. The earphone cushions should be inspected for cracks, crevices, and stiffness. Inspect the face of the audiometer to make sure there are no loose or misaligned dials.

1. Emits the signal at the level and frequency it claims to be producing
2. Delivers that signal only to the place (that is, a specific earphone) to which it is directed
3. Produces the signal, free of extraneous noises or unwanted by-products of the test signal

For details on inspection and checking the operation of an audiometer, see Chapters 22 and 23. The American National Standards Institute has published specifications (ANSI S3.6-1969 R1973) that establish allowable tolerances for the performance of audiometers. Reliable audiometer manufacturers produce audiometers that meet these specifications. Unfortunately, once an audiometer is built to meet these standards of performance, there is no assurance that it will remain in perfect working order for any definite period of time. As a matter of fact, because audiometers and the earphones are sensitive, they are very susceptible to unintentional abuse, which can result in faulty operation. Periodic calibration checks are required to assure accuracy and an overall high level of performance of the instrument.

Most industrial firms do not have the electronic instruments required to check the calibration of their audiometer. Distributors of audiometers usually provide a calibration service.

An audiometer can be compared to a watch. Just as a watch can run fast or slow, an audiometer can produce a signal that is too strong or weak. It can also develop extraneous sounds such as clicks, static, and after-ring.

Each day the audiometric technician should make both a visual and an auditory inspection of the instrument (see Figure 27–6). Once each month, a biological check should be performed to monitor the output level. The results of this check should be recorded and kept on file. A full cleaning, servicing, and a complete electroacoustic factory calibration should be made at least once a year, and ideally twice a year if many audiometric tests are being conducted. The company doing the annual check should place a dated and signed seal or notation on the audiometer verifying that it meets ANSI specifications.

Level of the pure tone

Clearly, if the measured output level of the pure tone produced by the audiometer does not correspond to the hearing threshold attenuator dial setting, the hearing thresholds obtained with this instrument will be incorrect (see Figure 27–7). If the audiometer hearing threshold level dial reads 40 dB when the subject first reports hearing the tone, but the hearing threshold level reaching the subject's ear is actually 50 dB, the audiogram will show the subject to have better hearing than is actually the case. The opposite situation (tone at 40 dB, hearing threshold dial set at 50 dB) can also occur.

Clicks and noises

If the audiometer produces a click or some other extraneous sound simultaneously with the desired test tone, the subject being tested may respond to the unwanted sound rather than the test tone. Even more critical to the subject, however, is the distraction and confusion that an unwanted sound can create. The subject being tested may give

676

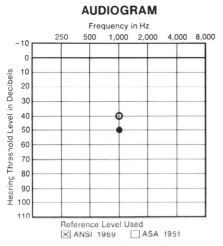

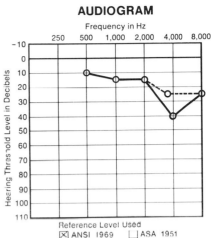

FIG. 27–7.—The hollow circle represents the apparent hearing threshold
level obtained at an attenuator dial
setting of 40 dB. The actual measured output level, shown as a solid
circle, at this frequency was found to
be 50 dB. An audiogram taken using
this instrument would show the subject to have better hearing than is
actually the case. The actual output
level of the pure tone produced by
the audiometer should correspond
with the hearing attenuator dial setting.

FIG. 27–8.—Shown as the dotted line
is an audiogram taken with an audiometer that produces a tone signal
of 3500 Hz with the frequency selector dial set at 4000 Hz. The solid
line shows an audiogram taken on an
accurate audiometer that produces a
tone signal at 4000 Hz when set to
4000 Hz. The audiogram taken with
the defective audiometer may show
better hearing than the one taken
with the accurate audiometer. Inaccurate frequency output can lead to
unreliable test results.

false responses because he hears a click and thinks he hears the test
tone signal. The unwanted sounds can be produced by a faulty spring
in the tone presenter switch, a worn stop-post in the switch, or some
other electrical malfunction or defect in the audiometer. Whatever the
cause, the industrial audiometric technician should be constantly alert
for this type of defect, and have it corrected immediately.

Earphone cords

A faulty earphone cord or connection at the earphone is likely to
produce an intermittent or erratic test signal. The tone may be reach-

ing the ear perfectly with the subject's head in one position, but if he turns his head, the tone may stop completely. This erratic response by the subject can cause considerable confusion for the technician, especially if he has never experienced this problem. In fact, it could conceivably result in an audiogram showing a marked hearing loss, when, in fact, the person has good hearing.

Inaccurate frequency

Inaccurate frequency output could lead to unreliable test results. That is, if one audiometer produces an actual signal of 3500 Hz with the frequency selector dial set at 4000 Hz, and another audiometer produces 4000 Hz when set to 4000 Hz, an audiogram obtained with the first (defective) audiometer may show better hearing than the one obtained with the other audiometer. This could occur if a person's hearing is defective at 4000 Hz, but good or considerably better at 3500 Hz (see Figure 27–8). The audiometric technician using the defective audiometer would think he was testing the subject's hearing at 4000 Hz, but actually would be presenting him with a tone 500 Hz lower in frequency. Thus, the hearing test results using the defective audiometer would show the subject's threshold for 3500 Hz instead of 4000 Hz. The audiogram obtained with the second (accurate) audiometer would show the subject's actual hearing threshold level at 4000 Hz.

Crosstalk

Crosstalk between earphones can occur and present problems to the audiometric technician. Crosstalk occurs when a test signal intended for just one earphone is heard in the opposite earphone. To check for crosstalk, disconnect the cord from one earphone, and present the pure tone signal to that earphone while listening for the pure tone signal in the opposite earphone (see Figure 27–9). If, under these circumstances, a pure tone signal is heard in one earphone while the test signal is directed to the opposite phone, there is crosstalk in the system. The person performing this check must have good hearing because crosstalk is usually of very low intensity. Also, the hearing level dial should be set to levels less than 60 dB. Some audiometers may have crosstalk at high intensity levels, but this condition is not unacceptable. The reason is that if one ear has a loss greater than 60 dB and the opposite ear has very good hearing, either a shadow curve would be suspected or masking would be used. (See Chapter 28, the section on Unilateral hearing loss, for an explanation of this situation.)

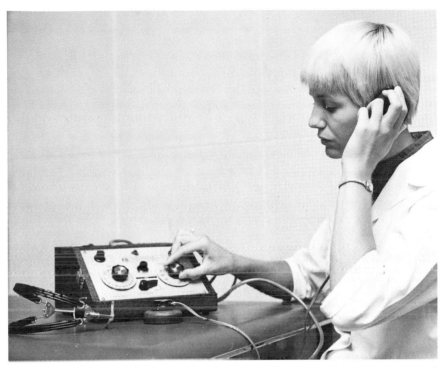

Fig. 27–9.—To check for crosstalk, disconnect the phone from the headband and listen for the pure tone while presenting the signal to the opposite earphone.

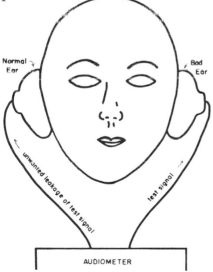

Fig. 27–10.—A schematic drawing showing crosstalk or leakage of a test signal for the intended earphone to the opposite phone. This situation could result in an audiogram showing two normal ears, where, in fact, one ear has a hearing loss.

679

Crosstalk at low hearing levels, however, should be corrected or repaired as soon as possible after detection.

Ideally, in hearing threshold measurements, when a test signal is delivered to one earphone, no signal should be in the opposite phone, except, of course, a masking stimulus when desired. Figure 27–10 illustrates the problem described here. This leakage or crosstalk can be so low in intensity that the average adult ear may not detect the unwanted leakage. However, some young adults have thresholds somewhat better than 10 dB. Therefore, if crosstalk is present, one could obtain absolutely normal thresholds for both ears in a case where one ear actually has a mild to moderate loss in acuity, because the good ear could have been stimulated when the test tone was delivered to the poorer ear.

Summary

Consistent attention to the test environment and the accuracy of an audiometer should not be taken lightly. The validity and reliability of hearing tests rest heavily upon a constantly quiet test room, and a properly functioning audiometer.

There are numerous reasons for an audiometer to go out of calibration—such as dropped earphones, overheating (leaving the audiometer turned on after covering it with a dust protector), exposure to excessive dust, excessive jarring, and normal aging of the electronic components. Whenever an audiometer or the earphones have been subject to unintentional abuse, it should be established that the equipment is still in calibration and working properly.

When audiometer problems are detected, arrangements for competent service should be made to correct the situation as quickly as possible. It is important that audiometric technicians recognize that audiometers cannot be used forever without service. In any case, an audiometer should be returned for calibration to the manufacturer or to an authorized service representative once a year. The importance of maintaining the audiometer in proper calibration cannot be overemphasized. The effectiveness of an industrial hearing conservation program is directly dependent upon accurate audiometry.

Chapter Twenty-Eight
Testing Problems: The Subject and the Technician

By Carl R. Harford, Ph.D.

Both the audiometric technician and the individual being tested can make voluntary or involuntary errors that result in invalid audiometric tests. It is the responsibility of the technician to be aware of the steps to take to reduce errors or avoid problems.

Four major factors can affect the accuracy of audiometric tests: (a) the test environment, (b) the audiometer, (c) the audiometric technician, and (d) the subject. Previous chapters discussed the pitfalls and problems involving the test environment and audiometric test equip ment. This chapter discusses the common pitfalls and problems induced by the audiometric technician and the subject.

Problems created by the audiometric technician

As a general rule, an audiogram should be obtained without delay. That is, changes from one frequency and one intensity to another should be made quickly, but not carelessly, so as to keep the time of the test as brief as possible. The longer the test, the more likely that the subject and the technician will become fatigued, and the test results will be less accurate.

Rushing through the audiometric test so rapidly that accurate thresholds are not obtained must be avoided. The audiometric technician should understand that some subjects take longer to respond than others. Sufficient time must be given to each subject to respond to the test signal. Faster and more definitive responses can be obtained if the subject understands what he is expected to do, and if the directions given by the technician are concise and explicit.

New, inexperienced technicians, who have not received careful training, are likely to use poor testing techniques that can have a detrimental influence on the responses from the person being tested. The following are the most common of these bad habits.

Instructing poorly

Poor instructions to the subject can induce false responses. The technician should carefully instruct the subject that no matter how soft he hears the test tone, he should raise his finger or push the button on the handheld switch. Some individuals prefer to hear the signal loudly before responding. The subject may reason: "Well, I'm not really sure about that one. I want to wait until I hear the signal a little louder before I indicate that I heard it." Prevent this attitude in the subject by telling him at the beginning of the audiometric test to respond, no matter how soft or faint the sound appears.

The technician should be explicit with instructions. Do not tell the subject more than he needs to know to perform the task. On the other hand, be careful not to underinstruct by leaving out some important points. (Refer to Chapter 24, Determination of Hearing Thresholds, for suggested instructions to the subject.)

Giving clues

When presenting tones to the subject, the technician should be fairly impassive in his facial expressions. The technician should never reveal that a response is expected or appear disappointed when he does not get one. Presenting the signal and then looking up at the subject as if to ask if he has heard the tone or giving other visual clues that the tone is "on" should be avoided. This is poor audiometric technique, and frequently, the subject will respond even though he does not hear the tone. In any event, if the subject learns to associate a head turn with the presentation of the test tone, he could become very confused if he ultimately senses the head turn, but does not hear the tone. This could induce false positive responses, that is, the subject indicates that he heard the tone when he really did not.

Another common error an inexperienced technician may make is to take his hand off the attenuator (hearing threshold level) dial each time he presents the tone. (There seems to be a common tendency to develop this undesirable habit.) The technician's hand should remain

Fig. 28–1.—A typical two-room audiometric test set-up. Note that the technician is able to see a profile of the subject's face, but the subject is not viewing the technician. The rationale for this arrangement is presented in Chapter 25.

on the hearing-level dial. Hand movement, just like a head turn, often can be perceived by the subject.

With some audiometers, when depressing the tone-presenter switch, the technician must be particularly careful not to press too hard or let it spring back too quickly so that it makes a clicking sound. This poor technique may induce responses to the click rather than to the pure tone presented at the earphone. By experimenting, it can be discovered that it is not necessary to depress the presentor switch so far that there is a click before the tone is emitted. The audiometer can be operated without producing a click, provided the technician learns how far he needs to press the presenter switch to elicit a tone. This technique should be viewed as a temporary solution to the problem of a noisy presenter switch. Obviously, this defect warrants repair. Of course, a

HEARING CONSERVATION DATA CARD NO._____

TYPE OF AUDIOGRAM

REFERENCE AND/OR
PRE-EMPLOYMENT ☐
RECHECK ☐
OTHER_____

A. IDENTIFICATION

LAST NAME	MIDDLE	FIRST	SEX	DATE OF BIRTH		
			MALE	DAY	MO.	YR.
			FEMALE			

SOCIAL SECURITY NUMBER	COMPANY NUMBER

B. CURRENT NOISE-EXPOSURE

JOB TITLE OR NUMBER	DEPARTMENT OR LOCATION	TIME IN JOB		
		NONE	MOS.	YRS.

NOISE-EXPOSURE

STEADY NOISE	IMPULSE NOISE	PERCENT TIME NOISE ON	EMPLOYEES ESTIMATE OF OWN HEARING
CONTINUOUS ☐	CONTINUOUS ☐	10 20 30 40 50	GOOD
INTERMITTENT ☐	INTERMITTENT ☐	60 70 80 90 100	FAIR
			POOR

☐ 1951 ASA   C. AUDIOGRAM   ☐ 1969 ANSI

TIME SINCE MOST RECENT NOISE-EXPOSURE	DURATION OF MOST RECENT NOISE-EXPOSURE
0-20 MIN 1 HR 4-7 HRS 1 DA	0-20 MIN 1 HR 4-7 HRS
21-50 MIN 2-3 HRS 8-16 HRS 2-3 DAS	21-50 MIN 2-3 HRS 7+ HRS
4+ DAS	

AGE	DATE OF AUDIOGRAM	DAY OF WEEK	TIME OF DAY	EAR PROTECTION
				WAS EAR PROTECTION WORN? YES NO

RIGHT EAR								LEFT EAR							
250	500	1000	1500	2000	3000	4000	6000	250	500	1000	1500	2000	3000	4000	6000

D. PREVIOUS NOISE-EXPOSURE AND MEDICAL HISTORY

PREVIOUS EMPLOYMENT (LAST 3 JOBS)

TYPE OF WORK	FOR WHOM	HOW LONG

HISTORY

HEAD INJURY (WITH UNCONSCIOUSNESS) ☐
HEARING LOSS IN FAMILY (BEFORE AGE 50) ☐
TINNITUS FOLLOWING NOISE-EXPOSURE   R.   L

RECORD ANY COMMENTS SUBJECT MAKES ABOUT HEARING

STATUS

PERFORATIONS OF DRUMHEAD   R   L
DRAINAGE FROM EAR   R   L
MALFORMATION OF EAR   R   L

TECHNICIAN_____

PHYSICIAN_____

☐ 1951 ASA   **AUDIOGRAM**   ☐ 1969 ANSI

TIME SINCE MOST RECENT NOISE-EXPOSURE				DURATION OF MOST RECENT NOISE-EXPOSURE		
0-20 MIN	1 HR	4-7 HRS	1 DA	0-20 MIN	1 HR	4-7 HRS
21-50 MIN	2-3 HRS	8-16 HRS	2-3 DAS	21-50 MIN	2-3 HRS	7+ HRS
			4+ DAS			

AGE	DATE OF AUDIOGRAM	DAY OF WEEK	TIME OF DAY	EAR PROTECTION
				WAS EAR PROTECTION WORN? YES NO

RIGHT EAR								LEFT EAR							
250	500	1000	1500	2000	3000	4000	6000	250	500	1000	1500	2000	3000	4000	6000

☐ 1951 ASA   **AUDIOGRAM**   ☐ 1969 ANSI

TIME SINCE MOST RECENT NOISE-EXPOSURE				DURATION OF MOST RECENT NOISE-EXPOSURE		
0-20 MIN	1 HR	4-7 HRS	1 DA	0-20 MIN	1 HR	4-7 HRS
21-50 MIN	2-3 HRS	8-16 HRS	2-3 DAS	21-50 MIN	2-3 HRS	7+ HRS
			4+ DAS			

AGE	DATE OF AUDIOGRAM	DAY OF WEEK	TIME OF DAY	EAR PROTECTION
				WAS EAR PROTECTION WORN? YES NO

RIGHT EAR								LEFT EAR							
250	500	1000	1500	2000	3000	4000	6000	250	500	1000	1500	2000	3000	4000	6000

Adapted from Guide for Conservation of Hearing in Noise, Committee on Conservation of Hearing of the American Academy of Ophthalmology and Otolaryngology.

FIG. 28–2.—An illustration of a form used to record information for industrial hearing conservation programs.

noisy presenter switch is not a problem in those testing situations where the subject is in a booth and the audiometer and technician are located outside, as illustrated in Figure 28–1.

Presenting the signal

The presentation time of the test signal should be not too long or not too short. An optimum length for the tone is about one to two seconds. To develop this timing, the technician might try saying to himself, "one thousand one" each time the signal is presented. After a while, this reminder is no longer necessary. Of course, the signal should be presented long enough so that the subject can decide that he actually heard it and respond.

When presenting a series of test tones, a rhythmic signal pattern

685

should be avoided, that is, do not present the tones in a regular time sequence. Break the pattern so as not to get a false response at a very critical point. If the tone presentation is not varied, the subject is prone to anticipate a tone and give a response when in fact he did not hear the sound.

Presenting the test signal for a sustained period of time is not necessary, because the ear is most sensitive to the initial portion of the tone. If a person cannot hear the test tone after one or one and one-half seconds, he will not hear it if it is left on any longer.

On the other hand, do not present the test signals too briefly. In the time it takes for the subject being tested to make the decision that he heard the tone, it is gone, and he may fail to respond. The technician cannot depend on always having cooperative subjects. He cannot expect that the subject will report each time, "Oh, I just heard that one, even though it is gone now." In fact, most subjects will not raise a finger or push a response button because the test signal came and went too fast. Then the technician may become confused by erratic responses. When the responses are inconsistent, the technician should check the length of his presentation of the test signals.

Keeping records

A uniform and concise method of recording pertinent information in addition to audiometric data is desirable. The person being tested should be identified by name, sex, social security and badge number, job location, age, and birth date (see Figure 28–2). In addition to the audiometric technician's name and signature, the make, model and serial number, and the standard to which the audiometer is calibrated should appear on the audiogram. The date, including the year, of each audiometric test is important, especially when several tests have been made on the same person.

A cursory examination of the external canal and eardrum may be performed. If so, the type of ear examination will vary, depending upon company policy. At the preemployment examination, the audiometric technician should note on the form any history of ear disease, military history, and previous noise exposure, as well as any family history of hearing loss (see Figure 28–3). A brief interim history may suffice for followup checks. If a marked abnormality or acute infection is present, the technician should refer the individual to a physician for an examination. Of course, this history may have been collected as part of a required physical examination, but it is very wise to double check those

686

EMPLOYEE		S.S. NUMBER		AGE	DATE	

OCCUPATIONAL HISTORY (LIST LAST JOB HELD ON LINE 1.)			DEPT.			
	EMPLOYER	CITY	DUTIES	LENGTH OF SERVICE	NOISE EXPOSURE	EAR PROTECTORS
1.					__YES __NO	__YES __NO
2.					__YES __NO	__YES __NO
3.					__YES __NO	__YES __NO

MILITARY SERVICE	TIME SERVED	BRANCH	__ARMY __NAVY __MARINES __ AIR FORCE	EXPOSURE TO GUNFIRE AND NOISE __YES __NO

M
E
D
I
C
A
L

CHECK IF YOU HAVE HAD THE FOLLOWING:

___DIZZINESS ___EARACHES ___FREQUENT COLDS ___MUMPS ___RINGING EARS ___RUNNING EARS
___ALLERGY ___MEASLES ___SCARLET FEVER OTHER _____

SERIOUS MEDICAL AILMENTS	SURGERY (TYPE OF OPERATION)	INDUSTRIAL INJURIES OR DISEASE

H
I
S
T
O
R
Y

HEAD INJURIES	HEARING IMPAIRMENT IN FAMILY	WORKERS EVALUATION OF HEARING GOOD ___ FAIR ___ POOR

EDUCATON (HIGHEST GRADE COMPLETED)	PREVIOUS HEARING TEST (DATE AND CO.)	DATE LAST WORKED

ADDITIONAL INFORMATION (HOBBIES, ETC.)

COMMENTS	SIGNATURE
	M.D.

Fig. 28–3.—An example of the type of personal history that is desirable for an effective audiometric program. The subject should be identified by name, sex, social security (and badge) number, age, and birth date.

items specific to hearing and the ear prior to a hearing test.

It is also desirable to record audiometric test conditions, such as location and type of test booth. It is especially wise to record a serial number or some other identification specific to the audiometer used for each test. Such factors as the elapsed time interval between the employee's last noise exposure and the audiometric test are important.

To reduce the possibility of error, the hearing threshold levels should be entered on the records exactly as shown on the audiometer hearing-level dial setting.

A variety of methods for recording audiometric data are in use at the present time. Perhaps the most common record is the audiogram,

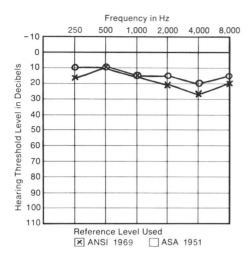

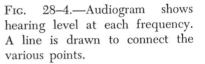

Fig. 28–4.—Audiogram shows hearing level at each frequency. A line is drawn to connect the various points.

where the hearing threshold at each frequency is recorded on a graph (see Figure 28–4). For air-conduction tests, it is customary to record the hearing threshold for the right ear as a circle (○), and to connect the circles with a continuous red line. The results for the left ear are recorded using an X and connected with a continuous blue line. (This has been pointed out in previous chapters.)

The disadvantage of this type of graph is that if many audiograms are taken of the same employee, the files become bulky. Further, it is more time consuming to compare one audiogram with another than to simply compare numbers.

Another method of recording and storing audiometric data is to prepare a serial audiogram sheet for each individual. A sample is shown in Figure 28–5. The hearing threshold level, in decibels, is recorded at each test frequency. Each succeeding audiometric test for the same employee is recorded below the previous entry on the same form, so that a serial record over a period of years can be compared.

It is important to record every hearing threshold level test even if there is considerable variation.

The information appearing on audiograms should *not* be altered. Care should be taken when this information is transcribed from one form to another. Audiograms are important written records of a person's hearing. Records of hearing thresholds should never be erased if on a repeat check the previous hearing threshold is found to differ

688

AUDIOMETRIC DATA SHEET

Medical Department

Name: Last	First	Middle	Department and or Insurance No.	Plant

Previous Significant Noise Exposure	Plant	Type of Work	Duration of Exposure
Civilian			

Military (type of exposure)...

History of Ear Disease ....

Family History of Deafness (relationship)..

Description of Ear Canal and Drum ....

Date ....

(The above information to be filled in at the time of employee's initial test. Attach additional pertinent data to this sheet.)

HEARING TESTS

HEARING LEVEL Left Ear Test Frequencies									HEARING LEVEL Right Ear Test Frequencies									TECHNICIAN'S SIGNATURE AND DATE	REMARKS
250	500	1000	1500	2000	3000	4000	6000	8000	250	500	1000	1500	2000	3000	4000	6000	8000		

From Industrial Noise Manual, *2d ed.,* American Industrial Hygiene Association.

FIG. 28–5.—Another type of audiometric form in which each succeeding audiometric test for the same employee is recorded below the previous entry, so that a serial record over a period of years can be compared.

689

Fig. 28–6.—A common problem encountered in industrial audiometry—the subject who thinks he hears the sound, but he's not sure. He may show this uncertainty by wiggling his finger or by the expression on his face.

from the latest test. Every audiogram should include the date and the name of the technician. The initials alone may be too vague for future identification of the technician.

Problems created by the subject

The accuracy of the responses of the subject is one of the most important ingredients in valid and reliable audiometry. If the subject is uncomfortable, overanxious, disinterested, ill, preoccupied, or easily distracted, he is likely to give spurious and inconsistent responses. It is extremely important to evaluate and reevaluate the physical and emotional status of the subject throughout a hearing test. Keep in mind that the technician is heavily dependent upon the accuracy of the subject's responses for obtaining a good hearing test. If the subject is easily excitable or nervous, he can start "hearing things," and this can

Fig. 28-7.—The technician should monitor the recording of the automatic audiometer. Irregularities, especially during the first minute of testing, are relatively common and may indicate that the instructions have not been understood.

lead to great frustration on the part of the technician as well as the subject.

The motivation of the employee and his attitude towards the audiometric test can affect the test results. A transfer or promotion, contingent upon the hearing-test results, may motivate the individual to perform better or worse.

An individual with a language problem, or one who is a slow learner, may require more time than the average to complete a hearing test. The technician should be prepared for such cases and adjust the testing schedule accordingly rather than treat all cases the same and be left with poor results for those subjects that truly need more time than the average person.

Occasionally, a subject's responses are so inconsistent that an accurate hearing test is not obtainable in the time period allotted. Rather than delay other persons who may be waiting to be tested, it is much wiser to recall this employee at a time when he can receive more individual attention. Do not report a vague general threshold when accurate responses are not attainable. It is better to reschedule the subject for another test when more time can be allotted, or seek audiological consultation.

Although not exhaustive, the conditions listed in the following sections can lead to inaccurate audiometric results. They can usually be rectified upon proper management by the technician or an audiologist.

Inconsistent responses

Perhaps the most common problem encountered during hearing tests is the subject who is undecided as to whether or not he hears the test tone signal. He will often manifest this uncertainty by wiggling his finger or pushing the response button intermittently causing the response light to flash off and on quickly (see Figure 28–6). This problem can often be solved by telling the subject, "Respond only when you hear the sound, no matter how faint it is. If you think you hear it, but you are not sure, don't do anything." The subject should not be encouraged to guess and to respond even if he thinks he hears a sound. This approach will encourage spurious responses.

If the technician is outside the audiometric test room or booth, he may need to go inside the booth to present these supplemental instructions. Some audiometers have a communication circuit that allows the technician to talk into the earphones ordinarily used for the presentation of the pure tone test signals. This arrangement makes supplemental instructions easier and quicker when the subject is in a soundproof room and the technician is seated outside.

Some subjects may say that they just hear the signal barely, faintly, or softly. Discourage such verbal responses, because the vocal sounds of the subject mask the audiometer signal coming into their ears, and the subject must get back into a listening "set" or mood after he has spoken. Verbal responses lengthen the time of an audiometric test, because more time must be allowed for the subject to respond between the presentation of the tone. Otherwise, there would be nothing wrong with verbal responses.

Tinnitus

A subject may have a ringing or some other type of noise in his ears. The technical term for this condition is *tinnitus*. A ringing-type of tin-

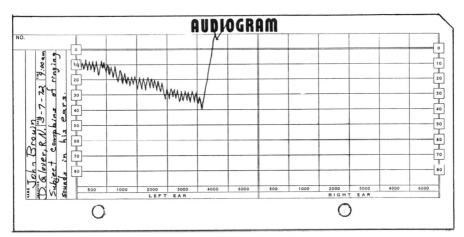

FIG. 28–8.—An audiogram of a subject taking a hearing test on an automatic audiometer. His smooth tracings follow the normal pattern and then reverse and go off scale as the subject follows the sounds of his tinnitus.

nitus can create difficulty for the subject to distinguish the sounds in his ears from the test tones. When a person reports that he cannot determine whether the sounds are coming from the audiometer or from his ears, he is reporting that he has tinnitus. Of course, the presence of tinnitus is usually noted on the history. This inability to distinguish the audiometer tones from his self-generated internal sounds is more likely to occur at the higher frequency test tones (2000, 3000, 4000, and 6000 Hz). Not all persons with ringing tinnitus have difficulty hearing a pure tone. However, the informed and experienced technician should anticipate this problem and be prepared to deal with it effectively.

Tinnitus is common in persons working in industrial situations where excessive noise levels are present. One of the symptoms of noise induced hearing loss is a ringing-type tinnitus. Tinnitus can complicate the audiometric test and induce inconsistent responses. Also, it is one of the reasons to monitor an automatic audiometer test closely (see Figure 28–7). A subject being tested with an automatic audiometer may trace his threshold very well in the region where he does not have tinnitus. He may be tracing a loss that appears like that shown in Figure 28–8. The smooth tracings follow the normal pattern, and then start to drop into the hearing-loss range. Then, the course of the tracing

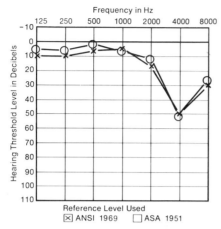

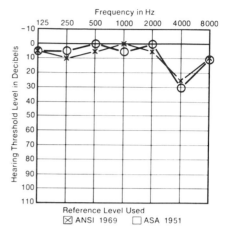

Fig. 28–9.—An audiogram showing a marked drop in hearing sensitivity at the higher frequencies. This drop could result from a temporary threshold shift due to recent noise exposure.

Fig. 28–10.—An audiogram taken on the same individual whose audiogram is shown in Figure 28–9. This audiogram, however, was taken after a 16-hour rest period away from noise exposure.

reverses as he starts to follow the sound of his tinnitus. He indicates that he hears the test tone by depressing the response switch until reaching the lower limit of the audiometer signal output. The technician can suspect that either the subject fell asleep with his finger on the handheld switch, or he is tracing the sound of his tinnitus.

The problem of tinnitus inducing inconsistent responses can be often rectified by giving supplemental instructions. Most tinnitus is fairly continuous—it may last for several minutes, hours, or even be permanent. Pathological tinnitus is not a sound that is on for one second and then off for one or two seconds. If the subject's responses are inconsistent because of tinnitus, it may be helpful to reinstruct as follows, "The audiometer sounds come and go rather quickly." Then, increase the hearing threshold level dial (attenuator) setting so that the subject can hear the audiometer signals very clearly. Then tell him, "My signals sound like this." He should be advised that *his* sound is on longer or all the time. The subject should be advised to respond only to the shorter sounds.

In the case of manual audiometry, the subject should be told to put his finger up when he hears the sound, and keep it up as long as he hears it. We deliberately do not tell the subject to do this in a routine hearing test, because it would be just another task that he would have to remember, and possibly confuse him.

One task at a time is enough for many subjects. Of course, under the conditions of these modified instructions, the technician must vary the length of time that he presents the test tone. The technician will be able to judge the accuracy of the subject's responses by comparing them to the length of the tone presentation.

Temporary threshold shift (TTS)

Measuring a temporary threshold shift in the subject that has been caused by recent noise exposure should be avoided. If a person has been exposed to excessive noise within approximately 14–16 hours before a hearing test, any loss that is measured may be due to a temporary threshold shift rather than a permanent shift in hearing sensitivity (see Figure 28–9). One cannot be certain if this threshold represents an actual hearing loss, or if it is due to the recent noise exposure. This hearing loss could be, at least in part, a temporary threshold shift (TTS). Upon rest away from the noise exposure, his hearing could, in fact, recover to a better level (see Figure 28–10).

What can be done about this? Under ideal conditions, an audiometric test should be made only after a person has had at least a 14 hour rest from exposure to excessive noise. This can present a management problem such as having 20 people lined up for a test each morning or at the beginning of a shift. Consequently, the person at the end of the line may show up on his job an hour or two late.

One solution to his problem is to issue hearing protectors to those workers scheduled to be tested on a particular day. Each person should be advised to wear the hearing protectors as soon as he enters the plant, and not to remove them until the audiometric test later on that same shift. If the subject is exposed to a noisy off-the-job environment, such as shooting at an indoor pistol range, motorcycling, or private flying, the technician can request that he abstain from these activities or wear hearing protection for his off-the-job noise exposures after he leaves work and again the following morning prior to the hearing test (see Figure 28–11).

Preparation for an examination is not a new concept. If a person is entering a hospital for a checkup or a test, he is often placed on a special diet the day before his admission. In the same manner, a

Courtesy Bausch & Lomb.

FIG. 28–11.—An employee whose hearing is to be tested should be instructed to wear hearing protection devices as soon as he enters the plant until the time of his test.

routine prehearing test procedure can be established for every person who works in a noisy area. The necessity for repeat audiometry will be less if this approach is used. An audiogram showing a typical noise induced notch may be the result of TTS unless it can be established that the worker was not exposed to hazardous noise levels for a period of 14 hours preceding the test.

Collapsing ear canals

Some ear canals will collapse when earphones are placed on them; this is a very well-known phenomenon in audiology and otology. Under such circumstances, the subject being tested would show a false hearing loss. Unfortunately, one cannot always tell by looking at an ear canal whether or not it is susceptible to collapsing. Collapsed ear canals may produce test results showing a 30 or 40 dB hearing loss, particularly for the higher frequencies (see Figure 28–12). Although some subjects'

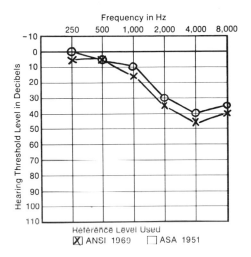

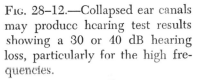

Fig. 28–12.—Collapsed ear canals may produce hearing test results showing a 30 or 40 dB hearing loss, particularly for the high frequencies.

audiograms will show a hearing loss, a clinical test will indicate normal hearing because a technique was used in the clinic that keeps the canal open during the test.

Keep in mind that with some subjects the ear canal can collapse, and if this is the case, recognize how it may bias the hearing test. Actually, the clinic should report back that this person has this problem. Hearing clinics do enough tests, both under earphones and by loudspeakers, to recognize this sort of problem and know how to deal with it appropriately. This problem of collapsing ear canals is cited here not with the intent of suggesting that the audiometric technician should deal with the problem directly. It is mentioned because some employees may show a hearing loss on tests taken at the plant, but the audiogram from the clinic may show normal hearing and indicate that the patient had collapsed canals when earphones were used.

Unilateral hearing loss

A person with good hearing in one ear and very poor hearing in the other ear presents a special kind of testing problem, which can lead to inaccurate results. When testing the person's bad ear, the pure tone is likely to cross through the head, via vibration, and be received by the good ear before the threshold of the poor ear can be obtained (see Figure 28–13). This is referred to as "lateralization" or "crossover" of the test tone, and the end result is actually an audiogram that reflects thresh-

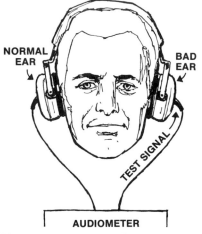

FIG. 28–13.—In this diagram, a 1000 Hz test tone at 50 dB is being delivered to the bad ear; the tone crosses over via vibration and stimulates the receptors in the good (or normal) ear. The crossover may occur at intensities lower than those necessary to elicit a response in the bad ear.

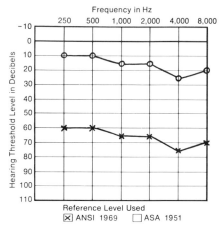

FIG. 28–14.—This audiogram shows a shadow curve of the thresholds for the poorer ear. The apparent thresholds for the bad ear reflect the hearing thresholds for the good ear. The lateralization of the test tone can occur at a difference of 50 dB, between the good ear threshold and the level delivered to the bad ear.

olds for the better ear tested twice; once through the test earphone over the good ear and again through the test earphone over the bad ear. The thresholds for the bad ear reflect what is often called a "shadow curve" with an audiometric configuration similar to that of the good ear (see Figure 28–14).

The crossover or lateralization of the test tone usually occurs at a difference of 50 to 55 dB between the good ear threshold and the level delivered to the bad ear. However, this crossover can occur at a *difference* as low as 40 dB between the good ear and the level fed to the poor ear, although this is unusual. It has been suggested that 50 dB be used as the danger crossover level. That is, if a threshold in the good ear is at least 50 dB better than the reported threshold in the poor ear, at the same frequency, the technician must question the possibility that the poor ear threshold is false.

The possibility of a false threshold for the bad ear of a person with a unilateral loss can be checked by "masking" the good ear. The simplest

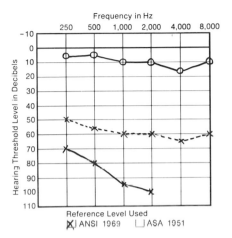

Frequency in Hz

Reference Level Used
☒ ANSI 1969   ☐ ASA 1951

FIG. 28–15.—Thresholds for the poorer ear will vary depending on whether or not masking was used. As shown here, the dotted line connects the apparent thresholds for the left (bad) ear resulting in a shadow curve of the right (better) ear. The real thresholds for the left ear are shown as the bottom solid curve.

explanation of masking in hearing testing is the creation of an artificial hearing loss in the good ear. (This is very similar to covering the right eye while testing the vision of the left eye. However, in hearing tests, a random-type noise is often used to cover the good ear.)

The industrial audiometric technician ordinarily is not expected to apply masking for those relatively few cases where it would be necessary. However, this does not preclude the fact that the average audiometric technician working in an industrial setting should be aware of the possibility of obtaining a shadow curve or false thresholds for the poorer ear of a person with a unilateral hearing loss. In any event, when the thresholds for the poor ear are at least 50 dB worse than those for the good ear, it is possible that the thresholds for the poor ear are worse than the audiogram suggests. If this individual is referred to a hearing clinic for more sophisticated testing, do not be surprised if the clinical audiometric tests show a greater loss for the poor ear than tests taken without masking (see Figure 28–15).

The malingerer

After a technician has had considerable experience testing hearing, he will develop a feeling for normal or average performance in the hearing-testing situation. Chapter 26 provides basic information in this area and describes how persons with varying degrees of hearing loss will behave in regular conversational situations.

A useful audiovisual aid in understanding the effects of various types of hearing impairment is the set of slides and recording *How They Hear*

TABLE 28-A
CAUSES OF AUDIOMETRIC CHANGES

Physical Variables

Improper placement of earphones
Ambient noise levels in test room
Equipment variables, such as accuracy of attenuator steps, type of earphone
cushions, hum, noise, etc

Physiological Variables

Age and sex
Pathology of the auditory organs
General health of subject
Temporary threshold shift
Tinnitus and other head noises

Psychological Variables

Motivation of subject
Momentary fluctuations of attention
Attitude toward the test situation
Personality attributes
Intellectual factors
    Comprehension of instructions
    Experience in test taking of any sort
Response conditions
    Type of response required of subject, i.e., button pressing, finger raising, ver-
        bal response, etc

Methodological Variables

Testing technique used
Time interval between successive tests
Instructions to subjects
Order of presentation of frequencies

NOTE: The factors listed have been demonstrated to influence audiometric measurements. Not all of these factors are of concern in the industrial hearing conservation program. The list is presented to illustrate the potential complexity of audiometry. (This information is adapted from *Guide for Conservation of Hearing in Noise,* published by the Committee on Conservation of Hearing of the American Academy of Ophthalmology and Otolaryngology.)

(see Chapter 33, Sources of Information) which simulates various types of hearing loss. The technician may find this record helpful in gaining a better understanding of the consequences of various types of hearing loss on communication.

The technician should be alert for the subject who shows a moderate-to-severe bilateral loss on a pure tone audiometric test, but who has little difficulty understanding the test instructions or normal conversation. In a clinical situation, in the regular practice of audiology and otolaryngology, a small percentage of individuals are encountered who, for one reason or another, do not report their true thresholds on a routine hearing test. There is no reason to suspect that a malingerer will not be encountered occasionally in an industrial hearing testing program. Experience to date indicates that malingering has not been found to be a problem in industrial hearing-conservation programs. Nevertheless, the audiometric technician should be on the alert for such cases and deal with them according to the advice of medical or audiological consultation.

The audiometric technician in industry is not expected to deal directly with the problem of malingering. However, the technician should be alert for such a problem, qualify the audiogram, and recommend appropriate referral for more detailed and sophisticated hearing tests.

Summary

The technician and the subject can introduce problems into the hearing testing situation that can lead to invalid and unreliable audiometric records (see Table 28-A). Most of these can be avoided or controlled by a well-trained and competent audiometric technician. As was indicated in previous chapters in Part Five—Industrial Audiometry, even the most accomplished audiometric technician will have his share of invalid audiometric tests. The intent of this chapter is simply to point out some of the most common problems that can be avoided, or at least recognized and dealt with appropriately. The typical audiometric technician is not prepared to cope with the last three problems cited— collapsing ear canals, unilateral hearing loss, and malingering. These are included because they are real and may well be encountered by anyone who engages in the measurement of hearing. These problems are best handled by professional audiologists who have specific training for a career in the measurement of hearing and evaluation of auditory communication problems. However, it would be negligent not to mention and briefly explain these more complex problems to alert the technician to the very real fact that some cases are likely to present complex problems, which require referral outside the industrial setting.

This chapter describes guidelines to follow when measuring hearing acuity to assure more accurate audiometry. Adherence to recommended procedures should help in obtaining valid and reliable hearing test results a higher percentage of the time.

It is highly recommended that industrial audiometric technicians be properly qualified by attendance at a training course (or its equivalent), which follows the guidelines recommended by the Intersociety Committee or other recognized institutions. Technicians should also be encouraged to attend a refresher course six months to a year from the date of the original training course.

This book will emphasize again and again that there is no substitute for direct, practical instruction in developing the skills required for the determination of valid air-conduction, pure tone hearing thresholds.

Part 6 Industrial Hearing-Conservation Programs

THE BLANK STAMPING CO
ANYTOWN, U.S.A.

SUBJECT: Industrial Hearing-Conservation Program

TO: All Employees

FROM: Company President

Steps are being taken to put into practice a hearing-conservation program that will provide protection against noise hazards wherever they might occur, in or around the work areas. In order for this program to be effective, everyone must cooperate.

To make certain that this program receives proper attention, an administrator has been appointed who will be responsible for the entire program. His duties and responsibilities include the following:

1. Appoint and select specialists or other qualified persons to assist in carrying out the details of the program.

2. Coordinate the efforts of all departments involved, such as industrial hygiene, safety, medical, engineering, and purchasing, as well as supervisory personnel.

3. Inspect the work areas throughout the organization. In conjunction with specialists and department heads, designate areas as hazardous to hearing. Personal hearing-protective devices will be required in those areas until such time as the noise levels can be reduced to a safe level.

4. The administrator of the hearing-conservation program will be responsible for recommending the proper type of personal hearing-protective device to be worn—based upon:
   a. The overall noise level.
   b. The character and composition of the noise.
   c. The period of time for which the hearing protector is to be worn.
   d. The attenuation and physical characteristics of hearing protective devices.

5. The administrator of the hearing-conservation program will develop and coordinate an educational program that has as its main objective the education of all employees as to the purpose and the benefits of an industrial hearing-conservation program.

6. All personnel will be informed about the abatement procedures to be followed until such time that excessive noise levels are reduced or eliminated.

FIG. 29–1.—An example of a policy statement that can be issued by a company inaugurating a hearing-conservation program. Such a statement should have the complete endorsement of management.

704

Chapter Twenty-Nine

Effective Hearing-Conservation Programs

Julian B. Olishifskl, P.E.

An effective industrial hearing-conservation program includes: (a) the prevention of noise induced occupational hearing loss, (b) minimizing annoyance or speech interference complaints by employees or residents of the surrounding community, (c) compliance with governmental noise regulations, and (d) control of the company's worker's compensation cost for occupational hearing loss. The purpose of this chapter is to provide information which can be helpful in establishing an effective hearing-conservation program.

Industry's prime concern with noise arising from industrial operations is the potential harmful effect on an employee's hearing ability. Prolonged exposure to excessive noise levels can cause a noise induced hearing loss. Permanent hearing loss caused by such exposure is now a recognized occupational hazard and is compensable in many states.

When an employee is exposed to excessive noise levels, usually the first physical effect that can be documented is a temporary hearing loss. This is a natural occurrence and most likely results from the hearing mechanism's attempt to protect itself from the potentially harmful environment. If the employee is transferred to a quiet environment, the hearing sensitivity should return to its preexposure level.

Absence of pain when exposed to high noise levels should not be construed to mean the absence of risk of sustaining hearing loss. Pain may be felt by a person in his ear if he is exposed to noise levels in the order of 130 dB, SPL; a temporary hearing loss, however, may be produced at considerably lower noise levels. Pain and annoyance are not reliable indicators of potential noise hazards. Consequently, the decision to initiate a hearing-conservation program should not be influenced by the presence or absence of these symptoms.

Almost every individual has experienced difficulty in conducting a conversation in a high-noise environment. In addition to speech interference high-level background noise can interfere with the reception of other important sounds, such as audible warning devices. The degree of speech interference depends on such factors as the ambient sound level, knowledge of the type of information expected, and the distance between the two persons attempting to communicate.

The degree of speech interference can be used as a rough guide to the potential noise hazard of an environment. If two people stand at arm's length from each other (assuming each has good hearing) and cannot carry on a normal conversation because of the background noise, then the possibility exists that continuous exposure of the ears to this environment over a period of time can produce permanent hearing damage.

An effective hearing-conservation program is one that prevents permanent hearing loss as a result of exposure to noise while on the job. With respect to existing worker's compensation laws, an effective hearing-conservation program is one that limits the amount of compensable hearing loss in the frequency range over which normal hearing is necessary in order to communicate.

An effective hearing-conservation program is required if an employee's noise exposure exceeds current limits as defined in the Occupational Safety and Health Act (see Appendix A-2). Increasing attention is being given to industrial noise problems in view of existing state and federal legislation covering noise exposure and compensation claims for noise induced hearing loss.

General procedures for carrying out an industrial hearing-conservation program are: (a) evaluation of noise exposures and the classification of operations or areas as to degree of hazard, (b) control of hazardous noise exposures by noise reduction if feasible, (c) use of personal hearing-protective devices such as earplugs or muffs wherever the noise cannot be or has not yet been, adequately controlled by administrative or engineering measures, (d) measurement of hearing ability of applicants for employment and employees exposed to harmful noise levels, (e) consideration of noise exposures in the planning of new operations and the purchase of new equipment or machinery.

An effective hearing-conservation protection program requires careful planning, diligent execution, and consideration of the following basic elements:

1. Declaration of policy
2. Program administration
3. Noise survey and analysis
4. Education, motivation, and training
5. Controls—engineering, administrative, personal hearing protection
6. Audiometry

Establishing a hearing conservation program

Declaration of policy

The details for carrying out an effective hearing-conservation program may be assigned, but the responsibility for the basic policy cannot be delegated. An overall policy statement should be prepared that will state clearly the objectives to be achieved if employee cooperation and participation are to be obtained (see Figure 29–1).

The goals and objectives of the hearing-conservation program should be discussed with all levels of management. Full cooperation between management and labor is essential for an effective hearing-conservation program. Management, from top level on down through first-line supervision, must believe in, support, and understand the reasons and objectives of the program. Employees must understand that the primary purpose for the program is to protect their hearing.

Top management must set the hearing-conservation policy and should declare the following:

1. Plant, project, and design engineers are to be responsible for the engineering control or reduction of noise at its source.

2. Suppliers must be informed that new equipment meet the company specification on allowable noise levels.

3. The supervisor must enforce the wearing of personal hearing protection in areas where needed.

The personnel or industrial relations manager can provide advice as to the policies and procedures which should be established to obtain supervisory and employee support of the program. The nurse and the physician to whom she reports are usually responsible for the audiometric testing program and the keeping of accurate medical records.

707

Program administration

The four main elements in the administration of an effective hearing-conservation program are: (*a*) identification of the noise problem, (*b*) evaluation of the extent of the noise hazard, if any, (*c*) determination of the proper corrective action, and (*d*) the implementation of the decision to minimize excessive noise exposure.

Good communication is required to make these elements work as an effective system. Communications can mean a number of things, including: (*a*) establishing a communications network that extends throughout all levels of the organization, (*b*) providing a focal point within the organization for hearing-conservation program-related matters, and (*c*) communicating with all levels of supervision and the employees to make sure that they understand the firm's policy.

For greater effectiveness, the responsibility for the administration of a hearing-conservation program should be assigned to one individual. Central authority and responsibility are necessary to assure overall co-ordination and direction. Hearing-conservation programs will vary according to the needs of particular industries and may involve specialists such as industrial audiometric technicians, nurses, safety professionals, industrial hygienists, acoustical engineers, audiologists, and physicians. In any case, overall responsibility for the entire program should rest with a single individual if the program is to achieve optimum results.

The decision as to who this individual should be is usually based upon the extent of the noise hazard, the size of the company and the nature of its operations. In companies with more than 500 workers or with scattered operations in many locations, the administrator of the program may be the vice president, general manager, safety director, nurse, director of industrial relations, industrial hygienist, or medical director. Some firms have hired an audiologist for this job.

In smaller companies, where there are less than 500 workers, the administrator of the hearing-conservation program may be the plant manager, superintendent, or other top-ranking official charged with the responsibility for industrial relations or personnel.

Responsibility. The practice of assigning the responsibility for the program to the highest possible ranking official demonstrates to the employees management's interest in the success of the program. On the other hand placing the direction of the hearing-conservation program under an individual with little official authority is evidence of the lack of interest on the part of management. The program would have little chance to function effectively under such an arrangement.

The administrator of the hearing-conservation program is ordinarily responsible for the following activities:

1. Formulating, administering, and making necessary changes in the hearing-conservation program.

2. Preparing regular, annual, monthly, or weekly reports on the status of the program.

3. Acting in an advisory capacity on all matters pertaining to hearing conservation as required for the guidance of management (superintendents, foremen, and supervisors) and departments—such as purchasing, engineering, safety, medical, industrial relations, and personnel.

4. Supervising or closely cooperating with the training supervisor in the training of employees on the proper use of personal hearing-protective devices.

5. Directing the activities of the staff so that the hearing-conservation program will be conducted efficiently.

6. Making certain the program complies with federal, state, or local laws or ordinances.

7. Establishing requirements for hearing-protective devices to be used by employees.

The amount of assistance needed by the administrator of the program depends upon the size and operating policies of the company, the nature and type of operations, and the degree of noise hazard. The success of the industrial hearing-conservation program requires the cooperative efforts of the engineering, medical, industrial hygiene, safety, purchasing, and maintenance departments and the supervisor, as well as the individual employee.

Engineering department. Engineering personnel should consider possible noise exposures when new equipment is ordered or new facilities planned. In selecting new equipment, noise levels and cost of noise control measures that will be added later should be considered.

The greatest noise reduction will come from quieter equipment, whether purchased or renovated to meet allowable noise standards. Noise from a particular process or machine is a function of the noise produced by the unit itself and of the acoustical properties of the area in which it is located. Noise is produced by: mechanical equipment such as compressors, pumps, electrical equipment motors, transformers, heating, ventilating and air conditioning equipment fans and blowers and structural vibrations.

The engineering group has an important role in the control of occupational noise hazards. Their responsibilities are to:

• Develop engineering concepts using established engineering principles to control noise at the source and to prevent unnecessary exposure to excessive noise levels.

• Notify the medical, industrial hygiene, and safety organizations whenever new potentially noisy operations or processes are planned.

Medical department. The responsibilities of the medical staff in a hearing-conservation program are:

• To examine and recommend the placement of new employees whose hearing ability meets the minimum job requirements.

• To cooperate with other departments in the development of adequate, effective measures to prevent excessive noise exposure.

• To examine periodically those employees who are exposed to hazardous noise levels.

• To restrict employees from further exposure to excessive noise on a medical basis whenever warranted by findings of a medical examination.

The **industrial hygiene department** may be responsible for monitoring the noise levels in the work environment and assisting the appropriate departments in making sure that a safe noise environment is provided and maintained. Noise surveys should be made as frequent as necessary, to keep abreast of noise exposure conditions.

The industrial hygiene personnel also have the following responsibilities:

• To advise the appropriate authority of the potential noise hazard arising from any current or proposed process or operation

• To recommend controls that will minimize employee exposure to excessive noise

• To assist the foreman or supervisor in educating employees on practices, precautions, and procedures to be followed

• To review present and proposed noise control practices and to make sure that they are in accordance with established standards

• To survey noise levels of all new installations before they are turned over to operating personnel

Safety department. Safety personnel have a responsibility to maintain an adequate stock of approved personal hearing protectors, issue or arrange for issue of such protectors to employees of the type and size prescribed by the medical department, keep records, observe the physical condition of personal hearing protectors in use and arrange for repairs or replacements as necessary, and continually encourage (through education of supervisors and employees) the regular wearing of hearing devices in designated areas.

The safety organization plays an integral part in the overall industrial hearing-conservation program. Its responsibilities are:

• To coordinate the educational, engineering, supervisory, and enforcement activities related to the program

• To assist the supervisor in teaching his employees rules, regulations, and procedures related to the program

• To conduct surveys to make sure that proper practices and procedures are being followed

• To recommend changes in rules, regulations, and procedures to keep pace with technological advancements

The **purchasing department** has a responsibility to make sure that only equipment approved by the engineering, industrial hygiene, safety, medical, and other departments is purchased for use in the company.

On orders for equipment, purchasing personnel should include a standard clause requesting noise information for use by those originating the purchase action for evaluation of the equipment (see Chapter 19, Sound Level Specifications). The purchasing department also should include any specific noise limits or other requirements on orders or inquiries as directed by the operating department.

Noise criteria must be developed to meet specific requirements based upon a thorough evaluation of the situation. These criteria must then be specified in the purchase order.

Maintenance department. Once noise levels have been reduced to acceptable levels, the maintenance staff is responsible for keeping them there. Worn gears, bad bearings, unbalanced fans, corroded mufflers, unlubricated fittings, and vibrating pipes all contribute to higher noise levels. An awareness of these sources by each member of the maintenance staff is essential if the effectiveness of the noise control program is to be sustained.

Changes in noise levels can often be correlated with changes in vibration levels. A change in vibration level on a machine can also be an early indication of required maintenance.

Supervisor's responsibilities. The supervisor's responsibilities are:

• To instruct employees periodically on precautions, procedures, and practices to be followed in order to minimize excessive exposure to noise

• To make sure that the required warning signs are posted in all areas where high noise levels exist

• To inform the engineering, industrial hygiene, and safety organizations of any operation or condition which appears to present a noise hazard to employees

• To make sure that employees are furnished with the proper personal-hearing protective equipment, and enforce wearing such equipment

• To consult with safety, industrial hygiene, engineering, and medical personnel when necessary

• To assist during a noise survey by:

1. Notifying surveyor of all potentially dangerous areas under the supervisor's responsibility and accompanying him or providing a guide while the surveyor is in the area.

2. Assisting the surveyor in determining the equivalent daily noise dose for each employee whenever required.

If the first-line supervisor feels that his area has a noise hazard then a request should be sent to the next level of supervision requesting that the area be evaluated.

The supervisor should refer all questions that he cannot promptly answer to his immediate superior or to the administrator of the hearing-conservation program.

Employee's responsibilities. Each employee is responsible for contributing his part towards the success of the industrial hearing-conservation program. This should include:

• Notifying his supervisor immediately when conditions or practices are changed and result in increased noise levels

• Observing all safety rules and making maximum use of all prescribed personal hearing-protective equipment

Noise survey analysis

Preliminary noise survey

A hearing-conservation program should start with a preliminary plant-wide noise level survey using appropriate sound level measuring equipment to locate operations or areas where workers may be exposed to hazardous noise levels.

A decision will have to be made whether to purchase sound level measuring equipment and train personnel to use it, or to contract the work out to an outside firm. The extent of the noise problem, the size of the plant, and nature of the work will affect this decision. In most plants, noise surveys are made by a qualified engineer, audiologist, technician, industrial hygienist, or safety professional.

A noise survey should be made of those work areas where it is difficult to communicate by speech in normal tones. Other indications for performing a noise survey are if people, after being exposed to high noise levels during their work shift, notice that speech and other sounds are muffled for several hours or if they develop head noises or ringing in the ears.

As a general guideline for conducting a noise survey, the information recorded should be sufficient to allow another individual to take the report, use the same equipment, locate the various measurement locations, and finally reproduce the measured and/or recorded data.

The preliminary noise survey normally does not define the noise environment in depth and therefore should not be used to determine employee exposure time and other details. The preliminary noise survey simply supplies sufficient data to determine if a noise problem exists and to indicate to what extent it exists.

713

Detailed noise survey

From the preliminary noise survey, it is relatively easy to determine specific locations requiring more detailed study and attention. A detailed noise study should then be made at each of the locations to determine the employee's time-weighted average exposure.

The purpose of a detailed noise survey is to (a) obtain specific information of the noise levels existing at each employee's work station, (b) develop guidelines for establishing engineering and/or administrative controls, (c) define areas where hearing protection will be required, and (d) determine those work areas where audiometric testing of employees is desirable and/or required. In addition, detailed noise survey data can be used to develop engineering control policies and procedures, and to determine compliance with specific company, state, or federal requirements. (See Chapter 14 and Appendix A.)

An effective hearing-conservation program always starts with the question, "Does a noise problem exist?" The answer must not be based on simply the subjective feeling that the problem exists, but on the results of a careful technical definition of the problem. Answers to the following questions must be obtained:

- How noisy is each work area?

- What equipment or process is generating the noise?

- Which employees are exposed to the noise?

- How long are they exposed?

The time that an employee spends at each of his work stations (if he has more than one) is determined by several means. Questioning of first-line supervisors can provide basic job function information concerning the type of noise levels in working areas and the percentage of time the worker spends in each of the areas. Production records can be examined and on-sight evaluation can provide information as to the extent of the noise problem.

The noise survey should be made using a general purpose sound level meter that meets American National Standard S1.4–1971, *Specifications for Sound Level Meters, Type 1 or 2*, and should be set for A-scale slow response. (See Chapter 6.)

Exposure may also be measured by a dosimeter which is equivalent in precision and accuracy to a sound level meter. Noise surveys should be conducted at least annually; they should also be conducted when changes in equipment or operations are made that would affect workers' noise exposure. The survey should include the noise levels obtained in the work area, along with information related to the noise source, number of people exposed, and the level of exposure.

Measurements of noise exposure should be taken at approximate ear level. No worker should be exposed to steady-state or interrupted steady-state sound levels exceeding the maximum listed in the current standard (see Appendix A). Information should include the name of the individual making the noise survey, along with the date, location, and time that the noise measurements were made. The serial numbers of the sound level meters and calibrators are also essential for compliance records.

Educational program

If the results of the detailed noise survey indicate that it is necessary to implement a hearing-conservation program, then an educational program should be developed. The main objective should be the education of all exposed and affected company personnel in the procedures to be followed in minimizing noise exposure.

To be considered adequate, an educational program should include most of the following topics:

1. Explaining to all employees the need and benefits of the hearing-conservation program

2. Giving special training to management, supervisors, and employees

3. Training employees in the proper use and care of personal hearing-protection devices

4. Explaining the audiometric testing procedure and interpretation of results according to guidelines developed by management

5. Using special handouts, posters and bulletin boards, meeting time, and company magazines (if they exist) to *remind* and *inform* employees of the continuing need to protect hearing and to inform them of successes achieved to date (see Figure 29–2).

6. Giving training to all new employees prior to the issuing of hearing protection and initial audiometric tests

A series of four letters to employees is shown in Figures 29–3 through 29–6. The letters are designed to elicit employee cooperation for the hearing-conservation program. Each letter is numbered to correspond

Fig. 29–2.—National Safety Council posters such as these serve to remind and inform the employees of the need for hearing protection.

with the step of the program in which it is to be distributed.

There are a number of ways in which the letters may be distributed—mailing them to employees at their home, enclosing them in pay envelopes, or personally distributing them to employees at their work stations.

The letters point out to the employees what their company is doing to protect their hearing—and why. The letters offer a rationale. They

(*Text continues on page 721.*)

Dear Employee:

(Company Name) has just launched a comprehensive hearing conservation program for our employees. I am writing you to explain the details, because we need your cooperation — and participation — to make the program a success.

Our new hearing conservation program is designed to safeguard your hearing while you are on the job. We're working to minimize the likelihood of hearing damage caused by occupational noise.

Though the idea is simple, the program is fairly complex, and it has several different elements:

1. The first step is to pinpoint any sources of excessive noise in our company by making sound level measurements. In the near future, you will see specially trained staffers at work making these measurements. They will be using electronic measuring instruments called sound level meters. In addition, you may be asked to carry a pocket-sized instrument (called an audio dosimeter) around with you for one workday. This device measures the actual "noise dose" that you receive while you do your job.

2. When feasible, (Company Name) will implement engineering and/or administrative controls to reduce employee exposure to excessive noise. In simpler terms, "engineering controls" are modifications to equipment, while "administrative controls" involve changes in employee work schedules or procedures.

3. Because engineering and/or administrative changes take time to develop, affected employees will be issued ear plugs or ear muffs to provide interim protection. These devices will also be used in situations where engineering or administrative controls are not practical and/or not completely effective.

4. As part of our hearing conservation program, we are establishing a program of periodic audiometric testing (hearing tests). These simple tests will help us make sure that the other program elements are doing their job to protect your hearing. Even more important, the tests can signal developing hearing loss which may be caused by many different factors — early enough to do something about it.

In a nutshell, this is our new program. Your hearing is one of your most precious "natural resources," and I urge you to learn as much as you can about hearing conservation. Your supervisor will be able to answer your questions, and to explain how the program affects you personally.

During the next few months, I'll be writing to you again to explain specific elements of the program in greater detail. I am sure that you will discover — as I have — that hearing conservation is both significant and worthwhile . . . an important safety effort that merits your interest and cooperation.

Yours sincerely,

Fig. 29–3.—The first letter explains to the employees the purpose of hearing conservation and should be distributed at the beginning of the program.

Courtesy Mine Safety Appliances Co., A Program for Hearing Conservation. *Copyright 1975 MSA. Used with permission.*

Dear Employee:

The first phase of our hearing conservation program is well under way. As you probably know, we are now conducting a "noise survey" of all *(Company Name)* facilities. Our goal is to track down the sources of occupational noise loud enough to be considered potentially hazardous.

Your supervisor has explained to you how loud occupational noise can — over the years — cause gradual, permanent hearing loss. Our new program is designed to protect your hearing from damage.

Broadly speaking, there are several different ways to measure the loudness of noise and the total "noise dose" received by workers exposed to loud noise. In the days ahead, you will see specially trained staffers using hand-held sound level meters (or other electronic instruments) to monitor noise levels throughout our facilities. They are searching for the sources of excessive noise.

Once these sources have been pinpointed, the next step is to find out if any *(Company Name)* employees are exposed to noise for a long enough period each working day to suffer hearing damage. In some cases, this will be determined by making time/motion noise studies during specific workshifts; in other cases, selected employees will be asked to carry a small electronic device called an "audio dosimeter" around with them during a full working day. Both of these procedures enable us to calculate the "noise dose" received by employees as they go about their jobs.

Our noise survey is really the heart of our hearing conservation efforts since it gives us the information we need to take corrective action in the future.

If you are asked to participate in the noise survey — perhaps by carrying a dosimeter — I encourage you to cooperate fully. Our hearing conservation program has but one goal: to protect and safeguard one of your most precious "natural resources"— hearing.

Yours sincerely,

FIG. 29–4.—The second letter familiarizes the employees with the procedures of noise surveys.

Copyright 1975 MSA.

Dear Employee:

(Company Name) will soon institute company-wide audiometric testing (hearing testing) of all employees as part of our new hearing conservation program.

Periodic audiometric testing accomplishes three important things:

1. The first audiometric test of a current or incoming employee provides our company physician with a "baseline" measurement (called an audiogram) of hearing ability. Subsequent tests can be compared with the baseline to verify that our hearing conservation program is doing its job by preventing noise-induced hearing loss.

2. Periodic audiometric testing provides an "early warning" of developing hearing disability . . . usually while there is still time to do something about it. Many factors other than excessive occupational noise can cause hearing loss. A few examples are noisy off-the-job activities, mechanical damage to the inner ear, and infections and diseases of the inner ear.

3. In rare cases, audiometric testing will uncover employees who should not work in noisy locations by virtue of prior hearing loss. Here, the risks of additional damage — and eventual deafness — are great.

Audiometric testing is a simple procedure. You will be seated in a quiet location or test booth and asked to wear a set of headphones. Then, you will listen to a series of audio tones and indicate — by raising your hand or touching a button — when you hear each tone in each ear. The complete test takes less than 10 minutes.

If you normally work in a noisy location, you may be asked to wear ear muffs on the day of your hearing test. The muffs will prevent your ears being "fatigued" by the noise . . . a circumstance that might cause a false indication of hearing loss.

As you are probably well aware, we are very serious about our hearing conservation efforts here at *(Company Name)*. We are determined that no employee will suffer needless hearing damage as a result of exposure to excessive occupational noise. Audiometric testing is a cornerstone of our program.

Yours sincerely,

Fig. 29–5.—The third letter explains the purpose and importance of audiometric testing.

Copyright 1975 MSA.

719

Sent only to employees who must wear hearing protectors

Dear Employee (or Dear Mr. XXXXX):

I have just been informed that you are one of *(Company Name's)* employees who will be asked to wear personal hearing protectors (ear muffs, ear plugs) during your workday. I am writing you to explain the importance of these simple devices, and to encourage you to wear them faithfully.

Your hearing protectors are designed to reduce the level of noise reaching your inner ear. If you cooperate by wearing them properly, they can provide virtually total protection against the damage to hearing that can be done by excessive occupational noise. In effect, ear plugs or ear muffs are your "first line of defense" against noise-induced hearing loss.

As you know, we recently conducted a noise survey throughout our facilities. Comprehensive monitoring with the latest electronic noise-measuring instruments determined that you receive an excessive "noise exposure" during your workday. It is possible that long-term exposure to this excessive occupational noise — over a period of many years — might bring on hearing loss.

Broadly speaking, *(Company Name)* is issuing hearing protectors to selected employees in two specific circumstances:

1. As an interim protection measure while engineering or administrative controls are being developed that will reduce employee exposure to excessive noise, and

2. As a long-term protective measure when engineering or administrative controls aren't feasible or aren't completely effective.

Your supervisor will explain why you have been asked to wear hearing protectors.

(Company Name) management has instructed your supervisor to make certain that all employees who receive protectors use them as directed. Hearing protectors are an integral part of our hearing conservation program — an important part! In fact, management has a clearly defined responsibility under Federal government regulations to make certain that employees use the ear muffs or ear plugs that they have been issued. Incidentally, under the same laws, the employee has a similar responsibility to wear the supplied protectors.

But, rules and regulations are only of secondary importance. What really matters is the safeguarding of one of your most precious "natural resources" . . . your sense of hearing. It's as simple as that.

Yours sincerely,

Fig. 29–6.—The fourth letter builds employee's motivation to use hearing protectors correctly.

Copyright 1975 MSA.

explain, rather than dictate. For this reason, they are helpful motivational tools.

In addition, the company employee newspaper can be used to cover the entire hearing-conservation program, from its inception to its implementation. The employees can be kept informed of each step of the program. If there is no employee publication, the articles shown in Figures 29–7 through 29–10 can be (a) used as an insert in employee's paycheck envelopes, or (b) distributed personally to the employees, or (c) posted on bulletin boards.

If the need exists for the development and implementation of a program, it is necessary to inform the employee, because if the employee does not understand what is happening and why, he quite often jumps to some unfortunate conclusions. Unnecessary fears may develop concerning job security. These false judgments can be greatly reduced if a well-organized educational program is initiated at the very beginning of the hearing-conservation program.

There are many educational materials available to assist in informing both the employer and the employee of the advantages of a hearing-conservation program. These include films, posters, leaflets, and recordings, which may be obtained from insurance carriers, manufacturers of equipment, and private sources such as the National Safety Council.

Control measures

After making noise surveys to determine the level, composition, and exposure time, a comparison can be made with government regulations and available damage-risk criteria to evaluate the hazard and consider what type of control measures are needed.

The daily noise dose of an employee is simply the actual duration of noise exposure (hours) divided by the exposure time (hours) permitted by the OSHA (or other regulatory) standard. Where actual noise exposures exceed those prescribed, steps should be taken to reduce noise levels for employees working in those areas.

There are three approaches:

1. The environment and the machines can be redesigned so that the overall noise level is reduced—noise reduction at the source.
2. There are many operations where the exposure of employees to the noise can be controlled administratively, without modifying the noise, by changing production schedules or rotating jobs so that exposure times are within safe limits.

(*Text continues on page 727.*)

ARTICLES FOR A COMPANY PUBLICATION

A series of four articles for a company employee publication that complements the letters to the employees is shown in Figures 29–7 through 29–10. The articles, based on the same subjects as the letters, explain the different phases of the hearing-conservation program. They tell the employees how the program benefits their health and safety. Keeping the employees informed is a key to building employee motivation.

Courtesy Mine Safety Appliances Co., A Program for Hearing Conservation. *Copyright 1975 MSA. Used with permission.*

FIG. 29–7.—*First article.*

COMPANY INITIATES HEARING-CONSERVATION PROGRAM

Within the next few days, XYZ Company will be launching a comprehensive hearing-conservation program for workers in all departments. Why? Because your company is concerned about the long-term effects that exposure to harsh on-the-job noises may have. Research shows that excessive occupational noise is more than just a daily annoyance. When listened to over a long period of time, it can cause permanent—irreparable—damage to your hearing.

Our new hearing-conservation program is designed to eliminate this hazard from your workday. Its primary goal is to safeguard your hearing while you are on the job.

This comprehensive program takes into account recent revisions to federal noise standards, and should do much to make our company a healthier and more pleasant place to work.

The program, to be successful, needs your support. Once you know the facts about job-related hearing loss, we're sure you'll be convinced that the long-term benefits of the program far outweigh any inconveniences.

The first step to protect your hearing will be to conduct a complete noise survey in all departments. Survey personnel will be in work areas shortly to measure and record noise levels to which you are subjected during an average working day.

If these noise surveys turn up any potential "noise hazards" in various departments, a number of measures will be undertaken to safeguard the hearing of employees working in these areas.

The first of these will be the implementation of appropriate engineering and/or administrative controls to reduce noise levels below the hazard level. Engineering controls are designed to eliminate or reduce the noise itself and include such things as machine enclosures, mufflers, baffles, or other sound damping supports. Administrative controls, on

722

the other hand, involve the arranging for employees to work in noisy locations less time each day by staggering work shifts or periodically shutting down noisy equipment.

While these engineering and/or administrative controls are being worked out, employees in affected areas will be provided with hearing-protective devices such as ear plugs and ear muffs to lower the intensity of the sound reaching the ears. Hearing-protective devices will also be issued to those who work in areas where engineering and/or administrative controls are not feasible.

As part of our hearing-conservation program, hearing tests (and periodic retesting) will be given to both present and new employees. The purpose of this testing will be to verify the effectiveness of our hearing-conservation program in preventing job-related hearing losses. This part of the program will be under the supervision of our company physician, Dr. ——.

Our program also calls for the periodic resurveying of noise levels in all departments and of employee noise exposures to make sure the program is effective and to verify that engineering and/or administrative controls are working.

That, in a nutshell, is a rundown of the basic parts of our hearing-conservation program. We'll be dealing with the subject in greater detail in future articles. Survey procedures will be explained in depth as will the company's audiometric testing program.

These articles will detail the problems of industrial noise. They will explain the purpose and operation of the new hearing conservation program, a program designed to pinpoint and eliminate noise hazards which might exist in the work environment.

Fig. 29–8.—Second article

NOISE SURVEYS
NOW UNDERWAY
IN ALL DEPARTMENTS

Survey personnel are now fanning out to various departments to measure the loudness of the montage of noises to which XYZ Company employees are exposed during an average work day. This activity signals the beginning of our hearing-conservation program. The first and most important step to the elimination of "noise hazards" in the workplace is the identification of high noise areas.

Surveyors are using sound level meters to record noise levels at all work stations as well as other locations visited by employees during their daily job routine.

Specifically, the surveyors are looking for noise levels which are potentially injurious to hearing. What is a potentially injurious noise level? Or, stated more simply, how loud a noise is too loud? There is no easy answer because the damage done by noise depends not only on intensity (or loudness), but also on the length of time which a person is exposed to it. This basic fact of hearing means that a particular noise level may be relatively safe for brief exposure times, yet hazardous for long-duration, cumulative exposures.

Second article—continued.

The reason can be found deep within the inner ear, inside the tiny hearing organ, called the cochlea. This fluid-filled canal contains thousands of tiny hairs that are connected to sensitive nerve cells. When a sound wave strikes the ear, the hairs inside the cochlea are set in motion, and nerve impulses stream from this organ through the auditory nerve to the brain. The sense of hearing depends on the willingness of these little hairs—and their nerve cells—to respond to sound waves.

Strong sound waves dull the ability of the hairs and nerve cells to respond. A brief barrage of loud noise can trigger temporary "nerve deafness" that will last a few hours or days until the fatigued nerve cells recover; repeated, long-term bombardment by intense sound waves can cause permanent deterioration of the cells.

The resulting hearing loss is irreparable, and cannot be eased by a hearing aid. Hearing damage begins with an increasing inability to hear high frequency sounds; a frequent accompanying symptom is ringing in the ear. Continued exposure to excessive noise over several years may eventually bring on total hearing loss.

The potential harm a particular noise level can do to your hearing depends on the loudness of that noise and the length of time you hear it. Federal lawmakers recognized this and incorporated into legislation time limits for exposure to noises of various intensities.

A person may be subjected to noises of several different levels during a typical work day. Measuring these noise levels and the length of time to which a person is exposed to them is difficult with an ordinary sound level meter. So as part of our hearing-conservation program, we'll be using a small electronic instrument called an audio dosimeter to measure the "noise dose" of employees in work areas thought to contain a noise hazard.

The audio dosimeter looks like an old-fashioned hearing aid, and is designed to be worn for an entire work shift. The dosimeter has a microphone which picks up noise and a small memory cell which keeps track of the daily noise dose. It will tell us at a glance if an employee's daily cumulative exposure to noise exceeds the allowable limit set by law.

If an employee's noise dose is excessive, we'll begin at once to develop engineering and/or administrative controls to deal with the problem. In the interim period or if such controls are not feasible, employees receiving more than the maximum noise dose per day will be fitted with personal hearing-protective devices.

Accurate measurements of noise levels in all work areas are the basis for an effective hearing-conservation program. They are necessary to assure the effectiveness of all other noise control activities, and can mean the difference between success and failure of our hearing-conservation program.

With that in mind, we ask your support and cooperation when survey personnel begin their activities in your area.

Fig. 29–9.—*Third article.*

AUDIOMETRIC TESTING PROGRAM NOW UNDERWAY

It has been several months since XYZ Company initiated its hearing-conservation program. In that time, the company has identified the potential noise hazards in various work stations. It has also determined which employees are receiving excessive "doses" of loud noise during their work shifts and has undertaken steps to reduce length of time to which these workers are exposed to them.

Now it's time to begin gaging the effectiveness of these various activities through a program of periodic audiometric testing. These audiometric tests—or hearing tests—will be conducted on all present and incoming employees to verify that our hearing-conservation program is, in fact, working to prevent hearing damage resulting from loud on-the-job noise.

Here at XYZ, audiometric tests will be given at least annually. However, more frequent testing may be undertaken for employees working in extremely noisy environments or for those workers using hearing-protective devices (ear muffs or ear plugs).

One word about audiometric testing: you can't "fail" a hearing test! The purpose of these hearing tests is to help make sure that your hearing is being protected in the working environment—and, more important, to warn of hearing problems while there is still time to take corrective action.

Just what is an audiometric test and how is it conducted? Basically, an audiometric test is a measure of the lowest-level sound you can hear at each of a series of different frequencies (or pitches). The results of the test are plotted on a graph called an audiogram, which is simply a "picture" of your hearing ability at the time the test was performed.

An audiometric test is painless and takes about 15 minutes from start to finish. Testing here at XYZ will be administered by (*name of company nurse*) under the supervision of (*company physician's name*).

Your audiometric test will be scheduled at least 14 hours after your last exposure to loud on-the-job noise. This procedure will be followed to allow your ears to recover from any excessive noise which might unfairly distort your audiogram. You may be asked to wear ear muffs on the day of your test to further protect your ears from loud noise.

When you arrive for the test, you'll be asked to sit in a soundproof enclosure which looks much like a telephone booth. You'll also be asked to don a pair of earphones, connected to a piece of equipment (called an audiometer) located outside the booth.

The audiometer is nothing more than an electronic instrument which generates sound signals and feeds them into the earphones. (*Name of company nurse*) will introduce sounds of different frequencies and volumes (typically seven in all) into each of your ears and plot your audiogram as you respond to the tones. This audiogram will show how your hearing compares to normal hearing. It will not, however, reveal a pattern of developing hearing loss.

Such losses can only be detected when a number of audiograms are

725

Third article—continued.

made and compared over a length of time. That's why XYZ has initiated an extensive annual audiometric testing program. These tests will help the company make sure that our hearing-conservation program is doing its job—protecting your sense of hearing.

FIG. 29–10.—*Fourth article.*

XYZ EMPLOYEES TO GET HEARING PROTECTORS

One day soon, you may be the proud wearer of a set of "personal hearing-protective devices." This long-winded label is the official name of ear plugs, ear muffs, and similar products designed to protect your hearing while you are on the job. If you are issued a set, feel free to call them anything you like—but be sure to wear them faithfully!

Your new ear plugs or ear muffs are a vital part of our hearing-conservation program. Used properly, they will help safeguard one of your most precious natural resources—your sense of hearing—from the damage that can be done by noise. Here are the details:

Our hearing-conservation program is in full swing. As you know, the program began a while back with an effort to locate "noise hazards." The first steps were to measure the loudness of noise throughout work areas, and to determine which employees may receive an excessive "dose" of loud noise during their work shifts. Now, we are studying the possibility of redesigning excessively noisy equip-

ment or of enclosing noisy equipment with sound absorbing enclosures. Also, we are considering selective re-arrangement of work schedules as a method of reducing exposure time to loud on-the-job noise.

But, engineering and administrative changes take time to implement. And, sometimes, they just aren't practical. This is where personal hearing protectors come into the picture. Ear muffs or ear plugs provide positive—effective—protection against industrial noise. They stand guard in front of your inner ear, to reduce the intensity of incoming sound waves to a safe level. If you do your part and wear them as directed, your personal hearing-protective devices will virtually eliminate any chance of long-term hearing damage caused by excessive noise.

The set of hearing protectors that you receive will be specially selected to cope with the level of excess noise at your work station, as determined by precision electronic noise measurements. If you are issued ear plugs, they will be carefully fitted to your ear canals; if you receive ear muffs, you will be shown how to adjust them for maximum comfort.

We won't kid you—both ear plugs and ear muffs take a bit of getting used to, just like eye glasses or a new pair of shoes do. They might feel odd at first—maybe even a little uncomfortable. After a few days, though, they will seem like old friends. If you don't believe it, ask someone who wears them now. He'll probably tell you that:

• Because hearing protectors reduce the level of all sound waves reaching your ear, sounds other than noise (such as voices) will also come across "muffled." This is perfectly normal, and proves that the devices are doing their job.

• The maintenance of ear plugs or ear muffs is a snap. Just keep them clean (use mild soap and water) and periodically check for wear and damage.

• If your ear muffs feel uncomfortable after you've worn them for a few days, they probably are adjusted wrong. The head piece should hold the ear cups firmly—*but not tightly*—in place over your ears.

• Your plugs should never be "lent" to another person. They have been fitted to *your* ear canals, and may not work properly for someone else.

• Ear plugs are made of a soft, rubbery material that is harmless to your ear. You can wear ear plugs safely all day.

If you are issued a set of ear plugs or ear muffs, your supervisor will insist that you wear them whenever you are exposed to potentially harmful noise. He's been given this responsibility by management to make sure that your hearing is protected while on-the-job. But, your cooperation is the key to success. After all, it's your hearing that's on the line.

---

3. Where it is impossible or not feasible to reduce noise to a comfortable or safe level, individuals can be provided with hearing protective devices.

The most desirable method of controlling a noise problem is to eliminate the noise at the source. This generally means modification of existing equipment and structures or the specification at the design stage of maximum permissible noise levels of new machinery and equipment.

Engineering

The most desirable course of action when starting a hearing-conservation program centers around the concept of applying engineering principles to reduce the noise levels to safe limits. Application of known principles of noise control usually can reduce any noise to any desired degree. However, economic considerations, and/or operational necessities may make the application impractical.

The most desirable method of controlling a noise problem is to eliminate the noise at the source. This generally means (a) modification of existing equipment and structures or (b) the specification at the design stage of maximum permissible noise levels of new machinery and equipment. Without going into the details of either approach (see Chapters 15, 16, 17, and 19), there are a few salient points which should be mentioned. Engineering controls are all procedures other than administrative or personal protection that reduce the sound level either at the source or in the hearing zone of the workers. The following are examples of applying engineering principles to reduce noise levels:

1. Maintenance
   a. Replacement or adjustment of worn and loose or unbalanced parts of machines
   b. Lubrication of machine parts and use of cutting oils
   c. Properly shaped and sharpened cutting tools

2. Substitution of machines
   a. Larger, slower machines for smaller, faster ones
   b. Step dies for single-operation dies
   c. Presses for hammers
   d. Rotating shears for square shears
   e. Hydraulic for mechanical presses
   f. Belt drives for gears

3. Substitution of processes
   a. Compression for impact riveting
   b. Welding for riveting
   c. Hot for cold working
   d. Pressing for rolling or forging

4. The driving force of vibrating surfaces may be reduced by
   a. Reducing the forces
   b. Minimizing rotational speed
   c. Isolating

5. The response of vibrating surfaces may be reduced by
   a. Damping
   b. Additional support
   c. Increasing the stiffness of the material
   d. Increasing the masses of the vibrating members
   e. Changing size to change resonance frequency

6. The radiation from the vibrating surfaces can be reduced by
   a. Reducing the radiating area
   b. Reducing overall size
   c. Perforating surfaces

7. Reduce sound transmission through solids by using
    a. Flexible mountings
    b. Flexible sections in pipe runs
    c. Flexible-shaft couplings
    d. Fabric sections in ducts
    e. Resilient flooring

8. Reducing sound produced by fluid flow
    a. Intake and exhaust mufflers
    b. Fan blades designed to reduce turbulence
    c. Large, low-speed fans for smaller, high-speed fans
    d. Reduce velocity of fluid flow (air)
    e. Increase cross section of streams
    f. Reduce the pressure
    g. Reduce air turbulence

9. Reducing noise by reducing its transmission through air
    a. Use of sound absorptive material on walls and ceiling in working areas
    b. Use of sound absorption along transmission path
    c. Complete enclosure of individual machines
    d. Use of baffles
    e. Confining high noise machines to insulated room

10. Isolating operator by providing a relatively sound-proof booth for the operator or attendant

Some of the noise control measures described here can be made quite inexpensively by plant personnel. Other controls require considerable expense and highly specialized technical knowledge to obtain the required results. It is strongly recommended that the services of competent acoustical engineers be obtained in planning and carrying out engineering noise control programs.

The possibility of excessive plant noise levels should be considered at the planning stage. Vendors supplying machinery and equipment should be advised that noise levels will be considered in the selection process. Suppliers should be asked to provide information on the noise levels of equipment. Noise specifications have been used successfully to obtain quiet equipment. Engineering specifications for new equipment should include a requirement for noise performance. If purchasers of industrial equipment demand quieter machines, designers will be forced to give consideration to this requirement (see Chapter 19).

It is not enough to specify for a particular machine that the sound pressure level at the operators' station shall be 90 dBA or less. If another identical machine is placed nearby, the sound level produced by the two machines can be 93 dBA at the operators' station.

To estimate the effect of a given machine on the total work environment, it is necessary to know the sound power which this machine produces. If there is no operator's work station in the immediate vicinity, the sound power specification may be sufficient; if, however, there is an operator in the near-sound field, more information is generally needed.

Objectionable noise levels, produced as a by-product of manufacturing operations, are found in almost every industry. Practical noise control measures are not easy to develop and few ready-made solutions are available. Unfortunately, a standard technique or procedure that will cover all or even most situations cannot be presented here. The same machine, process, or noise source in two different locations may present two entirely different problems, and be solved in two entirely different ways.

Administrative controls

There are many operations where the exposure of employees to the noise can be controlled administratively, without modifying the noise, by merely changing production schedules or rotating jobs so that exposure times are within safe limits. This includes such actions as transferring employees from a job location with high noise level to a job location with a lower one if this procedure would make the employee's daily noise exposure acceptable.

Administrative controls also refer to scheduling machine operating times so as to reduce the noise exposure. For example, if an operation is performed only one 8 hr day per week and the operator is over exposed for that one day, it might be possible to reduce the operation to one-half day (4 hours, two days per week) and the employee then would not be over exposed.

Employees who are particularly susceptible to noise can be transferred to work in a less noisy area. Transferring employees has its limitations as personnel problems can be caused due to loss of seniority and prestige, and lower productivity and pay.

Administrative controls may also be thought of as any administrative decision that results in lower sound levels or less exposure to the employee. This includes the formulation and implementation of purchase procedures that specify maximum noise exposure levels at the operator's position. Everything else being equal, the quietest machine or piece of equipment should be purchased.

Personal hearing protection

Pending engineering control, employee exposure to noise can be reduced by the mandatory use of hearing-protective devices. Either insert or circumaural devices are satisfactory within their limitations. It should be recognized that there are some ear canals into which a device should not be inserted. The inserts should be the correct size to fit the specific ear canal. They should be issued only by a person who has been properly trained to measure the ear canal and to recognize the contra-indications.

Once management has made the decision that hearing protectors should be worn, the success of such a program depends largely upon the method of initiation and proper indoctrination of supervisory personnel, as well as workers. Supervisory personnel must also wear ear protection in the areas where it is required. It can be extremely difficult to obtain compliance of employees if supervisors do not set an example by wearing their hearing protectors in noisy areas.

Some companies have found it very helpful to meet with the employees or their representatives to thoroughly review the contemplated activities and reach an understanding on the various problems involved. These include plant locations where ear protection will be provided or required, and state and federal safety regulations on noise.

The next step is the indoctrination and education of employees. This can usually be done best in small groups of about 25 people or less. A short explanation concerning the purpose and benefits to be derived through the wearing of hearing protectors is essential before discussing the details of how the devices will be issued, fitted, and replaced. (See Figure 29–11.)

The educational campaign should impress on the employee as well as all levels of supervision two important facts:

1. Excessive exposure to noise can cause hearing loss.

2. Until noise can be reduced to safe levels, wearing of hearing protectors in the form of earplugs or earmuffs is mandatory.

In one company, the medical director, noise control engineer, and safety director formulated an indoctrination program to be given to

(*Text continues on page 734.*)

A Typical Indoctrination Program For Industrial Hearing Conservation

Here's how one company introduces the various aspects and importance of its hearing conservation program. The person in charge of the program meets with new employees and conducts the session as follows:

Introduction

The purpose of this meeting is to familiarize you with our program for the conservation of hearing. We will present briefly the medical, industrial hygiene, and safety aspects of our program and outline the roles each of us play to make the program effective.

Following these brief discussions, we will show a film, which will give you further understanding regarding the important factors involved in noise control. An opportunity will be available during the latter portion of this session for you to ask questions, and for us to discuss any points that may not be clear to you.

This company has been outstanding in its leadership in the development of health and safety programs for its employees. We have accomplished a great deal with many safety devices, and now progress is being made in preventing hearing fatigue — the conservation of hearing as related to industrial noise.

The Ear — And How We Hear

The ear is the organ of hearing and is the most complex of the special sense organs. It serves as a source of communication and protection. It is divided into three main sections:

The Hearing Tests

Hearing losses are found by testing with an electronic apparatus, called an audiometer. This instrument measures the sensitivity of hearing. Regular, routine hearing tests enable us to measure any changes in your hearing ability.

The audiometric test consists of finding the faintest sound that you can hear at six different test frequencies. The point at which you can just hear each test frequency is marked on a card made for this purpoes. These points are connected by a line, and the resultant graph of hearing is called an audiogram.

Ear Protectors

Scientific research shows that prolonged exposure to excessive noise can harm the delicate hearing mechanism. Ear protectors—such as ear plugs or ear muffs—will reduce the noise before it reaches the ear drum.

It is our company policy for all employees in certain noisy areas to wear ear protectors, just as it is to use other safety equipment.

Types of Ear Protectors Available

1) Ear plugs or inserts are designed to fill the ear canal. Some are made of a soft pliable material, which is molded by hand to fit the ear canal. Others are solid, made of a rubber-like material, which is both easy to insert and comfortable to wear.

- External Ear • Middle Ear • Inner Ear

Each section has a particular job to perform. The outer ear collects and directs sound energy into the hearing mechanism. When the sound waves strike the ear drum, it vibrates and sets into motion three small bones, which are located in the middle ear. These bones carry sound as vibrations to the inner ear. There it is changed into electrical impulses and is transmitted to the brain for interpretation.

Types of Hearing Impairments

1) The ear canal may be plugged by wax, a foreign body, or by infection so that sound cannot reach the ear drum.

2) The middle ear may be injured by an explosion, or infection after a cold, or the small bones may become affected so that they cannot carry the vibrations to the inner ear.

3) The inner ear and its sensitive nerve endings may be damaged by diseases, childhood infections, head injuries, certain drugs, and age.

4) Prolonged exposure to excessive noise is now known to cause nerve damage.

In cases of external and middle ear conditions, a physician can often cure them. This is not always true of inner ear damage. Once the nerve endings are damaged, there is no way of restoring them, and they will not respond to treatment—medical or surgical.

Ear Examinations

Ear examinations are conducted by the company physician and his trained medical assistants.

2) Ear muffs are designed to cover the external ear. They are made of plastic ear cups with rubber-like pads to insure a good seal around the ears. They are held to the ears by a band.

Some employees prefer one type of ear protection to another. The important thing is that you wear properly fitted protectors of whichever is most comfortable for you.

The nurse will fit you and show you how to wear your ear protectors properly. Once you are fitted properly, you will always be able to tell whether you are wearing them correctly.

In a quiet room, ear protectors will muffle sounds; in a noisy work area, sounds such as speech and warning signals can be heard with ear protectors worn because the protectors reduce the background noise.

Points To Remember

1) The best ear protector is the one that is properly fitted and worn

2) Good protection depends on a snug fit. A small leak can destroy the effectiveness of the protector.

3) Ear plugs tend to work loose as a result of talking or chewing, and they must be reseated from time to time during the working day. Never put soiled ear plugs into your ears. Wash the plugs at least once a day with soap and water. With proper care, ear plugs will last for several months.

4) Don't try to adjust an ear protector yourself. If, after wearing for a couple of days, your ear protectors are uncomfortable, see the company nurse for adjustment or replacement. Whenever ear protectors become worn, stiff, or lose their shape, they should be replaced.

Fig. 29-11.

supervisors and employees before ear protectors were issued. This educational program lasted about 45 minutes and was given on company time.

The medical director explained how a person hears and the harmful effect of excessive noise exposure on the hearing mechanism.

The noise control engineer discussed problems associated with reducing noise at the source and what the engineers had done and were doing to reduce noise exposure.

The safety engineer emphasized that employees would be required to wear earplugs or muffs for their own protection—just as they are required to wear safety goggles, respirators, and other items of personal protective equipment. Periodic inspections would be made to determine compliance. Enforcement for wearing of ear plugs or muffs would be the responsibility of supervision.

Employees were then given an opportunity to examine various types of ear protectors. A question-and-answer period followed, giving employees an opportunity to ask questions about the hearing-conservation program. At the conclusion, each was given a small pocket folder summarizing the need of ear protection and giving suggestions on wear and care of such devices (see Figure 29–12).

In issuing insert-type hearing-protection devices, the following considerations are important:

• The employees' ear canals must be examined for possible wax buildup. If the eardrum is not visible, then it may be necessary for the physician to remove the wax buildup. (See Figure 29–13.)

• Earplugs must be properly fitted. All ear canals are not the same size. Not only does the canal size differ between different employees but some employees can have different sizes for the left and right ear canals.

If the plug is too large it will exert too much pressure on the ear canal and be very uncomfortable. On the other hand, the earplug must exert sufficient pressure on the ear canal to make and maintain a good seal.

Records of the type and size ear protectors that are issued to each employee should be maintained by the medical or safety department. The employees should be urged to report to their supervisor any damage or lost earplugs without delay. This information could then be given to the medical or safety department and a duplicate pair could then be delivered to the employee's supervisor with a minimum of delay.

LET'S REVIEW THE FACTS

1. It is necessary for employees in certain noisy areas to wear ear protectors.
2. Prolonged exposure to excessive noise can harm the delicate hearing mechanism.
3. Ear protectors such as ear plugs or ear muffs will reduce the noise before it reaches the ear drum.
4. Your job assignment will determine whether you should wear ear plugs (inserts) or muffs (covers).
5. Speech and warning signals can be fully heard with ear protectors in noisy shop areas.

WEAR YOUR EAR PROTECTORS

1. The nurse will fit them and instruct you how to wear them.
2. Wear them for short periods to start and gradually increase the wearing time. After a few days you will be able to wear them all day with minimum discomfort.

Suggested Wearing Time Schedule

	A.M.		P.M.
1st day =	30 minutes	—	1 hour
2nd day =	1 hour	—	1 hour
3rd day =	2 hours	—	2 hours
4th day =	3 hours	—	3 hours
5th day =	all day — all day thereafter		

3. If after five days the ear protectors feel uncomfortable, come in and see the nurse in the Company hospital.
4. Ear protectors should be replaced when they become worn, stiff or lose their shape.
5. If ear protectors are misplaced, a new pair should be obtained without delay.
6. Never put soiled ear plugs into your ears. Wash the ear plugs at least once a day with soap and water.
7. With proper care, ear plugs should last for several months and ear muffs should last for several years.

OTHER POINTS TO REMEMBER

1. The best ear protector is the one that is properly fitted and worn.
2. Good protection depends on a snug fit. A small leak can destroy the effectiveness of the protection.
3. Ear plugs tend to work loose as a result of talking or chewing, and they must be re-seated from time to time during the working day.
4. If ear plugs are kept clean, skin irritations and other reactions should not occur.

YOUR HEARING IS PRICELESS

PROTECT IT

Fig. 29–12.—Here is a sample of a card that one company issues to all of its employees who are required to wear some form of ear protective device. It highlights the care and use of the protectors. The National Safety Council publishes much motivational material.

The employees must be properly instructed in the proper care and use of hearing protection. Hearing protectors do not last forever, as some would like to assume when budgets are considered. Some existing types have been known to shrink one complete size in a relatively short period of time and, what may be more important, become very hard. Policy and procedures for replacing lost or destroyed hearing protection devices should be written down and explained to the employee. A simple handout should be formulated and issued. Areas where ear protectors are required should be clearly indicated by the posting of signs.

During safety meetings, the supervisor should review the need to wear personal hearing protection, as well as the benefits to the employee if he wears the protection provided. The employee should be informed that if he wears his protection properly while on the job, his hearing will be better when he leaves work at the end of the day.

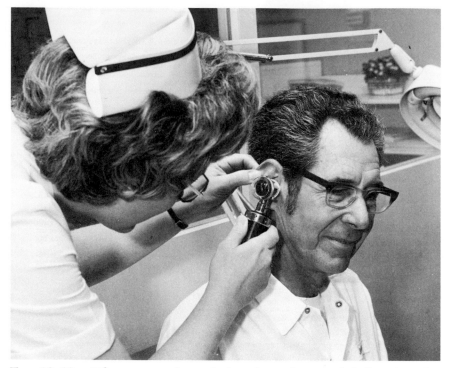

Fig. 29–13.—When an employee is issued earplugs, and before inserting them, he should have his ears examined for possible wax buildup.

Audiometric program

Audiometric testing is an essential part of an effective hearing-conservation program because test results can be used to determine the effectiveness of the noise reduction or engineering control measures.

Whether noise exposure control is by administrative means or by personal protectors, serial (successive) audiograms at appropriate intervals can be used to detect early changes in the employee's hearing threshold before the changes become irreversible. The audiogram is a graphic representation of an individual's hearing acuity. It is a plot of hearing threshold level as a function of frequency. Details are given in Part Five—Industrial Audiometry.

Audiometry has two objectives: (a) to detect changes in the hearing acuity of an individual, and (b) to detect trends in hearing of an exposed group. Conclusions about the general noise environment should not be based upon changes in the hearing of a single individual because of the wide variation in individual susceptibility to noise. Conclusions can, however, be drawn from the average changes, or lack of them, in a group of employees exposed to the same noise environment.

An effective industrial audiometric program should include consideration of the following factors:

- Medical surveillance

- Qualified personnel

- Suitable test environment

- Calibrated equipment

- Adequate records

Medical surveillance

Medical surveillance is essential for a hearing-testing program to serve its dual purpose of detecting hearing loss and providing valid records for compensation claims. Although many smaller companies do not have a medical department, they can, however, satisfy the general requirement of medical surveillance by using part-time medical consultants.

Noise susceptible workers are those employees who suffer handicapping hearing losses more quickly than do their companions under equivalent noise exposures. These workers constitute the group in which claims for compensation are most likely to arise and for whom the risk of hearing damage is likely to be greatest.

Audiometric examinations can be used to detect an individual's susceptibility to noise. By comparing audiograms taken at intervals, it is possible to determine the effect of a noisy working environment on a person's hearing ability. If the hearing acuity continues to decrease, the worker should be transferred to a less noisy job.

The preemployment examination should also provide a detailed history covering the applicant's prior occupational experience and a personal record of illnesses and injuries. For applicants who will work in noisy work environments, the history should include noise exposures in previous jobs as well as in military service. The medical phase of the inquiry should embrace frequency of earache, ear discharge, ear injury, surgery (ear or mastoid), head injury with unconsciousness, ringing in the ears, hearing loss in the immediate family, the use of drugs, and

737

history of allergy and toxic exposures. A standard form can be devised for this purpose.

Qualified personnel

Audiometric tests should be administered by a qualified individual such as a specially trained nurse, an audiologist, or a certified audiometric technician. A certified audiometric technician is an individual who can show evidence of the satisfactory completion of a course of training that meets, as a minimum, the guidelines established by the Intersociety Committee on Audiometric Technician Training or of certification by the Council for Accreditation of Occupational Hearing Conservation. (See Appendix D.)

The audiometric technician's duties are to perform baseline pure tone air-conduction threshold tests. Systematic supervision and encouragement of the industrial audiometric technician by the physician, audiologist, or other qualified person in charge of the audiometric program is recommended to maintain the high motivation required for good audiometric testing. The supervision should include periodic review of the testing procedures used by the audiometric technician to make sure that they conform to established procedures.

Not only is a special technical skill required for conducting industrial hearing tests, it is equally important that the audiometric technician remain alert and observant during the tests. Peculiarities in the audiogram or any unusual behavior on the part of the person being tested should signal the possibility of audiometer malfunction, failure to comprehend instructions, or an uncooperative attitude on the part of the person being tested.

Suitable test environment

According to OSHA standards, hearing measurements must be made in a test room or booth that conforms to the requirements of American National Standard S3.1–1960 (R1971), *Criteria for Background Noise in Audiometer Rooms*. It must be sufficiently quiet within the enclosure so that external noises do not interfere with the employee's perception of the test sounds. This usually requires a special, sound treated enclosure.

Hearing-test rooms should be located away from outside walls, elevators, and locations with heating and plumbing noises. If the background noise levels in the test area do not exceed the sound levels allowed by the standard, the background noise will not affect the hearing test results. The hearing test booth or room may be either a prefabricated unit or one that is built on the premises. Doors, gaskets, and other parts of the room or booth, which may deteriorate, warp, or crack, should be carefuly inspected periodically.

In addition to proper acoustical characteristics, the booth or room should allow for ease of access and egress, and be provided with good, comfortable ventilation and lighting. The audiometric technician should be able to sit outside the room or booth but able to see the interior through a window.

In order to select the proper room, it is necessary to make a noise survey at the proposed test location. Noise levels at each test frequency should be measured and recorded, using an octave-band sound level meter. The audiometric booth selected must have sufficient noise attenuation to reduce the background noise levels present at each test frequency so as not to exceed those listed in the OSHA requirements (see Appendix A)

Calibrated equipment

Audiometers must conform to American National Standard S3.6 (1969), *Specifications for Audiometers,* for limited range, pure tone audiometers. Two basic types of audiometers are available: the automatic recording audiometer and the manually operated audiometer. These were both described in detail earlier in this manual.

The audiometer should be subjected to a biological check, preferably before each day's use of the instrument. The biological check should be made by testing the hearing of a person with a known and stable audiometric curve. The monthly check should include movement and bending of cord, wire, and lead, knob turning, switch actuating, and button pushing to make sure that there are no sounds other than the test tones.

An exhaustive electronic calibration of the audiometric test instrument should be made at least every year by a repair and calibration facility that has the specialized equipment and skilled technical personnel necessary for this work (Figure 29–14). A certificate of the instrument calibration should be kept with the audiometer at all times.

Records

A permanent record (audiogram) should be kept of each audiometric test, and should include the following information:

1. The employee's name and social security number

2. The employee's job location

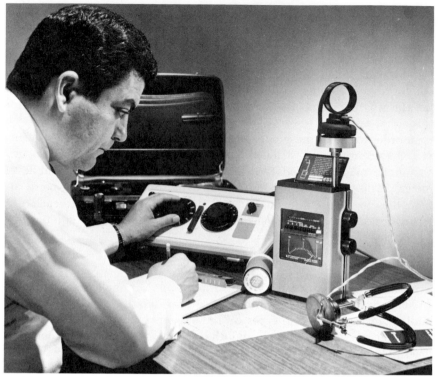

Courtesy General Radio Co.

FIG. 29–14.—A skilled technician and specialized equipment is necessary for the proper calibration of automatic audiometric equipment.

3. The examiner's name and signature

4. The date and time of the test

5. The model, make, and serial number of the audiometer

6. The date of calibration of the audiometer.

These audiometric test records should be kept for at least the duration of employment plus five years. Sometime in the future, they may be the basis for a settlement of a claim. For instance, the preplacement audiogram may show that the employee had a hearing loss at the time of employment with the company, or the separation audiogram may free the company from a claim that an employee sustained a loss while employed with the company, when it is shown that the loss occurred later.

The periodic audiograms are a profile of the employee's hearing acuity. Any change from previous audiograms should be explored. It may be that the employee's hearing protectors are not adequate or properly worn.

Preemployment audiometric examinations, together with the history as outlined previously, should be recorded on the preemployment examination form. Periodic followup examinations should be recorded on a type of form which allows for ease in comparing the results of successive hearing tests.

If the hearing test results are to be valid, it is necessary that the subject have a rested ear. This means the employee should not have been exposed to work place noise levels in excess of 80 dBA for the 14 hours preceding the hearing test. To meet this requirement, the use of hearing-protective devices that reduce the effective noise exposure below this level is considered satisfactory.

Audiometric tests should be made as frequently as specified by the company's regular or consulting physician or audiologist. An effective audiometric testing program should include the following:

• Within 90 days of employment, a baseline audiogram for all new employees who are exposed to sound pressure levels equal to one half or more of the permissible value (that is, an exposure equivalent to 85 dBA for a full 8 hour shift).

• Retesting of the employee yearly if exposed to sound levels at or above 85 dBA.

• Obtaining, if possible, a termination audiogram when the employee leaves the company.

The audiometric testing program should be both practical and feasible. In the case of small companies where the total number of employees to be tested is few in number, it would be impractical to purchase a booth and audiometer. It would be far more economical to consider a mobile audiometric testing service or to refer these few employees to a local, properly equipped and staffed hearing center or to an otologist for an audiometric examination.

Summary

Occupational noise exposure regulations require that protection against the effects of noise exposure shall be provided when the sound levels exceed certain minimums.

As summarized in this chapter an effective hearing-conservation program consists of:

1. Noise measurement and analysis

2. Noise control measures

3. Hearing evaluation

The effectiveness of a hearing-conservation program depends upon the cooperation of employers, supervisors, employees, and others concerned. Management's responsibility in this type of program includes noise measurements, initiation of noise control measures, provision of hearing protection equipment where it is required, and informing employees of the benefits to be derived from a hearing-conservation program.

It is the employee's responsibility to make proper use of the protective equipment provided by management. It is also the employee's responsibility to observe any rules or regulations in the use of equipment to minimize the noise level exposure.

Chapter Thirty

Role of the Nurse in Hearing Conservation

By Donna M. Reichmuth, R.N., C.O.H.N.

The occupational health nurse plays a key role in an industrial hearing-conservation program, because the success of the program depends to a great extent upon her competence, skills, and enthusiasm. In addition to audiometric testing, fitting of hearing protection, and keeping records, the nurse in the smaller company often has significant responsibilities for the implementation, coordination, and continuing administration of the hearing-conservation program.

Physician staffing at smaller companies or plants is frequently limited to a few hours per week, and more often than not, a physician is available only on an on-call basis. The smaller company physician's main concerns are usually trauma treatment and employment examinations. This is not to say that physicians have no interest in prevention and control of occupational hearing loss; they usually welcome and encourage a hearing-conservation program and appreciate the direct involvement of an adequately trained occupational health nurse.

In large companies or plants with full-time medical directors and multiple medical personnel, hearing-conservation programs are usually implemented and administered by the medical director with the approval and support of management. In large industries the nurse's responsibilities are usually limited to audiometry, fitting hearing protection, and recordkeeping.

Planning and implementation of a program

Many occupational health hazards, once they are properly identified and quantified, are rather simple to control. This is not true, however, for hazardous noise. To rush into a hearing-conservation program and attempt to control or eliminate a complex problem such as noise often results in a compromise that will, at best, only control some noise exposures and, at the worst, lead management into the false belief they have the noise exposure under control. Therefore, the time spent in methodically and systematically planning the program will yield the best results. If the program is well planned, it should fit into the daily routine of the medical department without compromising good hearing conservation or the other necessary department activities.

Management approval and support

There is general agreement that occupational safety or medical programs require the approval and total support of top management. All components of the program are essential to achieve its goals of: (a) conservation of employees' hearing, and (b) control of the dollar losses that may be spent paying for job-related hearing loss. The concerned and knowledgeable occupational health nurse has both an opportunity and a responsibility to demonstrate her knowledge and professional interest and must be prepared to discuss with management not only what is required for an effective program, but why these requirements must be met.

Enlightened management needs to know and deserves to know what must be done, why it must be done, what it can expect in the benefits to be derived, and what the cost is expected to be. Without this information presented in a logical, informative manner, management can hardly be expected to actively support and provide the funds required for an effective hearing-conservation program.

Since top management must approve and support the hearing-conservation program, it should establish a policy based upon the advice of medical personnel. The format of a written policy is as varied as the industry for which it is developed. Generally speaking, management should make a statement setting forth the goals of the hearing-conservation program, to prevent loss of hearing among workers exposed to noise, to provide screening audiometry to assist in the early detection of hearing loss from other causes, and to comply with state and federal statutory requirements. (See also Chapters 12 and 29.)

Fig. 30-1.—The hearing conservation program is a team effort and the nurse should communicate freely with the other team members.

It will be necessary to determine what type of audiometer would best meet the company's particular needs, that is, whether manual or automatic—the volume of testing is the usual determinant. Judgment should also be made concerning the type of testing booth. For an average plant, about $3000 may be required to fund the program. This figure would include a manual audiometer, testing booth, record forms, and training costs for one nurse, although this would vary according to the individual company's requirements.

The team approach

Responsibility for the total effort in hearing conservation should be shared among medical personnel, the safety director, the personnel department management and supervisors, and the engineering department. The nurse should be prepared to communicate freely with the other team members. (See Figure 30-1.)

If the company has a safety director, he will probably have direct responsibility for engineering controls, machine modifications, and periodic measurements of noise. The nurse should have copies of the noise measurements in order to estimate the degree of hazard in the

745

placement of certain employees. The frequency of audiometric tests and type of hearing protection needed would therefore depend on the noise exposure measurements.

Because it is difficult for the nurse to repeatedly tour the hazardous noise areas to determine whether employees are wearing their hearing protection properly, the ultimate responsibility for proper use of any protective device has to rest with the first-line supervisors. Because of their management responsibility, supervisors should also inform employees of the dates they are scheduled for audiograms, after receiving these lists from the nurse.

Before implementing a hearing-conservation program, it would be advisable to have a meeting attended by management, medical personnel, supervisors, and others with responsibility in order to explain the total hearing-conservation program, the reason for the program, and the responsibility of each team member.

This initial meeting would also be an excellent opportunity to demonstrate the various types of hearing protection and to demonstrate fitting as well as the appearance of properly worn hearing protection. Audiograms should be obtained for all team members. Those tested should be informed of the results, fitted with protection, and referred for further evaluation where indicated. Members of the group should agree to wear hearing protection whenever they are in a hazardous noise environment, primarily for purposes of setting a good example rather than the probability of incurring hearing loss.

Otologic history and ear inspection

Many occupational health nurses are responsible for completing medical and occupational histories and some portions of the preplacement or preemployment examination prior to the physician's final evaluation. The otologic history should also be included. An interview technique can be used here, with the nurse further explaining those questions the applicant or employee may not readily understand.

The question is frequently raised regarding the proper time to perform an ear inspection or examination and whether it is a proper nursing function.[1] In hearing-conservation programs, an otoscopic inspection is used to note the presence of impacted cerumen, dermatitis, or other inflammation of the external ear; to note the appearance of the eardrum; and to determine if there is gross anatomical deformity in the external ear. These problems can have an effect on the audiogram and make the fitting of insert-type ear protection difficult, if not impossible.

In those situations where it is necessary to obtain an audiogram before the physician's examination of job applicants, the nurse can, after careful instruction and demonstration by the physician, perform a limited otoscopic assessment to note and record the presence or absence of abnormalities. If abnormalities are found then the employee should be referred to a physician so he can fully evaluate the nurse's preliminary observations.

There may be a question concerning the propriety of having the nurse remove impacted cerumen. An audometric test would probably not be valid if there is significant impaction. Consequently, this question should also be resolved by the physician, and if he agrees that the nurse can remove it, then she should be carefully instructed.

Professional supervision

The specific roles of physicians and audiologists in the hearing conservation program are amply covered in Chapters 31 and 32.

If there is a common problem, in smaller companies at least, it is the absence of otologists, otolaryngologists, or audiologists to provide overall professional direction and guidance for the nurse.

The adequately trained nurse/technician may well have the knowledge to review an audiogram and determine whether the configuration is normal or abnormal. But she should not be expected to independently make decisions such as acceptability for employment, or transfer to less hazardous environments. She needs continuing guidance and advice from experts in hearing loss. During the initial planning of the hearing conservation program the nurse, with support from the company physician, should make a serious effort to convince management of the need for continuing expertise. Most company physicians readily understand this and will support it.

Audiograms and scheduling

The success of an effective hearing-conservation program is dependent to a great extent upon the validity and reliability of the audiometric data obtained by the nurse or technician responsible for the hearing threshold measurement of employees.

There are two major aspects to the preplacement physical examination. The first is to establish a medical history for the prospective employee including hereditary diseases, medications, previous work exposures, military service, and hobbies. The second is to place the employee within the organization according to his physical and medical limitations.

In the matter of job placement, one of the most important features is the audiometric test. This baseline hearing test serves as a point of reference for future hearing tests and also determines the employee's initial hearing threshold at the time of employment. The real purpose of a hearing-conservation program is to prevent hearing loss. If a preexisting hearing impairment is discovered, the employee should be referred to a physician so that he can receive adequate diagnosis and medical care, if indicated.

The noise level in the audiometric testing area should be checked and records maintained of both the calibration of the audiometer and the noise measurement of the booth. If the audiometer is sent out for calibration, it should be rechecked upon return. The earphones must be calibrated with the machine. (See Chapter 23 for a discussion of audiometer calibration.)

Preplacement audiometry

The first large group that should have initial audiograms and hearing protection (if necessary) is new employees at the time of the preplacement examination. It is in this group where the company is in the best possible position to make appropriate job assignments in terms of test results and, if necessary, to defend at a later time against alleged claims for hearing loss that may have occurred elsewhere. This test is also the only valid baseline from which changes in hearing thresholds can be evaluated at a later date.

Audiometry for present employees

If the fundamental goal of the program is to conserve hearing among all employees then everyone exposed should be tested at the outset.

Statutes will likely require eventual testing of all workers exposed. However, some industries may decide to defer testing this group until the statute is in effect and the noise level above which testing is required is defined. Others, aware that statutes already require ear protectors for all employees presently considered at risk, are completing baseline audiograms prior to the initial fitting of protectors. Since they know the threshold hearing levels of those who have been exposed for several months or years, they are in a better position to defend against possible claims for further hearing loss, assuming the protection is adequate and used properly.

Workers' compensation statutes vary among the states and this fact has influenced some industries' decisions regarding baseline audiograms on the current work force. (See Chapter 13.)

Because of the many variables it may be difficult for the nurse to advise management on the proper timing to test present employees. Again, if the goal is to conserve hearing among all those at risk, these employees should be tested at the outset.

Periodic audiometry

Unless there is an early plan to schedule periodic audiometric tests on a continuing, predetermined schedule, the nurse will find it necesssary to complete large numbers in a few days or a few weeks to meet these schedules. At other times there will be relatively few tests scheduled. This can create problems in the medical unit and in company production efforts. The same could be said for the baseline audiograms completed on the present work force.

One solution is to determine what hours and days during the nurse's work week would be most available. Hopefully, the tests can be scheduled after the start of the shifts.

Employees are then scheduled on the days selected sometime during the month of their respective birthdays. This should result in a continuous even flow of employees for periodic audiograms from all departments, thereby preventing both significant disruptions in production schedules and in other necessary activities in the medical department.

The supervising physician or audiologist should make the decision regarding appropriate intervals for periodic audiometry. Aside from those required, he will need current noise level readings in all areas of exposure so that adjustments in the test intervals may be made as indicated. Another factor that will influence his decision is the advisability of maintaining close followup of employees whose test results indicate a possibility of increasing hearing thresholds. (See Appendix A-3 and Chapter 13.)

Although there is agreement now that, provided new employees wear maximum hearing protection properly on the day of the scheduled audiogram, tests can be made at any convenient time during working hours, it would probably still be advisable to schedule these early in the work shift. The problem here is to have assurance the employees to be tested are wearing maximum protection properly—the supervisor's responsibility.

The proposed OSHA standard requires a 14 hour period away from noise above 80 dBA and the wearing of ear protection on the day of the test that reduces sound reaching the ear to below 80 dBA.

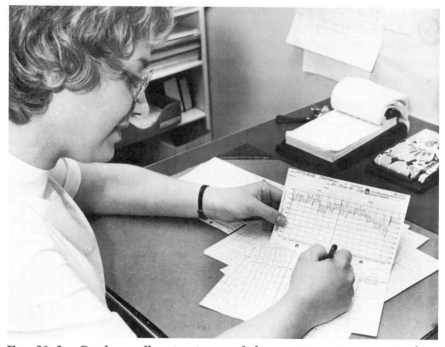

Fig. 30–2.—Good recordkeeping is one of the most important aspects of an audiometric testing program.

Maintaining audiometric skills

From the beginning, management must be aware of the problems that can arise if anyone other than an adequately trained individual is allowed to complete audiograms. It may be practical in a one-nurse firm or plant to have an alternate person trained to complete the tests in the nurse's absence. However, under no circumstances should the nurse train the other person. In fact, the Intersociety Committee, in its *Guide for Training of Industrial Audiometric Technicians,* recommends a specific training program (see Appendix D-2).

Other factors make this inadvisable. Few nurses would have sufficient broad knowledge to teach audiometric technique, and to make a judgment concerning that person's technique. Audiometric tests completed by an individual with less than acceptable skills may serve as the basis for later claims for hearing loss. Trial lawyers are aware that concerned professional societies have established definite minimum requirements for industrial audiometric training.

If it is not possible to have a trained alternate available in the company or plant, then appropriate outside facilities should be sought. However, these outside facilities should meet the same rigid conditions as those necessary in the firm or plant.

It may be necessary to defer preplacement audiograms in the nurse's absence unless an alternate is trained or there is a suitable outside facility. If the only solution is to defer audiograms, new employees assigned to work in a hazardous noise environment must then be informed that permanent employment is contingent on the satisfactory completion of the preplacement audiogram when the nurse returns. They should be required to wear continuous effective protection until that audiogram can be completed.

Hearing-conservation records

Good recordkeeping by the nurse is one of the most important aspects of any medical program (Figure 30-2). In an industrial hearing-conservation program it is vital. A good record includes an otologic history which lists previous noise exposure, job classification, and audiological record.

Audiometric record forms are available from manufacturers of audiometers. These are usually 4 × 5 in. file cards. Some contain limited space for otologic histories as well. With automatic audiometers, there is little opportunity to select an initial form acceptable for the individual company as the test form is made for insertion in the instrument. (See forms in Chapter 8 and Part Five.)

As is the case with other medical documents, all recordings should be clear, concise, accurate, and free of erasures. Extra care is warranted in recording precise hearing thresholds at all frequencies. Only established symbols and ink colors should be used. All audiograms should be signed in ink by the testor and, when personally reviewed by the physician or audiologist, signed by him as well.

If audiograms are kept in the individual employee's health folder, they often create considerable storage problems as they accumulate yearly, and they can be easily lost from the folder. For this reason, it may be more practical to maintain all individual audiograms in a separate, smaller file and to transcribe the test results chronologically on one log-type form for each employee. This form would include the date of the audiogram, the test results at various frequencies for each ear and other

CUMULATIVE AUDIOMETRIC RECORD

NAME _____

SOCIAL SECURITY NO. _____

Date and hours on job today	Job Location	Audiometer – Make, Model, Serial Number	Calibration Reference ASA 51 or ANSI 69	Threshold Levels (dB)														Comments and Technician's Signature
				Right Ear							Left Ear							
				500 Hz	1000 Hz	2000 Hz	3000 Hz	4000 Hz	6000 Hz	8000 Hz	500 Hz	1000 Hz	2000 Hz	3000 Hz	4000 Hz	6000 Hz	8000 Hz	

FIG. 30–3.

pertinent or required information that may influence the interpretation. (See Figure 30–3.)

This type of cumulative record also provides the nurse and the physician with ready information regarding the stability of the employee's hearing level. All test results should be kept during the entire period of employment and ten years after termination in the event there are inquiries or claims filed for hearing loss at a later date. (Check latest OSHA requirements.)

Records of all calibrations of audiometers, whether biologic, electroacoustic, or exhaustive, and all records of special service, repairs or instrument replacement should likewise be retained for ten years. The same practice should apply regarding records or affidavits of background noise measurements and service on the testing booth. Records of periodic noise measurements or noise analyses should be kept indefinitely. In smaller companies where these measurements might be done by an outside agency, it would seem appropriate for the nurse to keep this information as well.

Nurses, by virtue of their initial education and training, have a deep respect for the significance of records. It seems appropriate, therefore, that they assume responsibility for all documents applicable to the hearing-conservation program.

Hearing protection

The continual wearing of hearing protection by employees is considerably dependent on the counseling skills of the nurse and the example and concern of other team members. Experience seems to indicate, however, that complete compliance is rarely achieved unless the use of hearing protection is mandatory and employees understand this work rule. In some companies continued employment is contingent on an employee's compliance.

Employee motivation

For purposes of this discussion, *motivation* denotes the counseling effort directed to the individual employee by the nurse and *education* denotes any program directed to groups of employees.

The best time to fit hearing protection, explain its purpose, and motivate the employee to use it properly is immediately after taking the initial audiogram (see Figure 30–4). This may be the best opportunity the nurse will have to discuss the matter with the particular employee on a one-to-one basis.

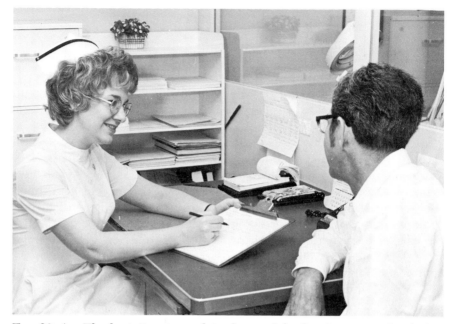

Fig. 30–4.—The best time to explain the need for hearing protection is immediately after the audiogram has been taken.

The nurse has already set the scene for this counseling effort by her explanation and completion of the initial audiogram. She has a captive audience at this point and the hearing test result can be used to motivate the employee to wear adequate hearing protection. If the test indicates a normal hearing level, the emphasis should be on preserving it. If the results indicate hearing losses in the higher frequencies, the nurse should emphasise the need to preserve the remaining hearing ability through diligent and proper use of hearing protection. The same advice would apply for those employees with preexisting hearing loss that is documented by their audiograms.

There is general agreement today that any patient, client, or employee who submits to any medical test or examination, whether completed on his request or at the request of a third party, has a right to know the results.[2,3] Since the employee will be informed of them anyway, it seems reasonable to use those results as a tool in the motivation effort. Further assistance or reinforcement, on an individual basis, should be provided as needed by the company physician or the outside specialist to whom selected employees are referred for diagnostic evaluation.

Fitting the protectors

The nurse should become familiar with the different types of protectors, so that she can easily demonstrate the need for a proper fit and the proper method of cleaning them. It is a good idea for the nurse herself to wear different types for short trial periods in order to become familiar with them and so be able to better understand the problems of those who must wear them on the job.

The employee at the outset should be given a choice of hearing protectors, and after having made a selection, the nurse should fit the protection and at the same time give instructions on its proper maintenance and use. (See Figures 30–5 and 30–6.) The employee should be invited to return to the nurse in one week to report on the success or difficulty in wearing hearing protection, and sooner if necessary.

In most instances, when the wearing of hearing protection is mandatory and a worker consistently refuses to wear the offered protection, the foreman or some other representative of management should take the necessary measures of enforcement. But the nurse's role in this part of the program is extremely important. She should have the knowledge, the ability, and most importantly, the time to talk to an individual worker about his reasons for not wearing his hearing protection. What is the problem? Is there another kind that he would rather try? It is up to the nurse to convince the individual of the necessity of wearing hearing protectors for his welfare.

This aspect of the hearing-conservation program requires extreme patience and understanding from the nurse. It is unrealistic to expect an employee to passively adjust to the continuous wearing of hearing protection unless he understands, in terms meaningful to him, the necessity for hearing protection and the serious problem he may have if he doesn't wear it.[4]

Another useful tool for explaining test results and fitting hearing protection is a simple anatomical drawing of the auditory mechanism. The nurse can briefly explain the anatomy and physiology of the ear and indicate the position of a properly fitted protective device. (See Figure 30–7.)

Some manufacturers of audiometers, hearing protection, and hearing

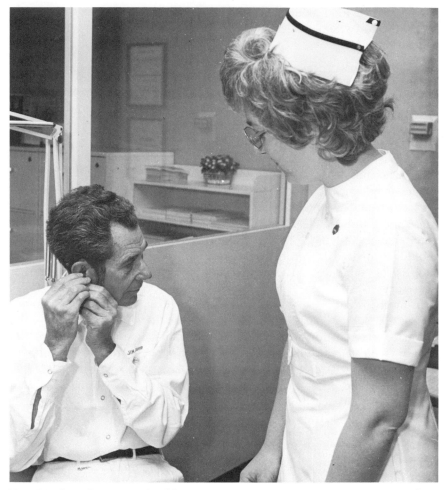

Fig. 30–5.—The nurse should instruct each person on the proper use of an earplug. Unless it is worn correctly, it will not provide adequate attenuation.

aids have excellent drawings; most of them are available for the asking. They, as well as some insurance companies, also have employee educational pamphlets available.

Group education

On a group basis several effective programs can be developed and

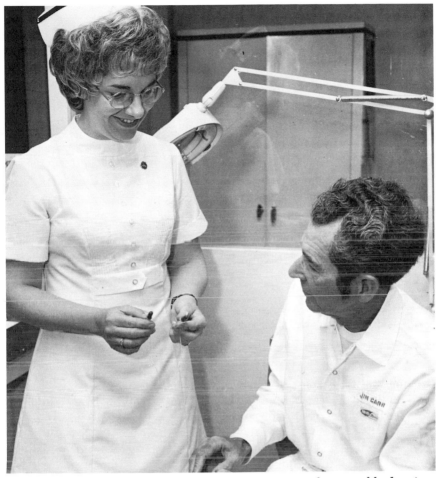

Fɪɢ. 30–6.—Often, if a person is offered a variety of acceptable hearing-protective devices to try, he will be more receptive to wearing the one he believes is most comfortable to him.

these would only be limited by the imagination of the nurse, the time available, and the approval of management.

There are many educational materials available to assist in informing both the employer and the employee of the advantages of an industrial hearing-conservation program. Included are films, posters, leaflets, and recordings, which may be obtained from insurance carriers, manufactur-

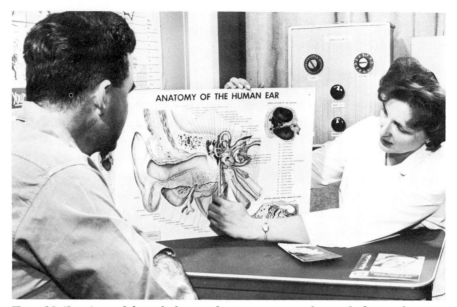

Fɪɢ. 30–7.—A useful tool for explaining test results and fitting hearing protectors is a simple anatomical drawing of the ear.

ers of equipment, and private sources such as the National Safety Council. (See Chapter 33.)

In both large and small industries, an effective program could include running one or more of the excellent films available. (See listing in Chapter 33.) Some film producers also have available large companion posters and pamphlets that can be posted nearby or handed out immediately after film showings to reinforce the film message. This type of group education, however, should never be considered a substitute for the initial one-on-one conference between the nurse and the employee or physician and the employee.

A word of caution is in order. If the decision is made to use films for employee education, the nurse and at least one other team member should screen these films before showing. Some are more properly used for medical personnel, and others are for management supervision. No film or other audiovisual can be expected to answer all questions employees may have about industrial hearing loss and their individual roles in its prevention. The nurse should attend all film showings and be available to answer employees' questions. Where other members of the team are available, including the company physician or audiologist, they can contribute as well.

Motivation and education in any aspect of safety and disease prevention is a never ending effort. This is especially true when dealing with the insidious nature of hearing loss from continued exposure to hazardous noise. Motivation and education programs can succeed when the nurse, the physician, management and supervisors believe in them and demonstrate their belief by their constant efforts.

Summary

The effectiveness of any hearing-conservation program is dependent on the degree of sound planning of that program and the sustained interest of the persons responsible for its administration. In those companies or plants having only one or two nurses and a part time or on-call physician, the effective program will also be dependent on the nurse's extending her function beyond audiometry, hearing protection, and recordkeeping. Unless the nurse accepts and carries out this additional responsibility, with the support of other team members, there will likely be a considerable expenditure of time, money and effort with poor or marginal results. The nurse may well be the only individual in the plant with any degree of familiarity with all the components of an adequate hearing-conservation program.

It will be necessary for the nurse, with the support of the company physician, to inform management of the broad as well as specific requirements of the program. Management can then adopt a statement of policy, provide the funding, and assign team responsibilities to carry out the policy.

It is of extreme importance that the nurse and company physician convince management that it should have an otologist, otolaryngologist, or audiologist actively involved in the program.

No company's or individual's interests are served if the only evidence of a hearing-conservation program is the occasional but inconsistent use of hearing protection and the accumulation of hundreds of audiometric tests with no effort to interpret these tests and place employees in jobs based on the interpretations.

A comprehensive hearing-conservation program should achieve its goal —the conservation of hearing. Other benefits both management and medical personnel have every right to expect are: compliance with statutory requirements, reduction of dollar losses that would likely be spent in compensation for hearing loss, improved employee productivity and morale, and, certainly, renewed respect from management and employees for the nurse's total effort.

References

1. Seedor, M. M. *The Physical Assessment.* New York, N.Y.: Teachers College Press, 1974. pp. 42–47.
2. Coffee v. McDonnell-Douglas Corp. 8 Cal. 3d 551 (1973).
3. Vela v. Wise. Cal. Super. Ct., Contra Costa Co., Docket No. 116083 (Nov. 1, 1973).
4. Maas, Roger. "Personal Hearing Protection: The Occupational Health Nurse's Challenge and Responsibility." *Occupational Health Nursing,* XXVII:25–27 (May 1969).

Chapter Thirty-One
Role of the Audiologist in Hearing Conservation

By Hayes A. Newby, Ph.D.

This chapter deals with the role of the audiologist in an effective industrial hearing-conservation program. One purpose of this chapter is to describe how the audiologist can assist the other professionals, who are concerned with the particular aspects of the hearing-conservation program. The many ways an audiologist can directly contribute to the success of an industrial hearing-conservation program are also pointed out.

Definitions

What is an audiologist? In the broadest sense, an *audiologist* is one who studies hearing—both normal and disordered. In a narrower sense, an audiologist is a professionally qualified individual who is concerned with the assessment of hearing and the rehabilitation of individuals with impairment of auditory function. The American Speech and Hearing Association defines audiologist simply as "an individual who holds the Association's certificate of clinical competence in audiology."

Audiology

The field of audiology is a broad one that can be approached from many aspects and to which varied specialists contribute their knowledge and skills. So far as these specialists make contributions to this field, they are properly designated as audiologists, even though audiology may

761

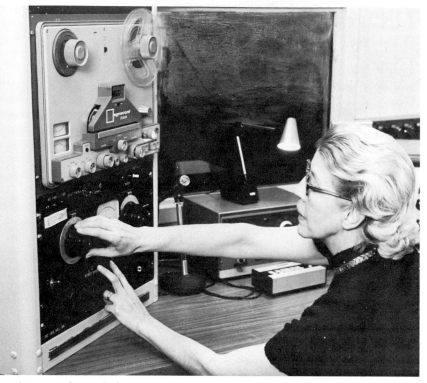

Fig. 31–1.—The audiologist must have knowledge of the phyiscal properties of sound stimuli. The measurement of hearing impairment requires accurate and dependable instrumentation.

be a secondary interest with them. Generally, however, the term "audiologist" is reserved for the individual whose primary interest is in the identification and measurement of hearing loss and the rehabilitation of those with hearing impairments. Usually his training has been academic rather than medical, although there are a few notable exceptions. Most audiologists are products of university graduate training programs in speech and hearing, and many have their Ph.D. degree in audiology.

Speaking figuratively, *audiology* is the offspring of two parents: speech pathology and otology. Speech pathology deals with the evaluation and nonmedical management of individuals who suffer from disorders in oral language. *Otology,* one division of otorhinolaryngology (ear, nose, and throat), is concerned with the diagnosis and medical/surgical treatment of individuals who have an ear disease or disorders of the peripheral mechanism of hearing. For the sake of clarification, remember that the otologist is "ear-oriented," whereas the audiologist is "hearing-oriented."

Speech pathology

Speech pathology is primarily a nonmedical specialty, whereas otology is purely a medical specialty. The two fields—speech pathology and otology—were combined during World War II in the so-called aural rehabilitation centers established by the armed forces to provide assistance to hearing-impaired service personnel.

The care and rehabilitation of hearing impaired people requires the closest teamwork between medical and nonmedical specialists. Some nonmedical persons recruited for work in the aural rehabilitation centers were former teachers of the deaf, but for the most part they were individuals whose training and experience had been in the field of speech pathology and speech science.

For years, speech pathologists had assumed responsibility for working with the speech problems of hard-of-hearing children and adults. Now in the aural rehabilitation centers, they extended their responsibilities to the development of tests of hearing function, selection of hearing aids, and the development of various rehabilitative techniques, which extended far beyond speech correction. Thus, through the cooperative efforts of the two disciplines of speech pathology and otology, a new field of specialization was created— *audiology.*

Physics

Two branches of the field of physics are also related to audiology: acoustics and electronics. Because hearing disorders represent an inability to respond normally to acoustic stimulation, the audiologist must have knowledge of the physical properties of sound stimuli. The measurement of the hearing function requires accurate and dependable instrumentation (see Figure 31–1). The audiologist, regardless of his basic orientation, has had to develop knowledge and understanding of electronics as it is applied to problems of evaluating and rehabilitating those with auditory impairments. The audiologist has to become knowledgeable to some extent, in acoustics and electronics (see Chapters 22, 23, and 25). Audiology, therefore, is a field that springs from many sources and draws upon a variety of skills and background.

Fig. 31–2.—The audiologist, regardless of his basic orientation, has to develop a knowledge and an understanding of electronics as it is applied to the evaluation of human auditory impairment.

Experimental work

The highly proficient audiologist would be a combination of speech pathologist, gerontologist, psychiatrist, psychologist, physicist, electronics engineer, and educator (see Figure 31–2). In addition to the clinical aspects of the field a new branch called experimental audiology is highly important, because from information gained experimentally come improved techniques for clinical application.

Physiology and psychology

In considering experimental work and research, the work of the physiologist must receive recognition. In audiology, the physiologist is concerned with problems of how people hear. On his work are based the principles of medical and surgical care of individuals with hearing impairment and also the principles of preventive medicine, or hearing conservation.

In addition to the physiologist, the experimental psychologist contributes to the field of audiology through research in the psychological processes of hearing and in the area of psychoacoustics. And, of course, all the specialists mentioned, in connection with the field of clinical audiology, are concerned also with research and experimentation in the areas of their principal orientation.

Otological consultation

At some stage in the audiological investigation, an otological consultation is necessary. It cannot be assumed that every case of impaired hearing among employees in a noisy industry is causally related to the work environment. For example, if an employee has a conductive-type hearing impairment, obviously some cause other than continued exposure to noise is responsible. All employees who manifest a conductive hearing loss should have an otological examination to determine whether their hearing can be improved through medical or surgical procedures. All employees with a sensori-neural loss should also be examined in order to rule out such conditions as tumors, Meniere's syndrome, drug ototoxicity, and the array of other causes of inner ear and nerve impairment.

Even when a sensori-neural loss appears to be noise induced, a closer look may show that the employee's hobby may be more responsible for his hearing loss than his work environment. For example, on week ends he may be an avid trapshooter or an electric guitarist in a rock group. Employees who are found to have hearing impairments that are capable of being treated by medical or surgical procedures should be advised to seek treatment in the interest of conserving their hearing.

The primary purpose in discovering hearing impairments that are not related to noise exposure, so far as the employer is concerned, is the employer's protection from future claims that noise exposure on the job has been the primary agent in producing an employee's hearing impairment—when in fact there may have been an entirely different etiology.

Establishing a system through which the hearing of all new employees is routinely tested is certainly desirable, for the reasons cited. It may not, however, be economically feasible for an employer to provide hearing tests for all new employees. However, reference audiograms should be obtained for all new employees who will be working in en-

Fig. 31–3.—An audiologist, by training and experience, is qualified to organize, administer, and monitor industrial audiometric programs.

vironments that have been identified as being potentially hazardous to hearing.

Industrial audiology

Because of interest in patient care and concern with prevention of handicapping hearing impairment, the audiologist can contribute substantially to the success of industrial hearing-conservation programs, either as a full-time employee or as a consultant.

By training and experience, the industrial audiologist is the professional person especially well qualified to organize and supervise audiometric evaluations of employees and to organize and administer the program of monitoring audiometry (see Figure 31–3). The audiologist can specify how and when tests are to be given. He should personally intervene whenever increasing hearing loss is discovered and help decide whether the affected employee should be removed from a work environment that may harm his hearing and that may result in a compensation claim.

Since noise induced hearing impairment has become a compensable disability, the area of hearing conservation has assumed increased economic importance to industry and to the insurance companies that write industrial workers' compensation insurance. The principle of apportionment of liability holds an employer liable for only that portion of a claimant's hearing loss that was incurred during his period of work for that employer, provided the employer can prove that the claimant had a preexisting hearing loss at the time he started his employment. It follows, then, *that for the employer to protect himself against being held liable for all of a claimant's hearing loss, he must perform a preemployment (initial) hearing test on each employee* (see Figure 31–4).

Baseline audiograms

The initial baseline audiogram is the "reference" with which future audiograms are compared to determine what changes, if any, have occurred in an employee's hearing levels during his period of employment (see Chapters 21 through 26). The decision as to where to place a new employee may hinge, to some extent, upon the preplacement audiogram. If the initial audiogram of a new employee shows evidence of noise induced hearing impairment, the audiologist should caution the employer that the employee has already demonstrated a susceptibility or a greater-than-normal sensitivity to noise. The audiologist can often determine whether or not there is a functional or nonorganic component in an employee's hearing impairment. For more details see Chapter 7.

Monitoring audiometry

Employees working in noisy environments should be tested periodically to determine whether or not there has been a change in their hearing level. The periodic retests, referred to as *monitoring audiometry,* should be performed after at least 14 hours off the job or away from noise exposure. Any change in hearing levels from the reference audiogram after 48 hours away from noise exposure should be regarded as a persistent threshold shift.

The detection of a persistent threshold shift should alert management to the need for taking action to prevent further deterioration of an em-

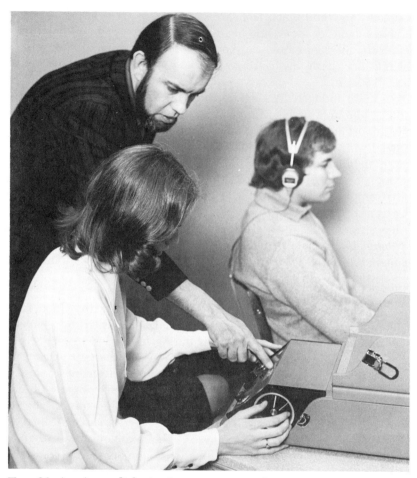

FIG. 31–4.—An audiologist is supervising a hearing test given by an audiometric technician to a new employee.

ployee's hearing. This action may be in the form of insisting that employees use available hearing-protective devices, reducing excessive noise levels at their source through engineering control methods, or limiting the employee's exposure time. In extreme cases, removal of the employee from the noisy environment in which he has been working is an alternate way of protecting employees with a persistent threshold shift.

As in the case of reference audiograms, monitoring audiometry also serves the purpose of discovering changes in hearing levels that may be caused by something other than noise exposure. Whenever monitoring audiometry reveals significant differences from the reference audiogram, there should be an audiological and otological evaluation of the employee whose hearing has been affected to determine, wherever possible, the cause for the change in hearing levels that has occurred.

Valid tests

To be valid, threshold testing requires a controlled sound environment and testing equipment that is properly calibrated (see Chapters 27 and 28). An audiologist can advise on the location of hearing test rooms, determine the acceptability of ambient noise levels within the rooms, choose appropriate audiometric equipment, and keep check on the calibration and operating efficiency of the audiometric equipment.

The accuracy of audiometric measurements depends heavily on the skill of the technician. In most cases, hearing tests will be conducted by industrial nurses or by technicians who may have only a minimal background in audiometry. An audiologist can select and supervise the audiometric technicians and monitor their work to make sure they follow acceptable practices that result in satisfactory intertest reliability.

Human values

The industrial audiologist is deeply committed to employee care. In addition to preventing financial loss to the employer due to compensation claims, the audiologist is also deeply involved in the human values of hearing conservation. He should investigate any hearing impairment discovered in order to determine the nature of the hearing problem. This means interviewing the employee to obtain any relevant history and administering special audiometric tests.

Hearing protection

The audiologist should be concerned that personal hearing protectors are available for all employees who work in noisy environments. It is not enough merely to have hearing protection available; to be effective, hearing protectors must be properly worn (see Chapter 20).

Employers must wage a constant campaign to persuade employees to wear protectors, and audiologists should do everything in their power to reinforce this effort. When monitoring audiometry indicates the beginning of noise induced hearing impairment, the audiologist should emphasize to the worker the importance of protecting his hearing from further damage (see Figure 31–5). Indeed, properly worn hearing protectors should be made a condition of the employee's continuing involvement in a noisy work environment.

Audiologists can help supervisory personnel who are responsible for managing hearing-conservation programs by advising on the adequacy of various protectors and by checking individual employees to make sure that their protectors fit properly, and that the employees are knowledgeable about proper care of protectors.

Expert witness

One important advantage to the employer in having an audiologist on the staff is the availability of an expert witness to testify as to the accuracy and validity of hearing measurements. When hearing loss claims are heard before compensation boards and in courts of law, key factors are (*a*) expert testimony as to the state of the claimant's hearing at the time of his initial employment and (*b*) the interpretation of any special tests administered by the audiologist or under his supervision—for example, tests for functional hearing problems. Because the results of audiometric examinations are potential medicolegal documents, great care should be taken in selecting, training, and supervising the industrial audiometric technicians.

Noise surveys

The industrial audiologist should be capable of conducting noise surveys, although in most situations other personnel—for example, industrial hygienists, acoustical or safety engineers—will be charged with this responsibility (see Chapter 6). Noise measurement is an art that requires not only technical proficiency in handling equipment, but also considerable judgment in determining where measurements should be taken and what kinds of noise analyses are needed.

Noise measurement should be attempted only by individuals who are thoroughly acquainted with the uses and limitations of the sound measuring equipment used and who have adequate background in making noise surveys.

Regardless of who does the noise survey the audiologist should be thoroughly familiar with the results showing the levels of industrial noise that employees are exposed to in various locations. Knowing the physical characteristics (work cycles, intensity, impact, etc.) and exposure time of this noise helps the audiologist make accurate judgments about the degree of hazard and the minimum requirements for a suitable

FIG. 31–5.—An audiologist is shown explaining the function of muff-type hearing protectors to employees.

hearing-conservation program.

Research

The industrial audiologist is involved in continuing research into such matters as adequacy of damage-risk criteria, problems of communication in noisy environments, effects of certain hearing loss patterns on the social adequacy of hearing, and comparative effectiveness of various hearing protective devices. By comparing the exposure times at various noise levels with changes in the audiograms of employees, the audiologist can gather valuable research data.

Audiologists also can promote industrial hearing conservation by making people aware of the hazards of noise and what they can do to reduce noise exposures to acceptable levels. Audiologists can interpret audiograms in terms of the problems that hearing-impaired persons face. The audiologist can act as a teacher and consultant to employers, employee groups, and the general public. He should be prepared to speak and write on such topics as susceptibility of the sense of hearing to damage from noise exposure, social and economic problems of noise induced hearing impairment, principles of noise control, and the value of hearing protectors. He should help prepare audiovisual aids such as movies, video tapes, and slides, illustrating all aspects of industrial hearing conservation.

Hearing centers

It is important that those individuals in management charged with the responsibilities of administering an industrial hearing-conservation program have some knowledge of the services available at hearing centers.

Modern hearing centers offer a variety of services, but basically their purpose is to provide hearing evaluation and rehabilitative services for individuals with hearing impairments. The term *hearing center,* as it appears here, refers to any nonprofit professional agency that offers evaluative and rehabilitative services to individuals with hearing impairments.

The services of a hearing center may generally be divided into two main classifications: evaluative services and rehabilitative or training services.

Evaluative services

Under the evaluative classification are such services as differential audiometry and hearing aid selection.

Differential audiometry refers to the measurement of a person's hearing and an analysis of his hearing problems for the purpose of assisting the otologist toward proper medical diagnosis. Differential audiometry is concerned with establishing a hearing loss as being organic or functional, and in determining the relation between air conduction and bone conduction as an indication of whether the loss is conductive or sensorineural in nature. The determination of the site of the lesion in sensorineural loss is an important part of differential audiometry. In the case of contemplated surgery on the middle ear, it is important for the otologist to have an accurate assessment of the patient's cochlear reserve.

Hearing aid selection. When hearing centers were first developed, their most important service was hearing aid selection. This service is

FIG. 31–6.—An audiologist can provide a professionally objective explanation of the results of an employee's hearing test

still one universally offered by hearing centers, regardless of their type or sponsorship. With so many makes and models of hearing aids commercially available at a wide range of prices, the hard-of-hearing person in need of a hearing aid is bewildered as to what criteria to use in making a choice. He requires professionally objective consideration of his problem, which the average hearing aid dealer is not usually equipped to give him. The hearing center is the place where the employee-patient can obtain the professional objectivity he deserves (see Figure 31–6).

Rehabilitative services

Those employees who cannot be helped by medical or surgical care, or whose hearing loss is of a permanent, nonreversible nature, can be placed in a rehabilitation program. This includes determination of the need for and selection of an individual hearing aid, speech reading (lip reading), auditory training, speech training as needed, and psychological and vocational guidance and counseling.

Counseling is a necessary and important part of the rehabilitation of the hard-of-hearing adult. Not infrequently, counseling must precede all other aspects of rehabilitation. Before an individual can benefit from instruction in speech reading, auditory training, or speech training, he must be motivated to want to help himself. Persuading him to accept a hearing aid, for example, may be a trying ordeal for the employee, the employee's family, and the audiologist. Rehabilitative work with an adult is rewarding, however, for once the employee has accepted the situation and decided that he should make an effort to help himself, he can make rapid progress under instruction.

The hard-of-hearing person who presents the greatest challenge to the audiologist is the one with a typical sensori-neural impairment, who has such good hearing in the low frequencies that he may not be able to utilize a hearing aid successfully, but whose hearing for the higher speech frequencies is so impaired that he has a severe speech-discrimination problem. This is the person who says, "I can hear you, but I don't understand what you are saying." He can usually benefit little from auditory training and will have to rely primarily on speech (lip) reading.

This individual, for whom so little can be done in the way of compensating for his hearing impairment, may become bitter. If so, the audiologist and the otologist are relatively powerless to help him. This individual observes other hard-of-hearing people deriving satisfactory results from hearing aids, and he resents his inability to secure similar assistance. He does not wear a hearing aid, therefore, he bears no outward sign of being handicapped, and people with whom he comes in casual contact do not realize that they must make special efforts when speaking to him.

The audiologist can only hope that such a person will eventually arrive at an acceptance of his disability and a thankfulness that he still has some hearing abilities. The best prognosis for this type of individual is when he admits freely to all with whom he comes in contact that his hearing is faulty. It is necessary for him to develop a "thick skin" and to take the initiative in benefitting himself by telling other people how they can communicate with him most effectively.

There is a tendency for the hard-of-hearing person to become withdrawn and psychologically depressed. The audiologist must make every effort through counseling the employee and, if necessary, through counseling the employee's family, to keep him from withdrawing from social contacts. Somehow he must be brought to the realization that, although

Fig. 31–7.—Some audiologists conduct courses and seminars in industrial audiometry, so that a pool of qualified individuals is available for placement in industrial hearing conservation programs. These courses should be accredited by the Council for Accreditation in Occupational Hearing Conservation.

his hearing loss is an inconvenience, it is not a tragedy.

A hearing impaired employee is the same individual he was before his hearing problem developed, and there is no need for him to abandon his former activities and interests because of the change in his hearing ability. Some adjustments may be necessary in his vocational and avocational pursuits, but they are usually of a minor character and need cause him no great concern. Employees who have major problems of psychological adjustment should be referred for psychiatric or psychological guidance.

Education

Audiologists who teach in university programs should organize courses and seminars in industrial audiology and should recruit students so that a pool of industrial audiologists becomes available for placement in industry (see Figure 31–7). University training programs in the past have paid too little attention to industrial hearing conservation and career possibilities for audiologists in industry.

As new university programs develop, campus placement officers should aggressively seek job openings for graduates. One way to meet the demand for industrial audiologists is for university audiologists (and nonuniversity audiologists, too, for that matter) to make themselves available to industry as consultants.

Consultation

By functioning as a consultant, an audiologist can keep abreast of industrial problems and demonstrate to employers the business value of audiological services. Audiologists serving as consultants to industry must be prepared to cope with a wide variety of situations, but they should not attempt to handle problems that are beyond the range of their expertise. Sometimes the application of common sense and the practice of common courtesy can be an effective means of dealing with the public relations aspects of an industrial noise control problem.

It is the responsibility of the audiologist to inform industry of the benefits to be derived from the utilization of his special abilities and demonstrate how he can function within the organization by proving that he has something of value to offer an employer. What this is depends on the needs of a particular industry and the experience, skills, and capabilities of the individual audiologist. It is difficult to define precisely the role of the audiologist in industrial hearing-conservation programs. The individual audiologist defines his own role, which should be shaped by his personal expertise, experience, and abilities rather than by the label "audiologist."

Environmental aspects

In addition, there is a real need for increased involvement of audiologists in community noise problems. Industrial noise is a specific aspect of the larger issue of environmental noise. Emphasis today is on citizen involvement in protection of the environment. Who are better qualified to speak out on matters of community noise than audiologists with their orientation toward preservation of hearing and prevention of hearing impairment?

It is said that noise levels in cities are increasing at the rate of one decibel each year. Because the progression is logarithmic, this means a doubling of noise levels every three years. Alarmists claim that by the end of the century America will be a nation of the deaf. While this conclusion is certainly debatable, it is clear that the tranquillity of our existence is threatened, if not our hearing acuity. Audiologists have a responsibility to society to work for reversal of this trend toward a noisier civilization.

The public must be made aware that the technology exists for quieting the environment, must be persuaded to demand quieter products, and must be willing to pay for them. Audiologists should contribute their expertise to betterment of the environment, in addition to practicing their professional skills to care for those whose hearing is impaired by whatever cause.

This chapter has been adapted from material presented at the Workshop on Industrial Hearing Conservation, held at Chicago, October 4–8, 1971, and sponsored by the National Association of Hearing and Speech Agencies—and from material appearing in the author's book *Audiology*, with permission of the original publisher, Appleton-Century-Crofts, New York.

Bibliography

Newby, H. A. *Audiology*. 3rd ed., New York: Appleton-Century-Crofts, 1972. (Englewood Cliffs, N.J.: Prentice-Hall, Inc.)

The National Association of Hearing and Speech Agencies. *Proceedings, Workshop on Industrial Hearing Conservation*, 1971. Silver Springs, Md.: National Association of Hearing and Speech Agencies, 1971.

Chapter Thirty-Two

Role of the Physician in Hearing Conservation

By Joseph Sataloff, M.D.

The principal purpose of this chapter is to define in a practical and comprehensive form the role of the physician in an industrial hearing-conservation program. Initiating and maintaining such programs require the active cooperation and participation of individuals in the medical, engineering, managerial, and other professional disciplines. It is also the intention of this chapter to foster a greater awareness on the part of nonmedical personnel of the vital importance of the early recognition of hearing impairment so that the employee may have the best opportunity for treatment or rehabilitation.

The physician's counsel is essential in an industrial hearing-conservation program in shaping proper attitudes, defeating prejudices, clearing up misunderstandings, and gaining employee acceptance of the idea of hearing conservation.

Social and economic factors

Noise induced hearing loss is one of the most challenging problems confronting industrial medicine, because no other physical disability can affect the personality of the patient so adversely.

It may not be possible to restore an employee's hearing to normal or to improve it to a nonhandicapping level, but it is always possible to mitigate the psychological impact. The employee with a hearing problem can be referred to specialists who can teach a person to hear more effectively with the hearing he has left.

Hearing conservation is basically a long-range educational process. Supervisors and foremen must deal with the hearing-conservation problem squarely and forcefully and bring it to the attention of the employees through individual discussions and meetings.

Physicians can be called upon at these meetings to explain the function of the hearing mechanism and the effects of noise exposure in simple terms. The employees should be informed that occupational hearing loss can be prevented if proper hearing protection is worn. It should be emphasized to employees that it is their hearing which can be damaged and that their cooperation in wearing hearing protectors is required.

Numerous factors have combined in recent years to make the prevention of noise induced hearing loss of increasing concern to physicians. The increasing use of higher speed and more complex industrial machinery has vastly increased the number of employees exposed to intense noise levels. As a result the incidence of occupational hearing loss may be expected to rise markedly.

Compensation aspects

Prior to 1948, hearing loss, caused by exposure of employees to industrial noise, was not considered important enough to be included in workers' compensation statutes in the majority of states (see Chapter 13). When compensation for hearing loss was provided, it was included incidentally in the sections pertaining to explosions and injuries. The statute usually specified, as a condition for financial recovery, that the hearing loss must be total in one ear or both ears.

Little or no mention was made in the compensation statutes of partial hearing losses due to prolonged exposure to excessive occupational noise. Perhaps the chief reason for this omission was that lawmakers were not aware of the far-reaching consequences of hearing loss. Since it did not seem to cause loss of wages or loss of earning power directly, legislators felt hearing loss did not warrant being considered as an occupational disease.

Legislation

A series of legal precedents were established in New York in 1948 and later in Wisconsin and Missouri that made partial loss of hearing resulting from exposure to industrial noise compensable though no wages might have been lost. Many other states, as well as the federal government, have now made partial hearing loss due to noise exposure compensable.

Some states have passed laws defining partial hearing loss as a specific disease: *occupational deafness.* In other states where hearing loss is not specifically included under worker's compensation or in special laws, redress for industrial hearing loss still can be obtained by recourse to common law. Large sums of money have been awarded by jury verdicts to workers suing employers in this manner.

The possibility of claims for compensation and the increased incidence of hearing impairment are not the sole factors motivating management's present interest in the problem of industrial noise induced hearing loss. Management and labor, as well as physicians, are genuinely concerned about employee health. The threat of noise induced hearing loss to the well-being of people generally and to skilled industrial workers in particular is receiving increasing recognition. The very serious repercussions that hearing loss may have on the individual and the damage it may inflict on his personality also are becoming more widely understood (see Chapter 9).

Medical aspects of hearing loss

Physicians are being called upon to assume vital roles in advising employers on medical problems arising from excessive noise exposure and to conduct hearing-conservation programs as part of their duty as company physicians. They are being requested to determine the course and degree of hearing impairment occurring in employees—especially in industrial compensation cases.

Audiometry

General practitioners should develop the audiometric skills required to determine the hearing threshold levels of their patients. This is especially essential in communities where there are no otolaryngologists. In larger cities where it is customary for general practitioners to refer patients with a hearing loss to an otolaryngologist and other specialists, it is advisable for the physician to have a clear understanding of audiometry, so that he may understand the significance of the reports sent to him by the specialist.

For the majority of patients who complain of hearing loss, the history, the ear examination, the tuning fork tests, and air- and bone-conduction audiometry provide sufficient information to make a tentative diagnosis. This information tells the physician approximately how much hearing is lost, what frequencies are affected, and even helps to determine whether the loss is conductive or sensori-neural in nature.

Since air-conduction audiometry measures only thresholds of hearing of pure tones, it provides only limited information. More sophisticated equipment is needed to detect certain phenomena such as recruitment, tone decay, tonal distortion, and the patient's ability to localize sound or to understand speech in a noisy environment. To obtain this information, special hearing tests have been devised. These rarely are given by general practitioners but many such tests are performed in an otologist's office or in hospital or university hearing centers. Though industrial physicians may not perform these tests, they should know when they are indicated, and they should be able to interpret the results as an aid in evaluating a patient's complete hearing status.

If a physician is to provide a professional judgment on the cause and degree of hearing impairment, he must have the best scientific information readily available in this highly specialized field. The physician will be required to make a specific diagnosis on the probable cause of the hearing impairment. He must be able to differentiate a hearing loss produced by exposure to noise from hearing losses of other sensory and sensori-neural etiologies (see Figure 32-1, next page).

Effect of prolonged exposure to intense noise

In its early stage, prolonged exposure to intense noise affects the outer hair cells of the organ of Corti and produces a slight dip in the hearing threshold level at about 4000 Hz. Actually, the hearing loss affects a range from 3000 to 4000 Hz, but audiometric measurements do not always cover this frequency spectrum. At this early stage the hearing loss rarely exceeds 40 dB. As the exposure to intense noise continues, the inner hair cells also become affected, and the degree of hearing impairment increases. As long as only the hair cells are damaged, the impairment is classified as sensory. With continued exposure to intense noise, the impairment becomes worse, the supporting cells in the cochlea also may become damaged, and subsequently the nerve fibers themselves become affected. The impairment then falls into the sensori-neural classification. As a matter of fact, most cases of occupational hearing loss that physicians see in their offices are of the sensori-neural type, because patients generally seek medical attention only after their losses have become handicapping. By this time the nerve fibers usually have become involved.

Referral

If a substantial hearing loss is found in an employee who is not ex-

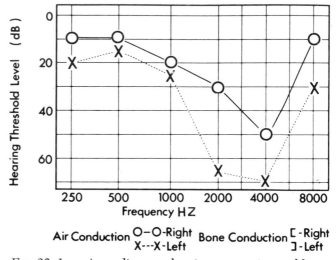

FIG. 32–1a.—An audiogram showing a sensori-neural hearing loss as a result of prolonged exposure to noise.

posed to excessive noise levels at his work location he should be referred (usually at his own expense) to his family physician for a diagnosis and prognosis. People with a conductive hearing loss usually can have their hearing improved by proper medical or surgical care. Employees with impaired hearing should be urged to obtain proper care to have their hearing restored so they can live better lives socially and vocationally. The serious psychological effect which hearing impairment can have upon the personality makes it imperative that management concern itself with the detection and prevention of all types of hearing loss, not only that caused by occupational noise exposure.

Types of hearing loss

Although in many cases it takes a competent otologist to diagnose the cause of a hearing loss, it is advisable that individuals responsible for the management of an industrial hearing-conservation program have a working knowledge of the difference between a conductive- and a sensori-neural-type hearing loss.

Conductive. If the site of damage causing the hearing loss is situated

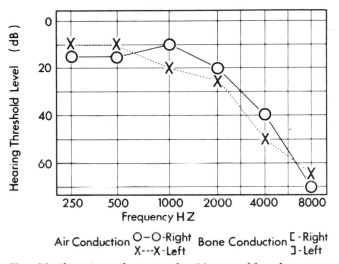

FIG. 32–1b.—An audiogram of a 62 year old male executive showing the typical pattern of sensori-neural deafness, primarily as a result of age—presbycusis.

in the region from the stapes footplate towards the outer ear, the hearing loss is called conductive.

Sensori-neural. If the site of damage is located inward from the stapes footplate, it is called sensori-neural. Conductive hearing loss is generally correctable; sensori-neural deafness is rarely correctable.

Impacted wax or foreign material in the ear canal, perforated eardrums, fluid in the middle ear, dislocation or deterioration of the middle ear bones (ossicles), fixation of the stapes bone, and infection of the outer ear or middle ear, can cause a conductive-type hearing loss.

Long exposure to excessive noise causes sensori-neural not conductive hearing impairment, but noise is by no means the only cause of sensori-neural impairment. Other common causes are congenital conditions, indiscriminate use of drugs (such as dihydrostreptomycin, quinine, neomycin, and kanamycin), severe head trauma, explosions, viral infection, blood disease, vascular spasms, Meniere's disease, aging, and others. Any of these conditions can produce a hearing profile similar to that caused by a long period of exposure to intense noise.

The typical audiometric pattern of a sensori-neural hearing impairment is a high-frequency loss, but the same pattern can be found in a conductive-type hearing loss. For example, by merely placing a few drops of heavy oil in the external ear canal a high-frequency loss will occur that is obviously conductive, not sensori-neural. It is apparent then, that one cannot diagnose a hearing loss on the basis of the shape of the audiogram alone. It is necessary to refer the patient to an otologist for a diagnosis.

One major reason for the need to obtain a diagnosis in certain individuals before they are hired is that, no matter what is determined to be a satisfactory criterion for hiring personnel with a preexisting hearing loss, applicants (especially for noisy jobs) who have a progressive type of nerve deafness should not be hired. The presence of this type of hearing loss should be determined prior to employment; detecting it in an employee long after he is hired obviously is too late. It is not always possible to diagnose every case of progressive nerve deafness, but competent otologists can do so in most instances.

Factors in hearing loss

How does a physician make a diagnosis that a sensori-neural hearing loss is due to prolonged exposure to intense noise and rule out other possible causes?

Presbycusis

There are points of distinction between early occupational hearing loss and presbycusis. In the former, the audiometric pattern displays a dip at 4000 Hz, with better hearing of pure tones in the higher frequencies. Testing at 8000 Hz shows fairly good residual hearing.

In presbycusis, the audiometric pattern shows that the higher frequencies almost invariably are more depressed than the lower ones (See Figure 32–1b). Furthermore, complete and continuous recruitment is present in early cases of occupational hearing loss but not in presbycusis. A diagnosis of occupational hearing loss requires a proven history of many months' exposure to a damaging noise level. A diagnosis of premature presbycusis in patients below 45 years of age should be made with great caution. There are so many more likely causes than the questionable diagnosis of "premature presbycusis."

It is not always possible to distinguish the more advanced cases of noise induced hearing loss, that is, cases of the sensori-neural type, from cases of presbycusis. Evidence of damaging noise exposure must exist. If, in addition, the patient hears pure tones better at 8000 Hz than those at lower frequencies, the principal cause of hearing loss is probably exposure to intense noise rather than aging. The presence of continuous recruitment also favors a diagnosis of noise induced hearing loss, provided that the patient has worked in a very noisy area.

Age

The subject's age is a vital diagnostic criterion, because a marked hearing loss in a young man would not be expected without other corroborating items (such as heredity, ototoxic drugs, and Meniere's disease, for example). Another point of distinction is that noise induced hearing loss does not seem to progress when noise exposure stops, whereas presbycusis generally progresses, though very slowly. Of course, it may take many months or years to obtain this information.

Unilateral deafness

The likelihood that prolonged noise exposure to both ears may be the cause of hearing loss in only one ear, while the hearing in the other ear remains practically normal, conflicts with available facts. Patients with unilateral total or partial hearing loss who work in noisy industries may claim that their problem was produced by many years' exposure to intense noise. Because many physicians are aware of a link between noise and hearing loss, some are likely to assume such a connection. However, all the evidence reported in the literature by competent investigators agrees that a sudden severe hearing loss in only one ear cannot result from prolonged exposure to intense industrial noise that is heard in both ears. Sudden acoustic trauma is a different story entirely.

Even if severe unilateral hearing loss develops over a period of several weeks, the probability of its being caused by prolonged exposure to intense industrial noise is almost negligible. Because both ears are almost equally sensitive in an individual, and industrial noise usually reaches both ears with almost equal intensity, it is unlikely that one ear could be seriously affected and the other not at all.

Individual susceptibility

Another complicating factor in the diagnosis of noise induced hearing loss is the varying degree of individual susceptibility to noise exposure. It is a common experience that two individuals working on the same job and exposed to the same noise level will not have the same degree of hearing damage. Often one employee has little hearing loss, whereas another working at the same job suffers some impairment at the higher frequencies that can reasonably be attributed to his exposure to noise. This subject of individual susceptibility requires much clarification.

Hearing conservation program

Noise standards

The 1970 Occupational Safety and Health Act (see Appendix A) specifies that 90 dBA is the maximum continuous noise level to which any unprotected employee may be exposed for eight hours daily over a period of many years (see Chapter 12, Federal Regulations and Guidelines). The Labor Department is making determined efforts to see that industries comply with these regulations so that noise induced hearing loss can be prevented. The 90 dBA criterion is a realistic level at which to institute a hearing-conservation program. This 90 dBA noise level is not intended for use in medico-legal situations or to be applied to individual cases.

The data upon which the 90 dBA noise standard was based have substantial shortcomings and are not sufficiently accurate to be used for a national standard, though they may be the best available. More extensive and valid data are needed to provide a standard that is accurate and practical.

Those individuals responsible for industrial hearing-conservation programs should not be misled into believing that the chief interest in preventing hearing loss comes solely from regulations promulgated under the Occupational Safety and Health Act. Management should realize that hearing loss, no matter what its cause, is a serious threat to the well-being of all employees. The industrial hearing-conservation program should be established not just to satisfy compliance officers, or only to protect exposed workers, but also to detect and possibly help all workers with hearing impairments.

Modern industrial hearing-conservation programs should be comprehensive, well-planned, properly guided humane efforts to determine the hearing status of all employees and to protect their hearing from being affected by environmental noise or disease.

Noise induced hearing loss

It is fortunate that the first indication of noise induced hearing loss is usually a decrease in hearing acuity for 4000 Hz. This tone is not significantly involved in the understanding of speech (speech discrimination) and hence the loss is not subjectively detectable, but its occurrence on the audiogram gives warning of impending future losses for sounds common to the speech range. Preventive measures should be undertaken before the speech-range frequencies become affected.

No problem in preventive medicine lends itself more readily to solution than noise induced hearing loss. By establishing effective hearing-conservation programs now, it will be possible in the forthcoming years to prevent noise induced handicapping hearing impairment in practically all segments of our working population.

The solution to the problem depends not so much on merely recording good noise measurements and audiometry but rather on effective noise control measures and enlightened use of personal hearing-protective devices wherever noisy machinery cannot be quieted immediately.

Top management should be informed how serious a handicap hearing loss can be and the adverse effect it has upon an employee's life style. A thoughtful, informed executive should quickly realize that management has much at stake in preventing hearing loss among employees. The industrialist has a special responsibility in this area because hearing loss produced by habitual occupational exposure to intense noise is irreversible; it is not curable.

Criteria for program

The initiation of a hearing-conservation program should be considered if excessive noise levels are present in working areas and employees have any of the following complaints.

- Difficulty in communicating by speech in working areas

- Head noises or ringing in the ear due to on-the-job noise exposures

- Complaints of muffling of speech sounds after exposure to on-the-job noise

Absence of pain when working in high noise levels should not be construed to mean absence of sustaining hearing loss. Pain may be felt in the ear during exposure to noise levels of the order of 130 dB. Noise induced hearing loss, however, may be produced at considerably lower noise levels. Pain and annoyance are not reliable indicators of potential noise induced hearing loss. The decision to initiate a hearing-conservation program should not be influenced by the presence or absence of these symptoms.

787

Outline of program

An industrial hearing-conservation program to be considered adequate must consist of at least three parts.

Determination of noise exposure of employees should be made in terms of sound level (in dBA), type of noise, duration, and time distribution or work cycle of noise exposure.

Control of noise exposure where warranted should be performed by engineering or administrative control measures or personal protection. When the noise level at the source cannot be sufficiently reduced, a combination of control methods may be required to conserve the employee's hearing.

Measurement of hearing should include preemployment or preplacement hearing tests and routine periodic follow-up tests. It is desirable that hearing be tested systematically during routine physical examinations. Hearing tests are as important and as necessary as tests of vision, or any of the other tests which accompany a physical examination.

Determination of noise exposure

All work areas should have noise measurements taken by experienced personnel using approved, calibrated equipment (see Chapters 4, 5, and 6). A general rule of thumb is that if the noise is loud enough for employees to have to shout to one another to be heard, or if they complain of the noise being far too loud, in all likelihood a hazardous noise level is present.

The primary purpose in making accurate sound measurements of the various noise levels is to classify broadly which noisy jobs are potentially hazardous to the hearing of exposed personnel.

Control of noise exposure

Areas where personnel are exposed to noise exceeding maximum permissible levels should be delineated as "noise hazardous" and noise control measures should be instituted or ear protectors required (see Chapters 15, 16, and 17). The preferred method of control of harmful noise exposure is through engineering procedures.

Realistically, in many noisy industries, the cost of reducing existing noise levels is prohibitive and the gradual replacement of noisy equipment will take many years. Reducing the number of hours of exposure of an employee to high noise levels (administrative control) is another method and is helpful during the period of machinery changeover.

When an industrial machine produces sound at an intensity level that may be harmful to hearing, the best procedure is to try to reduce the noise at its source by engineering control measures. This is usually accomplished by design engineers who have experience or training in acoustics. (See Part Four—Control of Noise.)

Hearing protection. Where engineering or administrative control measures cannot be used, hearing protectors must be issued to workers in noisy areas. Under ideal conditions, a hearing protector can attenuate airborne sound by as much as 50 dB (see Figure 20–15, page 546). In actual practice, however, the attenuation realized by a protector is generally not more than 30 dB.

Sound may reach the inner ear by air conduction by way of the external ear canal, and when sound intensity exceeds a certain critical level, it may pass directly through the bones of the skull—that is by bone conduction. In normal ears, the sound energy in the midfrequency range that reaches the inner ear by bone conduction alone is about 50 dB weaker than that which reaches it by air conduction. Employees with conductive hearing losses have some degree of built-in protection against damaging noise reaching the inner ear by air conduction.

Education. Personnel responsible for the hearing-conservation program must initiate an educational plan that will motivate employees to wear hearing protection. Filmstrips, posters, and discussions are helpful in convincing employees of the importance of wearing protectors in noisy areas. Employees should be informed about the possible handicaps that may confront those who fail to use the protection provided. Workers must also learn how hearing protectors are to be worn and cared for (see Chapter 20).

Continual reinforcement of hearing-conservation programs is vital to their long-term success. The direction of the program must set a high level, especially on a humane and medical basis, to provide the program with a proper foundation. The OSHAct in many ways makes it easier for industry to motivate employees to use hearing protectors where needed.

Measurement of hearing

An essential phase of any hearing-conservation program is the measurement of the hearing thresholds of employees, both before employment and routinely thereafter. This is particularly true from the medicolegal aspect, since measurements of hearing levels taken years after the employee's exposure to noise has occurred are of little value as a defense against a claim for occupational hearing loss.

The hearing-test program, generally, is the responsibility of the industrial physician, though he himself may not do the testing. That duty may be assigned to the plant nurse or to any well-trained person who has taken a comprehensive course in industrial audiometry, such as that recommended by the Council for Accreditation in Occupational Hearing Conservation (CAOHC).

Because industrial audiometry appears to be disarmingly simple, many audiograms taken in the past have been found to be inaccurate. For this reason a group of representatives from professional organizations concerned with hearing conservation formed an accreditation council and established guidelines for the certification of personnel to perform industrial audiometry.

The accreditation council specifies the course content and personnel needed to train technicians to obtain accurate audiograms in industrial hearing-conservation programs. The accreditation council guidelines emphasize the use of calibrated equipment in an adequately quiet room using an approved technique and keeping good records. Subsequent to certification, the accreditation council also recommends that followup refresher courses be taken by the technician to help maintain valid and reliable audiometric technique.

The technician should make certain that the employee's audiogram has been evaluated by a physician or an industrially oriented otologist. The need for this type of otologic evaluation is especially obvious if the data are to be used for research in the establishment of damage-risk criteria.

In self-recording or automatic audiometry, the technician should monitor the test as it is in progress. In some cases, depending on the response of the subject, the constant presence of the technician may be necessary.

Assuming proper calibration of test equipment, responses on repeat testing should agree within 5 to 10 dB of the initial test. Failure to do so would indicate the need for further and more comprehensive evaluation by qualified otological and audiological personnel.

Dubious appearing audiometric responses can generally be immediately recognized by an alert and experienced audiometric technician. Persons giving such responses should be scheduled for repeat testing. If the first audiogram was by the self-recording technique, the second should be performed by manual technique.

It may be better to postpone a hearing test than to perform audiometry inaccurately. Hearing tests, to be useful in medicolegal situations, must be carried out by expertly trained technicians interested in their work, using calibrated equipment in a suitable test room, and using approved techniques.

Supervision. The overall hearing testing program should be supervised by a physician or an audiologist thoroughly familiar with the special problems encountered in performing hearing tests in the occupational environment. Small industrial plants, in which only a few persons need to be tested, may find it advantageous to make an agreement with a local consultant to assume general supervision of the audiometric program and to work with management in interpreting audiograms and making individual recommendations. The management of a large industrial corporation will find it worthwhile to train a responsible person to undertake the hearing tests and to provide approved equipment and satisfactory test room facilities.

Preexisting hearing loss

An employee's preemployment audiogram establishes his baseline hearing status before he is exposed to noise in his new job. It has been estimated that about 20 percent of the applicants for noisy industrial jobs already have some hearing loss at the time they apply for employment. The importance of securing an accurate initial hearing threshold record for all new employees is obvious.

Instances are certain to arise in which so many new employees are hired in one day that it is impossible to test them all before they report for work. For this and other reasons it may be advisable to perform preplacement audiometry in addition to preemployment testing. This means that initial hearing tests are performed on employees when they are assigned to a job that has been classified as potentially hazardous to hearing.

The importance of the initial preemployment or preplacement audiogram as a document of legal importance should be considered carefully.

Applicants with a hearing handicap

One of the most important decisions that a physician may be called on to make in examining a prospective worker is whether or not to recommend a worker for employment if he has a preexisting hearing handicap. In general, it is inadvisable to adopt a blanket policy not to hire individuals with mild, high-tone hearing losses. To do so would deprive industry of many skilled workers and create an unwarranted labor scarcity.

Since there are many jobs which can be adequately performed by an individual who is totally deaf, there is not, and should not be, any generally accepted standard for hearing below which an individual is not employable.

An important factor in this decision will be the safety of the applicant and of the people with whom he will work. For example, for a job that requires operating or working near dangerous machinery, it is essential that an applicant have good hearing.

Another important factor in the decision will be the applicant's ability to communicate. It is reasonable to require that an applicant be capable of easily perceiving and comprehending the instructions necessary in his particular work. It is also reasonable to expect that he should be able to communicate easily with his supervisors and associates when necessary. Communication may be written, or in other forms, when it cannot be oral.

With the increasing trend to make loss of hearing from noise compensable, many employers are reluctant to hire individuals who have already sustained some hearing loss, because of the possibility of subsequent claims. Even if no valid claim appears probable, some employers would prefer to avoid the potential administrative burden of processing and adjudicating any claims.

The medical benefits provided by the employer, such as paid sick leave or medical expense insurance, may also be a factor in the decision. When such benefits are provided, it may be necessary to determine the origin of the hearing loss. The potential effect on employee benefit costs may vary greatly, from being negligible for an individual with stable congenital deafness, to being enormous for an individual whose hearing loss is a symptom of a brain tumor or other serious underlying disease.

Employees with preexisting sensori-neural impairment have not been shown to be more sensitive to the effects of noise than employees with normal hearing. Industry should feel free to hire such employees if the following concepts are kept in mind:

• The company should be able to assure itself and the applicant that, if he uses adequate hearing protection provided by the company, his hearing will not sustain any substantial additional damage due to the environmental noise exposure at his job.

• Preexisting hearing loss should be clearly defined in worker's compensation legislation (state and federal) to protect the employer.

• The applicant's hearing level should be sufficient to meet the demands of his job. The employer should not create a safety hazard as a result of impaired hearing, or as a result of excessive attenuation of his hearing acuity when wearing hearing protectors.

The physician should give careful consideration to the following questions:

1. Will the applicant's hearing be further damaged by on-the-job exposure? If the indications are positive, then the applicant should not be hired for that job unless proper hearing protection can be provided and adequate supervision maintained to make sure that the hearing protectors are properly worn.

2. Is the applicant's hearing loss now at the point at which a small degree of further loss will place him in the handicapped classification and make him a compensation problem? If the answer is "yes," the individual should be advised to seek employment in a low-noise industry and to conserve the hearing ability he has left.

3. Is the prospective employee so highly skilled that he is essential in the job under consideration, and is the risk of further hearing damage unavoidable because of his vocation? In such a case, industry can take a calculated risk. It would be only good sense to hire him, but he should be provided with the very best possible hearing protection for his work and adequately supervised and monitored to make sure that he wears it correctly.

There is no doubt that industry must continue to hire applicants with some degree of preexisting hearing loss for noisy jobs, but every possible effort must be made to protect employees from further unprotected on-the-job exposure to high-noise levels.

Preemployment hearing tests also serve to protect the employer from unwarranted claims. If the employer does not have a valid and reliable baseline hearing threshold measurement documenting the preexisting hearing loss, the employer is likely to find himself on the losing side of the litigation process.

Monitoring audiometry

Annual retests are usually performed on all workers in critical noise areas. It is important to determine meaningful changes in hearing thresholds and to seek out the cause. This is an excellent way to determine if hearing protectors are being worn properly, for there should be no changes in hearing due to noise exposure if protectors are effectively worn.

Employees not in critical noise areas should be retested every two years. A natural question to raise here is, "Why should we retest these employees?" The obvious answer is that the philosophy of the program should be management's concern for all personnel. Hearing can change for reasons other than noise, and these changes should be identified. In any case, to single out only those employees exposed to noise is certain to bring accusations of employer self-interest and general uncooperativeness in the hearing-conservation program.

An audiogram taken at termination of employment is a self-protection feature for management. Some states portion out liability for compensation payments to "previous employers," so it behooves these previous employers to know what hearing capability the worker had when he left.

Summary

It is the industrial physician's responsibility to interpret and evaluate the various medical data relating to each employee. The physician should determine suitability of a new job applicant for employment in a noisy environment, the presence of a preexisting hearing loss or a pathological disorder of the ear.

Proper job placement of workers, matching their physical capacities to the physical demands and exposure of jobs, is an established function of the industrial physician. Exposure of workers to industrial noise involves application of the same principles. The employee with a known hypersensitivity to noise should not be further exposed.

The individual employee with a partial hearing loss due to excessive noise exposure should be protected against further hearing loss by proper job assignment, by the use of adequate engineering controls, by the use of personal protective devices or other feasible means.

If the cause of the employee's hearing loss is nonoccupational, the worker should be urged to see his personal physician for prompt medical care. If the noise exposure at the employee's work station is determined to be a causative factor, every possible effort should be made to prevent any further deterioration of this employee's hearing.

Industry must be encouraged to provide places of employment that are not hazardous in terms of excessive noise levels, but it must be borne in mind that some high-tone hearing losses occur in the general population that is not exposed to noise, and that some high-tone hearing loss must be considered to be part of the unavoidable risk of living in an industrialized society.

Industry has come to appreciate that hearing loss, no matter what its cause, is a serious threat to the well-being of all the population. A hearing-conservation program should be established not just to satisfy compliance officers or to protect the exposed workers, but to detect and possibly help all workers with hearing impairments.

Bibliography

American Academy of Ophthalmology and Otolaryngology, *Guide for Conservation of Hearing in Noise,* Rev. Rochester, Minn.: American Academy of Ophthalmology and Otolaryngology, 1969.

Gosztonyi, R. E.; Vassallo, L. A.; and Sataloff, J. "Audiometric Reliability in Industry," *Archives of Environmental Health,* Vol. 18, June 1969.

Sataloff, J. and Vassallo, L. A. "Hearing Conservation," *Industrial Medicine,* Vol. 42, No. 2, p. 23.

Sataloff, J.; Vassallo, L. A.; and Menduke, H. "Hearing Loss From Exposure to Interrupted Noise," *Archives of Environmental Health,* Vol. 18, June 1969.

Sataloff, J.; Vassallo, L. A.; Valloti, J. M.; and Menduke, H. "Long-Term Study Relating Temporary and Permanent Hearing Loss," *Archives of Environmental Health,* Vol. 13, Nov. 1966.

Sataloff, J.; Vassallo, L. A.; and Menduke, H. "Occupational Hearing Loss and High Frequency Thresholds," *Archives of Environmental Health,* Vol. 14, June 1967.

Sataloff, J. *Hearing Loss.* Philadelphia, Pa.: J. B. Lippincott Company, 1966.

796

Chapter Thirty-Three — Sources of Information

Julian B. Ollshifski, P. E., and Robert Pedroza

The safety professional is frequently confronted with the need for highly specialized information on noise control and audiometry. Knowledge of the work, publications, and services available from trade associations, public agencies, and service organizations would be very helpful. Because there are numerous groups and their charters of responsibility are varied, this chapter classifies the various sources and defines their functions.

Consultants

Professional consultants and privately owned or endowed laboratories are available on a fee basis for concentrated studies of a specific noise problem or for a plant-wide or company-wide survey, which may be undertaken to identify and catalog individual environmental noise exposures. This chapter does not list consultants; for names, contact the professional society in the specific field of interest.

Selection of correct and adequate sound measurement, audiometric, and acoustical equipment and materials as well as hearing protection devices should be made in cooperation with the recommendations of a qualified industrial hygiene engineer, audiologist, otologist, or acoustical consultant. On first thought, this may seem like an unnecessary expense that can be trimmed quickly from the budget; however, qualified advice from professionals in the field can prove to be a financially sound investment.

Some consultants have more experience in the engineering aspects,

797

while others are more competent in hearing testing. The acoustical engineer can find the areas of excessive and unwanted sound and recommend ways either to reduce it or eliminate it. As discussed in previous chapters, this can be done by substitution of machinery, changing the path of the sound wave and reducing the exposure time.

A consultant in hearing testing and protection specifies audiometric testing equipment and procedures and supervises company personnel who administer the tests to the employees. The consultant can evaluate audiometric tests, aid in the selection of hearing-protective devices, and determine, with the advice of engineers, the proper area for testing hearing.

A qualified professional is needed to check the audiogram of all employees and advise on followup procedures or hearing protection for those employees whose job requires it.

The ideal hearing-conservation program is set up with the recommendations of both types of consultants. It should be noted that not all companies require the same degree of consultation.

Motivation and training

Convincing the employee of the importance of wearing proper hearing-protective devices and procedures for noise control can sometimes prove difficult. Included in this chapter is a selection of films, slide shows, and printed material that can be used in administering a hearing-conservation program.

Some of the material includes basic information on how the human ear operates in both safe and unsafe situations. Employees who are well informed about the hazards of excessive or damaging noise will usually be more willing to seek adequate protection.

Organizations

Many associations are concerned with industrial noise problems. For example, many of the industrial sections of the National Safety Council have an occupational health or industrial hygiene subcommittee that can be called upon for assistance.

The safety professional from the insurance company should not be overlooked as being a good source of information on noise control and audiometry. His work requires him to visit many plants, and his experience and acquaintance with other safety professionals can be extremely helpful.

Scientific and technical societies that can help with noise problems or hearing conservation are listed in this section. Some are prepared to provide consultation service to nonmembers; they should have a wealth of available technical information.

Another group of associations come under the broad heading of "trade associations." They are concerned with furthering the aims of their field of productive enterprise, including health preservation of employees and the public. These associations have trained personnel, cooperating committees, and publications that can be extremely helpful.

Service organizations

American Insurance Association (AIA)
Engineering and Safety Service
85 John Street, New York, N.Y. 10038

The American Insurance Association is an advisory organization serving a large number of companies in the property-liability insurance field. It is a multi-line organization designed to help its members' and subscribers' companies meet many of the diverse problems confronting the insurance business today. Embodied in the Association are the traditions, experience, and accomplishments of the National Board of Fire Underwriters, the Association of Casualty and Surety Companies, and the former American Insurance Association.

The Association's Engineering and Safety Service develops and publishes recommended safety programs and procedures, codes and ordinances for buildings, and the control of fire hazards. It prepares and issues bulletins to subscribing insurance companies and to public officials on matters of special interest in all manner of accident and fire prevention. It studies special hazards in specific industries and recommends safety measures.

American Mutual Insurance Alliance
20 North Wacker Dr., Chicago, Ill. 60606

The American Mutual Insurance Alliance is a national organization of leading mutual property-casualty insurance companies. Through its Accident and Fire Prevention Department, the Alliance makes a concerted effort to reduce accidents, fires, and other loss-producing incidents. Under the guidance of its Loss Control Advisory Committee, sound safety engineering, industrial hygiene, fire protection engineering, and other loss prevention principles are promoted. Major activities include the dissemination of information on safety subjects, conduct of specialized training courses for member company personnel, sponsorship of research, cooperation in the development of safety standards, development of visual aids, and the publication of technical and promotional safety literature. A catalog of safety materials is available without charge.

Industrial Health Foundation, Inc.
5231 Centre Ave., Pittsburgh, Pa. 15232

The foundation, a nonprofit research association of industries, advocates industrial health programs, improved working conditions, and bettered human relations.

The foundation maintains a staff of physicians, chemists, engineers, biochemists, and medical technicians at Mellon Institute, the foundation headquarters.

Activities fall into three major categories:

• To give direct professional assistance to member companies in the study of industrial health hazards and their control.

• To assist companies in the development of health programs as an essential part of industrial organization.

• To contribute to the technical advancement of industrial medicine and hygiene through investigation, research, and allied activities.

The foundation holds an annual meeting of members, conferences of member company specialists, and special conferences on problems common in a particular industry.

The following publications are issued:

Industrial Hygiene Digest, monthly.
Transactions of annual meetings.
Special bulletins prepared by advisory committees.
Pamphlets on current interest subjects.

National Safety Council (NSC)
425 North Michigan Ave., Chicago, Ill. 60611

The National Safety Council is the largest organization in the world devoting its entire effort to prevention of accidents. It is nonprofit and nonpolitical. Its staff members work as a team with more than 2000 volunteer officers, directors, and members of various conferences and committees to develop and maintain accident prevention material and programs in specific areas of safety. These areas include industrial, traffic, home, recreational, and public.

Recognizing that industry's safety problems often require specialized treatment, the Council has divided its industrial effort into sections; each is administered by its own executive committee, nominated and elected from the membership within that industry.

The Council sponsors the National Safety Congress in Chicago in the fall of each year. In terms of program, it is one of the largest conventions held anywhere. Proceedings of each Congress are published in *Transactions of the National Safety Congress*.

Professional societies

Academy of Rehabilitative Audiology (ARA)
Rochester School for the Deaf
1545 St. Paul St., Rochester, N.Y. 14621

Individuals who hold graduate degrees in audiology, language pathology, education of the deaf or allied fields, and who have at least five years' experience in rehabilitative or educational programs for persons with hearing impairment. Membership is by invitation. Provides forum for exchange of ideas in audiology; fosters professional education, research, and interest in programs for hearing handicapped persons.

Acoustical Society of America (ASA)
335 East 45th St., New York, N.Y. 10017

Members are physicists, engineers covering fields of electroacoustics, ultrasonics, architectural acoustics, physiological and psychological acoustics, music, noise and vibration control. Committees: Architectural Acoustics; Engineering Acoustics; Musical Acoustics; Noise; Physical Acoustics; Psychological and Physiological Acoustics; Shock and Vibration; Speech Communication; Underwater Acoustics.

The following publication is issued monthly: *Journal of the Acoustical Society of America*.

American Academy of Occupational Medicine (AAOM)
P.O. Box 27003, Richmond, Va. 23261

Members are physicians who devote full time to some phase of occupational medicine. Promotes maintenance and improvement of the health of industrial workers. Publishes the *Archives of Environmental Health* (monthly).

American Academy of Ophthalmology and Otolaryngology (AAOO)
15 Second St. S.W., Rochester, Minn. 55901

Professional society of medical doctors specializing in ophthalmology, otolaryngology, plastic surgery, and bronchoesophagology (diseases of the eye, ear, nose and throat). Committees: Hearing and Equilibrium; Insurance; Laryngeal and Voice Physiology; Otolaryngic Pathology; Prosthetic Devices.

American Association of Industrial Nurses (AAIN)
79 Madison Ave., New York, N.Y. 10016

Registered professional nurses employed by business and industrial firms, nurse educators, nurse editors, nurse writers and others interested in industrial nursing. Publishes: *Occupational Health Nursing*, monthly.

American Conference of Governmental Industrial Hygienists (ACGIH)
P.O. Box 1937, Cincinnati, Ohio 45201

A professional society of persons employed by official governmental units responsible for full-time programs of industrial hygiene, educators and others conducting research in industrial hygiene. Devoted to the development of administrative and technical aspects of worker health protection. Functions mainly as a medium for the exchange of ideas and the promotion of standards and techniques in industrial health. Publications: *Transactions of Annual Meeting* (annual); also publishes manuals, guides, studies, etc.

American Council of Otolaryngology (ACO)
1100 17th St. N.W., Washington, D.C. 20036

Individual physicians practicing otolaryngology. National, state, regional and local societies of otolaryngology. Purposes are: "to represent through a national organization the specialty of otolaryngology; to create a data base which will describe the professional activities, the training programs, the health care programs, the geographic distribution and the needs in manpower both professional and paramedical; to act as a clearinghouse for the specialty."

American Industrial Hygiene Association (AIHA)
66 South Miller Rd., Akron, Ohio 44313

Professional society of industrial hygienists. To promote the study and control of environmental factors affecting the health and well-being of industrial workers. Committees: Aerosol Technology; Aerospace; Air Pollution; Analytical Chemistry; Engineering; Hygienic Guides; Noise; Radiation; Respiratory Protective Equipment; Technical Publications; Toxicology. Publications: *The American Industrial Hygiene Association Journal*, monthly; also publishes several guides and manuals on problems related to industrial hygiene.

American Institute of Mining, Metallurgical, and Petroleum Engineers
345 East 47th St., New York, N.Y. 10017

The AIME promotes the arts and sciences connected with the economic production of useful minerals and metals and the welfare of those employed in this work.

Publications are *Mining Engineering, Journal of Metals,* and *Journal of Petroleum Technology,* all issued monthly; and *Transactions of the Society of Mining Engineers* (quarterly), *Transactions of the Metallurgical Society* (bimonthly), and *Society of Petroleum Engineers Journal* (bimonthly).

American Medical Association (AMA)
535 North Dearborn St., Chicago, Ill. 60610

The American Medical Association is a professional society for physicians. Its Council on Occupational Health is concerned with the protection and improvement of the health of the nation's working population through promotion of occupational health programs.

American Occupational Medical Association (AOMA)
150 North Wacker Dr., Chicago, Ill. 60606

Professional society of medical directors and plant physicians specializing in industrial medicine and surgery. Sponsors an Occupational Health Institute to advance education in occupational medicine and industrial health. Conducts program of evaluation, approval and certification of medical services in industry. Publishes *Journal of Occupational Medicine,* monthly.

American Society of Heating, Refrigerating and Airconditioning Engineers (ASHRAE)
345 East 47th St. New York, N.Y. 10017

Professional society of heating, ventilating, refrigeration, and air

803

conditioning engineers. Presents awards. Carries out a number of research programs in cooperation with universities and research laboratories on such subjects as human and animal environmental studies, effects of air conditioning, quality of inside air, heat transfer, flow processes, cooling processes. Publications: (1) *Journal*, monthly; (2) *Handbook*, annually; (3) *Transactions*, annually; (4) research reports, codes and engineering standards, preprints, and bulletins.

American Society of Mechanical Engineers (ASME)
345 East 47th St., New York, N.Y. 10017

Professional society of mechanical engineers. Membership also includes student members in 203 student sections. Conducts research; develops boiler and pressure vessel, and power test codes; serves as sponsor for the American National Standards Institute in developing safety codes and standards for equipment.

Publications: (1) *Applied Mechanics Reviews*, monthly; (2) *Mechanical Engineering*, monthly; (3) *Transactions* (*Journals of: Power; Industry; Heat Transfer; Applied Mechanics; Lubrication Technology; Dynamic Systems, Measurement and Control; Fluids Engineering; Engineering Materials and Technology; Pressure Vessel Technology*), quarterly.

American Society of Safety Engineers (ASSE)
850 Busse Highway, Park Ridge, Ill., 60068

The American Society of Safety Engineers is an organization of individual safety professionals dedicated to the advancement of the safety profession and to foster the well-being and professional development of its members.

The Society is actively pursuing its Professional Development Programs including the development of curricula for safety professionals, accreditation of degree programs, member education courses, additional publications, and defining research needs and communications to keep safety practitioners current. In addition, the Society has increased its participation and activity in national government affairs.

American Speech and Hearing Association (ASHA)
9030 Old Georgetown Rd., Bethesda, Md. 20014

Professional society of specialists in speech pathology and audiology. Presents annual awards. Publications: (1) ASHA, monthly; (2) Journal of Speech and Hearing Disorders, quarterly; (3) Journal of Speech and Hearing Research, quarterly; (4) Language, Speech and Hearing Services in Schools, quarterly; (5) Reports and monographs.

Committee on Noise as a Public Health Hazard
6570 AMRL (BBA)
Wright-Patterson AFB, Dayton, Ohio 45433

A committee of the American Speech and Hearing Association. Monitors research dealing with the effects of noise on health, including hearing and the body's autonomic system, and noise as a public nuisance and irritant. Holds periodic seminars to disseminate research findings.

Institute of Electrical and Electronics Engineers (IEEE)
345 East 47th St., New York, N.Y. 10017

Engineers and scientists in electrical engineering, electronics, and allied fields; membership includes 16,000 students. Holds numerous meetings and special technical conferences. Technical Activities Groups: Aerospace and Electronic Systems; Antennas and Propagation; Audio and Electroacoustics; Broadcast and Television Receivers; Broadcasting; Circuit Theory; Communication Technology.

Institute of Environmental Sciences (IES)
940 East Northwest Highway, Mr. Prospect, Ill. 60056

Engineers, scientists, and management people engaged in the simulation of the natural environment and the environments induced by equipment operation (for example, missiles, rockets, satellites, ships, aircraft, ground vehicles, nuclear radiation installations, etc.) and the testing of men, materials, and equipment in the simulated environments.

Committees: Acoustics; Bio-environmental Engineering; Corrosion; Data Acquisition and Reduction; Electromagnetics; Environmental Pollution; Handbook; IMET; Marine Environment; Nuclear Radiation; Shock and Vibration; Solar Radiation; USNC IEC/TC-SO. Divisions: Ecological Science; Physical Science; Standards and Practices; Test and Technology Applications; Training and Education. Publications: *Journal of Environmental Sciences*, bimonthly.

Institute of Noise Control Engineering (INCE)
P.O. Box 2167, Morristown, N.J. 07960

Scientific, technical and educational organization of individuals having an interest in noise control technology. Primarily professional engineers concerned with noise reduction in industry, buildings, transportation, products, appliances, community and noise instrumentation, standards, measurements, and legislation. Publication, *Noise Control Engineering*, is the technical journal of INCE and complements "Noise/News" which carries announcements of activities in the field.

Instrument Society of America (ISA)
400 Stanwix St., Pittsburgh, Pa. 15222

Scientific, technical, and educational organization dedicated to advancing the knowledge and practice related to the theory, design, manufacture, and use of instruments and controls in science and industry. Conducts conferences and symposia; develops standards; publishes and disseminates information.

Publications: (1) *Automation and Remote Control* (translation of Russian journal) monthly; (2) *Industrial Laboratory* (translation of Russian journal), monthly; (3) *Instrumentation Technology*, monthly; (4) *Measurement Technique* (translation of Russian journal), monthly; (5) *Instruments and Experimental Techniques* (translation of Russian journal), bimonthly; (6) *ISA Instrumentation Index*, quarterly; (7) *ISA Transactions*.

National Association of Hearing and Speech Agencies (NAHSA)
814 Thayer Ave., Silver Spring, Md. 20910

Otologists, social workers, school teachers and administrators, speech and hearing therapists, hard of hearing people, and others interested in the prevention of deafness, conservation of hearing, and rehabilitation of the hard of hearing.

Publications: *Hearing and Speech News,* bimonthly; also publishes pamphlets on audiology, hard of hearing children, rehabilitation, hearing aids, hearing tests, lipreading, local hearing agencies.

Society of Automotive Engineers (SAE)
2 Pennsylvania Plaza, New York, N.Y. 10001

Professional society of engineers in field of self-propelled ground, flight, and space vehicles; engineering students are enrolled in special affiliation. To promote the arts, sciences, standards, and engineering practices related to the design, construction, and utilization of self-propelled mechanisms, prime movers, components thereof, and related equipment. Publications: (1) *Journal of Automotive Engineering,* monthly; (2) *Handbook* (book of standards), annual; (3) *Proceedings,* annual; (4) *SAE Consultants,* annual; (5) *Transactions,* annual; also publishes *Advances in Engineering Series, Technical Progress Series,* technical papers, and other publications dealing with aerospace industry practices.

Society for Experimental Stress Analysis (SESA)
21 Bridge Square, Westport, Conn. 06880

Engineers and scientists in industry, consulting activity, government, and academic life, and others interested in the techniques employed in the measurement of stresses and strains as applied to metals and other materials. Publications: *Experimental Mechanics,* monthly; *Proceedings,* semiannual; also publishes handbooks, manuals and other material.

Trade associations

Air-conditioning and Refrigeration Institute (ARI)
1815 North Fort Myer Dr., Arlington, Va. 22209

Composed of manufacturers of air conditioning, warm air heating, and commercial and industrial refrigeration equipment, components, parts, accessories, and allied products. Develops and establishes equipment and application standards and certifies performance of certain industry products.

Air Moving and Conditioning Association (AMCA)
30 West University Dr., Arlington Heights, Ill. 60004

Composed of manufacturers of air moving and conditioning equipment for ventilation, heating and disposal of dust, and other waste materials. Conducts research on improvement of methods of utilization; develops standard codes for fans, steam and hot water unit heaters, air conditioning units, etc. Operates testing laboratory and performance certification program for fans and other air moving devices.

American Foundrymen's Society (AFS)
Golf and Wolf Rds., Des Plaines, Ill. 60016

This association serves as a clearinghouse for information in the foundry industry, including safety, hygiene and air pollution control. Many publications available through the society are of interest.

American Gas Association (AGA)
1515 Wilson Blvd., Arlington, Va. 22209

The association, through its Accident Prevention Committee, serves as a clearinghouse and in an advisory capacity to persons responsible for employee safety and to safety departments of its member companies. Its purposes are to study accident causes, recommend corrective measures, prepare manuals, and disseminate information to the gas industry that will help reduce employee injuries.

American Gear Manufacturers Association (AGMA)
1330 Massachusetts Ave. N.W., Washington, D.C. 20005

Composed of manufacturers of gears, geared speed changers, and related equipment for sale and for own use; also manufacturers of gear cutting and checking equipment.

American Iron and Steel Institute (AISI)
1000 16th St. N.W., Washington, D.C. 20036

This Institute represents 95 percent of the steel producers in the United States. Special committees on Safety, Industrial Hygiene and Industrial Health meet quarterly. Regional safety committees in the eastern, midwest and western sections of the United States also meet quarterly to provide informational forums for steel plant safety personnel.

American Mining Congress (AMC)
1200 18th St. N.W., Washington, D.C. 20036

Membership of the congress is from coal, metal, and nonmetal mining companies. This association has a Safety Committee organized within its Coal Division. The monthly *Mining Congress Journal* regularly carries articles and news items concerned with safety.

American Paper Institute
260 Madison Ave., New York, N.Y. 10016

The American Paper Institute conducts a broad safety education and informational service for the paper and pulp industry by means of a monthly safety letter covering topics of current interest. The institute also issues, to all participating companies, a "Safety Reference Desk Book" which contains information on subjects ranging from the organization of a safety department to technical advances in noise control.

American Petroleum Institute (API)
1801 K St. N.W., Washington, D.C. 20006

The objective of the Safety and Fire Protection Services of the American Petroleum Institute is to reduce the incidence of accidental occurrences, such as injuries to employees and the public, damage to property, motor vehicle accidents, and fires.

American Pulpwood Association (APA)
605 Third Ave., New York, N.Y. 10016

This association fosters study, discussion, and action programs to guide and help the pulpwood industry in growing and harvesting pulpwood raw material for the pulp and paper industry. The safety and training program of the Association is served through six regional Technical Divisions, each one of which has a safety and training committee.

Anti-Friction Bearing Manufacturers Association (AFBMA)
60 East 42nd St., New York, N.Y. 10017

Manufacturers of antifriction bearings, balls and rollers used in antifriction bearings. For the advancement of bearing standardization. Sponsors: Annular Bearing Engineers Committee (ABEC); Roller Bearing Engineers Committee (RBEC); Ball Manufacturers Engineers Committee (BMEC). Publications: *Standards for Ball and Roller Bearings and Steel Balls and Maintenance Manual.*

Associated General Contractors of America, Inc. (AGCA)
1957 E St. N.W., Washington, D.C. 20006

This association of contractors specializes in the building, highway, railroad, and heavy construction fields. All areas of the United States are served by the association's chapters, which carry on their own programs and render assistance to their members.

Association of Home Appliance Manufacturers (AHAM)
20 North Wacker Dr., Chicago, Ill. 60606

Comprised of companies manufacturing over 90 percent of the major appliances and a majority of the portable appliances produced in the United States each year. Associate members manufacture related products and component parts. Functions include: consumer affairs; representing the industry, especially in the area of government relations;

reporting industry statistics; developing appliance standards; certification of products.

Can Manufacturers Institute, Inc. (CMI)
1625 Massachusetts Ave. N.W., Washington, D.C. 20036

Can Manufacturers Institute, the manufacturing trade association for companies who make all-metal cans out of tin plate, black plate, terne plate, aluminum, or a combination thereof, acts as a clearinghouse of information on the metal can industry in the United States.

Compresed Air and Gas Institute (CAGI)
122 East 42nd St., New York, N.Y. 10017

Composed of manufacturers of equipment using compressed air.

Compressed Gas Association, Inc. (CGA)
500 Fifth Ave., New York, N.Y. 10036

The major purpose of the Compressed Gas Association is to provide, develop, and coordinate technical activities in the compressed gas industries, in the interest of safety and efficiency. Most of the work of the association is done by more than 30 technical committees, made up of representatives of member companies who are highly qualified technically in their respective areas:

Graphic Arts Technical Foundation
4615 Forbes Ave., Pittsburgh, Pa. 15213

This foundation is a coordinating organization. Members are large national and local printing and allied trade associations.

Gray and Ductile Iron Founders' Society, Inc.
Cast Metals Federation Bldg., Rocky River, Ohio 44126

The Gray and Ductile Iron Founders' Society is a trade association which represents gray and ductile iron foundries in the United States, Canada, and Mexico. The Society's Safety Committee convenes at least once a year to discuss in-plant safety programs for foundries.

Industrial Safety Equipment Association, Inc.
2425 Wilson Blvd., Arlington, Va. 22201

This association is composed of manufacturers and distributors of industrial safety equipment. Of outstanding importance is the broad representation of its members on numerous American National Standards Institute standards committees engaged in promulgation of industry-wide standards of performance for specific types of personal protective equipment.

Manufacturing Chemists' Association, Inc. (MCA)
1825 Connecticut Ave. N.W., Washington, D.C. 20009

The association supports a Safety and Fire Protection Committee composed of safety directors selected from its member companies. This committee meets four to six times a year and develops chemical safety information for use by member companies, state and federal health organizations, and the public. It also holds an annual safety workshop for those interested in chemical safety.

National Association of Manufacturers (NAM)
1776 F St. N.W., Washington, D.C. 20006

The NAM safety activities are carried on under the aegis of its Employee Health and Safety Committee which has a dual function: (*a*) promoting sound health and safety policies and programs in industry; and (*b*) working with the federal government to assure that present regulation of health and safety practices in industry and proposals for new legislation are realistic from industry's viewpoint.

National Coal Association (NCA)
1130 17th St. N.W., Washington, D.C. 20036

The association has a Department of Safety, which assists its members in the promotion of safety and accident prevention through cooperative, supervisory, educational, and research endeavors. Cooperation is extended to federal and state governments and private safety organizations.

National Electrical Manufacturers Association (NEMA)
155 East 44th St., New York, N.Y. 10017

Manufacturers of equipment and apparatus used for the generation, transmission, distribution and utilization of electric power, such as electrical machinery, motors, transportation, communication and lighting equipment. Develops product standards covering such matters as nomenclature, ratings, performance, testing and dimensions; participates in developing *National Electrical Code, National Electrical Safety Codes* and advocates their acceptance by state and local authorities.

National Fluid Power Association (NFPA)
P.O. Box 49, Thiensville, Wis. 53092

Manufacturers of components such as pumps, valves, cylinders, etc., used in transmitting power by means of a fluid (gas or liquid) under pressure (hydraulic or pneumatic); used in industrial material-handling, automotive, railway, aircraft, marine, agricultural, and other machinery. Activities include technical standards.

National Machine Tool Builders' Association (NMTBA)
7901 Westpark Dr., McLean, Va. 22101

Manufacturers of power-driven machines, not portable by hand, that are used to shape or form metal by cutting, impact, pressure, electrical techniques, or a combination of such processes. Seeks to improve methods of producing and marketing machine tools; promotes research and development in the industry. Participates in programs of standardization of design and performance of machine tools and components. Serves as a clearinghouse for technical aspects of the industry.

National Petroleum Refiners Association (NPRA)
1725 DeSales St. N.W., Washington, D.C. 20036

Primarily a service organization, the association services the petroleum refining industry and at the same time, serves as a clearinghouse for new ideas and developments in refining technology.

Portland Cement Association (PCA)
5420 Old Orchard Rd., Skokie, Ill. 60078

This organization is devoted to research, educational, and promotional activities to extend and improve the use of portland cement. The Accident Prevention Department of the association develops and administers the safety program.

The Society of the Plastics Industry, Inc. (SPI)
250 Park Ave., New York, N.Y. 10017

The Society of the Plastics Industry, Inc., is the national commercial trade association serving all segments of the plastics industry. SPI functions through divisions representing functional or product interest and committees representing the comprehensive interests of the entire industry.

Governmental agencies

State or Local

It is most important to have a good working knowledge of the state agencies responsible for the enforcement of safety and health laws. Contact the proper groups in the specific labor department or other agency and find out how the various boards, divisions, and services function.

Much difficulty can be avoided if one is familiar with the labor legislation and safety and health codes under which he is working. Codes and laws vary widely in the different states and provinces, and those who have safety jurisdiction in plants in a number of places should understand these differences.

The names and addresses of such state, commonwealth, or territorial agencies with which the safety professional may need to communicate, are available in an up-to-date listing of health authorities in Public Health Publication No. 75, "Directory of State, Territorial, and Regional Health Authorities," for sale by the U.S. Government Printing Office.

Federal

There is an overwhelming amount of information available from the federal government concerning all aspects of safety and health, environmental problems, pollution, statistical data, and other industry problems.

Occupational Safety and Health Administration (OSHA)

A new national policy was established on December 29, 1970, when President Richard M. Nixon signed into law the Occupational Safety and Health Act of 1970 (Public Law No. 91–596; 84 Statutes at Large 1593; 29 United States Code 655.

To assist in carrying out its responsibilities, OSHA has established ten regional offices and a number of area offices. The primary mission of the regional office chief, known as the Assistant Regional Director, is to supervise, coordinate, evaluate and execute all programs of OSHA in the region. Assisting the Assistant Regional Director are Associate Assistant Regional Directors for (a) training, (b) technical support, and (c) state and federal programs.

Area offices have been established within each region, each headed by an Area Director. The mission of the Area Director is to carry out the compliance program of OSHA within designated geographic areas. The area office staff carries out its activities under the general supervision of the Area Director with guidance of the Assistant Regional Director, using policy instructions received from the national headquarters. The real action for implementing the enforcement portion of the OSHAct is carried out by the area offices.

The locations of the Washington headquarters and regional offices are listed in the Directory at the end of this section.

Occupational Safety and Health Review Commission (OSHRC)

The Occupational Safety and Health Review Commission (OSHRC) is a quasijudicial board of three members appointed by the President and confirmed by the Senate. The Commission is an independent agency of the Executive Branch of the U.S. Government and is not a part of the Department of Labor. The principal function of the Commission is to adjudicate cases resulting from an enforcement action initiated against an employer by OSHA when any such action is contested by the employer or by his employees or their representatives.

National Institute for Occupational Safety and Health (NIOSH)

The National Institute for Occupational Safety and Health (NIOSH) was established within the HEW under the provisions of the OSHAct. Administratively, NIOSH is located in HEW's Center for Disease Control. NIOSH is the principal federal agency engaged in research, education, and training related to occupational safety and health.

The primary functions of NIOSH are to (*a*) develop and establish recommended occupational safety and health standards, (*b*) conduct research experiments and demonstrations related to occupational safety and health, and (*c*) conduct educational programs to provide an adequate supply of qualified personnel to carry out the purposes of the OSHAct.

Research and related functions. Under the OSHAct, NIOSH has the responsibility for conducting research for new occupational safety and health standards. NIOSH develops criteria for the establishment of such standards. Such criteria are transmitted to OSHA which has the responsibility for the final setting, promulgation, and enforcement of the standards.

Directory of Federal Agencies Concerned With the Occupational Safety and Health Act

National headquarters

OSHA

Occupational Safety and Health Administration, U.S. Department of Labor, Department of Labor Building, 14th Street and Constitution Ave. N.W., Washington, D.C. 20210

NIOSH

National Institute for Occupational Safety and Health, U.S. Department of Health, Education, and Welfare, Parklawn Building, 5600 Fishers Lane, Rockville, Md. 20852

BLS

Bureau of Labor Statistics, U.S. Department of Labor, 441 G St. N.W., Washington, D.C. 20212

OSHRC

Occupational Safety and Health Review Commission, 1825 K St. N.W., Washington, D.C. 20006

Regional offices

Region I

Connecticut
Maine
Massachusetts
New Hampshire
Rhode Island
Vermont

OSHA

18 Oliver St.
Boston, Mass. 02110

NIOSH

John F. Kennedy Federal Building
Room 1401-B-3, Government Center
Boston, Mass. 02203

BLS

John F. Kennedy Federal Building
Room 1603-A, Government Center
Boston, Mass. 02203

Region II

New York
New Jersey
Puerto Rico
Virgin Islands

OSHA

1515 Broadway (1 Astor Plaza)
New York, N.Y. 10036

NIOSH

26 Federal Plaza
New York, N.Y. 1007

BLS

1515 Broadway
New York, N.Y. 10036

Region III

Delaware
District of Columbia
Maryland
Pennsylvania
Virginia
West Virginia

OSHA
3535 Market St. (15220 Gateway
   Center)
Philadelphia, Pa. 19104

NIOSH
3535 Market St.
Philadelphia, Pa. 19104
(P.O. Box 13716, Philadelphia, Pa.
   19101)

BLS
3535 Market St., Room 14100
Philadelphia, Pa. 19104
(P.O. Box 13309, Philadelphia, Pa.
   19101)

Region IV

Alabama
Florida
Georgia
Kentucky
Mississippi
North Carolina
South Carolina
Tennessee

OSHA
1375 Peachtree St. N.E., Suite 587
Atlanta, Ga. 30309

NIOSH
50 Seventh St. N.E.
Atlanta, Ga. 30323

BLS
1371 Peachtree St. N.E., Suite 540
Atlanta, Ga. 30309

Region V

Illinois
Indiana
Michigan
Minnesota
Ohio
Wisconsin

OSHA
300 South Wacker Dr., Room 1201
Chicago, Ill. 60606

NIOSH
300 South Wacker Dr.
Chicago, Ill. 60606

BLS
300 South Wacker Dr., 8th Floor
Chicago, Ill. 60606

Region VI

Arkansas
Louisiana
New Mexico
Oklahoma
Texas

OSHA
1512 Commerce St. (Texaco Building)
7th Floor
Dallas, Texas 75201

NIOSH
1100 Commerce St., Room 8-C-53
Dallas, Texas 75202

BLS
1100 Commerce St., Room 6-B-7
Dallas, Texas 75202

Region VII

Iowa
Kansas
Missouri
Nebraska

OSHA
823 Walnut St.
(Waltower Building), Room 300
Kansas City, Mo. 64106

NIOSH
601 East 12th St.
Kansas City, Mo. 64106

BLS
911 Walnut St.
(Federal Office Building), 10th Floor
Kansas City, Mo. 64106

Region VIII

Colorado
Montana
North Dakota
South Dakota
Utah
Wyoming

OSHA
1961 Stout St.
(Federal Building), Room 15010
Denver, Colo. 80202

NIOSH
19th and Stout St.
(Federal Building), Room 9017
Denver, Colo. 80202

BLS
911 Walnut St.
(Federal Office Building), 10th Floor
Kansas City, Mo. 64106

Region IX

Arizona
California
Hawaii
Nevada

OSHA
450 Golden Gate Ave.
(Federal Building), Room 9470
San Francisco, Calif. 94102
(P.O. Box 36017, San Francisco,
Calif. 94102)

NIOSH
50 Fulton St.
(Federal Office Building), Room 254
San Francisco, Calif. 94102

BLS
450 Golden Gate Ave. (Federal
Building)
San Francisco, Calif. 94102
(P.O. Box 36017, San Francisco,
Calif. 94102)

Region X

Alaska
Idaho
Oregon
Washington

OSHA
506 Second Ave.
(Smith Tower Building), Room 1808
Seattle, Wash. 98104

NIOSH
1321 Second Ave. (Arcade Building)
Seattle, Wash. 98101

817

BLS
450 Golden Gate Ave. (Federal
    Building)
San Francisco, Calif. 94102
(P.O. Box 36017, San Francisco,
    Calif. 94102)

Canadian departments, associations, and boards

In all provinces of Canada there is a Workmen's Compensation Board or Commission. Some of these handle accident prevention directly. In other provinces there are provisions similar to Section 110 of the Quebec Workmen's Compensation Act, which stipulates "that industries included in any of the classes under Schedule I may form themselves into an Association for accident prevention and formulate rules for that purpose."

Furthermore, all provinces have legal safety requirements that are administered by the Department of Highways, Department of Labor, and the Department of Mines. These sources can be contacted by writing to the deputy minister of the department located in the capital of each province.

Technical literature

The intent of this section is to provide brief descriptions of the basic reference books and publications as well as knowledge of the work and functions of standards organizations having a high degree of interest in industrial noise control and hearing conservation.

Journals

IEEE Transactions on Audio and Electroacoustics. Institute of Electrical and Electronic Engineers, Inc. 345 East 47th St., New York, N.Y. 10017

The Journal of the Acoustical Society of America. Acoustical Society of America, 335 East 45 St., New York, N.Y. 10017

Journal of the Audio Engineering Society. Audio Engineering Society, Room 929 Lincoln Building, 60 East 42 St., New York, N.Y. 10017

Journal of Auditory Research. The C. W. Skilling Auditory Research Center, Inc., Box M, Groton, Conn. 06340

Journal of Sound and Vibration. Academic Press Inc., 111 Fifth Ave.,

New York, N.Y. 10003

"Noise/News." Institute of Noise Control Engineering, P.O. Box 2167, Morristown, N.J. 07960

Noise Control Engineering. Institute of Noise Control Engineering P.O. Box 2167, Morristown, N.J. 07960

Noise Regulation Reporter. Bureau of National Affairs, Inc., 1231 25th St. N.W., Washington, D.C. 20037

Journal of Speech and Hearing Disorders. American Speech and Hearing Association, 9030 Old Georgetown Road, Bethesda, Md. 20014

Sound and Vibration. Acoustical Publications, Inc., 27101 East Oviatt Road, Bay Village, Ohio 44140

OTHER PUBLICATIONS

There are a number of other publications that frequently contain articles dealing with noise and audiometry.

American Industrial Hygiene Association Journal. AIHA, 66 South Miller Road, Akron, Ohio 44313

Annals of Occupational Hygiene. Pergamon Press, Maxwell House, Fairview Park, Elmsford, N.Y. 10523

Archives of Otolaryngology. American Medical Association, 535 North Dearborn, Chicago, Ill. 60610

Industrial Hygiene News Report. Flournoy & Associates, 1845 West Morse Ave., Chicago, Ill. 60626

Job Safety and Health. Superintendent of Documents, U.S. Government Printing Office, Washington, D.C. 20402

Job Safety and Health Report. Business Publishers, Inc., P.O. Box 1067, Blair Station, Silver Spring, Md. 20910

Journal of Occupational Medicine. American Occupational Medical Association, 150 North Wacker Drive, Chicago, Ill. 60606

Occupational Health Nursing. American Association of Industrial Nurses, 79 Madison Ave., New York, N.Y. 10016

Occupational Safety and Health Reporter. Bureau of National Affairs, Inc., 1231 25th St. N.W., Washington, D.C. 20037

OSHA Up-To-Date. National Safety Council, 425 North Michigan Ave. Chicago, Ill. 60611

BIBLIOGRAPHY

Industrial Noise—A Selected Bibliography (*1963–1973*). Pittsburgh, Pa.: Industrial Health Foundation, Inc. 1973.

This bibliography of more than 1100 journal articles covers industrial noise and its measurement, effects on workers, and measures for control. Articles are in the English language; a subject index is included. (116 pp. paperbound.)

Basic texts

A suggested reference library includes the following basic and specialized books and pamphlets. Annotations are given for some of them.

GENERAL

American Industrial Hygiene Association. *Industrial Noise Manual,* 2nd ed. Detroit, Mich., 1966.

This manual is intended as a guide for industrial hygienists who are involved in implementing a hearing-conservation program.

This material is aimed at those who are new to hearing conservation and is written in a clear, easy-to-understand style. A great deal of technical information is imparted with a minimum use of mathematics.

The manual focuses upon: (*a*) the physical measurement of noise; (*b*) the medical evaluation of persons exposed to it; and, (*c*) the control of noise exposure.

The manual presents information on the physics of sound and follows through with discussions of noise measuring instruments and noise analysis.

The manual examines the means of controlling noise sources by using engineering methods. The legal aspects and liabilities associated with industrial hearing loss are then covered.

Beranek, L. L. *Acoustics.* New York, N.Y.: McGraw-Hill Book Co., 1954.

This book provides the material necessary for an engineer or scientist who wishes to practice in the field of acoustics and who does not intend to confine his efforts to theoretical matters.

This book is basically a text inasmuch as it is an outgrowth of a course in acoustics that the author taught to seniors and first-year graduate students in electrical engineering and communication physics. This text is divided into 13 chapters comprising 32 parts.

The basic wave equation and some of its more interesting solutions are discussed in detail in the first part of the text. The radiation of sound, components of acoustical systems, microphones, loudspeakers, and horns are treated in sufficient detail to allow the serious student to enter into electroacoustical design.

There is an extensive treatment of such important problems as sound enclosures, methods for noise reduction, hearing, speech intelligibility, and psychoacoustic criteria for comfort, for satisfactory speech intelligibility, and for pleasant listening conditions.

Bragon, C. R. *Noise Pollution—The Unquiet Crisis.* Philadelphia, Pa.: University of Pennsylvania Press, 1971.

This book is written not only for the concerned citizen, but for all those who can, and must, take immediate and effective action in our unquiet crisis: Urban planners, architects, hospital administrators, public health officials, transportation executives, lawyers, realtors, sound engineers, manufacturers of transportation equipment and household appliances, and community leaders.

The author inventories the noise levels of automobiles, buses, subways, airplanes, household appliances, and children's toys in numerous charts and tables, and relates these facts to the measurable social, physical, and psychological damage they do to human beings.

Broadbent, D. E. *Perception and Communication.* New York, N.Y.: Pergamon Press, 1958.

Brown, B., and Goodman, J. E. *High-Intensity Ultrasonics.* London, England: Iliffe Books Ltd., 1965.

Committee on the Problems of Noise. *Noise—Final Report.* London, England: Her Majesty's Stationery Office, 1964.

This book examines the nature of the sources and the effects of the problem of noise and advises what further measures can be taken to mitigate it. The authors believe that an important feature of the task was (a) to try to define (wherever possible) quantitative levels of noise which could become statutory limits; or (b) where statutory limits were not desirable or could not be laid down at present, to suggest levels which would serve as guides to what would be reasonable.

Crawford, A. E. *Ultrasonic Engineering,* New York, N.Y.: Academic Press, 1955.

Doelle, L. L. *Environmental Acoustics.* New York, N.Y.: McGraw-Hill Book Co., 1972.

This volume of applied acoustics has been specifically written to meet the needs of people whose work involves environmental noise control but who lack specialized training in acoustics.

Although intended for the architect and the architectural student, the book will also be useful to engineers, interior designers, builders, contractors, developers, and in general anyone whose occupation involves him in environmental acoustics. The acoustical expert, however, will find no new material here. What makes this book particularly easy to understand (regardless of level) is (*a*) a minimum of mathematical and physical details and (*b*) simple and practical recommendations with ample references to actual installations.

Ford, R. D. *Introduction to Acoustics.* Amsterdam, Netherlands: Elsevier, 1970.

Harris, C. M. *Handbook of Noise Control.* New York, N.Y.: McGraw-Hill Book Co., 1957.

This general reference work covers every aspect of noise and how to control it. Forty-six experts, each contributing information on his special topic, provide the authority and range that make this the complete reference book that it has become over the years. There are some areas that should be updated, but they do not detract from the value of the book. This book is definitely not for the beginner. It is a highly technical work geared to the practicing engineer.

The nature of noise, its measurement, and techniques of its control in buildings, industry, transportation, and the community are all discussed in detail. Also included are fundamentals, practical methods and engineering techniques, and useful reference data needed in connection with such a variety of subjects, such as, how noise affects man's efficiency, hearing losses, speech communication, vibration and its control, and uses of various materials in controlling industrial noise.

Particular stress has been given to the solution of noise problems by the use of examples taken from existing installations.

Hewlett-Packard Company. *Acoustics Handbook.* Palo Alto, Calif.: Hewlett-Packard Company, 1968.

Hosey, A. D., and Powell, C. H. *Industrial Noise—A Guide to its*

Evaluation and Control. Washington, D.C.: U.S. Government Printing Office, 1967.

This publication was originally designed as a supplement to the industrial noise course sponsored by the Public Health Service, and as a guide and reference for those who could not attend the classes and as a refresher for those already somewhat familiar with the subject.

The material in this book is aimed at engineers and industrial hygienists new to the field of occupational noise. The intent is to provide the basic procedures and techniques of recognition, evaluation, and control of noise in the work situation.

This manual presents lecture outlines and reference materials developed for use in one-week training courses in industrial noise for industrial hygienists.

Kinsler, L. E., and Frey, A. R. *Fundamentals of Acoustics*, New York, N.Y.: John Wiley & Sons, Inc., 1962.

Morse, P. M., and Ingard, K. Uno. *Theoretical Acoustics*. New York, N.Y.: McGraw-Hill Book Company, 1968.

The object of this book is to present the salient facts regarding acoustical phenomena in both descriptive and mathematical form to students and specialists in the field, and to provide an extended series of illustrations of the way mathematical physics can aid in the understanding of acoustics.

This book is a graduate-level text which should be useful to physical scientists and engineers who wish to learn about new developments in acoustical theory. Although on the graduate level, anyone familiar with calculus and vector analysis should be able to understand the mathematical techniques used.

Subject coverage in this book is restricted primarily to generation, propagation, absorption, reflection, and scattering of compressional waves in fluid media, in the distortion of such waves by viscous and thermal effects as well as by solid boundaries, and in their coupling through the induced vibration of walls and transmission panels. Material on the acoustics of moving media, on plasma acoustics, on nonlinear effects, and on the interaction between light and sound is also included.

Olson, H. F. *Acoustical Engineering*, New York, N.Y.: Van Nostrand, Reinhold Co., 1957.

Parkin, P. H., and Humphreys, H. R. *Acoustics, Noise and Buildings,* 3rd ed. London, England: Faber, 1969.

Richardson, E. G. and Meyer, E., eds., *Technical Aspects of Sound,* Vol. III, Amsterdam, Netherlands: Elsevier Publishing Company, 1962.

Seto, W. W. *Theory and Problems of Acoustics.* New York, N.Y.: McGraw-Hill Book Co., 1971.

This book is designed primarily to supplement standard tests in physical or applied acoustics in the belief that numerous solved problems constitute one of the best means for clarifying and fixing in mind basic principles.

Written at the senior undergraduate level, this book should be of considerable value to the physics and engineering student who is interested in the science of sound and its applications. The practicing engineer can also make frequent reference to the book for its numerical solutions of many realistic problems in the area of sound and vibration. Essential to the use of this book are knowledge of the fundamental principles of mechanics, electricity, strength of materials, and undergraduate mathematics including calculus and partial differential equations.

Sataloff, J., and Michael, P. *Hearing Conversation.* Springfield, Ill.: Charles C Thomas, 1973.

In developing effective hearing-conservation programs that will comply with current rules and regulations, many challenging problems arise. This book is intended to cover all these vital topics on a level that will be useful for engineers, executives, hearing testers, industrial hygienists, nurses, otologists, physicians, and safety personnel.

This book provides a basic, practical view of the physics of sound, how we hear, and the causes of auditory defects (needed to understand the problem of occupational deafness). In addition, it provides basic and practical instruction in hearing measurement, noise control, and other factors needed in the development of effective hearing conservation programs. Extra-auditory effects of noise are covered from the point of view of community reaction.

Stephens, R. W. R., and Bate, A. E. *Acoustics and Vibrational Physics.* New York, N.Y.: St. Martin's Press, 1966.

Suri, R. L. *Acoustics Design and Practice,* Vol. 1. Bombay, India:

Asia Publishing House, 1963.

The book deals with various aspects of the subject of noise. The acoustical design section covers all types of auditoria (assembly and conference halls, committee rooms, etc.) and also various types of studios for broadcasting, film, and television fields. It also deals with the problems of noise due to machines and ventilating systems and their control in industrial and nonindustrial buildings. The book besides covering the aspects of design devotes considerable space to the methods and techniques of various acoustical measurements. Exhaustive discussion has been included on actual measurements which provides first-hand information. The book also lays strong emphasis on the practical aspects of the problem and contains numerous illustrations, diagrams, and tables to serve as a design guide. The author embodies his ideas on design and technique and the background of his experience and knowledge over a long period with a foremost broadcasting organization. Teachers and students should find this book useful. It should also serve as a good reference book to architects and engineers interested in the subject of architectural acoustics.

Taylor, R. *Noise.* Harmandsworth, Middlesex, England: Penguin Books, 1970.

Ward, W. D., and Fricke, J. E., eds. *Noise as a Public Hazard.* Washington, D.C.: The American Speech and Hearing Association, 1969.

A conference on noise as a public hazard was held in an effort to present the best evidence available bearing on the general question: To what extent is noise a public health hazard? The conference was aimed at practicing professionals in the fields of medicine, psychology, industrial hygiene, engineers, city government, safety, and audiometry.

Papers in each of five panels, by different authors, were presented: effects of noise on man, industrial noise and the worker, noise in the community, special problems of recent technological development, and community noise control.

Wood, A. B. *A Textbook of Sound.* New York, N.Y.: Dover Publications, 1965.

SOUND MEASUREMENT

Arnold, R. R., Hill, H. C. and Nichols, A. V. *Modern Data Processing.* New York, N.Y.: John Wiley & Sons, Inc., 1969.

Beranek, L. L. *Acoustic Measurements.* New York, N.Y.: John Wiley & Sons, Inc., 1949.

Burroughs, L. *Microphones: Design and Application.* Plainview, N.Y.: Sagamore Publishing Co., Inc., 1974.

Because this book is basically for reference, it is organized for rapid referral. The author begins with a precise description of the various types of microphones and then pinpoints the exact situation where each type is most efficient. Real applications, with accounts of studios and theaters of various sizes, outdoor locations, various performing groups, etc., are described.

The intended audience is that group of persons who is involved in audio operations. Also, the book is based on a series of lectures given over a period of years by the author. Consequently, although the material itself is of a technical nature, it is presented in a conversational, easy-to-follow, step-by-step style that allows even the non-technical person to glean useful knowledge. More than 200 illustrations and diagrams clarify his explanations.

Cooper, G. R., and McGillem, D. C. *Methods of Signal and System Analysis.* New York, N.Y.: Holt, Rinehart and Winston, Inc., 1967.

Davenport, W. B., Jr., and Root, W. L. *Random Signals and Noise.* New York, N.Y.: McGraw-Hill Book Co., 1958.

Davis, D. *Acoustical Tests and Measurements.* Indianapolis: H. W. Sams, 1965.

Hueter, T. F., and Bolt, R. H. *Sonics.* New York, N.Y.: John Wiley & Sons, Inc., 1955.

Keast, D. N. *Measurement in Mechanical Dynamics.* New York, N.Y.: John Wiley & Sons, Inc. 1967.

King, A. J. *The Measurement and Suppression of Noise.* New York, N.Y.: Chapman & Hall, 1965.

Because noise is in the no man's land between physics and engineering, the inhabitant of this land must be able to converse intelligently with his neighbors on either side. It is hoped that the book will (a) act as an interpreter to enable engineers to appreciate at least something of the physical principles involved in noise problems and (b) help them tackle the problems logically with a good chance of success.

Kuo, F. F., and Kaiser, J. F. *System Analysis by Digital Computer,* New York, N.Y.: John Wiley & Sons, Inc., 1966.

Lange, H. F. H. *Correlation Techniques.* New York, N.Y.: Van Nostrand Reinhold Co., 1967.

Lathi, B. P. *Signals, Systems and Communication.* New York, N.Y.: John Wiley & Sons, Inc., 1965.

Lee, Y. W. *Statistical Theory of Communication.* New York, N.Y.: John Wiley & Sons, Inc., 1960.

Lord, P., and Thomas, F. L., eds. *Noise Measurement and Control.* London, England: Heywood & Co., Ltd., 1963.

Mackenzie, G. W. *Acoustics.* London, England: Focal Press, 1964.

Peterson, A. P. C., and Gross, E. E., Jr. *Handbook of Noise Measurement.* Concord, Mass.: General Radio Co., 1972.

The intent of the authors is to provide a basic handbook on sound measurement and instrumentation for the person who is new to the responsibility of making such measurements and complying with the existing noise exposure limits.

The authors endeavor to clarify the terms and definitions used in sound measuring, describe the measuring instruments and their use, aid the prospective user in selecting the proper equipment for the measurements that must be made, and show how the measurements can be interpreted to solve some typical problems.

Chapters include: what are noise and vibration; what noise and vibration do and how much is acceptable; hearing programs in industry; instrumentation for noise and vibration measurement; what noise and vibration measurements should be; techniques, precautions, and calibrations; noise and vibration control. Some case histories and nine very informative appendices are included.

Rettinger, M. *Acoustics, Room Design and Noise Control,* New York, N.Y.: Chemical Publishing Co., 1968.

CONTROL

Bell, L. H. *Fundamentals of Industrial Noise Control.* Trumbull, Conn.: Harmony Publishing Co.,

827

This is a practical textbook written for the plant or safety engineer who has the task of reducing plant noise to comply with OSHA regulations. It is a comprehensive, easy-to-use guide to industrial noise control with many case histories and illustrations from the machine tool, manufacturing, power plant, plastics, food processing, packaging, and petrochemical industries.

Beranek, L. L., ed. *Noise Reduction.* New York, N.Y.: McGraw-Hill Book Co., 1960.

The purpose of this book is to present the fundamentals of noise reduction to engineers who are not specialists in acoustics. This book attempts to lead the reader by gradual steps from the beginning of the subject into the more advanced aspects. Individuals with a noise problem should be able to find some assistance. The text contains numerical examples of solved problems and frequent comparisons of measured and calculated data.

The material is divided into the following four parts: sound waves and their measurement, fundamentals underlying noise and vibration control, criteria for noise and vibration control, and practical noise control.

Beranek, L. L. *Noise and Vibration Control.* New York, N.Y.: Mc-Graw-Hill Book Company, 1971.

Here is a detailed guide for those who have no special training in acoustics but who are concerned with the design of buildings, factories, power plants, air conditioning systems, farm and road equipment, and the like.

The material, written by various experts in the field, is geared to the practicing engineer who has a noise problem to solve. The text is graded somewhat in technical level starting with the fundamentals of sound and continuing through to the so-called criteria for noise control.

The organization of the text is well thought out beginning with the fundamentals of sound in free space outdoors and in small enclosures or rooms; this is followed by the methods for measuring and analyzing noise and vibration, methods for selecting design criteria, and methods for choosing materials structures, mufflers, and vibration isolators to quiet a product or building. The three principal aspects of a project for control are spelled out: source, path, and receiver.

This book abounds in graphs, diagrams, and tables; it has a comprehensive index and concise, well-planned appendices, all of which should be of great value to the reader.

Close, P. P. *Sound Control and Thermal Insulation of Buildings.* New York, N.Y.: Reinhold Publishing Corp., 1966.

This book is intended primarily to cover the fundamentals of essential design data and product information including application details. An endeavor has been made to simplify the presentation of the subject matter and to avoid complicated theory and formulas without compromising the essential facts. However, it is obviously impossible to avoid the use of a certain amount of technical language and terminology.

The sound control section deals with acoustic problems common to various types of buildings and includes:

• Auditorium acoustics involving good hearing conditions for speech and music

• Noise quietening in offices, industrial buildings, schools, hospitals, restaurants, and other buildings

• Noise control in buildings used for residential purposes such as homes, apartments, motels, hotels and institutions

• Control of machinery vibration and noise—a problem common to all types of buildings

At the end of each chapter, there is a rather thorough list of references that the reader may consult, if necessary.

Crocker, M. S., ed. *Noise and Vibration Control Engineering.* Lafayette, Ind.: Purdue University, 1971.

Proceedings of the Purdue Noise Control Conference held at Purdue University, in July 1971. The conference consisted of 9 sessions, 5 of which are of special interest to industrial hygienists and design engineers: machinery noise (8 papers); industrial noise criteria and control (7 papers); vibration control and biodynamics (8 papers); noise and vibration control (9 papers); noise in buildings (10 papers). The papers are reproduced in full, and the volume is supplemented by an extensive bibliography listing all the references given in the papers.

The section on machinery noise contains papers on the following subjects: control of noise from machinery, noise from large centrifugal compressors, control valve noise, gear noise, hermetic refrigeration compressor acoustics, estimating the performance of wall structures used for controlling machinery noise, quieting hydraulic components, and ways of vibration machine noise reduction.

Duerden, C. *Noise Abatement.* New York, N.Y.: The Philosophical Library, 1971.

The author has provided a book especially tailored for those who desire or who have need to know about noise and noise control, but do not have a technical background.

Noise Abasement is written in a simple, easy-to-read style, with a minimum of acoustic terminology and no math to speak of. It may be regarded as a textbook which will introduce the subject to the student, local government officials, industrialists, and architects.

Of the nine chapters, eight deal directly with noise and control measures: introduction to sound, properties of sound, sound level meter, planning—industry and transport, investigating a complaint, noise control, and practical examples. Chapter 9 (Legislation) and five appendices deal with the various British acts passed in an attempt to define and control environmental noise.

The principal drawback to this book is that its British author cites examples from his personal experience which is of course largely confined to England. As such, it is somewhat difficult for the reader on this side of the ocean to relate.

Hines, W. A. *Noise Control and Industry.* London, England. Business Publications Ltd., 1966.

Its specific purpose is to serve as a guide for all who are responsible for building projects of any kind—from factories and commercial buildings to administrative blocks and hospitals. The general contractor of the building will naturally rely on the architect and other specialists for the actual design and the construction of the project.

To discharge these responsibilities insofar as acoustic problems are concerned, it is necessary to have a good general background and knowledge of the theory and practice of noise control. Armed with this information, the owner of a building will at least know the questions to ask specialists, be in a position to evaluate their answers and so be able to make correct policy decisions on the basis of the advice received, leaving the specialist to carry out the detailed implementation.

Rosenblith, W. A., et al. *Handbook of Acoustic Noise Control,* Vol. 2. Noise and Man. Dayton, Ohio: Wright Air Development Center, 1963.

Sharland, I. *Woods Practical Guide to Noise Control.* Colchester,

England: Woods of Colchester Ltd., 1972.

This introductory textbook on noise control in 7 chapters covers: the fundamentals of noise (how it occurs, its structure, and units and definitions used to describe it); hearing and noise acceptability; measurement of acoustic properties; calculation of noise levels in given situations; calculation of noise in ventilatng systems; principles of noise control; noise control in practice. Appendices include: approximate values of absorption coefficients for various internal finishes to walls, floors, etc.; representative values of airborne sound-reduction indices for some common structures; a glossary of acoustical terms; and various conversion tables.

Thumann, A., and Miller, R. K. *Secrets of Noise Control.* Atlanta, Ga.: The Fairmont Press, 1974.

This book demonstrates the use of noise control techniques through the use of job simulation experience. Each experience presents problems as encountered in the field along with illustrated solutions. Although not a highly technical book, the subject matter is presented in such a manner that the reader should have some technical background in order to get the most from this book.

A broad range of topics is covered from the point of view of a noise control engineer. The nine chapters include sound propagation indoors and out, how to establish noise criteria, how to make sound measurements, a review of math (particularly logarithms), analyzing noise reports for management, silencing equipment, and how to meet your noise objectives.

Warring, R. H. *Handbook of Noise and Vibration Control.* Morden, Surrey, England: Trade and Technical Press Ltd., 1970.

A reference book including much new material and covering: psychoacoustics; sound fields, sound propagation; noise measurement techniques and instruments; torsional vibration; legal aspects; noise and vibration control; silencers, acoustic absorption materials, insulation, partitions, enclosures, windows, glazing; vibration generators; various types of buildings; different types of machinery (prime movers, road vehicles, electrical machines, pumps, pipework, heating and ventilation systems, pneumatic tools, etc). Conversion and other useful tables.

Yerges, L. F. *Sound, Noise, and Vibration Control.* New York, N.Y.: Van Nostrand Reinhold Company, 1969.

This book has a limited and specific purpose—to provide enough of the fundamentals of sound and vibration and their control to permit the professional to feel comfortable about the subject. This book is intended as a working guide for the professional. However, the technically informed person can easily understand the material. Math is not used extensively.

This book is divided into three basic parts: (1) basic theory of sound and vibration necessary for a real understanding of the effect of this form of energy on people and the environment in which they live and work; (2) broad, general principles of sound and vibration control, including the types of materials, systems, and constructions used for this purpose; and (3) important data, organized into tables, detailed drawings and sketches, and checklists for easy reference.

In general, the information includes well-established and widely used practices and procedures, universally available materials, and construction methods distilled from long experience in the field.

Effects on Man

American Medical Association. *Guides To the Evaluation of Permanent Impairment.* Chicago, Ill.: AMA, 1971.

The ad hoc Committee on Medical Rating of Physical Impairment was created by the Board of Trustees of the American Medical Association in September 1956. The Committee was directed to establish a series of practical guides for the rating of physical impairment of the various body systems.

With the assistance of outstanding consultants, a series of thirteen "Guides to the Evaluation of Permanent Impairment" was developed and published in the Journal of the American Medical Association.

With the publication in this single volume of the entire series of updated "Guides," the AMA is providing authoritative material to assist physicians and others in discharging a responsibility to their patients, clients, or applicants who are seeking benefits from the various agencies and programs serving the disabled.

American Mutual Insurance Alliance. *Background for Loss of Hearing Claims.* Chicago, Ill.: American Mutual Insurance Alliance, 1964.

This monograph describes, in nonscientific language, the factors involved in compensation claims for occupational loss of hearing. The objective has been to focus attention upon the key points essential to a basic understanding of the subject. Ample references have been provided for those who may desire to delve more deeply into the legal and legislative background or the medical, scientific, and technical considerations which influence the course of legal and legislative developments.

This material is designed primarily for claims and legal personnel. It may also be of some value to other groups as background information on various aspects of noise and loss of hearing.

No attempt has been made to be all-inclusive. This monograph is in no sense a claims manual or procedural guide, nor does it attempt to indicate the legal status of claims for loss of hearing other than in states where specific legislation has been enacted or precedents established. Every effort has been made to present a factual and objective summary of the situation.

American National Standards Institute. *The Effects of Shock and Vibration on Man*, Report S3. W39. New York, N.Y.: ANSI, 1974.

This report, a review of the effects of shock and vibration on man, covers a very broad scope. It deals with three problems: (1) the determination of the structure and properties of the human body considered as a mechanical as well as a biological system; (2) the effects of shock and vibration forces on this system; and (3) the protection required by the system under various exposure conditions and the means by which this protection is to be achieved. Standards Committee on Bioacoustics, S3, recommends dissemination of the report as pertinent background and reference material on human vibration research, until such time as more definitive guidance on permissible vibration exposures can be provided through a standard.

Ballantyne, J. *Deafness*, 2nd ed. Baltimore, Md.: Williams and Wilkins, 1970.

Bender, R. E. *The Conquest of Deafness.* Cleveland, Ohio: Case Western Reserve University.

Burns, W., and Murray, J. *Noise and Man.* London, England, 1968.

This introduction to the subject of noise and its effect is intended for audiologists, engineers, industrial and general physicians, otologists, physicians and health physicists, safety officers, psychologists, public health officers and administrators, and deals with: the physical properties of sound; types of sound; measurement of sound; the mechanism of hearing, normal hearing and its measurement; deafness; disturbance, annoyance and effect on health; measures to reduce interference effects; temporary effects of noise on hearing; permanent effects of noise on hearing; preservation of hearing; aircraft noise; impulse noise, intense noise, sonic boom and vibration; and present and future industrial and community noise.

Burns, W., and Robinson, D. W. *Hearing and Noise in Industry.* London, England: Her Majesty's Stationery Office, 1970.

The objective of this book is to establish a quantitative relationship between industrial noise and the resultant impairment of hearing. The aim was to do this in as broad a way as possible within the limits of a field of study of predetermined duration so that the results would be generally applicable to industry. The hearing acuity of more than 750 people employed in a wide variety of occupations and exposed daily to noise for periods up to 50 years forms one half of the main body of the experimental material. Data on the noise exposure forms the complementary half.

The absolute protection of every person's hearing through universal noise level restriction has to be accepted as an impractical target at present, but it has become possible to state precisely in terms of direct noise measurements or forecasts the degree of risk inherent in prolonged exposure to the noise environment.

Davis, H., and Silverman, S. R. *Hearing and Deafness.* New York, N.Y.: Holt, Rinehart, and Winston, Inc., 1970.

This book was written for the deaf and the hard of hearing and for their families, their parents, their teachers and their friends. It was also written for physicians, educators, social workers, and for all who are concerned with the conservation or improvement of hearing or with the approach to normal living for those who have suffered either complete or partial hearing loss. It was written to answer the questions that are continually being asked about the nature of hearing and the problems posed by partial or complete loss of hearing. Many professionals who are concerned primarily with these problems have been so besieged with questions over the years that writing this book is perhaps partly a gesture of self-defense.

DeCarlo, L. *The Deaf.* Englewood Cliffs, N.J.: Prentice-Hall, Inc. 1964.

Glorig, A. *Noise and Your Ear.* New York, N.Y.: Grune & Stratton, 1958.

Griffith, J. *Persons With Hearing Loss.* Springfield, Ill.: Charles C Thomas, 1969.

Gulick, L. W. *Hearing Physiology and Psychophysics.* New York, N.Y.: Oxford University Press, 1971.

This textbook was written for students of sensory physiology and sensory psychology. The author selected those facts necessary to the development of concepts and principles in the hope that coherence would be gained and the student's learning thereby advanced. Although limited to a matter deemed essential to an appreciation of the fundamentals of hearing, the treatment is sufficient in detail to prepare a careful reader for subsequent advanced study in specialized topics. This book is appropriate for use by undergraduate and graduate students in the study of hearing and sensory processes. Since this is a textbook, there are several points that distinguish its character: First, an historical account of the evolution of major hypotheses is provided. Second, concepts and principles are considered as more important than the results of any single experiment except when the validity of the concept or principle is called into question. Third, special attention is given to the auditory theories and the extent to which the major ones are able to handle the facts of hearing as they are now known. Finally, relationships between physiological and psychophysical data are examined.

Littler, T. S. *The Physics of the Ear,* New York, N.Y.: Pergamon Press, 1965.

Kryter, K. D. *The Effects of Noise on Man.* New York, N.Y.: Academic Press, 1970.

This book was written for persons who are interested in the protection of the health and well-being of people exposed to sound.

Arranged in four parts, Parts I and II include fundamental definitions of sound, its measurement, and concepts of the basic functioning and attributes of the auditory system and speech communications. Part III is devoted to man's nonauditory system responses and includes information about the effects of noise on such things as work performance, sleep, feeling of pain, vision, and blood circulation. Tolerable limits of noise with respect to its effects on man's auditory and nonauditory systems are suggested in various places. Part IV is a summary.

Mykebust, H. R. *The Psychology of Deafness.* New York, N.Y.: Grune and Stratton, 1964.

O'Neill, John J. *The Hard of Hearing.* Englewood Cliffs, N.J.: Prentice-Hall, 1964.

Robinson, D. W. *Occupational Hearing Loss.* New York, N.Y.: Academic Press, 1971.

This book is the result of a conference held in Teddington, England, in 1970. It was called to bring together researchers into the quantitative relationship between industrial noise exposure and its harmful effects on the normal and pathological ear, and those representatives of other disciplines concerned with the practical implementation of industrial hearing-conservation measures and of hearing-loss compensation.

For the most part, the information in this book is intended for a technical audience. That does not mean to say that only technical persons can profit from reading this book. However, it is difficult for those who are not technically trained to appreciate the ramifications of much that is contained in the book.

Because this book is, basically, the proceedings of the conference, there are 18 papers or chapters, each by a different author and expert. The topics are divided into three sections: field studies, practical aspects, and assessing hearing loss. What makes this book more than just a collection of papers presented at a conference is that it contains an excellent index and bibliography. In addition, at the end of each section, there is a discussion of the papers presented.

Sataloff, J. *Hearing Loss.* Philadelphia, Pa.: J. B. Lippincott Company, 1966.

In practical but comprehensive form, this book presents the principles and procedures for determining the causes of hearing loss, including the otologic history, otologic examination, and hearing tests. This book has been written with a special awareness of the need of busy general practitioners, industrial physicians, internists, pediatricians, and others (whose practices intersect only incidentally with problems of hearing loss and deafness) to have available a handy reference volume that would enable them to be of greater service to their patients. In addition, medical students, audiologists, and other paramedical personnel who lack the time to search through the literature for essential background information may also find this book helpful.

Much of what appears in this book will seem elementary to specialists in otolaryngology. This definitely is a book for professionals, such as those already listed. There is hardly a page without some sort of audiogram or related illustration. Although it is almost essential to have a good medical background in order to fully appreciate this book, a layman can read it to advantage, especially if he has a good medical dictionary.

Industrial physicians should find this book helpful in deciding whether hearing loss among personnel is related to industrial noise, head injury, or nonoccupational causes. In addition, this book should help answer a patient's most urgent question, "Can anything be done to restore my hearing?"

Stevens, S. S., and Warshofsky, F. *Sound and Hearing.* New York, N.Y.: Time, Inc., 1965.

Welch, B. L., and Welch, A. S., ed. *Physiological Effects of Noise.* New York, N.Y.: Plenum Press, 1970.

VanItallie, Phillips H. *How to Live With a Hearing Handicap.* New York, N.Y.: Eriksson, 1963.

Wallenfels, Herman G. *Hearing Aids for Nerve Deafness.* Springfield, Ill.: Charles C Thomas, 1971.

AUDIOLOGY

Emerick, L. *A Workbook in Clinical Audiometry.* Springfield, Ill.: Charles C Thomas, 1971.

Engelberg, M. W. *Audiological Evaluation for Exaggerated Hearing Level.* Springfield, Ill.: Charles C Thomas, 1970.

Fletcher, H. *Speech and Hearing in Communication.* New York, N.Y.: Van Nostrand Reinhold Co., 1953.

Glorig, A. *Audiometry Principles and Practices.* Baltimore, Md.: Williams & Wilkins Co., 1965.

The audiometrist, whether he be otologist, audiologist, or technician, should be thoroughly familiar with the equipment and its use and should be experienced in testing hearing. The Committee on Conservation of Hearing of the American Academy of Ophthalmology and Otolaryngology has long felt the need for a book for practicing otolaryngologists, resident in otolaryngology, and participants in home study courses. The original task was undertaken to satisfy this need, but as preparations continued it became increasingly obvious that such a volume would be useful to audiologists in training as well as those in practice; industrial physicians and nurses, pediatricians and general practitioners; special education teachers, psychologists, safety professionals, and school audiometrists.

Graham, A. B., ed. *Sensori-neural Hearing Processes and Disorders.* Boston, Mass.: Little, Brown and Co., 1967.

Hirsh, J. *The Measurement of Hearing.* New York, N.Y.: McGraw-Hill Book Co. 1952.

Jerger, J. J., ed. *Modern Developments in Audiology.* New York, N.Y.: Academic Press, 1963.

Katz, J., ed. *Handbook of Clinical Audiology.* Baltimore, Md.: Williams & Wilkins Co., 1972.

The current state of the science art of clinical audiology is presented by 32 contributors. The following sections are surveyed: basic evaluation; differential diagnostic evaluation (cochlear versus retrocochlear, central auditory function, nonorganic loss and other special procedures); hearing and auditory preceptual study of children; hearing aids; and aural rehabilitation. This book attempts to be thorough, with an eye to future trends in research and treatment.

Langenbeck, B. *Textbook of Practical Audiometry.* Baltimore, Md.: Williams & Wilkins Company, 1965.

This book is intended as an introduction to the art of audiometry for the beginner and for the advanced students and as a systematic presentation of the diagnostic possibilities. This book contains examination methods which have proved their value in routine use. There are certain modifications in the technique of individual tests practiced elsewhere which may be justified. These have been indicated. Many new illustrations have been added. The audiograms represent typical findings that are to be considered as examples of many similar observations and documents.

Newby, H. A. *Audiology.* New York, N.Y.: Appleton-Century-Crofts,

1972.

Because it was written primarily as a text for the beginning student in audiology, this book presents an overview of the entire field of audiology. This book can serve as a valuable reference for safety professionals who have program responsibilities in industrial hearing conservation. The material is well organized, and the many charts and tables included in the text will assist the reader in better understanding the subject matter. Included is a well-prepared subject and name index.

O'Neill, J. J., and Oyer, H. J. *Applied Audiometry.* New York, N.Y. Dodd, Mead, & Co., 1966.

Rose, D. E., ed. *Audiological Assessment.* Englewood Cliffs, N.J.: Prentice-Hall, 1971.

Rosenblith, W. A., ed. *Sensory Communication.* New York, N.Y.: Wiley and the M.I.T. Press, 1961.

Sanders, D. A. *Aural Rehabilitation.* Englewood Cliffs, N.J.: Prentice-Hall, 1971.

Tobias, J. V., ed. *Foundations of Modern Auditory Theory.* Vol. I. New York, N.Y.: Academic Press, 1970.

Travis, L. E., ed. *Handbook of Speech Pathology and Audiology.* New York, N.Y.: Appleton-Century-Crofts, 1971.

Ventry, R.; Chaiklin, J. B.; and Dixon, R. E. *Hearing Measurement, A Book of Readings.* New York, N.Y.: Appleton-Century-Crofts, 1971.

Zemlin, W. R. *Speech and Hearing Science.* Englewood Cliffs, N.J.: Prentice-Hall, 1968.

VIBRATION

Bendat, J. S., and Piersol, A. G. *Measurement and Analysis of Random Data.* New York, N.Y.: John Wiley and Sons, 1966.

Blake, M. P., and Mitchell, W. S. *Vibration and Acoustic Measurement Handbook.* New York, N.Y.: Spartan Books, 1972.

This book provides a practical approach to go along with theoretical material that is already available. An effort is made to consider always the practical problem and the industrial need and to address the maintenance engineer, the informed technician, or anyone who is charged with the task of using vibration and acoustic measurements. The first 11 chapters cover the fundamental aspects of acoustics and vibration together with some of the more useful methods of vibration isolation. The next eight chapters are devoted entirely to examples drawn from the field of machine maintenance. In this field, vibration measurements cover the entire gamut from safety to economy of operation. Seven more chapters are devoted to miscellaneous measuring examples and measuring instrumentation. Finally, a chapter is devoted to standards and general information that should prove most valuable in the industrial situation.

Crandell, S. H., and Mark, W. D. *Random Vibrations in Mechanical Systems.* New York, N.Y.: Academic Press, 1963.

Crede, C. E. *Shock and Vibration Concepts in Engineering Design.* Englewood Cliffs, N.J.: Prentice-Hall, Inc., 1965.

Den Hartog, J. P. *Mechanical Vibrations.* New York, N.Y.: McGraw-Hill Book Co., 1961.

Jacobsen, L. S., and Ayre, R. S. *Engineering Vibrations.* New York, N.Y.: McGraw-Hill Book Co., 1958.

Harris, C. M., and Creded, C. E. *Shock and Vibration Handbook.* New York, N.Y.: McGraw-Hill Book Co., 1961.

Macduff, J. N., and Curreri, J. R. *Vibration Control.* New York, N.Y.: McGraw-Hill Book Co., 1958.

The intent of this book is to provide the basic theory essential to understanding and comprehending the fundamentals of the subject. The book prepares the student for future study and development by pointing out limitations of the elementary theory and indicating problems and methods for advanced study and application.

The material has been selected so as to have as broad a coverage of the field as possible, ranging from elementay vibration theory through such specific studies as rotor balancing, vibration isolation, steady-state and transient response, and sound control problems.

Morrow, C. T. *Shock and Vibration Engineering.* New York, N.Y.: John Wiley and Sons, Inc., 1963.

Myklestead, N. O. *Fundamentals of Vibration Analysis,* New York, N.Y.: McGraw-Hill Book Co., 1956.

Snowden, J. C. *Vibration and Shock in Damped Mechanical Systems.* New York, N.Y.: John Wiley and Sons, 1968.

Thomson, W. T. *Vibration Theory and Applications.* Englewood Cliffs, N.J.: Prentice-Hall Inc., 1965.

Van Santen, G. W. *Mechanical Vibrations.* Philips Technical Library. Amsterdam, Netherlands: Elsevier Publishing Co., 1953.

Wilson, W. K. *Vibration Engineering.* London, England: Charles Griffin, 1959.

Wilcox, J. B. *Dynamic Balancing of Rotating Machinery.* London, England: Sir Isaac Pitman & Sons Ltd., 1967.

Standards and specifications groups

American National Standards Institute
1430 Broadway, New York, N.Y. 10018

The American National Standards Institute coordinates and administers the federated voluntary standardization system in the United States, which provides all segments of the economy with national consensus standards required for their operations and for protection of the consumer and industrial worker. It also represents the nation in international standardization efforts through the International Organization for Standardization (ISO), the International Electrotechnical Commission (IEC), and the Pacific Area Standards Congress (PASC).

ANSI is a federation of some 1000 national trade, technical, professional, labor, and consumer organizations, government agencies, and individual companies. It coordinates the standards development efforts of these groups and approves the standards they produce as American National Standards when its Board of Standards Review determines that a national consensus exists in their favor.

American Society for Testing and Materials
1916 Race St. Philadelphia, Pa. 19103

ASTM is the world's largest source of voluntary consensus standards for materials, products, systems and services. There are currently more than 4800 ASTM standards.

ASTM membership is drawn from a broad spectrum of individuals, agencies, and industries concerned with materials. Members include engineers, scientists, researchers, educators, testing experts, companies, associations and research institutes, governmental agencies and departments (federal, state, and municipal), educational institutions, and libraries.

The society also publishes standards for atmospheric sampling and analysis, fire tests of materials and construction, methods of testing building construction, nondestructive testing, fatigue testing, radiation effects, pavement skid resistance, protective equipment for electrical workers, and others.

These constitute basic reference materials for the safety professional who will frequently be confronted with ASTM standards.

Other standards groups

In addition to the American National Standards Institute, many governmental and other agencies have established specifications used by safety professionals. Many industries through their trade associations have also established either (a) codes covering operations in their own plants or (b) safe practices to be followed in the use of their products.

Standards

ANSI

The following standards in acoustics and mechanical shock and vibration can be purchased from the American Standards Institute.

S1.1–1960 (R1971), *Acoustical Terminology* (Including Mechanical Shock and Vibration) (Agrees with ISO R131) (ISO R16 and IEC 50–08)

S1.2–1962 (R1971), *Physical Measurement of Sound, Method for* (Partially revised by S1.13–1971 and S1.21–1972)

S1.4–1971, *Sound Level Meters, Specification for* (IEC 123)

S1.6–1967 (R1971), *Preferred Frequencies and Band Numbers for Acoustical Measurements* (Agrees with ISO R266)

S1.7–1970, *Sound Absorption of Acoustical Materials in Reverberation Rooms, Method of Test for* (ASTM C423–66)

S1.8–1969, *Preferred Reference Quantities for Acoustical Levels*

S1.10–1966 (R1971), *Calibration of Microphones, Method for the* (IEC 327)

S1.11–1966 (R1971), *Octave, Half-Octave, and Third-Octave Band Filter Sets, Specification for* (IEC 225)

S1.12–1967 (R1972), *Laboratory Standard Microphones, Specifications for*

S1.13–1971, *Sound Pressure Levels, Methods for the Measurement of* (Partial revision of S1.2–1962 (R1971)

S1.21–1972, *Sound Power Levels of Small Sources in Reverberation Rooms, Methods for the Determination of* (partial revision of S1.2–1962)

S2.2–1959 (R1971), *Calibration of Shock and Vibration Pickups, Methods for the*

S2.3–1964 (R1970), *High-Impact Shock Machine for Electronic Devices, Specifications for*

S2.4–1960 (R1971), *Specifying the Characteristics of Auxiliary Equipment for Shock and Vibration Measurements, Method for*

S2.5–1962 (R1971), *Specifying the Performance of Vibrating Machines, Recommendations for*

S2.6–1963 (R1971), *Specifying the Mechanical Impedance of Structures, Nomenclature and Symbols for*

S2.7–1964 (R1971), *Balancing Rotating Machinery, Terminology for*

S2.8–1972, *Resilient Mountings, Guide for Describing the Characteristics of*

S2.10–1971, *Analysis and Presentation of Shock and Vibration Data, Methods for*

S2.11–1969 (R1973), *Calibrations and Tests for Electrical Transducers Used for Measuring Shock and Vibration, Selection of*

S2.14 1973, *Performance of Shock Machines, Methods for Specifying the*

S2.15–1972, *Design, Construction, and Operation of Class HI (High Impact) Shock-Testing Machine for Lightweight Equipment, Specification for the* (Revision and redesignation of Z24.17–1955 (R1966))

S3.1–1960 (R1971), *Background Noise in Audiometer Rooms, Criteria for*

S3.2–1960 (R1971), *Monosyllabic Word Intelligibility, Method for Measurement of*

S3.3–1960 (R1971), *Electroacoustical Characteristics of Hearing Aids, Methods for Measurement of* (IEC 118)

S3.4–1968 (R1972), *Procedure for the Computation of Loudness of Noise* (ISO R357)

S3.5–1969 (R1973), *Methods for the Calculation of the Articulation Index*

S3.6–1969 (R1973), *Audiometers, Specifications for* (IEC 177)

S3.7–1973, *Coupler Calibration of Earphones, Method for* (Revision and redesignation of Z24.9–1949)

S3.8–1967 (R1971), *Hearing Aid Performance, Method of Expressing*

S3.13–1972, *Artificial Head-Bone for the Calibration of Audiometer Bone Vibrators*

S3.19–1974, *Measurement of the Real-Ear Protection of Hearing Protectors and Physical Attenuation of Earmuffs, Method for the*

S4.1–1960 (R1972), *Mechanically-Recorded Lateral Frequency Records, Methods of Calibration of* (58 IRE 19.S1: IEEE Std 192–1958)

S5.1–1971, *Measurement of Sound from Pneumatic Equipment, Test Code for the*

S6.1–1973, *Qualifying a Sound Acquisition System* (SAE RP J184–1973)

S6.2–1973, *Exterior Sound Level for Snowmobiles* (SAE RP J192–1973)

S6.3–1973, *Sound Level for Passenger Cars and Light Trucks* (SAE RP J986a–1973)

S6.4–1973, *Computing the Effective Perceived Noise Level for Flyover Aircraft Noise, Definitions and Procedures for* (SAE ARP 1071–June 1972)

The following will be redesigned as S standards as they are revised:

Z24.21–1957 (R1971), *Specifying the Characteristics of Pickups for Shock and Vibration Measurement, Method for*

The following are recommendations of the International Organization for Standardization (ISO) and are available from ANSI:

R16–1955, *Standard Tuning Frequency* (*Standard Musical Pitch*) (S1.1–1960)

R31–Part VII–1965, *Quantities and Units of Acoustics*

R131–1959, *Expression of the Physical and Subjective Magnitudes of Sound or Noise* (Agrees with S1.1–1960) (Including Supplement R357–1963)

R140–1960, *Field and Laboratory Measurements of Airborne and Impact Sound Transmission*

R226–1961, *Normal Equal-Loudness Contours for Pure Tones and Normal Threshold of Hearing Under Free Field Listening Conditions*

R266–1962, *Preferred Frequencies for Acoustical Measurements* (Agrees with S1.6–1967)

R354–1963, *Measurement of Absorption Coefficients in a Reverberation Room*

R357–1963, *Expression of the Power and Intensity Levels of Sound or Noise* (Supplement to R131–1959) (S3.4–1968)

R362–1964, *Measurement of Noise Emitted by Vehicles*

R389–1964, *Standard Reference Zero for the Calibration of Pure-tone Audiometers,* including Addendum 1–1971

Addendum 1–1971 *Additional Data in Conjunction with the 9-A Coupler,* sold separately

R454–1965, *Relation Between Sound Pressure Levels of Narrow Bands of Noise in a Diffuse Field and in a Frontally-Incident Free Field for Equal*

Loudness

R495–1966, *General Requirements for the Preparation of Test Codes For Measuring the Noise Emitted by Machines*

R507–1970, *Procedure for Describing Aircraft Noise Around an Airport*

R532–1966, *Method for Calculating Loudness Level*

R717–1968, *Rating of Sound Insulation for Dwellings*

R1761–1970, *Monitoring Aircraft Noise Around an Airport*

The following are standards of the International Electrotechnical Commission (IEC); they are available from ANSI:

50–08 (1960) *International Electrotechnical Vocabulary, Group 08: Electro-Acoustics* (S1.1–1960)

118 (1959) *Recommended Methods for Measurements of the Electro-Acoustical Characteristics of Hearing Aids* (S3.3–1960)

123 (1961) *Recommendations for Sound Level Meters* (S1.4–1961)

124 (1960) *Recommendations for the Rated Impedances and Dimensions of Loudspeakers*

126 (1973) *IEC Reference Coupler for the Measurement of Hearing Aids Using Earphones Coupled to the Ear by Means of Ear Inserts*

177 (1965) *Pure Tone Audiometers for General Diagnostic Purposes* (S3.6–1969)

178 (1965) *Pure Tone Screening Audiometers* (S3.6–1969)

179 (1973) *Precision Sound Level Meters*

184 (1965) *Methods for Specifying the Characteristics of Electro-Mechanical Transducers for Shock and Vibration Measurements*

222 (1966) *Methods for Specifying the Characteristics of Auxillary Equipment for Shock and Vibration Measurement*

225 (1966) *Octave, Half-Octave, and Third-Octave Band Filters Intended for the Analysis of Sounds and Vibrations* (S1.11–1966)

263 (1968) *Scales and Sizes for Plotting Frequency Characteristics*

303 (1970) *IEC Provisional Reference Coupler for the Calibration of Earphones Used in Audiometry*

327 (1971) *Precision Method for Pressure Calibration of One-Inch Standard Condenser Microphones by the Reciprocity Technique*

American Society for Testing and Materials (ASTM)

Test Methods

C367–57, *Strength Properties of Prefabricated Architectural Acoustical Materials*

845

C384–58, *Impedance and Absorption of Acoustical Materials by the Tube Method*

C423–66, *Sound Absorption of Acoustical Materials in Reverberation Rooms* (S1.7–1970)

C522–69, *Airflow Resistance of Acoustical Materials*

C643–69, *Painting Ceiling Materials for Acoustical Absorption Tests*

RECOMMENDED PRACTICES

E90–70, *Laboratory Measurement of Airborne Sound Transmission Loss of Building Partitions*

E336–67T, *Measurement of Airborne Sound Insulation in Buildings*

C636–69, *Installation of Metal Ceiling Suspension Systems for Acoustical Tile and Lay-In Panels*

DEFINITIONS

C634–69, *Acoustical Tests of Building Constructions and Materials*

American Society of Heating, Refrigerating and Air-Conditioning Engineers (ASHRAE)

36–62, *Measurement of Sound Power Radiated from Heating, Refrigerating and Air Conditioning Equipment*

36A–63, *Method of Determining Sound Power Levels of Room Air Conditioners and Other Ductless, Through-the-Wall Equipment*

36B–63, *Method of Testing for Rating the Acoustic Performance of Air Control and Terminal Devices and Similar Equipment*

36R–Feb 1970, *Method of Testing for Sound Rating Heating, Refrigerating and Air Conditioning Equipment*

Instrument Society of America (ISA)

RP37.2, *Guide for Specifications and Tests for Piezoelectric Acceleration Transducers for Aerospace Testing* (1964)

S37.10, *Specification and Tests for Piezoelectric Pressure and Sound-Pressure Transducers* (1969)

Institute of Electrical and Electronic Engineers (IEEE)

IEEE 85, *Airborne Noise Measurements on Rotating Electric Machinery* (1965)

IEEE 219, *Loudspeaker Measurements* (ANSI S1.5–1963)

IEEE 258, *Methods of Measurement for Close-Talking Pressure Type Microphones* (January 1965)

Society of Automotive Engineers (SAE)

SAE COMMITTEE A-21, AIRCRAFT NOISE MEASUREMENT

AIR 817, *A Technique for Narrow Band Analysis of a Transient*

AIR 852, *Methods of Comparing Aircraft Takeoff and Approach Noises*

ARP 865a, *Definitions and Procedures for Computing the Perceived Noise Level of Aircraft Noise*

ARP 866, *Standard Values of Atmospheric Absorption as a Function of Temperature and Humidity for Use in Evaluating Aircraft Flyover Noise*

SAE SOUND LEVEL COMMITTEE

J6a, *Ride and Vibration Data Manual*

J336, *SAE Recommended Practice, Sound Level for Truck Cab Interior*

J366, *SAE Recommended Practice, Exterior Sound Level for Heavy Trucks and Buses*

J377, *SAE Standard, Performance of Vehicle Traffic Horns*

J672a, *SAE Standard, Exterior Loudness Evaluation of Heavy Trucks and Buses*

J919, *SAE Recommended Practice, Measurement of Sound Level at Operator Station*

J952a, *SAE Standard, Sound Levels for Engine Powered Equipment*

J986a, *SAE Standard, Sound Level for Passenger Cars and Light Trucks*

J994, *SAE Recommended Practice, Criteria for Backup Alarm Devices*

Air Conditioning and Refrigeration Institute (ARI)

ARI 443–66, *Room Fan-Cool Air Conditioner*

ARI 270–67, *Sound Rating of Outdoor Unitary Equipment*

Air Diffusion Council (ADC)

AD–63, *Measurement of Room-to-Room Sound Transmissions Through Plenum Air Systems*

1062–R2, *Equipment Test Code*

Air Moving and Conditioning Association (AMCA)

Bulletin 300–67, *Test Code for Sound Rating*

Bulletin 301–65, *Method of Publishing Sound Ratings for Air Moving Devices*

Bulletin 302–65, *Application of Sone Loudness Ratings for Non-Ducted Air Moving Devices*

Bulletin 303–65, *Application of Sound Power Level Ratings for Ducted Air Moving Devices*

Publication 311–67, *AMCA Certified Sound Ratings Program for Air Moving Devices*

American Gear Manufacturers Association (AGMA)

295.02, *AGMA Standard Specification for Measurement of Sound in High Speed Helical and Herringbone Gear Units,* November 1965

The Anti-Friction Bearing Manufacturers Association, Inc.

AFBMA Standard, Section 13, Roller Bearing Vibration and Noise

Association of Home Appliance Manufacturers

Standard RAC-2-SR, *Room Air Conditioner Sound Rating,* January 1971

Compressed Air and Gas Institute (CAGI)

(See ANSI S5.1–1971)

National Electrical Manufacturers Association (NEMA)

SM33–1964, *Standards Publication, Gas Turbine Sound and Its Reduction*
MG1, *Motors and Generators*
Section 4.3.2, *Methods of Measuring Machine Noise*
TR1–1963, *Transformers, Regulators and Reactors* (Sections 9–04 and 9–05)

National Fluid Power Association (NFPA)

NFPA T3.9–70.12, *Method of Measuring Sound Generated by Hydraulic Fluid Power Pumps*

National Machine Tool Builders Association (NMTBA)

Noise Measurement Techniques

Educational

Training courses

Numerous short courses in noise and vibration control and audiometry are offered by universities, service organizations, and a few private firms. Specific information regarding any course can be obtained by writing to the sponsoring agency.

Center for Continuing Education
Appalachian State University
Boone, N.C. 28607

Speech and Hearing Clinic
Arizona State University
Tempe, Ariz. 85281

B&K Instruments, Inc.
5111 West 164th St.
Cleveland, Ohio 44142

Bolt Beranek and Newman, Inc.
50 Moulton St.
Cambridge, Mass. 02138

Department of Speech
Bowling Green State University
Bowling Green, Ohio 43403

Continuing Education in Engineering
University Extension
University of California
2223 Fulton St.
Berkeley, Calif. 94720

The Center for Professional Advance-
ment
P.O. Box 997
Somerville, N.J. 08876

Colby College
Waterville, Maine 04901

Colby College Institute
1721 Pine St.
Philadelphia, Pa. 19103

Continuing Education Services
University of Colorado School of
Nursing
4200 East Ninth Ave.
P.O. Container C-287
Denver, Colo. 80220

Dept. of Otolaryngology and
Maxillofacial Surgery
Medical Center
University of Cincinnati
Cincinnati, Ohio 45267

Engineering Society of Detroit
100 Farnsworth
Detroit, Mich. 48202

University of Hartford
Division of Continuing Education
200 Bloomfield Ave.
West Hartford, Conn. 06117

HCI, Inc.
Hearing Conservation Div.
810 East State St.
Rockford, Ill. 61104

Industrial Audiometry Institute
1530 N. Sycamore Ave.
Fullerton, Calif. 92631

IIT Research Institute
10 West 35th St.
Chicago, Ill. 60616

Institute on Noise Control, Inc.
3456 Altonah Road
Bethlehem, Pa. 18017

School of Medicine
University of Kansas
39th and Rainbow
Kansas City, Kans.

Office of Continuing Education
College of Engineering
University of Kentucky
Lexington, Ky. 40506

Kresge Hearing Research Laboratory
of the South
School of Medicine
Louisiana State University
1100 Florida Ave., Building 164
New Orleans, La. 70119

Communications Disorder Clinic
Department of Communication
Disorders
University of Massachusetts
Amherst, Mass. 01002

Dept. of Audiology and Speech
Pathology
Memphis State University
807 Jefferson Ave.
Memphis, Tenn. 38105

Engineering Summer Conferences
Chrysler Center, North Campus
University of Michigan
Ann Arbor, Mich. 48105

Extension Division
University of Missouri—Rolla
511 West 11th St.
Rolla, Mo. 65401

Department of Safety
School of Public Services
Central Missouri State University
Warrensburg, Mo. 64093

Safety Training Institute
National Safety Council
425 North Michigan Ave.
Chicago, Ill. 60611

Division of Continuing Education
North Carolina State University
P.O. Box 5125
Raleigh, N.C. 27607

Center for Continuing Education
Northeastern University
Huntington Ave.
Boston, Mass. 02115

Hearing Clinic
Northwestern University
Frances Searle Building
Evanston, Ill. 60201

Audiology Section
Department of Otolaryngology
The Ohio State University
3024 University Hospitals Clinic
456 Clinic Drive
Columbus, Ohio 43210

The University of Oklahoma
1700 Asp Ave.
Norman, Okla. 73069

Industrial Programs Service
School of Continuing Studies
Old Dominion University
Norfolk, Va. 23508

Continuing Education
The Pennsylvania State University
J. Orvis Keller Conference Center
University Park, Pa. 16802

School of Mechanical Engineering
Purdue University
West Lafayette, Ind. 47907

Spectral Dynamics Corp.
P.O. Box 671
San Diego, Calif. 92112

School of Continuing Studies
University of Toronto
119 St. George St.
Toronto, Ontario M5SIA9
Canada

Special Programs Office
Wells House
Union College
Schenectady, N.Y. 12308

Speech Pathology and Audiometry
University of Utah
1201 Behavioral Science Bldg.
Salt Lake City, Utah 84112

Bill Wilkerson Hearing and Speech
    Center
Vanderbilt University
1114 19th Ave. South
Nashville, Tenn. 37212

The Vibration Institute
5401 Katrine
Downers Grove, Ill. 60515

Medical College of Virginia
Virginia Commonwealth University
1109 East Clay St.
P.O. Box 157
Richmond, Va. 23219

Continuing Engineering Education
The George Washington University
Washington, D.C. 20052

School of Continuing Education
P.O. Box 1099
Washington University
St. Louis, Mo. 63130

Department of Audiology
School of Medicine
Wayne State University
261 Mack Blvd.
Detroit, Mich. 48201

Office of Continuing Professional
    Education
West Coast University
440 Shatto Place
Los Angeles, Calif. 90020

Dept. of Speech Pathology and
    Audiology
Western Michigan University
Kalamazoo, Mich. 49001

Department of Engineering
University of Wisconsin, Extension
432 North Lake St.
Madison, Wis. 53706

Health Sciences Unit
University of Wisconsin, Extension
600 West Kilbourn Ave.
Milwaukee, Wis. 53203

Audiovisuals

Films can be very effective when used within the context of a hearing conservation program. Following are some sources and films:

American Academy of Ophthalmology and Otolaryngology
1966 Inwood Road
Dallas, Tex. 75235

Audiometric Techniques. Motion picture, sound, 20 minutes.

Describes the audiometric technique (not the standard Inter-Society Committee technique, however)

The Ear and Noise. Motion picture, sound, color, 10 minutes.

Demonstrates control measures for conservation of hearing; for management.

Ear Protection and Noise. Motion picture, sound, 10 minutes.

Stresses the need for personal hearing protection for those exposed to high noise levels. Touches on noise control and audiometry. Especially valuable for supervisors and foremen.

The Effects of Noise on Man. Motion picture, sound, 25 minutes.

Describes the need and method of hearing-conservation measures and how noise can be combatted through engineering control and research. Film is excellent overview for people who are as yet unfamiliar with the problem.

851

American Optical Corporation
Safety Products Division
Southbridge, Mass. 01550

The Sound of Sound. Motion picture, sound, color, 16 minutes.

Film focuses on industrial noise; its insidious attack on hearing; the permanence of a noise related hearing loss. Industrial workers speak candidly of their occupational deafness—their loneliness and frustration. These men avidly support the film's basic message, "Keep the hearing you've got. Wear the proper hearing protection."

Bray Studios, Inc.
630 Ninth Ave.
New York, N.Y. 10036

Protect Your Hearing. Motion picture, sound, color, 15 minutes.

For industrial and general audiences, the film uses both live action and animation to explain the ear's vulnerability and how to protect it.

David Clark Company
360 Franklin St.
Worcester, Mass. 01604

Protect Your Hearing. Motion picture, sound, 16 minutes.

Film stresses hearing protection but shows all aspects of hearing conservation in the industrial setting. Excellent single teaching film for foremen and supervisors.

Educational Program Development
Naval Health Sciences
Education and Training Command
National Naval Medical Center
Bethesda, Md. 20014

Have You Heard? Motion picture, color, sound, 26 minutes.

A motivational and information film for all Navy personnel describing the hazards of high intensity noise found in military and industrial environments which can lead to progressive and permanent hearing loss. The film shows the effect hearing impairment can have in work and family relationships, illustrates the pathology of hearing loss from high intensity noise through the destruction of hair cells in the organ of Corti, and prescribes various noise control measures to reduce these hazards.

Medical Aspects of High Intensity Noise: Ear Defense. Motion picture, sound, black & white, 21 minutes.

Points out the hazards associated with high noise levels produced by jet aircraft and other noisy equipment found ashore and aboard ship. Describes the nature of noise, its effect on hearing and various devices that are used for the protection of hearing.

Medical Aspects of High Intensity Noise: General Effects. Motion picture, sound, black & white, 21 minutes.

Explains the increasingly serious hazards of high intensity noise; describes the nature of noise and some of its physiological and psychological effects; and gives examples of sounds of extreme intensity approximating conditions found near jet aircraft, artillery, and other noise-producing equipment.

Role of the Medical Department in Hearing Conservation. Motion picture, sound, color, 28 minutes.

Demonstrates in real-life situations the elements of the Navy Medical Department's program of hearing conservation: Noise Measurement and Analysis, Engineering Control, Audiometry, Use of Protective Devices, and Education. Emphasized prevention of hearing loss by means of periodic audiograms and the wearing of hearing-protective devices.

Encyclopaedia Britannica Educational Corporation
425 North Michigan Ave
Chicago, Ill. 60611

The Ear and Hearing. Motion picture, sound, color, 10 minutes.

Shows functioning of the hearing mechanism; has actual shots of the middle ear. Shows effects of injuries and disease on hearing.

Federal Aviation Administration
Film Library, AC-44.5
P.O. Box 25082
Oklahoma City, Okla. 73125

Can We Have a Little Quiet, Please? Motion picture, sound, color, 14½ minutes.

This documentary illustrates how government and industry are cooperating to reduce aircraft smoke emissions and noise (particularly around airports), and describes technical improvements that have been made to jet engines and sound abatement procedures.

Ford Motor Company
Medical Department
Dearborn, Mich. 48121

Hearing Conservation. Filmstrip and tape, 25 minutes.

Describes all phases of hearing conservation; stresses ear anatomy and personal hearing protection.

General Motors Corporation
Public Relations—Film Library
General Motors Building
Detroit, Mich. 48202

A Sound Motion Picture About Decibels. Motion picture, sound, color, 18 minutes.

Explores the sea of sound around us and, through the use of simple experiments, illustrates why some sounds please while others annoy. Industrial and vehicular noise pollution are investigated and some of the continuing research being conducted in these areas is revealed.

Gordon Stowe and Associates
P.O. Box 233
Northbrook, Ill. 60062

How They Hear. 33⅓-LP record and 35mm color slides with script.

Vivid audio demonstration of how a person with a hearing loss hears under various circumstances. Mild, moderate, and severe losses are demonstrated.

Alfred Higgins Productions, Inc.
9100 Sunset Blvd.
Los Angeles, Calif. 90069

Who Stole the Quiet Day. Motion Picture, sound, color, 15 minutes.

The purpose of this film is (*a*) to make the viewer aware that noise can be a health hazard and that he should take precautions against it, and (*b*) to explain how this damage occurs and to delineate ways a person can protect himself. Animated diagrams show how the ear converts sound waves into nervous impulses and how impairment can take place. An audiologist explains the first signs of damage. Then an otolaryngologist presents data that indicate that noise has a deleterious effect on blood pressure, the heart, and animals exposed to noise. Ways of protecting hearing are described. For the lay audience.

International Film Bureau, Inc.
332 South Michigan Ave.
Chicago, Ill. 60604

Listen While You Can—Dangerous Noise I. Motion picture, sound, color, 21 minutes.

Film stresses the importance of evaluating noise before it produces noise and demonstrates the type of hearing loss sustained from working in areas where noise is a hazard. The structure of the ear is explained.

Hearing Conservation—Dangerous Noise II. Motion picture, sound, color, 22 minutes.

Film stresses the importance of evaluating noise before it produces hearing loss. Graphs are plotted to show the noise levels on the worker who uses no ear protection and on the worker using earmuffs and earplugs.

Medical Aspects and Hearing Conservation. Motion picture, sound, color, 22 minutes.

Geared to audiences concerned with the medical, administrative, and technical management aspects of noise reduction.

Mines Accident Prevention Association of Ontario
1399 Hammond St.
North Bay, Ontario, Canada

Noise and Hearing. Motion picture, sound, color, 3 films 25 minutes each.

Part 1 deals with hearing and the effects of noise; Part 2 with the measurement of noise; Part 3 with the control of noise, using laboratory examples.

Modern Talking Picture Service, Inc.
(all major metropolitan areas)

To Conserve and Protect. Motion picture, sound, color, 14½ minutes.

As narrator James Mason says, "Noise pollution, if allowed to go unchecked, will rob millions of us of our God-given gift—the ability to hear." This important film covers the different aspects of noise pollution—its causes, bad effects on human beings, and what can and must be done to conserve and protect our precious ability to hear.

National Safety Council
425 North Michigan Ave.
Chicago, Ill. 60611

Hear What You Want To Hear. Motion picture, sound, color, 10 minutes.

Employees can actually participate in this film. The clever technique encourages use of hearing-protection devices.

It's Your Hearing. 30 2 x 2 color slides with script.

Explains how long exposure to excessive noise can harm hearing. Slides advise machine operators and others who work in high-noise areas to get good hearing protection and describe a number of types of hearing protectors.

Price Filmmakers
3491 Cahuenga Blvd.
Hollywood, Calif. 90028

An Approved Technique for Pure-Tone, Air-Conduction Audiometry. Motion Picture, sound, color, 14 minutes.

Technique for measuring hearing threshold of employees is explained in instructional steps. Why such testing is important and the need for standardizing techniques are also emphasized.

Ear Protection in Noise. Motion picture, sound, color, 12 minutes.

Seeks to make audiences aware of the problems of excessive noise.

Hearing—The Forgotten Sense. Motion picture, sound, color, 20 minutes.

Film describes noise and how it affects the ear. It discusses compensation for impaired hearing and describes a simple hearing test.

It Takes Two. Motion picture, sound, 18 minutes.

This film is specifically directed to the "why" and "how" of personal hearing protection. Uses case history approach to introduce a program of personal hearing protection. For nurses, supervisors, and foremen.

Pyramid Film Producers
P.O. Box 1048
Santa Monica, Calif. 90406

Sound Off. Motion picture, sound, color, 10 minutes.

Film graphically demonstrates the proper safeguards for hearing protection for the worker.

U.S. Department of the Interior
Mine Enforcement and Safety Administration
4800 Forbes Ave.
Pittsburgh, Pa. 15213

Mining Noise Hazards. Motion picture, sound, color, 21 minutes.

How we hear; the function of the outer, middle, and inner ear; and the effects of noise exposure on the ear are defined. Shown are actual underground mining scenes, on-the-job noise surveys, and laboratory tests and analysis. Preventive methods and recommendations are made.

I. C. Webb Company
301 North Madison St.
Rockford, Ill. 61110

Listen While You Can. Motion picture, sound, 15 minutes.

Covering noise level assessments, engineering control, audiometric examination, and methods of personal hearing protection, this film is excellent for safety directors, supervisors and foremen, and top management. British locale is no drawback for American audiences.

Willson Products Division
2nd & Washington Streets
Reading, Pa 19603

For Good Sound Reasons. Motion picture, sound, color, 17 minutes.

Hearing protection with emphasis on use of hearing equipment. Film dramatically shows effects of hearing loss on one family.

Beltone Electronics Corporation
Beltone Building
4201 West Victoria Street
Chicago, Il 60646

How We Hear. Filmstrip and 33⅓ LP record.

Very effective demonstration of the human hearing process. Shows how sound energy is converted to neural impulses.

Appendices

Appendix A Federal Occupational Noise Regulations

1. Walsh-Healey Regulations

This regulation will be found in Title 41—Public Contracts and Property Management, *Code of Federal Regulations,* Chapter 50.

§ 50–204.10. OCCUPATIONAL NOISE EXPOSURE

(a) Protection against the effects of noise exposure shall be provided when the sound levels exceed those shown in Table I of this section when measured on the A scale of a standard level meter at slow response. When noise levels are determined by octave band analysis, the equivalent A-weighted sound level may be determined as follows:

Octave band sound pressure levels may be converted to the equivalent A-weighted sound level by plotting them on the illustrated graph and noting the A-weighted sound level corresponding to the point of highest penetration into the sound level contours. This equivalent A-weighted sound level, which may differ from the actual

TABLE I
Permissible Noise Exposure*

Duration per day, hours	Sound level Slow response dBA
8	90
6	92
4	95
3	97
2	100
1½	102
1	105
½	110
¼ or less	115

* When the daily noise exposure is composed of two or more periods of noise exposure of different levels, their combined effect should be considered, rather than the individual effect of each. If the sum of the following fractions: $C_1/T_1 + C_2/T_2 \ldots C_n/T_n$ exceeds unity, then, the mixed exposure should be considered to exceed the limit value. C_n indicates the total time of exposure at a specified noise level, and T_n indicates the total time of exposure permitted at that level.

861

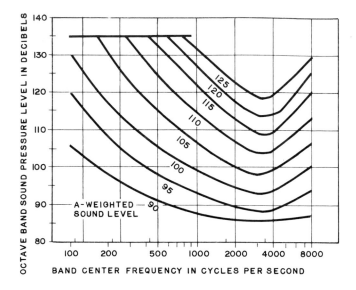

A-weighted sound level of the noise, is used to determine exposure limits from Table I.

(b) When employees are subjected to sound exceeding those listed in Table I of this section, feasible administrative or engineering controls shall be utilized. If such controls fail to reduce sound levels within the levels of the table, personal protective equipment shall be provided and used to reduce sound levels within the levels of the table.

(c) If the variations in noise levels

involve maxima at intervals of one second or less, it is to be considered continuous. In such cases, where the duration of the maxima are less than one second, they shall be treated as of one-second duration.

(d) In all cases where the sound levels exceed the values shown herein, a continuing, effective hearing-conservation program shall be administered.

Exposure to impulsive or impact noise should not exceed 140 dBC peak sound pressure level fast response.

2a. Occupational Safety and Health Act of 1970

After almost three years of contro versy and legislative battle, President Richard M. Nixon signed into law the Occupational Safety and Health Act (Public Law No. 91–596) on December 29, 1970.

Here's a summary of the major provisions:

Purpose

"To assure so far as possible every working man and woman in the nation safe and healthful working conditions . . . ."

Coverage

The act is applicable to all businesses engaged in interstate commerce—except for the mining and railroad industries, which are covered by other laws. Special provisions are made for federal and state government employees.

Employer duties

Each employer "shall furnish to each of his employees employment and a place of employment, which are free from recognized hazards that are causing, or are likely to cause, death or serious physical harm to his employees," and shall comply with the occupational safety and health standards and all rules pursuant to the act, except where an approved state plan is in effect.

Promulgation of standards

The Secretary of Labor has the full authority to promulgate (as soon as practicable but not later than two years after the effective date of the act) any national occupational safety or health consensus standard or any established federal standard, unless he determines that such a standard would not result in improved safety or health, and he need not hold hearings for the promulgation of such standards. A national consensus standard requires, among other things, that the standard has been adopted by a nationally recognized standards-producing organization under procedures that afforded opportunity for diverse views to be considered and that interested and affected persons have reached substantial agreement.

The secretary has the authority to promulgate, modify, or revoke any occupational safety or health standard using informal hearing procedures of the Administrative Procedure Act.

The secretary must provide for "emergency temporary standards," which are to take effect immediately upon publication in the *Federal Register,* if he determines that employees are exposed to "grave danger" from exposure to substances or agents determined to be toxic or physically harmful or from any new hazard, and that such emergency standard is necessary to protect employees from such danger. Such standards are to remain in effect until superseded by a standard promulgated under the procedures prescribed in this Act. Any affected employer may apply for variance from such standards promulgated or may file a petition challenging the validity of such standards.

Inspection and investigation

Inspections and investigations are to be made by the Secretary of Labor or the Secretary of Health, Education, and Welfare. Subject to regulations by the Secretary of Labor, both employer and employee representatives have the right to accompany an inspector during his physical examination of the plant. Employees may request an inspection if they believe an alleged violation threatens physical harm or that an "imminent danger" exists. Such inspections may be denied where the secretary determines that no reasonable grounds exist.

If the secretary determines that an employer has violated the "general duty" provision, promulgated standards or rules, or the records requirement, the secretary must issue a citation in writing for violation with reasonable promptness—but not later than six months following the occurrence of any violation. If the secretary arbitrarily or capriciously fails to seek judicial relief to counteract "imminent danger," any employee who may be injured by reason of such failure may bring action against the secretary to compel him to take such action.

Enforcement

If the secretary issues a citation for violation, he must notify the employer by certified mail of the penalty, if any, proposed to be assessed, and he must advise the employer that he has 15 working days within which to notify the secretary that he wishes to contest the citation or proposed assessment of penalty. If the employer notifies the secretary that he intends to contest a citation or proposed assessment of penalty, the secretary must immediately advise the Occupational Safety and Health Review Commission. Likewise, if any employee files a notice with the secretary alleging that the time set by the secretary for abatement of the violation is unreasonable, the secretary must immediately advise the commission. The commission, consisting of three presidential appointees, after opportunity for a hearing, shall adjudicate the dispute. The commission's orders become final 15 days after issuance unless stayed by court order. Uncontested citations shall be deemed to be a final order of the commission and shall not be subject to judicial review.

Judicial review

Any "person," which includes corporations or the secretary, adversely affected or aggrieved by an order of the commission may obtain a review of such order in a U.S. Court of Appeals, if sought within 60 days of the order's issuance. The proceeding will not operate as a stay of the commission's order unless so directed by the court.

Variances

Employers may obtain variances from a standard for one-year periods, renewable to a maximum of three years, on a showing of inabilities to meet standards because of unavailability of personnel or equipment or time to construct or alter facilities. However, the employer must also have a program for achieving compliance while taking all available safeguarding steps in the interim.

Imminent danger

A plant or location can be shut down because of "imminent danger" only by a court order. "Imminent danger" is defined as *a condition or practice involving a danger that could reasonably be expected to cause death or serious physical harm* immediately or before the imminence of such danger can be eliminated through the enforcement procedures otherwise provided by act.

Penalties

Civil penalties are provided up to $1000 for each violation—where they are not of a serious nature, such penalty is discretionary—and for each day in which a final order is violated. A penalty of up to $10,000 is provided for each willful or repeated violation of employer duties (described earlier). Criminal penalties are set for willful violations resulting in death.

State-federal relationships

The act places all jurisdiction regarding occupational safety and health under its terms in the federal government, except for those occupational safety and health issues for which no federal standard is in effect. A state can assume jurisdiction by submitting a state plan that is approved by the Secretary of Labor. The secretary may have dual jurisdiction in some respects for at least three years after approval of the state plan.

The act provides for grants to the states up to 90 percent of the total cost to assist them in identifying their needs and responsibilities in the area of occupational safety and health, in developing state plans, in developing plans for data collection, and for experimental and developmental projects. Likewise, grants up to 50 percent of the total cost are provided to assist the states in administering and enforcing programs for occupational safety and health contained in approved state plans.

Effect on other laws

Enforcement procedures of the new act become immediately applicable to the Walsh-Healey Public Contracts Act. Standards promulgated under the Walsh-Healey Public Contracts Act, the Service Contract Act, the National Foundation on Arts and Humanities Act, and the Longshoremen's and Harbor Workers' Compensation Act will be superseded when corresponding standards, which the secretary determines will be "more effective," are issued under the new Act.

Contractors of federal or federally financed contracts are subject to penalties of both the new Act and the 1969 Construction Safety Act. Congressional intent was stated that there be a single set of standards applicable to both categories of contractors.

Records requirement

Each employer will be required to maintain and make available such records as the secretary, in cooperation with the Secretary of HEW, may prescribe as appropriate for the enforcement of the act for developing information regarding the causes and prevention of occupational accidents and illnesses. Such rules may include provisions requiring employers to con-

duct periodic inspections (but not to determine or report their own state of compliance). The secretary must prescribe regulations requiring employers: (a) to maintain accurate records of work-related deaths, injuries, and illnesses involving medical treatment, loss of consciousness, restriction of work or motion, or transfer to another job but not minor injuries requiring only first aid treatment; and (b) to maintain records of employee exposure to potentially toxic materials or harmful physical agents. Certain information must be provided to the employee in the latter instance.

National Institute for Occupational Safety and Health

A national institute is created within the Department of Health, Education, and Welfare (HEW), and it is authorized, among other things, "to develop and establish recommended safety and health standards." Where feasible, HEW functions under the act are to be delegated to the institute.

National Commission on State Workers' Compensation Laws

This newly established commission is to be composed of 15 members appointed by the President. The commission is authorized to conduct a comprehensive study and evaluation of state workers' compensation laws to determine if such laws provide an adequate, prompt, and equitable system of compensation for injury or death arising out of, or in, the course of employment.

National Advisory Committee on Occupational Safety and Health

The Secretary of Labor and the Secretary of HEW must appoint such a committee consisting of 12 members —four of whom are to be designated by the Secretary of HEW—and composed of representatives of management, labor, occupational safety and occupational health professions, and the public. The committee is to advise, consult with, and make recommendations to both secretaries on matters relating to administration of this act.

Research

The act provides for research by HEW relating to occupational safety and health including, but not limited to, psychological factors involved, criteria dealing with toxic materials and harmful physical agents, and effects of low-level exposure to materials and processes on the potential for illness. At least annually, HEW shall publish a list of all known toxic substances and the concentrations at which such toxicity is known to occur.

Training

The Secretary of HEW is to conduct, directly or by grants or contracts: education programs to provide an adequate supply of qualified personnel to carry out the purposes of the act; informational programs on the importance of and proper use of safety and health equipment. The Secretary of Labor is to provide for the establishment and supervision of programs for the education and training of employers and employees with respect to effective means of preventing occupational injuries and illnesses.

Administrative matters

A new post, designated as Assistant Secretary of Labor for Occupational Safety and Health, is created.

Effective date

The act took effect April 28, 1971.

2b. Proposed Requirements for Occupational Noise Exposure

The following requirements and procedures were proposed by the Occupational Safety and Health Administration. They were published in Vol. 39, No. 207, of the *Federal Register* on October 24, 1974.

The background reasoning that led to them was described in Chapter 12, Federal Regulations and Guidelines. When finally promulgated, they will become part of Title 29, *Code of Federal Regulations*, Part 1910.

§ 1910.95 OCCUPATIONAL NOISE EXPOSURE

(a) Application and purpose

This section applies to occupational noise exposures in employments covered in this part. The purpose of this standard is to establish requirements and procedures that will minimize the risk of permanent hearing impairment from exposure to hazardous levels of noise in workplaces.

(b) Definitions

"Administrative controls" means any procedure which limits daily noise exposure by control of the work schedule. Hearing protectors do not constitute administrative controls.

"Assistant Secretary" the Assistant Secretary of Labor for Occupational Safety and Health, U.S. Department of Labor, or his designee.

"Audiogram" a graph or table of hearing level as a function of frequency that is obtained from an audiometric examination.

"Baseline audiogram" the first audiogram taken during employment with the current employer.

"Certified audiometric technician" an individual who meets the training requirements specified by the Intersociety Committee on Audiometric Technician Training (American Industrial Hygiene Association Journal, 27:303–304 (May–June 1966) or who is certified by the Council of

Accreditation in Occupational Hearing Conservation.

"Daily noise dose" (D) the cumulative noise exposure of an employee during a working day.

"dBA" (decibels—A-weighted)—a unit of measurement of sound level corrected to the A-weighted scale, as defined in ANSI S1.4-1971, using a reference level of 20 micropascals (2×10^{-5}) Newtons per square meter).

"Director" the Director, National Institute for Occupational Safety and Health, U.S. Department of Health, Education, and Welfare, or his designee.

"Engineering control" any design procedure that reduces the sound level.

"Hearing level" the amount, in decibels, by which the threshold of audibility for an ear differs from the standard audiometric reference level.

"Peak sound pressure level" the peak instantaneous pressure expressed in decibels, using a reference level of 20 micropascals.

"Workplace sound level" the sound level measured at the employee's point of exposure.

"Impulse or impact noise"—a sound with a rise time of not more than 35 milliseconds to peak intensity and a duration of not more than 500 milliseconds to the time when the level is 20 dB below the peak. If the impulses recur at intervals of less than one-half second, they shall be considered as continuous sound.

"Significant threshold shift" an average shift of more than 10 dB at frequencies of 2000, 3000, and 4000 Hz relative to the baseline audiogram in either ear.

TABLE G–16a

Sound Level (dBA)	Time Permitted (hours-minutes)	Sound Level (dBA)	Time Permitted (hours-minutes)
85	16–0	101	1–44
86	13–56	102	1–31
87	12–8	103	1–19
88	10–34	104	1–9
89	9–11	105	1–0
90	8–0	106	0–52
91	6–58	107	0–46
92	6–4	108	0–40
93	5–17	109	0–34
94	4–36	110	0–30
95	4–0	111	0–26
96	3–29	112	0–23
97	3–2	113	0–20
98	2–50	114	0–17
99	2–15	115	0–15
100	2–0		

(c) Permissible exposure limits

(1) *Steady state noise—single level.*

(i) The permissible exposure to continuous noise shall not exceed an eight hour time-weighted average of 90 dBA with a doubling rate of 5 dBA. For discrete permissible time and exposure limits, refer to Table G-16a, which is computed from the formula in paragraph (c)(1)(ii) of this section.

(ii) Where Table G-16a does not reflect actual exposure times and levels, the permissible exposure to continuous noise at a single level shall not exceed a time amount "T" (in hours) computed by the formula:

$$T = \frac{16}{2^{[0.2(L-85)]}}$$

where "L" is the workplace sound level measured in dBA on the slow

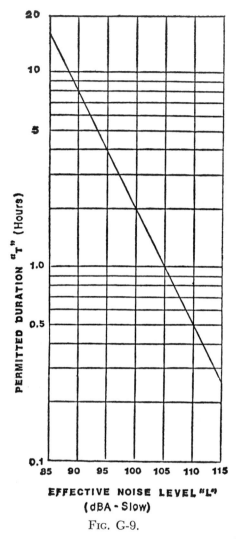

EFFECTIVE NOISE LEVEL "L"
(dBA - Slow)

Fɪɢ. G-9.

not exceed a daily noise dose "D" of unity (i) where "D" is computed by the formula:

$$D = \frac{C_1}{T_1} + \frac{C_2}{T_2} + \cdots + \frac{C_n}{T_n}$$

where C is the actual duration of exposure (in hours) at a given steady state noise level; and T is the noise exposure limit (in hours) for the level present during the time C, computed by the formula in paragraph (c)(1) (ii) of this section.

(3) *Maximum steady state noise level.* Exposures to continuous noise shall not exceed 115 dBA, regardless of any value computed in paragraphs (c)(1) or (c)(2) of this section.

(4) *Impulse or impact noise.*

(i) Exposures to impulse or impact noise shall not exceed a peak sound pressure level of 140 dB.

(ii) Exposures to impulses of 140 dB shall not exceed 100 such impulses per day. For each decrease of 10 dB in the peak sound pressure level of the impulse, the number of impulses to which employees are exposed may be increased by a factor of 10.

(d) Monitoring

(1) *Duty.* Each employer shall determine if any employee is exposed to a daily noise level dose of 0.5 or above, and shall determine if any employee is exposed to impulse or impact noise in excess of the exposure permitted by paragraph (c) (4) of this section. Such determinations shall be made:

(i) At least annually, and

(ii) Within 30 days of any change or modification of equipment or process, or other workplace or work practice modifications affecting the noise level.

scale of a standard sound level meter. The relationship between time and sound level is depicted in Figure G-9.

(2) *Steady state noise—two or more levels.* Exposures to continuous noise at two or more levels may

870

(2) *Procedure.* If determinations made pursuant to paragraph (d)(1) of this section reveal any employee exposure to a daily noise dose of 0.5 or above, or exposure to impulse or impact noise in excess of that permitted by paragraph (c)(4) of this section, the employer shall:

(i) Identify all employees who may be so exposed;

(ii) Measure the exposure of the employees so identified; and

(iii) Make all noise level measurements with the microphone of the sound measuring instrument at a position which most closely approximates the noise levels at the head position of the employee during normal operations.

(3) *Equipment.*

(i) Measurements of steady state noise exposures shall be made with a sound level meter conforming as a minimum to the requirements of ANSI S1.4-1971, Type 2, and set to an A-weighted slow response or with an audiodosimeter of equivalent accuracy and precision. The unit of measurement shall be decibels re 20 micropascals A-weighted.

(ii) Measurements of impulse or impact noise exposures shall be made with a sound level meter conforming as a minimum to the requirements of the ANSI S1.4-1971, Type 1 or Type 2, with a peak hold capability or accessory. For peak hold measurements, the rise time of the instrumentation shall be not more than 50 microseconds. The decay rate for the peak hold feature shall be less than 0.05 decibels per second. The unit of measurement shall be decibels peak sound pressure level re 20 micropascals.

(4) *Calibration of equipment.* An acoustical calibrator accurate to within plus or minus one decibel shall be used to verify the before and after calibration of the sound measuring instrument on each day noise measurements are taken.

(5) *Observation of monitoring.*

(i) *Duty.* The employer shall give employees or their representatives an opportunity to observe any monitoring of the noise levels in the workplace which is conducted pursuant to this section.

(ii) *Notification of employee right.* Written notice of the opportunity to observe the monitoring required by this section shall be prominently posted in a place regularly visited by affected employees and where notices to employees are usually posted. The employer shall take steps to insure that this notice is not altered, defaced, or covered by other material.

(A) The notice shall be posted at least three working days before monitoring is scheduled to occur.

(B) The notice shall list the time and place where monitoring will take place.

(C) The employer may require the employee or the employee representative to give advance written notification of intent to observe such monitoring.

(iii) *Exercise of opportunity to observe monitoring.*

(A) When observation of the monitoring of the workplace for noise levels requires entry into an area where the use of personal protective devices is required, the employer shall provide and the observer shall use such equipment and comply with all other applicable safety procedures.

(B) Observers shall be given an explanation of the procedure to be followed in measuring the workplace noise level.

(C) Observers shall be permitted, without interference with persons performing the monitoring, to:

(1) Visually observe all steps related to the collecting and evaluating of the noise level data that are being performed at the time;

(2) Record the results obtained; and

(3) Have a demonstration of the calibration function tests of the monitoring equipment when the calibrations are performed at the worksite before monitoring; where the calibrations are not performed at the worksite, the techniques shall be explained.

(e) Methods of compliance.

(1) Whenever employees are exposed to workplace sound levels exceeding those permitted by paragraph (c) of this section, engineering and administrative controls shall be utilized to reduce employee noise exposure to within permissible limits, except to the extent that such controls are not feasible. If such controls fail to reduce sound levels to within the permissible limits of paragraph (c) of this section, they shall be used to reduce the sound levels to the lowest level feasible and shall be supplemented by personal protective equipment in accordance with paragraph (f) of this section to further reduce the noise exposure to within permissible limits. Where the engineering and administrative controls which have been implemented do not reduce the sound levels to within the permissible limits of paragraph (c) of this section, the employer shall continue to develop and implement engineering and administrative controls as they become feasible.

(2) A program shall be established and implemented to reduce exposures to within the permissible exposure limit, or to the greatest extent feasible, solely by means of engineering controls. Written plans for such a program shall be developed and furnished upon request to authorized representatives of the Assistant Secretary and the Director.

(3) Exception. Hearing protectors may be provided to, and used by an employee to limit noise exposures in lieu of feasible engineering and administrative controls if the employee's exposure occurs on no more than one day per week.

(f) Hearing protectors

(1) Hearing protectors shall be provided to, and used by: (i) Employees receiving a daily noise dose between 0.5 and 1.0 (a daily noise dose of 0.5 is equivalent to an eight hour time weighted exposure of 85 dBA) if their audiograms show any significant threshold shift;

(ii) Employees who receive noise exposures in excess of the limits prescribed in paragraph (c) of this section: (A) During the period required for the implementation of feasible engineering and administrative controls; (B) in instances where engineering and administrative controls are feasible only to a limited extent; or (C) in instances where engineering and administrative controls have been shown to be infeasible.

(2) Hearing protectors shall reduce employee noise exposure to within the limits prescribed in paragraph (c) of this section.

(3) Procedures shall be established and implemented to assure proper issuance maintenance, and training in the use of hearing protectors.

(g) **Hearing conservation**

(1) *General.* (i) A hearing conservation program shall be established and maintained for employees who:

(A) Receive a daily noise dose equal to or exceeding 0.5; or

(B) Are required to wear hearing protectors pursuant to paragraph (f) of this section.

(ii) The hearing conservation program shall include at least an annual audiometric test for affected employees at no cost to such employees.

(iii) If no previous baseline audiogram exists, a baseline audiogram shall be taken within 90 days for each employee (A) who receives a daily noise dose of 0.5 or above; or (B) who is required to wear hearing protectors pursuant to paragraph (f) of this section.

(iv) Each employee's annual audiogram shall be examined to determine if any significant threshold shift in either ear has occurred relative to the baseline audiogram.

(v) (A) If a significant threshold shift is present, the employee shall be retested within one month.

(B) If the shift persists:

(1) Employees not having hearing protectors shall be provided with them in accordance with paragraph (f) of this section;

(2) Employees already having hearing protectors shall be retrained and reinstructed in the use of hearing protectors.

(3) The employee shall be notified of the shift in hearing level.

(2) *Audiometric testing.*

(i) Audiometric tests shall be administered by a certified audiometric technician or an individual with equivalent training and experience.

(ii) Audiometric tests shall be preceded by a period of at least fourteen hours during which there is no exposure to workplace sound levels in excess of 80 dBA. This requirement may be met by wearing hearing protectors which reduce the employee noise exposure level to below 80 dBA.

(iii) Audiometric tests shall be pure tone, air conduction, hearing threshold examinations, with test frequencies including as a minimum, 500, 1000, 2000, 3000, 4000 and 6000 Hz and shall be taken separately for each ear.

(iv) The functional operation of the audiometer shall be checked prior to each period of use to ensure that it is in proper operating order.

(v) Equipment, calibration and facilities shall meet the specifications set forth in the Appendix.

(h) **Information and warnings**

(1) *Signs.* Clearly worded signs shall be posted at entrances to, or on the periphery of, areas where employees may be exposed to noise levels in excess of the limits prescribed in paragraph (c) of this section. These signs shall describe the hazards involved and required protective actions.

(2) *Notification.* Each employee exposed to noise levels which exceed the limits prescribed in paragraph (c) of this section shall be notified in writing of such excessive exposure within 5 days of the time the employer discovers such exposure. Such notification shall inform the affected employee of the corrective action being taken.

(i) **Records**

(1) *Noise exposure measurements.*

(i) The employer shall keep an accurate record of all noise exposure measurements made pursuant to paragraph (d) of this section.

(ii) The record shall include the following information:

(A) name of employee, social security number and daily noise dose;

(B) location, date, and time of measurement and levels obtained;

(C) Name of person making measurement;

(D) Type, model and date of calibration of measuring equipment.

(iii) These records shall be maintained for a period of at least five years.

(2) *Audiometric tests.*

(i) The employer shall keep an accurate record of all employee audiograms taken pursuant to paragraph (g) of this section.

(ii) The record shall include the following information:

(A) Name of employee and social security number;

(B) Job location of employee;

(C) Date of the audiogram;

(D) The examiner's name and certification;

(E) Model, make and serial number of the audiometer; and

(F) Date of the last calibration of the audiometric test equipment.

(iii) These records shall be maintained for the duration of the affected employee's employment plus 5 years.

(3) *Calibration of audiometers.* (i) The employer shall keep an accurate record of all audiometer calibrations required to be made pursuant to paragraph (g) of this section and the Appendix.

(ii) The record shall include the following information:

(A) Type of calibration;

(B) Date performed; and

(C) All measurements obtained.

(iii) These records shall be maintained for a period of 5 years.

(4) *Access to records.* (i) All records required to be maintained by this section shall be made available upon request to authorized representatives of the Assistant Secretary and the Director.

(ii) Records of noise exposure measurements required to be maintained by this section shall be made available to employees and former employees and their designated representatives.

(iii) Employee audiometric data required to be maintained by this section shall be made available upon written request to the employee or former employee.

(j) References

(1) ANSI S1.4-1971, American National Standard Specification for Sound Level Meters, S1.4-1971, American National Standards Institute, 1430 Broadway, New York, New York 10018.

(2) ANSI S3.6-1969—American National Standard Specifications for Audiometers, S3.6-1969—American National Standards Institute, 1430 Broadway, New York, New York 10018.

(3) ANSI S1.11-1971—American National Standard Specification for Octave, Half-Octave, and Third Octave Band Filter Sets S1.11-1966 (Reaffirmed 1971), American National Standards Institute, 1430 Broadway, New York, New York 10018.

(4) ANSI Z24.22-1957 (R1971)— American National Standard Method for the Measurement of the Real-Ear Attenuation of Ear-Protectors at Threshold, Z24.22-1957, American National Standards Institute, 1430 Broadway, New York, New York, 10018.

(5) American Industrial Hygiene Association Journal, 27:303–304

(May–June 1966). American Industrial Hygiene Association, 66 S. Miller Road, Akron, Ohio 44313.

(6) Council for Accreditation in Occupational Hearing Conservation, 1619 Chestnut Avenue, Haddon Heights, New Jersey 08035.

(7) ISO R389-1964—International Organization for Standardization Recommendation R389-1964, Standard Reference Zero for the Calibration of Pure Tone Audiometers, including Addendum 1-1970. Available from the American National Standards Institute, 1430 Broadway, New York, New York 10018.

AUDIOMETRIC EQUIPMENT AND FACILITIES

(1) *Audiometric test rooms.* Rooms used for audiometric testing shall not have sound pressure levels exceeding those in Table G–16b

TABLE G–16b

Maximum Allowable Sound Pressure Levels for Audiometer Rooms

Octave band center frequency (Hz)	500	1000	2000	4000	8000
Sound pressure level (dB)	40	40	47	52	62

when measured by equipment conforming to the requirements of ANSI S1.4-1971, Type 1 or Type 2, and ANSI S1.11-1971.

(2) *Audiometric measuring instruments.*

(i) Instruments used for measurements required in paragraph (g) of this section shall be of the discrete frequency type which meet the requirements for limited range pure tone audiometers prescribed in ANSI S3.6-1969.

(ii) In the event that pulsed tone audiometers are used, they shall have a tone on-time of at least 200 milliseconds.

(iii) Self-recording audiometers shall comply with the following requirements:

(A) The chart upon which the audiogram is traced shall have lines at positions corresponding to all multiples of 10 dB hearing level within the intensity range spanned by the audiometer. The lines shall be equally spaced and shall be separated by at least ¼ inch. Additional gradations are optional. The audiogram pen tracings shall not exceed 2 dB in width.

(B) It shall be possible to set the stylus manually at the 10 dB gradation lines for calibration purposes.

(C) The slewing rate for the audiometer attenuator shall not be more than 6 dB/sec except that an initial slewing rate greater than 6 dB/sec is permitted at the beginning of each new test frequency, but only until the second subject response.

(D) The audiometer shall remain at each required test frequency for 30 seconds (±3 seconds). The audiogram shall be clearly marked at each change of frequency and the actual frequency change of the audiometer shall not deviate from the frequency boundaries marked on the audiogram by more than ±3 seconds.

(E) For audiograms taken with a self-recording audiometer, it must be possible at each test frequency to place a horizontal line segment parallel to the time axis on the audiogram, such that the audiometric tracing crosses the line segment at least six times at that test frequency. At each test frequency the threshold shall be the average of the midpoints of the tracing excursions.

(3) *Audiometer calibrations.*

(i) A biological calibration shall be made at least once each month and shall consist of testing a person having a known stable audiometric curve that does not exceed 25 dB hearing level at any frequency between 500 and 6,000 Hz and comparing the test results with the subject's known baseline audiogram, and

(ii) If the results of a biological calibration indicate hearing-level differences greater than 5 dB at any frequency, if the signal is distorted, or there are attenuator or tone switch transients, then the audiometer shall be subjected to a periodic calibration.

(iii) A periodic calibration shall be performed at least annually. The accuracy of the calibrating equipment shall be sufficient to assure that the audiometer is within the tolerances permitted by ANSI S3.6-1969. The following measurements shall be performed:

(A) With the audiometer set at 70 dB hearing threshold level, measure the sound pressure levels of test tones using a National Bureau of Standards Type 9A coupler, for both earphones and at all test frequencies.

(B) At 1000 Hz, for both earphones, measure the earphone decibel levels of the audiometer for 10 dB graduations in the range 70 to 10 dB hearing threshold level. This measurement may be made acoustically with a National Bureau of Standards Type 9A coupler or electrically at the earphone terminals.

(C) Measure the test tone frequencies between 500 and 6000 Hz with the audiometer set at 80 dB hearing threshold level, for one earphone.

(D) A careful listening test, more extensive than that required for biological calibration shall be made in order to ensure that the audiometer displays no evidence of distortion, unwanted sound, or other technical problems.

(E) The functional operation of the audiometer shall be checked to ensure that it is in proper operating order.

(iv) An exhaustive calibration shall be performed at least every five years. This shall include testing at all settings for both earphones. The test results shall demonstrate that the audiometer meets specific requirements stated in the applicable sections of ANSI S3.6-1969 as listed below:

(A) [Sections 4.1.2 and 4.1.4.3] Accuracy of decibel level settings of all test tones,

(B) [Section 4.1.2] Accuracy of test tone frequencies,

(C) [Section 4.1.3] Harmonic distortion of test tones,

(D) [Section 4.5] Tone-envelope characteristics, i.e., rise and decay times, overshoot, "Off" level,

(E) [Section 4.4.2] Sound from second earphone,

(F) [Section 4.4.1] Sound from test earphone, and

(G) [Section 4.4.3] Other unwanted sound.

For the convenience of the reader, the referenced portions of American National Standard *Specifications for Audiometers*, S3.6-1969, follow.

(A) *4.1.4.3 Accuracy of Sound Pressure Levels.* The sound pressure produced by an earphone as referred to the standard reference level shall not differ from the indicated value of sound pressure level at any reading of the hearing threshold level dial by more than 3 dB at the indicated frequencies of 250 to 3000 Hz inclusive, by more than 5 dB at frequencies above or below this range. Measurements for compliance with this requirement may be made by combining an acoustical measurement of sound pressure level at a 70 dB setting with voltage measurements at other settings.

(A) and (B) *4.1.2 Accuracy of Tone Frequencies.* Each of the above frequencies generated by the audiometer shall be within three percent of the indicated frequency, except as noted in 4.3.

(C) *4.1.3 Purity of Tones.* The sound pressure level of any harmonic of the fundamental shall be at least 30 dB below the sound pressure level of the fundamental. Measurements for compliance with this requirement shall be made at the fundamental frequencies of 4.1.1 and the output levels shown in Table 1.

(D) *4.5 Tone Switch.* A tone switch is required on manual wide range and limited range audiometers, and is optional on other pure tone audiometers. The tone switch is for optional presentation of the tone signal to the subject by the operator, and its operation shall be such as to establish and eliminate the tone without producing audible transients or extraneous frequencies. The time required for the sound pressure level to rise from -20 dB to -1 dB re its final steady value shall not be less than 0.02 second and not more than 0.1 second, and the time required for the sound pressure

level to decay by 20 dB shall be not less than 0.005 second and not more than 0.1 second.

The steady output voltage level in the "Off" position shall be at least 50 dB below its steady voltage in the "On" position, or at least 10 dB below that corresponding to the Zero Hearing Threshold Level, whichever of these two levels is the higher.

The operation of the tone switch shall not cause the output voltage level operating the earphone to attain at any time a value more than one dB above its steady state.

The tone switch shall be of the "normally off" type, although it is recommended that it be also adapted to "normally on" operation. It may also be arranged to operate automatically.

(E) *4.4.2 Sound from Second Earphone.* Any signal from the earphone not being used for test purposes shall have a sound pressure level at least 10 dB below the hearing threshold reference level, except that it need not be more than 70 dB below the signal from the "on" earphone.

(F) *4.4.1 Sound from Test Earphone.* Any sound from the test earphone, other than the desired test signal, shall not be of such a magnitude as to affect the threshold judgment of any properly instructed subject.

(G) *4.4.3 Other Unwanted Sound.* Any sound due to the operation of audiometer controls during the actual listening test, or to radiation from the audiometer, shall be inaudible at each setting of the hearing threshold level dial up to and including 50 dB. The test for this requirement shall be made by a subject having normal hearing, wearing a pair of disconnected earphones and located at the recommended test position, the electrical output of the audiometer being absorbed in a resistive load equal to the impedance of the earphone at 1000 Hz.

NOTE: This limitation on noise from controls would apply to any noise which could furnish a clue which would influence the test results. It is not intended to apply to a mechanism such as a detent on the frequency switch, any noise from which would occur when the subject is not actually being tested.

3. NIOSH Criteria Document

The Occupational Safety and Health Act of 1970 emphasized the need for standards to protect the health of workers exposed to an ever-increasing number of potential hazards at their workplace. To provide relevant data from which valid criteria and effective standards can be deduced, the National Institute for Occupational Safety and Health, in 1972, published criteria for a recommended standard on occupational exposure to noise. This was identified as Health Services and Mental Health Administration (HSM) document No. 73–11001.

The NIOSH Criteria Document consists of seven parts:

Except for Parts I and VI, most of the other material is duplicated in other portions of this Manual. Therefore, only these two parts, and the specific references, illustrations, and tables that apply to them, are included in this Appendix.

I. Recommendations for a noise standard

The National Institute for Occupational Safety and Health (NIOSH) recommends that employee exposure to noise in the workplace be controlled by requiring compliance with the standard set forth in the following sections. Control of employee exposure to the occupational limits stated and adherence to the precautionary procedures prescribed will improve the protection of the working population from incurring noise induced hearing loss that could impair their abilities to understand everyday speech. Such control and adherence at the workplace is believed sufficiently effective to reduce also the possibility of other forms of occupational injury and illness related to noise.

This standard is amenable to techniques that are valid, reproducible, and presently available. It will be reviewed and revised as additional information becomes available.

Section 1—Applicability

The provisions of this standard are applicable to occupational noise exposures at places of employment and are intended to apply for all noise even though additional controls may be necessary for certain specific types of noise, such as some impact and impulsive noise. For the purposes of this standard the noise exposure is determined for an 8 hour workday.

Section 2—Definitions

As used in this standard, the term:

(a) *Administrative control* means any procedure that limits daily exposure to noise by control of the work schedule.

(b) *Audiogram* means a graph or table obtained from an audiometric examination showing hearing level as a function of frequency.

(c) *Baseline audiogram* means an audiogram obtained from an audiometric examination that is preceded

by a period of at least 14 hours of quiet.

(e) *Audiometer setting* means a setting on an audiometer corresponding to a specific combination of hearing level and sound frequency.

(f) *Daily Noise Dose* means that value for D derived from the equation:

$$D = \frac{C_1}{T_1} + \frac{C_2}{T_2} + \cdots + \frac{C_n}{T_n}$$

where C_1, C_2 . . ., C_n are the actual durations of exposure for an employee at the various noise levels, T_1, T_2, . . ., T_n are the respective duration limits obtained from Figure I-1 and D is the Daily Noise Dose.

(g) *dBA-Slow* means the unit of measurement of sound level indicated by a sound level meter conforming as a minimum requirement to the American National Standard *Specification for Sound Level Meters*, ANSI S1.4-1971 Type S2A, when used for A-weighted sound level, slow response.

(h) *Engineering control* means any procedure other than administrative control that reduces the sound level either at the source of the noise or in the hearing zone of the employees.

(i) *Hearing level* means the amount, in decibels, by which the threshold of audibility for an ear differs from a standard audiometric threshold.

(j) *Environmental noise level* means the noise level in dBA-Slow as measured in accord with Section 3(c).

(k) *Effective noise level* means (1) for employees not wearing ear

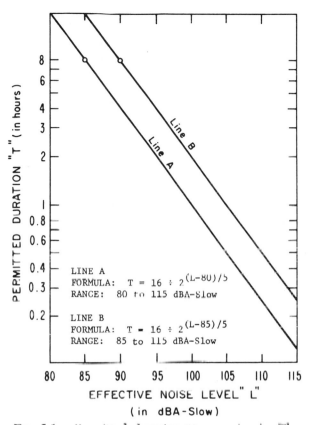

FIG. I 1. Permitted duration *vs.* noise level. The indicated duration limits which exceed 8 hr are to be used only for purposes of computing Daily Noise Dose and are not to be regarded as defining noise exposure limits for work days which exceed 8 hr.

protectors, the environmental noise level; (2) for employees wearing ear protectors, the result of subtracting the dBA reduction, R, for the ear protectors (determined as specified in Appendix I-A) from the measured environmental noise level. Effective noise level is expressed in units of dBA-Slow.

(*l*) *Noise exposure* means a com-

bination of effective noise level and exposure duration.

Section 3—Occupational environment

(*a*) The unit of measurement shall be *dBA-Slow*.

(*b*) Daily Occupational Noise Exposure

(*i*) Occupational noise exposure

shall be controlled so that no worker shall be exposed in excess of the limit described as line B in Figure I-1. New installations shall be designed with noise control so that the noise exposure does not exceed the limits described as line A in Figure I-1. For noise exposures consisting of two or more periods of exposure at different levels, the Daily Noise Dose, D, shall not exceed unity. Line A or line B, as applicable, shall be used in computing the Daily Noise Dose.

(*ii*) It is recommended that the limit described as line A become effective for all places of employment after a time period determined by the Secretary of Labor in consultation with the Secretary of Health, Education, and Welfare. This delay in effective date for all places of employment is believed necessary to permit the Department of Labor to conduct an extensive feasibility study.

(*iii*) At no time shall any worker be exposed to effective noise levels exceeding 115 dBA-Slow.

(*c*) Measurements

(*i*) Compliance with the permitted daily noise exposures defined by Section 3(*b*) shall be determined on the basis of measurements made with a sound level meter conforming as a minimum to the requirement of the American National Standard *Specification for Sound Level Meters,* S1.4 (1971) Type S2A, and set to use an A-weighted slow response.

(*ii*) All measurements shall be made with the sound level meter at a position which most closely approximates the noise levels at the head position of the employee during normal operations.

(*iii*) An acoustical calibrator accurate within plus or minus one decibel shall be used to calibrate the sound level meter on each day that noise measurements are taken.

Section 4—Medical

(*a*) Medical surveillance in the form of an audiometric testing program shall be provided by the employer when the Daily Noise Dose, D, equals or exceeds the limits specified in Section 3(*b*), and for all employees whose occupational noise exposure is controlled by personal protective equipment.

(*b*) The audiometric testing program required by (*a*) above shall conform to the following schedule:

(*i*) A baseline audiogram for each employee who is initially assigned or reassigned to work subject to conditions stated in (*a*) of this section shall be taken within 30 days of assignment to such employment, in the sixth year of such employment, and once every sixth year thereafter. It is recognized that some delay in implementation of this requirement may be necessary for employers with a small work force.

(*ii*) A baseline audiogram should be taken for each employee presently assigned to work subject to conditions stated in (*a*) of this section at the time of effective date of this regulation, in the sixth year, and once every sixth year thereafter.

(*iii*) In addition an audiogram, not necessarily baseline, for all exposed employees should be taken every second year.

(c) Each audiogram shall contain (1) employee's name or identifying number, (2) employee's job location, (3) significant aural medical history of the employee, (4) the examiner's name and signature, (5) the date and time of test, (6) serial number of the audiometer, and (7) last exposure to high level noise: number of hours since exposure; type of exposure; and noise level, if known.

(d) Each employee's audiogram shall be examined to determine whether it indicates for either ear any threshold shift (higher threshold), that equals or exceeds 10 dB at 500, 1000, 2000, or 3000 Hz, or 15 dB at 4000 or 6000 Hz as evidenced by a comparison of that audiogram with the employee's most recent baseline audiogram and with his initial baseline audiogram as corrected to his current age by the method described in Appendix I-B. If either comparison indicates a shift as described above:

(i) refer the employee for appropriate medical evaluation,

(ii) if the employee needs personal protective equipment or devices, ensure that he has the appropriate effective equipment and that he is instructed in the proper use and care of the equipment, and

(iii) if the audiogram was not a baseline audiogram, take a baseline audiogram within sixty days.

(e) Audiometric tests shall be pure tone, air-conduction, hearing threshold examinations, with test frequencies including 500, 1000, 2000, 3000, 4000, and 6000 Hz and shall be taken separately for the right and left ears.

(i) The tests shall be conducted in a room whose ambient noise levels conform to all requirements except that part concerning octave bands whose center frequencies are less than 250 Hz of the American National Standard *Criteria for Background Noise in Audiometer Rooms*, ANSI S3.1-1960 (R1971), when measured by equipment conforming to American National Standard *Specification for Sound Level Meters*, ANSI S1.4-1971 Type 2 and American National Standard *Specification for Octave, Half-Octave, and Third-Octave Band Filter Sets*, ANSI S1.11-1966 (R1971).

(ii) The tests shall be administered using an audiometer which conforms to the requirements for limited range pure tone audiometers prescribed by the American National Standard *Specifications for Audiometers*, ANSI S3.6-1969, and which is of the discrete frequency type. If a pulsed tone audiometer is used, the on-time of the tone shall be at least 200 milliseconds. The instrument used in the testing shall be either a manual audiometer, or a self-recording audiometer which is subject to the following additional restrictions:

(1) The chart upon which the audiogram is traced shall have printed lines at positions corresponding to all multiples of 10 dB hearing level within the intensity range spanned by the audiometer. The lines shall be equally spaced and shall be separated by at least ¼ inch. Additional graduations are optional. The pen which traces the audiogram shall

have a fine point so that the tracing shall not exceed 2 dB in width.

(2) It shall be possible to disable the stylus drive mechanism so that the stylus can be manually set at the 10 dB graduation lines for calibration purposes.

(3) The slewing rate for the audiometer attenuator shall be 6 dB/sec or less except that an initial slewing rate greater than 6 dB/sec is permitted at the beginning of each new test frequency, but only until the second subject response.

(4) The audiometer shall remain at each required test frequency for 30 seconds (±3 seconds). The audiogram shall be clearly marked at each change of frequency and the actual frequency change of the audiometer shall not deviate from the frequency boundaries marked on the audiogram by more than ±3 seconds.

(5) If an audiogram fails to pass the following criteria, the subject shall be retested.

At each test frequency it must be possible to place a horizontal line segment parallel to the time axis on the audiogram, such that the audiometric tracing crosses the line segment at least six times at that test frequency.

(*iii*) The audiometer shall be maintained in calibration in accordance with the provisions of Appendix I-C.

Section 5—Work practices

When employees are employed under conditions where noise exposures would exceed the limits prescribed in Section 3(*b*), administrative or engineering controls shall be utilized to reduce exposures to within those limits.

Section 6—Warning notice

(*a*) A warning sign shall be appropriately located at entrances to and/or the periphery of, areas where there exists sustained environmental noise at or in excess of the limit prescribed in Section 3(*b*).

(*b*) The notice shall consist of the following:

WARNING

NOISE AREA

MAY CAUSE HEARING LOSS

Use Proper Ear Protection

Section 7—Personal protective equipment

(*a*) If noise exposures to which employees could be exposed exceed the limits specified, personal protective equipment (*i.e*, ear protectors) shall be provided by the employer to be used in conjunction with an audiometric testing program, as specified in Section 4, subject to the following requirements:

(*i*) The use of personal protective equipment to prevent occupational noise exposure of the employer in excess of the prescribed limits is authorized only until engineering and administrative controls and procedures can be implemented to maintain the occupational noise exposures within prescribed limits.

(*ii*) Any ear protector used by an employee shall reduce the effective noise level to which he is exposed so that his noise exposure is within the limits prescribed in Section 3(*b*).

(*iii*) Insert-type protectors shall be fitted by a person trained in this procedure.

(*iv*) Inspection procedures to assure proper issuance, maintenance, and use of personal protective equipment shall be established by the employer.

(*b*) The employer shall provide training in the proper care and use of all personal protective equipment.

Section 8—Apprisal of employees of hazards from noise

Each worker exposed to noise shall be apprised of all hazards, relevant symptoms, and proper conditions and precautions for working in noisy areas. The information shall be kept on file and readily accessible to the worker at all places of employment where the noise levels equal or exceed the limits prescribed in Section 3(*b*).

Section 9—Monitoring and record-keeping requirements

(*a*) Employers will be required to maintain records of:

(*i*) environmental exposure monitoring for a period of 10 years.

(*ii*) all audiograms for a period of 20 years.

(*iii*) all audiometric calibration data for a period of 20 years.

(*b*) When exposure times of less than 8 hours per day are required in a specific work area or ear protection is used to meet the exposure limits, records of the method of control shall be maintained.

Appendix I-A—Determination of dBA reduction R for ear protectors

1. The pure tone attenuation versus frequency characteristics of the ear protector (normally supplied by the manufacturer) shall have been determined in accordance with the American National Standard *Method for Measurement of the Real-Ear Attenuation of Ear Protectors at Threshold,* ANSI Z24.22-1957. Let Q_1, Q_2, . . ., Q_7 be defined (in dB) as follows:

Q_1 = attenuation at 125 Hz, plus 16.2 dB

Q_2 = attenuation at 250 Hz, plus 8.7 dB

Q_3 = attenuation at 500 Hz, plus 3.3 dB

Q_4 = attenuation at 1000 Hz

Q_5 = attenuation at 2000 Hz, minus 1.2 dB

Q_6 = average of attenuation at 3000 and 4000 Hz, minus 1.0 dB

Q_7 = average of attenuations at 6000 and 8000 Hz, plus 1.1 dB

2. The following procedure shall be used to determine the dBA reduction R of the ear protector when used for an occupational noise whose octave-band sound pressure levels have been measured.

Let L_1, L_2, L_3, L_4, L_5, L_6, and L_7 denote the octave band levels in dB at 125, 250, 500, 1000, 2000, 4000, and 8000 Hz respectively; and let L_A denote the dBA-Slow level of the noise. Then the dBA reduction as connected is given by

$$R = L_A - 10 \log S - 10.0$$

where

S = antilog $(0.1 \times [L_1 - Q_1]$ + antilog $(0.1 \times [L_2 - Q_2])$ + antilog $(0.1 \times [L_3 - Q_3])$ + antilog $(0.1 \times [L_4 - Q_4])$ + antilog $(0.1 \times [L_5 -$

Q_5]) + antilog $(0.1 \times [L_6 - Q_6])$ + antilog $(0.1 \times [L_7 - Q_7])$

The "−10.0" correction term is to account for possible noise spectrum irregularities and noise leakage which might be caused by long hair, safety glasses, head movement, or various other factors.

3. If the octave-band levels of the noise are not known, then the dBA reduction R may be computed simply as

$$R = -10 \log S - 3.0$$

where

S = antilog $(-0.1 \times Q_1)$ + antilog $(-0.1 \times Q_2)$ + antilog $(-0.1 \times Q_3)$ + antilog $(-0.1 \times Q_4)$ + antilog $(-0.1 \times Q_5)$ + antilog $(-0.1 \times Q_6)$ + antilog $(-0.1 \times Q_7)$

This calculation is approximate, and is based upon the assumption that the octave-band levels are equal. For most types of noise it will give results close to those obtained by the more accurate method of (2) above.

EXAMPLE:

Typical Pure Tone Attentuation Characteristics of an Ear Protector

125	250	500	1000	2000	Hz
24	21	23	29	30	dB

	3000	4000	6000	8000	Hz
	35	31	29	27	dB

Thus $Q_1 = 40.2$; $Q_2 = 29.7$; $Q_3 = 26.3$; $Q_4 = 29.0$; $Q_5 = 28.8$; $Q_6 = 31.0$; $Q_7 = 29.1$

If the octave-band noise levels are not known, then

$$R = -10 \log S - 3.0$$

where

S = antilog $(-4.02$ + antilog (-2.97) + antilog (-2.63) + antilog (-2.90) + antilog (-2.88) +

antilog (-3.10) + antilog (-2.91)

or $S = 0.00811$

So $R = -10 \log 0.0081 - 3.0 = 20.9 - 3.0 \cong 18$ dBA

Now suppose the ear protector is to be used in an area with an environmental noise level of 95 dBA, for which the octave-band noise levels are as follows:

125	250	500	1000
99	94	94	90

2000	4000	8000	Hz
84	82	75	Octave-band level

In this case the dBA reduction is

$$R = L_A = 10 \log S - 10.0$$

Where S = antilog $(9.9 - 4.02)$ + antilog $(9.4 - 2.97)$ + antilog $(9.4 - 2.63)$ + antilog $(9.0 - 2.90)$ + antilog $(8.4 - 2.88)$ + antilog $(8.2 - 3.10)$ + antilog $(7.5 - 2.91)$. So $S = 11,090,000$

Thus $R = 95.0 - 10 \times 7.05 - 10.0 \cong 85.0 - 70.5$

So $R = 14.5$ dBA

Appendix I-B—Method for Correcting Initial Baseline Audiograms for Age

Age corrections to initial baseline audiograms shall be made in the following manner:

For each audiometric test frequency:

1. Determine from Table B-1 or B-2 the age correction values for the employee
   (a) for the age at which the most recent audiogram was taken and
   (b) for the age at which the initial baseline audiogram was taken.

TABLE B-1

Age Corrections Values to be Used for Age Correction of Initial Baseline Audiograms for Males

Age (years)	Audiometric Test Frequencies (Hz)					
	500	1000	2000	3000	4000	6000
20 or younger	10	5	3	4	5	8
21	10	5	3	4	5	8
22	10	5	3	4	5	8
23	10	5	3	4	6	9
24	10	5	3	5	6	9
25	11	5	3	5	7	10
26	11	5	4	5	7	10
27	11	5	4	6	7	11
28	11	6	4	6	8	11
29	11	6	4	6	8	12
30	11	6	4	6	9	12
31	12	6	4	7	9	13
32	12	6	5	7	10	14
33	12	6	5	7	10	14
34	12	6	5	8	11	15
35	12	7	5	8	11	15
36	12	7	5	9	12	16
37	13	7	6	9	12	17
38	13	7	6	9	13	17
39	13	7	6	10	14	18
40	13	7	6	10	14	19
41	13	7	6	11	15	20
42	14	8	7	11	16	20
43	14	8	7	12	16	21
44	14	8	7	12	17	22
45	14	8	7	13	18	23
46	14	8	8	13	19	24
47	14	8	8	14	19	24
48	15	9	8	14	20	25
49	15	9	9	15	21	26
50	15	9	9	16	22	27
51	15	9	9	16	23	28
52	15	9	10	17	24	29
53	16	9	10	18	25	30
54	16	10	10	18	26	31
55	16	10	11	19	27	32
56	16	10	11	20	28	34
57	16	10	11	21	29	35
58	17	10	12	22	31	36
59	17	11	12	22	32	37
60 or older	17	11	13	23	33	38

TABLE B-2

Age Corrections Values to be Used for Age Correction of Initial Baseline Audiograms for Females

Age (years)	Audiometric Test Frequencies (Hz)					
	500	1000	2000	3000	4000	6000
20 or younger	15	7	4	3	3	6
21	16	7	4	4	3	6
22	16	7	4	4	4	6
23	16	7	5	4	4	7
24	16	7	5	4	4	7
25	16	8	5	4	4	7
26	16	8	5	5	4	8
27	17	8	5	5	5	8
28	17	8	5	5	5	8
29	17	8	5	5	5	9
30	17	8	6	5	5	9
31	17	8	6	6	5	9
32	17	9	6	6	6	10
33	18	9	6	6	6	10
34	18	9	6	6	6	10
35	18	9	6	7	7	11
36	18	9	7	7	7	11
37	18	9	7	7	7	12
38	18	10	7	7	7	12
39	19	10	7	8	8	12
40	19	10	7	8	8	13
41	19	10	8	8	8	13
42	19	10	8	9	9	13
43	19	11	8	9	9	14
44	20	11	8	9	9	14
45	20	11	8	10	10	15
46	20	11	9	10	10	15
47	20	11	9	10	11	16
48	20	12	9	11	11	16
49	21	12	9	11	11	16
50	21	12	10	11	12	17
51	21	12	10	12	12	17
52	21	12	10	12	13	18
53	21	13	10	13	13	18
54	21	13	11	13	14	19
55	22	13	11	14	14	19
56	22	13	11	14	15	20
57	22	13	11	15	15	20
58	22	14	12	15	16	21
59	22	14	12	16	16	21
60 or older	23	14	12	16	17	22

2. Subtract the values found in (*a*) from the values found in (*b*).

3. Add the difference found in (2) to the employee's initial baseline audiogram to obtain the initial baseline audiogram corrected for age.

EXAMPLE:

Employee is 56 years old and male. His initial baseline audiogram was taken at age 26 and his hearing levels at that age were as follows:

Hz	500	1000	2000	3000	4000	6000
Left ear	5	0	10	5	10	10
Right ear	10	0	5	0	5	15

Enter Table B-1 at age 56 and at age 26 and subtract.

Hz	500	1000	2000	3000	4000	6000
Age 56	16	10	11	20	28	34
Age 26	11	5	4	5	7	10
Difference	5	5	7	15	21	24

Add the differences to his initial baseline audiogram to obtain his corrected initial baseline audiogram as follows:

Hz	500	1000	2000	3000	4000	6000
Left ear	10	5	18	20	31	34
Right ear	15	5	13	15	26	39

Appendix C—Procedures for Calibration of Audiometers

The accuracy of an audiometer shall be determined by (1) a biological calibration, (2) a periodic calibration, and (3) an exhaustive calibration.

A. A biological calibration shall be made at least once each month and shall consist of (1) testing a person having a known stable audiometric curve that does not exceed 25 dB hearing level at any frequency and comparing the test results with the known curve, and (2) registering the subject's response to distortions and unwanted sounds from the audiometer. If the results of a biological calibration indicate hearing level differences greater than ±5 dB at any frequency, if the signal is distorted, or if there are attenuator or tone switch transients, then the audiometer shall be subjected to a periodic calibration within thirty days.

B. A periodic calibration shall be performed at least annually or as indicated by results of a biological check and shall include the following:

(1) Set audiometer to 70 dB hearing threshold level and measure sound pressure levels of test tones using an NBS-9A-type coupler, for both earphones and at all test frequencies.

(2) At 1000 Hz, for both earphones measure the earphone decibel levels of the audiometer for 10 dB settings in the range 70 to 10 dB hearing threshold level. This measurement may be made acoustically with a 9A coupler, or electrically at the earphone terminals.

(3) Measure the test tone frequencies with the audiometer set at 70 dB hearing threshold level, for one earphone only.

(4) In making the measurements in (1) through (3) above, the accuracy of the calibrating equipment shall be sufficient to prove that the audiometer is within the tolerances permitted by ANSI S3.6-1969.

(5) A careful listening test, more extensive than that required in the

biological calibration, shall be made in order to ensure that the audiometer displays no evidence of distortion, unwanted sound, or other technical problems.

(6) General function of the audiometer shall be checked, particularly in the case of a self-recording audiometer.

(7) All observed deviations from required performance shall be corrected.

C. An exhaustive calibration shall be performed at least every five years. This shall include testing at all settings for both earphones. The test results must prove unequivocally that the audiometer meets for the following parameters the specific requirements stated in the applicable sections of ANSI S.3-1969 as noted in parenthesis.

(1) Accuracy of decibel level settings of test tones (Sections 4.1.4.1 and 4.1.4.3).

(2) Accuracy of test tone frequencies (Section 4.1.2).

(3) Harmonic distortion of test tones (Section 4.1.3).

(4) Tone-envelope characteristics, i.e., rise and decay times, overshoot, "off" level (Section 4.5).

(5) Sound from second earphone (Section 4.4.2).

(6) Sound from test earphone (Section 4.4.1).

(7) Other unwanted sound (Section 4.4.3).

* * *

VI. Development of the standard

Attempts at limiting human exposure to noise have been based on damage-risk criteria. The purpose of such criteria is to define maximum permissible levels of noise for stated durations which, if not exceeded, would result in an acceptably small effect on hearing levels over a working lifetime of exposure.

Previous damage risk criteria

(a) Damage risk criteria before 1950

Early efforts at determining the maximum safe level of exposure relied heavily on overall levels of sound pressure. A listing of criteria developed prior to 1950 is presented in Table VIII. As may be seen from the table, there was, even at that time, quite a diversity of opinion with regard to the limit of safe exposure to noise. Estimates ranged from a low of 75 dB SPL[67] to a high of 100 dB SPL.[68-71] This situation was further complicated in 1945 by Goldner's suggestion that a nominal daily exposure for at least two years to a noise having an overall sound pressure level of 80 dB could be hazardous to hearing.[72]

In tracing the possible sources of error in these pre-1950 criteria, Kryter[73] suggested that one problem inherent in most of the studies was the high ambient noise levels which characterized the hearing testing environments. It was thought that such high ambient noise levels could account for an overestimation of the degree of hearing loss by as much as 10 to 15 dB. Probably the greatest source of error, however, was the fact that exposures were characterized us-

CRITERIA DOCUMENT TABLE 8
DAMAGE-RISK CRITERIA PRIOR TO 1950

Author	Overall Sound Pressure Level (dB)		
	Safe	Borderline	Harmful
McKenzie (1934)			90
Rosenblith (1942)	75–80		
Bunch (1942)		80–90	
McCoy (1944)	80–85	90–100	110–130
Davis (1945)		100	115–120
Goldner (1945)			80
Schweishmer (1945)		80–90	
MacLaren (1947)		100	
Fowler (1947)		100	
Canfield (1949)	80		100–110
Grave (1949)	90		
Guild (1950)	<90 dB above hearing threshold		

Adapted from Jonco (Reference 130).

ing overall sound pressure level and no other factors.

(b) Damage risk criteria since 1950

It was apparent by 1950 that proposed limits must consider, in addition to intensity, other physical dimensions and characteristics of noise exposure. In 1953, Rosenblith and Stevens[74] published an extensive document entitled *Noise and Man* in which they delineated the following variables important to the development of damage-risk criteria:

- Measurement of spectral distribution (noise spectrum).

- Determination of the temporal characteristics of exposure (noise duration).

- Identification of a protection goal (biologic response).

In the discussion which follows, selected damage-risk criteria listed in Table IX, will be compared and contrasted with respect to the above variables. The table represents a compilation of most criteria developed between 1950 and 1971, and where appropriate, criteria expressed in octave-band levels have been converted to equivalent dBA. For purposes of performing these conversions, a "pink" noise spectrum (*i.e.*, equal sound pressure level in each octave band), typical of many common industrial noises, was assumed.

(c) Criteria based on octave-band levels

Beginning with Kryter[73] in 1950, concern shifted from measurement of noise based solely on overall sound pressure to measurements which are

(*Text continues on page 897.*)

CRITERIA DOCUMENT TABLE 9

DAMAGE RISK CRITERIA FOR 5 TO 8 HOUR EXPOSURES AS PROPOSED FROM 1950–1971

Author and Year	Ref. No.	Basis of Criteria (notes explained at end of table)	Protection Goal	Actual or Computed* Octave-Band SPL								Actual or Computed** dBA
				20–75	75–150	150–300	300–600	600–1200	1200–2400	2400–4800	4800–9600	
Kryter (1950)	73	No "critical band"[1] > 85 dB SPL (re: 0.0002 MB)	No PTS or TTS	81	85	90	93	96	97	96	95[2]	88**
		No "critical band" > 85 dB SPL (re: MAF)	No PTS or TTS	125	115	108	101	100	92	87	102[2]	94**
Hardy (1952)	131	100 sones[3] per octave	Upper limit above which definite hazard to hearing exists	115	112	108	106	104	95	91	102	98**
		50 sones per octave	Lower limit below which no hazard to hearing exists	104	100	97	95	92	87	85	95	92**
Rosenblith, Stevens (1953)	74	Octave-band SPL with respect to the sensitivity of the ear broadband noise	Prevention of permanent damage due to noise	110	102	97	95	95	95	95	95	102**
Rosenblith, Stevens (1953)	74	Same as above except for pure tones and critical bands of noise	Prevention of permanent damage due to noise	100	92	87	85	85	85	85	85	Not Applicable

Author and Year	Ref. No.	Basis of Criteria (notes explained at end of table)	Protection Goal	Actual or Computed* Octave-Band SPL								Actual or Computed** dBA
				20–75	75–150	150–300	300–600	600–1200	1200–2400	2400–4800	4800–9600	
Lindman (1955)	132	Interpolation between sound pressure of sorting octaves⁴ and allowance for less sensitivity in lower frequencies	Protects most, but not all persons with unprotected ears	113	105	100	90	90	85	85	85	92**
CFR 160–3 (1956)	108	Octave-band levels at or above which ear protection must be used	Preservation of hearing of 15 dB or better at the frequencies 500, 1000, 2000 HZ	—	—	95	95	95	95	95	—	102**
		Octave-band levels at or above which the use of ear protection is recommended	Same as above	—	—	85	85	85	85	85	—	92**
AAOO (1957)	133	Octave-band SPL at these bands most likely to have an effect on the frequency listed in a protection goal	Protect man's hearing for speech (i.e., losses at 500, 1000, 2000 Hz)	—	—	—	—	85	85	—	—	92**

CONTINUATION (2) OF TABLE 9

Author and Year	Ref. No.	Basis of Criteria (notes explained at end of table)	Protection Goal	Actual or Computed* Octave-Band SPL								Actual or Computed** dBA
				20–75	75–150	150–300	300–600	600–1200	1200–2400	2400–4800	4800–9600	
Jones, Church (1960)	134	Octave-band SPL	Allowable weekly exposure dose, determining when hearing conservation is mandatory	100	91	87	86	85	85	85	92	92**
ISO (1961)	80	Octave band levels; primary emphasis on those with center frequency 500, 1000, 2000, NR Curve 85	Protect against TTS_5 or PTS_5 greater than 12 dB at 500, 1000, 2000 for 50% of the persons exposed	102	95	91	87	85	82	80	79	86**
Kryter (1963 and 1965)	135 136	Octave-band levels Broadband noise	Protect against normal ears producing TTS_2 of 10 dB at 1000 Hz, 15 dB at 2000 Hz, & 20 dB at 3000 Hz.	—	98	92	89	86	85	85	86	92**
Kryter (1963 and 1965)	135 136	Narrow-band levels	Protect against normal ears producing TTS_2 of 10 dB at 1000 Hz, 15 dB at 2000 Hz, and 20 dB at 3000 Hz	—	93	87	84	81	80	80	81	Not applicable

Author and Year	Ref. No.	Basis of Criteria (notes explained at end of table)	Protection Goal	Actual or Computed* Octave-Band SPL								Actual or Computed** dBA
				20–75	75–150	150–300	300–600	600–1200	1200–2400	2400–4800	4800–9600	
AAOO (1964)	137	Octave-band levels encompassing "speech frequencies"	Prevention of hearing loss in those people who are "normally" susceptible at the frequencies 500, 1000, 2000 Hz	—	—	—	85	85	85	—	—	92**
CHABA (1966)	107	Octave-band levels	No permanent or temporary loss greater than 10 dB at 1000 Hz, 15 dB at 2000 Hz, and 20 dB at 3000 Hz in 50% of the people exposed	—	98	92	89	86	85	85	86	98**
		Narrow-band levels			92	88	84	81	80	80	81	Not applicable
		Pure tones		—	92	88	84	81	80	80	81	Not applicable
Intersociety (1970)	27	dBA	An increase of 10 percentage points (10 more people per 100) in the number of people who develop hearing impairment by retirement age due to exposure									90

CONTINUATION (3) OF TABLE 9

Author and Year	Ref. No.	Basis of Criteria (notes explained at end of table)	Protection Goal	Actual or Computed* Octave-Band SPL								Actual or Computed** dBA
				20– 75	75– 150	150– 300	300– 600	600– 1200	1200– 2400	2400– 4800	4800– 9600	
British Occupational Hygiene Society (1971)	87	Noise immision based on dBA and total duration of exposure	Protect 99% of the exposed population from developing an average NIPTS of 40 dB or average hearing level of 48 dB for the frequencies 0.5, 1, 2, 3, 4, and 6 KHz									90
Kryter (1970)	88	Octave-band level	Maximum allowable TTS or PTS for 75% of those exposed limited to 0 dB below 2 KHz and 10 dB above 2 KHz	91	83	78	73	68	61	52	53	65

* Damage risk criteria not given in octave-band levels, but computed by author referenced by number following OBL 4800–9600 Hz.

** Computed, assuming a "pink" noise spectrum (equal energy in each octave band).

[1] Critical band— ". . is that frequency band of sound, being a portion of a continuous-spectrum noise covering a wide band, that contains sound power equal to that of a simple (pure) tone centered in the critical band and just audible in the presence of the wide-band noise" (Reference 4).

[2] From Eldredge, D. H. (Reference 91).

[3] Sone— ". . a unit of loudness. By definition, a simple tone of frequency 1000 Hz, 40 decibels above a listener's threshold, produces a loudness of 1 sone. The loudness of any sound that is judged by the listener to be n times that of the 1-sone tone is n sones" (Reference 138).

[4] Levels selected by Z24-X–2 sorting octaves (Reference 138).

[5] Average hearing level at 500, 1000, and 2000 Hz of 15 dB re ASA (1951) or 25 dB re ANSI (1969) (References 15 and 95).

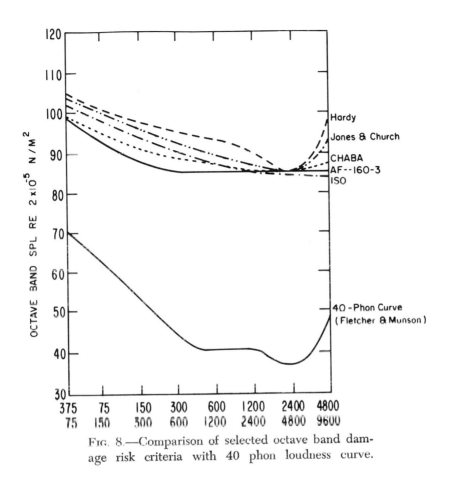

FIG. 8.—Comparison of selected octave band damage risk criteria with 40 phon loudness curve.

more indicative of the response of the hearing mechanism. Consistent with this thinking, several modern damage-risk criteria have emphasized limit setting by frequency bands, usually one octave in width. Two lines of evidence were responsible for this shift in thought. First of all, data on minimum audible field sensitivity[75] and measurements of equal loudness[76] indicated that the ear was not equally sensitive at all frequencies. It was found that the ear is most sensitive to acoustic stimuli in the frequency range 2000 to 4000 Hz, and less sensitive to frequencies both below and above this range. Shown in Figure 8 are several damage risk criteria (DRC) developed between 1952 and 1966. For comparison, the 40 phon equal-loudness curve[76] is presented in the lower part of this figure. As may be seen from this figure, although the DRC differ in estimates of safe sound pressure level per octave, they all weight the spectrum similarly.

The second major impetus for measurement of noise based on octave band analysis came from research which indicated that, at least for most audiometric frequencies, the amount of threshold shift observed (either temporary or permanent) was closely related to the frequency or spectrum of the stimulus. Results of "stimulation deafness" (temporary threshold shift) studies indicated that for pure tone stimuli the maximum shift in hearing appears to be about one-half octave above the frequency of total stimulation.[77-79] Similar findings were reported for octave bands of noise and broadband noise by Davis et al.,[77] Kylin,[29] and Ward.[30] However, for these latter stimuli there was some difference of opinion as to the exact location of maximum effect. Davis et al.[77] and Kylin[29] suggested that the maximum effect occurs one-half octave to one octave above the center frequency of the octave band, whereas more recently Ward[30] found that the maximum change in hearing occurs one-half octave to one octave above the upper cutoff frequency of the noise.

Prior to 1956, damage risk criteria set as a goal for protection (see the later section Protection goal) the prevention of hearing loss at all frequencies. This necessitated assessment of the noise at each octave band. After this time, however, much more qualified protection goals were established (usually protection of loss in the so-called "speech frequencies") such that only knowledge of the sound pressure in certain critical octave bands (not to be confused with aural critical bands) was required in order to assess the risk of noise exposure to hearing. This approach characterized the damage-risk criteria developed by the Air Force in 1956, The American Academy of Ophthalmology and Otolaryngology in 1957, the International Standards Organization in 1961, and the American Academy of Ophthalmology and Otolaryngology in its revision of the 1957 criteria in 1964 (see Table IX).

The procedure for rating noise hazards by this method consists of measuring the octave-band levels in the critical octaves, and then comparing the measured levels with damage-risk contours. This is best exemplified by the use of the "Noise rating" curves developed by the International Standards Organization.[80] The octave-band levels of the noise are measured and then compared with the noise rating curve (Figure 9).* The highest curve which is exceeded by the level of these bands yields the noise rating number (N). For this particular scheme, a noise rating of 85 was suggested as the protection criterion.

The use of A-weighted sound level

Since the publishing of the first Intersociety *Guidelines for Noise Exposure Control*,[81] a relatively new approach, A-weighted sound level measurement, has become a popular measure for assessing overall noise hazard. As stated in Part III, the

---

* This will be found in Chapter 11, Figure 11–1.

weighting on the A-scale approximates the 40-phon equal loudness contour (Figure 8). Use of the A-weighting is thought, therefore, to assure the rating of noises in a reasonably similar manner as would the human ear.

Several studies have been conducted in order to evaluate the efficacy of using A-weighted sound levels in rating hazardous exposures to noise. In a study of 580 industrial noises, Botsford[82] showed that the A-weighted sound level indicated the hazard to hearing as accurately as did limits expressed as octave-band sound pressure levels in 80 percent of the cases and was slightly more conservative than octave band measures in 16 percent of the noises. Passchier-Vermeer[26] found that, except in one noise condition, sound level in dBA was as accurate as Noise Rating (NR) in estimating noise induced hearing loss. In a study of hearing loss in 759 subjects, Robinson[83] concluded that the error incurred from using dBA in predicting hearing level was within ±2 dB, even for noises ranging in slope from +4 dB/octave to −5 dB/octave. A recent study[84] found that even though dBA perhaps discounted too much low-frequency energy, in all cases but one it predicted TTS$_2$ resulting from exposures to noises of different spectra (slopes of −6 dB/octave, 0 dB/octave, and 6 dB/octave) as well as, or better than, other noise rating schemes which employed spectral measurements in octave bands.

As a result of its simplicity and accuracy in rating hazard to hearing, the A-weighted sound level was adopted as the measure for assessing noise exposure by the American Conference of Governmental Industrial Hygienists (ACGIH)[85] and by an Intersociety Committee[27] consisting of representatives from the American Academy of Occupational Medicine, American Academy of Ophthalmology and Otolaryngology, ACGIH, Industrial Hygiene Association, and the Industrial Medical Association. A-weighted sound level measurement was also adopted by the U.S. Department of Labor as part of the Occupational Safety and Health Standards[86] and by the British Occupational Hygiene Society in its Hygiene Standard for Wide-Band Noise.[87]

In keeping with the several precedents which have been established for its use in rating the hazard resulting from industrial noise exposure, and because it has been shown to be a reasonably accurate measure of such hazard, the A-weighted sound level measurement has been recommended for use in rating noise hazard in the Recommended Standard.

Protection goal

The limit of noise exposure that is established ultimately depends upon the degree of hearing which is to be protected and the number of persons in an exposed population to be protected. If a very strict protection criterion is contemplated such that no person exposed to noise will develop hearing loss at any frequency, the maximum permissible noise level

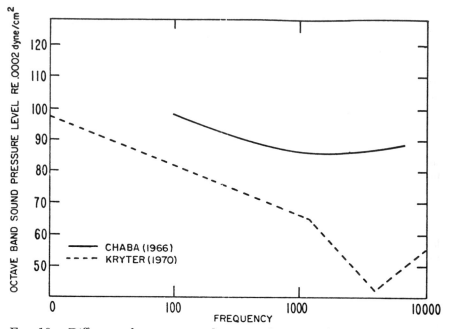

Fig. 10.—Difference between two damage risk criteria based on protection goal recommended by CHABA (1966) and Kryter (1970).

governing a daily, or near-daily, exposure would be quite low. Conversely, if the protection goal were to permit a certain amount of hearing loss in a small percentage of workers over a working lifetime, then the permissible exposure level would be raised accordingly. For example, Figure 10 compares the permissible levels of exposure for an eight-hour day recommended by the National Academy of Sciences, National Research Council Committee on Hearing, Bioacoustics and Biomechanics (NAS-NRC) CHABA Working Group 46 with the damage risk criterion recently proposed by Kryter[88] for the same amount of exposure. Although both criteria are based upon either the same or similar types of data, the damage risk level is much higher in the CHABA criterion than in that proposed by Kryter. The major reason for this difference is that CHABA established as its protection goal attainment of no more than 10 dB of permanent threshold shift at 1000 Hz, 15 dB at 2000 Hz, and 20 dB at 3000 Hz in 50 percent of the people exposed to noise; whereas, Kryter set as his protection goal attainment of "0" dB of threshold shift at the frequencies 2000 Hz and below, and 10 dB of shift in the frequencies above 2000 Hz in 75 percent of the people exposed to noise.

The problem is further illustrated by a comparison of the protection criteria developed by the Intersociety in the *Guidelines for Noise Exposure Control* with the Hygiene Standard for Wide Band Noise[87] developed by the British Occupational Hygiene Society. Both standards established 90 dBA as the limit for a near daily 8 hour-per-day continuous exposure. However, as the following quotations indicate, there is quite a difference of opinion as to how much protection is actually afforded by 90 dBA:

• "In the population exposed to 90 dBA to age 50 through 59, the amount of impairment is increased 10 percentage points (ten more persons per 100 exposed) as compared to the population with no occupational exposure." (Intersociety, 1970)

• "A noise emission of 105 dB (equivalent to 90 dBA for a working lifetime) is acceptable exposure on the basis that no more than 1 percent of exposed persons will experience handicap due to noise after lifetime exposure." (British Occupational Hygiene Society, 1071)

The difference here, as in the previous example, follows from a difference in the definition of the protection goal, specifically, the definition of hearing impairment or hearing handicap. The first criteria (Intersociety, 1970) adopted the AAOO-AMA definition of hearing impairment.[15] This definition states that hearing impairment begins as the average hearing level at 500, 1000, and 2000 Hz exceeds 15 dB re ASA S3.6-1951 (25 dB re ANSI S3.6-1969). Conversely,

the British Occupational Hygiene Standard defined as its "low fence" of impairment an average permanent noise induced threshold shift (not to be confused with hearing level) of 40 dB in the six frequency range 0.5 to 6.0 KHz for 30 years of exposure. Recently, Robinson[89] computed hearing impairment risk values on the British data using the AAOO-AMA definition of hearing impairment. His figures indicate that near-daily 8 hour exposure to continuous noise at a level of 90 dBA for 40 years would result in an increase in hearing impairment of between 13 to 15 persons per 100, depending upon the incidence figure of the non-noise exposed control population used for comparison. This "risk" value is comparable to the one presented by the Intersociety Committee[27] but about 6 to 8 percent below the International Standards Organization value for the same exposure.[90]

The question of how much hearing should be protected and in what percentage of the people hearing losses of certain magnitudes should be permitted has long been an issue of much controversy. The ultimate decision, according to Eldredge,[91] must be based on social and humane values.

Historically, the most common protection goal has been one directed at the preservation of hearing for speech. Direct measures for evaluating hearing for speech have been, and are being, developed. These tests generally fall into two classes: (1) those which measure the threshold of speech or the ability to hear speech, and (2) those which measure discrimination, or the ability to understand speech. Although speech tests have been widely accepted for use in aural diagnostics, several objections have been raised as to their use and validity in industrial testing. These are: (1) speech test items are sometimes unfamiliar to the listener; (2) speech tests frequently measure the size of one's vocabulary as well as hearing impairment for speech; (3) several speech tests or different forms of a single test designed to measure the same speech hearing function may yield different results; and (4) considerable training is required on the part of the examiner to administer and score speech tests. It has become, therefore, a common practice to measure pure tone sensitivity and relate hearing levels at certain specific frequencies to the ability to hear and understand speech.

In 1929, Fletcher[92] proposed what has now become known as his "Point 8" formula whereby the ability to hear everyday speech was estimated by multiplying the hearing levels at 500, 1000, and 2000 Hz by 0.8 and then computing the average over these three frequencies. The major contribution of this formula was the introduction of the concept that hearing loss for speech could be estimated by the average hearing levels at what has now become known as the "speech frequencies"—500, 1000, and 2000 Hz.

The American Medical Association[93] in 1947 recommended that hearing loss for speech be determined by the pure tone hearing losses at 500, 1000, 2000, and 4000 Hz. The four frequencies were given a weighting in accordance with what was presumed to be the importance of each frequency in hearing for speech (i.e., 15 percent at 500 Hz, 30 percent at 1000 Hz, 40 percent at 2000 Hz, and 15 percent at 4000 Hz). This guideline further suggested that hearing loss for speech does not begin until the weighted-average hearing loss equaled 10 dB, and total loss for speech hearing occurred when the loss at 500 Hz reached 90 dB or the losses at the other three frequencies reached 95 dB.

In a later article which reviewed the assumptions in computing hearing loss for speech, the AMA[94] made the following observations and recommendations:

• The 1947 formula was inadequate for calculating hearing loss for speech in sensorineural hearing loss. (This is particularly interesting in that the method used today for computing hearing loss for speech, developed by the AAOO in 1959 and accepted by the AMA in 1961, eliminated the most sensitive indicator of sensorineural hearing loss (i.e., losses at 4000 Hz.))

• Everyday communication should be the basis for evaluation of hearing disability.

• Losses greater than 15 dB (re ASA, 1951 Zero Audiometric standard) at 500, 1000, and 2000 Hz are abnormal and usually noticeable by the individual in everyday communications. Furthermore, a loss greater than 30 dB at 4000 Hz can be considered abnormal.

A new formula was developed by the Subcommittee on Noise of the American Academy of Ophthalmology and Otolaryngology (AAOO). This formula was subsequently adopted by the AAOO Committee on Conservation of Hearing[95] in 1959 and by the American Medical Association[15] in 1961. The bases of this formula are explained by the following excerpts taken from the *Guides to the Evaluation of Permanent Impairment,* published by the American Medical Association.[15]

"Estimated hearing level for speech is the simple average of hearing levels at the three frequencies of 500, 1000, and 2000 cycles per second (cps).

"Ideally, hearing impairment should be evaluated in terms of ability to hear everyday speech under everyday conditions. The ability to hear sentences and to repeat them correctly in a quiet environment is taken as satisfactory evidence for correct hearing of everyday speech. Because of present limitation of speech audiometry, the hearing loss for speech is estimated from measurements made with a pure tone audiometer. For this estimate, the simple average of the hearing levels at the three frequencies 500, 1000, and 2000 cps is recommended.

"In order to evaluate the hearing impairment, it must be recognized that the range of impairment is not nearly so wide as the audiometric range of human hearing. Audiometric zero, which is presumably the average normal threshold level, is not the point at which impairment begins. If the average hearing level at 500, 1000, and 2000 cps is 15 dB or less, usually no impairment exists in the ability to hear everyday speech under everyday conditions."

The only major change in this formula from 1959 to the present time has been the result of the change in audiometric reference for hearing level (HL). The 15 dB average hearing level at 500, 1000, and 2000 Hz referenced to the 1951 ASA standard[96] corresponds to a 25 dB average hearing level at the same frequencies according to the recent reference pressure adopted by the American National Standards Institute.[97]

On the basis of the results of recent research which has investigated the relationship between pure tone hearing loss and hearing loss for speech, a slightly different definition of "hearing impairment" has been adopted for the purposes of this document. Simply stated, hearing impairment for speech communication begins when the average hearing level at 1000, 2000, and 3000 Hz exceeds 25 dB re ANSI (1969). The principal reasons for this definition are as follows:

(1) The basis of hearing impairment should be not only the ability to hear speech, but also the ability to understand speech.

(2) The ability to hear sentences and repeat them correctly in quiet is *not* satisfactory evidence of adequate hearing for speech communication under everyday conditions.

(3) From the two reasons above, the ability to understand speech under everyday conditions is best predicted on the basis of the hearing levels at 1000, 2000, and 3000 Hz.

(4) The point at which the average of hearing losses in the stated three frequency range of 1000 to 3000 Hz begins to have a detrimental effect on the ability to understand speech is 25 dB re ANSI (1969).

With reference to (1) *Determination of hearing impairment,* the ability to "hear" speech, measured in terms of the lowest intensity at which a listener can barely identify speech materials, provides little information concerning communication difficulties under everyday conditions. As Sataloff[98] states, "It [occupational deafness] implies the presence of obvious difficulties in hearing speech. Actually, the difficulty more often lies not so much in 'hearing' speech as in 'understanding' it." Furthermore, Davis and Silverman[99] observed that ". . . a man with severe high-tone nerve deafness (as is seen in occupational noise induced hearing loss) will always fail to hear certain sounds and will never make a perfect articulation score. On the other hand, the same man may hear some words, the easy low-frequency words, as well as anyone else does. He may have a normal *threshold* for speech."

This issue is further clarified if one compares the "typical" clinical picture of a person having a conductive hearing loss versus a person having a sensorineural hearing loss resulting from noise exposure. Both cases would be expected to have elevated speech reception thresholds (a measure of hearing for speech); however, in the case of the conductive loss, speech discrimination (measure of understanding) would be approximately the same as that for a person having normal hearing, provided that the presentation level is sufficiently above the speech reception threshold level. The person with occupational hearing loss (sensori-neural), on the other hand, would have relatively poor discrimination scores, and the effect of raising the presentation level to higher levels often serves to reduce the articulation score[100] (see example in Figure 11). In applying the AAOO-AMA formula to the cases shown in Figure 11, it is possible that both would be rated identically in terms of hearing impairment, yet the sensori-neural case has much more difficulty in understanding speech than does the conductive case. It is apparent, therefore, that the formula applied to compute hearing impairment should consider discrimination ability and that the pure tone frequencies used in the formula should be highly correlated with this later function.

With reference to (2) *speech communication under everyday conditions,* it has been assumed by the AAOO-AMA formula that the "ability to hear sentences and repeat them correctly in a quiet environment is taken as satisfactory evidence for correct hearing of everyday speech."[15,95] According to Kryter[88] this definition of everyday speech employs a type of speech material and a listening condition which is not indicative of ev-

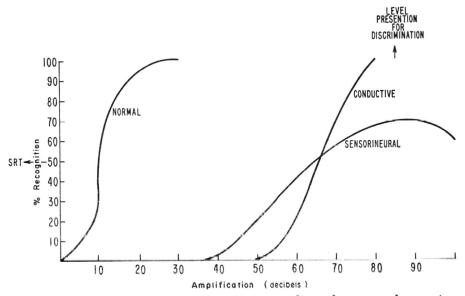

FIG. 11.—Articulation indices representing normal, conductive, and sensori-neural cases. (From Hood, 1971.)

eryday conditions and one which is "least likely to show any impairment in the deafened person."

Actually, everyday communication is placed under a wide variety of environmental stresses. Estimates of the amount of time that everyday speech is distorted range from a conservative figure of 50 percent[100] up to about 100 percent.[101] Furthermore, everyday speech rarely takes the form of complete sentence communications; thus, the number of speech cues available for accurate speech perception under everyday conditions is greatly reduced.[88] From this discussion, it may be concluded that an appropriate predicting scheme for determination of hearing impairment must include some consideration for an actual daily communication environment rather than some optimum condition as suggested by the AAOO AMA.

With reference to predicting ability to understand speech on the basis of heavy levels at (3) *the pure tone average at 1000, 2000, and 3000 Hz,* results of several studies indicate that hearing levels at these three frequencies predict hearing loss for speech under mild conditions of distortion better than the three frequency average at 500, 1000, and 2000 Hz. Mullins and Bangs,[102] investigating the relationship between speech discrimination and several indices of hearing loss, found that the pure tone hearing losses at 2000 and 3000 Hz

905

had the highest correlation with speech discrimination. Harris, Haines, and Meyers[103] studied the effect that speeded speech had on discrimination in subjects with high-frequency sensori-neural hearing loss. They concluded that a nearly normal audiogram at 3000 Hz was essential for high sentence intelligibility if the speech material is distorted by increasing the speech rate. It was further concluded that once hearing losses progressed to include 2000 Hz, the effect on discrimination of speeded speech was quite devastating.

Kryter, Williams, and Green,[17] in a study of the effects of background noise on speech discrimination, found that in 114 adult male soldiers who had varying degrees of sensori-neural hearing loss, threshold levels at 2000, 3000, and 4000 Hz correlated best with speech discrimination loss. They concluded, however, that the average hearing loss at 1000, 2000, and 3000 Hz should be used to predict speech-hearing loss since this average represented a "reasonable compromise" for the results of the various studies which have dealt with the topic.

In a comparison of normal-hearing subjects and subjects with sensori-neural hearing losses on several different measures of hearing acuity, Ross et al.[104] found that in the hearing impaired group: (1) speech discrimination scores in quiet tended to be poorer as the losses at 2000 and 4000 Hz increased, and (2) neither pure tone threshold at 500 Hz nor speech reception threshold levels were related to speech discrimination in quiet.

Furthermore, in 1965 Harris[16] conducted an investigation to explore the effects of audiometric losses on discrimination scores for speech which was mildly and severely distorted. The results of this study indicated the frequency regions of greatest impact on intelligibility were somewhat different, depending upon the severity of the distortion. However, Harris concluded that ". . . the region 2000 Hz and below is inadequate for predicting intelligibility of speech in noise, and that a point of vanishing returns is reached by adding anything beyond 3000 Hz."

Recently, Acton[105] investigated the effect of different signal-to-noise ratios on speech discrimination in a group of industrial workers who had incurred characteristic noise induced hearing losses. Results indicated that a significant loss in speech intelligibility occurred when high-frequency hearing loss involved the 2000 Hz audiometric test frequency, and quite profound effects upon intelligibility once the loss had progressed to 1 KHz. In another recent investigation[106] of speech discrimination in industrial employees, it was found that hearing level at 2000 Hz had the highest correlation with speech discrimination (0.769, P 0.0001) under the most favorable condition of signal-to-noise ratio (S/N = +10).

In summary, it is evident that in order to accurately assess hearing loss for speech under everyday conditions by means of pure tone hearing loss, a modification in the three frequency average recommended by the AAOO and the AMA is warranted. Such a

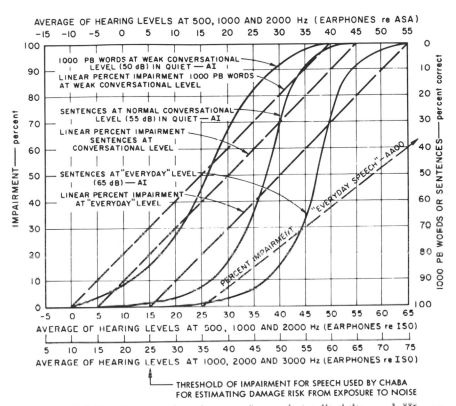

AVERAGE OF HEARING LEVELS AT 500, 1000 AND 2000 Hz (EARPHONES re ASA)

FIG. 12.—Relation between impairment of speech intelligibility and HL, as calculated by AI and as proposed by AAOO. (From Kryter, 1970.)

modification should include the elimination of 500 Hz from the formula, and the addition of 3000 Hz in its place.

With reference to (4) *Level of beginning hearing impairment for speech,* it would appear that an average hearing level of 25 dB re ANSI (1969) at 1000, 2000, and 3000 Hz signals the beginning of speech communication difficulties in everyday situations. In a comprehensive review of the topic of hearing impair-

ment, Kryter[88] constructed several curves (see Figure 12) which related pure tone hearing level average to speech impairment for various samples of speech presented at different levels in quiet. As may be seen from this figure, the AAOO-AMA definition of impairment (average HL at 500, 1000, and 2000 Hz of 25 dB re ANSI (1969)) allows for negligible impairment for sentences presented at an "everyday" level or normal conversational level, and only 15 percent impairment in the perception of isolated words presented at the weak conversational level.

Kryter, Williams, and Green[17] found that in subjects with sensorineural hearing losses, a dramatic change in perception of speech occurred as the average hearing level at 1000, 2000, and 3000 Hz shifted from approximately 18 dB re ANSI (1969) to about 31 dB re ANSI (1969). Corresponding to these shifts in average hearing level, sentence intelligibility in a mild background of noise (S/N = +5) dropped from 90 to 78 percent whereas PB word intelligibility, with slightly less noise (S/N = +10), decreased from 75 to 58 percent.

Results of the study conducted by Acton[105] concerned with speech intelligibility in a group of industrial workers indicated that a significant, although slight, shift in speech intelligibility (compared with normals) occurred when the hearing level (group mean) at 2000 Hz had reached 25.3 dB. At this point, the average hearing level at 1000, 2000, and 3000 Hz was 25 dB re ANSI (1969).

Temporal characteristics of exposure

The damage-risk criteria in Table IX are specifically concerned with limits of safe exposure to continuous noise for five hours or more. It has long been recognized that the ear can tolerate greater amounts of energy provided that the exposure time is limited.[73,74,107] Furthermore, research indicates that noises which are interrupted on a regular or irregular basis are much less hazardous to hearing.[109-111]

The decision as to how much noise can be tolerated for daily short-duration continuous exposures and interrupted exposures ultimately depends upon how the ear integrates noise over time. Probably the two most popular theories on how the ear responds to such stimulation are the equal energy and the equal pressure rules.

The equal energy rule states that equal quantities of *acoustic energy* entering the ear canal are equally injurious, regardless of how they are distributed in time. This rule dictates that, as exposure time doubles, the level of noise must be reduced by 3 dB in order to maintain an equal degree of hazard. The equal pressure rule, on the other hand, hypothesizes that the ear integrates noise on a *pressure*, rather than an energy basis. Such a rule maintains that for each doubling of the exposure time the level of noise must be reduced by 6 dB to maintain an equal degree of hazard.

Research attempting to determine which rule is appropriate has generally been inconclusive. Spieth and Trittipoe,[112] investigating the effects of high-level, short duration exposures in human subjects, found that two different exposures would produce the same TTS if one exposure were 6 dB lower and twice the duration of the other. Conversely, Ward[113] recently found that four separate exposure conditions, equated in terms of equal energy, all caused about the same amount of temporary threshold shift in chinchillas. However, he cautioned that his findings were only applicable to continuous exposures and not to intermittent exposures.

Variables that are germane to interrupted exposures but do not play a significant role in limiting hazard from short-term continuous exposures further complicate the problem of how the ear responds and integrates noise over time. One such variable is the "acoustic" or "middle ear" reflex. When the ear is exposed to loud noise, the middle ear muscles contract, thus altering the impedance of the middle ear. This reflex, which serves to attenuate the noise reaching the inner ear, adapts out or disappears quickly if the noise is continuous and relatively unchanging over time. However, if the noise level varies considerably or is interrupted on a regular or irregular basis, then the reflex is sustained.

A second variable which plays an important role in reducing the hazard of interrupted noises relative to short-term continuous noises concerns the off-time of the exposure cycles. Depending upon the overall level of the noise and the nature of the relationship between on-time and off-time, a considerable reduction in the degree of temporary hearing threshold shift may be observed.

To date, the only empirical data available on permanent hearing losses resulting from intermittent exposures comes from a study of iron ore miners conducted by Sataloff et al.[114] Their findings indicated that intermittent noises had to be some 15 dB more intense than continuous noises to cause the same additional hearing impairment in men ages 30 to 50 years. Although this evidence confirms the general notion that intermittent exposures are less hazardous than continuous steady-state exposures of the same duration and noise level, the applicability of this rule to other schedules of intermittency must await further investigation.

Since 1960, several damage risk criteria have been proposed to limit exposure to intermittent noise.[82,107,115] For the most part, these criteria, like the rules for assessing intermittent noise exposure discussed next, have been based predominantly upon evidence collected from studies of temporary threshold shift.

At least three different rules have been proposed in order to assess the hazard of exposures to intermittent noise. The first of these rules, developed by Ward et al.[28] was called the "on-fraction" rule. This rule states that the amount of temporary threshold shift resulting from a given intermittent exposure can be determined on the basis of noise level and average on-fraction (the time the noise is on, divided by the total duration of exposure). This procedure assumes that levels below 75 dB SPL are not hazardous to hearing; thus, the amount of on-time is taken as the total time the noise is above 75 dB SPL. In a critical test of the on-fraction rule, Selters and Ward[111] found that this rule was invalid when the regular on-off times exceeded two minutes.

For burst durations longer than

CRITERIA DOCUMENT TABLE 10

ACCEPTABLE EXPOSURES TO NOISE IN dBA AS A
FUNCTION OF THE NUMBER OF OCCURRENCES PER DAY.

Daily Duration		Number of times the noise occurs per day						
Hours	*Min*	*1*	*3*	*7*	*15*	*35*	*75*	*160 up*
8		90	90	90	90	90	90	90
6		91	93	96	98	97	95	94
4		92	95	99	102	104	102	100
2		95	99	102	106	109	114	
1		98	103	107	110	115		
	30	101	106	110	115			
	15	105	110	115				
	8	109	115					
	4	113						

To use the table, select the column headed by the number of times the noise occurs per day; read down to the average sound level of the noise and locate directly to the left in the first column the total duration of noise permitted for any 24 hour period. It is permissible to interpolate if necessary. Noise levels are in dBA.

From Guidelines for Noise Exposure Control, 1970

two minutes, a second rule has been suggested. This second rule, developed by Ward et al.,[109] is called the "exposure equivalent" rule. According to the concept of exposure equivalency, the amount of hearing change observed at the end of the day may be computed as follows:

1. Calculate the amount of TTS resulting from the exposure to the first bursts of noise.

2. Using generalized recovery curves, compute the residual TTS remaining at the end of the "off-time."

3. Determine how much exposure (time) to the noise causing the initial TTS in (1) above is necessary to cause the residual TTS.

4. Add the time in (3) above to the time of the subsequent noise burst and predict the TTS₂ at the end of the second exposure.

5. Repeat steps (2), (3), and (4) for each cycle in the daily exposure.

The essential feature of this approach is that residual TTS is translated into exposure time.

One of the crucial assumptions of the "exposure-equivalent" rule is that the course of recovery from TTS is independent of the type of noise that produces the TTS. In a recent article, Ward[33] has presented data that question the validity of this assumption. It appears that intermitttent exposure

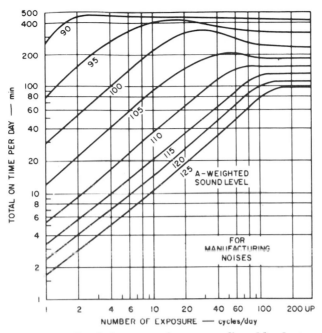

FIG. 13.—Total duration of a noise allowable during an 8 hr day as a function of the number of periodic interruptions. An exposure cycle is completed each time the A-weighted sound level decreases to or below 89 dB. (From Botsford, 1967.)

to high-level, high-frequency noise causes a considerable delay in the recovery of TTS relative to intermittent low-frequency exposures.

A third approach in determining hazard from interrupted noise has been to determine the total on-time of the noise, regardless of how the noise bursts are distributed in time, and to consider the intermittent exposure in terms of an equivalent continuous exposure. This approach attempts to take into consideration the reduced hazard of interrupted noise by adjusting the rule which relates noise level and exposure duration. Although possibly not as scientifically rigorous as the previously mentioned

procedures, the "equivalent continuous" rule is not constrained by the assumption concerning the regularity of exposure cycles which is basic to the other rules.

Intermittent noise exposure criteria based upon the first and/or second rules include those developed by Glorig, Ward, and Nixon,[115] CHABA Working Group 46,[107] and Botsford.[32] Botsford's[82] intermittency criteria reflect a simplification and consolidation of the CHABA continuous exposure, long-burst intermittent, and short-burst intermittent contours into one general figure relating dBA level, total on-time (noise level above 89 dBA), and number of exposure cycles (see Figure 13). The limits of intermittent exposure expressed in these contours (shown in Table X) have recently been adopted by the Second Intersociety Committee.[27] Similar limits have been adopted as part of a revision of the German document concerned with assessment of industrial noise in working areas.[116]

Recent research designed to investigate the efficacy of the limits proposed in Table X have generally shown that the limits do not accurately predict risk to hearing, at least insofar as temporary threshold shift is concerned. In a laboratory study[117] designed to evaluate selected exposure conditions from Table X, it was found that (1) the table permits concentrations of noise exposure within an 8-hour workday than can cause excessive amounts of temporary threshold shift, and (2) the conditions did not yield equal effects on hearing, thus not affording equal protection. Conversely in a study of forestry employees[118] it was found that although the noise exposures were rated as hazardous according to Table X, the audiometric results indicated that the exposures did not pose a risk to hearing.

Considerably more data must be collected to evaluate present criteria which attempt to designate safe levels of exposure to intermittent noise. Furthermore, additional research is needed to define the relationship of exposure level and duration. Until such information is made available, a change in the present 5 dB rule for halving or doubling of exposure time and a change in the assessment of intermittent noise in terms of equivalent continuous exposure is unwarranted.

One variable which does warrant alteration concerns the lower level or "off" level of noise in intermittent exposures. The designation of such a level implies (1) noises below this level do not of themselves cause any significant temporary or permanent hearing threshold shift, and (2) in combination with intermittent high levels of noise, optimum recovery may take place between noise bursts.

Various noise "cut-off" levels have been suggested. As mentioned previously, Glorig, Ward, and Nixon,[115] based on results of continuous noise exposure on temporary threshold shift, designated 75 dB SPL in any octave band as the level at which no TTS_2 would develop. The CHABA Working Group 46,[107] on the other hand, suggested that the "off" level was frequently dependent. For example, the safe level of exposure for the octave band 300–600 Hz was seen to be 89 dB SPL, whereas it was approximately 85 dB SPL for octave band 1200–2400 Hz.

Recently, Botsford[82] computed a dBA equivalent from the octave-band damage-risk criteria developed by CHABA. The results of this computation suggested that the "off level" based upon one-third octave or octave-band sound pressure level will, in many cases, be below the level designated by Botsford (particularly in the case of strong narrow-band components in the noise). Both the CHABA and Botsford criteria do not appear to be in accord with the intended meaning of a safe intermittent level in that present data suggest that there is a significant increase in the proportion of the population having hearing impairment in those groups exposed to continuous noise levels at and slightly below 85 dBA as compared with a non-noise exposed population.

Two lines of evidence suggest that the lower limit of interrupted exposure is considerably below the levels mentioned above. In a review of much of the available TTS and PTS data, Kryter[88] stated that a level of 65 dBA would cause "(a) no more temporary threshold shift than 0 dB for frequencies up to 2000 Hz and 10 dB for frequencies above 2000 Hz, measured two minutes after initial exposure for the average normal ear, and (b) a like amount of permanent noise-induced threshold shift following 20 years of nearly eight hours of daily exposure to noise in the hearing of no more than 25 percent of the population." Furthermore, results of a study[119] that investigated interrupted exposures using three different quiet levels indicated that the interval level of 57 dBA had a significant effect on the resultant TTS_2, TTS_{30}, and 30 minute recovery rate when compared with 67 dBA and/or 77 dBA interruption levels. It was concluded that recovery from intermittent noise exposure is maximized in quiet levels below 67 dBA.

It would appear from the foregoing discussion that a level of approximately 65 dBA meets the requirements of criteria established for a true "off-level" for intermittent exposure.

Support of the standard

To comply with the protection goal of the NIOSH standard (see Part I), hearing impairment for an individual is considered to occur when the average of hearing threshold levels at the three audiometric frequencies 1000, 2000, and 3000 Hz for both ears exceeds 25 dB (thresholds re ANSI S3.6 (1969)). As described next, NIOSH noise and hearing study data relevant to hearing impairment were analyzed, and the incidence of hearing impairment of noise exposed employee groups was compared with that of unexposed employee groups of comparable age and work experience. For the purposes of this part, noise exposed employees are those exposed to 80 dBA-Slow to 102 dBA-Slow and non-noise exposed employees are those exposed to less than 80 dBA-Slow. These comparisons resulted in the risk values applicable to the NIOSH standard (incidence of hearing impairment of exposed group minus incidence of hearing impairment of unexposed group).

Data collected from 1968 to 1971 by NIOSH, represented the steelmaking, paper bag processing, aluminum processing, quarrying, printing, tunnel police, wood working, and trucking employees included in 13 noise and hearing surveys. Audiometric data from non-noise exposed employees were collected in 12 of these 13 surveys. The audiometric data were analyzed using the current "fence" of hearing handicap, 25 dB average hearing threshold level at 0.5, 1, and 2 kHz (thresholds re ANSI

CRITERIA DOCUMENT TABLE 11
DISTRIBUTION OF NIOSH DATA OVER NOISE EXPOSURE LEVEL, AGE, AND EXPERIENCE

Age Groups (in yrs.)	17–27	28–35	36–45	46–54	55–70
Number of Workers	228	292	287	215	150
Experience Groups (in yrs.)	0–1	2–4	5–10	11–20	21–41
Number of Workers	133	154	308	314	263
Exposure Groups* (in dBA-Slow)	<80	80–84	85–89	90–94	95–102
Number of Workers	380	51	387	314	40

* In the data analysis, noise exposure levels were not grouped.

S3.6–1969, as well as the fence appropriate to this document, 25 dB average hearing threshold level at 1, 2, 3 kHz (thresholds re 1969 audiometric zero). The total sample of more than 4000 audiograms, however, could not be used to represent a qualitative measure of hearing loss. Employees not exposed to a specified continuous noise level in dBA-Slow over their working lifetime and those with abnormal hearing levels as a result of their medical history and a variety of otological problems were eliminated from the sample. Thus, 1172 audiograms were used which represented 792 noise exposed and 380 nonnoise exposed employees. The distribution of employees with respect to noise exposure, age, and experience is listed in Table XI.

The audiometric van used for the hearing tests was capable of testing six individuals at one time. All employees were tested before the beginning of their work shift, and, due to scheduling problems, the number of employees in a test session ranged from one to six. When less than six employees were present at a testing session, an attempt was made to randomize the assignment of audiometers. It was also necessary to use headphones with otocups to properly shield the employees from the possible effects of interference caused from hearing the other test tones in the van. However, it was found from the results of two independent studies in the NIOSH laboratory that there was no significant difference in measured thresholds between headphones fitted with otocups and those fitted with standard MX-41/Ar-type ear cushions.

Before data analysis could be done, it was necessary to check the calibration data accumulated during the respective survey. Calibration of the audiometers used to take the audiograms was usually performed before and after each survey. The data were corrected where necessary to the appropriate values given in the American National Standard *Specifications for Audiometers*, ANSI S3.6–1969.

Used for purposes of data analysis

were the three-frequency averages mentioned above in the definitions of hearing impairment. HLI ($\overline{0.5, 1, 2}$) and HLI ($\overline{1, 2, 3}$) are used to denote these averages performed over both ears. (HLI stands for "hearing level index.")

The samples were grouped into age and experience ranges to assure equal numbers per cell and a consistent spread of the data across the various dBA levels.

The following lists the steps made in the data analysis:

(1) Hearing level indices for 87 and 94 dBA noise exposed individuals were grouped into 31 samples for three-way cross-classification with respect to dBA level, age group, and experience group. The data were transformed by taking natural logarithms, and the resulting variances of log HLI ($\overline{0.5, 1, 2}$) and log HLI ($\overline{1, 2, 3}$) were computed for each sample. For each of the two dBA levels, Bartlett's tests for homogeneity of variances were performed over all age and experience combinations. Separate tests were performed for HLI ($\overline{0.5, 1, 2}$) and HLI ($\overline{1, 2, 3}$) average noise indices. Of the four Bartlett's tests, three showed no suggestion of nonhomogeneity of variance, but the fourth was significant at the 0.05 probability level. However, only one atypical variance was found within the "nonhomogeneous" group, and this was believed to be caused by an improbable combination of purely random variations and not indicative of a real elevation of variability for the cell in question. Thus, the conclusions were that variability of log

HLI ($\overline{0.5, 1, 2}$) and log HLI ($\overline{1, 2, 3}$) for replicate subjects was stable over all cells defined by the cross-classification.

(2) Fifth-degree orthogonal polynomial regression curves were fitted to log HLI *vs.* dBA for each age and experience cell using data for *all* dBA levels. Significance tests for nullity of regression coefficients were performed. For most of the curves which exhibited any significant trend, a straight line fitted the data within the limits of unexplained variability. In several cases, fourth or fifth degree coefficients showed significance, but examination of the plotted points revealed these to be artifacts due to clustering of the dBA levels for those plots, i.e., too few levels of the independent variable so that the polynomial tended to "fit the random errors."

(3) Histograms of pooled deviations of log HLI values from the respective regression lines for HLI ($\overline{0.5, 1, 2}$) and HLI ($\overline{1, 2, 3}$) were constructed by fitting normal distribution curves. Chi-square goodness-of-fit tests were performed. The tests revealed that the log HLI deviations from the means were normally distributed over the full range of variability to a very significant degree of approximation as shown in Figures 14 and 15. Means were found to be zero, and pooled variances were calculated for use in later stages of the analysis.

(4) Regression lines for different age groups within an experience level were tested for parallelism, and in every case, the lines were found to be

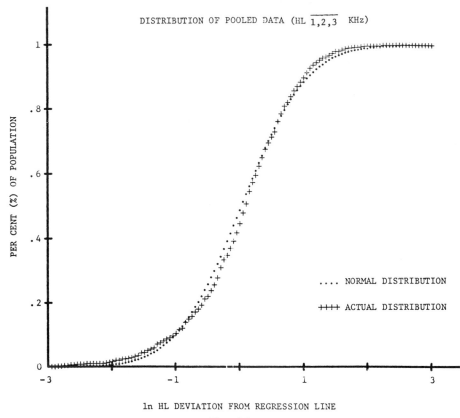

DISTRIBUTION OF POOLED DATA (HL $\overline{1,2,3}$ KHz)

.... NORMAL DISTRIBUTION

++++ ACTUAL DISTRIBUTION

ln HL DEVIATION FROM REGRESSION LINE

Fig. 14.

parallel within the limits of error in the slope estimates. Pooled slopes were calculated, and the intercepts were revised to reflect the small differences between the separate and pooled slopes. Families of parallel lines were plotted. Tests for coincidence of sets of parallel lines were then made by the method of covariance analysis. This revealed significant difference at the 0.01 probability level in all cases.

(5) Regression lines for different

experience levels within an age group were not found to be parallel, and, for each age group, the intercepts were compared by means of Student's t-tests. The "intercepts" were defined as ordinates of the regression lines at a dBA of 79, which represented the control group exposed to less than 80 dBA. These regression lines were found to be significantly different families of nonparallel lines from common intercepts.

(6) For each age and experience

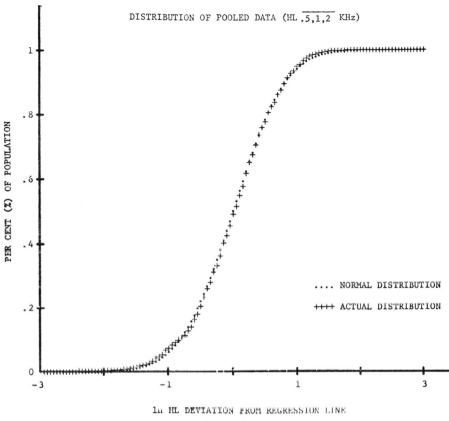

DISTRIBUTION OF POOLED DATA (HL .5,1,2 KHz)

.... NORMAL DISTRIBUTION

++++ ACTUAL DISTRIBUTION

In HL DEVIATION FROM REGRESSION LINE

FIG. 15.

combination, the normal distribution of pooled variation in replicate subjects was distributed about the regression line with its zero mean centered at the ordinate of the line. This model was then used to calculate a predicted percentage of subjects whose hearing levels exceeded a "fence." Thus, such percentages could be tabulated as a function of dBA for each age and experience category. Furthermore, risk values were then derived as the percentage

difference between employees exposed to noise levels 80 dBA or greater and those exposed to less than 80 dBA (Tables XII and XIII).

This analysis indicates that the 85 dBA-Slow noise limit for an 8 hour day, in conjunction with the medical program prescribed in the standard, will improve the protection of the working population from hearing loss that could impair their abilities to understand everyday speech. The reliability of the analysis is evidenced

CRITERIA DOCUMENT TABLE 12

DEPENDENCE OF HEARING IMPAIRMENT ON AGE, EXPERIENCE, AND NOISE EXPOSURE—HLI (0.5, 1, 2)

| | Noise Exposure in dBA-Slow | | | | | |
	80*	80	85	90	95	100
Experience: 2–4 years						
Age (in years)						
17–27	1.3	1.5	2.4	3.9	6.0	9.0
28–35	3.2	3.5	5.5	8.2	11.9	16.6
36–45	4.9	5.3	8.0	11.6	16.2	—
46–54	9.1	9.8	14.0	19.2	25.4	32.6
Experience: 5–10 years						
Age (in years)						
17–27	1.3	1.5	2.8	4.9	—	—
28–35	3.3	3.7	6.2	10.0	15.2	22.0
36–45	5.0	5.5	9.0	13.8	20.2	28.2
46–54	9.3	10.2	15.4	22.3	30.6	—
Experience: 11–20 years						
Age (in years)						
28–35	3.3	3.8	6.8	11.5	—	—
36–45	5.0	5.7	9.7	15.7	23.6	33.3
46–54	9.4	10.4	16.6	24.7	34.6	45.7
55–70	20.0	21.7	31.0	41.8	—	—
Experience: 21–41 years						
Age (in years)						
36–45	5.2	6.0	11.7	20.4	32.2	—
46–54	9.6	10.9	19.3	30.8	44.6	59.0
55–70	20.4	22.6	34.9	49.0	63.3	75.9

* Non-noise exposed.

by homogeneity of the variance and normality of the population distributions. In other words, the evaluation is repeatable and is representative of a random sample.

Comparison of NIOSH data with other published data

Three analyses comparable to the NIOSH analysis use a definition of hearing impairment different from that used in the NIOSH standard. In order to compare NIOSH data with these analyses, the NIOSH data was analyzed using the following definition: hearing impairment is considered to occur when the average of the hearing threshold levels at the audiometric frequencies 500, 1000, and 2000 Hz for both ears, HLI (0.5, 1, 2), exceeds 25 dB (thresholds re ANSI S3.6–1969). Again, risk is defined as the additional incidence of hearing impairment of noise exposed worker

918

CRITERIA DOCUMENT TABLE 13
DEPENDENCE OF HEARING IMPAIRMENT ON AGE, EXPERIENCE, AND NOISE EXPOSURE—HLI (1, 2, 3)

	Noise Exposure in dBA-Slow					
	80*	80	85	90	95	100
Experience: 2–4 years						
Age (in years)						
17–27	1.4	1.6	2.7	4.4	6.8	10.2
28–35	7.4	8.0	11.8	16.7	22.8	29.9
36–45	8.3	9.0	13.1	18.3	24.7	—
46–54	16.9	18.0	24.4	31.7	39.9	48.5
Experience: 5–10 years						
Age (in years)						
17–27	1.5	1.8	4.0	8.0	—	—
28–35	7.7	8.8	15.7	25.5	37.7	51.3
36–45	8.7	9.8	17.2	27.5	40.1	53.0
46–54	17.5	19.4	30.3	43.3	57.0	—
Experience: 11–20 years						
Age (in years)						
28–35	7.9	9.1	17.6	29.7	—	—
36–45	8.8	10.2	19.2	31.9	47.2	62.9
46–54	17.8	20.0	32.9	48.3	64.0	77.6
55–70	27.6	30.4	45.7	61.6	—	—
Experience: 21–41 years						
Age (in years)						
36–45	8.7	9.8	17.2	40.0	—	—
46 54	17.5	19.4	30.2	43.2	56.9	69.9
55–70	27.3	29.6	42.7	56.5	69.7	80.6

* Non-noise exposed.

groups when compared with that of equivalent nonnoise exposed groups, or the difference between the two incidences.

NIOSH risk data for retirement age groups are compared in Tables XIV, XV, and XVI with the following sets of risk data: (1) that used by the American Conference of Governmental Industrial Hygienists,[85] the OSHA Federal Standard,[86] as well as the Intersociety Committee;[27] (2) that used by the International Organization for Standardization (ISO); and (3) that developed by the National Physical Laboratory (U.K.).[89] In all cases, the age grouping and sound levels are similar to those of the NIOSH data.

The Intersociety Committee, composed of representatives from the American Academy of Occupational Medicine, American Academy of Ophthalmology and Otolaryngology,

CRITERIA DOCUMENT TABLE 14
COMPARISON OF RISK* FOR RETIREMENT AGE POPULATIONS
AS DETERMINED BY INTERSOCIETY COMMITTEE AND NIOSH

	dBA	80	85	90	95	100
	Total Percent Impaired	23	26	33	43	56
Intersociety**	Normal Percent Impaired	22	22	22	22	22
	Risk	1	4	11	21	34
	Total Percent Impaired	11	19	31	45	59
NIOSH***	Normal Percent Impaired	10	10	10	10	10
(Age 46–54)	Risk	1	9	21	35	49
	Total Percent Impaired	23	35	49	63	76
NIOSH***	Normal Percent Impaired	20	20	20	20	20
(Age 55–70)	Risk	3	15	29	43	56

* Where impairment is defined as average threshold level in excess of 15 dB re ASA 1951 (25 dB re ANSI (1969)) at 500, 1000, 2000 Hz.
** Age group 50–59, assumes monotonic growth of exposure with age.
*** Age groups 46–54 and 55–70, respectively, experience 21–41 years. (See Table XI)

American Conference of Governmental Industrial Hygienists, Industrial Hygiene Association, and Industrial Medical Association, in 1970, published an analysis similar to the NIOSH analysis. It studied a combination of several noise and hearing studies[120-124] in order to determine risk from noise exposure. There are several features of this analysis, however, which differ from that by NIOSH.

First, most of the Intersociety data consisted of hearing levels for only the right ear. Although the right ear may statistically be better than the left, both ears were used in the NIOSH analysis in order to obtain a more realistic incidence of hearing impairment since a person hears with both ears, not one. This same feature

of the Intersociety analysis is discussed by Botsford[125] who determined that the use of the average of the two ears produces a higher risk factor.

Also, the Intersociety data were not separated into experience groups within each age group. The NIOSH analysis found that work experience ranged from 0 to 40 years in the older age groups, and thus, it was necessary to classify employees by experience as well as by age.

Moreover, some of the studies used in the Intersociety analysis used Speech Interference Level (SIL—the average of octave band levels with center frequencies 500, 1000, and 2000 Hz) as the measure of exposure in analyzing the noise levels encountered by the employees. NIOSH considers this unsatisfactory since the

CRITERIA DOCUMENT TABLE 15

COMPARISON OF RISK* FOR RETIREMENT AGE POPULATIONS AS DETERMINED BY INTERNATIONAL ORGANIZATION FOR STANDARDIZATION AND NIOSH

		Age 50 years				
	dBA	80	85	90	95	100
ISO**	Total Percent Impaired	14	22	32	45	50
	Normal Percent Impaired	14	14	14	14	14
	Risk	0	8	18	31	44
NIOSH***	Total Percent Impaired	11	19	31	45	59
(Age 46–54)	Normal Percent Impaired	10	10	10	10	10
	Risk	1	9	21	35	49
		Age 60 Years				
	dDA	80	85	90	95	100
ISO**	Total Percent Impaired	33	43	54	62	74
	Normal Percent Impaired	33	33	33	33	33
	Risk	0	10	21	29	41
NIOSH***	Total Percent Impaired	23	35	49	63	76
(Age 55–70)	Normal Percent Impaired	20	20	20	20	20
	Risk	3	15	29	43	56

* Where impairment is defined as average threshold level in excess of 15 dB re ASA 1951 (25 dB re ANSI(1969)) at 500, 1000, 2000 Hz.
** Ages 48 and 58 years, respectively, experience is equal to 18 years.
*** Age groups 46–54 and 55–70, respectively, experience is 21 41 years.

conversion of SIL to dBA is generally inaccurate and is based on tenuous assumptions.

Finally, the Intersociety analysis used the noise exposed populations from a variety of different studies with one non-noise exposed population and one "general" population (including both noise exposed and non-noise exposed individuals) for their composite determination of risk. Furthermore, the different investiga-

tions used in this analysis were each unique with respect to screening (or excluding) criteria, audiometric equipment, and data analysis. The NIOSH study used a nonnoise exposed population which consisted of a pool of employees similar in these respects to each other and to the noise exposed population under study.

Thus, the Intersociety analysis differs from that of NIOSH in several characteristics—use of one ear only,

921

CRITERIA DOCUMENT TABLE 16
COMPARISON OF RISK* FOR RETIREMENT AGE POPULATION
AS DETERMINED BY ROBINSON AND NIOSH

		Age 50 Years			
	dBA	87	92	97	102
Robinson**	Total Percent Impaired				
	a) thresholds re: 97 British controls	3	8	17	33
	b) thresholds re: +10 dB correction	16	26	40	59
	Normal Percent Impaired				
	a) thresholds re: 97 British controls	1	1	1	1
	b) thresholds re: +10 dB correction	3	3	3	3
	Risk				
	a) thresholds re: 97 British controls	3	8	17	33
	b) thresholds re: +10 dB correction	13	23	37	56
NIOSH***	Total Percent Impaired	24	36	50	65
	Normal Percent Impaired	10	10	10	10
	Risk	14	26	40	55

* Where impairment is defined as average threshold level in excess of 15 dB re ASA 1951 (25 dB re ANSI(1969)) at 500, 1000, 2000 Hz.
** Based on 30 years exposure. Risk computed by Robinson[87] using a fence of 25 dB re ANSI (1969).
*** Age group 46–54, experience is 21–41 years.

nonseparation of experience groups, use of SIL in noise levels, and use of a dissimilar composite population. Some of these characteristics tend to produce lower risk values and considerably more uncertainty than the NIOSH analysis, as evidenced in Table XIV.

Another study whose analysis determined risk is published in ISO Recommendation R1999 (1971).[90] This analysis differs from the NIOSH analysis in three ways. The first is that only the right ear was used. The second is that no separation of age groups into work experience groups was done. The third is that no screening for otological abnormalities was done in the ISO study. On the other hand, the entire sample of data used in this analysis is homogeneous in that all members of the sample were taken from one comprehensive examination.[126] The lack of otological screening has some effect on incidence of hearing impairment for both the noise exposed and the nonnoise exposed groups, but, when risk is calculated by subtracting the two incidences, the effect is essentially cancelled. Thus the NIOSH risk values are very similar to the ISO values, as evidenced in Table XV.

Another study, by the British National Physical Laboratory,[127] developed an equation for calculating hearing levels of the populations exposed to noise. This equation was

CRITERIA DOCUMENT TABLE 17
COMPARISON OF NIOSH RISK VALUES FOR TWO DEFINITIONS OF HEARING IMPAIRMENT

	Age 46–54 Experience 21–41					
	dBA	80	85	90	95	100
HLI (0.5,1,2)	Total Percent Impaired	11	19	31	45	59
	Normal Percent Impaired	10	10	10	10	10
	Risk	1	9	21	35	49
HLI (1,2,3)	Total Percent Impaired	19	30	43	57	70
	Normal Percent Impaired	18	18	18	18	18
	Risk	1	12	25	39	52

	Age 55–70 Experience 21–41					
	dBA	80	85	90	95	100
HLI (0.5,1,2)	Total Percent Impaired	23	35	49	63	76
	Normal Percent Impaired	20	20	20	20	20
	Risk	3	15	29	43	56
HLI (1,2,3)	Total Percent Impaired	30	43	56	70	81
	Normal Percent Impaired	27	27	27	27	27
	Risk	3	16	29	43	54

used by Robinson[89] to develop risk tables for various groups and noise levels.

In comparing the British risk values with those of the NIOSH, shown in Table XVI, it can be seen that the British risk values are much lower. The nature of this discrepancy is difficult to determine; however, it may result from the severity of the British screening for otological abnormalities and previous noise exposure. It is also possible that the reason for the discrepancy is the baseline, or reference level, used in this analysis. The British used a baseline (which they considered to be audiometric zero), determined by a non-noise exposed industrial group of people 18–25 years of age, which was actually lower than audiometric zero (thresholds re ANSI 1969 (or ISO R389)). It has been found, however, in many United States studies[5,21,22,126,128,129] including the NIOSH analysis, that the average hearing threshold level over the audiometric frequencies 500, 1000, and 2000 Hz (HLI(0.5, 1, 2)) is 5–10 dB (thresholds re ANSI S3.6–1969) for nonnoise exposed employees 18–25 years of age, which is approximately 10 dB higher than that of the 97 nonnoise controls used by the British. Thus, if the British data are used to calculate risk with a 10 dB correction, which brings the baseline of their data into coincidence with the baseline appropriate to the protection goal of this standard and which is representative of the baseline found in occupational environments in many U.S. studies, then the risk values using the British data are, in fact, very similar to those found in both the NIOSH and ISO risk tables, as shown in Table XVI.

The *Hygiene Standard for Wide-Band Noise* of the British Occupational Hygiene Society[87] is based on assumptions radically different from those of the NIOSH standard. As mentioned previously, the British consider hearing impairment to occur when the average hearing loss at the audiometric frequencies 500, 1000, 2000, 3000, 4000, and 6000 Hz for both ears exceeds 40 dB [(threshold re ISI R389–1964)] (48 dB minus 8 dB for presbycusis or aging effects). This 48 dB "fence" is comparable to an HLI ($\overline{0.5, 1, 2}$) of approximately 39 dB for thresholds re ANSI S3.6–1969. Such a high fence is not in line with the protection goal of the NIOSH standard.

Effect of hearing impairment definition on risk

The NIOSH standard was based on risk calculated using the definition of hearing impairment as the condition when the average of the hearing threshold levels at the three audiometric frequencies 1000, 2000, and 3000 Hz, HLI ($\overline{1, 2, 3}$) for both ears exceeds 25 dB (thresholds re ANSI S3.6-1969). Another definition was used to compare the NIOSH risk data with other data. This definition

was that hearing impairment for an individual is considered to occur when the average of the hearing threshold levels for the audiometric frequencies 500, 1000, and 2000 Hz, HLI ($\overline{0.5, 1, 2}$) for both ears exceeds 25 dB re ANSI S3.6-1969. Some of the NIOSH risk values calculated using both definitions are shown in Table XVII. Although the incidences of hearing impairment are higher for the definition using HLI ($\overline{1, 2, 3}$), the risks due to noise are, in fact, quite similar. Thus, even though the two definitions reflect the incidence of hearing impairment in the population differently, the different definitions have little effect when risk is calculated.

Comparison of the NIOSH standard with other standards

The present federal standard for occupational noise exposure,[86] which is based on the same data as that of the Intersociety Committee, ACGIH, and Walsh-Healey Public Contract Act mentioned above, differs in several respects from that of the NIOSH standard, and the analysis shows lower risk than does NIOSH for the same noise levels. Indeed, industrial employee data more recent than the Intersociety data, published as ISO R1999,[90] has shown trends comparable to those of the NIOSH analysis. Thus, the 85 dBA-Slow noise exposure level for a nominal 8 hour day should allow no more than an increase of 10 to 15 percentage points in the incidence of hearing impairment, as compared to the non-noise exposed population. (This statistic is for employees aged 50 to 65 years, having a minimum of 20 years' noise exposure.)

The recommended occupational exposure level of 85 dBA for an 8 hour day will be applicable to all newly designed installations six months after the effective date of the standard. However, the level of 85 dBA is not applicable to established installations until such time as determined by the Secretary of Labor in consultation with the Secretary of Health, Education, and Welfare. Such a provision was necessary because of the lack of sufficient available evidence upon which to determine a reasonable time period for the development of technologically feasible methods to meet the 85 dBA level.

VII. List of references

4. *Acoustical Terminology.* American National Standards Institute, N.Y. S1.1–1960 (R1971).

15. "Guides to the Evaluation of Permanent Impairment—Ear, Nose, Throat and Related Structures." *J. Am. Med. Assoc.*, Vol 177, pp 489–501, 1961. [Published in 1971 in *Guides to the Evaluation of Permanent Impairment.*—Ed.]

16. Harris, J. D. "Pure Tone Hearing Acuity and the Intelligibility of Everyday Speech." *J. Acoust. Soc. Am.*, Vol. 37, pp. 824–830, 1965.

17. Kryter, K. D., Williams, C.; and Green, D. M. "Auditory Acuity and Perception of Speech." *J. Acoust. Soc. Am.*, Vol. 34, pp. 1217–1223.

26. Passchier-Vermeer, W. "Hearing Loss Due to Exposure to Steady-State Broadband Noise." Rep. No. 35, Institute for Pub. Health Eng., The Netherlands, 1968.

27. "Guidelines for Noise Exposure Control." *Sound and Vibration*, Vol. 4, pp. 21 24, 1070.

28. Ward, W. D., Glorig, A.; and Sklar, D. L. "Temporary Threshold Shift from Octave-Band Noise: Applications to Damage Risk Criteria." *J. Acoust. Soc. Am.*, Vol. 31, pp. 522–528, 1959.

29. Kylin, B. "Temporary Threshold Shift and Auditory Trauma Following Exposures to Steady-State Noise." *Acta Otolaryngol.*, Vol. 51, Suppl. No. 152, 1960.

30. Ward, W. D. "Damage-Risk Criteria for Line Spectra." *J. Acoust. Soc. Am.*, Vol. 34, pp. 1610–1619, 1962.

32. Nixon, J. C. and Glorig, A. "Noise-Induced Permanent Threshold Shift at 2000 and 4000 cps." *J. Acoust. Soc. Am.*, Vol. 33, pp. 904–908, 1961.

33. Ward, W. D. "Temporary Threshold Shift and Damage Risk Criteria for Intermittent Noise Exposures." *J. Acoust. Soc. Am.*, Vol. 48, 1961.

67. Rosenblith, W. A. "Industrial Noise and Industrial Deafness." *J. Acoust. Soc. Am.*, Vol. 13, p. 220, 1942.

68. McCoy, D. W. "Industrial Noise—Its Analysis and Interpretation for Preventative Treatment." *J. Ind. Hyg. and Toxicol.*, Vol. 26, p. 120, 1944.

69. Davis, H. A. "Protection of Workers Against Noise." *J. Ind. Hyg. and Toxicol.*, Vol. 27, p. 56, 1945.

70. MacLaren, W. R., and Chaney, A. L. "An Evaluation of Some Factors in the Development of Occupational Deafness." *Ind. Med.*, Vol. 16, p. 109. 1947.
71. Fowler, E. P., Jr. "Medical Aspects of Hearing Loss," *Hearing and Deafness,* H. Davis, ed., 1947.
72. Goldner, A. "Occupational Deafness." *Arch. Otolaryng.*, Vol. 42, p. 407, 1945.
73. Kryter, K. D. "The Effects of Noise on Man." *J. Speech and Hear. Dis.,* Monograph Suppl. 1, 1950.
74. Rosenblith, W. A., and Stevens, K. N. *Handbook of Acoustic Noise Control— Noise and Man.* USAF, WADC Tech. Rept. No. 52.204, Vol. II, 1953.
75. Sivian, L. J., and White, S. D. "On Minimum Audible Fields." *J. Acous. Soc. Amer.*, Vol. 4, p. 288, 1933.
76. Fletcher, H. and Munson, W. A. "Loudness, its Definition, Measurement, and Calculation." *J. Acous. Soc. Am.*, Vol. 5, p. 82, 1933.
77. Davis, H.; Morgan, C. I.; Hawkins, J. E., Jr.; Galambos, R.; and Smith, F. W. "Temporary Deafness Following Exposure to Loud Tones and Noise." *Acta Otolaryng.*, Suppl. 88, 1950.
78. Perlman, H. B. "Acoustic Trauma in Man; Clinical and Experimental Studies." *Arch. Otolaryng.*, Vol. 34, p. 409, 1946.
79. Ruedi, L., and Furrer, W. "Physics and Physiology of Acoustic Trauma." *J. Acous. Soc. Am.*, Vol. 18, p. 409, 1946.
80. "Rating Noise with Respect to Hearing Conversation, Speech Communication, and Annoyance." International Organization for Standardization, ISO Tech. Comm. 43, 1961.
81. "Guidelines for Noise Exposure Control." *Am. Ind. Hyg. Assoc. J.*, Vol. 28, p. 418, 1967.
82. Botsford, J. H. "Simple Method for Identifying Acceptable Noise Exposures." *J. Acous. Soc. Am.*, Vol. 42, p. 810, 1967.
83. Robinson, D. W. "Relations Between Hearing Loss and Noise Exposure" in *Hearing and Noise in Industry,* W. Burns and D. W. Robinson, eds., Appen. 10, Her Majesty's Stationery Office, London, 1970.
84. Cohen, A.; Anticaglia, J. R.; and Carpenter, P. "Temporary Threshold Shift in Hearing from Exposure to Different Noise Spectra at Equal dBA Level." *J. Acous. Soc. Am.*, Vol. 51, p. 503, 1972.
85. "Threshold Limit Values of Physical Agents." Am. Conf. of Gov't. Ind. Hyg., Cincinnati, 1970.
86. "Occupational Safety and Health Standards." Occup. Safety and Health Admin., DOL, Vol. 36, p. 105, 1971.
87. "Hygiene Standard for Wide-Band Noise." Brit. Occup. Hyg. Soc., Pergamon Press, Elmsford, N.Y., 1971.
88. Kryter, K. D. *The Effects of Noise on Man.* Academic Press, N.Y., 1970.
89. Robinson, D. W. "Estimating the Risk of Hearing Loss due to Continuous Noise." In *Occupational Hearing Loss,* R. W. Robinson, ed. Academic Press, N.Y.,, 1970.
90. *Assessment of Occupational Noise Exposure for Hearing Conservation Purposes.* International Organization for Standardization (ISO R1999), 1971.
91. Eldredge, D. H. *The Problems of Criteria for Noise Exposure.* Armed Forces—NRC Committee on Hearing and Bio-Acoustics, 1960.

92. Fletcher, H. *Speech and Hearing.* D. Van Nostrand Co., Inc., N.Y., 1929.

93. "Tentative Standard Procedure for Evaluating the Percentage Loss of Hearing Medicolegal Cases." Council on Physical Med., *J. Am. Med. Assoc.,* Vol. 133, p. 396, 1947.

94. "Principles for Evaluating Hearing Loss." *J. Am. Med. Assoc.,* Vol. 157, p. 1408, 1955.

95. "Guides for the Evaluation of Hearing Impairment." *Trans. Am. Acad. Ophthal. Otolaryngol.,* 1961.

96. *Audiometers for General Diagnostic Purposes.* Am. Standards Institute, N.Y., S3.6–1969.

98. Sataloff, J. *Hearing Loss.* J. B. Lippincott Co., Philadelphia and Toronto, 1966.

99. Davis, H., and Silverman, S. R. *Hearing and Deafness.* Holt, Rinehart and Winston, Inc., N.Y., 1963.

100. Hood, J. D. "Discussion of Papers, Section III." In *Occupational Hearing Loss,* D. W. Robinson, ed. Academic Press, N.Y., 1971.

101. Niemeyer, W. "Speech Discrimination in Noise-Induced Deafness." *Intern. Audiol.,* Vol. 6, p. 42, 1967.

102. Mullins, C. J. and Bangs, J. L. "Relationships between Speech Discrimination and Other Audiometric Data." *Acta Otolaryng.,* Vol. 47, p. 149, 1957.

103. Harris, J. D.; Haines, H. L.; and Myers, C. K. "The Importance of Hearing at 3 Kc for Understanding Speeded Speech." *Laryng.,* Vol. 70, No. 2, p. 131, 1960.

104. Ross, M.; Huntington, D.; Newby, H. A.; and Dixton, R. F. *Speech Discrimination of Hearing-Impaired Individuals in Noise—Its Relationship to Other Audiometric Parameters.* USAF Tech. Doc. Rept. No. 62–130, 1963.

105. Acton, W. I. "Speech Intelligibility in a Background Noise and Noise Induced Hearing Loss." *Ergonomics,* Vol. 13, No. 5, p. 546, 1970.

106. Lindeman, H, E, "Relation Between Audiological Findings and Complaints by Persons Suffering from Noise-Induced Hearing Loss." *J. Am. Indus. Hyg. Assoc.,* Vol. 32, p. 447, 1971.

107. Kryter, K. D. W.; Miller, J. D.; and Eldredge, D. H. "Hazardous Exposure to Intermittent and Steady-State Noise." *J. Acoust. Soc. Am.,* Vol. 39, No. 3, p. 451, 1966.

108. "Hazardous Noise Exposure." Air Force Regulation 160–3, Dept. of the Air Force, 1956.

109. Ward, W. D.; Glorig, A.; and Sklar, D. L. "Temporary Threshold Shift Produced by Intermittent Exposure to Noise." *J. Acous. Soc. Am.,* Vol. 31, p. 791, 1959.

110. Ward, W. D. "Studies on the Aural Reflex II: Reduction of Temporary Threshold Shift from Intermittent Noise by Reflex Activity; Implications for Damage-Risk Criteria." *J. Acous. Soc. Am.,* Vol. 34, p. 234, 1962.

111. Selters, W. and Ward, W. D. "Temporary Threshold Shift with Changing Duty Cycle." *J. Acous. Soc. Am.,* Vol. 34, p. 122, 1962.

112. Spieth, W. and Trittipoe, W. J. "Intensity and Deviation of Noise Exposure and Temporary Threshold Shifts." *J. Acous. Soc. Amer.,* 30, 710, 1958.

113. Ward, W. D. and Nelson, D. A. "On the Equal-Energy Hypothesis Relative to

Damage-Risk Criteria in the Chinchilla." In *Occupational Hearing Loss,* D. W. Robinson, ed. Acad. Press, N.Y., 1971.

114. Sataloff, J.; Vassello, L.; and Menduke, H. "Hearing Loss from Exposure to Interrupted Noise." *Arch. of Env. Health,* Vol. 18, p. 972, 1969.

115. Glorig, A.; Ward, W. D.; and Nixon, J. "Damage-Risk Criteria and Noise-Induced Hearing Loss." *Arch. Otolaryng.,* Vol. 74, p. 413, 1961.

116. "The Assessment of Industrial Noise in Working Areas in Reference to Hearing Damage," VD12058 Part 2, Germany, 1970.

117. Schmidek, M.; Margolis, M.; and Henderson, T. L. "Evaluation of Proposed Limits for Intermittent Noise Exposures with Temporary Threshold Shift as a Criterion." (To be published)

118. Schmidek, M. and Carpenter, P. "Study of Chain Saw Operators: Nature of Intermittent Noise Exposure and Associated Damage Risk to Hearing." (To be published)

119. Schmidek, M.; Margolis, B.; and Henderson, T. L. "Effects of the Level of Noise During Interruptions on Temporary Threshold Shift." (To be published)

120. Gallo, R., and Glorig, A. "Permanent Threshold Shift Changes Produced by Noise Exposure and Aging," *Am. Ind. Hyg. Assoc. J.,* Vol. 25, pp. 237–245, 1964.

121. AIHA Noise Committee, *Industrial Noise Manual,* Sec. Ed., Chapter 7, Am. Ind. Hyg. Assoc., Detroit, 1966.

122. Glorig, A., and Nixon, J. "Distribution of Hearing Loss in Various Populations," *Annals of Otology, Rhinology, and Laryngology,* Vol. 69, p. 497, 1960.

123. Glorig, A.; Ward, D.; and Nixon, J. "Damage-Risk Criteria and Noise Induced Hearing Loss." *Arch. Otolaryng.,* Vol. 74, p. 413, 1961.

124. Unpublished data available to the Intersociety Committee, 1968.

125. Botsford, J. H. "Prevalence of Impaired Hearing and Sound Levels at Work." *J. Acous. Soc. Am.,* Vol. 45, pp. 79–82, 1969.

126. Unpublished data available to NIOSH, 1968.

127. Burns, W. and Robinson, D. W. *Hearing and Noise in Industry,* Her Majesty's Stationery Office, London, 1970.

128. Glorig, A. and Nixon, J. "Hearing Loss as a Function of Age." *Laryngoscope,* Vol. LXXII, No. 11, 1962.

129. Unpublished data available to NIOSH, 1971.

130. Jones, H. "Noise Standards." In *Industrial Noise: A Guide to Its Evaluation and Control.* A. D. Hosey and C. H. Powell, eds., U.S.P.H.S. Pub. No. 1572, 1967.

131. Hardy, H. C. "Tentative Estimate of a Hearing Damage Risk Criterion for Steady-State Noise." *J. Acous. Soc. Am.,* Vol. 24, p. 756, 1952.

132. Lindman, M. W. "Noise Level Limits for the Avoidance of Deafness in Shipboard Machinery Spaces." Washington: USN Bu Ships Rep. No. 271-N-72, 1955.

133. "Guide for Conservation of Hearing in Noise." *Am. Acad. of Ophthal. and Otolaryng.,* 1957.

134. Jones, A. R. and Church, F. W. "A Criterion for Evaluation of Noise Exposures." *J. Am. Indus. Hyg. Assoc.,* Vol. 21, No. 6, p. 481, 1960.

135. Kryter, K. D. "Exposure to Steady-State Noise and Impairment of Hearing." *J. Acous. Soc. Am.,* Vol. 35, No. 10, p. 1515, 1963.
136. Kryter, K. D. "Damage-Risk Criterion and Contours Based on Permanent and Temporary Hearing Loss Data." *J. Am. Ind. Hyg. Assoc.,* Vol. 26, No. 1, p. 34, 1965.
137. "Guide for Conservation of Hearing in Noise." *Am. Acad. of Ophthal. and Otolaryngol.,* Suppl., 1964.
138. *The Relations of Hearing Loss to Noise Exposure.* Am. Standards Assoc., N.Y., Z24-X-2 Committee Report, 1954.
139. Draft Proposal for "Hearing Levels of Non-Noise Exposed People at Various Ages." ISO/TC, 43/SC, 1 (Netherlands, 1) 57E.

4. Coal Mining Act

This regulation will be found in Title 30—Mineral Resources, *Code of Federal Regulations,* Chapter 1, Subchapter O.

§ 70.500 Definitions.

As used in this Subpart F, the term:

(a) "dBA" means noise level in decibels, as measured with the A-weighted network of a standard sound level meter using slow response;

(b) "Noise exposure" means a period of time during which the noise level is 90 or more dBA;

(c) "Multiple noise exposure" means the daily noise exposure is composed of two or more different noise levels;

(d) "Noise level" is the average dBA during a noise exposure; and,

(e) "Qualified persons" means, as the context requires, an individual deemed qualified by the Secretary and designated by the operator to make tests and examinations required by this Act.

§ 70.501 Requirements.

Every operator of an underground coal mine shall maintain the noise levels during each shift to which each miner in the active workings of the mine is exposed at or below the permissible noise levels set forth in Table I of this subpart.

EXAMPLE: If a noise is recorded to be 110 dBA then exposure shall not exceed 30 minutes during an 8-hour shift.

§ 70.502 Computation of multiple noise exposure.

The standard will be considered to have been violated in the case of multiple noise exposure where such exposure totals exceed one as computed by adding the total time of exposure at each specified level (C_1, C_2,

C_3 etc.) divided by the total time of exposure permitted at that level (T_1, T_2, T_3). Thus,

$$\frac{C_1}{T_1} + \frac{C_2}{T_2} + \frac{C_3}{T_3} \text{ must not exceed 1.}$$

EXAMPLE I: Exposure of 2 hours at 92 dBA and 1 hour at 100 dBA during an 8-hour shift.

$$\frac{\text{Total minutes of noise exposure at dBA level}}{\text{Total minutes of permissible noise exposure at dBA level}}$$

$$\frac{120 \text{ min.}}{360 \text{ min.}} + \frac{60 \text{ min.}}{120 \text{ min.}}$$

$$= \tfrac{2}{6} + \tfrac{1}{2} = \tfrac{2}{6} + \tfrac{3}{6} = \tfrac{5}{6}$$

The sum of the fractions does not exceed one; hence the exposure for the shift would not violate the standard.

EXAMPLE II: Exposure of 3 hours at 95 dBA and 1 hour at 100 dBA during an 8 hour shift.

$$\tfrac{3}{4} + \tfrac{1}{2} = \tfrac{3}{4} + \tfrac{2}{4} = \tfrac{5}{4}$$

The sum of the fractions exceeds one; hence the exposure for the shift would violate the standard.

§ 70.503 Noise level measurements; general.

Every coal miner operator shall take accurate readings of the noise levels to which each miner in the active workings of the mine is exposed during the performance of the duties to which he is normally assigned.

§ 70.504 Noise level measurements; by whom done.

The noise level measurements required by this Subpart F shall be taken by, or as directed by, a person who has met the minimum requirements set forth in § 70.504–1, and has been certified by the Administrator, Mining Enforcement and Safety Administration as qualified to take noise level measurements as prescribed in this Subpart F;

§ 70.504–1 Persons qualified to measure noise levels; minimum requirements.

The following persons shall be considered qualified to take noise level measurements as prescribed in this Subpart F:

(a) Any person who has been certified by the Mining Enforcement and Safety Administration as an instructor in noise measurement training programs;

(b) Any person who has satisfactorily completed a noise training course conducted by the Administration and has been certified by the Bureau as a qualified person; and,

(c) Any person who has satisfactorily completed a noise training course approved by the Administration and has been certified by the Administration as a qualified person.

§ 70.504–2 Certification of qualified persons by the Bureau of Mines.

Upon a satisfactory showing that a person has met the minimum requirements set forth in § 70.504–1, the Mining Enforcement and Safety Administration shall certify that such person has the ability and capacity to conduct tests of the noise levels in a coal mine and to report and certify the results of such tests to the Secretary and the Secretary of Health, Education, and Welfare.

§ 70.505 Noise level measurement equipment.

(a) Noise level measurements shall be taken only with instruments which are approved by the Bureau of Mines as permissible electric face equipment under the provisions of Part 18 of this chapter (Bureau of Mines, Schedule 2G), and which meet the operational specifications of the American National Standards Institute for Sound Level Meters S1.4–1971 (Type S2A).

(b) Noise level measurement equipment shall be set to operate with the A-weighted network and slow response and shall be acoustically calibrated in accordance with the manufacturer's instructions before, during and after each shift on which such equipment is used.

§ 70.506 Noise level measurement procedures.

(a) Noise level measurements shall be made at locations where the noise is typical of that entering the ears of the miner whose exposure is under consideration.

(b) Five measurements shall be made for each type of noise exposure producing operation to which the miner under consideration is exposed.

(c) Each measurement shall be made by observing the A-scale readings for 30 seconds and recording the noise level.

(d) The average of the five noise level measurements shall be considered as the noise level measurement which is representative of the operation.

(e) Where different and distinct noise levels occur at various phases of an operation, noise level measurements shall be made in accordance with this section for each distinct phase.

(f) The noise levels and the estimated length of time the miner is exposed to each level during a normal work shift shall be reported for the operation. The range of the five noise level measurements used in paragraph (d) of this section shall also be reported.

§ 70.507 Initial noise level survey.

On or before June 30, 1971, each operator shall:

(a) Conduct, in accordance with this subpart, a survey of the noise levels to which each miner in the active workings of the mine is exposed during his normal work shift; and,

(b) Report and certify to the Mining Enforcement and Safety Administration, and the Department of Health, Education, and Welfare, the results of such survey using the Coal Mine Noise Data Report, Figure 1. Reports shall be sent to:

Division of Automatic Data Processing, Mining Enforcement and Safety Administration, Building 53, Denver Federal Center, Denver, Colo. 80225.

(a) At intervals of at least every 6 months after June 30, 1971, but in no case shall the interval be less than 3 months, each operator shall conduct, in accordance with this subpart, periodic surveys of the noise levels to which each miner in the active workings of the mine is exposed and shall report and certify the results of such surveys to the Mining Enforcement and Safety Administration, and the Department of Health, Education, and Welfare, using the Coal Mine Noise Data Report Form. Reports shall be sent to:

Division of Automatic Data Processing, Mining Enforcement and Safety Administration, Building 53, Denver Federal Center, Denver, Colo. 80225.

(b) Where no A-scale reading recorded for any miner during an initial

FIGURE 1
(Submit one form for each miner)
COAL MINE NOISE DATA REPORT

Date:...............Mine ID No.:.......

Section ID No.:.....Miner's SSA No.:.....

Occupation:...........................

Actual production—tons this shift:.......

Type of mining:

    Development.........................

    Retreat.............................

Method of mining:

    Continuous..........................

    Conventional........................

    Longwall............................

    Other...............................

Equipment in operation:

    Electric.............................

    Pneumatic...........................

    Other...............................

    Voltage......Pressure-p.s.i.......

    a.c. or d.c............

Total horsepower......................

Description of equipment (make, model
No., order No., etc.):.................
....
....

Seam conditions: Name of seam:.......
....

Coal height—inches:......

Average width of place:...........

Type of roof (sandstone, slate, etc.):.....
....

Hearing protective device used?
Yes......No......

Type and model number of sound level
meter:...............................

Check if section will be closed before next
sampling: Yes...... No......

☐ Initial survey ☐ Periodic survey ☐ supplementary survey

Signature of qualified person:..........
....
Coal Mine Noise Data Report

Date:..........   Mine ID No.:.......
List noise level measurements for each
level of exposure.

Operation (loading, tram- ming, etc.)	*Noise level dBA average*	*Range (high and low readings)*	*Cumulative exposure- minutes*

Mantrip...............................

Signature of qualified person:..........

933

or periodic noise level survey exceeds 90 dBA, the operator shall not be required to survey such miner during any subsequent periodic noise level survey required by this section: *Provided, however,* That the name and job position of each such miner shall be reported in every periodic survey and the operator shall certify that such miner's job duties and noise exposure levels have not changed substantially during the preceding 6-month period.

§ 70.509 Supplemental noise level survey; reports and certification.

(a) Where the certified results of an initial noise level survey conducted in accordance with § 70.507, or a periodic noise level survey conducted in accordance with § 70.508, show that any miner in the active workings of the mine is exposed to a noise level in excess of the permissible noise level prescribed in Table I, the operator shall conduct a supplemental noise level survey with respect to each miner whose noise exposure exceeds this standard. This survey shall be conducted within 15 days following notification to the operator by the Bureau of Mines to conduct such survey.

(b) Supplemental noise level surveys shall be conducted by taking noise level measurements in accordance with § 70.506, however, noise level measurements shall be taken during the entire period of each individual operation to which the miner under consideration is actually exposed during his normal work shift.

(c) Each operator shall report and certify the results of each supplemental noise level survey conducted in accordance with this section to the Bureau of Mines and the Department of Health, Education, and Welfare using the Coal Mine Noise Data Report Form to record noise level readings taken with respect to all operations during which such measurements were taken.

(d) Supplemental noise level surveys shall, upon completion, be mailed to:

Division of Automatic Data Processing, Mining Enforcement and Safety Administration, Building 53, Denver Federal Center, Denver, Colo. 80225.

§ 70.510 Violation of noise standard; notice of violation; action required by operator.

(a) Where the results of a supplemental noise level survey conducted in accordance with § 70.509 show that any miner in the active workings of the mine is exposed to noise levels which exceed the permissible noise levels prescribed in Table I, the Secretary shall issue a notice to the operator that he is in violation of this subpart.

(b) Upon receipt of a Notice of Violation issued pursuant to paragraph (a) of this section, the operator shall:

(1) Institute promptly administrative and/or engineering controls necessary to assure compliance with the standard. Such controls may include protective devices other than those devices or systems which the Secretary or his authorized representative finds to be hazardous in such mine.

(2) Within 60 days following the issuance of any Notice of Violation of this subpart, submit for approval to a joint Mining Enforcement and Safety Administration-Health, Education, and Welfare committee, a plan for the administration of a continuing, effec-

934

TABLE I
PERMISSIBLE NOISE EXPOSURES

Duration per day (hours)	Noise level (dBA)
8	90
6	92
4	95
3	97
2	100
1½	102
1	105
¾	107
½	110
¼ or less	115

tive hearing conservation program to assure compliance with this subpart, including provision for:

(i) Reducing environmental noise levels;

(ii) Personal ear protective devices to be made available to the miners;

(iii) Preemployment and periodic audiograms.

(3) Plans required under subparagraph (2) of this paragraph shall be submitted to:

Assistant Administrator, Coal Mine Health and Safety, Mining Enforcement and Safety Administration, Department of the Interior, Washington, D.C. 20240.

SUBPART D—NOISE STANDARD

§ 71.300 Noise standard; general requirements.

Each operator of an underground coal mine and each operator of a surface coal mine shall, during each shift, maintain the noise level to which each miner in each surface installation and at each surface worksite is exposed at or below the maximum noise exposure level prescribed in Subpart F, Part 70 of this Subchapter O.

§ 71.301 Measurement of noise levels.

Each operator shall measure the noise level to which each miner is exposed in each surface installation and at each surface worksite in the manner prescribed in Subpart F, Part 70, of this Subchapter O.

§ 71.302 Initial noise level survey.

On or before November 13, 1974, each operator shall:

(a) Conduct, in accordance with this subpart, a survey of the noise levels to which each miner in each surface installation and at each surface worksite is exposed during his normal work shift; and,

(b) Report and certify to the Mining Enforcement and Safety Administration and the Department of Health, Education, and Welfare, the results of such survey using the Coal Mine Noise Data Report. (See Figure 1, Part 70 of this subchapter.) Reports shall be sent to:

Division of Automatic Data Processing, Post Office Box 25407, Building 41, Denver Federal Center, Denver, Colo. 80225.

935

§ 71.303 Periodic noise level survey.

(a) At intervals of at least every 6 months, after November 13, 1974, each operator shall conduct periodic surveys of the noise levels to which each miner in each surface installation and at each surface worksite is exposed and shall report and certify the results of such surveys to the Mining Enforcement and Safety Administration and the Department of Health, Education, and Welfare, using the Coal Mine Noise Data Report Form. The interval between each survey shall not be less than 3 months. Reports shall be sent to:

Division of Automatic Data Processing, Post Office Box 25407, Building 41, Denver Federal Center, Denver, Colo. 80225.

(b) Where no A-scale reading recorded for any miner during an initial or periodic noise level survey exceeds 90 dBA, the operator shall not be required to survey such miner during any subsequent periodic noise level survey required by this section: *Provided, however,* That the name and job position of each such miner shall be reported in every periodic survey and the operator shall certify that such miner's job duties and noise exposure levels have not changed substantially during the preceding 6-month period.

§ 71.304 Supplemental noise level survey; reports and certification.

(a) Where the certified results of an initial noise level survey conducted in accordance with § 71.302 or a periodic noise level survey conducted in accordance with § 71.303 indicate that any miner may be exposed to a noise level in excess of the permissible noise level, the operator shall conduct a supplemental noise level survey with respect to each miner whose noise exposure exceeds this standard. This survey shall be conducted within 15 days following notification to the operator by the Mining Enforcement and Safety Administration to conduct such survey.

(b) Supplemental noise level surveys shall be conducted by taking noise level measurements in accordance with § 70.506 of this Subchapter O; however, noise level measurements shall be taken of each individual operation to which the miner under consideration is actually exposed during his normal work shift and the duration of each such exposure shall be recorded.

(c) Each operator shall report and certify the results of each supplemental noise level survey conducted in accordance with this section to the Mining Enforcement and Safety Administration and the Department of Health, Education, and Welfare using the Coal Mine Noise Data Report Form to record noise level readings taken with respect to all operations during which such measurements were taken.

(d) Supplemental noise level surveys shall, upon completion, be mailed to:

Division of Automatic Data Processing, Post Office Box 25407, Building 41, Denver Federal Center, Denver, Colo. 80225.

§ 71.305 Violation of noise standard; notice of violation; action required by operator.

(a) Where the results of a supplemental noise level survey conducted in accordance with § 71.304 indicate that any miner is exposed to noise levels which exceed the permissible noise levels, the Secretary shall issue a notice to the operator that he is in violation of this subpart.

(b) Upon receipt of a notice of violation issued pursuant to paragraph (a) of this section, the operator shall:

(1) Institute, promptly, administrative and/or engineering controls necessary to assure compliance with the standard. Such controls may include protective devices other than those devices or systems which the Secretary or his authorized representative finds to be hazardous in such mine.

(2) Within 60 days following the issuance of the first notice of violation of this subpart, submit for approval to a joint Mining Enforcement and Safety Administration/Health, Education, and Welfare committee, a plan for the administration of a continuing, effective hearing conservation program to assure compliance with this subpart, including provision for:

(i) Reducing environmental noise levels;

(ii) Personal ear protective devices

to be made available to the miners;

(iii) Preplacement and periodic audiograms.

(iv) Those administrative and engineering controls that it has instituted to assure compliance with the standard.

(3) Plans required under subparagraph (2) of this paragraph shall be submitted to:

Division of Automatic Data Processing, Post Office Box 25407, Building 41, Denver Federal Center, Denver, Colo. 80225.

(c) Within 30 days following the issuance of any subsequent notice of violation of this subpart, the operator shall submit in writing:

(i) a statement of the manner in which the plan is intended to prevent the violation or

(ii) a revision to its plan to prevent similar future violations.

NOTE.—The incorporation by reference provision in this document was approved by the Director of the Federal Register on July 6, 1973.

937

Appendix B Federal Environmental Noise Regulations

1. Noise Control Act of 1972

Public Law 92–574; enacted by Congress, October 18, 1972; signed by the President, October 27, 1972.

SHORT TITLE

SECTION 1. This Act may be cited as the "Noise Control Act of 1972".

FINDINGS AND POLICY

SEC. 2. (a) The Congress finds—
(1) that inadequately controlled noise presents a growing danger to the health and welfare of the Nation's population, particularly in urban areas;
(2) that the major sources of noise include transportation vehicles and equipment, machinery, appliances, and other products in commerce; and
(3) that, while primary responsibility for control of noise rests with State and local governments, Federal action is essential to deal with major noise sources in commerce control of which require national uniformity of treatment.
(b) The Congress declares that it is the policy of the United States to promote an environment for all Americans free from noise that jeopardizes their health or welfare. To that end, it is the purpose of this Act to establish a means for effective coordination of Federal research and activities in noise control, to authorize the establishment of Federal noise emission standards for products distributed in commerce, and to provide information to the public respecting the noise emission and noise reduction characteristics of such products.

DEFINITIONS

SEC. 3. For purposes of this Act:
(1) The term "Administrator" means the Administrator of the Environmental Protection Agency.
(2) The term "person" means an individual, corporation, partnership, or association, and (except as provided in sections 11(e) and 12(a)) includes any officer, employee, department, agency, or instrumentality of the United States, a State, or any

938

political subdivision of a State.

(3) The term "product" means any manufactured article or goods or component thereof; except that such term does not include—

(A) any aircraft, aircraft engine, propeller, or appliance, as such terms are defined in section 101 of the Federal Aviation Act of 1958; or

(B)(i) any military weapons or equipment which are designed for combat use; (ii) any rockets or equipment which are designed for research, experimental, or developmental work to be performed by the National Aeronautics and Space Administration; or (iii) to the extent provided by regulations of the Administrator, any other machinery or equipment designed for use in experimental work done by or for the Federal Government.

(4) The term "ultimate purchaser" means the first person who in good faith purchases a product for purposes other than resale.

(5) The term "new product" means (A) a product the equitable or legal title of which has never been transferred to an ultimate purchaser, or (B) a product which is imported or offered for importation into the United States and which is manufactured after the effective date of a regulation under section 6 or section 8 which would have been applicable to such product had it been manufactured in the United States.

(6) The term "manufacturer" means any person engaged in the manufacturing or assembling of new products, or the importing of new products for resale, or who acts for, and is controlled by, any such person in connection with the distribution of such products.

(7) The term "commerce" means trade, traffic, commerce, or transportation—

(A) between a place in a State and any place outside thereof, or

(B) which affects trade, traffic, commerce, or transportation described in subparagraph (A).

(8) The term "distribute in commerce" means sell in, offer for sale in, or introduce or deliver for introduction into, commerce.

(9) The term "State" includes the District of Columbia, the Commonwealth of Puerto Rico, the Virgin Islands, American Samoa, Guam, and the Trust Territory of the Pacific Islands.

(10) The term "Federal agency" means an executive agency (as defined in section 105 of title 5, United States Code) and includes the United States Postal Service.

(11) The term "environmental noise" means the intensity, duration, and the character of sounds from all sources.

FEDERAL PROGRAMS

SEC. 4. (a) The Congress authorizes and directs that Federal agencies shall, to the fullest extent consistent with their authority under Federal laws administered by them, carry out the programs within their control in such a manner as to further the policy declared in section 2(b).

(b) Each department, agency, or instrumentality of the executive, legislative, and judicial branches of the Federal Government—

(1) having jurisdiction over any property or facility, or

(2) engaged in any activity resulting, or which may result, in the emission of noise, shall comply with Federal, State, interstate, and local requirements respecting control and abatement of environmental noise to

the same extent that any person is subject to such requirements. The President may exempt any single activity or facility, including noise emission sources or classes thereof, of any department, agency, or instrumentality in the executive branch from compliance with any such requirement if he determines it to be in the paramount interest of the United States to do so; except that no exemption, other than for those products referred to in section 3(3)(B) of this Act, may be granted from the requirements of sections 6, 17, and 18 of this Act. No such exemption shall be granted due to lack of appropriation unless the President shall have specifically requested such appropriation as a part of the budgetary process and the Congress shall have failed to make available such requested appropriation. Any exemption shall be for a period not in excess of one year, but additional exemptions may be granted for periods of not to exceed one year upon the President's making a new determination. The President shall report each January to the Congress all exemptions from the requirements of this section granted during the preceding calendar year, together with his reason for granting such exemption.

(c)(1) The Administrator shall coordinate the programs of all Federal agencies relating to noise research and noise control. Each Federal agency shall, upon request, furnish to the Administrator such information as he may reasonably require to determine the nature, scope, and results of the noise-research and noise-control programs of the agency.

(2) Each Federal agency shall consult with the Administrator in pre-scribing standards or regulations respecting noise. If at any time the Administrator has reason to believe that a standard or regulation, or any proposed standard or regulation, of any Federal agency respecting noise does not protect the public health and welfare to the extent he believes to be required and feasible, he may request such agency to review and report to him on the advisibility of revising such standard or regulation to provide such protection. Any such request may be published in the Federal Register and shall be accompanied by a detailed statement of the information on which it is based. Such agency shall complete the requested review and report to the Administrator within such time as the Administrator specifies in the request, but such time specified may not be less than ninety days from the date the request was made. The report shall be published in the Federal Register and shall be accompanied by a detailed statement of the findings and conclusions of the agency respecting the revision of its standard or regulation. With respect to the Federal Aviation Administration, section 611 of the Federal Aviation Act of 1958 (as amended by section 7 of this Act) shall apply in lieu of this paragraph.

(3) On the basis of regular consultation with appropriate Federal agencies, the Administrator shall compile and publish, from time to time, a report on the status and progress of Federal activities relating to noise research and noise control. This report shall describe the noise-control programs of each Federal agency and assess the contributions of those programs to the Federal Government's overall efforts to control noise.

Identification of Major Noise Sources; Noise Criteria and Control Technology

Sec. 5. (a)(1) The Administrator shall, after consultation with appropriate Federal agencies and within nine months of the date of the enactment of this Act, develop and publish criteria with respect to noise. Such criteria shall reflect the scientific knowledge most useful in indicating the kind and extent of all identifiable effects on the public health or welfare which may be expected from differing quantities and qualities of noise.

(2) The Administrator shall, after consultation with appropriate Federal agencies and within twelve months of the date of the enactment of this Act, publish information on the levels of environmental noise the attainment and maintenance of which in defined areas under various conditions are requisite to protect the public health and welfare with an adequate margin of safety.

(b) The Administrator shall, after consultation with appropriate Federal agencies, compile and publish a report or series of reports (1) identifying products (or classes of products) which in his judgment are major sources of noise, and (2) giving information on techniques for control of noise from such products, including available data on the technology, costs, and alternative methods of noise control. The first such report shall be published not later than eighteen months after the date of enactment of this Act.

(c) The Administrator shall from time to time review and, as appropriate, revise or supplement any criteria or reports published under this section.

(d) Any report (or revision thereof) under subsection (b)(1) identifying major noise sources shall be published in the Federal Register. The publication or revision under this section of any criteria or information on control techniques shall be announced in the Federal Register, and copies shall be made available to the general public.

Noise Emission Standards for Products Distributed in Commerce

Sec. 6 (a)(1) The Administrator shall publish proposed regulations, meeting the requirements of subsection (c), for each product—

(A) which is identified (or is part of a class identified) in any report published under section 5(b)(1) as a major source of noise,

(B) for which, in his judgment, noise emission standards are feasible, and

(C) which falls in one of the following categories:

(i) Construction equipment.

(ii) Transportation equipment (including recreational vehicles and related equipment).

(iii) Any motor or engine (including any equipment of which an engine or motor is an integral part).

(iv) Electrical or electronic equipment.

(2)(A) Initial proposed regulations under paragraph (1) shall be published not later than eighteen months after the date of enactment of this Act, and shall apply to any product described in paragraph (1) which is identified (or is a part of a class identified) as a major source of noise in any report published under section 5(b)(1) on or before the date

941

of publication of such initial proposed regulations.

(B) In the case of any product described in paragraph (1) which is identified (or is part of a class identified) as a major source of noise in a report published under section 5(b)(1) after publication of the initial proposed regulations under subparagraph (A) of this paragraph, regulations under paragraph (1) for such product shall be proposed and published by the Administrator not later than eighteen months after such report is published.

(3) After proposed regulations respecting a product have been published under paragraph (2), the Administrator shall, unless in his judgment noise emission standards are not feasible for such product, prescribe regulations, meeting the requirements of subsection (c), for such product—

(A) not earlier than six months after publication of such proposed regulations, and

(B) not later than—

(i) twenty-four months after the date of enactment of this Act, in the case of a product subject to proposed regulations published under paragraph (2)(A), or

(ii) in the case of any other product, twenty-four months after the publication of the report under section 5(b)(1) identifying it (or a class of products of which it is a part) as a major source of noise.

(b) The Administrator may publish proposed regulations, meeting the requirements of subsection (c), for any product for which he is not required by subsection (a) to prescribe regulations but for which, in his judgment, noise emission standards are feasible and are requisite to protect the public health and welfare. Not earlier than six months after the date of publication of such proposed regulations respecting such product, he may prescribe regulations, meeting the requirements of subsection (c), for such product.

(c)(1) Any regulation prescribed under subsection (a) or (b) of this section (and any revision thereof) respecting a product shall include a noise emission standard which shall set limits on noise emissions from such product and shall be a standard which in the Administrator's judgment, based on criteria published under section 5, is requisite to protect the public health and welfare, taking into account the magnitude and conditions of use of such product (alone or in combination with other noise sources) the degree of noise reduction achievable through the application of the best available technology, and the cost of compliance. In establishing such a standard for any product, the Administrator shall give appropriate consideration to standards under other laws designed to safeguard the health and welfare of persons, including any standards under the National Traffic and Motor Vehicle Safety Act of 1966, the Clean Air Act, and the Federal Water Pollution Control Act. Any such noise emission standards shall be a performance standard. In addition, any regulation under subsection (a) or (b) (and any revision thereof) may contain testing procedures necessary to assure compliance with the emission standard in such regulation, and may contain provisions respecting instructions of the manufacturer for the maintenance, use, or repair of the product.

(2) After publication of any proposed regulations under this section,

the Administrator shall allow interested persons an opportunity to participate in rulemaking in accordance with the first sentence of section 553(c) of title 5, United States Code.

(3) The Administrator may revise any regulation prescribed by him under this section by (A) publication of proposed revised regulations, and (B) the promulgation, not earlier than six months after the date of such publication, of regulations making the revision; except that a revision which makes only technical or clerical corrections in a regulation under this section may be promulgated earlier than six months after such date if the Administrator finds that such earlier promulgation is in the public interest.

(d)(1) On and after the effective date of any regulation prescribed under subsection (a) or (b) of this section, the manufacturer of each new product to which such regulation applies shall warrant to the ultimate purchaser and each subsequent purchaser that such product is designed, built, and equipped so as to conform at the time of sale with such regulation.

(2) Any cost obligation of any dealer incurred as a result of any requirement imposed by paragraph (1) of this subsection shall be borne by the manufacturer. The transfer of any such cost obligation from a manufacturer to any dealer through franchise or other agreement is prohibited.

(3) If a manufacturer includes in any advertisement a statement respecting the cost or value of noise emission control devices or systems, such manufacturer shall set forth in such statement the cost or value attributed to such devices or systems by the Secretary of Labor (through the Bureau of Labor Statistics). The

Secretary of Labor, and his representatives, shall have the same access for this purpose to the books, documents, papers, and records of a manufacturer as the Comptroller General has to those of a recipient of assistance for purposes of section 311 of the Clear Air Act.

(e)(1) No State or political subdivision thereof may adopt or enforce—

(A) with respect to any new product for which a regulation has been prescribed by the Administrator under this section, any law or regulation which sets a limit on noise emissions from such new product and which is not identical to such regulation of the Administrator; or

(B) with respect to any component incorporated into such new product by the manufacturer of such product, any law or regulation setting a limit on noise emissions from such component when so incorporated.

(2) Subject to sections 17 and 18, nothing in this section precludes or denies the right of any State or political subdivision thereof to establish and enforce controls on environmental noise (or one or more sources thereof) through the licensing, regulation, or restriction of the use, operation, or movement of any product or combination of products.

AIRCRAFT NOISE STANDARDS

SEC. 7. (a) The Administrator, after consultation with appropriate Federal, State, and local agencies and interested persons, shall conduct a study of the (1) adequacy of Federal Aviation Administration flight and operational noise controls; (2) adequacy of noise emission standards on new and existing aircraft, together with recommendations on the retrofitting and phaseout of existing aircraft; (3) implications of identifying and achieving levels of cumulative noise exposure around airports; and (4) additional measures available to airport operators and local governments to control aircraft noise. He shall report on such study to the Committee on Interstate and Foreign Commerce of the House of Representatives and the Committees on Commerce and Public Works of the Senate within nine months after the date of the enactment of this Act.

(b) Section 611 of the Federal Aviation Act of 1958 (49 U.S.C. 1431) is amended to read as follows:

"CONTROL AND ABATEMENT OF AIRCRAFT NOISE AND SONIC BOOM

"SEC. 611. (a) For purposes of this section:

"(1) The term 'FAA' means Administrator of the Federal Aviation Administration.

"(2) The term 'EPA' means the Administrator of the Environmental Protection Agency.

"(b)(1) In order to afford present and future relief and protection to the public health and welfare from aircraft noise and sonic boom, the FAA, after consultation with the Secretary of Transportation and with EPA, shall prescribe and amend standards for the measurement of aircraft noise and sonic boom and shall prescribe and amend such regulations as the FAA may find necessary to provide for the control and abatement of aircraft noise and sonic boom, including the application of such standards and regulations in the issuance, amendment, modification, suspension, or revocation of any certificate authorized by this title. No exemption with respect to any standard or regulation under this section may be granted under any provision of this Act unless the FAA shall have consulted with EPA before such exemption is granted, except that if the FAA determines that safety in air commerce or air transportation requires that such an exemption be granted before EPA can be consulted, the FAA shall consult with EPA as soon as practicable after the exemption is granted.

"(2) The FAA shall not issue an original type certificate under section 603(a) of this Act for any aircraft for which substantial noise abatement can be achieved by prescribing standards and regulations in accordance with this section, unless he shall have prescribed standards and regulations in accordance with this section which apply to such aircraft and which protect the public from aircraft noise and sonic boom, consistent with the considerations listed in subsection (d).

"(c)(1) Not earlier than the date of submission of the report required by section 7 (a) of the Noise Control Act of 1972, EPA shall submit to the FAA proposed regulations to provide such control and abatement of aircraft noise and sonic boom (including control and abatement through the exercise of any of the FAA's regulatory authority over air commerce or transportation or over aircraft or airport operations) as EPA determines is necessary to protect the public health and welfare. The FAA shall consider such proposed regulations submitted by EPA under this paragraph and shall, within thirty days of the date of its submission to the FAA, publish the proposed regulations in a notice of proposed rulemaking. Within sixty days after such publication, the FAA shall commence a hearing at which interested persons shall be afforded an opportunity for oral (as well as written) presentations of data, views, and arguments. Within a reasonable time after the conclusion of such hearing and after consultation with EPA, the FAA shall—

"(A) in accordance with subsection (b), prescribe regulations (i) substantially as they were submitted by EPA, or (ii) which are a modification of the proposed regulations submitted by EPA, or

"(B) publish in the Federal Register a notice that it is not prescribing any regulation in response to EPA's submission of proposed regulations, together with a detailed explanation providing reasons for the decision not to prescribe such regulations.

"(2) If EPA has reason to believe that the FAA's action with respect to a regulation proposed by EPA under paragraph (1)(A)(ii) or (1)(B) of this subsection does not protect the public health and welfare from aircraft noise or sonic boom, consistent with the considerations listed in subsection (d) of this section, EPA shall consult with the FAA and may request the FAA to review, and report to EPA on, the advisability of prescribing the regulation originally proposed by EPA. Any such request shall be published in the Federal Register and shall include a detailed statement of the information on which it is based. The FAA shall complete the review requested and shall report to EPA within such time as EPA specifies in the request, but such time specified may not be less than ninety days from the date the request was made. The FAA's report shall be accompanied by a detailed statement of the FAA's findings and the reasons for the FAA's conclusions; shall identify any statement filed pursuant to section 102(2)(C) of the National Environmental Policy Act of 1969 with respect to such action of the FAA under paragraph (1) of this subsection; and shall specify whether (and where) such statements are available for public inspection. The FAA's report shall be published in the Federal Register, except in a case in which EPA's request proposed specific action to be taken by the FAA, and the FAA's report indicates such action will be taken.

"(3) If, in the case of a matter described in paragraph (2) of this subsection with respect to which no statement is required to be filed under such section 102(2)(C), the report of the FAA indicates that the proposed regulation originally submitted by EPA should not be made, then EPA may request the FAA to file a supplemental report, which shall be published in the Federal Register within such a period as EPA may specify (but such time specified shall not be less than ninety days from the date the request was made), and which shall contain a comparison of (A) the environmental effects (including those which cannot be avoided) of the action actually taken by the FAA in response to EPA's proposed regulations, and (B) EPA's proposed regulations.

"(d) In prescribing and amending standards and regulations under this section, the FAA shall—

"(1) consider relevant available data relating to aircraft noise and sonic boom, including the results of research, development, testing, and evaluation activities conducted pursuant to this Act and the Department of Transportation Act;

"(2) consult with such Federal,

945

State, and interstate agencies as he deems appropriate;

"(3) consider whether any proposed standard or regulation is consistent with the highest degree of safety in air commerce or air transportation in the public interest;

"(4) consider whether any proposed standard or regulation is economically reasonable, technologically practicable, and appropriate for the particular type of aircraft, aircraft engine, appliance, or certificate to which it will apply; and

"(5) consider the extent to which such standard or regulation will contribute to carrying out the purposes of this section.

"(e) In any action to amend, modify, suspend, or revoke a certificate in which violation of aircraft noise or sonic boom standards or regulations is at issue, the certificate holder shall have the same notice and appeal rights as are contained in section 609, and in any appeal to the National Transportation Safety Board, the Board may amend, modify, or reverse the order of the FAA if it finds that control of abatement of aircraft noise or sonic boom and the public health and welfare do not require the affirmation of such order, or that such order is not consistent with safety in air commerce or air transportation."

(c) All—

(1) standards, rules, and regulations prescribed under section 611 of the Federal Aviation Act of 1958, and

(2) exemptions, granted under any provision of the Federal Aviation Act of 1958, with respect to such standards, rules, and regulations, which are in effect on the date of the enactment of this Act, shall continue in effect according to their terms until modified, terminated, superseded, set aside, or repealed by the Administrator of the Federal Aviation Administration in the exercise of any authority vested in him, by a court of competent jurisdiction, or by operation of law.

LABELING

SEC. 8. (a) The Administrator shall by regulation designate any product (or class thereof)—

(1) which emits noise capable of adversely affecting the public health or welfare; or

(2) which is sold wholly or in part on the basis of its effectiveness in reducing noise.

(b) For each product (or class thereof) designated under subsection (a) the Administrator shall by regulation require that notice be given to the prospective user of the level of the noise the product emits, or of its effectiveness in reducing noise, as the case may be. Such regulations shall specify (1) whether such notice shall be affixed to the product or to the outside of its container, or to both, at the time of its sale to the ultimate purchaser or whether such notice shall be given to the prospective user in some other manner, (2) the form of the notice, and (3) the methods and units of measurement to be used. Sections 6(c)(2) shall apply to the prescribing of any regulation under this section.

(c) This section does not prevent any State or political subdivision thereof from regulating product labeling or information respecting products in any way not in conflict with regulations prescribed by the Administrator under this section.

IMPORTS

SEC. 9. The Secretary of the Treasury shall, in consultation with the Administrator, issue regulations to carry out the provisions of this Act with respect to new products imported or offered for importation.

PROHIBITED ACTS

SEC. 10. (a) Except as otherwise provided in subsection (b), the following acts or the causing thereof are prohibited:

(1) In the case of a manufacturer, to distribute in commerce any new product manufactured after the effective date of a regulation prescribed under section 6 which is applicable to such product, except in conformity with such regulation.

(2)(A) The removal or rendering inoperative by any person, other than for purpose of maintenance, repair, or replacement, of any device or element of design incorporated into any product in compliance with regulations under section 6, prior to its sale or delivery to the ultimate purchaser or while it is in use, or (B) the use of a product after such device or element of design has been removed or rendered inoperative by any person.

(3) In the case of a manufacturer, to distribute in commerce any new product manufactured after the effective date of a regulation prescribed under section 8(b) (requiring information respecting noise) which is applicable to such product, except in conformity with such regulation.

(4) The removal by any person of any notice affixed to a product or container pursuant to regulations prescribed under section 8(b), prior to sale of the product to the ultimate purchaser.

(5) The importation into the United States by any person of any new product in violation of a regulation prescribed under section 9 which is applicable to such product.

(6) The failure or refusal by any person to comply with any requirement of section 11(d) or 13(a) or regulations prescribed under section 13(a), 17, or 18.

(b)(1) For the purpose of research, investigations, studies, demonstrations, or training, or for reasons of national security, the Administrator may exempt for a specified period of time any product, or class thereof, from paragraphs (1), (2), (3), and (5) of subsection (a), upon such terms and conditions as he may find necessary to protect the public health or welfare.

(2) Paragraphs (1), (2), (3), and (4) of subsection (a) shall not apply with respect to any product which is manufactured solely for use outside any State and which (and the container of which) is labeled or otherwise marked to show that it is manufactured solely for use outside any State; except that such paragraphs shall apply to such product if it is in fact distributed in commerce for use in any State.

ENFORCEMENT

SEC. 11. (a) Any person who willfully or knowingly violates paragraph (1), (3), (5), or (6) of subsection (a) of section 10 of this Act shall be punished by a fine of not more than $25,000 per day of violation, or by imprisonment for not more than one year, or by both. If the conviction is for a violation committed after a first conviction of such person under this subsection, punishment shall be by a fine of not more than $50,000 per day of violation, or by imprisonment for not more than two years, or by both.

(b) For the purpose of this section, each day of violation of any paragraph of section 10(a) shall constitute a separate violation of that section.

(c) The district courts of the United States shall have jurisdiction of actions brought by and in the name of the United States to restrain any violation of section 10(a) of this Act.

(d)(1) Whenever any person is in violation of section 10(a) of this Act, the Administrator may issue an order specifying such relief as he determines is necessary to protect the public health and welfare.

(2) Any order under this subsection shall be issued only after notice and opportunity for a hearing in accordance with section 554 of title 5 of the United States Code.

(e) The term "person," as used in this section, does not include a department, agency, or instrumentality of the United States.

CITIZEN SUITS

SEC. 12. (a) Except as provided in subsection (b), any person (other than the United States) may commence a civil action on his own behalf—

(1) against any person (including (A) the United States, and (B) any other governmental instrumentality or agency to the extent permitted by the eleventh amendment to the Constitution) who is alleged to be in violation of any noise control requirement (as defined in subsection (e)), or

(2) against—

(A) The Administrator of the Environmental Protection Agency where there is alleged a failure of such Ad-

ministrator to perform any act or duty under this Act which is not discretionary with such Administrator, or

(B) the Administrator of the Federal Aviation Administration where there is alleged a failure of such Administrator to perform any act or duty under section 611 of the Federal Aviation Act of 1958 which is not discretionary with such Administrator.

The district courts of the United States shall have jurisdiction, without regard to the amount in controversy, to restrain such person from violating such noise control requirement or to order such Administrator to perform such act or duty, as the case may be.

(b) No action may be commenced—

(1) under subsection (a)(1)—

(A) prior to sixty days after the plaintiff has given notice of the violation (1) to the Administrator of the Environmental Protection Agency (and to the Federal Aviation Administrator in the case of a violation of a noise control requirement under such section 611) and (ii) to any alleged violator of such requirement, or

(B) If an Administrator has commenced and is diligently prosecuting a civil action to require compliance with the noise control requirement, but in any such action in a court of the United States any person may intervene as a matter of right, or

(2) under subsection (a)(2) prior to sixty days after the plaintiff has given notice to the defendant that he will commence such action.

Notice under this subsection shall be given in such manner as the Administrator of the Environmental Protection Agency shall prescribe by regulation.

(c) In an action under this section, the Administrator of the Environmental Protection Agency, if not a party, may intervene as a matter of right. In an action under this section respecting a noise control requirement under section 611 of the Federal Aviation Act of 1958, the Administrator of the Federal Aviation Administration, if not a party, may also intervene as a matter of right.

(d) The court, in issuing any final order in any action brought pursuant to subsection (a) of this section, may award costs of litigation (including reasonable attorney and expert witness fees) to any party, whenever the court determines such an award is appropriate.

(e) Nothing in this section shall restrict any right which any person (or class of persons) may have under any statute or common law to seek enforcement of any noise control requirement or to seek any other relief (including relief against an Administrator).

(f) For purposes of this section, the term "noise control requirement" means paragraph (1), (2), (3), (4), or (5) of section 10(a), or a standard, rule, or regulation issued under section 17 or 18 of this Act or under section 611 of the Federal Aviation Act of 1958.

RECORDS, REPORTS, AND INFORMATION

SEC. 13. (a) Each manufacturer of a product to which regulations under section 6 or section 8 apply shall—

(1) establish and maintain such records, make such reports, provide such information, and make such tests, as the Administrator may reasonably require to enable him to determine whether such manufacturer has acted or is acting in compliance with this Act.

(2) upon request of an officer or employee duly designated by the Administrator, permit such officer or employee at reasonable times to have access to such information and the results of such tests and to copy such records, and

(3) to the extent required by regulations of the Administrator, make products coming off the assembly line or otherwise in the hands of the manufacturer available for testing by the Administrator.

(b)(1) All information obtained by the Administrator or his representatives pursuant to subsection (a) of this section, which information contains or relates to a trade secret or other matter referred to in section 1905 of title 18 of the United States Code, shall be considered confidential for the purpose of that section, except that such information may be disclosed to other Federal officers or employees, in whose possession it shall remain confidential, or when relevant to the matter in controversy in any proceeding under this Act.

(2) Nothing in this subsection shall authorize the withholding of information by the Administrator, or by any officers or employees under his control, from the duly authorized committees of the Congress.

(c) Any person who knowingly makes any false statement, representation, or certification in any application, record, report, plan, or other document filed or required to be maintained under this Act or who falsifies, tampers with, or knowingly renders inaccurate any monitoring device or method required to be maintained under this Act, shall upon conviction be punished by a fine of not more than $10,000, or by imprisonment for not more than six months, or by both.

RESEARCH, TECHNICAL ASSISTANCE, AND PUBLIC INFORMATION

SEC. 14. In furtherance of his responsibilities under this Act and to complement, as necessary, the noise-research programs of other Federal agencies, the Administrator is authorized to:

(1) Conduct research, and finance research by contract with any person, on the effects, measurement, and control of noise, including but not limited to—

(A) investigation of the psychological and physiological effects of noise on humans and the effects of noise on domestic animals, wildlife, and property, and determination of acceptable levels of noise on the basis of such effects;

(B) development of improved methods and standards for measurement and monitoring of noise, in cooperation with the National Bureau of Standards, Department of Commerce; and

(C) determination of the most effective and practicable means of controlling noise emission.

(2) Provide technical assistance to State and local governments to facilitate their development and enforcement of ambient noise standards, including but not limited to—

(A) advice on training of noise-control personnel and on selection and operation of noise-abatement equipment; and

(B) preparation of model State or local legislation for noise control.

(3) Disseminate to the public information on the effects of noise, acceptable noise levels, and techniques for noise measurement and control.

DEVELOPMENT OF LOW-NOISE-EMISSION PRODUCTS

SEC. 15. (a) For the purpose of this section:

(1) The term "Committee" means the Low-Noise-Emission Product Advisory Committee.

(2) The term "Federal Government" includes the legislative, executive, and judicial branches of the Government of the United States, and the government of the District of Columbia.

(3) The term "low-noise-emission product" means any product which emits noise in amounts significantly below the levels specified in noise emission standards under regulations applicable under section 6 at the time of procurement to that type of product.

(4) The term "retail price" means (A) the maximum statutory price applicable to any type of product; or (B) in any case where there is no applicable maximum statutory price, the most recent procurement price paid for any type of product.

(b)(1) The Administrator shall determine which products qualify as low-noise-emission products in accordance with the provisions of this section.

(2) The Administrator shall certify any product—

(A) for which a certification application has been filed in accordance with paragraph (5)(A) of this subsection;

(B) which is a low-noise-emission product as determined by the Administrator; and

(C) which he determines is suitable for use as a substitute for a type of product at that time in use by agencies of the Federal Government.

(3) The Administrator may establish a Low-Noise-Emission Product Advisory Committee to assist him in determining which products qualify as low-noise-emission products for purposes of this section. The Committee shall include the Administrator or his designee, a representative of the National Bureau of Standards, and representatives of such other Federal agencies and private individuals as the Administrator may deem necessary from time to time. Any member of the Committee not employed on a full-time basis by the United States may receive the daily equivalent of the annual rate of basic pay in effect for grade GS-18 of the General Schedule for each day such member is engaged upon work of the Committee. Each member of the Committee shall be reimbursed for travel expenses, including per diem in lieu of subsistence as authorized by section 5703 of title 5, United States Code, for persons in the Government service employed intermittently.

(4) Certification under this section shall be effective for a period of one year from the date of issuance.

(5)(A) Any person seeking to have a class or model of product certified under this section shall file a certification application in accordance with regulations prescribed by the Administrator.

(B) The Administrator shall publish in the Federal Register a notice of each application received.

(C) The Administrator shall make determinations for the purpose of this section in accordance with procedures prescribed by him by regulation.

(D) The Administrator shall conduct whatever investigation is necessary, including actual inspection of the product at a place designated in regulations prescribed under subparagraph (A).

(E) The Administrator shall receive and evaluate written comments and documents from interested persons in support of, or in opposition to, certification of the class or model of product under consideration.

(F) Within ninety days after the receipt of a properly filed certification application the Administrator shall determine whether such product is a low-noise-emission product for purposes of this section. If the Administrator determines that such product is a low-noise-emission product, then within one hundred and eighty days of such determination the Administrator shall reach a decision as to whether such product is a suitable substitute for any class or classes of products presently being purchased by the Federal Government for use by its agencies.

(G) Immediately upon making any determination or decision under subparagraph (F), the Administrator shall publish in the Federal Register notice of such determination or decision, including reason therefor.

(c)(1) Certified low-noise-emission products shall be acquired by purchase or lease by the Federal Government for use by the Federal Government in lieu of other products if the Administrator of General Services determines that such certified products have procurement costs which are no more than 125 per centum of the retail price of the least expensive type of product for which they are certified substitutes.

(2) Data relied upon by the Administrator in determining that a product is a certified low-noise-emission product shall be incorporated

in any contract for the procurement of such product.

(d) The procuring agency shall be required to purchase available certified low-noise-emission products which are eligible for purchase to the extent they are available before purchasing any other products for which any low-noise-emission product is a certified substitute. In making purchasing selections between competing eligible certified low-noise-emission products, the procuring agency shall give priority to any class or model which does not require extensive periodic maintenance to retain its low-noise-emission qualities or which does not involve operating costs significantly in excess of those products for which it is a certified substitute.

(e) For the purpose of procuring certified low-noise-emission products any statutory price limitations shall be waived.

(f) The Administrator shall, from time to time as he deems appropriate, test the emissions of noise from certified low-noise-emission products purchased by the Federal Government. If at any time he finds that the noise-emission levels exceed the levels on which certification under this section was based, the Administrator shall give the supplier of such product written notice of this finding, issue public notice of it, and give the supplier an opportunity to make necessary repairs, adjustments, or replacements. If no such repairs, adjustments, or replacements are made within a period to be set by the Administrator, he may order the supplier to show cause why the product involved should be eligible for recertification.

(g) There are authorized to be appropriated for paying additional amounts for products pursuant to, and for carrying out the provisions of, this section, $1,000,000 for the fiscal year ending June 30, 1973, and $2,000,000 for each of the two succeeding fiscal years.

(h) The Administrator shall promulgate the procedures required to implement this section within one hundred and eighty days after the date of enactment of this Act.

JUDICIAL REVIEW; WITNESSES

SEC. 16. (a) A petition for review of action of the Administrator of the Environmental Protection Agency in promulgating any standard or regulation under section 6, 17, or 18 of this Act or any labeling regulation under section 8 of this Act may be filed only in the United States Court of Appeals for the District of Columbia Circuit, and a petition for review of action of the Administrator of the Federal Aviation Administration in promulgating any standard or regulation under section 611 of the Federal Aviation Act of 1958 may be filed only in such court. Any such petition shall be filed within ninety days from the date of such promulgation, or after such date if such petition is based solely on grounds arising after such ninetieth day. Action of either Administrator with respect to which review could have been obtained under this subsection shall not be subject to judicial review in civil or criminal proceedings for enforcement.

(b) If a party seeking review under this Act applies to the court for leave to adduce additional evidence, and shows to the satisfaction of the court that the information is material and was not available at the time of the proceeding before the Administrator of such Agency or Administration (as the case may be), the court may order such additional evidence (and evidence in rebuttal thereof) to be taken before such Administrator, and to be adduced upon the hearing, in such manner and upon such terms and conditions as the court may deem proper. Such Administrator may modify his findings as to the facts, or make new findings, by reason of the additional evidence so taken, and he shall file with the court such modified or new findings, and his recommendation, if any, for the modification or setting aside of his original order, with the return of such additional evidence.

(c) With respect to relief pending review of an action by either Administrator, no stay of an agency action may be granted unless the reviewing court determines that the party seeking such stay is (1) likely to prevail on the merits in the review proceeding and (2) will suffer irreparable harm pending such proceeding.

(d) For the purpose of obtaining information to carry out this Act, the Administrator of the Environmental Protection Agency may issue subpenas for the attendance and testimony of witnesses and the production of relevant papers, books, and documents, and he may administer oaths. Witnesses summoned shall be paid the same fees and mileage that are paid witnesses in the courts of the United States. In cases of contumacy or refusal to obey a subpena served upon any person under this subsection, the district court of the United States for any district in which such person is found or resides or transacts business, upon application by the United States and after notice to such person, shall have jurisdiction to issue an order requiring such person to appear and give testimony before the Administrator,

to appear and produce papers, books, and documents before the Administrator, or both, and any failure to obey such order of the court may be punished by such court as a contempt thereof.

RAILROAD NOISE EMISSION STANDARDS

SEC. 17. (a)(1) Within nine months after the date of enactment of this Act, the Administrator shall publish proposed noise emission regulations for surface carriers engaged in interstate commerce by railroad. Such proposed regulations shall include noise emission standards setting such limits on noise emissions resulting from operation of the equipment and facilities of surface carriers engaged in interstate commerce by railroad which reflect the degree of noise reduction achievable through the application of the best available technology, taking into account the cost of compliance. These regulations shall be in addition to any regulations that may be proposed under section 6 of this Act.

(2) Within ninety days after the publication of such regulations as may be proposed under paragraph (1) of this subsection, and subject to the provisions of section 16 of this Act, the Administrator shall promulgate final regulations. Such regulations may be revised, from time to time, in accordance with this subsection.

(3) Any standard or regulation, or revision thereof, proposed under this subsection shall be promulgated only after consultation with the Secretary of Transportation in order to assure appropriate consideration for safety and technological availability.

(4) Any regulation or revision thereof promulgated under this subsection shall take effect after such

period as the Administrator finds necessary, after consultation with the Secretary of Transportation, to permit the development and application of the requisite technology, giving appropriate consideration to the cost of compliance within such period.

(b) The Secretary of Transportation, after consultation with the Administrator, shall promulgate regulations to insure compliance with all standards promulgated by the Administrator under this section. The Secretary of Transportation shall carry out such regulations through the use of his powers and duties of enforcement and inspection authorized by the Safety Appliance Acts, the Interstate Commerce Act, and the Department of Transportation Act. Regulations promulgated under this section shall be subject to the provisions of sections 10, 11, 12, and 16 of this Act.

(c)(1) Subject to paragraph (2) but notwithstanding any other provisions of this Act, after the effective date of a regulation under this section applicable to noise emissions resulting from the operation of any equipment or facility of a surface carrier engaged in interstate commerce by railroad, no State or political subdivision thereof may adopt or enforce any standard applicable to noise emissions resulting from the operation of the same equipment or facility of such carrier unless such standard is identical to a standard applicable to noise emissions resulting from such operation prescribed by any regulation under this section.

(2) Nothing in this section shall diminish or enhance the rights of any State or political subdivision thereof to establish and enforce standards or controls on levels of environmental noise, or to control, license, regulate,·

or restrict the use, operation, or movement of any product if the Administrator, after consultation with the Secretary of Transportation, determines that such standard, control, license, regulation, or restriction is necessitated by special local conditions and is not in conflict with regulations promulgated under this section.

(d) The terms "carrier" and "railroad" as used in this section shall have the same meaning as such terms have under the first section of the Act of February 17, 1911 (45 U.S.C. 22).

MOTOR CARRIER NOISE EMISSION STANDARDS

SEC. 18. (a)(1) Within nine months after the date of enactment of this Act, the Administrator shall publish proposed noise emission regulations for motor carriers engaged in interstate commerce. Such proposed regulations shall include noise emission standards setting such limits on noise emissions resulting from operation of motor carriers engaged in interstate commerce which reflect the degree of noise reduction achievable through the application of the best available technology, taking into account the cost of compliance. These regulations shall be in addition to any regulations that may be proposed under section 6 of this Act.

(2) Within ninety days after the publication of such regulations as may be proposed under paragraph (1) of this subsection, and subject to the provisions of section 16 of this Act, the Administrator shall promulgate final regulations. Such regulations may be revised from time to time, in accordance with this subsection.

(3) Any standard or regulation, or revision thereof, proposed under

this subsection shall be promulgated only after consultation with the Secretary of Transportation in order to assure appropriate consideration for safety and technological availability.

(4) Any regulation or revision thereof promulgated under this subsection shall take effect after such period as the Administrator finds necessary, after consultation with the Secretary of Transportation, to permit the development and application of the requisite technology, giving appropriate consideration to the cost of compliance within such period.

(b) The Secretary of Transportation, after consultation with the Administrator shall promulgate regulations to insure compliance with all standards promulgated by the Administrator under this section. The Secretary of Transportation shall carry out such regulations through the use of his powers and duties of enforcement and inspection authorized by the Interstate Commerce Act and the Department of Transportation Act. Regulations promulgated under this section shall be subject to the provisions of sections 10, 11, 12, and 16 of this Act.

(c)(1) Subject to paragraph (2) of this subsection but notwithstanding any other provision of this Act, after the effective date of a regulation under this section applicable to noise emissions resulting from the operation of any motor carrier engaged in interstate commerce, no State or political subdivision thereof may adopt or enforce any standard applicable to the same operation of such motor carrier, unless such standard is identical to a standard applicable to noise emissions resulting from such operation prescribed by any regulation under this section.

(2) Nothing in this section shall diminish or enhance the rights of any State or political subdivision thereof to establish and enforce standards or controls on levels of environmental noise, or to control, license, regulate, or restrict the use, operation, or movement of any product if the Administrator, after consultation with the Secretary of Transportation, determines that such standard, control, license, regulation, or restriction is necessitated by special local conditions and is not in conflict with regulations promulgated under this section.

(4) For purposes of this section, the term "motor carrier" includes a common carrier by motor vehicle, a contract carrier by motor vehicle, and a private carrier of property by motor vehicle as those terms are defined by paragraphs (14), (15), and (17) of section 203(a) of the Interstate Commerce Act (49 U.S.C. 303(a)).

2. Public Health and Welfare Criteria for Noise

"Public Health and Welfare Criteria for Noise" is the Criteria Document published by the Environmental Protection Agency on July 27, 1973, as directed by the Noise Control Act of 1972 (see Appendix B–1). It provides a partial basis for the establishment by EPA of standards and regulations required by the Act. The Criteria Document reflects "the scientific knowledge most useful in indicating the kind and extent of all identifiable effects of noise on the public health and welfare which may be expected from differing quantities and qualities of noise." Following are summaries of each of the document's 12 Sections. The summaries were made in the "Noise Regulation Reporter," published by Bureau of National Affairs, Inc., Washington, D.C. 20037. Copyright 1974 by B.N.A., they are used with permission.

Types of noise affecting public health and welfare

Historical evidence shows that excessive noise has long been considered a menace to the public health and welfare. Over the past two centuries, industrial development has resulted in a steady increase in the extent of noise impact.

Noise can affect the ability to communicate or to understand speech and other signals. This may arise from either actual impairment of the hearing mechanism or as a result of intrusions of sounds such that the desired ones cannot be understood by the listener.

The physics of sound provide the appropriate background for the difficult task of assessing human response to noise. As sound waves travel over increasing distances, their energy diminishes proportionally, being spread over an ever-increasing area. Once the source ceases to be in motion, the movement of the air particles ceases and the sound waves usually disappear almost instantaneously.

Sound may be described scientifically in terms of three variables associated with the characteristics of waves. These are its amplitude (loudness), its frequency (pitch), and its duration (time). Sound intensity is the average rate of sound energy transmitted through a unit area. Frequency is the number of compressions and rarefactions of air molecules in a unit of time associated with a sound wave. The temporal nature of sound relates to the duration of its generation and presence. The variables of sound make sound measurement a complex problem.

Noise is frequently differentiated into ongoing and impulsive noise, to evaluate its effect on public health and welfare. Ongoing noise is further differentiated into steady-state, fluctuating, and intermittent noise.

Rating schemes for environmental community noise

The description of community noise must account for:

1. Those parameters of noise that have been shown to contribute to the effects of noise on man.
2. The variety of noises found in the environment.
3. The variations in noise levels that occur as a person moves through the environment.
4. The variations associated with the time of day.

Over the years, considerable effort has been expended to develop scales that reduce the dimensions of sound and perception into a one-number scheme. Much effort has been focused on combining measures of frequency content and overall level into a quantity proportional to the magnitude of sound as heard by a person. An example of this type of rating scheme is embodied in the sound level meter, although other rating schemes are reviewed as well. Others have described noise by a statistical approach that takes time into account. This is done by giving the complete curve depicting the cumulative distribution of sound levels. Finally, schemes designed to assess the effects of the constant high-level noise intrusion or the intermittent single-event noise intrusion are also reviewed. It is found that to date one measure of noise that appears to be emerging as one of the most important measures of environmental noise in terms of the effects of noise on man is the "energy mean noise level," L_{eq}, which by definition is the level of the steady state continuous noise having the same energy as the actual time-varying noise.

Annoyance and community response

Numerous techniques have been devised to measure annoyance, from a simple scale of annoyance level to complicated techniques involving social surveys. Laboratory studies of individual response to noise have helped isolate a number of the factors contributing to annoyance, such as the intensity level and spectral characteristics of the noise, duration, the presence of impulses, pitch, information content, and the degree of interference with activity.

Social surveys have revealed several factors related to the level of community annoyance. Some of these factors include:

1. Fear associated with activities of noise sources such as fear of crashes in the case of aircraft noise.
2. Socioeconomic status and educational level.
3. The extent to which community residents believe that they are being treated fairly.
4. Attitude of the community's residents regarding the contribution of the activities associated with the noise source to the general well-being of the community.
5. The extent to which residents of the community believe that the noise source could be controlled.

The highly convergent trend of the various investigations of annoyance and community response leads to the following conclusions:

• The degree of annoyance due to noise exposure expressed by the population average for a community is highly correlated to the magnitude of noise exposure in the community.

• Variations in individual annoyance or response, relative to the community average, are related to individual susceptibilities to noise; and these are highly correlated with definable personal attitudes about noise.

• The number of complaints about noise registered with the authorities is small compared to the number of people annoyed, or who wish to complain. However, the number of actual complaints is highly correlated with the proportion of people in the community who express high annoyance.

• The high correlation between those noise rating methods that account for the physical properties of noise exposure over a day's time suggests that the simplest acoustical measure that accounts for sound magnitude, frequency distribution, and temporal characteristics of sound over 24 hours is an adequate measure for noise exposure in communities.

Normal auditory function

Normal hearing is regarded as the ability to detect sounds in the audio-frequency range (16 Hz to 20 kHz) according to established standards or norms. This range varies little in human populations around the world. However, there is considerable individual variation in hearing ability. As a general rule, for example, women in industrial countries typically have better hearing than men.

In the normal auditory mechanism, sound is transmitted to the inner portion of the ear when sound vibrations imported to the eardrum are transported across the middle ear.

The stapedius and tensor tympani muscles, when contracting, increase the tension of the conductive mechanism and thereby reduce the amount of sound energy delivered to the inner ear. Since high intensity sound causes these contractions, the ear has a limited built-in protective device. However, there is enough of a lag between sound onset and muscle contraction, that a sudden impulse is not attenuated by the protective mechanism.

Hearing sensitivity normally diminishes with age, a condition known as presbycusis. Consequently, corrections for aging should be considered in examining data on hearing loss due to noise exposure.

Noise-induced hearing loss
Temporary and permanent shifts
In auditory threshold following
noise exposure

Ongoing noise has been proven to cause permanent hearing loss in industrial settings and among young people exposed to loud music over extended periods of time. Noise is also known to cause temporary hearing loss and ringing in the ears (tinnitus).

However, since there is a relative lack of information about the effect of shorter term intermittent or incomplete daily exposures, several theories have been postulated to relate noise exposure to hearing loss in these situations.

One theory that has been fairly widely used is the Equal-Energy Hypothesis, which postulates that hearing damage is determined by the total sound energy entering the ear on a daily basis.

Another theory suggests that the long term hazard is predicted by the average temporary threshold shift produced by daily noise exposures. There is evidence to support both of these theories within reasonable limits of extrapolation.

Impulsive noise (such as gunshots) has also been shown to cause damage. CHABA has recently developed a noise hazard numerical weighting system that takes into account such factors as intensity, duration, and number of noise impulses.

Averaging the NIPTS predictions over various industrial noise hazard prediction methods gives a fairly dependable measure of the hearing risk of noise exposed populations. Hearing damage has been noted as levels as low as 75 dBA after 10 years.

The only important factor in increasing hearing risk appears to be noise exposure, and artificial ear protection devices do appear to be of value in preventing damage. Neither sex-related nor cultural differences appear to significantly affect hearing risk due to noise-exposure.

It is evident from the noise exposure data that noise can damage hearing and can cause both NITTS and NIPTS. The relationship between noise exposure and hearing loss is well understood in industrial settings and in the case of high intensity impulsive sound (i.e. gunshots). However, in the case of fluctuating or intermittent noise, data is generally lacking and it is necessary to rely on data extrapolations to estimate effects.

Masking and speech interference

Speech interference is one aspect

of "masking"—an interaction of two acoustic stimuli whereby one of them changes the perceived quality of the other, shifts its apparent location or loudness, or makes it completely inaudible. Much information is available concerning the masking or fairly simple signals such as pure tones, noise bands and nonsense syllables by noises of various spectra; and general laws have been developed that will allow rather accurate prediction of whether or not a speech sound will be masked by a particular noise.

In describing speech interference, the noise concerned can be defined either in terms of its specific spectrum and level or in terms of any number of summarizing schemes. In addition to the average A-weighted sound level, the two most generally-used alternative methods of characterizing noises in respect to their speech-masking abilities are the articulation index (AI) and the speech interference level (SIL). The AI takes into account the fact that certain frequencies in the masking noise are more effective in masking than other frequencies. The SIL is more simplified, indicating only the average general masking capability of the noise. Since much speech is spoken at a reasonably constant level, it is possible to express many of the empirical facts about average speech communication in a single graph showing noise level, vocal effort, and distance.

Various factors enter into the degree of speech interference. Speech, age, and hearing of individuals affect communications. Children have less precise speech than adults do. Older listeners are more susceptible to interference from background noise.

Situational factors influence the degree of speech interference. In some contexts, the predictability of the message will decrease speech interference. Nonverbal communication and lipreading have the same effect. Spatial variables may facilitate or impede speech communication in noise. The exact characteristics of noise are important in predicting speech communication.

Effects of noise on autonomic nervous system functions and other systems

Noise can elicit many different physiological responses. However, no clear evidence exists indicating that the continued activation of these responses leads to irreversible changes and permanent health effects. Sound of sufficient intensity can cause pain to the auditory systems. Except for those persons with poorly designed hearing aids, such intense exposures should not normally be encountered in the non-occupational environment. Noise can also effect the equilibrium of man, but the scarce data available indicates that the intensities required must be quite high or similar to the intensities that produce pain.

Noise-induced orienting reflexes serve to locate the source of a sudden sound and, in combination with the startle reflex, prepare the individual to take appropriate action in the event danger is present. Apart from possibly increasing the chance of an accident in some situations, there are no clear indications that the effects are harmful since these effects are of short duration and do not cause long time body changes.

Noise can interfere with sleep; however, the problem of relating noise exposure level to quality of sleep is difficult. Even noise of a very moderate level can change patterns of sleep, but the determination of the significance of these changes is still an open question.

Noise exposure may cause fatigue, irritability, or insomnia in some individuals, but the quantitative evidence in this regard is unclear. No firm relationships between noise and these factors can be established at this time.

Noise exposure can be presumed to cause general stress by itself or in conjunction with other stressors. Neither the relationship between noise exposure and stress nor the threshold noise level or duration at which stress may appear has been resolved.

Noise exposure to moderate intensities likely to be found in the environment affects the cardiovascular system in various ways; however, no definite permanent effects on the circulatory system have been demonstrated. Noise of moderate intensities has been found to cause vasoconstriction of the peripheral areas of the body and pupillary dilation. Although several hypotheses exist, there is no evidence at this time that these reactions to noisy environments can lead to harmful consequences over a period of time. Speculations that noise might be a contributory factor to circulatory difficulties and heart diseases are not yet supported by scientific data.

Performance and work efficiency

Continuous noise levels above 90 dBA appear to have potentially detrimental effects on human performance, especially on what have been described as noise-sensitive tasks such as vigilance tasks, information gathering and analytical processes. Effects of noise on more routine tasks appear to be much less important, although cumulative degrading effects have been demonstrated by researchers. Noise levels of less than 90 dBA can be disruptive, especially if they have predominantly high-frequency components, are intermittent, unexpected, or uncontrollable. The amount of disruption is highly dependent on:

- The type of task
- The state of the human organism
- The state of morale and motivation

Noise does not usually influence the overall rate of work, but high levels of noise may increase the variability of the work rate. There may be "noise pauses" or gaps in response, sometimes followed by compensating increases in work rate. Noise is more likely to reduce the accuracy of work than to reduce the total quantity of work. Complex or demanding tasks are more likely to be adversely effected than are simple tasks. Since laboratory studies represent idealized situations there is a pressing need for field studies in real-life conditions.

961

Interaction of noise and other conditions or influences

Determination of how various agents or conditions interact with noise in producing a given effect requires three separate experiments measuring the effect produced by the noise alone, the effect produced by the other agent alone, and the effect produced by the joint action of the agent and the noise. These results indicate whether the joint effect is indifferent, additive, synergistic, or ameliorative.

Chemical agents may have a joint effect with noise. Ototoxic drugs that are known to be damaging to the hearing mechanism can be assumed to produce at least an additive effect on hearing when combined with noise exposure. There are instances in which individuals using medication temporarily suffer a hearing loss when exposed to noise, but there is no definitive data on the interactive effects of ototoxic drugs and noise on humans. Evidence linking exposure to noise plus industrial chemicals with hearing loss is also inconclusive.

The possibility of a reciprocally synergistic effect exists when noise and vibration occur together. Vibration is usually more potent than noise in effecting physiological parameters. There appears to be consensus that vibration increases the effect of noise on hearing.

Health conditions may interact with noise to produce a hearing loss. Mineral and vitamin deficiencies are one example but little research has been done on the effect of such deficiencies on susceptibility to noise. Another reasonable hypothesis is that illness increases an individual's susceptibility to the adverse effects of noise. However, as with the other hypotheses, conclusive evidence is lacking.

Infrasound and ultrasound

Frequencies below 16 Hz are referred to as infrasonic frequencies. Sources of infrasonic frequencies include earthquakes, winds, thunder, and jet aircraft. Man-made infrasound occurs at higher intensity levels than those found in nature. Complaints associated with infrasound resemble mild stress reactions and bizarre auditory sensations, such as pulsating and fluttering. It does not appear, however, that exposure to infrasound, at intensities below 130 dB SPL, present a serious health hazard.

Ultrasonic frequencies are those above 20,000 Hz. They are produced by a variety of industrial equipment and jet engines. The effects of exposure to high intensity ultrasound (above 105 dB SPL) also resemble those observed during stress. However, there are experimental difficulties in assessing the effects of ultrasound:

1. Ultrasonic waves are highly absorbed by air.

2. Ultrasonic waves are often accompanied by broad-band noise and by sub-harmonics. At levels below 105 dB SPL there have been no observed adverse effects.

Effects of noise on wildlife and other animals

Noise produces the same general types of effects on animals as it does on humans, namely: auditory, masking of communication, behavioral, and physiological.

As previously mentioned, the most observable effects of noise on farm and wild animals seem to be behavioral. Clearly, noise of sufficient intensity or noise of adverse character can disrupt normal patterns of animal existence. Exploratory behavior can be curtailed, avoidance behavior can limit access to food and shelter, and breeding habits can be disrupted. Hearing loss and the masking of auditory signals, as mentioned before, can further complicate an animal's efforts to recognize its young, detect and locate prey, and evade predators. Competition for food and space in an "ecological niche" results in complex interrelationships and, hence, a complex balance.

Many laboratory studies have indicated temporary and permanent noise induced threshold shifts. However, damage-risk criteria for various species have not yet been developed. Masking of auditory signals has been demonstrated by commercial jamming signals, which are amplitude and frequency modulated.

Physiological effects of noise exposure, such as changes in blood pressure and chemistry, hormonal balance, and reproductivity, have been demonstrated in laboratory animals and, to some extent, in farm animals.

But these effects are understandably difficult to assess in wildlife. Also, the amount of physiological and behavioral adaptation that occurs in response to noise stimuli is as yet unknown.

Considerable research needs to be accomplished before more definitive criteria can be developed. The basic needs are:

1. More thorough investigations to determine the point at which various species incur hearing loss.

2. Studies to determine the effects on animals of low-level, chronic noise exposures.

3. Comprehensive studies on the effects of noise on animals in their natural habitats. Such variables as the extent of aversive reactions, physiological changes, and predator-prey relationships should be examined.

Until more information exists, judgments of environmental impact must be made on existing information, however incomplete.

Effect of noise on structures

The three general types of effects of noise on material are: sonic boom effects, noise induced vibration, and sonic fatigue. These are secondary effects of noise on the health and welfare of man. Sound can also excite buildings to vibrate, which can cause direct effects on man.

The effects caused by sonic booms are the most significant from an environmental standpoint. Sonic booms of sufficient intensity not only can break windows, but can damage buildings structures as well. Nevertheless, as with noise in general, the intensity of sonic booms can be controlled to levels that are completely innocuous with respect to material or structures.

Noise induced vibration can cause noticeable effects on community windows near large rocket launch sites. Construction may also cause such effects, but such relationships are poorly defined at this time.

Sonic fatigue is a very real problem where material is used near intense sound sources. However, such considerations are normally the responsibility of a design engineer and do not cause environmental problems.

Building vibrations excited by impulse noise such as sonic booms or from low-frequency noise from aircraft or rockets can result in human reactions such as startle, discomfort or interference with some tasks. These effects occur primarily in the infrasound range and point toward the close relationship between sound and vibration. Criteria for human exposure to vibration are available but not discussed in this report.

3. Levels of Environmental Noise Requisite To Protect Public Health

"Information on Levels of Environmental Noise . . ." is the Levels Document published by the Environmental Protection Agency, Office of Noise Abatement and Control, as directed by the Noise Control Act of 1972. It appeared in March 1974. The document, says EPA, "is based on analyses, extrapolations and evaluations of the present state of scientific knowledge." Its contents do not constitute Agency regulations or standards. It "will provide state and local governments as well as the Federal Government and the private sector with an informational point of departure for the purpose of decision-making." The following excerpts from the introduction and other early portions were made by the Bureau of National Affairs, Inc., Washington, D.C. 20037, and are used with permission.

In order to provide adequately for the federal emission control requirement and to assure federal assistance and guidance to the state and localities, the Congress has established two separate but related requirements with regard to scientific information about health and welfare effects of noise. First, the Environmental Protection Agency was called upon to publish descriptive data on the effect of noise which might be expected from various levels and exposure situations. Secondly, the Agency is required to publish "information" as to the levels of noise "requisite to protect the public health and welfare with an adequate margin of safety."

Summary

The first requirement was completed . . . when the document "Public Health and Welfare Criteria for Noise" was published. The present document represents the second step.

After a great deal of analysis and deliberation, levels were identified to protect public health and welfare for a large number of situations. They are summarized in Table 1 according to the public health and welfare effect to be protected against, the requisite sound level, and the areas which are appropriate for such protection.

In order to identify these levels, a number of considerations and hypotheses were necessary, which are listed here:

1. In order to describe the effects of environmental noise in a simple, uniform and appropriate way, the best descriptors are the long-term equivalent A-weighted sound level (L_{eq}) and a variation with a nighttime weighting, the day-night sound level (L_{dn}).

2. To protect against hearing impairment:
   a. The human ear, when damaged by noise, is typically affected first at the 4000 Hz frequency.
   b. Changes in hearing level of less than 5 dB are generally not considered noticeable or significant.
   c. One cannot be damaged by sounds considered normally audible, which one cannot hear.
   d. Protecting the population up to a critical percentile (ranked according to decreasing ability to hear) will also protect those above that percentile (in view of consideration 2c above), thereby protecting virtually the entire population.

3. To correct for intermittency and duration in identifying the appropriate level to protect against hearing loss:
   a. The "equal energy hypothesis"
   b. The TTS [temporary threshold shift] hypothesis

4. To identify levels requisite to protect against activity interference:
   a. Annoyance due to noise, as measured by community surveys, is the consequence of activity interference.
   b. Of the various kinds of activity interference, speech interference is the one that is most readily quantifiable.

Explanation of Table 1

1. Briefly, $L_{eq(8)}$ represents the sound energy averaged over an 8 hour period while $L_{eq(24)}$ energy averages over a 24 hour period. L_{dn} represents the L_{eq} with a 10 dB nighttime weighting.

2. The hearing loss identified here represents annual averages of the daily level over a period of 40 years. (These are energy averages, not to be confused with arithmetic averages.)

3. Relationship of an $L_{eq(24)}$ of 70 dB to higher exposure levels.

EPA has determined that for purposes of hearing conservation alone, a level which is protective of that segment of the population at or below the 96th percentile will protect virtually the entire population. This level has been calculated to be an L_{eq} of 70 dB over a 24 hour day.

Given this quantity, it is possible to calculate levels which, when aver-

TABLE 1

SUMMARY OF NOISE LEVELS IDENTIFIED AS REQUISITE TO PROTECT PUBLIC HEALTH AND WELFARE WITH AN ADEQUATE MARGIN OF SAFETY

Effect	Level	Area
Hearing loss	$L_{eq(24)} \leq 70$ dB	All areas
Outdoor activity interference and annoyance	$L_{dn} \leq 55$ dB	Outdoors in residential areas and farms and other outdoor areas where people spend widely varying amounts of time and other places in which quiet is a basis for use.
	$L_{eq(24)} \leq 55$ dB	Outdoor areas where people spend limited amounts of time, such as school yards, playgrounds, etc.
Indoor activity interference and annoyance	$L_{dn} \leq 45$ dB	Indoor residential areas
	$L_{eq(24)} \leq 45$ dB	Other indoor areas with human activities such as schools, etc.

aged over given durations shorter than 24 hours, result in equivalent amounts of energy. For example, the energy contained in an 8 hour exposure to 75 dB is equivalent to the energy contained in a 24 hour exposure to 70 dB. For practical purposes, the former exposure is only equivalent to the latter when the average level of the remaining 16 hours per day is negligible (i.e., no more than about 60 dB* for this case).

An $L_{eq(8)}$ of 75 is considered an appropriate level for this particular duration because 8 hours is the typical daily work period. In addition, the 24 hour exposure level was derived from data on 8 hour daily exposures over a 40 year working life. In planning community noise abatement activities, local governments should bear in mind the special need of those residents who experience levels higher than $L_{eq(8)}$ at 70 on their jobs.

These levels are not to be construed as standards as they do not take into account cost or feasibility. Nor should they be thought of as discrete numbers, since they are described in terms of energy equiva-

---

* This is not to imply that 60 dB is a negligible exposure level in terms of health and welfare considerations, but rather that levels of 60 dB make a negligible contribution to the energy average of $L_{eq} = 70$ dB when an 8 hour exposure of 75 dB is included.

lents. As specified in this document, it is EPA's judgment that the maintenance of levels of environmental noise at or below those specified above, is requisite to protect the public from adverse health and welfare effects. Thus, as an individual moves from a relatively quiet home, through the transportation cycle, to a somewhat noisier occupational situation, and then back home again, his hearing will not be impaired if the daily equivalent of sound energy in his environment is no more than 70 dB. Likewise, undue interference with activity and annoyance will not occur if outdoor levels are maintained at an energy equivalent of 55 dB and indoor levels at 45 dB. However, it is always assumed throughout that environmental levels will fluctuate even though the identified energy equivalent is not exceeded. Likewise, human exposure to noise will vary during the day even though the daily "dose" may correspond well to the identified levels.

Environmental noise exposure

A complete physical description of a sound must describe its magnitude, its frequency spectrum, and the variations of both of these parameters in time. However, one must choose between the ultimate refinement in measurement techniques and a practical approach that is no more complicated than necessary to predict the impact of noise on people. The Environmental Protection Agency's choice for the measurement of environmental noise is based on the following considerations:

1. The measure should be applicable to the evaluation of pervasive long-term noise in various defined areas and under various conditions over long periods of time.

2. The measure should correlate well with known effects of the noise environment on the individual and the public.

3. The measure should be simple, practical and accurate. In principle, it should be useful for planning as well as for enforcement or monitoring purposes.

4. The required measurement equipment, with standardized characteristics, should be commercially available.

5. The measure should be closely related to existing methods currently in use.

6. The single measure of noise at a given location should be predictable, within an acceptable tolerance, from knowledge of the physical events producing the noise.

7. The measure should lend itself to small, simple monitors which can be left unattended in public areas for long periods of time.

These considerations, when coupled with the physical attributes of sound that influence human response, lead EPA to the conclusion that the magnitude of sound is of most importance insofar as cumulative noise effects are concerned. Long-term average sound level, henceforth referred to as equivalent sound level (L_{eq}), is considered the best measure for the magnitude of environmental noise to fulfill the above seven requirements.

Equivalent A-weighted sound level is the constant sound level that, in a given situation and time period, conveys the same sound energy as the actual time-varying A-weighted sound. The basic unit of equivalent sound levels is the decibel . . . and the symbol for equivalent sound level is L_{eq}. Two sounds, one of which contains twice as much energy but lasts only half as long as the other, would be characterized by the same equivalent sound level; so would a sound with four times the energy lasting one fourth as long. The relation is often called the equal-energy rule.

The following caution is called to the attention of those who may prescribe levels: It should be noted that the use of equivalent sound level in measuring environmental noise will not directly exclude the existence of very high noise levels of short duration. For example, an equivalent sound level of 60 dB over a 24 hour day would permit sound levels of 110 dB but would limit them to less than one second duration in the 24 hour period. Comparable relationships between maximum sound levels and their permissible durations can easily be obtained for any combination relative to any equivalent sound level.

In determining the daily measure of environmental noise, it is important to account for the difference in response of people in residential areas to noises that occur during sleeping hours as compared to waking hours. During nighttime, exterior background noises generally drop in level from daytime values. Further, the activity of most households decreases at night, lowering the internally generated noise levels. Thus, noise events become more intrusive at night, since the increase in noise levels of the event over background noise is greater than it is during the daytime.

Methods for accounting for these differences between daytime and nighttime exposures have been developed in a number of different noise assessment methods employed around the world. In general, the method used is to characterize nighttime noise as more severe than corresponding daytime events; that is, to apply a weighting factor to noise that increases the numbers commensurate with their severity. Two approaches to identifying time periods have been employed: one divides the 24 hour day into two periods, the waking and sleeping hours, while the other divides the 24 hours into three periods—day, evening, and night. The weighting applied to the non-daytime periods differs slightly among the different countries, but most of them weight nighttime activities by about 10 dB. The evening weighting, if used, is 5 dB.

An examination of the numerical values obtained by using two periods versus three periods per day shows that for any reasonable distribution of environmental noise levels, the two-period day and the three-period day are essentially identical; *i.e.*, the 24-hour equivalent sound levels are equal within a few tenths of a decibel. Therefore, the simpler two-period day is used in this document,

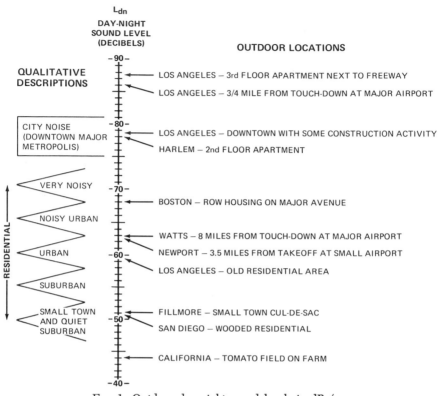

FIG. 1.—Outdoor day-night sound levels in dB (re 20 micropascals) at various locations.

with daytime extending from 7 A.M. to 10 P.M. and nighttime extending from 10 P.M. to 7 A.M. The symbol for the 15 hour daytime equivalent sound level is L_d; the symbol for the 9 hour nighttime equivalent sound level is L_n; and the day-night weighted measure is symbolized at L_{dn}.

The L_{dn} is defined as the A-weighted average sound level in decibels (re 20 micropascals) during a 24 hour period with a 10 dB weight-ing applied to nighttime sound levels. Examples of the outdoor present day (1973) day-night noise level at typical locations are given in Figure 1.

L_{eq} for the 24 hour average sound level to which an individual is exposed ($L_{eq(24)}$): This situation is related to the cumulative noise exposure experienced by an individual irrespective of where, or under what situation, this exposure is received. The long-term health and welfare effects of noise on an individual are related to the cumulative noise exposure he receives over a lifetime.

Relatively little is known concerning the total effect of such lifetime exposures, but dose-effect relations have been studied for two selected situations:

- The average long-term exposure to noise primarily in residential areas leading to annoyance reactions and complaints.

- The long-term effects of occupational noise on hearing, with the daily exposure dose based on an eight-hour work day.

An ideal approach to identifying environmental noise levels in terms of their effect on public health and welfare would be to start by identifying the maximum noise not to be exceeded by individuals. However, the noise dose that an individual receives is a function of lifestyle. For example, exposure patterns of office workers, factory workers, housewives, and school children are quite different. Within each group, the exposures will vary widely as a function of the working, recreational, and sleeping patterns of the individual. Thus, two individuals working in the same office will probably accumulate different total noise doses if they use different modes of transportation, live in different areas, and have different TV habits. However, detailed statistical information on the distribution of actual noise doses and the relationship of these doses to long-term health and welfare effects is still missing. Therefore, a realistic approach to this problem is to identify appropriate noise levels for places occupied by people as a function of the activity in which they are engaged, including a gross estimate of typical average exposure times.

From a practical viewpoint, it is necessary to utilize the wealth of data relating to occupational noise exposure, even though some of it must be interpreted in order to arrive at extrapolations upon which the identification of safe levels of daily (24 hour) exposures can be used.

Use of identified environmental noise levels

One of the purposes of this document is to provide a basis for judgment by states and local governments as a basis for setting standards. In doing so the information contained in this document must be utilized along with other relevant factors. These factors include the balance between costs and benefits associated with setting standards at particular noise levels, the nature of the existing or projected noise problems in any particular area, the local aspirations and the means available to control environmental noise.

In order to bring these factors together, states, local governments and the public will need to evaluate in a systematic manner the following:

1. The magnitude of existing or projected noise environments in defined areas as compared with the various levels identified in this document.

2. The community expectations for noise abatement with respect to existing or projected conditions.

3. The affected elements of the public and the degree of impact of present or projected environmental noise levels.

4. The noise sources, not controlled by federal regulations, that cause local noise problems.

5. Methods available to attack environmental noise problems (use limitations, source control through noise emission standards, compatible land use planning, etc.).

6. The costs inherent in reducing noise to certain levels and benefits achieved by doing so.

7. The availability of technology to achieve the desired noise reduction.

The levels of environmental noise identified in this report provide the basis of assessing the effectiveness of any noise abatement program. These noise levels are identified irrespective of the nature of any individual noise source. One of the primary purposes of identifying environmental noise levels is to provide a basis by which noise source emission regulations, human exposure standards, land use planning, zoning, and building codes may be assessed, as to the degree with which they protect the public health and welfare with respect to noise. Such regulatory action must consider technical feasibility and economic reasonableness, the scale of time over which results can be expected, and the specific problems of enforcement. In the process of balancing these conflicting elements, the public health and welfare consequence of any specific decision can be determined by comparing the resultant noise environment against the environmental noise levels identified in this report.

4. Community Noise Criteria

The following is taken from *Control of Community Noise*, published by American Foundrymen's Society, Des Plaines, Ill. 60016.

The Noise Control Act of 1972 (see Appendix B-1) requires that the Administrator of the Environmental Protection Agency (EPA) develop and publish criteria with respect to noise. These criteria are to reflect the scientific knowledge most useful in indicating the kind and extent of all identifiable effects of noise on the public health and welfare which may be expected from differing quantities and qualities of noise.

The criteria are descriptions of cause-and-effect relationships. Standards and regulations must take into account not only the health and welfare considerations described in the criteria, but also, as called for in the Noise Control Act of 1972, technology and cost of control.

The term "health and welfare," as used in the Noise Control Act, includes the physical and mental well-being of the human population. The term also includes other indirect effects, such as annoyance, interference with communication, loss of value and utility of property, and effects on other living things.

Noise ratings must be capable of easy measurement and yet correlate well with subjective response. For most noises, the decibel level read on the A-scale of a sound level meter (dBA) is adequate. For aircraft noise, a more complex rating scale must be used which includes consideration of the duration of the noise and the presence of a pure tone (engine whine). For still other noises (industrial and construction), some consideration must be given to the total daily exposure time of the noise, in addition to its dBA rating.

Factors

The acceptability of a noise in a given community depends on many factors. The range of possible response corresponds to a range in noise level of about ±10 dB. One reason for this large variation in response is that the annoyance produced in an individual is related most strongly to his attitudes and feelings regarding the noise source rather than the level of the noise itself.

The factors which strongly influence individual response can be illustrated by answering the following questions:

• Is the source of noise a necessary or valuable asset to the community?

• Is the noise necessary in view of the operations being performed?

• Are those in authority controlling the noise as much as possible?

• Does the noise interfere with social activities (such as a back yard barbecue)?

• Is the noise damaging to health?

• Does the source of noise create other environmental problems?

It is difficult to guess the answers to such questions in the minds of community residents. Community surveys by independent investigators can provide the information, but usually at considerable expense.

Description

An adequate description of community noise must account for the amplitude, frequency, duration, and variations in noise levels associated with the time of day at any given location. Descriptions of community noise should take into account the time and location variations in the noise environment throughout the community so that the descriptions are relevant to the effects of environmental noise on people, whether they are located indoors or outdoors. This chapter will describe some schemes that have been developed over the years to illustrate the techniques and problems involved, so as to facilitate an understanding of their application and use.

The ambient levels in the community are the levels produced by the composite of all noise sources present, including the facility under evaluation. If the latter does not increase the ambient levels significantly over those which would exist in the absence of the facility as a noise source, there is little likelihood of adverse community reaction.

Noise parameters

A noise parameter should indicate some description of the noise at a particular time and at a particular location. For example, the reading of a sound level meter as well as calculated loudness levels can be considered noise parameters.

Octave- and one-third octave-band levels are commonly used noise parameters. The sound pressure level can be filtered by standard filters with bandwidths one octave wide. One-third octave bands (subdividing the octave into three parts) are used for a more detailed analysis of noise. Since the sound at a particular time at a specific location is divided into frequency bands, this parameter gives a level for each frequency band.

In most instances the noise problem is twofold. It involves either constant high noise levels or the intermittent single-event noise intrusion. With the considerable increase in flyovers of jet aircraft, the latter type of problem has grown considerably over the years. Studies of jet aircraft noise have contributed significant data and insight on community annoyance and have stimulated the development of indices for assessing the cumulative effect of intrusive noises.

Noise ratings

A noise rating index averages noise effects over a longer period of time and/or larger area. A noise index attempts to present averages and, in general, is valid only for the particular situations for which it was developed. It gives methods for quantifying situations so that a data base can be constructed.

Because a variety of ratings are already in use, a reasonable familiarity with the application and the ratings is required to identify the correct one for a specific situation.

Rosenblith-Stevens Model

Rosenblith and Stevens developed, in the early 1950's a model for relating the probable community reaction to intrusive aircraft noise. The model included seven factors that were corrected for:

1. Magnitude of the noise.
2. Duration of the intruding noise.
3. Time of the year (winter/summer; windows opened or closed).
4. Time of day (night/day).
5. Outdoor noise level when the intruding noise is not present.
6. History of prior exposure to the noise.
7. Frequency components in the noise or its impulsive nature.

Noise criteria (NC) curves

In 1956, Beranek developed a set of noise criteria curves, a family of curves of sound pressure level versus octave-band frequencies that are shaped somewhat like equal loudness contours. When octave-band levels of indoor background noise at a work site are plotted on the NC curves, an NC value is assigned to the noise spectrum corresponding to the highest NC curve crossed. The rating has been applied primarily to background noise in buildings and it has been widely accepted for noise control in buildings.

Perceived noise level

Kryter, in the late 1950's, developed a scale of perceived intensity called the perceived noise level. Its units are in decibels and are often popularly referred to as PNdB. This scale represents the sound pressure level of an octave band of noise at 1000 Hz that would be judged equally noisy to the sound to be rated. "Equally noisy" means that in a comparison of sound, one would just as soon have one noise as the other at his home during the day or night. The observer was asked to compare noises on the basis of their acceptability or their "noisiness." The resulting judgments were found to be similar to those for loudness, but enough difference was noticed to give a somewhat different rating for various sounds. On the basis of these results, Kryter has set up a calculation procedure for "perceived noise level." Given octave-band data, the composite PNL can be computed.

Effective perceived noise level (EPNL)

Experiments conducted by Kryter and Pearson during the 1960's indicated that annoyance was influenced by the duration of an intruding sound and by the presence of strong pure tone components within an otherwise broadband signal. The latter situation is particularly noticeable under the approach path of aircraft equipped with turbojet engines.

The experiments indicated corrections to PNL based upon the duration and pure tone content of the flyover. The duration correction is basically an energy integration of the perceived noise for the time the level is within 10 dB of the peak. Thus EPNL is no longer a measure of a particular time and position but is a rating of a particular flyover.

This quantity is somewhat more exact than the A-weighting in relating man's perception of sound to the physical parameters of sound, particularly in the case of aircraft noise. For this reason, it has become a major element in the procedures utilized by the Federal Aviation Administration for the certification of aircraft noise. For most sounds, the effective perceived noise level exceeds the A-weighted noise level by 13 dB, depending primarily upon the amount of correction for pure tones.

The effective perceived noise level scale requires complex analysis and instrumentation to define a sound. Thus, it has not been utilized extensively, particularly since in most instances the simple A-weighted sound level appears to adequately describe environmental noise at a location at a given time, and does not require particularly complex instrumentation.

Noise and number index

Another rating for aircraft noise is called NNI, the noise and number index. It is based on perceived noise levels. It was developed to take into account the annoyance effect of the number of aircraft flyovers per day.

Two factors are used to compute NNI: (a) the peak noise value in PNdB of each aircraft flyover averaged over the number of aircraft, and (b) 15 log N where N is the number of aircraft heard in the specified period. Thus the rating incorporates the peak from an event and the number of events. The duration of a specific event is not used.

It is defined by the following relation. NNI = (Average peak perceived noise level) $+ 15(\log_{10}N) - 80$, where N is the number of aircraft per day or night. The value "80" is subtracted to bring the index to about 0 for conditions of no annoyance.

The "average peak perceived noise level" is obtained in the following way. The peak perceived noise level that occurs during the passage of each airplane is noted. These peak levels are then converted into equivalent power and averaged. This average value is then converted back into a level and used in the equation.

If the perceived noise level is approximated by the use of A-weighted sound levels, the 80 is reduced to about 67. This gives NNI = (Average peak A-level) $+ 15(\log_{10}N)$ − 67.

Noise exposure forecast (NEF)

The NEF accounts for both duration and pure tone content of each single event. The NEF rating is a technique in which noise contours are calculated for an airport to yield forecasts of the noise environment.

In the computation, the noise exposure at a point is estimated as the sound-pressure-squared level of the noise produced by an aircraft flying on a prescribed flight path. Nighttime operations are weighted as contributing approximately 17 times more to the overall noise exposure than daytime operations. The total NEF at a given ground point is determined by summing all of the individual contributions on an energy basis. Noise exposure forecasts are often used to forecast the effects of changes in airport operations or the addition of new runways.

Noise pollution level (NPL)

This measure is derived from two terms, one involving the average sound level (L_{eq}) of the noise and one involving the magnitude of the time variation of the noise level. The

L_{np} concept embodies some simple principles:

1. Other things being equal, the higher the noise level, the more the disturbance.
2. Other things being equal, the less steady the noise level, the greater its annoying quality.

Single event noise exposure level (SENEL)

The SENEL rating has been incorporated into California regulations to control noise around airports. Its purpose is to quantify the disturbance of a flyover into a single number related to total annoyance of the event so that limits can be established for individual flyovers. The SENEL rating squares and integrates the A-weighted sound level to obtain the energy equivalent of a single flyover as a single number. The rating is a direct parallel to EPNL without tone corrections.

Community noise equivalent level

California introduced the "community noise equivalent level" (CNEL). This rating represents the average noise level determined for a 24 hour period, with different weighting factors for noise levels occurring during the day, evening, and night period. Essentially, it is an L_{eq} for a 24 hour period with special corrections of 5 and 10 dB, respectively, for evening and nighttime. It is designed to account for the increased disturbance caused by noise events during the evening and the night.

The CNEL rating is the companion to SENEL for community noise. The energy average of A-weighted sound level is established for each hour during the day. These hourly averages are summed with weighting factors based upon the time of day.

Criteria for evaluation

The acceptability of a particular noise in a given community should be evaluated according to its effect on the ambient level and if this noise exceeds the level established by local noise regulations or ordinances. Specific noise requirements that are applicable to the area surrounding or adjoining the plant or activity being evaluated should be first in priority of consideration. The noise produced must be evaluated in the terms stated by the ordinance. Some older city ordinances, for example, specify noise level limits in octave bands, necessitating this type of analysis. Typical ranges of municipal ordinances in U.S. cities are: for residential zones, 43–56 dBA; for commercial/retail zones, 53–65 dBA; and for industrial zones, 58–70 dBA. These values are applicable to continuous steady broadband noise occurring in the daytime. Some ordinances require nighttime levels as much as 10 dB lower than specified daytime levels; this is the case mainly where the permissible daytime levels are relatively high.

The ambient noise level will not be significantly increased if the particular noise in question has a level at least 5 dB lower than the preexisting or background noise level due to all other sources.

Community noise surveys

Community noise surveys may nor-mally be made with the procedures and instrumentation described in this Manual. The use of an A-weighted sound level is recommended because this facilitates detailed exploration of the spatial and temporal characteristics of the noise. Contour techniques are often useful for defining "noise impact" areas in the community.

The complexity of a community survey program depends upon the noise environment itself and the purpose of the program. The latter may include (a) investigation of complaints, (b) site evaluation for new construction, and (c) current assessment of community noise. In general, enough measurement points in the community should be taken to obtain a consistent delineation of the noise pattern. At least some points should be selected well away from road traffic, since traffic noise tends to dominate in the vicinity of a roadway and sometimes produces misleading results.

Complaints

Complaints regarding noise do arise, in spite of efforts to maintain a satisfactory community noise environment. Although motives for complaints vary, with rare exceptions the complainant is disturbed by the noise and has some expectation that corrective action will be taken. If complaints are received, the prevailing noise conditions should be assessed. Every effort should be made to determine the precise nature of the annoyance, enlisting the cooperation of the complainant as much as possible. Often the real object of concern is a particular aspect of the noise that otherwise would be overlooked in the course of measurement and evaluation. If the source of annoyance can be removed, this should of course be done. In any case, all possible means should be employed to maintain good communications with the complainant.

The following specific suggestions for avoiding or dealing with community complaints are offered:

1. Control dust, odors, and other irritants, as well as noise as much as possible.
2. Deal frankly and constructively with all complaints.
3. Maintain a record of all complaints, investigate them fully, and record all actions taken.
4. Maintain good community relationships in general.
5. Make no promises regarding any type of corrective action unless it is certain that they can be kept.

Every effort should be made to establish and maintain good relationships with the community. Assuming that efforts are being made to control excessive noise in the community, these should be publicized, but care should be taken not to lose credibility by exaggeration in statements or promises.

5. Noise Pollution Control Regulations of the State of Illinois

Typical state regulations governing the emission of sound from property-line noise sources are given in this appendix. Officially, they are Chapter 8 of the Rules and Regulations of the Illinois Pollution Control Board, effective August 9, 1973.

Following these regulations are the "Measurement Techniques for Enforcement of Noise Pollution Control Regulations" used to determine compliance.

Chapter 8: Noise Regulations
Part 1—General Provisions

RULE 101: DEFINITIONS

Except as hereinafter stated and unless a different meaning of a term is clear from its context, the definitions of terms used in this chapter shall be the same as those used in the environmental protection act.

All definitions of acoustical termi-nology shall be in conformance with those contained in ANSI S1.1–1960, "Acoustical Terminology."

(a) **ANSI:** American National Standards Institute or its successor bodies.

(b) **Construction:** On-site erection, fabrication, installation, alteration, demolition or removal of any structure, facility, or addition thereto, including all related activities, including, but not restricted to, clearing of land, earthmoving, blasting and landscaping.

(c) **Daytime hours:** 7:00 A.M. to 10:00 P.M., local time.

(d) **dB(A):** Sound level in decibels determined by the A-weighting of a sound level meter.

(e) **Decibel (dB):** A unit of measure, on a logarithmic scale to the base 10, of the ratio of the magnitude of a particular sound pressure to a standard reference pressure, which, for purposes of this Chapter, shall be 20 micronewtons per square meter (μN/m²).

(f) **Existing property-line-noise-source:** Any property-line-noise-source, the construction or establishment of which commenced prior to the effective date of this Chapter. For the purposes of this sub-section, any property-line-noise source whose A, B or C land use classification changes, on or after the effective date of this Chapter, shall not be considered an existing property-line-noise-source.

(g) **Impulsive sound:** Either a single pressure peak or a single burst (multiple pressure peaks) for a duration less than one second.

(h) **New property-line-noise-source:** Any property-line-noise-source, the establishment of which commenced on or after the effective date of this Chapter.

(i) **Nighttime hours:** 10:00 P.M. to 7:00 A.M., local time.

(i) **Noise pollution:** The emission of sound that unreasonably interferes with the enjoyment of life or with any lawful business or activity.

(k) **Octave band sound pressure level:** The sound pressure level for the sound being measured contained within the specified octave band. The reference pressure is 20 micronewtons per square meter.

(l) **Person:** Any individual, corporation, partnership, firm, association, trust, estate, public or private institution, group, agency, political subdivision of this State, any other State or political subdivision or agency thereof or any legal successor, representative, agent or agency of the foregoing.

(m) **Preferred frequencies:** Those frequencies in hertz preferred for acoustical measurements which, for the purposes of this Chapter, consist of the following set of values: 20, 25, 31.5, 40, 50, 63, 80, 100, 125, 160, 200, 250, 315, 400, 500, 630, 800, 1000, 1250, 1600, 2000, 2500, 3150, 4000, 5000, 6300, 8000, 10,000, 12,500.

(n) **Prominent discrete tone:** Sound, having a one-third octave band sound pressure level which, when measured in a one-third octave band at the preferred frequencies, exceeds the arthmetic average of the sound pressure levels of the two adjacent one-third octave bands on either side of such one-third octave band by:

(a) 5 dB for such one-third octave band with a center frequency from 500 hertz to 10,000 hertz, inclusive. Provided: such one-third octave-band sound pressure level exceeds the sound pressure level of each adjacent one-third octave band, or;

(b) 8 dB for such one-third octave band with a center frequency from 160 hertz to 400 hertz, inclusive. Provided: such one-third octave-band sound pressure level exceeds the sound pressure level of each adjacent one-third octave band, or;

(c) 15 dB for such one-third octave band with a center frequency from 25 hertz to 125 hertz, inclusive. Provided: such one-third octave-band sound pressure level exceeds the sound pressure level of each adjacent one-third octave band.

(o) **Property-line-noise-source:** Any equipment or facility, or combination thereof, which operates within any land used as specified by Rule 201 of this Chapter. Such equipment or facility, or combination thereof, must be capable of emitting sound beyond the property line of the land on which operated.

(p) **SLUCM:** The "Standard Land Use Coding Manual" (1969, United States Government Printing Office) which designates land activities by means of numerical codes.

(q) **Sound:** An oscillation in pressure in air.

(r) **Sound level:** In decibels, a weighted sound pressure level, determined by the use of metering characteristics and frequency weightings specified in ANSI S1.4–1971, "Specification for Sound Level Meters."

(s) **Sound-pressure level:** In decibels, 20 times the logarithm to the base 10 of the ratio of the magnitude of a particular sound pressure to the standard reference pressure. The standard reference pressure is 20 micronewtons per square meter.

(t) **Unregulated safety relief valve:** A safety relief valve used and designed to be actuated by high pressure in the pipe or vessel to which it is connected and which is used and designed to prevent explosion or other hazardous reaction from pressure buildup, rather than being used and designed as a process pressure blowdown.

RULE 102: PROHIBITION OF NOISE POLLUTION

No person shall cause or allow the emission of sound beyond the boundaries of his property so as to cause noise pollution in Illinois, or so as to violate any provision of this Chapter or the Illinois Environmental Protection Act.

RULE 103: MEASUREMENT TECHNIQUES

Test procedures to determine whether emission of sound is in conformance with this Chapter shall be in substantial conformity with Standards and Recommended Practices established by the American National Standards Institute, Inc. (ANSI) and the Society of Automotive Engineers, Inc. (SAE), and the latest revisions thereof, including ANSI S1.1–1960, ANSI S1.6–1967, ANSI S1.8–1969, ANSI S1.2–1962, ANSI S1.4–1971—Type 1 Precision, ANSI S1.11–1966, ANSI S1.13–1971 Field Method, SAE J–184.

The Agency may adopt procedures which set forth criteria for the measurement of sound. Such procedures shall be revised from time to time to reflect current engineering judgment and advances in noise measurement techniques. Such procedures, and the revisions thereto, shall not become effective until filed with the Index Division of the Office of the Secretary of State as required by "An Act concerning administrative rules," approved June 14, 1951, as amended.

RULE 104: BURDEN OF PERSUASION REGARDING EXCEPTIONS

In any proceeding pursuant to this Chapter, if an exception stated in this Chapter would limit an obligation, limit a liability, or eliminate either an obligation or a liability, the person who would benefit from the application of the exception shall have the burden of persuasion that the exception applies and that the terms of the exception have been met. The

Agency shall cooperate with and assist persons in determining the application of the provisions of this Chapter.

RULE 105: SEVERABILITY

If any provision of these rules or regulations is adjudged invalid, or if the application thereof to any person or in any circumstance is adjudged invalid, such invalidity shall not affect the validity of this Chapter as a whole or of any part, sub-part, sentence or clause thereof not adjudged invalid.

Part 2—Sound Emission Standards and Limitations
For Property-Line-Noise-Sources

All terms defined in Part 1 of this Chapter which appear in Part 2 of this Chapter have the same definitions specified by Rule 101 of Part 1 of this Chapter.

RULE 201: CLASSIFICATION OF LAND ACCORDING TO USE

(a) **Class A Land**
Class A land shall include all land used as specified by SLUCM Codes 110 through 190 inclusive, 651, 674, 681 through 683 inclusive, 691, 711, 762, 7121, 7122, 7123 and 921.

(b) **Class B Land**
Class B land shall include all land used as specified by SLUCM Codes 397, 471 through 479 inclusive, 511 through 599 inclusive, 611 through 649 inclusive, 652 through 673 inclusive, 675, 692, 699, 7124, 7129, 719, 721, 722 except 7223 used for automobile and motorcycle racing, 723 through 761 inclusive except 7311 used for automobile and motorcycle racing, 769 through 790 inclusive, and 922.

(c) **Class C Land**
Class C land shall include all land used as specified by SLUCM Codes 211 through 299 inclusive, 311 through 396 inclusive, 399, 411 except 4111, 412 except 4121, 421, 422, 429, 441, 449, 460, 481 through 499 inclusive, 7223 and 7311 used for automobile and motorcycle racing, and 811 through 890 inclusive.

(d) A parcel or tract of land used as specified by SLUCM Code 81, 83, 91 or 922, when adjacent to Class B or C land may be classified similarly by action of a municipal government having zoning jurisdiction over such land. Notwithstanding any subsequent changes in actual land use, land so classified shall retain such B or C classification until the municipal

TABLE 1

Octave Band Center Frequency (hertz)	Allowable Octave-Band Sound Pressure Levels (dB) of Sound Emitted to any Receiving Class A Land from		
	Class C Land	Class B Land	Class A Land
31.5	75	72	72
63	74	71	71
125	69	65	65
250	64	57	57
500	58	51	51
1000	52	45	45
2000	47	39	39
4000	43	34	34
8000	40	32	32

government removes the classification adopted by it.

RULE 202: SOUND EMITTED TO CLASS A LAND DURING DAYTIME HOURS

Except as elsewhere in this Part 2 provided, no person shall cause or allow the emission of sound during daytime hours from any property-line-noise-source located on any Class A, B or C land to any receiving Class A land which exceeds any allowable octave band sound pressure level specified in Table 1, when measured at any point within such receiving Class A land, provided, however, that no measurement of sound pressure levels shall be made less than 25 feet from such property-line-noise-source.

RULE 203: SOUND EMITTED TO CLASS A LAND DURING NIGHTTIME HOURS

Except as elsewhere in this Part 2 provided, no person shall cause or allow the emission of sound during nighttime hours from any property-line-noise-source located on any Class A, B or C land to any receiving Class A land which exceeds any allowable octave band sound pressure level specified in Table 2, when measured at any point within such receiving Class A land, provided, however, that no measurement of sound pressure levels shall be made less than 25 feet from such property-line-noise-source.

RULE 204: SOUND EMITTED TO CLASS B LAND

Except as elsewhere in this Part 2 provided, no person shall cause or allow the emission of sound from any property-line-noise-source located on any Class A, B or C land to any receiving Class B land which exceeds any allowable octave band sound

TABLE 2

Octave Band Center Frequency (hertz)	Allowable Octave-Band Sound Pressure Levels (dB) of Sound Emitted to any Receiving Class A Land from		
	Class C Land	Class B Land	Class A Land
31.5	69	63	63
63	67	61	61
125	62	55	55
250	54	47	47
500	47	40	40
1000	41	35	35
2000	36	30	30
4000	32	25	25
8000	32	25	25

TABLE 3

Octave Band Center Frequency (hertz)	Allowable Octave Band Sound Pressure Levels (dB) of Sound Emitted to any Receiving Class B Land from		
	Class C Land	Class B Land	Class A Land
31.5	80	79	72
63	79	78	71
125	74	72	65
250	69	64	57
500	63	58	51
1000	57	52	45
2000	52	46	39
4000	48	41	34
8000	45	39	32

pressure level specified in Table 3, when measured at any point within such receiving Class B land, provided, however, that no measurement of sound pressure levels shall be made less than 25 feet from such property-line-noise-source.

RULE 205: SOUND EMITTED TO CLASS C LAND

Except as elsewhere in this Part 2 provided, no person shall cause or allow the emission of sound from any property-line-noise-source located on any Class A, B or C land to any receiving Class C land which exceeds any allowable octave band sound pressure level specified in Table 4, when measured at any point within such receiving Class C land, provided, however, that no measurement of sound pressure levels shall be made less than 25 feet from such property-line-noise-source.

RULE 206: IMPULSIVE SOUND

No person shall cause or allow the emission of impulsive sound from any property-line-noise-source located on any Class A, B or C land to any receiving Class A, B or C land which exceeds the allowable dB(A) sound level specified in Table 5, when measured at any point within such receiving Class A, B or C land, provided, however, that no measurement of sound levels shall be made less than 25 feet from the property-line-noise-source.

TABLE 4

Octave Band Center Frequency (hertz)	Allowable Octave Band Sound Pressure Levels (dB) of Sound Emitted to any Receiving Class C Land from	
	Class C Land	Class B Land and Class A Land
31.5	88	79
63	83	78
125	78	72
250	73	64
500	67	58
1000	60	52
2000	54	46
4000	50	41
8000	47	39

TABLE 5

Classification of Land on which Property-Line-Noise-Source is Located	Allowable dB(A) Sound Levels of Impulsive Sound Emitted to Designated Classes of Receiving Land			
	Class C Land	Class B Land	Class A Land	
			Daytime	Nighttime
Class A Land	57	50	50	45
Class B Land	57	57	50	45
Class C Land	65	61	56	46

RULE 207: PROMINENT DISCRETE TONES

(a) No person shall cause or allow the emisison of any prominent discrete tone from any property-line-noise-source located on any Class A, B or C land to any receiving Class A, B or C land, provided, however, that no measurement of one-third octave band sound pressure levels shall be made less than 25 feet from such property-line-noise-source.

(b) This rule shall not apply to prominent discrete tones having a one-third octave band sound pressure level 10 or more dB below the allowable octave band sound pressure level specified in the applicable table in Rules 202 through 205 for the octave band which contains such one-third octave band. In the application of this sub-section, the applicable table for sound emitted from any existing property-line-noise-source to receiving Class A land, for both daytime and nighttime operations shall be Table 1 (Rule 202).

RULE 208: EXCEPTIONS

(a) Rules 202 through 207 inclusive shall not apply to sound emitted from land used as specified by SLUCM Codes 110, 140, 190, 691, 7311 except as used for automobile and motorcycle racing, and 742 except 7424 and 7425.

(b) Rules 202 through 207 inclusive shall not apply to sound emitted from emergency warning devices and unregulated safety relief values.

(c) Rules 202 through 207 inclusive shall not apply to sound emitted from lawn care maintenance equipment and agricultural field ma-chinery used during daytime hours. For the purposes of this sub-section, grain dryers operated off the farm shall not be considered agricultural field machinery.

(d) Rules 202 through 207 inclusive shall not apply to sound emitted from equipment being used for construction.

(e) Rule 203 shall not apply to sound emitted from existing property-line-noise-sources during nighttime hours, provided, however, that sound emitted from such existing property-line-noise-sources shall be governed during nighttime hours by the limits specified in Rule 202.

RULE 209: COMPLIANCE DATES FOR PART 2

(a) Except as provided in Rules 209(f), 209(g), 209(i), and 209(j), every owner or operator of a new property-line-noise-source shall comply with the standards and limitations of Part 2 of this Chapter on and after the effective date of this Chapter.

(b) Except as otherwise provided in this Rule 209, every owner or operator of an existing property-line-noise-source shall comply with the standards and limitations of Part 2 of this Chapter on and after twelve months from the effective date of this Chapter.

(c) Every owner or operator of an existing property-line-noise-source who emits sound which exceeds any allowable octave band sound pressure level of Rules 202, 203, 204 or 205 by 10 dB or more in any octave band with a center frequency of 31.5 hertz, 63 hertz or 125 hertz shall comply with the standards and limitations of Part 2 of this Chapter on and after eighteen months from the effective date of this Chapter.

(d) Except as provided in Rules 209(f), 209(g) and 209(h), every owner or operator of an existing property-line-noise-source required to comply with Rule 206 of this Chapter shall comply with the standards and limitations of Part 2 of this Chapter on and after eighteen months from the effective date of this Chapter.

(e) Every owner or operator of an existing property-line-noise-source required to comply with Rule 207 of this Chapter shall comply with the standards and limitations of Part 2 of this Chapter on and after eighteen months from the effective date of this Chapter.

(f) Every owner or operator of Class C land now or hereafter used as specified by SLUCM Codes 852 and 854 shall have three years from the effective date of this Chapter to bring the sound from necessary explosive blasting activities in compliance with Rule 206, provided that such blasting activities are conducted between 8:00 A.M. and 5:00 P.M. local time, at specified hours previously announced to the local public.

(g) Every owner or operator of Class C land now and hereafter used as specified by SLUCM Code 4112 shall have three years from the effective date of this Chapter to bring the sound from railroad car coupling in compliance with Rule 206.

(h) Every owner or operator of Class C land on which forging operations are now conducted shall have three years from the effective date of this Chapter to bring sound from the impact of forging hammers into full compliance with the limits specified in Rule 206 for emissions to any receiving land.

(i) Every owner or operator of Class C land now and hereafter used as specified by SLUCM Code 291 shall comply with the standards and limitations of Part 2 of this Chapter on and after two years from the effective date of this Chapter.

(j) Every owner or operator of Class C land now and hereafter used as specified by SLUCM Code 7223 and 7311 when used for automobile and motorcycle racing shall comply with the standards and limitations of Part 2 of this Chapter on and after two years from the effective date of this Chapter.

"Measurement Techniques for Enforcement of Noise Pollution Control Regulations"

1.0 General

1.1 This Report, pursuant to Rule 103, Chapter 8, of the Noise Pollution Control Regulations, establishes the measurement techniques to be used by the Agency in determining compliance with the rules governing the emission of sound from property-line-noise-sources. It is recommended that persons making noise measurements employ the measurement techniques used by the Agency. The procedures and instrumentation specified herein do not establish limits on sound.

1.2 The procedures and instrumentation specified herein provide the methodology necessary to establish compliance with the rules governing the emission of sound from property-line-noise-sources.

1.3 ANSI Standards 1.4–1971, 1.11–1966 and 1.13–1971 listed in Rule 103, Chapter 8 of the Noise Regulations refer to specific sound measurement techniques and data gathering equipment. The sections of ANSI 1.4–1971, 1.11–1966 and 1.13–1971 applicable to this report are those dealing only with outdoor sound measurements (field method). No indoor measurements will be recorded for the purposes of determining compliance with Part 2, Chapter 8.

1.4 The remaining ANSI Standards referred to in Chapter 8, Rule 103, contain definitions, terminology or other acoustic data not directly related to measurement techniques but, nevertheles, of technical interest to those persons who may perform sound level measurement.

2.0 Personnel qualifications

2.1 Personnel conducting sound measurements shall have been trained and experienced in the current techniques and principles of sound measurement and in the selection and operation of sound measuring instrumentation.

3.0 Instrumentation

3.1 A sound level meter and octave band and one-third octave band filter set shall be used for the acquisition of data. These instruments shall conform with the following standards or their later revisions:

(a) American National Standards Institute (ANSI) S1.4–1971 Specification for Sound Level Meters,

Type 1 precision sound level meter.

(b) American National Standards Institute (ANSI) S1.11–1966 Specifications for Octave, One-Half Octave and One-Third Octave Band Filter Sets.

(c) American National Standards Institute (ANSI) S1.6–1967 Preferred Frequencies and Band Numbers for Acoustical Measurements.

(d) American National Standards Institute (ANSI) S1.8–1969 Preferred Reference Quantities for Acoustical Levels.

If a magnetic tape recorder or a graphic level recorder or other indicating device is used, the system shall meet the requirements of:

(a) Society of Automotive Engineers (SAE) Recommended Practice J184 Qualifying a Sound Data Acquisition System.

3.2 An anemometer and compass or other suitable devices shall be used to measure wind speed and direction in accordance with the manufacturer's recommended limits.

3.3 A thermometer suitable for measurement of ambient temperature shall be used in accordance with the manufacturer's recommended limits.

3.4 An hygrometer suitable for the measurement of relative humidity shall be used in accordance with the manufacturer's recommended limits.

3.5 A barometer suitable for the measurement of barometric pressure shall be used in accordance with the manufacturer's recommended limits.

4.0 Instrumentation Setup

4.1 Instruments shall be set up to conform to ANSI S1.13–1971 Methods for the Measurement of

Sound Pressure Level (Field Method), with the following additions:

(a) Connect microphone to the sound level meter with a manufacturer's specified (10 foot minimum) extension cable.

(b) Attach an appropriate wind screen to the microphone.

5.0 Data Acquisition and Operation

5.1 Following manufacturer's instructions and with the instrumentation set up as described in 4.0, the following steps will be taken to acquire and record the data:

(a) Check condition of power supply prior to recording of data.

(b Calibrate the instrumentation set up with an acoustic calibrator prior to recording data.

(c) Set meter to "fast" response and "A-weighting" network, or connect to external filter whenever octave or one-third octave analysis is being made. (Note: "Slow" response may be used if necesasry to stabilize the meter needle. If slow response is used it shall be so noted in the data record.)

(d) Record all pertinent atmospheric conditions, i.e., wind speed and direction, temperature, relative humidity, barometric pressure, and general weather conditions.

(e) Record instruments used (manufacturer, model and kit number).

(f) Record location of sound source of interest and sound measuring microphone locations relative to the sound source, and any unusual microphone positions.

(g) Calibrate the instrumentation setup with an acoustic calibrator following the recording of data.

(h) Check the power supply following the recording of data.

6.0 General Conditions

6.1 While measurements are being recorded, constant visual surveillance of extraneous sound sources should be made to insure that the measurements are of the sound being investigated.

6.2 If operator or instrument positions other than those specified in ANSI S1.13–1971 (Field Method) are employed, the reason and description shall be described in the recording of data.

7.0 Impulsive Sound

7.1 Impulsive sound shall be measured on the "A-weighted" scale with the meter set to "fast" response.

Appendix C Intersociety Committee Guidelines for Noise Control

This is the Intersociety Committee Report, revised 1970, that originally appeared in the *Journal of Occupational Medicine*, Vol. 12, pp. 276–281 (July 1970).

In the fall of 1967, a paper entitled *Guidelines for Noise Exposure Control* was published in various scientific journals.[1] This paper was the result of the deliberations of an ad hoc committee, composed of two members from each of five technical groups, who were assigned the responsibility of developing a reliable noise criteria for use in noise control and hearing conservation programs for industrial workers. The concluding statement in their paper was: "This document will be reviewed at intervals not exceeding three years and reaffirmed or revised as indicated by the current state of knowledge."

In the fall of 1968, a second ad hoc committee was formed to review the existing Guidelines as related to new scientific data which had been developed since the initial publication. The members of the second committee who revised the original Guidelines, and the organizations they represent, follow:

American Conference of
Governmental Industrial Hygienists
Herbert H. Jones
Floyd A. Van Atta, Ph.D.

American Academy of
Ophthalmology and Otolaryngology
Aram Glorig, M.D.
Meyer S. Fox, M.D.

American Academy of
Occupational Medicine
Edwin DeJongh, M.D.
Capt. N. E. Rosenwinkel, M.C., USN

American Industrial Hygiene
Association
Paul L. Michael, Ph.D.
James H. Botsford

Industrial Medical Association
Joseph Sataloff, M.D.
William L. Baughn, M.D.

It is believed that the following Guidelines will aid industrial management and official agency groups in establishing a common base for the development of noise control measures and an effective hearing conservation program.

Jack C. Radcliffe, Chairman
Second Intersociety Committee on
Guidelines for Noise Exposure Control

I. Foreword

Noise has long been recognized as one of the several causes of hearing loss. Exposure to high noise levels may cause temporary or permanent changes in hearing threshold level. Permanent hearing loss which impairs communication by speech is a handicap or impairment. Competent medical specialists have defined impairment as average hearing threshold level in excess of 15 dB, ASA-Z24.5 (1951), at 500, 1,000, and 2,000 Hz.[2,3] This definition is accepted for this document.

Noise-induced hearing loss increases with both the intensity of the noise and the duration of exposure. Generally, many years of exposure to high noise levels are required to produce significant permanent impairment in the exposed group; however, there will be marked differences in the hearing of individuals and in their response to noise. A portion or all of a hearing impairment may be due to causes other than noise exposure. These guidelines will be directed toward the prevention of that portion of the permanent hearing loss resulting from exposure to steady noise whether continuous or intermittent.

II. Objective

To provide practical guidelines for evaluating the hazard from noise exposures and for minimizing the development or aggravation of permanent hearing impairment resulting from prolonged exposure.

III. Occupational Hearing Loss Control Programs

The following procedures are necessary to accomplish the objective.

A. Evaluation of the Noise Hazard

The hazard to hearing produced by a given noise exposure depends on the intensity and frequency distribution of the noise and on the duration of the exposure. Each of these factors must be considered in determining which exposures are hazardous.

1. Noise Measurement

Continuous or intermittent steady noise is readily measured by standard instruments; impulsive noise measurement requires special procedures not considered here.[4,5]

(a) Instruments

All sound level meters and octave band analyzers used in noise determinations should meet the pertinent specifications of the American National Standards Institute.[6,7]

A critical level may be selected, by methods presented below, for initiation of noise exposure controls. If the A-scale reading is well above or well below the selected critical level for noise control, the exposure rating based on this measure may be used without validation. When A-scale readings lie within 3 or 4 decibels of such a level, the exposure rating should be verified by octave band methods.

(b) Survey Methods.

Surveys should be conducted by competent persons according to accepted practices.[4,8]

(1) Measurement:

Noise may be measured using either the A-scale of the sound level meter or octave band analysis. Levels measured by the sound level meter should be designated dBA. Where the octave band analysis is made, the equivalent dBA value for hazard rating may be determined from the curves of Fig. 1.[*] Noise levels measured at a wide variety of industrial operations have been pub-

lished,[9] and these may serve as useful guides until noise surveys can be made.

(2) Duration and Time Pattern:

The total time of the noise exposure and the distribution of exposure periods throughout the working day should be determined or estimated.

2. Hazard Rating.

(a) Continuous Exposure.

Several criteria for acceptable noise exposure have been proposed. The differences between these criteria result primarily from different definitions of acceptable incidence of hearing impairment. Table I and Fig. 2[**] have been prepared from a number of controlled studies which correlate noise exposure with incidence of hearing impairment. The first column of Table I indicates the steady noise levels to which the various groups were exposed in terms of the A-scale reading.

[*] FIG. 1.–Equivalent sound level contours. Octave band sound pressure levels may be converted to the equivalent A-weighted sound level by plotting them on this graph and noting the A-weighted sound level corresponding to the point of highest penetration into the sound level contours. This equivalent A-weighted sound level, which may differ from the actual A-weighted sound level of the noise, is used to determine exposure limits. (This illustration is identical to that in Appendix A-1.)

[**] FIG. 2.–Incidence of hearing impairment (average hearing threshold level in excess of 15 dB ASA at 500, 1000, and 2000 cps[1,2]) in the general population and in selected populations by age groups and by occupational noise exposure. (This illustration is identical to Figure 12–5 in Chapter 12.)

TABLE 1

INCIDENCE OF HEARING IMPAIRMENT IN THE GENERAL POPULATION AND IN SELECTED POPULATIONS BY AGE GROUPS AND OCCUPATIONAL NOISE EXPOSURE

A-weighted Sound Level (Continuous Noise) dB	Percentage of Population Having Impaired Hearing[a] By age groups				Source of Information[b]
	20–29	30–39	40–49	50–59	
Non-Noise	3	5	10	20	18
General Population	2	5	14	24	19
80	1	3	6	19	22
85	3	7	13	27	22
90	6	14	21	37	22
92[c]	3	9	15	28	19
95	10	22	32	47	22
96[b]	3	10	19	—	18
97	7	22	32	40	20
100	16	36	46	60	22
102	10	18	30	45	18
104	5	21	35	57	21
105	24	50	62	75	22

a Average hearing threshold level in excess of 15 dB at 500, 1,000, and 2,000 Hz.
b Refers to list of references at the end of this article. The population in each study group was as follows: (18) 2,282 non-noise; 1,834 exposed to 96 dBA; 666 exposed to 102 dBA; (19) 20,459 general population; 9,653 exposed at 92 dBA; (20) 400; (21) 174; (22) 6,585.
c Level estimated from octave band data.

The remaining columns of Table I show, for various age groups, the percentage of the groups having impaired hearing. The first line of the table shows the incidence of hearing impairment in a population having no exposure to injurious noise and no other explanation for observed hearing impairments. It is presumed that at least this minimal incidence of impairment will be found in any population and that the other groups may be regarded as exhibiting injurious effects of noise only if they show significantly higher rates of incidence. The information contained in Table I is presented graphically in Fig. 2.

The upper curve in Fig. 2 indicates that of 100 persons exposed to 85 dBA, about 26 will have impaired hearing when they reach the age group of 50 to 59. This compares to about 22 persons out of 100 with no occupational noise exposure. This is an increase of four persons per 100 population for the noise exposed group, or four percentage points. Because of the wide scatter of the data, so small a difference between groups cannot be attributed to differences in noise exposure with much certainty and, therefore, is not considered to be real or significant in the statistical sense.

In the population exposed to 90 dBA to age 50–59, the amount of impairment is increased 10 percentage points (ten more persons per 100 exposed) as compared to the population with no occupational exposure. The difference is probably greater than the limits of precision of the data.

With exposure to 95 dBA until age 50 to 59, the number of persons with impairment is increased by 22 percentage points (22 more persons per 100 working lives) 56 per cent of persons in the sixth decade will have no hearing impairment and 22 per cent would have some impairment from presbycusis alone irrespective of occupational noise exposure. Therefore, 78 per cent will probably be little affected by a lifetime occupational exposure to 95 dBA.

(b) Intermittent and Part-time Exposure.

The studies on which Table I and Fig. 2 are based dealt with men exposed to noise during a normal workday of eight hours duration. There are few long-term studies of the extent to which the risk of permanent hearing impairment may be reduced by shortening the daily duration of exposure, or by interrupting the exposure periodically.[10] Guidance comes mainly from studies based on temporary threshold shifts (TTS) resulting from various types of noise exposure. Results of TTS studies are summarized in Table II which may be used to estimate the effect of intermittency of noise exposure on risk of hearing impairment.

The information in Table II may be approximated by the simple rule that for each halving of daily exposure time the noise level may be increased by 5 dB without increasing the hazard of hearing impairment, as illustrated in Table III. The entries in Table II and Table III vary, for instance, the single occurrence at 15 minutes, because the 5 dB steps in Table III include an allowance for the number of occurrences ordinarily found in high-level noise.

(c) Maximum Recommended Exposures.

In view of the above ob-

TABLE 2
ACCEPTABLE EXPOSURES TO NOISE IN dBA
AS A FUNCTION OF THE NUMBER OF OCCURRENCES PER DAY

Daily Duration Hours	Min	Number of Times the Noise Occurs Per Day						
		1	3	7	15	35	75	160 up
8		90	90	90	90	90	90	90
6		91	93	96	98	97	95	94
4		92	95	99	102	104	102	100
2		95	99	102	106	109	114	
1		98	103	107	110	115		
	30	101	106	110	115			
	15	105	110	115				
	8	109	115					
	4	113						

To use the table, select the column headed by the number of times the noise occurs per day, read down to the average sound level of the noise, and locate directly to the left in the first column the total duration of noise permitted for any 24-hour period. It is permissible to interpolate if necessary. Noise levels are in dBA.

servations, the Committee considers a reasonable objective for hearing conservation is an environmental level of 90 dBA for steady-state noise with permissible increase of 5 dBA for each halving of the exposure time up to 115 dBA as shown in Table III.

When the daily noise exposure is composed of two or more periods of noise exposure of different levels, their combined effect should be considered, rather than the individual effect of each. If the sum of the following fractions:

$$\frac{C_1}{T_1} + \frac{C_2}{T_2} + \cdots \frac{C_n}{T_n}$$

exceeds unity, then, the mixed

exposure should be considered to exceed the threshold limit value. C_1 indicates the total time of exposure at a specified noise level, and T_1 indicates the total time of exposure permitted at that level.

TABLE 3
MAXIMUM RECOMMENDED EXPOSURE TO NOISE IN dBA FOR EIGHT HOURS AND LESS

Daily Exposure Time (Hours)	Sound Level (dBA)
8	90
4	95
2	100
1	105
½	110
¼ or less	115

Noise exposures of less than 90 dBA do not enter into the above calculation.[11]

(d) Limitations.

The methods of exposure rating proposed above apply only to groups, not to individuals. They cannot be used to determine whether an individual has or has not suffered a hearing loss resulting from noise exposure; medical evaluation is required for such a determination. These methods also do not apply to impulsive or impact noises, concerning which there are not sufficient data available for a reasonable assessment. There are some indications that impact noises which peak below 140 dBA sound pressure level are not harmful. Tentative definitions of impact noises and methods of measurement are given in a reference.[12]

B. Exposure Control Methods

When noise exposure exceeds the critical level selected for control, measures should be taken to (1) reduce the environmental noise levels, (2) reduce the duration of exposure, or (3) protect the exposed personnel by the use of ear protector devices.

1. Noise Reduction.

The most desirable exposure control method is to reduce noise to noninjurious levels both for the prevention of hearing loss and for other benefits which accrue. Application of known principles of noise control[4,5] usually can reduce any noise to any desired degree; however, economic considerations and/or operational necessities may make the application impractical. Where practical methods of reducing noise to safe levels have been developed, they should be adopted.

2. Reduction of Exposure Time.

Reduction of exposure time is seldom a practical method of reducing noise hazards in industry. Where the exposure can be limited and interspersed with recovery periods in noninjurious noise, Table II may be used as a guide.

3. Ear Protection.

Where it is not feasible to reduce environmental noise to acceptable levels, ear protectors are commercially available which are capable of reducing noise entering the ears to acceptable levels for most noise exposures encountered in industry.[13] Plugs inserted into the ear canals must form an airtight seal in order to obtain the full noise exclusion of which they are capable. In order to assure satisfactory protection, plugs should be fitted by a competent person. Muffs covering the entire ear are subject to fewer uncertainties of fit and positioning than ear plugs and usually are more dependable.

C. Planning Hazard-free Operations

Every effort should be made at the planning stage to minimize noise exposure. At this stage, a competent acoustical consultant can provide valuable services.

1. Noise Abatement in Engineering Design.

   Engineers and architects should consider potentially hazardous noise exposures in the design of buildings and machines, and in the layout of floor plans. They should incorporate architectural and production features such as isolation of noisy operations, sound-absorbing materials in construction, spacing of noisy machines, and utilize all other available means to minimize noise exposures to personnel.

2. Noise Rating Considerations in Purchasing Equipment.

   Consideration should be given to possible noise exposures when new equipment is ordered or new facilities planned. When hazardous noise exposures are likely to result from use of the equipment, under consideration, noise data should be obtained from suppliers so that realistic estimates of noise exposure can be made. In selecting equipment, noise should be given due consideration.

D. Audiometry

Hearing acuity of persons likely to be exposed to excessive noise should be determined by pure tone audiometry. Audiometry should be conducted under medical supervision according to the conditions and procedures suggested below.

1. Facilities.

   To insure accurate audiograms, the facility must meet the following minimum standards:
   (a) Test Room.

Audiograms should be obtained only in environments which meet the requirements of the American National Standards Institute for background noise.[14]
(b) Audiometer.

Audiometers should meet the specifications of the American National Standards Institute,[15] and should be maintained in calibration in accordance with recognized procedures.

2. Personnel.

Persons obtaining audiograms should be trained in air conduction audiometry either by formal course work at accredited educational institutions or by individual instruction provided by an audiologist or otologist. Such a course has been described in an Intersociety document.[16] Audiometry should be conducted under medical supervision.

3. Audiograms.

Preplacement audiograms should include hearing thresholds for both ears at frequencies of 500, 1000, 2000, 3000, 4000, and 6000 cps. Subsequent audiograms utilizing the same frequencies should be obtained as deemed necessary by the supervising physician. The frequency of follow-up audiograms will generally be related to the type and intensity of the noise exposure. The reference zero of the audiometer used should be noted on each audiogram.

IV. Review

This document will be reviewed at intervals not exceeding three years and reaffirmed or revised as indicated by the current state of knowledge.

References

1. "Guidelines for Noise Exposure Control."
   Am Ind Hygiene Assn J, 28:418–424 (Sept.–Oct.) 1967.
   Arch Environ Health, 15:674–678 (Nov.) 1967.
   J Occup Med, 9:571–575 (Nov.) 1967.
   Am Assn Ind Nurses J. Vol. 16, No. 5, pp 17–21, (May) 1968.
2. "Guides for the Evaluation of Hearing Impairment." *Trans Amer Acad Ophthalmology and Otolaryngology*, pp. 167–168 (March–April) 1959.
3. Committee on Medical Rating of Physical Impairment: "Guides to the Evaluation of Permanent Impairment; Eye, Nose, Throat and Related Structures." *J Amer Med Assoc*, 177:489 (Aug.) 1961.
4. AIHA Noise Committee: *Industrial Noise Manual.* Second Ed. American Industrial Hygiene Association, Detroit, Michigan; 1966.
5. Harris, C. M. (Ed.): *Handbook of Noise Control.* McGraw-Hill Book Co.; 1957.
6. American Standards Association: *Specifications for General Purpose Sound Level Meters SI .4–1961.* American Standards Association: New York, N.Y.; 1961.
7. American Standards Association: *Specification for the Octave Band Filter Set for the Analysis of Noise and Other Sounds Z24.10-1953.* American Standards Association: New York, N.Y.; 1953.
8. American Standards Association: *Method for the Physical Measurement of Sound S1.2-1962.* American Standards Association: New York, N.Y.; 1962.
9. Karplus, H. B., and Bonvallet, G. L.: "A Noise Survey of Manufacturing Industries." *Amer Ind Hyg Assoc Quarterly*, 14:235 (Dec.) 1953.
10. Sataloff, J.; Vassallo, L.; and Menduke, H.: "Hearing Loss From Exposure to Interrupted Noise." *AMA Arch Environ Health* 18:972, 1969.
11. American Conference of Governmental Industrial Hygienists, "Threshold Limit Values of Physical Agents Adopted by ACGIH for 1969."
12. Coles, R. R. R.; Garinther, G. R.; Hodge, D. C.; and Rice, C. G.: "Hazardous Exposure to Impulse Noise." *J Acoust Soc Amer*, 43:336, 1968.
13. Webster, J. C., and Rubin, E. R.: "Noise Attenuation of Ear-Protective Devices." *Sound* 1:34, 1962.
14. American Standards Association: *Criteria for Background Noise in Audiometer Rooms S3.1-1960.* American Standards Association: New York, N.Y.: 1960.
15. American Standards Association: *Specifications for Audiometers for General Diagnostic Purposes Z24.5-1951.* American Standards Association: New York, N.Y.: 1951.
16. Intersociety Committee on Industrial Audiometric Technician Training: "Guide for Training of Industrial Audiometric Technicians," *Am Ind Hyg Assn J*, 27:303–304 (May–June) 1966.
17. Subcommittee on Noise: "Guide for Conservation of Hearing in Noise." A supplement to the *Trans Amer Acad Ophthalmology and Otolaryngology* (Revised

1964). Published by Research Center Subcommittee on Noise: 327 Alvarado Street, Los Angeles, Calif.

18. AIHA Noise Committee: *Industrial Noise Manual.* Second Ed. Chapter 7, American Industrial Hygiene Association: Detroit, Michigan; 1966.

19. Glorig, A., and Nixon, J.: "Distribution of Hearing Loss in Various Populations." *Annals of Otology, Rhinology and Laryngology* 69:497 (June) 1960.

20. Gallo, R., and Glorig, A.: "Permanent Threshold Shift Changes Produced by Noise Exposure and Aging." *Amer Ind Hyg Assoc J*, 25:237, 1964.

21. Glorig, A.; Ward, D.; and Nixon, J.: "Damage Risk Criteria and Noise Induced Hearing Loss." *Arch Otolaryngology*, 74:413 (Oct.) 1961.

22. Unpublished data available to the Committee (1968).

Appendix D

Council for Accreditation in Occupational Hearing Conservation

Legislation to control excessive occupational noise exposures plus the requirement for maintaining audiometric records for those employees exposed to time-weighted average levels of more than 85 dBA (or who are required to wear personal hearing protection on the job) has created a new demand for properly trained industrial audiometric technicians. A uniform method for testing employees' hearing, throughout industry, is essential for accurate, valid, and reliable audiometry.

The proposed "OSHA Rules and Regulations for Noise Exposure"— as published in the October 24 1974 issue of the *Federal Register* (published in this manual as Appendix A-2b)—state that "OSHA believes that the audiometric testing program will detect any changes in hearing level in workers, so that the employer can adopt corrective action and inform employees before the changes become significant. In order for the results of such a testing program to be valid and meaningful, the audiometric environment and technique must be well standardized and stable over a sufficient number of years to represent a significant fraction of the employee's working life. It is also essential that these factors be reasonably identical from one employment to another. For these reasons, mandatory requirements are proposed for audiometric test rooms and the calibration of audiometers."

Certified audiometric technician

In 1972, as an outgrowth of an Intersociety Committee concerned with standards of competency in Industrial Audiometry and Hearing Conservation, the Council for Ac-

Hearing conservation

The proposed requirements and procedures for occupational noise exposure, as published in the October 24, 1974, issue of the *Federal Register* (see Appendix A-2) included the following:

A hearing conservation program shall be established and maintained for employees who:

• Receive a daily dose equal to or exceeding an eight hour, time-weighted exposure of 85 dBA;

• Are required to wear hearing protectors.

The hearing conservation program shall include at least an annual audiometric test for affected employees.

If no baseline audiogram exists, one shall be taken within 90 days for each employee affected.

Each employee's annual audiogram shall be examined to determine if any significant hearing threshold shift in either ear has occurred relative to the baseline audiogram.

If a significant threshold shift is present, the employee shall be retested within one month.

Audiometric tests

Audiometric tests shall be administered by a certified audiometric technician or an individual with equivalent training and experience.

The employer shall keep an accurate record of all employee audiograms.

These records shall be maintained for a period of five years.

Certified audiometric technician

A "certified audiometric technician" is an individual who meets the training requirements specified by the Inter-Society Committee on Audiometric Technician Training, or who is certified by the Council of Accreditation in Occupational Hearing Conservation.

creditation in Occupational Hearing Conservation was formed. Representatives of the following organizations unanimously approved the establishment of an independent and scientifically oriented Board for Industrial Hearing Conservation Technicians—American Association of Industrial Nurses, American Academy of Occupational Medicine, American Council of Otolaryngology, American Industrial Hygiene Association, American Speech and Hearing Association, Industrial Medical Association, and the National Safety Council.

The objective of this board is to set standards and establish training policies and methods for providing industry with technicians who will be able to conduct excellent hearing conservation programs in large and small plants.

A Certified Audiometric Technician will be an individual who meets the training requirements specified by the Intersociety Committee. (See *Guide for Training Industrial Audiometric Technicians*, Appendix D-2.)

The need for quality control for faculty as well as technicians was realized, so the Council for Accreditation in Occupational Hearing Con-

---

Occupational hearing conservation training programs criteria

- Minimum course of study is 20 hours (following Inter-Society Syllabus)

- A certified faculty member—Certified by the Council for Accreditation in Occupational Hearing Conservation, this faculty member is to be responsible for the balance of the staff, who should be experienced in occupational hearing conservation.

- There should be a ratio of one instructor for every six students.

- There should be a ratio of one audiometer for every three students—with ample time for practice on varied audiometers under supervision.

- Successful completion of the preferred study course will include a satisfactory final practical examination.

---

servation certifies both faculty and student. In order for the student to participate in a course eligible for certification, the course must be under the direction of a certified faculty member.

The purpose of the Council for Accreditation is to implement and conduct the program of certification of qualified faculty and students, and in connection therewith to:

- Establish standards for professional faculty in occupational hearing conservation, to certify professionals who meet such standards, and to maintain a roster of holders of Faculty Certificates to be known as the Directory of Certified Faculty for the Council for Accreditation in Occupational Hearing Conservation;

- Establish standards for technicians in occupational hearing conservation, to certify those who meet the standards, and to maintain a roster of

holders of certificates granted to be known as the Directory of Conservationists Certified by the Council for Accreditation in Occupational Hearing Conservation;

- Elevate and maintain the quality of occupational hearing conservation in American industry;

- Stimulate the development of improved standards and programs in the field of occupational hearing conservation.

Standard course of study

Perhaps the primary purpose and justification for offering courses for training industrial audiometric technicians is to standardize the techniques used in industrial audiometry to obtain hearing threshold levels. There is no substitute for direct practical instruction in developing the skills required for the determination of air-conduction, pure tone hearing thresholds.

A consistent technique, patterned upon a tried and proved basic procedure, should be used by industrial audiometric technicians. This will encourage consistent testing techniques from place to place and from one time to another. Inconsistent use of different techniques could lead to discrepant results and cause confusion in the interpretation of the audiometric findings.

A matter of skill and motivation

Persons of "normal" intelligence can learn the skills necessary for conducting the air conduction tests recommended for industrial audiometry. Many companies have utilized an industrial nurse to administer hearing tests, but a medical background is *not* a necessary requirement.

Experience has shown that technical skill alone is also no guarantee that reliable and valid audiograms will be obtained. The person to whom the task of audiometric testing is assigned often treats this aspect of the job as just another addition to an already over-burdened schedule. The technician may attempt novel test procedures to relieve the monotony. Other technicians may fall into poor testing practices without realizing it.

The technical skill required for conducting industrial hearing tests is minimal, but it is required that the technician remain alert and observant during the test.

Peculiarities in the audiogram being obtained or any unusual behavior on the part of the person being tested should signal the possibility of audiometer malfunction, failure on the part of the subject to comprehend the instructions, an uncooperative attitude on the part of the person being tested, or a failure on the part of the technician.

The use of the self-recording, automatic type of audiometer does not relieve the technician of the necessity of remaining alert and observant. The self-recording audiometer greatly reduces the amount of effort required in testing, but it still requires that the technician monitor the recording of the threshold as it is being recorded. For example, irregularities in the threshold record, especially during the first minute of testing, are relatively common and generally indicate that the instructions have not been understood. This problem is easily corrected either by repeating or by elaborating on the instruction, but this can be done only if the technician notices that something is wrong.

Systematic supervision and encouragement of the testing personnel by the physician or other *qualified* person in charge of the program is recommended to maintain the high motivation required for good audiometric testing. The supervision should include periodic review of the testing procedures used by the test technician to make sure that they conform to the established preferred procedures.

The industrial audiometric technician should have a basic knowledge about the anatomy and physiology of the hearing mechanism, elementary physics of sound, basic audiometry, and the application and function of personal hearing protective devices. The technician should also have an understanding of the basic vocabulary and terminology of industrial audiometry.

Under medical supervision, the industrial audiometric technician is responsible for measuring hearing thresholds of employees and obtaining valid, reliable audiograms by using appropriate equipment.

Conducting a valid test of an individual's hearing ability involves much more than the mere turning of dials and knobs. If the industrial audiometric technician fails to instruct the subject properly, to provide the proper test environment, and to understand what is taking place, invalid, inaccurate audiometry can result, because the audiograms are based on the technician's interpretation of the subject's responses to the tones produced by the audiometer.

The industrial audiometric technician should have a fairly good reading ability and a high interest in working with people and rendering services. The technician should have a moderate level of dexterity that will permit the operation of appropriate equipment, and should be able to work quickly, tactfully, and carefully. Technicians should be individuals who adapt to problem situations easily and who can work under close supervision.

Training

The recommended industrial audiometric technician training program consists of both an academic

---

Industrial audiometric technician duties and qualifications

- Be able to obtain a valid pure-tone, air-conduction audiogram, which necessitates instruction and practice in:
  1. Audiometric testing techniques,
  2. Audiometer care and operation,
  3. Calibration techniques,
  4. Identification of a suitable testing environment.
- Be able to identify medically questionable audiograms for the purpose of referral to the proper medical or paramedical services.
- Be able to fit and dispense personal hearing protection devices.
- Be able to present to management information pertaining to the development of an effective program, including:
  1. Cost analysis;
  2. Audiometric monitoring;
  3. Hearing protection.

---

portion and a work experience or practicum portion. Certified instructors in this program should be selected for their knowledge of industrial audiometry or related subjects, and their ability to present the material in an intensive, yet meaningful, manner; they should have a genuine interest in instructing industrial audiometric technicians.

The certified industrial audiometric

technician is an important individual who has successfully completed a training course that is based upon the guidelines recommended by the Intersociety Committee.

Objectives

Audiometric testing or measurement of hearing ability is an important element in an industrial hearing conservation program. In general, it will have two primary objectives:

• To serve as a means of control by checking upon the effectiveness of noise control measures and by identifying those individuals who are hypersusceptible to noise damage. (The amount of hearing loss produced by a given noise exposure varies from person to person, and it is vitally important to discover noise-susceptible individuals.)

• To provide a medical record of the worker's hearing ability—preemployment, periodic rechecks, and at termination. (Legal decisions involving considerable amounts of money may be based upon the audiometric records.)

The success of an effective hearing conservation program is dependent to a great extent upon the validity and reliability of the audiometric data obtained by the technician responsible for the hearing threshold measurement of employees.

For more information about certification and accreditation, including application forms, write to: Secretary, Council for Accreditation in Occupational Hearing Conservation, 1619 Chestnut Ave., Haddon Heights, N.J. 08035.

1. Accreditation as Faculty in Training Programs

There are two methods for achieving accreditation as faculty in training programs for industrial hearing conservationists. One requires a credential review and takes into account the background and experience of the applicant. The second employs a work-shop designed to acquaint the applicant with the content of training courses for occupational hearing conservationists (which will be acceptable to the Council for Accreditation), to show types of instruction that have proven to be successful in other training programs, to acquaint the applicant with national, state, and local legislation pertaining to industrial hearing conservationists, to introduce the applicant to the needs of industry and particularly to the needs of the occupational hearing conservationists in fulfilling their responsibility in industrial hearing conservation programs.

Both methods for accreditation require evidence of formal academic training in one of the disciplines related to hearing conservation in industry. Regardless of the method chosen by the person interested in accreditation as faculty, an application will have to be obtained from the Council for Accreditation in Occupational Hearing Conservation, and completed. The application and the applicant's credentials will be reviewed, and, upon approval of the council, accreditation will be awarded. The fee for accreditation is $30, payable to the Council for Accreditation in Occupational Hearing Conservation. For application, write: Secretary, CAOHC, 1619 Chestnut Ave., Haddon Heights, N.J. 08035.

Requirements

• Related disciplines as defined by the council include: Industrial Medicine; Otolaryngology; Industrial Nursing; Industrial Hygiene; Safety Engineering; Audiology. Evidence of this formal training requires that the applicant complete an educational biography, including information about the location of academic training, when training was taken, the academic major, and the degree granted. In addition, the applicant should have professional certification, or should be eligible for certification, in one or more of the following: Industrial Medicine; Otolaryngology; Industrial Hygiene; Safety Professional; Audiology; Acoustics Engineering; Occupational Health Nursing.

• Evidence of previous experience may be submitted either as of employment in industry in a position related to industrial hearing conservation, or of previous experience as faculty in industrial audiometric training programs.

(a) Applicant must submit evidence of full-time employment for one year in industry, associated with industrial medical or safety programs in a position related to industrial hearing conservation. Hearing conservation programs include activities such as hearing testing, fitting hearing protectors, noise surveys, noise control, otologic physical examinations, collecting auditory historical information, and keeping appropriate records. In addition to describing such employment and responsibilities, the applicant must submit a letter of support obtained from the employer. If experience is part time, such as a consultant for various industries, this experience should include a minimum

of one full day as a consultant for each month of the calendar year (12 full days). As with full-time employment, part-time activity must be in the area of hearing conservation. Applicant must list each employer and responsibilities as a consultant, and letters of support must be submitted. The part-time experience must have been obtained in the two-year period preceding the application.

(b) To qualify with previous experience as faculty, the applicant must have been involved in a minimum of four separate training courses over the past two years. The training courses must have followed the guidelines issued by the Inter-Society Committee for Training of Industrial Audiometric Technicians. Applicant must state where courses were taught, who sponsored courses, the number of students involved, and a description of responsibility in teaching the course.

• Applicants who meet the requirements for formal academic training, but who do not meet the required criteria of experience in industry or as faculty in previous training programs, can elect to attend a work-shop sponsored and approved by the Council for Accreditation. Participation in such work-shops is limited only to those who meet the formal academic training requirements. Upon completion of the work-shop, the person may apply to the council for accreditation as faculty.

2. Guide for Training Industrial Audiometric Technicians

Recommendations of the *Intersociety Committee on Industrial Audiometric Technician Training*, as published in *American Industrial Hygiene Association Journal*, Vol. 27, pp. 303–304 (May–June 1966).

In recognition of the need for reproducible and reliable audiometric testing in industrial programs for hearing conservation, the American Industrial Hygiene Association has joined with the American Speech and Hearing Association, the American Industrial Medical Association, and the American Industrial Nurses Association in the development of this guide for training of industrial audiometric technicians. A joint committee with representation from each of the four associations has promulgated the finished guide after long and careful consideration. This document now has the endorsement of each of the participating associations.

The American Industrial Hygiene Association is publishing this document for the benefit of our members and of the field of industrial hygiene. The use of this guide, where appropriate, is recommended. Requests for assistance or information relative to this guide may be directed to Paul L. Michael, Ph.D., 667 Franklin Street, State College, Pa. 16801, or to the offices of AIHA.

Resolution

The American Speech and Hearing Association Executive Council passed the following resolution:

Whereas, there is a need for hearing tests to be performed in industrial settings; and

Whereas, representatives of the American Industrial Medicine Association, the American Industrial Hygiene Association, the American Industrial Nurses Association and the American Speech and Hearing Association have been devising a syllabus for a course of study for audiometric technicians in industry; and

Whereas, it is desirable that ASHA assume a position of leadership in training audiometric technicians; therefore,

Resolved, that the Executive Council endorse in principle a two and a half day program of training for audiometric technicians in industry as outlined in the draft of *Guide for Training Technicians* in Industry, dated March 10, 1965.

Guide for training

In recognition of the need for a conservation of hearing program that will be practical and economically feasible for the managements of both large and small industries, a practical guide has been outlined for the industrial audiometric technician. This guide is recommended for use in training technicians to:

1. Perform pure tone air conduction audiograms for industry.

2. Implement an adequate hearing protection program.

3. Assist management in planning and carrying out a hearing conservation program in industry under adequate supervision. (Implicit in these recommendations is that one technician not train another.)

First day

Topic I—1 hour

A. Hearing conservation in industry —why?
   1. Importance of hearing
   2. Social aspects
   3. Economic and legal aspects

4. Health maintenance, physical examination

B. Objectives of training program
   1. Valid audiograms
   2. Effective ear protection
   3. Medical follow-up

C. The industrial audiometric technicians' responsibility and limitations in implementing hearing conservation programs

Topic II—1 hour

A. Basic discussion of how the ear functions (Anatomy and Physiology) Encyclopaedia Britannica Film, "The Ear and Hearing."

B. Causes of hearing problems and interpretations of audiograms

Topic III—1 hour

A. Physics of sound and its measurement

B. Practical demonstration of noise measurement with noise level meter and analyzer

Topic IV—1 hour

A. The Audiometer—What it is and how it works; its calibration and care—intensity limits of audiometer

Topic V—3½ hours

A. Instruction and preparation of the subject—standard audiometric technique (AIHA *Industrial Noise Manual,* October 1965, Chapter VIII)

B. Record keeping

C. Pitfalls

Supervised Audiograms—2½ hours

Each student performs audiograms on numerous people (all audiogram cards are kept by technician for discussion purposes; Topic VI)

Second day

Topic VI—3 hours

Review of audiograms performed and additional practice

Types of audiograms—their significance and interpretation—30 minutes

Practice—2 hours

Film "Hearing Conservation" (Produced by Ford Motor Company, Dearborn, Michigan)—30 minutes

Topic VII—1 hour

Medical-Legal Aspects

Topic VIII—3½ hours

Hearing Protection and Noise Control

Third day

Summation—morning

Question and answer period and general review

Written examination and review of papers

Continued practice in taking audiograms for those who need more confidence

Certificate of attendance

Text material

A copy should be obtained for each student, as well as a notebook.

1. *Industrial Noise Manual*—Second Edition. Published by American Industrial Hygiene Association, 14125 Provost Street, Detroit, Michigan 48227.

2. *Guide for Conservation of Hearing in Noise*. Research Center, Subcommittee on Noise in Industry, Aram Glorig, M.D., Director, 3851 Cedar Springs Road, Dallas, Texas.

Recommendations. That these courses be planned through

1. Centers of learning with speech and hearing clinics directed by otologists or audiologists with experience in industrial hearing conservation programs.

2. Enrollment be limited to 24 with one clinician to supervise each six trainees during practice.

A refresher course for at least one day to be given six months to a year from date of original course.

Personal qualifications of the trainee. Any nurse or technician who will be responsible for hearing measurement procedures and fitting the protective devices.

A detailed guide is being prepared to help instructors cover the most important features of each topic they discuss.

Representation

Roger Maas, Ed.D., Chairman, American Speech and Hearing Association

Paul Michael, Ph.D., American Industrial Hygiene Association

Anne Murphy, R.N., American Industrial Nurses' Association

Joseph Sataloff, M.D., American Industrial Medical Association

Appendix E   Threshold Limit Values for Noise for 1974

The American Conference of Government Industrial Hygienists for a number of years have recommended limits of exposure to chemical agents in the working environment by the setting of threshold limit values. During the past few years, a number of suggested limits of exposures for physical agents have been proposed by various organizations, but none of the limits have been accepted universally. Due to this lack of uniformity, the American Conference of Governmental Industrial Hygienists (ACGIH) in May 1967 established a Committee on Physical Agents. This Committee was directed to review the existing data on exposures of individuals to various physical agents and to recommend to the Conference safe limits of exposure.

In establishing any limit of exposure, many factors have to be considered. Among these are the types of data available and the validity of this data; methods of control of exposure and their feasibility; and of primary importance, the percentage of the group which will be protected by the established limits.

Various procedures have been suggested in the past for rating the hearing loss potential of noise. These have included "C" scale reading of a sound level meter, "A" scale reading of a sound level meter, average of the 500, 1000, and 2000 Hz octave bands, maximum limits in each of the three octave bands of 500, 1000, and 2000 Hz, and limits for each of the eight octave bands. After considering the merits of each system and their ease of application it was decided to use the "A" scale reading from the sound level meter. It must be pointed out that the "A" scale reading is used for hazard rating only, but if studies are made for the purpose of engineering control, then octave band analysis should be made of the noise.

After considering the above factors, the Committee decided to establish a limit of 90 dBA for an eight hours per day, forty hours per week exposure. Data indicates that this will protect about 90 per cent of the people exposed to this level for a normal working lifetime. As more exposure data becomes available and the cost of engineering controls are reduced, it would be desirable to revise the limit, if necessary, to protect a larger percentage of the exposed population.

For a number of years it was assumed that equal energy would produce equal damage to the ear. If this assumption were true, then each time the sound level is increased three decibels the exposure time should be reduced one-half. Laboratory data on temporary threshold shift and field data indicate that for the shorter exposure times the ear can tolerate more acoustical energy per day than for a continuous eight-hour exposure. Also laboratory data and very limited field data indicate that if the exposure is intermittent in nature, (rest periods between exposures) the ear can tolerate considerably more acoustical energy than for a single exposure to continuous noise. Considering these two factors, the limit is increased five decibels for each halving of the exposure time. Thus these limits are a compromise between the more conservative equal energy concept and the more liberal intermittent exposure concept.

At one time it was thought that limits of exposure for narrow bands of noise or pure tones should be 10 decibels lower than for broad band noise. This was then revised to only five decibels, but some of the latest data available indicated that even five decibels is too conservative. In the present limit no correction is made for pure tones or narrow bands of noise.

There is very little data available on the effects of exposure to impact or impulsive noise. Many factors possibly influence the effects; among them are: peak sound pressure level, rise time, decay time, repetition rate, time interval between impacts or impulses, number per day, and background sound pressure levels. It is known that exposure to a small number of 140 dB impulsive noises of short duration will produce a temporary threshold shift. Until additional data are available, a limit of 140 dB peak sound pressure level is recommended.

As additional data becomes available to the Committee it will be reviewed and, if necessary, revisions will be recommended to the Conference. These revisions are normally made at the annual meetings of the Conference which are held in May of each year.

1974 Committee members

Herbert H. Jones, USPHS, Chairman
Peter A. Breysse, University of Washington
Irving H. Davis, Michigan Dept. of Health
LCDR Joseph J. Drozd, USN
Dr. David A. Fraser, Univ. of North

Carolina
Maj. George S. Kush, USAF
Tom Cummins, Ontario Dept. of Health
Dr. Wordie H. Parr, NIOSH
David H. Sliney, U.S. Army
Dr. Robert N. Thompson, FAA
Thomas K. Wilkinson, USPHS
Eugene G. Wood, U.S. Dept. of Labor
Ronald D. Dobbin, NIOSH
Lt. Col. Robert T. Wangemann, U.S. Army

Any comments or questions regarding these limits should be addressed to Herbert H. Jones, Chairman, Threshold Limits Committee for Physical Agents, American Conference Governmental Industrial Hygienists, 1014 Broadway, Cincinnati 45202.

Preface
Physical agents

These threshold limit values refer to levels of physical agents and represent conditions under which it is believed that nearly all workers may be repeatedly exposed day after day without adverse effect. Because of wide variations in individual susceptibility, exposure of an occasional individual at, or even below, the threshold limit may not prevent annoyance, aggravation of a pre-existing condition, or physiological damage.

These threshold limits are based on the best available information from industrial experience, from experimental human and animal studies, and when possible, from a combination of the three.

These limits are intended for use in the practice of industrial hygiene and should be interpreted and applied only by a person trained in this discipline. They are not intended for use, or for modification for use, (1) in the evaluation or control of the levels of physical agents in the community, (2) as proof or disproof of an existing physical disability, or (3) for adoption by countries whose working conditions differ from those in the United States of America.

These values are reviewed annually by the Committee on Threshold Limits for Physical Agents for revisions or additions, as further information becomes available.

Notice of intent. At the beginning of each year, proposed actions of the Committee for the forthcoming year are issued in the form of a "Notice of Intent." This notice provides not only an opportunity for comment, but solicits suggestions of physical agents to be added to the list. The suggestions should be accompanied by substantiating evidence.

As legislative code. The Conference recognizes that the Threshold Limit Values may be adopted in legislative codes and regulations. If so used, the intent of the concepts contained in the Preface should be maintained and provisions should be made to keep the list current.

Reprint permission. This publication may be reprinted provided that written permission is obtained from the Secretary-Treasurer of the Conference and that this Preface be published in its entirety along with the Threshold Limit Values.

1013

Noise*

These threshold limit values refer to sound pressure levels that represent conditions under which it is believed that nearly all workers may be repeatedly exposed without adverse effect on their ability to hear and understand normal speech. The medical profession ([1,2]) has defined hearing impairment as an average hearing threshold level in excess of 25 decibels (ANSI S3.6-1969) at 500, 1000, and 2000 Hz, and the limits which are given have been established to prevent a hearing loss in excess of this value. These values should be used as guides in the control of noise exposure and, due to individual susceptibility, should not be regarded as fine lines between safe and dangerous levels.

Continuous or intermittent

The sound level shall be determined by a sound level meter, meeting the standards of the American National Standards Institute and operating on the A-weighting network with slow meter response. Duration of exposure shall not *exceed that shown in the table.*

These values apply to total time of exposure per working day regardless of whether this is one continuous exposure or a number of short-term exposures but does not apply to impact or impulsve type of noises.

When the daily noise exposure is composed of two or more periods of

---

* See Notice of Intended Changes, following.

Table F
PERMISSIBLE NOISE EXPOSURES

Duration per day Hours	Sound Level dBA*
8	90
6	92
4	95
3	97
2	100
1½	102
1	105
¾	107
½	110
¼	115**

* Sound level in decibels as measured on a standard level meter operating on the A-weighting network with slow meter response.
** No exposure in excess of 115 dBA

noise exposure of different levels, their combined effect should be considered, rather than the individual effect of each. If the sum of the following fractions,

$$\frac{C_1}{T_1} + \frac{C_2}{T_2} + \cdots \frac{C_n}{T_n}$$

exceeds unity, then, the mixed exposure should be considered to exceed the threshold limit value, C_1 indicates the total time of exposure at a specified noise level, and T_1 indicates the total time of exposure permitted at that level. Noise exposures of less than 90 dBA do not enter into the above calculations.

Impulse or impact noise

It is recommended that exposure to impulsive or impact noise should not exceed 140 decibels peak sound pressure level.

Notice of intended changes (for 1974)

These physical agents, with their corresponding values, comprise those for which either a limit has been proposed for the first time, or for which a change in the "Adopted" listing has been proposed. In both cases, the proposed limits should be considered trial limits that will remain in the listing for a period of at least one year. If, after one year no evidence comes to light that questions the appropriateness of the values herein, the values will be reconsidered for the "Adopted" list.

Noise

These threshold limit values refer to sound pressure levels and durations of exposure that represent conditions under which it is believed that nearly all workers may be repeatedly exposed without adverse effect on their ability to hear and understand normal speech. The medical profession has defined hearing impairment as an average hearing threshold level in excess of 25 decibels (ANSI-S3.6-1969) at 500, 1000, and 2000 Hz, and the limits which are given have been established to prevent a hearing loss in excess of this level. The values should be used as guides in the control of noise exposure and, due to individual susceptibility, should not be regarded as fine lines between safe and dangerous levels.

Continuous or intermittent

The sound level shall be determined by a sound level meter, conforming as a minimum to the requirements of the American National Standard Specification for Sound Level Meters, S1.4 (1971) Type S2A, and set to use the A-weighted network with slow meter response. Duration of exposure shall not exceed that shown in Table, "Threshold Limit Values."

These values apply to total duration of exposure per working day regardless of whether this is one continuous exposure or a number of short-term exposures but does not apply to impact or impulsive type of noise.

When the daily noise exposure is composed of two or more periods of noise exposure of different levels, their combined effect should be considered, rather than the individual effect of each. If the sum of the following fractions:

$$\frac{C_1}{T_1} + \frac{C_2}{T_2} + \cdots \frac{C_n}{T_n}$$

exceeds unity, then, the mixed exposure should be considered to exceed the threshold limit value, C_1 indicates the total duration of exposure at a specific noise level, and T_1 indicates the total duration of exposure permitted at that level. All on the job noise exposures of 80 dBA or greater shall be used in the above calculations.

Impulsive or impact noise

It is recommended that exposure to impulsive or impact noise should not exceed 140 decibels peak sound pressure level. Impulsive or impact noise is considered to be those variations in noise levels that involve maxima at intervals of greater than one per second. Where the intervals are less than one second, it should be considered continuous.

It should be recognized that the application of the TLV for noise will not protect all workers from the adverse effects of noise exposure. A hearing conservation program with audiometric testing is necessary when workers are exposed to noise at or above the TLV levels.

THRESHOLD LIMIT VALUES

Duration per day Hours	Sound Level dBA*
16	80
8	85
4	90
2	95
1	100
½	105
¼	110
⅛	115**

* Sound level in decibels as measured on a sound level meter, conforming as a minimum to the requirements of the American National Standard Specification for Sound Level Meters, S1.4 (1971) Type S2A, and set to use the A-weighted network with slow meter response.
** No exposure to continuous or intermittent in excess of 115 dBA

References

1. "Guides for the Evaluation of Hearing Impairment." *Transactions of the American Academy of Ophthalmology and Otolaryngology* (March–April, 1961).
2. "Guides to the Evaluation of the Permanent Impairment; Ear, Nose, Throat and Related Structures." *Journal of the American Medical Association 197:489* (August 1961).

Appendix F Glossary

This glossary defines words and terms used in industrial noise control and audiometry. Many definitions were taken from references listed in the Bibliography at the end of this appendix.

A

A-weighting network. See *Weighting network.*

Absorbents, diaphragmatic. Materials that flex under sound pressures and vibrate as a diaphragm, dissipating acoustic energy within their structure as heat and as mechanical energy of vibration.

Absorbents, sound. Materials that absorb sound readily. These are usually construction materials designed specifically for the purpose of absorbing acoustic energy.

Absorption, sound. See *Sound absorption.*

Absorption coefficient, α. For a surface, the ratio of the sound energy absorbed by a surface of a medium (or material) exposed to a sound field (or to sound radiation) to the sound energy incident on the surface. The stated values of this ratio are to hold for an infinite area of the surface. The conditions under which measurements of absorption coefficients are made must be stated explicitly. The absorption coefficient is a function of both angle of incidence and frequency. Tables of absorption coefficients usually list the absorption coefficients at various frequencies, the values being those obtained by averaging overall angles of incidence.

Absorption loss. That part of the transmission loss due to the dissipation or conversion of sound energy into other forms of energy (for example, heat), either within the medium or attendant upon a reflection.

Acceleration. A vector that specifies the time rate of change of velocity. Various self-explanatory modifiers, such as peak, average and rms, are often used. The time interval must be indicated over which the average (for example) was taken. Acceleration may be (*a*) oscillatory, in which case it may be defined by the acceleration amplitude (if simple harmonic) or the rms acceleration (if random); or (*b*) nonoscillatory, in which case it is designated "sustained" or "transient" acceleration. In simple harmonic motion, acceleration is the instantaneous value of the acceleration of point *P* along the diameter. In calculus, this is the derivative of velocity with respect to time, or the second derivative of displacement with respect to time.

Acceleration amplitude. See *Amplitude.*

Accelerometer. A transducer for converting acceleration into a proportional electrical or mechanical analog.

Accuracy. A statistical measure of the agreement between a quantitative observation and the observed phenomenon—the relationship of the measured value to the actual value.

Acoustic. Acoustical. The qualifying adjectives acoustic and acoustical mean containing, producing, arising from, actuated by, related to, or associated with sound. Acoustic is used when the term being qualified designates something that has the properties, dimensions, or physical characteristics associated with sound waves; acoustical is used when the term being qualified does not designate explicitly something that has such properties, dimensions, or physical characteristics. Usually the generic term is modified by acoustical, whereas the specific technical implication calls for acoustic. See also *Acoustics.*

• The following examples qualify as having the properties or physical characteristics associated with sound waves and hence would take Acoustic: impedance, inertance, load (radiation field), output (sound power), energy, wave, medium, signal, conduit, absorptivity, transducer.

• The following examples do not have the requisite physical characteristics and therefore take Acoustical: society, method, engineer, school glossary, symbol, problem, measurement, point of view, end-use device.

Acoustic dispersion. (*a*) The change of speed of sound with frequency in a nonhomogeneous medium. (*b*) Spreading of a sound, such as by a loudspeaker. See also *Distribution.*

Acoustic energy. The total energy of a given part of the transmitting medium minus the energy which would exist in the same part of the medium with no sound waves present. The energy added is the result of the sound vibrations.

Acoustic impedance. The acoustic impedance of a sound medium on a given surface lying in a wave front is the impedance obtained from the ratio of the sound pressure (force per unit area) on that surface by the flux (volume velocity, or linear velocity multiplied by the area) through the surface. When concentrated rather than distributed impedances are considered, the impedance of a portion of the medium is based on the pressure difference effective in driving that portion and the flux (volume velocity). The acoustic impedance may be expressed in terms of mechanical impedance divided by the square of the area of the surface considered. (Velocities in the direction along which the impedance is to be specified are considered positive.) See *Impedance.*

Acoustic neuroma. Tumor or growth on or of the auditory nerve.

Acoustic power. See *Sound power.*

Acoustic power level. See *Sound power level.*

Acoustic radiation pressure. A unidirectional, steady-state pressure exerted upon a surface exposed to an acoustic wave.

Acoustic reactance. The imaginary component of the acoustic impedance. See *Impedance.*

Acoustic refraction. See *Refraction.*

Acoustic resistance. The real component of the acoustic impedance.

Acoustic scattering. The irregular reflection, refraction, or diffraction of a sound in many directions.

Acoustic transmission system. An assembly of elements adapted for the transmission of sound. See also *Transmission coefficient.*

Acoustic trauma. A hearing injury produced by exposure to sudden intense acoustic energy, such as from blasts and explosions or from direct trauma to the head or ear. It should be thought of as one single incident relating to the onset of hearing loss.

Acoustical analysis. A detailed study of the use of the structure, the location and orientation of its spaces, and a determination of noise sources and the desirable acoustical environment in each usable area.

Acoustical environment. All of the factors, interior or exterior, which affect the acoustic conditions of the location, space, or structure under consideration.

Acoustical ohm. An acoustic resistance, reactance, or impedance has a magnitude of 1 acoustical ohm when a sound pressure of 1 micro-

bar produces a volume velocity of 1 cc/sec.

Acoustical tile. Acoustical absorbents produced in the form of sheets or units resembling tiles, usually 12 × 12 in. or multiples thereof.

Acoustical treatment. The use of acoustical absorbents, isolation, or any changes or additions to the structure to correct acoustical faults or improve the acoustical environment.

Acoustics. The science of sound, including its production, transmission, and effects. The acoustics of a room are those qualities that together determine its character with respect to sound and hearing. See also *Acoustic.*

Acoustics, architectural. The acoustics of buildings and structures.

Active transducer. A transducer whose output waves are dependent upon sources of power, apart from that supplied by any of the actuating waves, which are controlled by one or more of these waves. See also *Passive transducer.*

Acusis. Normal hearing ability.

Adaptation. A hearing impairment characterized by an inability to hear a tone at threshold level and at uniform intensity. Also called pathologic fatigue or abnormal tone decay.

Air bone gap. The difference in decibels between the hearing levels for sound at a particular frequency as determined by air conduction and bone conduction—threshold measurements.

Airborne sound. Sound transmitted through air as a medium. See also *Structure-borne sound.*

Air conduction. The process by which sound is conducted to the inner ear through the air in the outer ear canal utilizing the tympanic membrane and the ossicles as part of the pathway.

All-pass network. A network designed to minimize appreciable attenuation at any frequency.

AMA hearing impairment formula. The American Medical Association has established a formula for hearing impairment based on the average of the hearing levels of the three speech frequencies of 500, 1000, and 2000 Hz.

Ambient noise. The all-encompassing noise associated with a given environment, being usually a composite of sounds from many sources near and far. See also *Background noise.*

American Academy of Ophthalmology and Otolaryngology (AAOO). A professional organization of medical specialists who establish standards pertaining to vision and hearing.

Amplification, electronic. Increasing the intensity level of a sound signal by means of electronic circuitry.

Amplitude. The maximum displacement to either side of the normal or "rest" position of the molecules, atoms, or particles of the medium transmitting the vibration. The maximum value of displacement, velocity, or acceleration in simple harmonic motion is designated as the amplitude.

Amplitude of a periodic quantity. The maximum value of the quantity.

Anacusis. Total loss of hearing or inability to perceive sound.

Analog. If a first quantity or structural element plays a mutually similar role as a second quantity or structural element, the first quantity is called the analog of the second, and vice versa.

Analogy. Analogous. An analogy is a recognized relationship of consistent, mutual similarity between the equations and structures appearing within two or more fields of knowledge. Two or more quantities, or structural elements, from two or more fields of knowledge that play mutually similar roles in an analogy are said to be ANALOGOUS.

Analyzer. A combination of a sound filter system and a system for indicating the relative sound energy that is passed through the filter system. The filter is usually adjustable so that the signal applied to the filter can be measured in terms of the relative energy passed through the filter as a function of the adjustment of the filter response versus frequency characteristic. This measurement is usually interpreted as giving the distribution of energy of the applied signal as a function of frequency.

Anechoic room. An enclosure whose boundaries absorb effectively all the sound incident thereon, thereby affording essentially free-field conditions. Also called free-field room or dead room. See also *Free field.*

Annoyance. High-pitched (approximately 1500 Hz to 10,000 Hz) noise is more annoying than low-pitched noise. Also intermittent or irregular noise may be considerably more annoying than a steady noise of the same intensity.

ANSI. The American National Standards Institute, a standards-making body.

Antinode. A point, line, or surface in a standing wave where some characteristic of the wave field has maximum amplitude. Also called a LOOP. The appropriate modifier should be used before the word "antinode" to signify the type that is intended; that is, the DISPLACEMENT ANTINODE, VELOCITY ANTINODE, or PRESSURE ANTINODE.

Antinode microphone. A pressure-gradient microphone that discriminates against sound coming from a relatively distant source in favor of sound from a nearby source.

Antiresonance. For a system in forced oscillation, antiresonance exists at a point when any change, however small, in the frequency of excitation causes an increase in the response at this point. See also *Resonance.*

Antiresonance frequency. A frequency at which antiresonance exists.

Anvil. See *Ossicles.*

Aphasia. Loss by the brain of its ability to interpret the input received from the eyes and ears, and to send out directing impulses for speaking and writing. It is not due to a mechanical defect in the hearing, speech, or other organs.

Applied shock. Any excitation that, if applied to a system, would produce mechanical shock. The excitation may be either a force applied to the system or a motion of its support.

Articulation (percent articulation). Intelligibility (percent intelligibility. The percentage of the speech units spoken by a talker or talkers that is correctly repeated, written down, or checked by a listener or listeners, in a communications system. The word ARTICULATION is used when the units of speech material are meaningless syllables or fragments; the word INTELLIGIBILITY is used when the units of speech material are complete, meaningful words or phrases.

It is important to specify the type of speech material and the units into which it is analyzed for the purpose of computing the percentage. The units may be fundamental speech sounds, such as syllables, words, and sentences.

The percent articulation or percent intelligibility is a property of the entire communication system: talker, transmission equipment or medium, and listener. Even when attention is focused upon one component of the system (for example, a talker or a radio receiver), the other components of the system should be specified.

The kind of speech material used is identified by an appropriate adjective in phrases such as syllable articulation, individual sound articulation, vowel (or consonant) articulation, monosyllabic word intelligibility, discrete word intelligibility, discrete sentence intelligibility.

Artificial ear. A device that presents an acoustic impedance to the earphone equivalent to the impedance presented by the average human ear.

Artificial voice. A small loudspeaker mounted in a shaped baffle that is proportioned to simulate the acoustical constants of the human head. The artificial voice is used for calibrating and testing close-talking microphones.

ASA. American Standards Association, now called ANSI, American National Standards Institute.

Attenuate. To reduce in strength or force. In sound, to reduce the intensity as expressed in decibels.

Attenuation, sound. The reduction, expressed in decibels, of the sound intensity. For a signal confined to a channel or duct, the intensity values are averaged over the cross section. Reflected signals are disregarded in computing the attenuation.

Attenuator. The loudness-level control on the audiometer.

Audibility, threshold of. See *Threshold of audibility*.

Audible. Capable of being heard.

Audible range. The frequency range over which normal ears hear (approximately 20 Hz through 20,-000 Hz). See also *Ultrasonic* and *Infrasonic*.

Audio frequency. Any frequency corresponding to a normally audible sound wave. Audio frequencies range roughly from 20 to 20,000 Hz. The word audio may be used as a modifier to indicate a device or system intended to operate at audio frequencies, for example, audio amplifier.

Audiogram. A chart or table relating hearing level (for pure tones) to frequency. A record of hearing loss or hearing level measured at several different frequencies, usually 500 to 6000 Hz. The audiogram may be presented graphically or numerically. Also called a Pure tone, air-conduction audiogram or a Threshold audiogram.

Audiologist. A person trained in the specialized problems of hearing and deafness.

Audiometer, pure tone. An electroacoustical generator that provides pure tones of selected frequencies and of calibrated output, for the purpose of determining an individual's threshold of audibility. The units are generally of three types: Wide range which covers the major portion of the human auditory range in frequency and sound pressure level and is used primarily for clinical and diagnostic purposes; Limited range which covers frequencies between 500 Hz and 6000 Hz and levels from 10 dB to 70 dB; and the Narrow range which is even more restricted in frequency and sound pressure level than the limited-range audiometer.

Audiometer, speech. A unit which provides spoken material, either live or recorded, at controlled sound pressure levels. The same unit but more restricted in range of sound pressure levels is designated a Limited range speech audiometer.

Audiometric reference level. That sound pressure level (ASA, ISO, or ANSI) to which the audiometer is calibrated. A declared value, at a particular frequency, of the threshold of hearing for normal persons within a given age range, normally 18 to 25 years.

Audiometric technician. A person trained and qualified to administer audiometric examinations.

Audiometry. The science of measurement of hearing ability.

Auditory agnosia. Ability to perceive sound at the end organ with inability to interpret it centrally.

Auditory nerve. The eighth cranial nerve.

Auditory sensation area. The region enclosed by the curves defining the threshold of pain and the threshold of audibility.

Aural. Of or pertaining to the ear or hearing.

Aural critical band. That frequency band of sound that contains sound power equal to that of a simple (pure) tone centered in the critical band and just audible in the presence of the wide-band noise. "Just audible" means audible in a specified fraction of the trials. To be just audible in a wide-band continuous noise, the level of a simple tone in decibels must exceed the spectrum level of the continuous noise (at the same frequency) by 10 times the logarithm to the base 10 of the ratio of the critical bandwidth to unit bandwidth.

Aural harmonic. A harmonic generated in the auditory mechanism.

Auricle (pinna). The outer ear, including the opening to the ear canal.

Automatic (recording) audiometry. The method which allows the subject to manually control a mechanism, indicating on a graph when sound is first heard and when the same sound disappears.

Autonomic nervous system. The division of the vertebrate nervous system that regulates involuntary action, as of the intestines, heart, and glands, and comprises the sympathetic nervous system and parasympathetic nervous system.

Average speech power. The average value of the instantaneous speech power over a stated time interval.

Average value. Of a time function, is the numerical average of a large number of measurements along a time scale. Also called STATIC VALUE. Also see *Rectified average value, Root-mean-square value,* and *Peak value.*

B

B-weighting network. See *Weighting network.*

Background noise. The total of all sources of interference in a system used for the production, detection, measurement, or recording of a signal, independent of the presence of the signal. Ambient noise detected, measured, or recorded with the signal becomes part of the background noise.

Included in this definition is the interference resulting from primary power supplies (separately described as hum).

Baffle. A shielding structure or partition used to increase the effective length of the external transmission path between two points in an acoustic system as, for example, between the front and back of an electro-acoustic transducer.

Ballistics. A term used by sound engineers to describe the fluctuations of a meter needle due to inertia and momentum.

Band-center frequency. The designated (geometric) mean frequency of a band of noise or other signal. For example, 1000 Hz is the band center frequency for the octave band that extends from 707 Hz to 1414 Hz, or for the third-octave band that extends from 891 to 1123 Hz.

Band-pass filter. A wave filter that has a large insertion loss for one frequency band, neither the critical nor cutoff frequencies being zero or infinite.

Band-pressure level. The band-pressure level of a sound for a specified frequency band is the effective sound pressure level for the sound energy contained within the band. The width of the band and the reference pressure must be specified. The width of the band may be indicated by the use of a qualifying adjective; that is, octave-band (sound pressure) level, half-octave band level, third-octave band level, 50 Hz band level. If the sound pressure level is caused by thermal noise, the standard deviation of the band-pressure level will not exceed 1 dB if the product of the bandwidth in Hz by the integration time in seconds exceeds 20. See also *Octave-band pressure level.*

Bandwidth. When applied to a band-pass filter, bandwidth is determined by the interval of transmitted waves between the low and high cutoff frequencies.

Bar. See *Microbar.*

Barotrauma. Injury to the ear caused by a sudden alteration in barometric pressure.

Baseline. A measurement of parameters at a specific point in time in order to define existing conditions and especially variations with time.

Basic frequency. The basic frequency of an oscillatory quantity having sinusoidal components with different frequencies is the frequency of the component considered to be the most important. In a driven system, the basic frequency would, in general, be the driving frequency, and in a periodic oscillatory system, it would be the fundamental frequency.

Beam pattern. See *Directional response pattern.*

Beats. Beat frequency. Periodic variations that result from the superposition of two simple harmonic quantities of different frequencies f_1 and f_2. They involve the periodic increase and decrease of amplitude at the beat frequency $(f_1 - f_2)$.

Bel. A unit of sound level when the base of the logarithm is 10. Use of the bel is restricted to levels of quantities proportional to power. See also *Decibel.*

Bias. Consistent error, a fixed difference between the average, in a set of experimental results, and the true value.

Bidirectional microphone. A microphone whose response is a maximum for sound incident at two directions 180 degrees apart.

Bone conduction. The process by which sound is conducted to the inner ear through the cranial bones.

Bone-conduction test. A special test conducted by placing an oscillator on the mastoid process to determine nerve-carrying capacity or efficiency of the cochlea and the eighth cranial nerve leading to the brain.

Bone-conduction vibrator. Electromechanical transducer intended to produce the sensation of hearing by vibrating the bones of the head.

Broadband. A wide frequency range. Sound whose energy is distributed over a broad range of frequency (generally, more than one octave).

C

C-weighting network. See *Weighting network.*

Cardiovascular. Pertaining to heart and blood vessels.

Centile. See *Distribution, 3.*

Central tendency. Of a range of values of a variable (x), the general position of the distribution of the values indicated by the following measures:

(*a*) ARITHMETIC MEAN (AVERAGE): the sum of all the values divided by their number, denoted by $\bar{x} = (\Sigma x)/n$.

(*b*) MEDIAN: the value occupying the central position when the values are arranged in sequence from the least to the greatest.

(*c*) MODE: the most commonly occurring value of x.

Ceramic microphone. Microphone using a ceramic cartridge which exhibits piezoelectric properties. See *Piezoelectric material.*

Cerumen. Ear wax.

Characteristic impedance. The characteristic impedance of a medium is the ratio of the effective sound pressure at a given point to the effective particle velocity at the point in a free-plane progressive sound wave. The characteristic impedance is equal to the product of the density (ρ) times the speed of sound in the medium (c), for example, (ρc). See also *Impedance.*

Chart recorder. See *Graphic level recorder.*

Clipping (recording). Where the reproduced signal amplitude varies from the input signal effective amplitude by more than one dB, due to electronic compression or magnetic saturation.

Close-talking microphone. A microphone designed for use close to the mouth of the speaker.

Cochlea. The auditory part of the internal ear, shaped like a snail shell, containing the basilar membrane on which are distributed the end organs of the acoustic or eighth cranial nerve.

Coincidence. The condition (or frequency range) at which the velocity of the parallel component of the sound wave incident upon a panel equals the velocity of the shear wave in the panel.

Combination microphone. A combination of two or more similar or dissimilar microphones. For example, two oppositely phased pressure microphones acting as a gradient microphone, or a pressure microphone and a velocity microphone acting as a unidirectional microphone.

Communication. The signals or stimuli (or their transmission) which produce reactions, orient us in our environment, and furnish information on which to base decisions.

Community noise equivalent level (CNEL). Community noise equivalent level is a scale which takes account of all the A-weighted acoustic energy received at a point, from all noise events causing noise levels above some prescribed value. Weighting factors place greater importance upon noise events occurring during the evening hours (7:00 p.m. to 10:00 p.m.) and even greater importance upon noise events at night (10:00 p.m. to 6:00 a.m.). See Appendix B-4.

Complex quantity. A mathematical quantity containing imaginary elements such as *i* which indicates the square root of minus one.

Complex tone. A sound wave containing simple sinusoidal components of different frequencies. A complex tone is a sound sensation characterized by more than one pitch.

Compliance. (*a*) Roughly, the ease with which a panel of a material can be flexed by application of a force or pressure. Compliance is the reciprocal of stiffness. (*b*) To conform or adapt to applicable laws and standards.

Composite noise rating (CNR). A scale which takes account of the aircraft operations in quantifying the total noise environment. It was the earliest method for evaluating compatible land use around airports and is still in wide use by the Department of Defense in predicting noise environments around military airfields. See Appendix B-4.

Compressional wave. An elastic medium which causes an element of the medium to change its volume without undergoing rotation. A compressional plane wave is a longitudinal wave.

Condenser microphone. Microphone using a condenser (capacitor) as the transducer which converts the sound pressure to voltage.

Conductive hearing loss. See *Hearing loss.*

Confidence limits. The upper and lower values of the range over which a given percent probability applies. For instance, if the chances are 99 out of 100 that a sample lies between 10 and 12, the 99 percent confidence limits are said to be 10 and 12.

Consistent error. The fixed difference in a set of experimental results between the numerical average of the results and the true values.

Contact microphone. A microphone actuated by contact with the sound generator.

Continuous sound spectrum. The spectrum of wave components which are continuously distributed over a frequency region.

Continuous system, distributed system. A system that is considered to have an infinite number of possible independent displacements. Its configuration is specified by a function of a continuous spatial variable or variables in contrast to a discrete or lumped parameter system which requires only a finite number of coordinates to specify its configuration.

Correlation. A procedure for investigating the degree of association between two (or more) characteristics of a population.

Correlation coefficient. A numerical indication of the degree of association between two characteristics of a population, x and y. Complete dependence of one characteristic on the other yields unity; no dependence, zero. It may be positive, for direct relations, or negative, for inverse relations. The value is given by

$$ r = \frac{\Sigma(x - \bar{x})\ (y - \bar{y})}{\{\Sigma(x - \bar{x})^2\ \Sigma(y - \bar{y})^2\}^{\frac{1}{2}}} $$

Also called the PRODUCT-MOMENT CORRELATION COEFFICIENT.

Coulomb damping, dry friction damping. The dissipation of energy that occurs when a particle in a vibrating system is resisted by a force whose magnitude is a constant independent of displacement and velocity, and whose direction is opposite to the direction of the velocity of the particle.

Coupled modes. Modes of vibration that are not independent but that mutually influence one another because of energy transfer from one mode to the other.

Coupler. A device for acoustic loading of earphones. It has a specified arrangement of acoustic elements, and is provided with a microphone for measurement of the sound pressure, developed in a specified portion of the device.

Coupling. Any means of joining separated masses of any media so that sound energy is transmitted between them.

Covariation. Coincident variation.

Cranial nerves. There are 12. (*a*) OLFACTORY—sense of smell. (*b*) OPTIC—the eye. (*c*) OCULOMOTOR —eye movement. (*d*) TROCHLEAR —also eye movement. (*e*) TRIGEMINAL—feeling on the face and chewing. (*f*) ABDUCERS—lateral eye movement. (*g*) FACIAL—movement of face muscles. (*h*) VESTIBULAR COCHLEARIA—auditory nerve. (*i*) GLOSSOPHARYNGEAL—taste. (*j*) VAGUS—branches to voice organ, windpipe, gullet, heart, stomach, etc. (*k*) ACCESSORY—palate, etc. (*l*) HYPOGLOSSAL—also taste.

Crest factor. Ratio of peak-to-rms voltage values in a pulsed wave form.

Critical band. See *Aural critical band*.

Critical damping. The minimum viscous damping that will allow a displaced system to return to its initial position without oscillation.

Critical speed. A speed of a rotating system that corresponds to a resonance frequency of the system.

Crossover frequency. As applied to electric dividing networks, the frequency at which equal electric powers are delivered to each of the adjacent frequency channels when all channels are terminated in the loads specified.

Crosstalk. (*a*) Correlated signal produced in another channel (for example, print-through on magnetic tapes). (*b*) In audiometry, the presence of an undesired signal in the earphone not under test.

Cumulative distribution. See *Distribution, 2*.

Cycle. (*a*) The complete sequence of values of a periodic quantity that occurs during a period. (See also *Frequency*.) (*b*) The entire sequence of movement of a particle (during periodic motion) from rest to one extreme of displacement, back through rest position to the opposite extreme of displacement, and back to rest position.

Cycle per second (cps). A unit of frequency. The preferred terminology is hertz, abbreviated Hz.

Cyclical noise levels. Levels that change repetitiously during machine duty cycle.

D

Damage-risk criterion specifies the maximum allowable exposure to which people may be exposed if risk of hearing impairment is to be avoided. A damage-risk criterion may include in its statement a specification of such factors as time of exposure, noise intensity, and frequency, amount of hearing loss that is considered significant, percentage of the population to be protected, and method of measuring the noise.

Damped natural frequency. The frequency of free vibration of a damped linear system. The oscillation of a damped system may be considered periodic in the limited sense that the time interval between zero crossings in the same direction is constant, even though successive amplitudes decrease progressively. The frequency of the oscillation is the reciprocal of this time interval.

Damping. Any means of dissipating or attenuating vibrational energy within a vibrating medium. Usually the energy is converted to heat. Any influence which extracts energy from a vibrating system is known as damping. See also *Coulomb damping*, *Critical damping*, and *Viscous damping*.

Damping ratio. For a system with

1027

viscous damping, the ratio of actual damping coefficient to the critical damping coefficient.

Data recorder. A device used to preserve instantaneous data for subsequent analysis.

Dead room. A room that is characterized by an unusually large amount of sound absorption. See also *Anechoic room.*

Deaf. Unable to hear, hard of hearing.

Deafness. Loss of ability to hear without designation of degree of loss or cause. For the sake of clarity more precise terms are now preferred. See *Acusis, Anacusis, Hyposacusis, Dysacusis, Auditory agnosia, Presbycusis,* and *Diplacusis.*

Decay rate. See *Rate of decay.*

Decibel (dB). A nondimensional unit used to express sound levels. It is a logarithmic expression of the ratio of a measured quantity to a reference quantity. In audiometry, a level of zero decibels represents roughly the weakest sound that can be heard by a person with good hearing.

The decibel is a unit of level when the base of the logarithm is the tenth root of ten, and the quantities concerned are proportional to power.

Examples of quantities that qualify are power (any form), sound pressure squared, particle velocity squared, sound intensity, sound energy density, voltage squared. Thus the decibel is a unit of sound-pressure-squared level; it is common practice, however, to shorten this to sound pressure level because ordinarily no ambiguity results from so doing. The decibel is one tenth of a bel.

Where a weighted network filter is employed in making sound pressure measurements this is indicated by a suffix added to the unit symbol, for example:

• dBA. A sound level reading in decibels made on the A-weighted network of a sound level meter.

• dBC. A sound level reading in decibels made on the C-weighted network of a sound level meter. See also *Weighting network.*

Decile. See *Distribution 4.*

Degrees of freedom (DF). The effective number of independent values of a variable, equal to the actual number (n) minus the number of constraints. In calculating a mean, $DF = n$; in calculating the standard deviation of a sample, $DF = n - 1$; in calculating a correlation coefficient, $DF = n - 2$. The number of degrees of freedom of a mechanical system is equal to the minimum number of independent generalized coordinates required to define completely the positions of all parts of the system at any instant of time. In general, it is equal to the number of independent generalized displacements that are possible.

Difference limen. Differential threshold. The increment in a stimulus which is just noticed in a specified fraction of the trials. The relative difference limen is the ratio of the difference limen to the absolute magnitude of the stimulus to which it is related. Also called the JUST-NOTICEABLE DIFFERENCE.

Diffracted wave. A wave whose front has been changed in direction by an obstacle or other nonhomogeneity in a medium, other than by reflection or refraction.

Diffraction. That process that produces a diffracted wave. The distortion of a wave front caused by the presence of an obstacle in the sound field. Also, roughly, the ability of a sound wave to flow around an obstruction or through openings with little loss of energy. See also *Refraction*.

Diffuse sound field. A field in which the time average of the mean-square sound pressure is everywhere the same and the flow of energy in all directions is equally probable. See also *Time average* and *Mean-square value*.

Diffusion. Dispersion of sound within a space so that there is uniform energy density throughout the space.

Diplacusis. A difference of perception of sound by the two ears, either in time or in pitch, so that one round is heard as two.

Directional gain. Directivity index. The directional gain of a transducer, in decibels, is 10 times the logarithm to the base 10 of the directivity factor.

Directional microphone. A microphone whose response varies significantly with the direction of sound incidence.

Directional response pattern. Beam pattern. The directional response pattern of a transducer used for sound emission or reception is a description, often presented graphically, of the response of the transducer as a function of the direction of the transmitted or incident sound waves in a specified plane and at a specified frequency. The directional response pattern is often shown as the response relative to the maximum response.
A complete description of the direction response pattern of a transducer would require a three-dimensional presentation.

Directivity factor. (*a*) The directivity factor of a transducer used for sound emission is the ratio of the sound pressure squared, at some fixed distance and specified direction, to the mean-square sound pressure at the same distance averaged over all directions from the transducer. The distance must be great enough so that the sound appears to diverge spherically from the effective acoustic center of the sources. Unless otherwise specified, the reference direction is understood to be that of maximum response. (*b*) The directivity factor of a transducer used for sound reception is the ratio of the square of the open-circuit voltage produced in response to sound waves arriving in a specified direction to the mean-square voltage that would be produced in a perfectly diffused sound field of the same frequency and mean-square sound pressure.

Directivity index. In a given direction from a sound source, the difference in decibels between (*a*) the sound pressure level produced by the source in that direction, and (*b*) the space average sound pressure level of that source, measured at the same distance. It is defined as ten times the logarithm (base 10) of the ratio of the axial sound intensity to some reference intensity.

Disability of hearing. See *Percent impairment of hearing*.

Discrimination. The ability of the hearing apparatus to discern discrete, particular signals in a complex sound field.

Discrimination loss. Discrimination loss is the difference in percent between the normal discrimination score for the test and the score obtained for the ear under test.

Discrimination score for speech. The percent of items in an appropriate form of hearing test, usually monosyllabic words, that is correctly repeated, written down, or checked by the listener. This form of test is usually administered at an acoustic level well above the individual's threshold for speech. The normal value of discrimination (or articulation score) for each test must be determined empirically.

Dispersion. The spread of values of a variable about the mean or median.

Displacement. A vector quantity that specifies the change of position of a body or particle and is usually measured from the mean position or position of rest. In general, it can be represented by a rotation vector, a translation vector, or both.

Distortion. Any change in the transmitted sound which alters the character of the energy-frequency distribution within the signal so that the sound being received is not a faithful replica of the source sound.

Distributed system. See *Continuous system.*

Distribution. (*a*) The pattern of sound intensity levels within the space; also, the patterns of sound dispersion as the sound travels within the space. (*b*) Statistics: The manner in which values in a sample of a population are distributed about the arithmetic mean. Various types and descriptions of distributions are recognized.

1. Normal (gaussian) distribution: A distribution commonly found in biological systems, in which the mean, median, and mode coincide at the apex of a symmetrical bell-shaped curve of frequency of occurrence plotted against magnitude of the measurement.

2. Cumulative distribution: A presentation in which the variable is plotted horizontally with the percentage of occurrence of all values less than the abscissa plotted vertically. On a suitable graph paper, the points will fall upon a straight line if the distribution is gaussian.

3. Centile: A unit for defining the position in a series of values arranged in sequence from least to greatest, whereby each division is 1 percent of the total.

4. Decile: 10 centiles, usually used to designate the 10th and 90th centiles.

5. Quartile: 2.5 deciles, normally restricted to describing the 25th and 75th centiles.

6. Skewness: Asymmetry of a distribution.

7. Kurtosis: Symmetrical deviation from the gaussian distribution. An abnormally peaked distribution is called Leptokurtic; an abnormally flattened one, Platykurtic.

Divergence loss. That part of the transmission loss due to the divergence on spreading of the sound rays in accordance with the geometry of the system (for example, spherical waves emitted by a point source).

Dizziness. An imprecise term commonly used to describe various peculiar subjective symptoms such as faintness, giddiness, lightheadedness, or unsteadiness. See also *Vertigo.*

Doppler effect. The phenomenon evidenced by the change in the observed frequency of a wave in a transmission system caused by a time rate of change in the effective length of the path of travel between the source and the point of observation.

Dosimeter. A device worn on the person for determining the accumulated sound exposure with regard to level and time.

Drift. Instability. Changes of a property of a system with time.

Dry friction damping. See *Coulomb damping.*

Dubbing. The combining of two or more sources of sound into a complete recording, at least one of the sources being a recording. In common usage, the term is often applied to the duplication of a single recording.

Duration of shock pulse. The time required for the acceleration of the pulse to rise from some stated fraction of the maximum amplitude and to decay to this value.

Dynamic range. That region of measurement of an instrument or system between the noise floor and maximum permissible distortion point.

Dynamic vibration absorber. Tuned damper. A tuned damper is a device for reducing vibration of a primary system by the transfer of energy to an auxiliary resonant system which is tuned to the frequency of the vibration. The force exerted by the auxiliary system is opposite in phase to the force acting on the primary system.

Dysacusis. Any hearing impairment that is not primarily a loss of ability to perceive sound. Loss of discrimination for words, syllables, or phonemes, or loss of discrimination in terms of understanding of words or in terms of pitch. Also, pain or discomfort in ear from exposure to sound.

E

Ear protector. A device worn to reduce the passage of ambient noise into the auditory system. EARPLUGS are inserted in the external ear canal. EARMUFFS fit over the entire ear and snug against the head.

Earphone. Receiver. An electroacoustic transducer intended to be closely coupled acoustically to the ear. The term RECEIVER should be avoided when there is risk of ambiguity.

Echo. A wave that has been reflected or otherwise returned with sufficient magnitude and delay to be detected as a wave distinct from that directly transmitted.

Effective perceived noise level. See *Perceived noise level (PNL)*.

Effective sound pressure. Root-mean-square (rms) sound pressure. The effective sound pressure at a point is the rms value of the instantaneous sound pressures over a time interval at the point under consideration. In the case of periodic sound pressures, the interval must be an integral number of periods or an interval that is long compared to a period. In the case of nonperiodic sound pressures, the interval should be long enough to make the value obtained essentially independent of small changes in the length of the interval. The term "effective sound pressure" is frequently shortened to SOUND PRESSURE. See *Root-mean-square value.*

Efficiency. The efficiency of a device with respect to a physical quantity is the ratio of the useful output of the quantity to its total input.

Elastic medium. Any substance in which strain or deformation is directly proportional to stress or loading.

Electroacoustic transducer. A transducer that consists of a capacitor and depends upon interaction between its electric field and the change of its electrostatic capacitance. An example is the condenser or capacitor microphone.

Electroacoustics. Transforming sound waves into electric current (and vice versa) by means such as microphone, amplifiers, etc.

Electroencephalogram (EEG). A graphic recording of electric currents developed in the brain.

Electromechanical transducer. A transducer for receiving waves from an electric system and delivering waves to a mechanical system, or vice versa.

Electrostatic actuator. An apparatus constituting an auxiliary external electrode that permits the application of known electrostatic forces to the diaphragm of a microphone for the purpose of obtaining a primary calibration.

Environment. The surrounding conditions, influences, or forces to which a person is exposed.

Epithelium. The sheetlike tissue which covers and lines the surfaces of the body. It forms the outer layer or covering of the skin as well as the lining of the hollow organs, and the passages of the digestive, respiratory, and urinary systems.

Equal loudness contours. Graphical representation for rating the loudness of a sound. A rating by an observer of sound pressure levels at various frequencies as compared to a 1000 Hz tone at a given sound pressure level.

Equivalent network. A network that, under certain conditions of use, may replace another network.

Equivalent system. A system that may be substituted for another system for the purpose of analysis. Many types of equivalence are common in vibration and shock technology; for example: (a) equivalent stiffness, (b) equivalent damping, (c) torsional system equivalent to a translational system, and (d) electrical or acoustical system equivalent to a mechanical system.

Equivalent viscous damping. A value of viscous damping assumed for the purpose of analysis of a vibratory motion, such that the dissipation of energy per cycle at resonance is the same for either the assumed or actual damping force.

Etiology. The study of the cause of disease.

Eustachian tube. A tube approximately 2½ in. long leading from the back of the throat to the middle ear. It equalizes the pressure of air in the middle ear with that outside the eardrum.

Evoked response audiometry (ERA). It refers to the use of the electroencephalogram with a summing or averaging computer. This makes possible the evaluation of slight changes in electrical activity of the brain in response to a sound stimulus.

Excitation. Stimulus. Excitation is an external force (or other input) applied to a system that causes the system to respond in some way.

F

Facial nerve. See *Cranial nerves*.
Far field. See *Field*.
Fast response. A selectable mode

of operation of a sound level meter or analyzer in which the needle indicator has minimum damping and can therefore respond rapidly to changes in sound level.

Fatigue, auditory. See *Temporary threshold shift.*

Feedback. The return to the input of a part of the output of a system.

Fenestration. An operation in which the oval window, which has been made inoperative through a fixation of the stapes, is bypassed and a window created in the horizontal semicircular canal. The new opening is covered by a skin flap and now sound vibrations are able to reach the inner ear. Maximum improvement is to within 20 to 30 dB of normal.

Fidelity. The faithful reproduction of the source sound.

Field, far. That part of a sound field for which spherical divergence occurs; that is, SPL decreases by −6 dB for each doubling of distance. As a general rule, it is also considered as that part of a sound field which is beyond a distance of 3 to 4 times the largest dimension of the source or greater than the maximum wavelength of sound for the lowest frequency of interest. It is also sometimes possible to satisfy the far-field conditions over a limited region between the near field and the reverberant field, if the absorption within the enclosure is not too small so that the near field and the reverberant field merge.

Filter. A device for separating components of a signal on the basis of frequency. It allows components in one or more frequency bands to pass relatively unattenuated, and it attenuates greatly components in

other frequency bands. A WAVE FILTER is a transducer for separating waves on the basis of their frequency. It introduces relatively small insertion loss to waves of one or more frequency bands and relatively large insertion loss to waves of other frequencies. See also *Band-pass filter* and *Bandwidth.*

Flanking paths. Transmission paths which transmit acoustic energy around a sound barrier; paths which bypass the intended barrier.

Flat response. The characterization of microphone, instrument, or recorder having a sensitivity or response that is constant regardless of frequency.

Flutter. A rapid reflection or echo pattern between parallel walls, with sufficient time between each reflection to cause a listener to be aware of separate, discrete signals.

Flutter echo. A rapid succession of reflected pulses resulting from a single initial pulse.

Flux. Rate of transfer of fluid, particles, or energy across a given surface.

Focusing. Concentration of acoustic energy within a limited location in a room as the result of reflections from concave surfaces.

Force. That which changes the state of rest or motion of matter.

Forced oscillation. Forced vibration. The oscillation or vibration of a system is forced if the response is imposed by the excitation. If the excitation is periodic and continuing, the oscillation is steady-state. See also *Oscillation.*

Forcing frequency. The frequency associated with a harmonically varying force or motion acting upon a system.

Foundation. A structure that supports the gravity load of a mechanical system. It may be fixed in space, or it may undergo a motion that provides excitation for the supported system.

Free field. A free sound field is a field in a homogeneous, isotropic medium free from boundaries. In practice it is a field in which the effects of the boundaries are negligible over the region of interest. The actual pressure impinging on an object (for example, a microphone) placed in an otherwise free sound field will differ from the pressure which would exist at that point with the object removed, unless the acoustic impedance of the object matches the acoustic impedance of the medium.

Free-field room. A room in which essentially free-field conditions exist. See also *Anechoic room.*

Free-field voltage sensitivity. See *Sensitivity.*

Free oscillation. Free vibration. Free oscillation of a system is the oscillation of some physical quantity of the system when there are no externally applied driving forces. Such oscillation is maintained by the transfer of energy between elastic restoring forces and inertia forces. The oscillation may arise from initial displacements, velocities, or a force suddenly applied and withdrawn. FREE VIBRATION is the vibratory motion which takes place when an elastic system is displaced from its equilibrium position and released. See also *Oscillation.*

Free progressive wave. Free wave. A free progressive wave is a wave in a medium free from boundary effects. A free wave in a steady state can only be approximated in practice.

Frequency. The time rate of repetition of a periodic phenomenon. The frequency is the reciprocal of the period. In sound (as in electricity) the unit of measurement for frequency is the hertz (Hz) which equals one complete waveform (cycle) per second. See also *Hertz* and *Pitch.*

Frequency, infrasonic. Frequencies below the range of human hearing, usually below 20 Hz. See *Infrasonic frequency.*

Frequency, ultrasonic. Frequencies above the range of human hearing, usually above 20,000 Hz. See also *Ultrasonics.*

Frequency analyzer. Electrical apparatus capable of measuring the acoustic energy present in various frequency bands of a complex sound. See also *Analyzer.*

Frequency bands. A division of the audible range of frequencies into subgroups for detailed analysis of sound.

Functional hearing loss (psychogenic). A hearing impairment due to nonorganic causes, that is, psychological.

Fundamental. The component in a periodic wave corresponding to the fundamental frequency.

Fundamental frequency. The fundamental frequency (*a*) of a periodic quantity is equal to the reciprocal of the shortest period during which the quantity exactly reproduces itself, and (*b*) of an oscillating system is the lowest natural frequency. The normal mode of vibration associated with this frequency is known as the FUNDAMENTAL MODE.

Fundamental mode of vibration.

The fundamental mode of vibration of a system is the mode having the lowest natural frequency.

G

g. The acceleration produced by the force of gravity, which varies with the latitude and elevation of the point of observation. By international agreement, the value 980.665 cm/sec^2 = 386.087 in./sec^2 = 32.1739 ft/sec^2 has been chosen as the STANDARD ACCELERATION DUE TO GRAVITY.

Gaussian distribution. See *Distribution*, 1.

Gaussian random noise. See *Random noise*.

Graphic level recorder. A graphic recorder with a built-in detector so that dynamic waveforms can be applied directly to its input.

Grazing incidence. A microphone positioned so that its axis is perpendicular to a line from the microphone to the noise source. See also *Incidence of sound*.

H

Hair cells. Sensory receptors for sound stimuli located in the basilar membrane.

Hammer. *Malleus.* See *Ossicles.*

Hard of hearing. See *Deaf.*

Harmonic. Subharmonic. A sinusoidal quantity having a frequency that is an integral multiple of the frequency of a periodic quantity to which it is related. A subharmonic is a sinusoidal quantity having a frequency that is an integral submultiple of the fundamental frequency of a periodic quantity to which it is related.

Harmonic frequency. The frequency of a component of a periodic quantity is an integral multiple of the fundamental frequency. For example, a component whose frequency is twice the fundamental frequency is the second harmonic of that frequency.

Harmonic motion (simple). A periodic motion that has a single frequency or amplitude.

Harmonic series of sounds. A series in which each basic frequency in the series is an integral multiple of a fundamental frequency.

Hearing. The subjective response to sound and the nervous and cerebral operations which translate the physical stimuli into meaningful signals.

Hearing aid. An electronic device, fitted to the ear, which amplifies sound.

Hearing conservation. The prevention or minimizing of noise induced hearing loss through the use of hearing protection devices and the control of noise through engineering methods or administrative procedures.

Hearing disability. Actual or presumed inability to remain employed at full wages as a result of a hearing loss.

Hearing handicap. The disadvantage imposed by impairment sufficient to affect one's efficiency in the situation of everyday living.

Hearing impairment. A deviation or change for the worse in either structure or function, usually outside the normal hearing range.

Hearing level. A measured threshold of hearing at a specified frequency, expressed in decibels relative to a specified standard of normal hearing. The deviation in decibels of an individual's threshold from the zero reference of the audiometer.

Hearing loss. An increase in the threshold of audibility, at specific frequencies, as the result of normal aging, disease, or injury to the hearing organs. It is the symptom of reduced auditory sensitivity, synonymous with auditory impairment, when a specific cause can be ascribed. Also used, in a general sense, to describe the process of losing auditory sensitivity. Types of hearing loss are as follows:

1. CONDUCTIVE. A hearing loss originating in the conductive mechanism of the ear.
2. SENSORI-NEURAL. A hearing loss originating in the cochlea or the fibers of the auditory nerve. (Formerly called PERCEPTIVE DEAFNESS.)
3. NOISE INDUCED. A sensori-neural hearing loss attributable to the effects of noise.

Hearing loss (hearing level) for speech. The difference in decibels between the speech levels at which the average normal ear and the defective ear, respectively, reach the same intelligibility, often arbitrarily set at 50 percent.

Hearing threshold level (HTL). Amount (in decibels) by which the threshold of audibility for that ear exceeds a standard audiometric threshold.

Herpes zoster. Disease caused by a virus. It is characterized by the formation of small water blisters on the skin and inflammation of groups of nerve cells.

Hertz (Hz). Synonymous term for CYCLES PER SECOND. Most standardizing agencies have adopted hertz as the preferred unit of frequency.

High-frequency loss. Usually showing a hearing impairment, starting with 2000 Hz and beyond.

High-pass filter. A wave filter having a single transmission band extending from some critical or cutoff frequency, not zero, up to infinite frequency.

Histogram. A graphic representation of a frequency distribution in which the widths of the contiguous vertical bars are proportional to the class widths of the variable and the heights of the bars are proportional to the class frequencies.

Homogeneous. Being of a uniform structure or composition throughout.

Horn loudspeaker. A loudspeaker whose diaphragm is coupled to the external acoustic medium by means of a horn used as an impedance-matching and directivity-controlling device.

Hum. See *Background noise.*

Hyperacusis. Abnormal acuteness of hearing, due to increased irritability of sensori-neural mechanism.

Hypoacusis. Hearing impairment attributed to deficiency in the peripheral organs of hearing; may be conductive or sensori-neural.

I

Impact. A collision of one mass in motion with a second mass, which may be either in motion or at rest.

Impact noise. The noise resulting from the collision of two masses. See *Impulse noise.*

Impact noise reduction. A single number rating system which compares the impact isolation of a test specimen with a standard contour. (Also given in terms of a comparable impact insulation class. Strictly, only when tested in accordance with a specific test procedure.)

Impedance. A complex ratio related to the sound absorption or transmission characteristics of acoustical materials. Similar to electrical or mechanical impedance. Actually, the rate at which a given volume of any material can accept energy. An impedance is the ratio of two complex quantities whose arguments increase linearly with time and whose real (or imaginary) parts represent a force-like and velocity-like quantity respectively. Examples of force-like quantities are: force, sound pressure, voltage, temperature, electric field strength. Examples of the velocity-like quantities are: velocity, volume velocity, current, heat, flow, magnetic flux.

Impedance mismatch. The flow of acoustic energy along the path from source to receiver can be impeded by discontinuities which reflect the energy back toward the source. In other words, an impedance mismatch. Sound transmission in the open air can also be impeded. For example, the stack of an exhaust blower can be designed to provide the greatest reflection of acoustic fan-noise energy at its outlet, in order to minimize the radiation of blower noise from the stack. Acoustic filters and mufflers operate on this principle, although some mufflers may also include absorption in the transmission path.

Impulse. The product of a force and the time during which the force is applied; more specifically, the impulse is $\int_{t_1}^{t_2} F dt$ where the force F is time dependent and equal to zero before time t_1 and after time t_2.

Impulse noise. Impulse noises are usually considered to be singular noise pulses, each less than 1 second in duration, or repetitive noise pulses occurring at greater than 1 second intervals. Also defined as a change of sound pressure of 40 dB or more within 0.5 second.

Incidence of sound. The direction from which a sound wave approaches an object.

● GRAZING INCIDENCE. Describes the direction of sound incidence when sound is incident upon an object parallel to a specified surface of the object.

● NORMAL INCIDENCE. Describes the direction of sound incidence when sound is incident upon an object normal (perpendicular) to a specified surface of the object.

● RANDOM INCIDENCE. If an object is in a diffuse sound field, the sound is said to strike the object at random incidence.

Incus. See *Ossicles*.

Industrial audiometry. The use of audiometric procedures in an industrial hearing-conservation program.

Industrial hygiene. That science or art which devotes itself to the recognition, evaluation, and control of those environmental factors arising in or from the workplace which may cause impaired health or significant discomfort among workers or residents of the community.

Inertia. The tendency of a mass to resist any change in its state of motion or rest.

Infrasonic frequency. A frequency lying below the audio frequency range (about 20 Hz). The word INFRASONIC may be used as a modifier to indicate a device or system intended to operate at an infrasonic frequency. See also *Ultrasonic*.

Insertion loss. The difference in two sound pressure levels measured at the same point in space before and after a muffler is inserted at that point.

Instantaneous sound pressure. The instantaneous sound pressure at a point is the total pressure at that point minus the static pressure at that point.

Instantaneous speech power. The rate at which sound energy is being radiated by a speech source at any given instant.

Intelligibility. See *Articulation*.

Intensity. The sound intensity measured in a specified direction at a point is the average rate at which sound energy is transmitted through a unit area perpendicular to the specified direction at the point considered. Only in plane or spherical free progressive sound waves is the intensity I related to the average pressure p by the equation $I = p^2/\rho c$, where ρc represents the characteristic impedance of air. In general, however, there is no simple relation between sound pressure level and intensity level.

Intensity level. The intensity level, in decibels, of a sound is 10 times the logarithm to the base 10 of the ratio of the intensity of this sound to the reference intensity. The reference intensity shall be stated explicitly.

Internal noise. Electrical circuit noise. In sound level meters, especially vacuum-tube type, such noise must be taken into account at lower levels around 35 or 40 dB.

Interrupter switch. More properly called the PRESENTER SWITCH, because it permits the stimulus to be presented or cut off from the earphones but leaves the audiometer circuits in operation, thus making it possible to change the frequency and intensity level of the tone.

ISO. International Organization for Standardization. Publications are available through ANSI.

Isolation. A reduction in the capacity of a system to respond to an excitation by the use of a resilient support.

Isolation, sound. Materials or constructions (or the use of such materials or constructions) which resist the passage of sound.

Isolator. See *Vibration isolator*.

Isotropic. Exhibiting properties with the same values when measured in all directions.

J

Jerk. A vector which specifies the time rate of change of the acceleration of a particle; the third derivative of the displacement of the particle with respect to time.

Just audible. See *Aural critical band*.

Just-noticeable difference. See *Difference limen*.

K

Kilohertz (kHz). 1000 Hz.

Kurtosis. See *Distribution*, 7.

L

Laboratory standard microphone. A microphone that satisfies American National Standards Institute *Specification for Laboratory Standard Pressure Microphones*, S1.12–1967, and is therefore suitable for use as a secondary sound pressure reference standard.

Lapel microphone. A microphone adapted to positioning around the neck of the user. Also called a LAVALIERE MICROPHONE.

Leaks, sound. Any opening which permits airborne sound transmission.

Least squares. A criterion to establish the curve, or straight line of best fit, to points on a graph.

Level. In acoustics, the level of a quantity is the logarithm of the ratio of that quantity to a reference quantity of the same kind. The base of the logarithm, the reference quantity, and the kind of level must be specified. Examples of kinds of levels in common use are sound power level, sound-pressure-squared level, and sound intensity level.

The level as here defined is measured in units of the logarithm of a reference ratio that is equal to the base of logarithms. In symbols

$$L = \log_r(q/q_0)$$

where L = level of kind determined by the kind of quantity under consideration, measured in units of $\log_r r$.

r = base of logarithms and the reference ratio,

q = the quantity under consideration,

q_0 = reference quantity of the same kind.

Level above threshold. Sensation level. The level above threshold of a sound is the level of the sound in decibels above its threshold of audibility for the individual observer or for a specified group of individuals.

Level meter, sound. See *Sound level meter.*

Level recorder. See *Graphic level recorder.*

Limen. The threshold of physiological and psychological response. See also *Difference limen.*

Limited-range audiometer. See *Audiometer, pure tone.*

Limpness. Property of material where the material under stress does not recover from the deformation upon removal of stress.

Line spectrum. A spectrum whose components occur at a number of discrete frequencies.

Lineal. An old word, used to distinguish running feet from square or cubic feet. It is used here to distinguish to-and-fro vibration from torsional vibration. The words lateral and linear are invariably used instead of lineal.

Linear. In a mathematical sense this usually connotes a straight-line functional relationship such as $y = Cx$, where C is constant. In a sense of vibration classification, it connotes LINEAL.

Linear mechanical impedance. A complex ratio of force to velocity or relative velocity. The real part of a mechanical impedance is a mechanical reactance. If the force and velocity are measured at the same point, the ratio is designated as a DRIVING-POINT IMPEDANCE; if measured at different points, the ratio is designated as a TRANSFER IMPEDANCE.

Linear regression of one characteristic on another. A description of the manner in which one characteristic varies with respect to another, as described by a straight line fitted to the data by the method of least squares, the deviations being measured in the direction of the first characteristic. REGRESSION COEFFICIENT—the numerical value of the change that, according to the slope of the regression line, takes place in one characteristic for unit change in the other.

Lip microphone. A microphone adapted for use in contact with the lip.

Live room. A room characterized by an unusually small amount of sound absorption. See also *Reverberant room*.

Logarithm. The logarithm of a number is the exponent or that power of a number to which another number, the base, must be raised to yield the number first named. In symbols:

$$10^y = x \qquad \log_{10} x = y$$

Longitudinal wave. A wave in which displacement of the molecules in the medium is parallel with the direction of propagation of the wave.

Loop. See *Antinode*.

Loudness. Loudness is the intensive attribute of an auditory sensation, in terms of which sounds may be ordered on a scale extending from soft to loud. Loudness depends primarily upon the sound pressure of the stimulus, but also depends upon the frequency and waveform of the stimulus.

Loudness contour. A curve that shows the related values of sound pressure level and frequency required to produce a given loudness sensation for the typical listener.

Loudness level. The loudness level of a sound, in phons, is numerically equal to the median sound pressure level, in decibels, relative to 0.0002 microbar, of a free progressive wave of 1000 Hz frequency presented to listeners facing the source, which in a number of trials is judged by the listeners to be equally loud. The man-

ner of listening to the unknown sound, which must be stated, may be considered one of the characteristics of that sound.

Loudspeaker. Speaker. An electroacoustic transducer intended to radiate acoustic power into the air, the acoustic wave form being essentially equivalent to that of the electrical input.

Loudspeaker system. A combination of one or more loudspeakers and associated baffles, horns, and dividing networks arranged to work together as a coupling means between the driving electric circuit and the acoustic medium.

Low-pass filter. A wave filter having a single transmission band extending from zero frequency up to some critical or cutoff frequency, not infinite.

M

Machine envelope. The space occupied by a machine tool. The boundary of the machine envelope is a reasonably uniform line around the periphery of the machine, approximately 3 ft from machine components. The machine envelope extends up and over the top and down and under the bottom of the machine. For machines mounted on long runways, the machine envelope contains only active portions of the machine, excluding vacant portions of the runway.

Malingerer. One who pretends deafness or other hearing abnormalities to avoid work.

Malleus. See *Ossicles*.

Mask microphone. A microphone designed for use inside an oxygen or other respiratory face mask.

Masking. The amount or process

by which the threshold of audibility of a sound is raised by the presence of another (masking) sound. The unit customarily used is the decibel. The stimulation of one ear of a patient by controlled noise to prevent his hearing with that ear the tone or signal given to his other ear.

Mass. The quality of matter which permits it to resist acceleration; the quality of matter which produces the effect of inertia. See also *Force*.

Mass law. The approximately linear relationship between the sound insulation of a partition, expressed in decibels, and the logarithm of its weight per unit area.

Mastoiditis. An inflammation of the skull bone behind the ear.

Maximum sound pressure. The maximum sound pressure for any given cycle of a periodic wave is the maximum absolute value of the instantaneous sound pressure occurring during that cycle. In the case of a sinusoidal sound wave, this maximum sound pressure is also called the pressure amplitude.

Mean free path. The mean free path for sound waves in an enclosure is the average distance sound travels between successive reflections in the enclosure.

Mean square value. The value which results from converting the negative instantaneous values of any wave form to corresponding positive values by squaring the waveform.

Measurand. A physical quantity, property, or condition which is measured.

Meatus, auditory. The external opening of the ear canal.

Mechanical shock. Shock that occurs when the position of a system is significantly changed in a relatively short time in a nonperiodic manner. It is characterized by suddenness and large displacements, and develops significant internal forces in the system.

Mechanical system. An aggregate of matter comprising a defined configuration of mass, mechanical stiffness, and mechanical resistance.

Mechanical transmission system. An assembly of elements adapted for the transmission of mechanical power.

Mel. A unit of pitch. By definition, a simple tone of 1000 Hz frequency, 40 dB above a listener's threshold, produces a pitch of 1000 mels. The pitch of any sound that is judged by the listener to be n times that of a 1-mel tone is n mels.

Meniere's disease. The combination of deafness, tinnitus, nausea, and vertigo.

Microbar (Dyne per square centimeter). A unit of pressure commonly used in acoustics. One microbar is equal to 1 dyne per square centimeter. The term "bar" properly denotes a pressure of 10^6 dynes per square centimeter.

Microphone. An electroacoustic transducer that responds to sound waves and delivers essentially equivalent electric waves.

Microphone directivity. The variation in response of a microphone dependent on the direction of arrival of the sound wave.

Microphonics. Internal equipment noise caused by vibration.

Minima. The lowest value.

Modal number. In general, a vibratory system can be analyzed in terms of its normal modes. The modes may be arranged in a discrete sequence associated with a set of ordered integers which are called modal numbers.

Modal shape. One of the characteristic shapes of a vibrating body or system. It corresponds to a normal mode of vibration.

Mode of vibration. In a system undergoing vibration, a mode of vibration is a characteristic pattern assumed by the system. Two or more modes may exist concurrently in a multiple-degree-of-freedom system. See also *Fundamental Mode of Vibration.*

Modulation. The process or the result of the process whereby some characteristic of one wave is varied in accordance with some characteristic of another wave. Modulation is the variation in the value of some parameter characterizing a periodic oscillation. Thus, amplitude modulation of a sinusoidal oscillation is a variation in the amplitude of the sinusoidal oscillation.

Monaural hearing. Hearing with one ear only. See also *Monophonic.*

Monophonic. A one-channel audio system.

Mounting, resilient. Any mounting, attachment system, or apparatus which permits room surfaces or machinery to vibrate without transmitting all of the energy of vibration to the structure.

Mounting, tile. The method of attaching acoustical tile to the building structure or building surfaces.

Muffler. A special duct or pipe that impedes the transmission of sound by reducing the velocity of the air or gas flow (Dispersive type), by sound absorption (Dissipative), or by reflecting the sound back toward the source through a series of cavities and side chambers (Reac-

tive type).

Multiple-degree-of-freedom system. A system for which two or more coordinates are required to define completely the position of the system at any instant.

N

Narrow band. Applies to a narrow band of transmitted waves, with neither of the critical or cutoff frequencies of the filter being zero or infinite.

Natural frequency. The frequency of free oscillation of a system. For a multiple-degree-of-freedom system, the natural frequencies are the frequencies of the normal modes of vibration. In a damped system, the natural frequency is a quasi-frequency in that the motion is not periodic but is generally taken as the frequency at which the velocity reverses sign. The frequency at which a resiliently mounted mass would vibrate when set into vibration, under the influence of gravity alone with no added force or constraints. Often called Resonance frequency.

Near field. See *Field.*

Neper. A unit of level when the logarithm is on the Napierian base *e*. Examples of quantities that qualify are voltage, current, particle velocity, sound pressure.

Neural. Of, or relating to, or affecting a nerve or the nervous system.

Neuritis. Inflammation of a nerve.

Neuroma. A general term indicating a tumor on or of the nerve. On the basis of newer knowledge such tumors (or neoplasms) are now classified in more specific categories, for example, gangloneuroma, neurinoma, etc.

NIPTS. Noise induced permanent threshold shift. See also *Permanent threshold shift.*

NNI. The noise and number index based on perceived noise level. It is used for rating airplane fly-by noise.

Node. A point, line, or surface in a standing wave where some characteristic of the wave field has essentially zero amplitude. The appropriate modifier should be used before the word "node" to signify the type that is intended; for example, displacement, velocity, or pressure node. See also *Antinode.*

Noise. (*a*) Any undesired sound. Noise is any unwanted disturbance within a useful frequency band. (*b*) An erratic, intermittent, or statistically random oscillation.

If ambiguity exists as to the nature of the noise, a phrase such as ACOUSTIC NOISE or ELECTRIC NOISE should be used. Since these definitions are not mutually exclusive, it is usually necessary to depend upon context for the distinction.

See also *Background noise, Pink noise, Random noise, White noise,* and *Noise level.*

Noise and number index (NNI). A measure based on perceived noise level, and with weighting factors added to account for the number of noise events, and used (in some European countries) for rating the noise environment near airports. (See also *Perceived noise level.*)

Noise exposure. A generic term signifying the total acoustic stimulus applied to the ear over a period of time.

Noise exposure forecast (NEF). A scale (analogous to CNEL and CNR) which has been used by the federal government in land use planning guides for use in connection with airports. In the NEF scale, the basic measure of magnitude for individual noise events is the EFFECTIVE PERCEIVED NOISE LEVEL (EPNL), in units of EPNdB. This magnitude measure includes the effect of duration per event. See also *Community noise equivalent level* and *Composite noise rating.*

Noise floor. Electrical circuit noise within an instrument or system. See also *Internal noise.*

Noise immission. An index of the total noise energy incident on the ear over a specified period of time. Correlative of NOISE EMISSION.

Noise immission level (NIL). The A-weighted noise immission, expressed in decibels relative to a specified datum.

Noise induced hearing loss. This terminology is usually restricted to mean the slowly progressive inner ear hearing loss which results from exposure to noise over a long period of time as contrasted to acoustic trauma or physical injury to the ear. See also *Hearing loss.*

Noise isolation class (NIC). A single number rating derived in a prescribed manner from the measured values of noise reduction. It provides an evaluation of the sound isolation between two enclosed spaces that are acoustically connected by one or more paths.

Noise level. A level of noise, the type of which must be indicated by further modifier or context. The physical quantity measured, the instrument used, and the bandwidth or other weighting characteristic must be indicated.

For airborne sound, unless specified to the contrary, noise level is the WEIGHTED SOUND PRESSURE LEVEL, also called SOUND LEVEL; the weighting must be indicated.

Noise rating curves. An agreed set of empirical curves relating octave bands to a NOISE RATING (NR), which is numerically equal to the sound pressure level at the intersection of the curve with the ordinate at 1000 Hz. The noise rating of a given noise is found by plotting the octave-band spectrum on the same diagram and selecting the highest noise rating curve which the spectrum just penetrates.

Noise reduction (NR). A decrease of the sound pressure level at a specified observation point. Noise reduction is also used to designate the differences in sound pressure levels existing at two different locations at a single time, when the designated structures (such as baffles or partitions) are in position. The term NOISE CONTROL is meaningful only when noise control components and the points of observation are fully specified.

Noise reduction coefficient (NRC). A measure of the acoustical absorption performance of a material, calculated by averaging its sound absorption coefficients at 250, 500, 1000, and 2000 Hz, expressed to the nearest integral multiple of 0.05.

Nominal band-pass center frequency. The geometric mean of the nominal cutoff frequencies.

Nominal bandwidth. The nominal bandwidth of a filter is the difference between the upper and lower cutoff frequencies. The difference may be expressed in hertz as a percentage of the band-pass center frequency, or as the interval between the upper and lower nominal cutoffs in octaves.

Nominal microphone sensitivity. The frequency range in which the sensitivity is constant.

Nominal upper and lower cutoff frequencies. The nominal upper and lower cutoff frequencies of a filter are those frequencies above and below which the signal is rejected.

Nondirectional microphone. See *Omnidirectional microphone.*

Nonlinear damping. Damping due to a damping force that is not proportional to velocity.

Normal distribution. See *Distribution, 1.*

Normal incidence. Microphone positioned so it is pointing toward sound source. See also *Incidence of sound.*

Normal mode of vibration. In general, any composite motion of the system is analyzable into a summation of its normal modes. The characteristic pattern of motion typically consists of a space distribution, one part of which is negative in relation to the other part. Thus, at the same time that the particles in one part are moving outward in the positive direction from their positions of equilibrium, the particles in the other part are moving outward in the negative direction, and conversely. Vibration in a normal mode occurs at a natural frequency of the undamped system. The terms NATURAL MODE and CHARACTERISTIC MODE are synonymous with normal mode.

Normal threshold of audibility. At a given frequency, this is the value of the minimum sound level which produces an auditory sensation in a large number of young persons with normal ears.

Noy. A unit or term applied to the

divisions of a scale, comparable with the loudness scale of relative annoyance or noisiness of sound. A unit used in the calculation of perceived noise level.

NR. See *Noise rating curves.*

Nystagmus. An abnormal and involuntary movement of the eyes— side to side, up and down, or rotary.

O

Octave. (*a*) The interval between two sounds having a basic frequency ratio of two. (*b*) An octave is the pitch interval between two tones such that one tone may be regarded as duplicating the basic musical import of the other tone at the nearest possible higher pitch.

Octave band. A division of the audible range of frequencies into subgroups such that in each division the upper frequency limit is twice the lower limit. The center frequencies used to designate the octaves are twice the center frequency of the preceding octave band. ONE-THIRD OCTAVE BANDS: a split of an octave band into three equal parts for more detailed analysis of distribution of sound energy. See also *Spectrum.*

Octave-band analyzer. A device for measuring pressure levels of a sound for a frequency band corresponding to a specific octave.

Octave-band spectrum. See *Spectrum.*

Ohm. The unit of electrical resistance equal to the resistance of a circuit in which a potential difference of one volt produces a current of one ampere. OHM'S LAW states that the voltage is equal to the resistance (in ohms) times the current (in amperes): $E = IR$.

Omnidirectional microphone. A microphone whose response is essentially independent of the angle of arrival of the incident sound wave. It should be noted that omnidirectional here refers to elevation as well as to azimuth. In radio antenna practice this is not necessarily so.

Operator's normal position. The location of an operator with respect to a machine during the time it is operating. The position should be recorded for the particular machine installation. More than one position may be necessary.

Organ of Corti. An aggregation of nerve cells lying on the basilar membrane which senses vibrations that are transmitted to the brain, where they are interpreted as sound.

Oscillation. The variation, usually with time, of the magnitude of a quantity with respect to a specified reference when the magnitude is alternately greater and smaller than the reference. See also *Forced oscillation* and *Free oscillation.*

Ossicles, auditory. Any one of the small bones such as the malleus, incus, or stapes which forms a chain for the transmission of sound from the tympanic membrane to the oval window.

Otitis media. Inflammation and infection of the middle ear.

Otogenous. Originating within the ear especially from inflammation of the ear.

Otolaryngologist. A physician or surgeon specializing in the practice of OTOLOGY (ear disease), RHINOLOGY (nose disease), and LARYNGOLOGY (throat and larynx diseases).

Otolaryngology. The branch of medicine or surgery which deals with the diseases of the ear, the nose, and the throat and larynx. See also *Otolaryngologist.*

Otologist. A physician or surgeon who specializes in the diagnosis and treatment of the disorders and diseases of the ear.

Otology. The branch of medicine which deals with the diagnosis and treatment of the disorders and diseases of the ear.

Otomycosis. An infection due to fungus in the external auditory canal.

Otorrhea. A discharge or running from the ear; said especially of a discharge containing pus.

Otosclerosis. A condition marked by the growth of spongy bone around the delicate structures within the ear, particularly the stapes and oval window, resulting in a gradual loss of hearing.

Otoscope. An instrument, usually provided with a lens and an illuminating system, for inspecting the ear, especially the external auditory meatus and the eardrum.

Overload level of a component or system is that level at which operation ceases to be satisfactory as a result of signal distortion, overheating, etc.

P

P. Probability. See *Statistical significance.*

Paget's disease. A generalized skeletal disease of older persons of unknown cause, involving a thickening and softening of the bones, as in the skull and ear.

Pain, threshold of. A sound pressure level sufficiently high to produce the sensation of pain in the human ear (usually above 120 dBA).

Parabolic-reflector microphone. A microphone employing a parabolic reflector to improve its directivity.

Paracusis Willisii. The sensation of a deafened person indicating that he can hear better in a noisy area.

Partial. (*a*) A partial is a physical component of a complex tone. (*b*) A partial is a component of a sound sensation which may be distinguished as a simple tone that cannot be further analyzed by the ear and which contributes to the timbre of the complex sound.

Partial node. A partial node is the point, line, or surface in a standing wave system where some characteristic of the wave field has a minimum amplitude differing from zero. The appropriate modifier should be used with words "partial node" to signify the type that is intended; for example, DISPLACEMENT PARTIAL NODE, VELOCITY PARTIAL NODE, PRESSURE PARTIAL NODE.

Particle velocity. In a sound field, the velocity of a given infinitesimal part of the medium, with reference to the medium as a whole due to the sound wave.

Passive transducer. A transducer whose output waves are independent of any sources of power that are controlled by the actuating waves. See also *Active transducer.*

PB words. A measure of speech intelligibility.

Peak level. The maximum instantaneous level that occurs during a specified time interval. In acoustics, PEAK SOUND PRESSURE LEVEL is to be understood, unless some other kind of level is specified.

Peak sound pressure. For any specified time interval, this is the maximum absolute value of the instantaneous sound pressure in that interval. In the case of a periodic wave, if the time interval considered is a complete period, the peak sound pressure becomes identical with the maximum sound pressure.

Peak speech power. The maximum value of the instantaneous speech power within the time interval considered.

Peak-to-peak value. Of an oscillating quantity, this value is the algebraic difference between the extremes of the quantity.

Peak value. The extreme value of the quantity.

Perceived noise level (PNL). The level in decibels assigned to a noise by means of a calculation procedure that is based on an approximation to subjective evaluations of noisiness. Units are PNdB.

Percent impairment of hearing. Percent hearing loss. An estimate of a person's ability to hear. It is usually based, as the result of an arbitrary rule, on the pure tone audiogram. The specific rule for calculating percent hearing loss varies from state to state. Impairment refers specifically to a person's "illness or injury that affects his personal efficiency in the activities of daily living." Disability has the additional medicolegal connotation that an impairment reduces a person's ability to engage in gainful activity. Impairment is only a contributing factor to the disability. See also *Hearing loss.*

Percent intelligibility. See *Articulation.*

Perceptive deafness. See *Hearing loss.*

Perforation, eardrum. A puncture or rupture of the eardrum.

Period. The time required to complete one cycle; it is the reciprocal of the frequency. The period of a periodic quantity is the smallest increment of the independent variable for which the function repeats itself.

Periodic quantity. An oscillating quantity whose values recur for certain increments of the independent variable.

Permanent threshold shift (PTS). The component of threshold shift which shows no progressive reduction with a passage of time when apparent cause has been removed.

Phase shift. Change of frequency. See also *Doppler effect* and *Acoustic dispersion.*

Phon. A measure of loudness level (on a logarithmic scale) which compares the loudness of a sound to the loudness of a 1000 Hz tone of a given sound pressure level. See also *Loudness level.*

Piezoelectric material. A nonconducting crystal which exhibits the property of generating an electrical charge when mechanically stressed.

Piezoelectric transducer. A transducer that depends for its operation on the interaction between the electric charge and the deformation of certain asymmetric crystals having piezoelectric properties. Also called a crystal or ceramic transducer.

Pink noise. Noise whose spectrum level decreases with increasing frequency to yield constant energy per octave of bandwidth.

Pistonphone. A small chamber equipped with a reciprocating piston of measurable displacement that permits the establishment of a known sound pressure in the chamber.

Pitch. That attribute of auditory sensation in terms of which sounds may be ordered on a scale extending from low to high. Pitch depends primarily upon the frequency of the sound stimulus, but also depends upon the sound pressure and waveform of the stimulus.

Plane wave. A wave in which the fronts are everywhere parallel planes normal to the direction of propagation.

PNL. PNdB. See *Perceived noise level.*

Point source. See *Simple sound source.*

Population. The ensemble of values of a variable from which a sample is drawn.

Power (level) gain. The excess of the output power level, in decibels, over the input power level, in decibels. Ordinarily, the name of this quantity is shortened without ambiguity to POWER GAIN IN DECIBELS.

Power level. In decibels, 10 times the logarithm to the base 10 of the ratio of a given power to a reference power. The reference power must be indicated.

Power spectrum. See *Spectrum density.*

Precision. A measure of the resolution (the smallest detectable change) of an instrument or observation. Also how close the results compare to each other. See also *Accuracy* and *Repeatability.*

Presbycusis. Decline in hearing acuity that normally occurs with aging process.

Presenter switch. See *Interrupter switch.*

Pressure, acoustic. The instantaneous pressure at a point as a result of the sound vibration minus the static pressure at that point. The change in pressure resulting from the sound vibration.

Pressure amplitude. See *Maximum sound pressure.*

Pressure-gradient microphone. A microphone in which the electric output substantially corresponds to a component of the gradient (space derivative) of the sound pressure.

Pressure level, sound. See *Sound pressure level.*

Pressure microphone. A microphone in which the electric output substantially corresponds to the instantaneous sound pressure of the impressed sound wave.

Pressure spectrum level. The pressure spectrum level of a sound at a specified frequency is the effective sound pressure level for the sound energy contained within a band 1 Hz wide, centered at the specified frequency. Ordinarily this has significance only for sound having a continuous distribution of energy within the frequency range under consideration. The reference pressure should be explicitly stated.

Primitive period. Period. The primitive period of a periodic quantity is the smallest increment of the independent variable for which the function repeats itself. If no ambiguity is likely, the primitive period is simply called the PERIOD OF THE FUNCTION.

Probability. See *Statistical significance.*

Product-moment correlation coefficient. See *Correlation coefficient.*

Propagation velocity. See *Wave velocity.*

Prospective study. The form of serial study in which the first observation of hearing level occurs before any occupational noise exposure has been sustained.

Psychoacoustics. That branch of psychophysics dealing with acoustic stimuli.

Psychogenic deafness. Deafness originating in or produced by the mental reaction of an individual to his physical or social environment. Sometimes called FUNCTIONAL DEAFNESS or FEIGNED DEAFNESS.

Pulse. Cannot be exactly defined. A term more generic than impact and including impact and impulse. A term usually used in a more generic sense than pulse is excitation, which includes pulse and vibration. See *Excitation.*

Pulse rise time. The interval of time required for the leading edge of a pulse to rise from some specified small fraction to some specified larger fraction of the maximum value.

Pure tone. A sound wave, the instantaneous sound pressure of which is a simple sinusoidal function of the time. Audibly, it is characterized by a singleness of pitch.

Push-pull microphone. A microphone in which two like microphone elements actuated by the same sound wave operate 180 degrees out of phase.

Q

Q. Quality factor. A measure of the sharpness of resonance or frequency selectivity of a resonant vibratory system having a single degree of freedom, either mechanical or electrical.

Quality. Usually refers to the spectral distribution of acoustic energy.

Quartile. See *Distribution,* 5.

R

Random error. See *Residual variance.*

Random incidence. If an object is in a diffuse sound field, the sound is said to strike the object at random incidence. See also *Incidence of sound.*

Random noise. An oscillation whose instantaneous magnitude is not specified for any given instant of time. The instantaneous magnitude of a random noise is specified only by probability distribution functions giving the fraction of the total time that the magnitude, or some sequence of magnitudes, lies within a specified range.

A random noise whose instantaneous magnitudes occur according to the gaussian distribution is called GAUSSIAN RANDOM NOISE. See also *Distribution* and *Noise.*

Rank. The ordinal number of an item in a series which is arranged in sequence from the least to the greatest value of the items.

Rate of decay. The time rate at which the sound pressure level (or other stated characteristic) decreases at a given point and at a given time. A commonly used unit is decibels per second.

Reactive muffler. See *Muffler.*

Readout. The meter reading or oscillographic display or other final expression of information within the capability of the eye. Sometimes the term CALL-OUT is used in relation to the ear.

Real-time analyzer. An instrument for performing computations during the time of observation, so that the results are available for controlling the operation or process.

Receiver. The affected person or equipment. See also *Earphone*.

Recognition differential. For a specified aural detection system it is that excess of the signal level over the noise level presented to the ear which results in a 50 percent probability of detection of the signal. The bandwidth of the system, within which signal and noise are presented and measured, must be specified.

Recorder, magnetic. Equipment incorporating an electromagnetic transducer and means for moving a ferromagnetic recording medium relative to the transducer for recording electric signals as magnetic variation in the medium. The generic term "magnetic recorder" can also be applied to an instrument which has not only facilities for recording electric signal as magnetic variations, but also for converting such magnetic variations back into electrical variations (playback).

Recording channel. Either one of a number of independent recorders in a recording system or independent recording tracks on a recording medium. One or more channels may be used at the same time for covering different ranges of the transmitted frequency band, for multichannel recording, or for control purposes.

Recording system, multitrack. A recording system which provides two or more recording paths on a medium which may carry either related or unrelated recordings in common time relationship.

Recording system, sound. A combination of transducing devices and associated equipment suitable for storing sound in a form capable of subsequent reproduction.

Recruitment. The condition where faint or moderate sounds cannot be heard while at the same time there is little or no loss in the sense of loudness of loud sounds. The condition in which an individual perceives an abnormally rapid increase in loudness as the sound pressure goes up. Usually characteristic of severe sensorineural deafness.

Rectified average value. Determined by converting all negative values of the waveform to equivalent positive values and then averaging.

Reference levels. See *Levels* and also see *Audiometric reference levels*.

Reflecting surfaces. Room surfaces from which significant sound reflections occur; or special surfaces used particularly to direct sound throughout the space.

Reflection. The return from surfaces of sound energy not absorbed upon contact with the surfaces.

Refraction, acoustic. The process by which the direction of sound propagation is changed due to spatial variation in the speed of sound in the medium. See also *Diffraction* and *Scattering*.

Refraction loss. That part of the transmission loss due to refraction resulting from nonuniformity of the medium.

Regression coefficient. See *Linear regression*.

Relative velocity. The relative velocity of a point with respect to a reference frame is the time rate of change of a position vector of that point with respect to the reference frame.

Repeatability. The ability of a transducer to reproduce an output reading when the same measurand value is applied to it repeatedly, under the same conditions, and in the same direction. Also the ability to obtain exact duplicate measurements. See also *Accuracy* and *Precision*.

Residual noise level. The noise which exists at a point as a result of the combination of many distant sources, individually indistinguishable.

Residual variance. In general, the unexplained part of the total variance. In regressions, the mean squared deviation from the regression line, sometimes referred to loosely as regression variance.

Resolution. The magnitude of output step changes (expressed in percent of full scale output) as the measurand is continuously varied over the range. The smallest detectable change in a measurand is the THRESHOLD.

Resonance. When a system is acted upon by an external harmonic force whose frequency equals the natural frequency of the system, the amplitude becomes great and the system is said to be in a STATE OF RESONANCE. Displacement resonance exists between a body, or system, and an applied force if any small change in frequency of the applied force causes a decrease in the amplitude of displacement. The natural sympathetic vibration of a volume of air or a panel of material at a particular frequency as the result of excitation by a sound of that particular frequency. See also *Oscillation* and *Forced oscillation.*

Resonance frequency. Resonant frequency. A frequency at which resonance exists. See also *Natural frequency.*

Response. The response of a device or system resulting from an excitation (stimulus) under specified conditions. Modifying phrases must be prefixed to the term response to indicate what kinds of input and output are being utilized.

The response characteristic, often presented graphically, gives the response as a function of some independent variable such as frequency or direction. For such purposes it is customary to assume that other characteristics of input are held constant.

Retrospective study. Cross-sectional study. The study of hearing levels of persons who have been exposed to known noises for known periods of time.

Reverberation. (*a*) The persistence of sound in an enclosed space, as a result of multiple reflections after the sound source has stopped. (*b*) The sound that persists in an enclosed space, as a result of repeated reflection or scattering, after the source of the sound is stopped.

Reverberation chamber. An enclosure in which all the surfaces have been made as sound reflective as possible. Reverberation chambers are used for certain acoustical measurements.

Reverberation time. (t_{60}). For any given frequency, the time required for the average sound pressure level, originally in a steady state, to decrease 60 dB after the source is stopped. Usually the pressure level for the upper part of this range is measured and the result extrapolated to cover 60 dB.

Reverberation time, optimum. An empirically determined reverberation time, varying directly with room volume, which produces hearing conditions considered ideal by an average listening audience.

Risk. The percentage of persons who, as a result of exposure to a specified noise level, may be expected to sustain a noise induced hearing loss equal to or greater than a defined value.

Rms. See *Root mean square value.*

Room constant. Equal to the product of the average absorption coefficient of the surface and the total internal area of the room divided by the quantity one minus the average absorption coefficient. It is the name given to the expression $S\alpha/(1-\alpha)$, where S is the total area of the bounding surfaces of the room in square feet, and α is the average absorption coefficient of the sound absorptive surface present in the room.

Room shape. The configuration of enclosed space, resulting from the configuration, orientation, and arrangement of surfaces defining the space.

Room volume. The cubic feet (or cubic meters) of space enclosed by the room surfaces.

Root-mean-square sound pressure. See *Effective sound pressure.*

Root-mean-square value (rms). The square root of the arithmetic mean of the squares of a set of values.

S

Sabin. A measure of sound absorption of a surface; equivalent to perfect absorption having the dimensions of square feet. The METRIC SABIN has a dimension of square meters. See also *Absorption, sound.*

Scalar. Refers to quantities that have magnitude only. See *Vector.*

Scatter diagram. A graphical presentation, for example, relating two series of variables, in which a separate point indicates each pair of corresponding values.

Scattering. Acoustic scattering is the irregular reflection, refraction or diffraction of a sound in many directions. See also *Diffraction* and *Refraction.*

Self-induced (self-excited) vibration. The vibration of a mechanical system is self-induced if it results from conversion, within the system, of nonoscillatory excitation to oscillatory excitation.

Semicircular canals. Special organs of balance located in the inner ear. They have nothing to do with the hearing mechanism.

Sensation level. See *Level above threshold.*

Sensitivity. The sensitivity of a receiving transducer is the ratio of its electrical output to its mechanical input.

FREE-FIELD VOLTAGE SENSITIVITY (RECEIVING VOLTAGE RESPONSE). The free-field voltage sensitivity of a transducer used for sound reception is the ratio of the output open-circuit voltage to the free-field sound pressure in the undisturbed sound field. The frequency and direction of incidence must be specified.

The ratio may be expressed, for example, in microvolts per microbar.

Unless otherwise specified, the undisturbed free field is understood to mean a plane progressive wave.

Sensitivity level. The sensitivity (or response) level of a transducer, in decibels, is 20 times the logarithm to the base 10 of the ratio of the amplitude sensitivity to the reference sensitivity, where the amplitude is a quantity proportional to the square root of power. The kind of sensitivity and the reference sensitivity must be indicated.

Sensori-neural deafness. Decreased

sensitivity of the auditory mechanism in the cochlea or paralysis of the acoustic nerve or its center in the brain. This was formerly called INNER EAR IMPAIRMENT or PERCEPTIVE DEAFNESS.

Sensori-neural hearing loss. See *Hearing loss.*

Septum. A thin partition separating two cavities or soft masses.

Serial study. Longitudinal study. The observation of the hearing level of an individual at successive intervals of time.

Shear wave. A wave motion in which movement of the media is at right angles to direction of propagation of the wave.

Shock. Shock pulse. A substantial disturbance characterized by a rise and decay of acceleration in a short period of time. Shock pulses are normally displayed graphically as curves of acceleration as a function of time. The duration of a shock pulse is the time required for the acceleration of the pulse to rise from some stated fraction of the maximum amplitude and to decay back to this value.

Shock absorber. A device for the dissipation of energy to modify the response of a mechanical system to applied shock.

Shock isolator. Shock mount. A resilient support that tends to isolate a system from applied shock.

Shock machine. A device for subjecting a system to controlled and reproducible mechanical shocks.

Shock spectrum. A plot of the maximum acceleration experienced by a system as a function of its own natural frequency in response to an applied shock.

Signal. A signal is (*a*) a disturbance used to convey information;

(*b*) the information to be conveyed over a communication system.

Signal-to-noise-ratio (S/N). The ratio between the rms signal level within a specified frequency band and the rms level of the electrical system noise within the same band. This S/N ratio is generally expressed in decibel notation. This S/N ratio establishes the dynamic range of the system as a function of frequency.

Simple harmonic motion. The projection on a diameter of point P moving around the circumference of a circle with uniform angular velocity. The DISPLACEMENT AMPLITUDE is the radius of the circle; the VELOCITY AMPLITUDE is the peripheral velocity of point P on the circle; and the ACCELERATION AMPLITUDE is the centrifugal acceleration associated with the motion of point P on the circle.

Simple sound source. A source that radiates sound uniformly in all directions under free-field conditions.

Simple tone. See *Pure tone.*

Sine wave. A sound wave in which the sound pressure is a sinusoidal function of time.

Single-degree-of-freedom system. A system for which only one coordinate is required to define completely the configuration of the system at any instant. See also *Degrees of freedom.*

Sinus. A hollow in bone or other tissue. Also a channel for passage of blood or lymph.

Skewness. See *Distribution 6.*

Slow response. A selectable mode of operation of a sound level meter or analyzer in which the indicator has high damping and therefore slowly responds to change in sound level. This mode tends to provide an average reading.

Snubber. A device used to increase the stiffness of an elastic system (usually by a large factor) whenever the displacement becomes larger than a specified amount.

Sociocusis. Hearing loss resulting from the accumulative effects of daily nonindustrial noise exposure.

Sone. A unit of loudness. By definition, a simple tone of frequency 1000 Hz, 40 dB above a listener's threshold, produces a loudness of 1 sone. The loudness of any sound that is judged by the listener to be n times that of the 1 sone tone is n sones. The loudness scale is a relation between loudness and level above threshold for a particular listener. In presenting data relating loudness in sones to sound pressure level, or in averaging the loudness scales of several listeners, the thresholds (measured or assumed) should be specified.

A MILLISONE is equal to 0.001 sone.

Sonics. The technology of sound in processing and analysis. Sonics includes the use of sound in any non-communication process.

Sound. (*a*) The auditory sensation produced through the organs of hearing usually by vibrations transmitted in a material medium, commonly air. (*b*) Sound is an oscillation in pressure, stress, particle displacement, particle velocity, etc., in a medium with internal forces, for example, elastic, viscous), or the superposition of such propagated oscillations.

In case of possible confusion, the term "SOUND WAVE" or "ELASTIC WAVE" may be used for concept (*b*), and the term "SOUND SENSATION" for concept (*a*). Not all sound waves can evoke an auditory sensation; for example, ultrasound. The medium in which the sound exists is often indicated by an appropriate adjective; such as, airborne, waterborne, or structure-borne sounds.

Sound absorption. The change of sound energy into some other form, usually heat, in passing through a medium or on striking a surface. In addition, sound absorption is the property possessed by materials and objects, including air, of absorbing sound energy. See also *Attenuation.*

Sound absorption coefficient. In reference to a surface the fraction of incident sound energy absorbed or otherwise not reflected by the surface. Unless otherwise specified, a diffuse sound field is assumed.

Sound analyzer. A device for measuring the band-pressure level or pressure-spectrum level of a sound as a function of frequency.

Sound control. The application of the science of acoustics to the design of structures and equipment, to permit them to function properly and to create the proper environment for the activities intended.

Sound energy. In a given part of a medium, the total energy in this part of the medium minus the energy which would exist in the same part of the medium with no sound waves present.

Sound energy density. At a point in a sound field the sound energy contained in a given infinitesimal part of the medium divided by the volume of that part of the medium.

The terms INSTANTANEOUS ENERGY DENSITY, MAXIMUM ENERGY DENSITY, and PEAK ENERGY DENSITY have meanings analogous to the related terms used for sound pressure.

In speaking of average energy density in general, it is necessary to distinguish between the space average (at a given instant) and the time average (at a given point).

Sound energy flux. The average rate of flow of sound energy for one period through any specified area.

Sound energy flux density level. See *Intensity level*.

Sound field. A region containing sound waves.

Sound insulation. (*a*) The use of structures and materials designed to reduce the transmission of sound from one room or area to another or from the exterior to the interior of a building. (*b*) The degree by which sound transmission is reduced by means of sound insulating structures and materials.

Sound intensity, Sound energy flux density, Sound power density. The sound intensity in a specified direction at a point is the average rate of sound energy transmitted in the specified direction through a unit area normal to this direction at the point considered.

Sound level meter. A device to measure sound level or weighted sound level, constructed in accordance with the standard specifications for sound level meters set up by the American National Standards Institute (ANSI). The sound level meter consists of a microphone, an amplifier to raise the microphone output to useful levels, a calibrated attenuator to adjust the amplification to values appropriate to the sound levels being measured, and an instrument to indicate the measured sound level; optional weighting networks are included to adjust the overall frequency characteristic of the response.

Sound power level (L_w). The sound power level of a sound source, in decibels, is 10 times the logarithm to the base 10 of the ratio of the sound power radiated by the source to a reference power.

Sound power of a source. The total sound energy radiated by the source per unit time.

Sound pressure. The sound pressure at a point is the total instantaneous pressure at that point in the presence of a sound wave minus the static pressure at that point. See *Effective sound pressure*.

Sound pressure level (SPL). The sound pressure level, in decibels, of a sound is 20 times the logarithm to the base 10 of the ratio of the pressure of this sound to the reference pressure. The reference pressure shall be explicitly stated. The following referenced pressures are in common use: (*a*) 0.0002 microbar, or (*b*) 1 microbar. Reference pressure (*a*) is in general use for measurements concerned with hearing and with sound in air, while (*b*) has gained widespread acceptance for calibration of transducers and various kinds of sound measurements in liquids.

Unless otherwise explicitly stated, it is to be understood that the sound pressure is the EFFECTIVE (RMS) SOUND PRESSURE.

Sound probe. A device that responds to some characteristic of an acoustic wave (for example, sound pressure, particle velocity) and that can be used to explore and determine this characteristic in a sound field without appreciably altering that field.

Sound reduction between rooms. The sound reduction, in decibels, between two rooms is the amount by which the mean-square sound pressure level averaged throughout the source room exceeds the same level

room.
averaged throughout the receiving

Sound reflection coefficient. The sound reflection coefficient of a surface is the fraction of incident sound reflected by the surface. Unless otherwise specified, reflection of sound energy in a diffuse sound field is assumed.

Sound, speed of. At 20 C (68 F), sound travels 344 meters/second (1127 fps). The speed varies in direct proportion to the temperature; that is, the speed increases 0.607 meter/second per degree Celsius (1.1 fps per degree Fahrenheit).

Sound transmission class (STC). A single-number rating system which compares the sound transmission loss (STL) of a test specimen with a standard contour.

Sound transmission coefficient. See *Transmission coefficient.*

Sound transmission loss (STL). See *Transmission loss.*

Sound transmission loss of a partition. See *Transmission loss of a partition.*

Sound waves. Vibratory motion caused by a vibrating body which transmits a series of alternating compressions and rarefactions (expansions) of the elastic medium.

Speaker. See *Loudspeaker.*

Specific acoustic impedance. Unit-area acoustic impedance. The specific acoustic impedance at a point in the medium is the complex ratio of sound pressure to particle velocity.

Specific acoustic reactance. The imaginary component of the specific acoustic impedance.

Specific acoustic resistance. The real component of the specific acoustic impedance.

Spectrum. Spectrum is used to signify a continuous range of components, usually wide in extent, within which waves have some specified common characteristics; for example, audiofrequency spectrum.

• OCTAVE-BAND SPECTRUM. The ensemble of octave-band pressure levels comprising a sound. The audible spectrum is conventionally taken to embrace eight octaves. See also *Octave band.*

• ⅓ OCTAVE-BAND SPECTRUM. The ensemble of band-pressure levels in bands ⅓ octave wide comprising a sound.

Spectrum level. Spectrum density level. The spectrum level of a specified signal at a particular frequency is the level of that part of the signal contained within a band 1 Hz wide, centered at the particular frequency. Ordinarily this has significance only for a signal having a continuous distribution of components within the frequency range under consideration. The words "spectrum level" cannot be used alone, but must appear in combination with a prefatory modifier, such as pressure, velocity, voltage.

Speech audiometer. See *Audiometer, speech.*

Speech interference level (SIL). The speech interference level of a noise is the average, in decibels, of the sound pressure levels of the noise in the three octave bands of frequency 600–1200, 1200–2400, and 2400–4800 Hz, or 500, 1000, and 2000 Hz if preferred center frequencies are used.

Speech perception test. A measurement of hearing acuity by the administration of a carefully controlled list of words. The identification of correct responses is evaluated in terms of norms established by the average performance of normal listeners.

Speech reading. Also called LIP READING or VISUAL HEARING. The interpretation of movements of head, lips, and face as an aid to communication by speech.

Spherical wave. A wave in which the wave fronts are concentric spheres.

Spondee words. Standardized word list composed of two-syllable words equally stressed on both syllables.

Standard deviation. A numerical indication of the scatter of values of a variable around their mean (symbol σ). Specifically, it is the square root of the mean squared deviation, given by

$$\sigma = \{\Sigma(X - x)^2/n\}^{\frac{1}{2}}$$

In a normal distribution, a range of two standard deviations on each side of the mean will include 95.45 percent of all values, and three standard deviations on each side of the mean, 99.73 percent.

When the standard deviation is determined from a sample of values and not from the parent population (which is frequently unknown), the denominator within the square root should be taken in n − 1, due to the fact that one degree of freedom out of the total n is used to estimate the mean value x.

Standard error of the mean of a sample. The standard error is found by dividing the standard deviation of the items in the sample by the square root of their number. The mean of the population from which the sample was obtained is not likely to differ from the sample mean by more than plus or minus twice the standard error.

Standing wave. A periodic wave having a fixed distribution in space which is the result of interference of progressive waves of the same frequency and kind. Such waves are characterized by the existence of nodes or partial nodes and antinodes that are fixed in space.

Stapedectomy. A corrective operation for otosclerosis. The stapes can be removed and replaced by a prosthetic device.

Stapes. See *Ossicles.*

Static deflection. The deflection of the spring in an elastic system resulting from the dead weight of the supported load.

Static pressure. The static pressure at a point is the pressure that would exist at that point in the absence of sound waves.

Static value. See *Average value.*

Stationary wave. A standing wave in which the net energy flux is zero at all points. Stationary waves can only be approximated in practice.

Statistical significance. An expression of the degree of confidence which can be attached to the occurrence of an event. It is normally given a numerical value by stating the proportion of times on which the event could occur by chance. The conventional level of significance is that the occurrence could only arise by chance in 5 percent of cases; commonly designated probability, $P = 0.05$.

Steady state. That state said to have been reached by a system when the relevant variable of the system no longer changes as a function of time.

Steady-state noise. Noises that are continuous or that consist of impulses spaced less than one second apart shall be considered to be steady-state noises.

Steady-state vibration. After a vibrating system has been acted upon by a definite force for a sufficient time, it will follow a definite cycle of events described as steady-state vibration. Steady-state vibration exists in a system if the velocity of each particle is a continuing periodic quantity.

Stereophonic sound system. A sound system in which a plurality of microphones, transmission channels, and loudspeakers are arranged so as to provide a sensation of spatial distribution of the sound sources to the listener.

Stiffness. The ratio of change of force (or torque) to the corresponding change in translational (or rotational) displacement of an elastic element.

Stimulus. See *Excitation.*

Strain. The ratio of the elongation, ΔL, of a material subject to stress, to the original length of the material, L.

Strength of a sound source. The strength of a sound source is the maximum instantaneous rate of volume displacement produced by the source when emitting a wave with sinusoidal time variation. The term is properly applicable only to sources whose dimensions are small with respect to the wavelength.

Structure-borne sound. Sound energy transmitted through the solid media of the building structure.

Subharmonic. A sinusoidal quantity having a frequency which is an integral submultiple of the fundamental frequency of a periodic quantity to which it is related. For example, a wave, the frequency of which is half the fundamental fre-

quency of another wave, is called the SECOND SUBHARMONIC of that wave.

Subsonic. See *Infrasonic frequency.*

Supersonics is the general subject covering phenomena associated with speed higher than the speed of sound (as in the case of aircraft and projectiles traveling faster than sound). Supersonics was once used in acoustics synonomously with ultrasonics; however, because of the ambiguity its use is not recommended. See *Ultrasonics.*

Sweep rate. The time rate of change of frequency. A uniform sweep rate exists if the rate of change of frequency is constant.

T

Temporary threshold shift (TTS). The component of threshold shift which shows progressive reduction with the passage of time when the apparent cause has been removed. The temporary hearing loss suffered as the result of noise exposure, if all or part of the loss is recovered during an arbitrary period of time when one is no longer subjected to the noise.

Test space. An area where sound surveys are made.

Threshold of audibility. The threshold of audibility for a specified signal is the minimum effective sound pressure level of the signal that is capable of evoking an auditory sensation in a specified fraction of the trials. The characteristics of the signal, the manner in which it is presented to the listener, and the point at which the sound pressure level is measured must be specified.

Unless otherwise indicated, the ambient noise reaching the ears is assumed to be negligible. The threshold is usually given as a level in decibels. See also *Aural critical band*.

Threshold of feeling (or tickle) for a specified signal. The minimum sound level at the entrance to the external auditory canal which, in a specified fraction of the trials, will stimulate the ear to a point at which there is a sensation of feeling that is different from the sensation of hearing.

Threshold of hearing of a continuous sound. The minimum value of the sound pressure which excites the sensation of hearing.

Threshold of pain for a specified signal. The minimum effective sound pressure level of that signal which, in a specified fraction of the trials, will stimulate the ear to a point at which there is a sensation of pain that is distinct from a feeling of discomfort.

Timbre. That attribute of auditory sensation in terms of which a listener can judge that two sounds similarly presented and having the same loudness and pitch are dissimilar. Timbre depends primarily upon the spectrum of the stimulus, but it also depends upon the waveform, the sound pressure, the frequency location of the spectrum, and the temporal characteristics of the stimulus.

Tinnitus. A subjective sense of noises in the head or ringing in the ears for which there is no observable external cause.

Tone. (*a*) A sound wave capable of exciting an auditory sensation having pitch. (*b*) A sound sensation having pitch. See *Pure tone*.

Tone deafness. The inability to make a close discrimination between fundamental tones close together in pitch.

Tone decay, abnormal. See *Adaptation*.

Transducer. A device capable of being actuated by waves from one or more transmission systems or media and of supplying related waves to one or more other transmission systems or media. The waves in either input or output may be of the same or different types (electric, mechanical, or acoustic).

Transient. Transient motion. Any motion which has not reached or has ceased to be a steady state.

Transient vibration. Temporarily sustained vibration of a mechanical system. It may consist of forced or free vibration or both. Any motion in a vibrating system which occurs during the time required for the system to adapt itself from one condition to another is called transient vibration.

Transmissibility. The nondimensional ratio of the response amplitude of a system in steady-state forced vibration to the excitation amplitude. The ratio may be one of forces, displacements, velocities, or accelerations. See *Isolation*.

Transmission. The propagation of a vibration through various media.

Transmission coefficient. The sound-transmission coefficient of a partition is the fraction of incident sound transmitted through the partition. The angle of incidence and the characteristic of sound observed must be specified; for example, pressure amplitude at normal incidence.

Transmission loss. The ratio, expressed in decibels, of the sound energy incident on a structure to the sound energy which is transmitted. The term is applied both to building structures (walls, floors, etc.) and to air passages (muffler, ducts, etc.).

Transmission loss is the reduction in the magnitude of some characteristic of a signal, between two stated points in a transmission system. The characteristic is often some kind of level, such as power level; in acoustics the characteristic that is commonly measured is sound pressure level. Thus, if the levels are expressed in decibels, the transmission level loss is likewise in decibels.

It is imperative that the characteristic concerned (such as the sound pressure level) be clearly identified because in all transmission systems more than one characteristic is propagated.

Transmission loss of a partition (TL). The sound transmission loss of a partition is equal to the number of decibels by which sound incident on a partition is reduced in transmission through it. It is thus a measure of the airborne sound insulation of the partition. Unless otherwise specified, it is to be understood that the sound fields on both sides of the partition are diffuse.

Transverse wave. A wave in which the direction of displacement at each point of the medium is parallel to the wave front.

Trauma. An injury or wound brought about by an outside force.

Trend curves. A set of curves describing the central tendency, and sometimes the dispersion or growth with time.

Tuned damper. See *Dynamic vibration absorber.*

Tuning fork. Instrument used to produce sound waves by vibrating prongs.

Tympanum. The space between the drum membrane and the bony capsule of the inner ear. Also called the MIDDLE EAR.

U

Ultrasonics. The technology of sound at frequencies above the audio range.

Undamped natural frequency. In a mechanical system, the frequency of free vibration resulting from only elastic and inertial forces of the system.

Unidirectional microphone. A microphone that is responsive predominantly to sound incident from a single solid angle of one hemisphere or less.

Unit-area acoustic impedance. See *Specific acoustic impedance.*

Unwanted sound. NOISE; interfering sound, whatever its source or nature. See also *Noise.*

V

Variable-resistance transducer. A transducer that depends for its operation upon sound actuated variation in electrical resistance.

Variance. The mean squared deviation of the values of a variable from their mean. The square of the standard deviation.

Vascular. Pertaining to or involving the blood or lymph vessels.

Vector. A quantity that has both magnitude and direction. See *Scalar.*

Velocity. Velocity is a vector that specifies the time rate of change of displacement with respect to a reference frame.

Velocity microphone. A microphone in which the electric output substantially corresponds to the instantaneous particle velocity in the impressed sound wave.

Velocity of sound. See *Sound, speed of.*

Velocity shock. A mechanical shock resulting from a nonoscillatory change in velocity of an entire system.

Vertigo. The sensation of a person that objects are turning in circles around him. Sensation of movement causing unsteadiness.

Vestibular mechanism. The balance portion of the inner ear, consisting of the UTRICLE, the SACCULE and the three SEMICIRCULAR CANALS. See also *Semicircular canals.*

Vibration. Vibration is an oscillation wherein the quantity is a parameter that defines the motion of a mechanical system. See also *Fundamental mode of vibration* and *Mode of vibration.*

Vibration isolation. Any of several means of preventing transmission of sound vibrations from a vibrating body to the structure in which or on which it is mounted.

Vibration isolator. A resilient support that tends to isolate a system from steady-state excitation.

Vibration meter. An apparatus for the measurement of displacement, velocity, or acceleration of a vibrating body.

Viscous damping. The dissipation of energy that occurs when a particle in a vibrating system is resisted by a force that has a magnitude proportional to the magnitude of the velocity of the particle and direction opposite to the direction of the particle.

Voltage (level) gain. The excess of the output voltage level in decibels over the input voltage level in decibels. By reason of the properties of logarithms, it is also 20 times the common logarithm of the ratio of the output voltage to the input voltage.

Ordinarily the name of this quantity can be shortened without ambiguity to voltage gain in decibels.

Volume unit (VU). A unit for expressing the magnitude of a complex electric wave, such as that corresponding to speech or music. The volume in VU is equal to the number of decibels by which the wave differs from reference volume.

Volume velocity. The rate of alternating flow of the medium through a specified surface due to a sound wave.

W

Watch tick test. These tests and other crude testing devices, and WHISPER TESTS, are now outmoded by audiometric measurements.

Wave. A disturbance which is propagated in a medium in such a manner that at any point in the medium, the quantity serving as measure of disturbance is a function of the time, while at any instant the displacement at a point is a function of the position of the point. Any physical quantity that has the same relationship to some independent variable (usually time) that a propagated disturbance has, at a particular instant with respect to space, may be called a wave. Note: For a particular waveform, look under the characteristic; for example, for plane wave see *Plane wave.*

Wave filter. See *Filter.*

Wave front. (*a*) The wave front of a progressive wave in space is a continuous surface which is a locus of points having the same phase at a given instant. (*b*) The wave front of a progressive surface wave is a continuous line which is a locus of points having the same phase at a given instant.

Wave interference. The phenomenon which results when waves of the same or nearly the same frequency are superposed; it is characterized by a spatial or temporal distribution of amplitude of some specified characteristic that differs from that of the individual superposed waves.

Wave velocity. Propagation velocity. A vector quantity that specifies the speed and direction with which a sound wave travels through a medium.

Wavelength of a periodic wave in an isotropic medium is the perpendicular distance between two wave fronts in which the displacements have a difference in phase of one complete period.

Weighting curve. A curve describing the frequency response of a sound level meter.

Weighting network. An electrical network designed to be incorporated in a sound level meter such that the latter conforms to a specified weighting curve. The three networks are designated A, B, and C, according to ANSI S1.4–1971. See also *Sound level.*

White noise. A noise whose spectral density (or spectrum level) is substantially independent of frequency over a specified range. White noise need not be random. See also *Noise.*

Wideband. Applies to a wide band of transmitted waves, with neither of the critical or cutoff frequencies of the filter being zero or infinite. See also *Broadband.*

Wide-range audiometer. See *Audiometer, pure tone.*

Windscreen. A shield placed around a microphone. When strong enough the wind causes a turbulent airstream to be developed around the microphone which in turn causes the microphone diaphragm to move in a way similar to that produced by a high noise level.

Bibliography

American Industrial Hygiene Association (AIHA). *Industrial Noise Manual,* 2nd ed. Detroit, Mich., 1966.

American National Standards Institute (ANSI). *Acoustical Terminology,* S1.1–1960. New York, N.Y., 1960.

Blake, Michael P., and Mitchell, William S., eds. *Vibration and Acoustic Measurement Handbook.* Spartan Books, New York, N.Y., 1972.

Burns, W., and Robinson, D. W. *Hearing and Noise in Industry.* Her Majesty's Stationery Office, London, England, 1970.

Crede, Charles E. "Principles of Vibration Control." *Handbook of Noise.* McGraw-Hill Book Co., New York, N.Y., 1957.

Employers Insurance of Wausau, Safety and Health Services. *Guide for Industrial Audiometric Technicians,* 2nd ed. Wausau, Wis., 1967.

Instrument Society of America (ISA). *Nomenclature and Specification Terminology for Aerospace Test Transducers with Electrical Output,* RP37.1–1963. Pittsburgh, Pa., 1963.

Keast, David N. *Measurements in Mechanical Dynamics.* McGraw-Hill Book Co., New York, N.Y., 1967.

Petersen, Arnold P. G. and Gross, Ervin, E., Jr. *Handbook of Noise Measurement,* 6th ed. General Radio Co. West Concord, Mass., 1967.

Schmidt, J. E. *Paramedical Dictionary.* Charles C Thomas, Springfield, Ill., 1969.

Williams & Wilkins Co. *Stedman's Medical Dictionary,* 22nd ed. Baltimore Md., 1972.

Yerges, Lyle F. *Sound, Noise, and Vibration Control.* Van Nostrand Reinhold Co., New York, N.Y., 1969.

Appendix G Review of Mathematics

Significant figures

Measurements often result in what are called *approximate numbers* in contrast to *discrete counts*.

For example: The dimensions of a table are reported as 29.6 in. by 50.2 in. This implies that the measurement is to the nearest tenth of an inch and that the table is less than 50.25 in. and more than 50.15 in. in length. One can show the same thing for the width, using the symbolic notation:

29.55 in. < width < 29.65 in.

If, on the other hand, one knows the degree of precision of the measurement (say ± 0.03 in. or ± 0.08 in.), he may write

50.2 ± .03 or 50.2 ± .08

to indicate the degree of accuracy of the measurement of the length.

In reporting results, the number of significant digits that can be recorded is determined by the precision of the instruments used.

Rules

• In any approximate number, the significant digits include the digit that determines the degree of precision of the number and all digits to the left of it, except for zeros used to place the decimal.

• All digits from 1 to 9 are significant.

• All zeros that are between significant digits are significant.

• Final zeros of decimal numbers are significant. For example:

Number	Number of Significant Digits
0.0702	3
0.07020	4
70.20	4
7,002	4
7,020	3

Scientific notation

One case where it is difficult to determine the number of significant digits: Example: 7000. In general, it is considered to have only one significant digit. It is better to use scientific notation.

In standard scientific notation, one writes the number as a number between 1 and 10, in which only the significant digits are shown, multiplied by an exponential number to the base 10. For example:

Number		Number of Significant Digits
$5,320,000 = 5.32 \times 10^6$		3
$= 5.320 \times 10^6$		4
$= 5.3200 \times 10^6$		5
$0.00000532 = 5.32 \times 10^{-6}$		3

Addition and subtraction

The result must not have more decimal places than the number with the fewest decimal places. For example:

21.262	should be	21.3
23.74	should be	23.7
139.6	should be	139.6
184.602	should be	184.6

This can also be written as:

$$21.262 \pm 0.0005$$
$$23.75 \pm 0.005$$
$$139.6 \pm 0.05$$
$$184.612 \pm 0.0555$$

Multiplication and division

The result must not have more significant places than are possessed by the number with the fewest significant digits. For example:

$$(50.20)(29.6) = 1485.92$$
$$= 1490$$
$$= 1.49 \times 10^3$$

Logarithms

Logarithms are exponents. The logarithm of any number is the power to which a selected base must be raised to produce the number. The laws of exponents apply to logarithms.

The two equations $a^x = y$ and $x = \log_a y$ are two ways of expressing the same thing, i.e., the exponent applied to a to give y is equal to x. The number a is called the base of the system of logarithms.

Although any positive number greater than 1 can be used as the base of some system of logarithms, there are two systems in general use. These are the *common* or Briggs's system and the *natural* or Napierian system. In the common system the base is 10, while in the natural system the base is a certain irrational number $e = 2.71828 \ldots$.

Common logarithms

Common logarithms use the base 10 and are identified by the notation "log." The common logarithm of a number consists of a characteristic, which locates the decimal point in the number, and a mantissa, which defines the numerical arrangement of the number.

A bar over a characteristic indicates

(*Text continues on page 1068.*)

COMMON

N	0	1	2	3	4	5	6	7	8	9
0	....	0000	3010	4771	6021	6990	7782	8451	9031	9542
1	0000	0414	0792	1139	1461	1761	2041	2304	2553	2788
2	3010	3222	3424	3617	3802	3979	4150	4314	4472	4624
3	4771	4914	5051	5185	5315	5441	5563	5682	5798	5911
4	6021	6128	6232	6335	6435	6532	6628	6721	6812	6902
5	6990	7076	7160	7243	7324	7404	7482	7559	7634	7709
6	7782	7853	7924	7993	8062	8129	8195	8261	8325	8388
7	8451	8513	8573	8633	8692	8751	8808	8865	8921	8976
8	9031	9085	9138	9191	9243	9294	9345	9395	9445	9494
9	9542	9590	9638	9685	9731	9777	9823	9868	9912	9956
10	0000	0043	0086	0128	0170	0212	0253	0294	0334	0374
11	0414	0453	0492	0531	0569	0607	0645	0682	0719	0755
12	0792	0828	0864	0899	0934	0969	1004	1038	1072	1106
13	1139	1173	1206	1239	1271	1303	1335	1367	1399	1430
14	1461	1492	1523	1553	1584	1614	1644	1673	1703	1732
15	1761	1790	1818	1847	1875	1903	1931	1959	1987	2014
16	2041	2068	2095	2122	2148	2175	2201	2227	2253	2279
17	2304	2330	2355	2380	2405	2430	2455	2480	2504	2529
18	2553	2577	2601	2625	2648	2672	2695	2718	2742	2765
19	2788	2810	2833	2856	2878	2900	2923	2945	2967	2989
20	3010	3032	3054	3075	3096	3118	3139	3160	3181	3201
21	3222	3243	3263	3284	3304	3324	3345	3365	3385	3404
22	3424	3444	3464	3483	3502	3522	3541	3560	3579	3598
23	3617	3636	3655	3674	3692	3711	3729	3747	3766	3784
24	3802	3820	3838	3856	3874	3892	3909	3927	3945	3962
25	3979	3997	4014	4031	4048	4065	4082	4099	4116	4133
26	4150	4166	4183	4200	4216	4232	4249	4265	4281	4298
27	4314	4330	4346	4362	4378	4393	4409	4425	4440	4456
28	4472	4487	4502	4518	4533	4548	4564	4579	4594	4609
29	4624	4639	4654	4669	4683	4698	4713	4728	4742	4757
30	4771	4786	4800	4814	4829	4843	4857	4871	4886	4900
31	4914	4928	4942	4955	4969	4983	4997	5011	5024	5038
32	5051	5065	5079	5092	5105	5119	5132	5145	5159	5172
33	5185	5198	5211	5224	5237	5250	5263	5276	5289	5302
34	5315	5328	5340	5353	5366	5378	5391	5403	5416	5428
35	5441	5453	5465	5478	5490	5502	5514	5527	5539	5551
36	5563	5575	5587	5599	5611	5623	5635	5647	5658	5670
37	5682	5694	5705	5717	5729	5740	5752	5763	5775	5786
38	5798	5809	5821	5832	5843	5855	5866	5877	5888	5899
39	5911	5922	5933	5944	5955	5966	5977	5988	5999	6010
40	6021	6031	6042	6053	6064	6075	6085	6096	6107	6117
41	6128	6138	6149	6160	6170	6180	6191	6201	6212	6222
42	6232	6243	6253	6263	6274	6284	6294	6304	6314	6325
43	6335	6345	6355	6365	6375	6385	6395	6405	6415	6425
44	6435	6444	6454	6464	6474	6484	6493	6503	6513	6522
45	6532	6542	6551	6561	6571	6580	6590	6599	6609	6618
46	6628	6637	6646	6656	6665	6675	6684	6693	6702	6712
47	6721	6730	6739	6749	6758	6767	6776	6785	6794	6803
48	6812	6821	6830	6839	6848	6857	6866	6875	6884	6893
49	6902	6911	6920	6928	6937	6946	6955	6964	6972	6981
50	6990	6998	7007	7016	7024	7033	7042	7050	7059	7067
N	0	1	2	3	4	5	6	7	8	9

LOGARITHMS

N	0	1	2	3	4	5	6	7	8	9
50	6990	6998	7007	7016	7024	7033	7042	7050	7059	7067
51	7076	7084	7093	7101	7110	7118	7126	7135	7143	7152
52	7160	7168	7177	7185	7193	7202	7210	7218	7226	7235
53	7243	7251	7259	7267	7275	7284	7292	7300	7308	7316
54	7324	7332	7340	7348	7356	7364	7372	7380	7388	7396
55	7404	7412	7419	7427	7435	7443	7451	7459	7466	7474
56	7482	7490	7497	7505	7513	7520	7528	7536	7543	7551
57	7559	7566	7574	7582	7589	7597	7604	7612	7619	7627
58	7634	7642	7649	7657	7664	7672	7679	7686	7694	7701
59	7709	7716	7723	7731	7738	7745	7752	7760	7767	7774
60	7782	7789	7796	7803	7810	7818	7825	7832	7839	7846
61	7853	7860	7868	7875	7882	7889	7896	7903	7910	7917
62	7924	7931	7938	7945	7952	7959	7966	7973	7980	7987
63	7993	8000	8007	8014	8021	8028	8035	8041	8048	8055
64	8062	8069	8075	8082	8089	8096	8102	8109	8116	8122
65	8129	8136	8142	8149	8156	8162	8169	8176	8182	8189
66	8195	8202	8209	8215	8222	8228	8235	8241	8248	8254
67	8261	8267	8274	8280	8287	8293	8299	8306	8312	8319
68	8325	8331	8338	8344	8351	8357	8363	8370	8376	8382
69	8388	8395	8401	8407	8414	8420	8426	8432	8439	8445
70	8451	8457	8463	8470	8476	8482	8488	8494	8500	8506
71	8513	8519	8525	8531	8537	8543	8549	8555	8561	8567
72	8573	8579	8585	8591	8597	8603	8609	8615	8621	8627
73	8633	8639	8645	8651	8657	8663	8669	8675	8681	8686
74	8692	8698	8704	8710	8716	8722	8727	8733	8739	8745
75	8751	8756	8762	8768	8774	8779	8785	8791	8797	8802
76	8808	8814	8820	8825	8831	8837	8842	8848	8854	8859
77	8865	8871	8876	8882	8887	8893	8899	8904	8910	8915
78	8921	8927	8932	8938	8943	8949	8954	8960	8965	8971
79	8976	8982	8987	8993	8998	9004	9009	9015	9020	9025
80	9031	9036	9042	9047	9053	9058	9063	9069	9074	9079
81	9085	9090	9096	9101	9106	9112	9117	9122	9128	9133
82	9138	9143	9149	9154	9159	9165	9170	9175	9180	9186
83	9191	9196	9201	9206	9212	9217	9222	9227	9232	9238
84	9243	9248	9253	9258	9263	9269	9274	9279	9284	9289
85	9294	9299	9304	9309	9315	9320	9325	9330	9335	9340
86	9345	9350	9355	9360	9365	9370	9375	9380	9385	9390
87	9395	9400	9405	9410	9415	9420	9425	9430	9435	9440
88	9445	9450	9455	9460	9465	9469	9474	9479	9484	9489
89	9494	9499	9504	9509	9513	9518	9523	9528	9533	9538
90	9542	9547	9552	9557	9562	9566	9571	9576	9581	9586
91	9590	9595	9600	9605	9609	9614	9619	9624	9628	9633
92	9638	9643	9647	9652	9657	9661	9666	9671	9675	9680
93	9685	9689	9694	9699	9703	9708	9713	9717	9722	9727
94	9731	9736	9741	9745	9750	9754	9759	9763	9768	9773
95	9777	9782	9786	9791	9795	9800	9805	9809	9814	9818
96	9823	9827	9832	9836	9841	9845	9850	9854	9859	9863
97	9868	9872	9877	9881	9886	9890	9894	9899	9903	9908
98	9912	9917	9921	9926	9930	9934	9939	9943	9948	9952
99	9956	9961	9965	9969	9974	9978	9983	9987	9991	9996
100	0000	0004	0009	0013	0017	0022	0026	0030	0035	0039
N	0	1	2	3	4	5	6	7	8	9

Number (Anti-log)	Exponential Form	Common Logarithmic Form		
		Characteristic	Mantissa	Complete Log
0.0005	5×10^{-4}	-4	0.7	$\overline{4}.7$
0.05	5×10^{-2}	-2	0.7	$\overline{2}.7$
5.0	5×10^{0}	0	0.7	0.7
500.0	5×10^{2}	2	0.7	2.7
50000.0	5×10^{4}	4	0.7	4.7

a negative characteristic and a positive mantissa. The log may be written $\overline{4}.7$ or 6.7–10 or −3.3. The form −3.3 does not contain a characteristic and mantissa.

The integral part of a logarithm is called the *characteristic* and the decimal part is called the *mantissa*. In log 845, the characteristic is 2 and the mantissa is 0.9269. For convenience in constructing tables, it is desirable to select the mantissa as positive even if the logarithm is a negative number. For example, log ½ = −0.3010; but since −0.3010 = 9.6990 − 10, this may be written log ½ = 9.6990 − 10 with a positive mantissa. This is also the log of 0.5, which we should have looked up in the first place. The following illustration shows the method of writing the characteristic and mantissa:

log	8245	= 3.9162
log	824.5	= 2.9162
log	82.45	= 1.9162
log	8.245	= 0.9162
log	0.8245	= 9.9162−10
log	0.08245	= 8.9162−10

How to use logarithms

If the laws of exponents are rewritten in terms of logarithms, they become the *laws of logarithms*:

$$\log_a (x^n) = n \log_a x$$
$$\log_a \left(\frac{x}{y}\right) = \log_a x - \log_a y$$
$$\log_a (x^n) = n \log_a x$$

Logarithms derive their main usefulness in computation from the above laws, since they allow multiplication, division, and exponentiation to be replaced by the simpler operations of addition, subtraction, and multiplication, respectively.

How to use logarithm tables

In this appendix is a "four-place" table of logarithms. In this table, the mantissas of the logarithms of all integers from 1 to 999 are recorded correct to four decimal places, which is all one needs to work with decibels, which at best have three significant digits.

To find the logarithm of a given number, use the table as follows.

For example, find the logarithm of 63.5.

Glance down the column headed *N*

for the first two significant digits (63), and then along the top of the table for the third figure (5). In a row with 63 and under the column with 5 is found 8028. This is the mantissa. Adding the proper characteristic 1, the logarithm (or log) of 63.5 is 1.8028.

Conversely, one can find the number that corresponds to a given logarithm (the antilogarithm).

For example, find the number whose logarithm is 1.6355. The mantissa 6355 corresponds to the number in the table that is under the column with 2 and in the row with 43. Thus, the mantissa corresponds to the number 432. Since the characteristic is 1, the number whose logarithm is 1.6355 is 43.2.

Because in measuring sound we are concerned only with three significant digits, the number whose logarithm is 1.6360 would also be 43.2. The number whose logarithm is 1.6361 would be 43.3.

Decibel notation

If two sound intensities P_1 and P_2 are to be compared according to the ability of the ear to detect intensity differences, we may determine the number of decibels which expresses the relative value of the two intensities by

$$N_{db} = 10 \log_{10} \frac{P_1}{P_2}$$

where P_1 is greater than P_2.

The factor 10 comes into this picture because the original unit devised was the *bel*, which is the logarithm of

10 to the base 10 and which represents the audible difference between two powers one of which is 10 times as great as the other. The decibel is one-tenth of a bel, and, therefore, there are 10 times as many decibels in any expression involving the relation between two sound intensities as there are bels.

The decibel is a logarithmic unit. Each time the amount of power is increased by a factor of 10, we have added 10 decibels (abbreviated dB).

To determine the number of decibels by which two powers differ, we must *first determine the ratio of the two powers*, then we look up this ratio in a table of logarithms to the base 10 and then we multiply the figure obtained by a factor of 10.

If we want to find the relative loudness of 10,000 people who can shout louder than 100 people can, we use the following reasoning.

The logarithm (to the base 10) of any number is merely the number of times 10 must be multiplied by itself to be equal to the number. In the example here, 100 represents 10 multiplied by itself, and the logarithm of 100 to the base 10, therefore, is 2. For example, the number of decibels expressing the relative loudness of 10,000 people shouting compared with 100 is

$$N_{db} = 10 \log_{10} (10,000 \div 100)$$
$$= 10 \log_{10} 100$$
$$= 10 \times 2.0$$
$$= 20$$

Now let us see what happens if we double the number of people to 20,000.

$N_{db} = 10 \log_{10} (20{,}000 \div 100)$
$= 10 \log_{10} 200$
$= 10 \times 2.3010$
$= 23$ (rounded to significant digits)

It can be seen, therefore, that decibels are logarithm ratios. In their use in sound measurement, P_o (the usual reference level) is 20 micropascals or 0.0002 dynes/square centimeter, which approximates the "threshold of hearing," the sound that can just be heard by a young person with excellent hearing. This was described in Part I—Introduction and in Part II—Sound Measurement.

SELECTED COMMON ABBREVIATIONS

A	area	**ft**	foot, feet
abs	absolute	**ft lb**	foot pound
AC	air conduction	**fpm**	feet per minute
AI	articulation index	**fps**	feet per second
amb	ambient	**fps system**	foot-pound-second system
amp	ampere	**g**	gauss, gravity
atm	atmosphere	**gal**	gallon
BC	bone conduction	**gm**	gram
c/s	cycles per second (*see* hertz)	**hp**	horsepower
		HL	hearing level
cc, cm³	cubic centimeter	**hr**	hour
cfm	cubic feet per minute	**HTL**	hearing threshold level
cfs	cubic feet per second	**Hz**	hertz (cycles per second)
cg	centigram	**I**	sound intensity
cgs system	centimeter-gram-second system	**ID**	inside diameter
		IIC	impact insulator class
cm	centimeter	**in.**	inch
CNEL	community noise equivalent level	**INR**	impact noise rating
		j	jerk
CNR	composite noise rating	**kg**	kilogram
D	daily noise dose	**km**	kilometer
dB	decibel	**L**	sound level, loudness
dBA	decibel A-weighting	L_A	dBA or A-weighted sound level
dBC	decibel C-weighting		
DI	directivity index	L_{DN}	day/night average sound level
DRC	damage risk criteria		
EPNL	effective perceived noise level	L_{eq}	equivalent sound level
		L_L	loudness (phons)
f	frequency	L_{NP}	noise pollution level

L_{rms}	root-mean-square sound level
L_{10}	sound level exceeded 10 percent of the time
L_{50}	sound level exceeded 50 percent of the time
liq	liquid
log	logarithm (common)
ln	logarithm (natural)
m	meter
max	maximum
Mel	unit of pitch
mks system	meter-kilogram-second system
mph	miles per hour
mg	milligram
ml	milliliter
mm	millimeter
min	minute, minimum
NC	noise criteria
NIL	noise immission level
NIC	noise isolation class
NIHL	noise induced hearing loss
NEF	noise exposure forecast
NIPTS	noise induced permanent threshold shift
NNI	noise and number index
NR	noise reduction, noise rating
NRC	noise reduction coefficient
OB	octave band
OBA	octave band analyzer
OD	outside diameter
oz	ounce
P	sound power
p–p	peak-to-peak
Pa	pascal
PNC	preferred noise criteria curve
PNL	perceived noise level
PNLdB	perceived noise level (dB)
psi	pounds per square inch
psig	pounds per square inch (gage)
PTS	permanent threshold shift
PWL	sound power level
Q	quality factor
rms	root mean square
rpm	revolutions per minute
S	loudness
sec	second
SENEL	single event noise exposure level
SIL	speech interference level
SLM	sound level meter
S/N	signal-to-noise ratio
SPL	sound pressure level
STC	sound transmission class
STL	sound transmission loss
STP	standard temperature and pressure
T	time, noise exposure limit
TL	transmission loss
TTL	temporary threshold loss
TTS	temporary threshold shift
VU	volume units
α	absorption coefficient, sound-power reflection coefficient, transmission coefficient
β	elastic modulus
δ	phase angle between stress and strain
Δ	logarithmic resilience
ϵ	axial strain, strain
θ	temperature
λ	wavelength, positive number
μ	damping coefficient
ρ	density
σ	axial stress, standard deviation
τ	time, one-half wave period
ϕ	phase angle
ψ	damping capacity, attenuation constant
ω	natural radian frequency of a system, circular frequency, angular frequency, frequency (radians/second)

TABLE OF UNIT PREFIXES

Multiples and submultiples	Prefixes	Symbols
$1,000,000,000,000 = 10^{12}$	tera-	T
$1,000,000,000 = 10^9$	giga-	G
$1,000,000 = 10^6$	mega-	M
$1,000 = 10^3$	kilo-	k
$100 = 10^2$	hecto-	h
$10 = 10$	deka-	D
$0.1 = 10^{-1}$	deci-	d
$.01 = 10^{-2}$	centi-	c
$.001 = 10^{-3}$	milli-	m
$.000001 = 10^{-6}$	micro-	μ
$.000000001 = 10^{-9}$	nano-	n
$.000000000001 = 10^{-12}$	pico-	p

National Bureau of Standards

SIGNS AND SYMBOLS

$+$	plus, addition, positive		$\sqrt{}$	square root
$-$	minus, subtraction, negative		$\sqrt[n]{}$	nth root
$\pm$	plus or minus, positive or negative		a^n	nth power of a
$\mp$	minus or plus, negative or positive		$\log$, $\log_{10}$	common logarithm
$\div, /, —$	division		$\ln$, $\log_e$	natural logarithm
$\times, \cdot, ()()$	multiplication		e or ϵ	base of natural logs, 2.718
$()[]$	collection		π	pi, 3.1416
$=$	is equal to		$\angle$	angle
$\neq$	is not equal to		$\perp$	perpendicular to
$\equiv$	is identical to		$\parallel$	parallel to
$\cong$	equals approximately, congruent		n	any number
			$|n|$	absolute value of n
$>$	greater than		$\bar{n}$	average value of n
$\ngtr$	not greater than		a^{-n}	reciprocal of nth power
$\geqq$	greater than or equal to			of $a = \left\{ \dfrac{1}{a^n} \right\}$
$<$	less than			
$\nless$	not less than		$n°$	n degrees (angle)
$\leqq$	less than or equal to		n'	n minutes, n feet
$\because$	proportional to		n''	n seconds, n inches
$:$	ratio		$f(x)$	function of x
$\sim$	similar to		Δx	increment of x
$\propto$	varies as, proportional to		dx	differential of x
$\rightarrow$	approaches		Σ	summation of
∞	infinity		sin	sine
$\therefore$	therefore		cos	cosine
			tan	tangent

EXPONENTS—POWERS OF TEN

Extremely large or small numbers are conventiently expressed as exponential numbers using the base 10. The general laws of exponents apply.

$10^0 = 1$

$10^1 = 10$

$10^2 = 10 \times 10 = 100$

$10^3 = 10 \times 10 \times 10 = 1000$

$10^4 = 10 \times 10 \times 10 \times 10 = 10,000$

$10^5 = 10 \times 10 \times 10 \times 10 \times 10 = 100,000$

$10^6 = 10 \times 10 \times 10 \times 10 \times 10 \times 10 = 1,000,000$

$10^{-1} = \dfrac{1}{10} = 0.1$

$10^{-2} = \dfrac{1}{10^2} = \dfrac{1}{100} = 0.01$

$10^{-3} = \dfrac{1}{10^3} = \dfrac{1}{1000} = 0.001$

$10^{-4} = \dfrac{1}{10^4} = \dfrac{1}{10,000} = 0.0001$

Any number may be expressed as an integral power of 10, or as the product of two numbers one of which is an integral power of 10 (e.g., 300 = 3 × 10².

Examples: $22,400 = 2.24 \times 10^4$ $0.0454 = 4.54 \times 10^{-2}$

In the expression 10^5, the base *is 10 and the* exponent *is 5.*

In multiplication, exponents of like bases are added.

$$a^3 \times a^5 = a^{3+5} = a^8$$
$$10^2 \times 10^3 = 10^{2+3} = 10^5$$
$$10^7 \times 10^{-3} = 10^{7-3} = 10^4$$
$$8000 \times 2500 = (8 \times 10^3)(2.5 \times 10^3) = 20 \times 10^6 = 2 \times 10^7 \text{ or } 20,000,000$$

In division, exponents of like bases are subtracted.

$$\frac{a^5}{a^3} = a^{5-3} = a^2$$

$$\frac{10^2}{10^5} = 10^{2-5} = 10^{-3}$$

$$\frac{0.0078}{120} = \frac{7.8 \times 10^{-3}}{1.2 \times 10^2} = 6.5 \times 10^{-5} \text{ or } 0.000065$$

An expression with an exponent of zero is equal to 1.

$$a^0 = 1, \qquad 10^0 = 1, \qquad (4 \times 10)^0 = 1, \qquad 7 \times 10^0 = 7$$

A factor may be transferred from the numerator to the denominator of a fraction, or vice versa, by changing the sign of the exponent.

$$10^{-4} = \frac{1}{10^4} \qquad 5 \times 10^{-3} = \frac{5}{10^3} \qquad \frac{7}{10^{-2}} = 7 \times 10^2$$

The meaning of the fractional exponent is illustrated by the following:

$$10^{\frac{2}{3}} = \sqrt[3]{10^2} \qquad 10^{\frac{3}{2}} = \sqrt{10^3} \qquad 10^{\frac{1}{2}} = \sqrt{10}$$

To extract the square root, divide the exponent by 2. If the exponent is an odd number it should be increased or decreased by 1, and the coefficient adjusted accordingly. To extract the cube root, divide the exponent by 3. The coefficients are treated independently.

$$\sqrt{90,000} = \sqrt{9 \times 10^4} = 3 \times 10^2 \text{ or } 300$$

Conversion of Octave-Band Levels to A-Weighted Levels

A-weighted sound levels are widely used for noise ratings and it may be necessary to convert measured octave-band levels to the equivalent A-level.

This conversion is readily accomplished by means of the table below.

1. Add the correction numbers given in the table to each of the corresponding measured octave-band levels.

2. Following the rules for the addition of decibels, add up the corrected band levels to obtain the sum (see p. 198).

CORRECTION NUMBERS IN dB
To Be Applied to Octave-Band Levels
To Obtain Equivalent Levels For A-Weighted Levels

Band Center Frequency HZ	Weighting Used for Original Analysis	
	Flat	C
31.5	−37.0	−34.2
63	−24.6	−23.9
125	−15.1	−14.9
250	− 8.0	− 8.0
500	− 2.9	− 2.9
1000	0	0
2000	+ 1.2	+ 1.4
4000	+ 0.9	+ 1.8
8000	− 1.4	+ 1.9

Illustrative example of the conversion of octave-band levels to an A-weighted level.

Measured octave-band levels at a typical industrial noise source:

Band Center Frequency (Hz)	Band Level (dB)
31.5	78
63	76
125	78
250	82
500	81
1000	80
2000	80
4000	73
8000	65

Correction for A-weighting to obtain corrected level.

Band Center Hz	Band Level dB	Correction for A-Weighting	Corrected Level
31.5	78	−37	41
63	76	−25	51
125	78	−15	63
250	82	− 8	74
500	81	− 3	78
1000	80	0	80
2000	80	+ 1	81
4000	73	+ 1	74
8000	65	− 1	64
			85.5 dBA

Sum of corrected levels corresponds to an A-weighted level of 85.5 dBA.

Index

Acoustical (*continued*)
  energy, 298
  engineer, 383, 798
  engineer's zero, 609
  environment, 141, 1019
  materials; *see* Materials
  ohm, 1019
  seal, 538
  stimulus, 52
  tile, 404, 1019
  treatment, 1019
Acoustics, 51, 1019
  architectural, 1019
Active transducer, 1019
  see also Passive transducer
Acusis, 1019
Adaptation, 1019
Adaptive change, 274
  see also Loudness adaptation
Administrative controls, 35, 325, 445,
    730–731
Aerodynamic forces, 412, 416, 493
  see also Noise sources
Aerotitis barotrauma, 229
Aging, 229, 446
  see also Presbycusis
Air compressor, 409
Air conduction, 26, 553, 575, 1019
Air Diffusion Council, 847
Air ejection system, 417, 458–459
Air Force regulations, 298–300, 318
Air Moving and Conditioning Asso-
    ciation, 807, 847
Airborne, 412, 1019
Air-bone gap, 228, 258, 1019
Air-conditioning and Refrigeration In-
    stitute, 807, 847
Aircraft noise standards, 943–946
All-pass network, 1020
Ambient noise, 4, 505, 769, 1220
  levels, 75, 518, 974
American Academy of Occupational
    Medicine, 801

American Academy of Ophthalmol-
    ogy and Otolaryngology, 648,
    802, 809, 851, 1020
American Association of Industrial
    Nurses, 802
American Conference of Governmen-
    tal Industrial Hygienists, 365,
    802
American Council of Otolaryngology,
    802
American Foundrymen's Society, 807
American Gas Association, 808
American Gear Manufacturers Asso-
    ciation, 808, 848
American Industrial Hygiene Associa-
    tion, 802
American Institute of Mining, Metal-
    lurgical, and Petroleum Engi-
    neers, 803
American Insurance Association, 799
American Iron and Steel Institute,
    808
American Medical Association, 336
    803
  hearing impairment formula, 338,
    1020
American Mining Congress, 808
American Mutual Insurance Alliance,
    799
American National Standards Insti-
    tute, 513, 841, 1020
  American Standards Association,
    1021
American Occupational Medical As-
    sociation, 803
American Paper Institute, 808
American Petroleum Institute, 809
American Pulpwood Association, 809
American Society of Heating, Refrig-
    erating, and Airconditioning
    Engineers, 803
American Society of Mechanical En-
    gineers, 804

Audiogram (*continued*)
  termination, 794
  threshold of hearing, 567
  types of, 568
  see also Records
Audiological referral criteria, 636–637
Audiologist, 556, 738, 747, 797, 1022
  in community noise problems, 776
  as a consultant, 776–777
  definition of, 761
  experimental work, 764–765
  as an expert witness, 770
  in hearing-conservation program, 761–777
  hearing test supervision, 768
  industrial, 767–768
  as instructor, 775–776
  performing noise surveys, 770
  in research, 771–772
Audiology, 761–763
  industrial, 766–767
  psychology, 764–765
  physiology, 764–765
Audiometer, 25, 565, 566, 570–571, 628–641
  block diagram, 570
  calibrator; *see* Calibration
  characteristics of, 674–675
  clicks and noises, 676–677
  crosstalk, 678–680
  description of, 565
  face panel, 572
  periodic checks, 559–560
  power source, 571
  problems, 677–678
  schematic, 569
Audiometer, automatic, 577, 625
  advantages of, 634–635
  comparison with manual, 633–636, 638–641
  fixed-frequency type, 632
  operation of, 631
  sweep-frequency type, 631

Audiometer calibration; *see* Calibration
Audiometer care, 565–585
  attenuator noise, 583
  biological check, 583–584
  checklist, 577–583
  crosstalk, 583, 678–680
  daily calibration check, 589–591
  earphones, 578–580, 582
  factory check, 582
  problem areas, 577
Audiometer, continuous tone
  tone switching, 600–602
Audiometer, manual, 571, 745
  comparison with automatic audiometer, 633–636, 638–641
  operation, 565–585
    audiograms, 566–570
    bone-conduction device, 575
    ear-selector switch, 575
    frequency selector, 573–575
    half-octave capability, 574
    intensity control (attenuator), 573
    maximum intensity output, 575
    power switch, 571
    pure-tone level, 576
    transducer, 575, 577
    tone presenter, 571–573
    wide-range audiometer, 565
  power source, 571
  presenting the tone, 571
  test frequencies, 629
Audiometer noise, 582
Audiometer, types of
  automatic; *see* Audiometer, automatic
  Bekesy, 630, 638
  diagnostic, 588
  limited range, 566, 628
  manual; *see* Audiometer, manual
  pure-tone, air-conduction, 589–605, 625–644, 1022
  screening, 589

Ceramic microphone, 103, 139, 1025
  see also Microphones
Certification, 1000–1010
Certified audiometric technician, 738,
  868, 1000–1002
Cerumen, 533, 746, 1025
  impactions, 207
Charge amplifier, 486
Characteristic impedance, 1025
Chart recorder; *see* Graphic level re-
  corder
Cholesteatoma, 209
Circuit noise, 155
Clipping, 149, 1025
Clinical audiologist, 640
  see also Audiologist
Close-talking microphone, 1025
Coal Mining Act, 930–937
Cochlea, 14, 75, 215, 216, 218,
  1025
Coefficient
  absorption, 1017
  temperature, 103
  transmission, 400–402
Coil springs, 496, 498
Coincidence, 1025
Color perception, 268
Collapsing ear canals, 696–697
Combination microphone, 1025
Combined technique, 637
  see also Hearing threshold, deter-
  mination of
Combustion noise, 448–449
Committee on Hearing and Bio-
  Acoustics, 303, 900
Committee on Noise as a Public
  Health Hazard, 805
Communication, 1025
  gap, 635
  in a hearing-conservation program,
  708
Communication problems, 645, 659–
  664

Community noise
  ambient levels, 974
  annoyance, 957
  community response, 957
  complaints, 978–979
  criteria, 973–980
  noise parameters, 974
  problems, 776
  surveys, 978
Community noise equivalent level,
  977
Compensation; *see* Workers' Com-
  pensation
Complex quantity, 1025
Complex tone, 1025
Compliance, 1025
  records, 715
Complaints, symptomatic, 263
Composite noise rating (CNR), 1025
Compressed air, 449
Compressed Air and Gas Institute,
  810, 848
Compressed Gas Association, Inc.,
  810
Compressional wave, 1026
Compressors, 522
Condenser microphone, 106, 1026
Conductive hearing loss; *see* Hearing
  loss *and* Hearing impairment
Configuration; *see* Audiogram
Congenital deafness, 220, 792
Congenital deformities, 214
Confidence limits, 1026
Consistent error, 1026
Consonant discrimination, 249
Constant percentage analyzer, 115
Construction materials, 382
Construction Safety Act, 318
Consultants; *see* Professional societies
Contact microphone, 1026
Continuous sound spectrum, 1026
Continuous system, 1026
Continuous tone, 632

O

Vibration isolators, 391, 465, 478, 498
Vibration level, 170, 463, 464, 484, 712
Vibration measurement, 70–72, 118–121, 470, 481–493
  acceleration, 71–72, 484
  accelerometers, 485–486, 487, 493
  accessories, 488–489
  analyzers, 486, 488
  cable noise, 492
  calibrators, 489
  displacement, 70, 484
  in the field, 489–493
  ground loop connection, 493
  handheld probes, 491
  jerk, 72
  meters, 119, 120, 485, 1061
  mounting accelerometers, 489, 491, 493
  narrow-band analyzer, 472
  operating speeds, 468
  pickup system, 120, 472, 485
  preamplifier, 486
  recorder, 488
  setting up instruments, 492–493
  shock, 489
  special instruments, 467–468
  spectrum analysis, 483
  velocity, 70, 484
Vibratory motion, 468, 485
Vibrissae, 206
Viral infection, 783
Viscoelastic material, 392, 429
Viscous damping, 1061
Vocal effort; see Speech interference level
Voice coil, 577
Voltage amplifier, 486
Voltage (level) gain, 1061
Voltmeter, 108, 598
Volume unit (VU), 1061
Volume velocity, 1061
Vowel sounds, 661–662

W

Wage loss, 328
Waiting period, 341
Walls
  acoustical treatment, 442–443
  as a sound barrier, 439–440
  double and single compared, 441
  cinder block, 439
  double construction, 440–441
  single solid construction, 439–440
  sound reflection from, 441–443
  transmission loss, 439, 440
  see also Noise control
Walsh-Healey regulations, 9, 316–317, 335, 861–862
  see also Federal noise regulations
Watch tick test, 1061
Wave
  analyzer, 126
  compressional, 1026
  definition, 1061
  elastic, 43
  filter, 1061
  front, 1061
  interference, 1062
  motion, 39, 41, 230
  plane, 43
  reflecting, 456, 457
  spherical, 43, 44, 1057
  standing, 63, 142, 182, 193, 1057
  transmission through solids, 412
  velocity, 1062
Waveform, 58
Wavelength, 45, 47
  definition, 413, 1062
  diffraction, 47
Wax, 529, 783
  buildup, 736
  impactions, 207
  impregnated cotton, 530
Weighting curve, 1062
Weighting networks, 20, 55, 111, 1062